Proceedings of the 20th International Symposium on

SHOCK WAVES

Proceedings of the 20th International Symposium on

SHOCK WAVES

Volume I

Pasadena, California, USA July 1995

Editors

B Sturtevant

J E Shepherd

H G Hornung

Graduate Aeronautical Laboratories
California Institute of Technology

World Scientific
Singapore • New Jersey • London • Hong Kong

Published by

World Scientific Publishing Co. Pte. Ltd.

P O Box 128, Farrer Road, Singapore 912805

USA office: Suite 1B, 1060 Main Street, River Edge, NJ 07661

UK office: 57 Shelton Street, Covent Garden, London WC2H 9HE

This work relates to Department of Navy Grant N00014-95-1-0301 issued by the Office of Naval Research. The U.S. Government has a royalty-free license throughout the world in all copyrightable material contained herein.

This work is sponsored in part by the Defense Nuclear Agency under Grant DNA001-95-1-0007. This manuscript is submitted for publication with the understanding that the U.S. Government is authorized to reproduce and distribute reprints for Government purposes notwithstanding any copyright notions thereon.

British Library Cataloguing-in-Publication Data
A catalogue record for this book is available from the British Library.

PROCEEDINGS OF THE 20TH INTERNATIONAL SYMPOSIUM ON SHOCK WAVES (Vol. I)

ISBN 981-02-2593-8 (Set)
ISBN 981-02-2957- 7 (Vol. I)
ISBN 981-02-2958-5 (Vol. II)

Printed in Singapore.

Preface

The International Symposia on Shock Waves are the definitive meetings of the scientific community devoted to the study and use of shock-wave phenomena of all kinds. They take place every two years, usually at places close to centers of activity in the field. The format of the meetings has grown over the first thirty years of their existence, but the number of presentations has deliberately been held approximately constant since then. Because of steadily increasing activity in the field, this requires fairly stringent selection of the material to be presented, and consequently also leads to the high standard and good reputation that the proceedings enjoy. Although much of the work published there has also reached the archival literature in more detailed form, these volumes are a valuable resource because they bring the shock-wave information together in one place.

The broad range of topics covered by the presentations may be divided into applications such as combustion, high-speed flight, detonations, physical chemistry, medicine, and geology; the use and development of devices that produce shock waves, such as shock tubes and shock tunnels, with the associated technology advances including diagnostic techniques; and fundamental studies such as the refraction and reflection of shocks, their interaction with vortices, and chemical kinetics.

Four hundred fifty abstracts were submitted to the 20th Symposium. Each abstract was judged by three members of the Program Committee; 320 abstracts were selected for presentation, 165 as oral and 155 as poster presentations. The presentations included 12 memorable invited lectures. Two of these, the Paul Vieille and Otto Laporte lectures, were delivered by Paul Roth of Duisburg and Heinz Reichenbach of the Ernst Mach Institut in Freiburg, respectively. Both were high-quality and rich presentations that will be remembered as highlights of the Symposium.

Three hundred thirty four participants and 30 companions came to Pasadena from 18 countries: USA, 110; Japan, 57; Germany, 36; France, 32; Australia, 18; Russia, 17; UK, 16; Israel, 13; Canada, 11; China, 9; South Africa and Netherlands, 4 each; Korea, 2; and Belarus, Brazil, Czech Republic, India, and Poland, 1 each.

Two hundred fifty eight papers are published in the Proceedings. This number is smaller than the number of presentations at the Symposium for a number of reasons. For example, some authors had already published similar material elsewhere, as in the case of one of the invited talks (Kanamori), and some wished to publish in other places, where the usual limitation of the ISSW Proceedings to six pages did not constrain them.

The Proceedings is organized such that papers on related topics are grouped together, without making distinctions between poster and oral presentations. It was necessary to break the material into two volumes, which should, however, be seen as parts of a set. Accordingly, the page numbers do not start again in Volume II, but continue from the end of Volume I. For convenience, the Table of Contents in each volume includes the entries of both volumes. Similarly, complete author and subject indexes may be found at the end of each volume.

The papers here were selected from contributions by the Scientific Committee. We thank these individuals for their service to the shock-wave community:

T. Ahrens, R. Akbar, D. Baganoff, W. Behrens,
J. Bélanger, R.F. Benjamin, D. Bershader, R. Bonazza,
C. Brennen, M. Brouillette, A. Bruckner, J. Budzinski,
G. Candler, P. Cassady, J. Cates, B.M. Cetegen,
W. Christiansen, P. Colella, M.S. Cramer, E. Cummings,
J. Davis, J.M. Dewey, J.L. Ellzey, R.J. Emrich,
D. Frost, H. Glaz, J.J. Gottlieb, R.K. Hanson,
B. Hartenbaum, H. Hornung, J.W. Jacobs, J.H. Kiefer,
D. Knight, J.K. Knowles, J.H.S. Lee, P.V. Marrone,
J.R. Maus, S. McCahan, J.C. McDaniel, R. Menikoff,
R.B. Miles, P. Miller, S. Molder, F.K. Owen,
E.G. Puckett, D. Pullin, G. Ravichandran, V. Rupert,
A. Roshko, D.A. Russell, R. Samtaney, S.R. Sanderson,
D.W. Schwendeman, R.E. Setchell, S. Sharma, J.E. Shepherd,
M. Sichel, B. Sturtevant, A. Teodorczyk, P. Thibault,
W.P. Thompson, W. Tsang, T. Vreeland, G. Wilson,
N.J. Zabusky

The International Advisory Committee of the Symposium contributed many suggestions. We recognize these individuals for their assistance:

T.V. Bazhenova, G. Ben-Dor, D. Bershader, M. Brouillette,
R. Brun, F. Demmig, J.M. Dewey, L. Dumitrescu,
R. East, J.L. Ellzey, N.A. Fomin, B. Forestier,
B. Gelfand, J.J. Gottlieb, H. Grönig, Z. Han,
R.K. Hanson, L.F. Henderson, A. Hertzberg, F. Higashino,
R. Hillier, H. Honma, H. Hornung, L. Houas,
F. Houwing, O. Igra, V.K. Kedrinskii, Y.W. Kim,
A. Lifshitz, H. Matsui, B.E. Milton, H. Miyajima,
F. Obermeier, N.M. Reddy, H.W. Reichenbach, P. Roth,
D.A. Russell, J.E. Shepherd, B. Skews, R.J. Stalker,
J.L. Stollery, B. Sturtevant, K. Takayama, M.E.H. Van Dongen,
Z.A. Walenta, H.-Y. Yu

The following organizations contributed financial support essential to the operation of the Symposium:

Air Force Office of Scientific Research under Grant F49610-92-J-0110
Defense Nuclear Agency under Grant DNA001-95-1-0007
National Science Foundation under Grant CTS-9509360
Office of Naval Research under Grant N00014-95-1-0301
Division of Engineering and Applied Science,
California Institute of Technology
Seismological Laboratory at the California Institute of Technology
The Estate of Ralph D. Baker, M.S., Ph.D., Caltech, 1938

We gratefully acknowledge the efforts of Jamie Guthrie, Donna Shepherd, Joe Cates, James Quirk, the GALCIT, Mechanical Engineering and Geophysics graduate students, and Caltech staff in organizing the Symposium. We also wish to thank Marionne Kirk for performing the arduous task of copy editing the Proceedings.

The Executive Committee

H. G. HORNUNG	Chairman
T. J. AHRENS	
G. RAVICHANDRAN	
A. ROSHKO	
J. E. SHEPHERD	
B. STURTEVANT	Editor
K. CHEETHAM	Secretary

Contents

Volume I

Irving Glass Memorial Session

Hypersonic Flow

Supersonic Flow

Wave Reflection and Diffraction

Wave Refraction

Waves, General

Viscous Effects

Volume II

Chemical Physics

Combustion

Detonation

Plasma and Radiation

Multi-phase Flows

Condensed Matter

Technology and Diagnostics

Medical Applications

Volume I

Paul Vielle Lecture

Shock-tube kinetics in homogeneous and heterogeneous reaction systems

P. Roth
Institut für Verbrennung und Gasdynamik
Gerhard-Mercator-Universität Duisburg
D-47048 Duisburg, Germany

Abstract: The shock tube used as a high-temperature wave reactor has dominated high-temperature kinetics for more than 45 years. The nearly instantaneous heating to high temperatures, the accessible wide temperature and pressure ranges, and the diffusion-free reaction conditions are the main advantages of this technique for measuring rate coefficients at high temperatures. In this paper some applications of the shock-tube technique for kinetic studies in homogeneous and heterogeneous reaction systems will be discussed. The examples to be presented were obtained in the author's laboratory. They include thermally and photolytically induced chemical reactions, which were studied by applying different optical absorption techniques.

Key words: Shock-tube kinetics

1. Introduction

The modern applications of shock-tube technology for kinetic studies are characterized by two interrelated facts: very high dilution of the reactants in an inert bath gas (mostly argon) and high sensitivity of the optical diagnostics used. The big advantage of highly diluted reactants is that the endothermicity or exothermicity of the chemical reactions has no effect on the structure and stability of the wave. Furthermore, the number of elementary reactions, which determine progress in the shock-induced reaction system, is limited, thus facilitating the kinetic interpretation of measured properties. On the other hand, access to intermediate chemical species by the use of modern optical diagnostics (see Hanson 1995) provides an insight into details of chemical reactions which are highly influenced by radical species.

Compared to research on homogeneous chemical reactions in shock waves, which was strongly stimulated by the Göttingen researchers (see Wagner 1971), research on heterogeneous kinetics in shock waves is at an early stage. The reasons are that both the degree of reaction complexity and the difficulties in the diagnostics are significantly higher. A precondition for this type of experiment is that the solid phase must be homogeneously distributed in the gas phase. The kinetic problems interfere to a high degree with the two-phase flow difficulties. Detailed access to the adsorption-desorption surface reactions is not possible in shock-tube experiments so that rate parameters obtained are limited to global expressions. On the diagnostics side, sensitive optical methods for gas-phase properties like laser schlieren or laser absorption are significantly disturbed by the dispersed particles.

In the present paper, some examples of different types of shock-wave-induced chemical reactions all obtained in the author's laboratory will be illustrated: simple thermal decomposition reactions, thermally or photolytically initiated radical reactions, and reactions in dispersed systems (aerosols). In all cases, optical absorption ranging from vacuum ultra-violet (VUV) to the infrared region was used, together with the Lambert-Beer Law, to measure time-dependent concentrations of reactive species.

2. Thermal decomposition reactions

The classic application of shock-tube technology, which reached a certain state of completeness many years ago, is to study the above type of reaction by observing the relaxation towards chemical equilibrium after shock heating of the test gas. The decomposition of molecules A can in those cases be described by the Lindemann mechanism, which considers a collisional activation or deactivation followed by the dissociation:

$$\mathrm{A} + \mathrm{M} \underset{2}{\overset{1}{\rightleftharpoons}} \mathrm{A}^* + \mathrm{M}$$
$$\mathrm{A}^* \overset{3}{\rightarrow} \text{products}$$

M is the collision partner. Besides temperature, the progress in the disappearance of A depends on pressure. Two limiting cases are possible: at low pressure the collisional deactivation of A* is much rarer than its chemical reaction and the whole reaction obeys a second-order rate law (low-pressure limit). At high pressure, the rate of disappearance of A becomes independent of the concentration of the collision partner M and obeys a first-order rate law (high-pressure limit). The reality of the above type of reaction is normally much more complicated. The energy required for dissociation is transferred to the molecule by a large number of collisions which can have different efficiencies. Also the internal dynamics of the energy transfer is of importance, i.e., the coupling between vibrational relaxation and dissociation can play a significant role resulting in an apparent delayed dissociation.

The thermal decomposition of diatomic molecules proceeds for most applications under conditions close to the low-pressure limit. A good example is the dissociation of N_2. Because of its importance in hypersonic flows, its rate coefficient was determined in numerous studies, all following the disappearance rate of N_2. For this and other types of decomposition reactions it seems useful to measure the rate of appearance of the atomic dissociation product and to determine in this way the dissociation rate coefficient of the mother molecule. In the case of N_2, the low-pressure decomposition is simply:

$$N_2 + M \xrightarrow{4} N + N + M$$

It is not an easy task to follow N-atom formation under shock-tube conditions. Since the middle of the 1970s atomic resonance absorption spectroscopy (ARAS) for measuring H- and O-atom concentrations has been made available for shock-tube research (see Just 1981), which was later extended to other species like N, C, Si, and S atoms. The diagnostic method is basically a spectroscopic line-emission line-absorption technique. The respective atomic species have strong resonance absorption lines in the vacuum UV, which makes it possible to measure their ground-state concentration with very high sensitivity.

A schematic diagram of the experimental arrangement employed for ARAS is shown in Fig. 1 together with other devices, which will be of interest later on. Microwave power is coupled into a low-pressure flow of a He/N_2 mixture. The

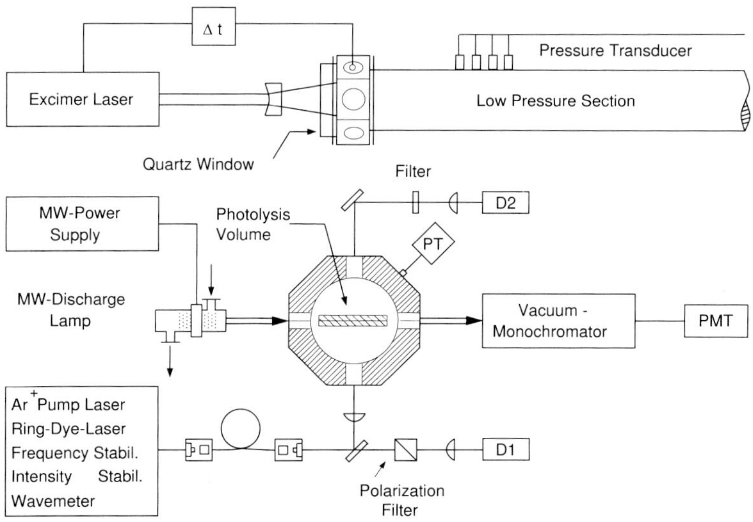

Figure 1. Experimental arrangement for measurements in a shock tube: atomic resonance absorption spectroscopy (ARAS), ring-dye-laser spectroscopy (RDLS), and laser photolysis (LP).

high-temperature plasma formed contains the NI triplet radiation at 119.9 nm, which is single-passed through the shock tube about 10 mm from the end wall.

A one-meter VUV monochromator separates the strongest of the N-atom lines from the radiation of the plasma lamp. N atoms formed in the shock tube during high-temperature thermal decomposition absorb the resonance radiation very effectively. The absorption measured by a solar-blind photomultiplier is related to the N-atom concentration. For a detailed evaluation of the signal calibration is needed, see Thielen and Roth (1986).

An example of a data trace for N absorption/concentration in a reflected shock experiment in a mixture of 200 ppm N_2/Ar at 5500 K and 1.43 bar is shown in Fig. 2 (left). The arrival time of the reflected shock is marked. After

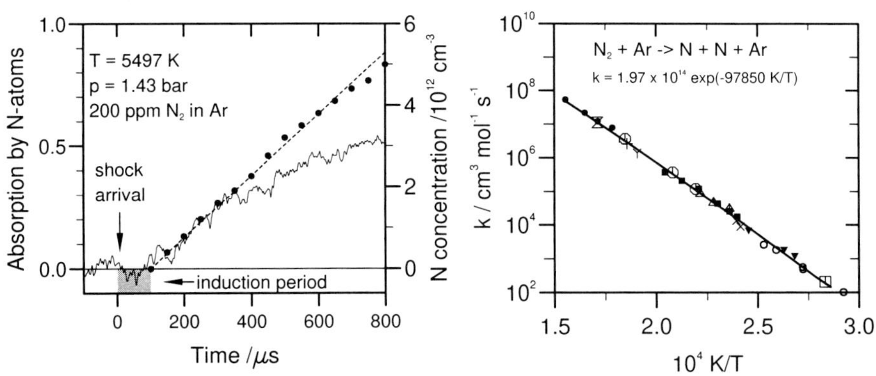

Figure 2. N-atom ARAS signal and corresponding concentration during thermal decomposition of N_2 (left). Rate coefficient of N_2 dissociation (right).

a certain delay the N-atom concentration increases linearly with time. From the N-atom slope the rate coefficient for the N_2 dissociation can be determined. Results of a series of shock-wave experiments in N_2/Ar mixtures are summarized in Arrhenius form in Fig. 2 (right). Because of the high sensitivity of the optical diagnostics, the low temperature end is very much extended compared to other experimental results, the data points are spread over several orders of magnitude, and the rms deviation is quite small.

The thermal decomposition of triatomic molecules like H_2O, COS, or N_2O is more complicated. They can dissociate immediately out of the ground state or can dissociate in a spin-forbidden way from an electronically excited state, which requires a singlet-triplet transition. Decomposition reactions of molecules with four or more atoms are more and more influenced by secondary and consecutive reactions, which must — even under high dilution conditions — be described with kinetic mechanisms.

3. Precursor-initiated radical reactions

The second type of elementary reaction is motivated by the strong interest in obtaining rate-coefficient data of radical–molecule or radical–radical reactions. A chemical precursor undergoing a very fast "priming" reaction after shock heating seeds the test gas with reactive intermediate species at known concentrations. The reaction of interest begins to proceed at a measurable rate only after the radical formation process is nearly finished. This idea of using the shock tube as a chemical reactor goes back to Gardiner et al. (1971). It is clear that the radical formation reaction must be fast compared to the reaction of interest. Through the years, many precursors were tested in our and in other laboratories. The application and successful use of the precursors depend on the detailed reaction conditions.

A good example for a precursor-initiated atom reaction, which also illustrates again the high sensitivity of ARAS, is the reaction of Si atoms with CO.

$$Si + CO \xrightarrow{7} SiO + C \qquad \Delta^R h = 276.1 \text{ kJ/mol}$$

This reaction was studied in mixtures containing 0.75 to 20 ppm SiH_4 and 0.2 to 5% CO diluted in Ar at temperatures 2720 K$\leq T \leq$ 5190 K using the Si-atom ARAS as given in Fig. 1. An individual example is represented in Fig. 3 (left)

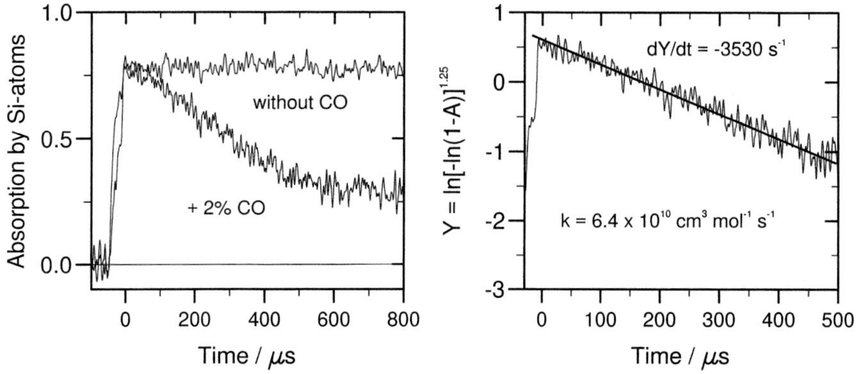

Figure 3. Si-atom absorption measured in mixtures of Ar and 1 ppm SiH_4 without and with CO addition, T = 3660 K, p = 0.8 bar (left). Evaluation of the absorption signal by the first-order method (right).

showing Si-atom absorption measurements in mixtures without and with CO. Immediately behind the reflected shock wave the precursor SiH_4 decomposes nearly instantaneously to form Si atoms (and H_2) at a constant level. In the mixture with CO, the silicon atoms in turn react as given in the above equation, resulting in a decrease in the Si absorption.

The kinetic evaluation of the above type of absorption signal is facilitated by the fact that the disappearance of Si proceeds under first-order conditions, because the CO concentration is practically constant.

$$d\ln[\mathrm{Si}]/dt = -\mathrm{k}_7[\mathrm{CO}]_0$$

The silicon atom concentration can, via the modified Lambert-Beer Law with a concentration exponent of n = 0.8, be directly related to the measured absorption. The above equation rewritten yields:

$$\frac{d\ln\{-\ln(1-\mathrm{A})^{1/n}\}}{dt} \equiv \frac{d\mathrm{Y}}{dt} = -\mathrm{k}_7[\mathrm{CO}]_0$$

with A being the measured fractional Si absorption. The diagram on the right-hand side of Fig. 3 shows the property Y as a function of the reaction time. The linear behavior of Y confirms the first-order assumption of reaction 7 and allows determination of the rate coefficient from the slope. A summary of all rate coefficients obtained is shown in the Arrhenius diagram of Fig. 4. It also

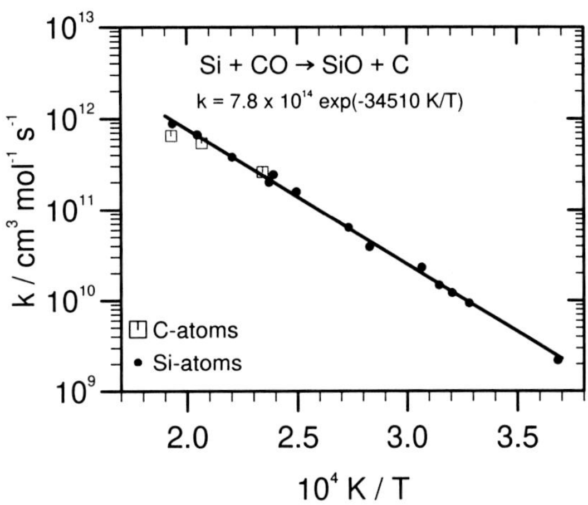

Figure 4. Rate coefficient of the reaction Si + CO ⟶ SiO + C.

includes results of C-atom measurements. The standard deviation of the data due to experimental scatter is, at ±10%, very low. Rate coefficients of several silicon reactions obtained in the author's laboratory are summarized in Table 1.

Table 1. Rate coefficients of various Si-containing reactions obtained by ARAS; $k_i = A_i \exp(-C_i/T)$ $\mathrm{cm^3\ mol^{-1}s^{-1}}$.

REACTION			A_i	C_i
Si + CO	$\longrightarrow$	SiO + C	7.8×10^{14}	34510
Si + CO_2	$\longrightarrow$	SiO + CO	6.0×10^{14}	9420
Si + NO	$\longrightarrow$	SiO + N	3.2×10^{13}	1775
Si + O_2	$\longrightarrow$	products	2.7×10^{14}	1765
Si + N_2O	$\longrightarrow$	SiN + NO	5.0×10^{14}	8100
	$\longrightarrow$	SiO + N_2	8.0×10^{13}	—

4. Photolytically initiated radical reactions

The method of generating radical species by photolysis and following their further reaction is an old and well-established technique in chemical kinetics, mostly used in room-temperature static cells. In the shock-wave community, flash lamps were first used to photolyze molecules in shock-heated gases. With the availability of high-intensity pulsed-UV lasers, which have a higher spectral intensity and a greater monochromaticity compared to flashlamps, these light sources were also used in shock-tube experiments, see e.g. Hanson (1995), Yoshimura et al. (1992). Compared to the chemically formed radicals, described in the previous section, pulsed-laser photolysis is an effective and instantaneous source allowing direct preparation of radical species widely independent of the properties of the shocked gases. An extension to low temperatures is therefore unproblematic. A disadvantage is that the energy transfer or the absorption coefficient during photolysis is not known in detail, rendering prediction of the identity and the energy states of the photofragments produced more difficult.

To illustrate a photolytically initiated radical reaction behind the reflected shock wave, I have chosen the reaction:

$$\mathrm{S + H_2 \xrightarrow{9} HS + H} \qquad \Delta^R h = 80.3\ \mathrm{kJ/mol}$$

for which no experimental high-temperature data are available. Sulfur atoms were photolytically produced by an excimer-laser pulse at 193 nm in mixtures of 25 or 30 ppm CS_2 and 1000 to 4000 ppm H_2 diluted in Ar. The laser was coupled into the measurement plane of the shock tube through an end plate made of quartz glass, see Fig. 1. The originally rectangular beam was expanded by a cylindrical lens allowing the illumination of the whole ARAS diagnostic path; for more details see Woiki and Roth (1993). The temperature range of the present experiments was 1260 to 1840 K. The progress in the chemical reaction studied was followed by applying $S(^3P)$ ARAS at 147.4 nm.

The absorption profile of a typical experiment is shown on the left-hand side of Fig. 5. The arrival of the reflected shock, which is associated with a

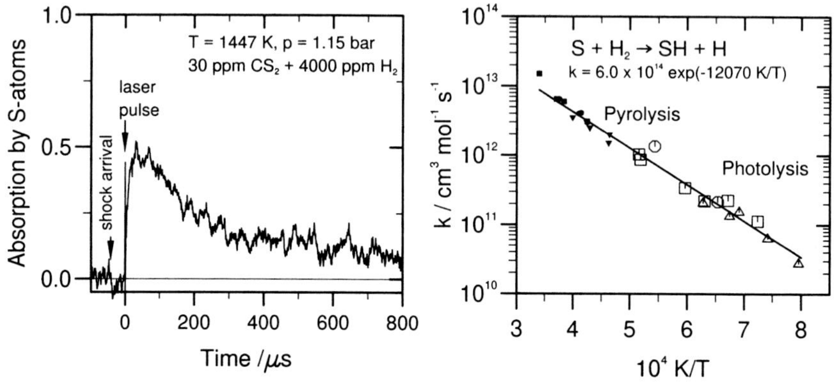

Figure 5. S-atom ARAS signal after photolysis of a shock-heated $CS_2/H_2/Ar$ mixture (left). Rate coefficient of the reaction $S+H_2 \longrightarrow HS + H$ (right).

temperature discontinuity does not initiate any measurable S-atom formation in the mixture. Only the laser photolysis pulse of about 13-ns duration, which was delayed with respect to the reflected wave by about 50 μs, causes a quasi-step-like peak in S-atom absorption. Subsequently the signal decreases due to S consumption by the above reaction with H_2. Data reduction could again be achieved by assuming first-order decay of S, as illustrated in the last section. Results of all the experiments are summarized in the Arrhenius diagram of Fig. 5, right-hand side. The data points can be approximated by the Arrhenius expression

$$k_9 = 6.0 \times 10^{14} \exp(-12070 \text{ K}/T) \text{ cm}^3\text{mol}^{-1}\text{s}^{-1}$$

In this diagram also the results of another series of experiments are included, where the S atoms were chemically formed by pyrolysis of COS. Both data sets agree quite nicely and illustrate the high potential of photolysis experiments at low temperatures, and the good applicability of pyrolysis-generated reactive species at high-temperature reaction conditions.

5. Thermally initiated reactions in dispersed systems

The reactions of solid particles homogeneously dispersed in a gas mixture at high temperature is of interest in many technical applications, for example, soot oxidation, pulverized fuel combustion, or SO_2 reduction by limestone particles. The degree of complexity of such heterogeneous reactions is much higher than in the case of gas-phase reactions, and the availability of kinetic data, even for global processes, is poor. The idea of applying the shock-tube technique to reactive aerosol systems seemed to suggest itself, because a dilute suspension of particles behaves essentially as a gas. The initially unreacted mixture can be heated by a shock wave thus initiating the post-shock chemical reactions, which must be recorded by optical methods. But there are some difficulties compared to experiments with gas mixtures:

- the preparation of the gas/particle mixture (aerosol) and its homogeneous distribution along the low-pressure part of the shock tube,
- the difficulties to be expected during shock-wave heating of the two-phase system,
- the necessity to measure the reactive surface of the aerosol system,
- the application of an optical diagnostic for gas-phase species in a particle-loaded system.

The first problem was solved by developing an expansion-wave-driven aerosol generator allowing dispersion of small amounts of powders, see Roth and von Gersum (1993). The second difficulty can be overcome by choosing sub-μm particles having low overall particle load. In this case the effect of particles on the fluid-flow properties is very weak, i.e. gas temperature and pressure increase quasi-instantaneously behind a shock wave to nearly constant levels. Because

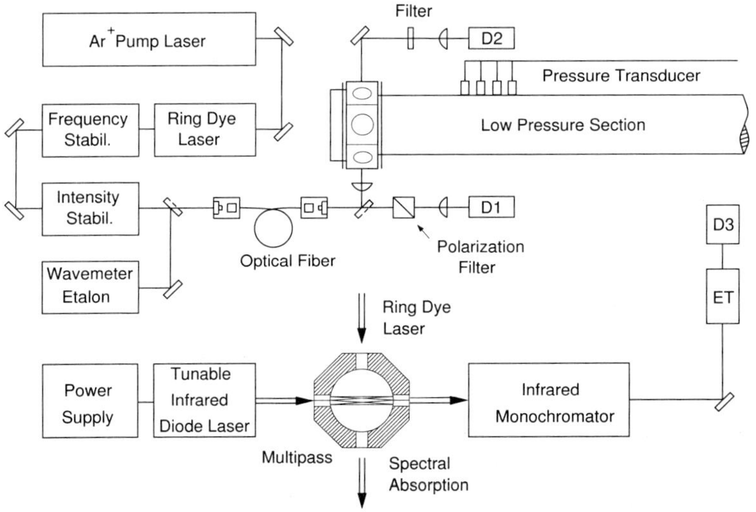

Figure 6. Experimental arrangement for measurement in a shock tube: ring-dye-laser spectroscopy (RDLS) and infrared-diode-laser spectroscopy (IRDS).

of the small size of the particles, the slip-flow region behind the shock wave, in which the heat and momentum transfer from the gas to the particle phase occurs, is very small compared to the extension of the heterogeneous reaction zone, i.e. the suspended particles are also nearly instantaneously heated up behind the shock wave to a constant level. If gas-phase reactants are either initially available in the mixture or chemically formed during a fast homogeneous rate process, their reactions with the hot solid particle surface can proceed. The rate of a heterogeneous reaction depends on, besides other surface properties, the size or the reacting area, i.e. the particles must be characterized with respect to these properties. This can be done by applying laser-light scattering or laser-light extinction at different wavelengths, which will not be discussed here; for details see Brandt and Roth (1989).

The remaining problem of the above list is to determine the progress of the heterogeneous reaction by measuring the formation of gas-phase products or the reduction of gas-phase educts. A well-suited technique is tunable IR-diode laser absorption spectroscopy, because interference by the particles is small, as light extinction by small particles is weak in the infrared region. The optical arrangement used is shown in Fig. 6. It consists of a pulsed, tunable IR-diode laser, the shock tube arranged as a multipass optical cell, a 0.5-m infrared monochromator for mode filtering, and an InSb detection system. The laser is pulsed with a frequency of 25 kHz and can additionally be tuned in each pulse over one or two lines thus producing a series of highly resolved absorption lines, from which time-dependent species concentration can be determined, see Brandt and Roth (1989).

An example of a thermally initiated reaction in a dispersed system is the oxidation of soot particles by O atoms. The expected oxidation products are CO and CO_2, but the only clearly detectable product in the time scale of the present experiments was CO, i.e. the oxidation proceeds in the present case mainly via the global heterogeneous reaction:

$$\mathrm{C_s + O \longrightarrow CO}$$

C_s represents the solid carbon in the soot particles. The O atoms were generated by the fast thermal decomposition of N_2O that was present in the initial gas/particle mixture. According to the above reaction, the rate of formation of CO is:

$$\frac{d[\mathrm{CO}]}{dt} = \alpha_0 Z_0 a_p \qquad Z_0 = \frac{1}{4}\,\overline{c_0}\,[\mathrm{O}]$$

In this equation α_0, Z_0, and a_p are the reaction probability of O atoms with the soot surface, the collision number of O atoms per time and unit area, and the reacting particle surface per unit volume of suspension, respectively. The collision number Z_0 can be calculated from the Hertz-Knudsen equation with $\overline{c_0}$ being the mean thermal velocity of O atoms. The desirable kinetic property is the reaction probability α_0, which can for known [O] be determined from measured properties [CO] and a_p.

A typical example of an individual shock-tube experiment in soot/N_2O/Ar mixture at 2490 K is shown in Fig. 7. The laser was tuned in a wavelength range close to the (0,1) P8 line of CO, the pulse frequency was about 25 kHz. The arrival of the shock front is marked. In the lower left part of Fig. 7 selected scans of the same experiment are shown in more detail with an enlarged time scale. The strong absorption lines in scan A were caused by spectral absorption of N_2O under pre-shock conditions. Beginning with the first post-shock scans, sharp spectral lines of CO appear showing increasing absorption. These illustrate the formation of an increasing amount of CO during the heterogeneous soot oxidation.

The reduction of the spectroscopic data in terms of CO concentration can be performed based on the Lambert-Beer Law. The necessary spectral-absorption coefficient depends on the line strength and the line-shape factor. The latter was determined by fitting a Voigt profile to the measured absorption line. Taking this into account, the sequence of absorption obtained can be transformed into a time-dependent CO concentration, as illustrated on the right-hand side of Fig. 7. From the initial slope together with the above equation or from computer simulation of this and other oxidation experiments, reaction probabilities α_0 were determined. A summary is given in the Arrhenius diagram of Fig. 8,

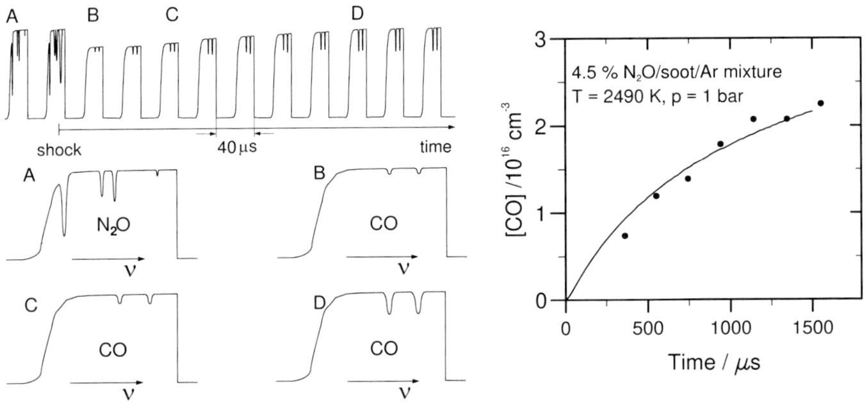

Figure 7. Example of pulsed-IR-diode-laser scans in a shock heated soot/N_2O/Ar mixture showing increasing CO spectral absorptions (left); corresponding CO concentration (right).

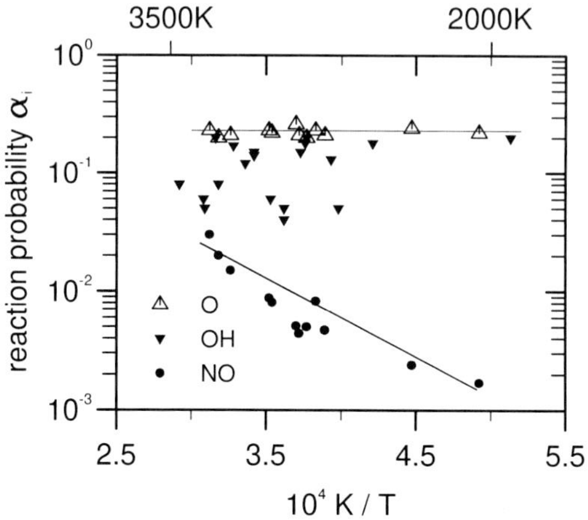

Figure 8. Reaction probabilities of soot oxidation by O, OH, and NO obtained from shock-tube experiments.

which also contains results of soot oxidation by OH and NO. The α_0 values obtained are nearly independent of temperature and can be approximated by a mean value of $\alpha_0 = 0.23$. For more details see Roth and von Gersum (1993).

6. Conclusion

Modern applications of shock-tube technology for kinetic studies in both homogeneous and heterogeneous reaction systems were illustrated in several examples. It could be shown that the trends are in the direction of studying the rate processes in highly diluted systems, partly as low as 1 ppm of reactant. The availability of highly sensitive optical, mostly laser-based diagnostics make such experiments possible. Care is required concerning the influence of impurities in the test gases as well as in the shock tubes, which are in our case constructed and used as ultra-high vacuum apparatuses. A problem can also be wall adsorption of strongly polar reactants in the highly diluted systems. Nevertheless the high-temperature wave reactor will also in the future be of great importance in high-temperature kinetics.

Acknowledgement. The financial support of the German Science Foundation is gratefully acknowledged.

References

Brandt O, Roth P (1989) A tunable IR-diode laser technique for measuring reaction rates of high-temperature aerosols. Comb and Flame 77:69.

Gardiner WC, McFarland M, Morinaga K, Takeyama T, Walkes BF (1971) The shock tube as a chemical reactor. In: Proceedings of the 8th Int. Shock Tube Symposium, Chapman and Hall, London, p. 23/1.

Hanson RK (1995) The rate of lasers in shock tube studies of chemical kinetics. In: Proceedings of the 19th Int. Shock Wave Symposium, R Brun and LZ Dumitrescu (eds), Springer-Verlag.

Just T (1981) Atomic resonance absorption spectrometry in shock waves. In: Shock Waves in Chemistry, A Lifshitz (ed), Marcel Dekker Inc., New York, p. 279.

Roth P, von Gersum S (1993) High temperature oxidation of soot particles by O, OH, and NO. In: Turbulence and Molecular Processes in Combustion, T Takeno (ed), Elsevier, Amsterdam, p. 149

Thielen K, Roth P (1986) N atom measurements in high-temperature N_2 dissociation kinetics. AIAA Journal 24:1102–1105.

Wagner HG (1971) Chemical reaction of shock waves. In: Proceedings of the 8th Int. Shock Tube Symp., Chapman and Hall, London, p. 4/1.

Woiki D, Roth P (1993) ARAS measurments on $S(^1D)$ formation in shock heated COS/Ar mixtures. Ber Bunsenges Phys Chem 97:996.

Yoshimura M, Koshi M, Matsui H, Kamia K, Hideaki U (1992) Non-arrhenius temperature dependance of the rate constant for the reaction $H+H_2S$. Chem Phys Lett 189:195–204.

Otto Laporte Lecture

120 years of shock-wave visualization — a fascinating activity

H. Reichenbach
Fraunhofer-Institut für Kurzzeitdynamik
Ernst-Mach Instach Institut
Freiburg, Germany

Abstract: After a brief chronology of Otto Laporte's life and a few personal remembrances, contributions are described of some pioneers in shock-wave visualization. August Toepler (1836–1912) observed shock waves created by electric sparks only with his eye. Ernst Mach (1838–1916), philosopher and originator of instationary gasdynamics and shock-wave research, found the irregular shock-wave reflection and invented various high-speed diagnostic methods. Cranz (1858–1946) and Schardin (1902–1965) contributed significantly to the development of shock visualization. From countless applications only a few are picked out to demonstrate the fascinating development to the state of the art nowadays.

Key words: Shock waves, Visualization, Historical report

1. Introduction

Assuming that not all participants know exactly who Otto Laporte was, in Table 1 a brief chronology is presented, quoted from a bibliographical memoir (Crane and Dennison 1979), where also many details of his life[1] are given. Otto Laporte is well known as an excellent spectroscopist. From his youth he was interested in this subject, starting when his father gave him a prism of an old periscope. Otto Laporte spoke German and English fluently, and he was also perfect in French and Japanese. Now and then he liked to apply these languages in the same sentence. He managed perfectly to mix in such a manner that nobody had any difficulty in understanding him.

With such a talent for languages, it is not surprising that Otto Laporte was given the task of interviewing German scientists concerning their research work during the Second World War. In this way he met Hubert Schardin who was Director of the Ballistics Institute in Berlin until 1945.

During the interview with Schardin, Otto Laporte did not act like an inquisitor — he was too much a scientist for that. On the contrary, he spoke passionately about shock waves and the manifold applications of shock tubes.

[1]I gratefully acknowledge the help of Prof. Y. W. Kim, Lehigh University, Bethlehem PA, who made the Otto Laporte bibliographical memoir available to me.

Table 1. Chronology: Otto Laporte.

1902	Born July 23, Mainz, Germany. Parents: Wilhelm Laporte, heavy artillery officer in the Imperial German Army; Anna Geyl
1923	Analyzed iron and vanadium spectra. Enunciated Laporte rule
1924	Ph.D. University of Munich, under Arnold Sommerfeld. Thesis: "Exact Treatment of Scattering of EM Waves by a Sphere"
1924-26	International Education Fellow; studied in Europe, Japan, and finally in the USA in the spectroscopy section of the National Bureau of Standards
1927	Assistant Professor
1935	Became naturalized US citizen
1944	Changed field of activity from atomic spectroscopy to fluid mechanics
1945	Promoted to full professor
1946	Assumed charge of shock-tube laboratory at University of Michigan
1949-50	Scientific Intelligence Analyst, US Army of Occupation, Heidelberg, Germany
1951	Initiated research on reflected shock waves
1954-55	Scientific attaché at US Embassy in Tokyo
1956	Cited by US State Department as instrumental in securing atomic-energy agreement between USA and Japan
1965	Chairman, Division of Fluid Dynamics, American Physical Society
1971	Died March 28, Ann Arbor, Michigan
1971	Elected, posthumously, to the National Academy of Sciences

As ballistician, Schardin was of course very well acquainted with shock waves and high-speed measurement techniques. Remember, together with his academic teacher and mentor, he developed the famous Cranz-Schardin Camera in 1928. Of course, Schardin immediately realized the advantage of combining a shock tube with a Cranz-Schardin Camera for shock wave research. Even today, nearly 50 years later, this combination is still an important tool in this field.

As shock-wave research flourished in Germany, France, and other European countries after the Second World War, Otto Laporte was one of those who gave a significant impetus to the development of shock-wave visualization.

If I dare now to reflect on the development of the last 120 or more years, I must restrict myself to a small, modest, and subjective selection. I apologize ahead of time if I cannot mention all important contributions in this field.

2. First visualizations of shock waves

Although Ernst Mach is now recognized as the pioneer and originator of instationary gasdynamics and shock-wave research, there was another scientist before him who carefully investigated shock-wave phenomena from spark discharges. Since no sensitive photographic plates were available at that time, he observed the phenomena in a *subjective* manner, namely, with his eye by means of a telescope. The name of this great scientist was August Toepler[2].

[2]P. Krehl from Ernst-Mach Institut, Freiburg, has investigated intensively the literature on Toepler's publications (see Krehl and Engemann 1995). I gratefully acknowledge Krehl's

2.1. August Toepler (1836–1912)

Toepler is best known as the originator of the schlieren method. But his contributions to shock-wave phenomena and other fields of physics and chemistry have fallen into oblivion. His pioneering work was at first not noticed by experts. He was forced to print it at his own expense in 1864. Besides the spherical pressure waves that he first called shock waves (Stoßwellen), he also investigated the reflection of shock waves.

Toepler was not only an excellent experimentalist but also a talented artist. The shock fronts he saw were afterwards drawn by hand using pen and ink. Figure 1 is a reproduction of a woodcut of such a drawing, published 1864. He found that the wave fronts reflected in such a way that it appeared as if a second source (c') was present and symmetrically orientated to the reflecting surface. He noticed that the reflected wave front met the incident front at the surface, but he also mentioned that in some tests the reflected wave seemed to *precede* the incident wave near the reflection point. What he saw was probably the pattern of a Mach reflection! Such a reflection was unimaginable for him, so he assumed that it must be an optical illusion. Toepler also studied shock refraction phenomena, and investigated gas combinations with sound velocities larger and smaller than air.

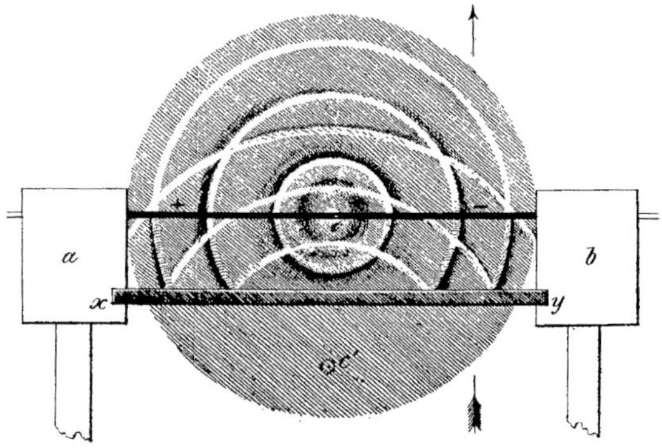

Figure 1. Reflection of spark-generated shock waves (woodcut, 1864).

Nowadays it sounds unbelievable that Toepler observed the instationary gasdynamic effects only with his eye. While he had unfortunately no appropriate method at his disposal for an *objective* registration, another scientist was much more successful in this respect.

---

help and information he gave me concerning Toepler's scientific contributions.

2.2. Ernst Mach (1838–1916)

Ernst Mach was an inventive scientist, not only skilled in experimental techniques, but also in theoretical interpretation of experimental observations. He invented various high-speed diagnostic methods. His life and his contributions to philosophy and physics are discussed in numerous publications (e.g., Merzkirch 1970; Blackmore 1972; Reichenbach 1983; Wolters 1987; Prosser and Folta 1988).

The discovery of an irregular reflection pattern at shock waves — today called Mach reflection — is certainly Mach's most important and spectacular contribution to fluid dynamics. Even today this phenomenon is not completely understood. The era of objective visualization started when E. Mach and J. Wosyka published in June 1875 (exactly 120 years ago) the result of their ingenious experiments. A spark discharge along a line, drawn with a conducting paint on a glass plate, was used as the source of a strong acoustic wave. The line

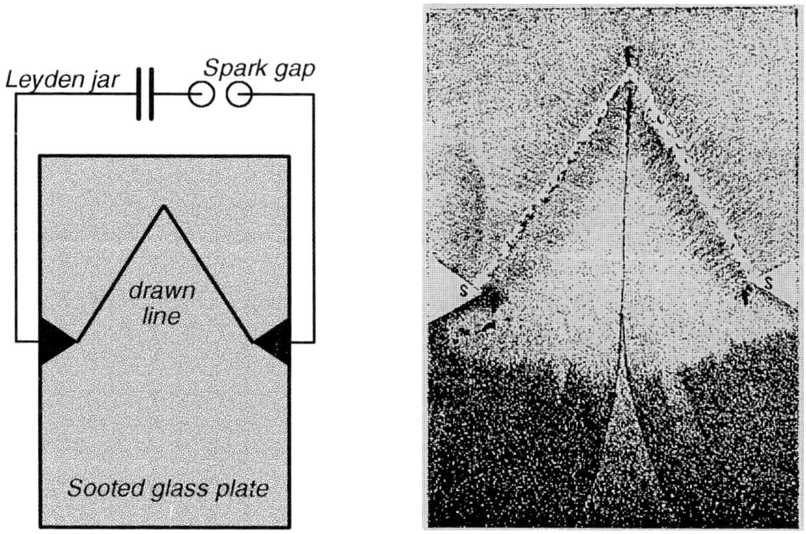

(a) (b)

Figure 2. Mach's study of the V-phenomenon (1875): (a) test arrangement; (b) visualization of Mach-V by means of a sooted plate.

formed an angle as illustrated in Fig. 2a. A Leyden jar capacitor was charged by an hand-operated electrostatic machine until the spark gap fired, thus closing the circuit. At discharge, two cylindrical shocks start out from the angular line. To visualize the intersection Mach applied the soot-plate technique developed by Antolik (1875). The glass plate with the drawn circuit was sooted. The soot was partly blown away by the shock waves and revealed their intersection, with the so-called Mach V into which the line of intersection bifurcated. This can be seen in Fig. 2b which is a reproduction of an original woodcut of an exposed soot plate.

Mach assumed immediately that a new wave must be the reason for the bifurcation. To visualize this wave, he used in another ingenious experiment two line systems in opposite positions. If this new wave exists, the soot must remain at rest at the intersection line inside the Mach Vs. That is exactly what the experiment showed. This wave is called today Mach wave or Mach stem. Only a few years later in 1878, Mach formulated a criterion for Mach reflection (Mach 1878). This formula corresponds to what we call today: the "Mechanical Equilibrium Condition." It was Ernst Mach who clearly emphasized that the experiment must be the basis and the touchstone of theory in natural science.

3. Progress in experimental techniques

Serious shock-wave research at the end of the last century was just barely possible, because the required experimental techniques were not available. New precise tools were necessary if Mach's idea of the control or validation mechanisms of experiments should be translated into action.

3.1. Carl Cranz (1858–1946)

Carl Cranz was one of those who fully accepted Mach's concept. In 1903 he took the Chair of Ballistics of the Military-Technical Academy in Berlin; he received enough funding to make his Institute a premier institution for more than three decades.

Figure 3, taken before 1914 (Cranz et al. 1914), gives proof of the excellent quality of the high-speed pictures. It shows a projectile propagating inside a tube with slots. The bow wave can escape through the openings. It was shown that the bow wave is the envelope of Huygens wavelets.

But the most famous technical development of Cranz and his student Hubert Schardin was the invention of a multiple-spark camera in 1928, today known as Cranz Schardin Camera. In the classical paper "Cinematography on Fixed Film with Extremely High Picture Frequency" numerous applications were treated; picture frequencies as high as 3 million frames per second were achieved (Cranz and Schardin 1929).

3.2. Hubert Schardin (1902–1965)

The work of Cranz was continued by his successor and former student Hubert Schardin. Like Cranz, he was an excellent scientist, with an ability to grasp the principle underlying complicated events and develop a simple approach for their study. Schardin studied for instance the question whether rarefaction shocks exist. When an evacuated container implodes, a bang is heard. Is this caused by the expansion wave or by the pressure wave propagating after the implosion? Schardin's set-up was indeed a shock tube where the outer atmosphere represented the pressure chamber. The shock-wave studies, undertaken so hopefully at first, were not followed up. They did not find a receptive audience at the time, since people had not yet realized the numerous applications of a shock tube. One must remember that in those days ballistic problems seemed more

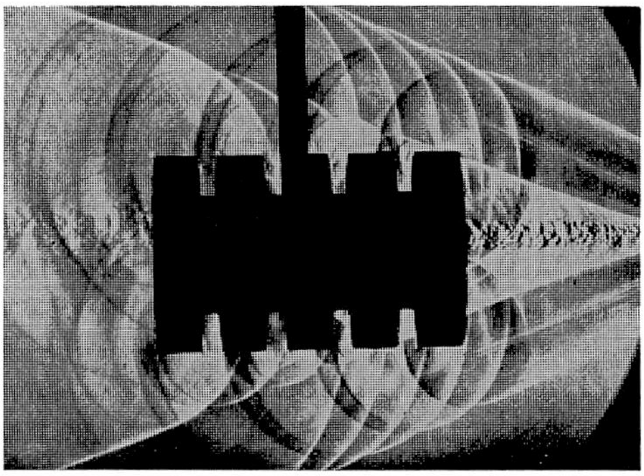

Figure 3. Projectile transversing a tube with slots (Cranz, before 1914).

important. In retrospect one could say that all the knowledge of visualization techniques was developed in Cranz's institute and available to Schardin in the thirties; only the idea of carrying out shock-wave experiments in a shock tube was lacking. As often happens in science, all information and methods are available, but a new impetus is necessary to create a new era.

4. Applications

Two military problems spawned rapidly increasing activities in shock-wave research: nuclear explosions and protection against their effects, and the forces and heat transfer acting on intercontinental missiles at reentry into the atmosphere. During and after the Second World War, experiments were performed to solve these problems and to understand the physics of the numerous new phenomena. As examples, I would like to mention Bleakney's group in Princeton and Glass's group in Toronto, and also Schardin in St. Louis (and later on in Freiburg). New visualization principles were developed, such as phase-contrast systems, differential interferometer, color-schlieren systems, and color interferometry (e.g. Holder and North 1952; Wolters 1956).

4.1. Reflection and diffraction

Let me present a few examples of reflection and diffraction. The wave pattern around a triangular prism is shown in Fig. 4. Most remarkable are the spiral vortices at the corners, Mach effects at the crossing diffracted shock fronts, rarefaction waves centered at the corners, and slip lines starting at the triple points. To understand the complicated wave interactions, a systematic study of simple cases was necessary, such as reflection of a plane shock front at oblique solid surfaces. Regular reflection and Mach reflection were expected. But at higher shock Mach numbers a new type of reflection appeared; the double Mach reflection, with two triple points and two slip lines. In the meantime various kinds of shock-wave reflection were isolated and transition criteria were formulated. Ben-Dor recently published a book dealing only with shock-wave reflection phenomena (Ben-Dor 1992). Is it not fascinating to find that there are so many solutions to such a simple problem as the reflection of a plane shock from a flat rigid surface? Even today it is not clear whether all possible solutions have been discovered (Ben-Dor and Takayama 1992).

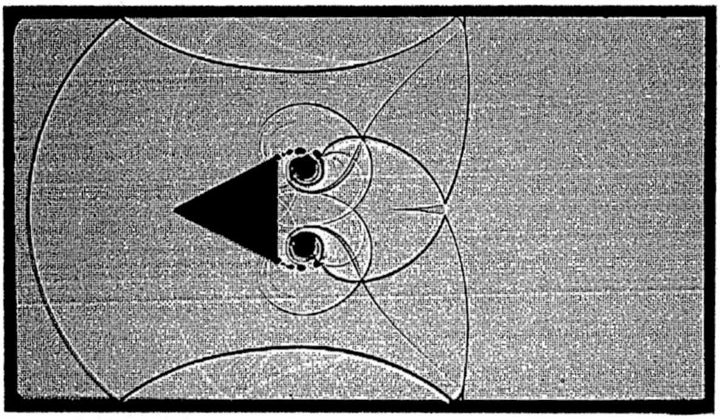

Figure 4. Diffraction around a triangular prism.

Very early in the fifties, investigations were started on the diffraction at corners and edges. Around that time, color films became available and color-schlieren systems were developed. (Holder and North 1952; North 1954). Schardin noticed that color-schlieren photography is an ideal method for visualizing diffraction and vortex formation at corners. One of his first attempts is shown in Fig. 5. It is a good example to demonstrate the advantages of color photography. Color interferometry, applied by us since the sixties, is also useful for visualizing diffraction phenomena.

Figure 5. Color-schlieren photograph: diffraction at a wedge (Schardin, about 1953).

4.2. Influence of surface properties

Surface properties (e.g., roughness, porosity, and elasticity) can influence the propagation and reflection of shock waves. Extensive tests were performed to investigate the physical mechanisms. If a shock wave propagates along a rough surface, it is reflected and diffracted at each roughness center. To visualize this effect, a surface was chosen with periodic grooves perpendicular to the propagation direction of the shock wave. The result is given in Fig. 6 for a groove depth of 2.5 mm. The elementary waves emanating from each step (representing two-dimensional roughness centers) overlap each other, causing distinct bending of the original vertical front, starting where the first elementary wave hits the incident shock front.

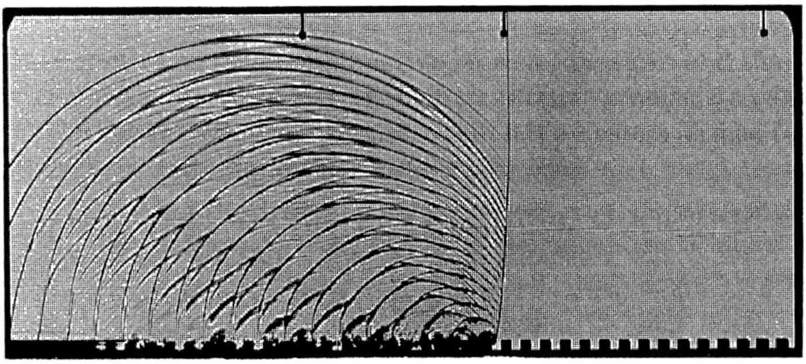

Figure 6. Shock propagation along a surface with periodic grooves; $M_S = 1.38$.

The same effect must also result from oblique reflections at a rough surface. This implies that, at the reflection point, the angle between shock front and the wedge surface does not exactly correspond to the wedge angle. The transition from regular to Mach reflection on rough surfaces must occur at smaller wedge angles than on a smooth surface. This is indeed observed. But no comprehensive theory has yet been found.

4.3. Precursors and decursors

Another field where further effects are involved is the shock-wave interaction with thermal layers, or layers with gases of different sound speeds. We have developed several methods to produce such layers in a shock tube: the most important ones are: (a) electric heating of the surface, (b) diffusion of a gas through a porous ceramic flat plate (Reichenbach and Kuhl 1987). The sound speed of helium for instance is much higher than that of air.

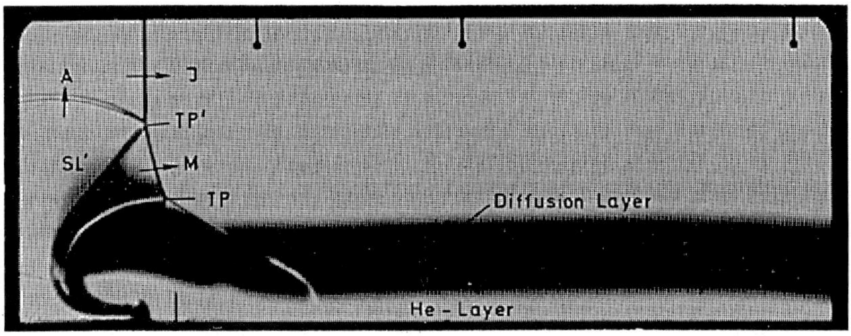

Figure 7. Precursor start-up period, $M_S = 1.77$.

If a shock wave interacts with such a layer a precursor develops. Figure 7 shows the start-up period of the interaction with a helium layer. Between the helium layer (white zone at the bottom) and the pure air in the upper part of the schlieren picture, a mixing layer is visible as a black zone. The start-up of the precursor proceeds through a transition phase to the final phase.

A completely different behavior occurs with a freon layer; a decursor develops. The sound speed of freon R12 is 150 m/s (at T = 293 K). Therefore a shock wave propagates more slowly in the Freon layer than in air, Fig. 8. This causes the shock front to be curved backward near the wall. The oblique shock compresses the layer and reflects from the floor as a regular reflection. The reflected shock reflects off the layer interface as a rarefaction wave which stops the compression of the layer. The rarefaction wave reflects back and forth within the layer, causing a series of expansion and compression waves that radiate from the layer. These waves strongly perturb the interface, which

Figure 8. Decursor formation, $M_S = 1.29$.

acquires an undulated shape and transitions rapidly to a turbulent boundary-layer flow. Wave-dynamics effects damp out and mixing processes begin to dominate. Coherent structures and striations appear (Kuhl et al. 1992). We call this a baroclinic boundary layer.

4.4. Shock-induced turbulent flow

When visualization techniques developed into a practical tool, investigations of shock-induced flow and interactions with small-scale structures increased rapidly. As only one example I choose the propagation of a planar shock through an aligned baffle system. Figure 9a shows a single-frame color interferogram in the infinite-fringe mode of the same shock-induced flow. The undisturbed region of this aligned 4-baffle system appears in a uniform color. In this case, the fringes are lines of constant density (isopycnics).

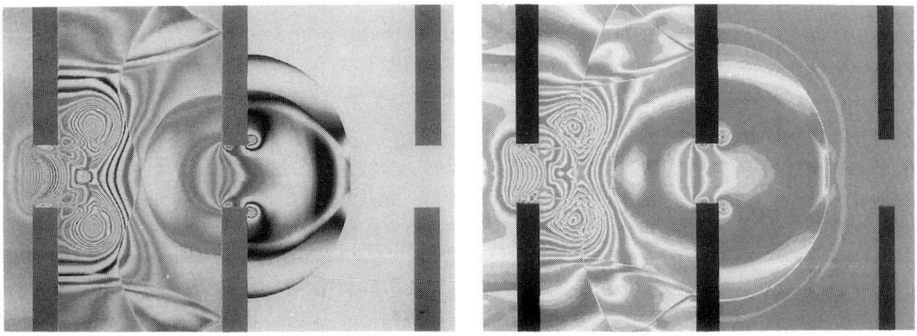

(a) (b)

Figure 9. Shock wave passing a baffle system: (a) experimental color interferogram; (b) numerical interferogram.

The last figure demonstrates a classical visualization method. Numerous improvements have been made in the last 50 years, and new principles were developed like phase-contrast and differential-interferometry methods. The advent of the laser in 1960 provided a revolution in optics, and opened up applications such as holographic visualization techniques. There is not enough time to describe these methods. Professor Takayama from Sendai for instance and many others have presented excellent results again and again. I would now like to pass onto another type of visualization that is applied more and more frequently nowadays.

5. Numerical simulation

As a consequence of the Nuclear Atmospheric Test Ban Treaty of 1962, multi-dimensional hydrocodes have been developed for shock-wave research, thus allowing visualization by numerical simulation. With sophisticated dataprocessing programs, it is possible to present the results of numerically simulated shock-wave phenomena in form of shadow-, schlieren- or interferograms etc. A new era of visualization possibilities has started.

Figure 9b demonstrates an example. The shock-induced flow in the baffle system under the same conditions as for the previously presented experiments was calculated with the Adaptive Mesh Refinement code AMR (Bell et al. 1994; Berger and Colella 1989). At first glance, it is not easy to distinguish between the experimental and the numerical interferogram. However, in small details differences can be detected, especially in the vortex region where some instabilities seem to be more apparent in the calculated interferogram. Altogether, the agreement between experiment and simulation is excellent.

Numerical methods also provide information on quantities which cannot be measured or visualized directly, like entropy, internal energy, vorticity etc. In order to avoid misinterpretations, it becomes important to validate computer codes, under the assumption that a good agreement between calculated and measured quantities also implies that non-directly measurable data are reproduced in a trustworthy manner. Experiments will therefore never become unnecessary, since an experiment is essentially nothing but an analog computer.

6. Final remarks

It was the aim of my presentation to trace in rough outline the historic development of shock-wave visualization. It is fascinating to see what has resulted from the scientific work of a few pioneers of our research discipline. Shock-wave research leads often to surprising results, and visualized effects do not only help create a better physical understanding, but sometimes also fulfill artistic and aesthetic ambitions. Also, for the future we can certainly assume further significant progress in shock-wave research by sophisticated visualization methods.

References

Antolik K (1875) Das Gleiten elektrischer Funken. Pogg Ann Phys Chem 1:14.

Bell JB, Berger M, Saltzman J, Welcome M (1994) Three-dimensional adaptive mesh refinement for hyperbolic conservation laws. SIAM J Stat Comp 15(1):127.

Ben-Dor G (1992) Shock Wave Reflection Phenomena, Springer, Berlin, Heidelberg, New York.

Ben-Dor G, Takayama K (1992) The phenomena of shock wave reflection — a review of unsolved problems and future research needs. Shock Waves 2:211.

Berger MJ, Colella P (1989) Local adaptive refinement for shock hydrodynamics. J Comp Phys 82:64.

Blackmore JT (1972) Ernst Mach — His Work, Life and Influence, Univ Calif Press, Berkley, Los Angeles, London.

Crane HR, Dennison DM (1979) Otto Laporte 1902–1971; A Bibliographical Memoir, Bibliogr Memoirs, Vol 50, Nat Acad of Sciences, USA.

Cranz C, Günther PA, Külp F (1914) Photographische Aufnahmen von sehr rasch verlaufenden Vorgängen, insbesondere von Schußvorgängen mittels Vorderbelichtung durch das Licht elektrischer Funken. Z.f.d. Gesamte Schieß-und Sprengstoffwesen 4:42–59.

Cranz C, Schardin H (1929) Kinematographie auf ruhendem Film und mit extrem hoher Bildfrequenz. Z f Physik 56:147.

Holder DW, North RJ (1952) A schlieren apparatus giving an image in colour. Nature 15:466.

Krehl P, Engemann S (1995) August Toepler — the first who visualized shock waves. Shock Waves 5:1.

Kuhl AL, Reichenbach H, Ferguson RE (1992) Shock interaction with dense-gas wall layers. In: Proc 18th Int Symp on Shock Waves, pp 159-166.

Mach E, Wosyka J (1875) Über einige mechanischen Wirkungen des elektrischen Funkens. Sitzungsber Akad Wiss Wien 72:44.

Mach E (1878) Über den Verlauf der Funkenwellen in der Ebene und im Raum. Sitzungsber Akad Wiss Wien 77:819.

Merzkirch W (1970) Mach's contribution to the development of gasdynamics. Boston studies in philosophy and science 6:42.

North RJ (1954) A colour schlieren system using multi-colour filters of simple construction. NPL/Aero 266.

Prosser V, Folta J (1988) Ernst Mach and the development of physics. In: Int Conf Prague, Karolinum (ed), (conference paper), Prague 1991.

Reichenbach H (1969) Contribution of H. Schardin to the theory and applications of the shock tube. Phys Fluids, Shock Tube Symp 12(5):I3–I8.

Reichenbach H (1983) Contribution of Ernst Mach to fluid mechanics. Ann Rev Fluid Mech 1–28.

Reichenbach H, Kuhl AL (1987) Techniques for creating precursors in shock tubes. In: Proc 16th Int Symp on Shock Waves, pp 847–853.
Reichenbach H (1992a) In the footsteps of Ernst Mach — a historical review of shock wave research at the Ernst Mach Institut. Shock Waves 2:65.
Reichenbach H (1992b) Hubert Schardin (1902–1965) — his life and his work. Proc 20th Int Congr on High Speed Photogr and Photonics.
Schardin H (1932) Bemerkungen zum Druckausgleichsvorgang in einer Rohrleitung. Phys Z 2:60.
Schardin H (1959) Carl Cranz als Mitbegründer der Kurzzeitphotographie. Proc 4th Int Congr on High Speed Photogr and Cinematogr.
Toepler A (1864) Beobachtungen nach einer neuen Methode — Ein Beitrag zur Experimentalphysik, Max Cohen & Sohn, Bonn.
Wolters G (1987) Mach I, Mach II, Einstein und die Relativitätstheorie - Eine Fälschung und ihre Folgen, Walter de Gruyter, Berlin, New York.
Wolters H (1956) Schlieren-, Phasenkontrast- und Lichtschnittverfahren. Handb Phys XXIV, Grundlagen der Optik, Springer, Berlin, Göttingen, Heidelberg.

Plenary Lectures

Transport and relaxation behind shock waves: old and new problems

R. Brun
Laboratoire IUSTI, MHEQ Université de Provence
Centre Saint Jérôme
13397 Marseille, Cedex 20 France

Abstract: An analysis of possible ways of modeling nonequilibrium flow behind shock waves is presented: the corresponding Navier-Stokes systems differ by their transport expressions and relaxation equations. Then, a general model is exposed which avoids misunderstandings and covers all nonequilibrium situations: particular emphasis is laid on vibrational nonequilibrium and examples of computation of transport terms are given. Furthermore, vibrational relaxation models and corresponding experimental results are examined, particularly the evolution of individual vibrational populations behind shocks with and without dissociation; models for the vibration-dissociation coupling are also presented. Finally, emission measurements of CN behind strong shock waves moving in the reconstituted Titan atmosphere of a free-piston shock tube are discussed.

Key words: Nonequilibrium flow models, Transport terms, Vibrational populations, Dissociation

1. Introduction

Transport and relaxation are two aspects of the same phenomenon, characterizing the nonequilibrium of parameters representing physico-chemical processes having a time scale larger than those of perturbations in the considered medium. The case of a shock wave, representing an intense, brief, and well-defined perturbation, allows one to analyze simply the evolution of these processes behind the shock, where the flow time and process time have the same order of magnitude. From a theoretical point of view, the modeling of these phenomena may give rise to different formulations and it is necessary to know the precise starting assumptions and definitions, particularly about the time scales, known or assumed, of the different processes. In the same way, the interpretation of experiments in nonequilibrium regions necessitates the use of more or less sophisticated models with results to be carefully examined.

At the macroscopic level in the continuum regime, the description of the flow behind shock waves is essentially based on the Euler or Navier-Stokes equations more or less simplified: Viscous Shock Layer (VSL), Parabolized Navier-Stokes (PNS), Boundary Layer (BL). These equations result from successive approximations of Boltzmann equations specific to different species and (or) quantum levels. If we except the N-moment methods and Gross-Jackson procedures, which are valid outside the collisional regime but difficult to apply, the most popular method of solving the Boltzmann equations remains the Chapman-Enskog method (CE) first established for pure structureless gases, then extended to mixtures and to gases with internal modes in nonequilibrium situations or eventually reactive (Brun et al. 1984). Based on an expansion of the distribution function in a series of a "small" parameter chosen in most but not all cases as the ratio of the relaxation translation-rotation time to a reference flow time, the method requires one to define the time scale of the different processes as compared to this reference time: depending on the considered cases, frozen, equilibrium, and nonequilibrium flows may be defined. The Chapman-Enskog expansion differs for each case and it is the same for the corresponding Navier-Stokes systems. There is also no matching between the solutions which therefore cannot describe a complete flow passing for example from a frozen to an equilibrium situation (shock) or the inverse (expansion). A general matching method or generalized Chapman-Enskog method (GCE) is proposed hereafter which can cover all situations, whatever the degree of nonequilibrium.

Applications to vibrational nonequilibrium are treated for pure gases (Brun 1988) and then for mixtures like N_2/O_2 (Pascal & Brun 1993). Furthermore, the collision integrals which unavoidably appear in the formal solutions may be expressed as functions of transport terms insensitive to the internal content of molecules (shear viscosity, mass diffusion) or of global phenomenological relaxation times (rotation, vibration) (Mason & Monchick 1962), so that explicit solutions are presented. Connected to the CE method, the problem of vibration-dissociation coupling is globally treated and allows one to explicitly compute the influence of the vibrational nonequilibrium on the dissociation rate constants.

As for the relaxation itself, the availability of CW lasers allows one to carry out absorption measurements of characteristic rotational-vibrational lines and to follow the evolution of vibrational populations (NO and CO) (Hanson et al. 1976). Thus, the non-Boltzmann character of these populations is evidenced and incubation times for high levels may be measured. A comparison of different models including TV and VV transitions is also presented without and with dissociation and popular models of vibration-dissociation coupling are discussed.

Finally, regarding chemical kinetics, some recent results obtained in the free-piston shock tube of Marseille are analyzed. They consist of emission measurements of CN lines behind an intense shock propagating in a $N_2/CH_4/Ar$ mixture (Titan atmosphere).

2. Theoretical models for transport

2.1. Boltzmann equation and Chapman-Enskog method

In a reactive flow, the Boltzmann equation for a component p and/or an internal level i, with i = ir, iv, where ir and iv are respectively rotational and vibrational levels, is symbolically writen:

$$\frac{d\,f_{\rm ip}}{d\,t} = J \tag{1}$$

where J is the collisional balance of the corresponding particles and f is the corresponding distribution function. The CE method consists in expanding f in a series of a small parameter ε and here, in stopping the expansion at the first order, that is $f_{\rm ip} = f^0_{\rm ip}\,(1+\phi_{\rm ip})$ where $f^0_{\rm ip}$ is the zeroth-order term and $\phi_{\rm ip}$ the perturbation brought at the first order. Then the collisions are classified in two types depending on their probability or macroscopically on the characteristic relaxation times, that is $\tau_{\rm I}$ and $\tau_{\rm II}$. Noting $\varepsilon_{\rm I} = \tau_{\rm I}/\vartheta$ and $\varepsilon_{\rm II} = \tau_{\rm II}/\vartheta$, where ϑ represents a characteristic flow time, (1) may be written under the nondimensional form:

$$\frac{d\,f^*_{\rm ip}}{d\,t^*} = \frac{1}{\varepsilon_{\rm I}}J^*_{\rm I} + \frac{1}{\varepsilon_{\rm II}}J^*_{\rm II} \tag{2}$$

In the continuum regime $\varepsilon_{\rm I} = \varepsilon \ll 1$ and represents the "small parameter." As for $\varepsilon_{\rm II}$, three cases are possible, depending on its order of magnitude. The first case, for which $\varepsilon_{\rm I} = 1/\varepsilon \ll 1$, may be left out, since it represents the frozen case, for which the processes of type II have no influence on the first two orders of f and therefore on the description of the flow. The two other cases: $\varepsilon_{\rm II} \ll 1$ and $\varepsilon_{\rm II} \approx 1$, correspond respectively to the so-called Weak Non-Equilibrium (WNE) and Strong Non-Equilibrium (SNE) cases, examined below.

2.2. WNE and SNE cases

In the first case ($\varepsilon_{\rm II} \ll 1$), the expansion allows one to obtain the following system:

$$J^0_{\rm I} + J^0_{\rm II} = 0 \qquad\qquad \frac{d\,f^0_{\rm ip}}{d\,t} = J^1_{\rm I} + J^1_{\rm II} \tag{3}$$

The corresponding zeroth-order solution is therefore an equilibrium solution (Euler level) and the first order one brings a weak nonequilibrium solution (Navier-Stokes level) giving transport terms in which collisions of type I and II are included. Thus, this is called the Weak Non-Equilibrium (WNE) case.

In the second case ($\varepsilon_{\rm II} \approx 1$), the system is written:

$$J^0_{\rm I} = 0 \qquad\qquad \frac{d\,f^0_{\rm ip}}{d\,t} = J^1_{\rm I} + J^0_{\rm II} \tag{4}$$

The zeroth-order solution is therefore a nonequilibrium solution and at the corresponding Euler level, relaxation (or kinetic) equations are present, as well as at the first-order Navier-Stokes level in which the transport terms include only collisions of type I. This is called the Strong Non-Equilibrium case (SNE).

The application of both methods has been made for polyatomic gases or mixtures possessing vibrational modes eventually in nonequilibrium. The extension to the case of chemical reactions is in principle similar and a particular example proposed hereafter allows us to treat the problem of vibration-dissociation interaction.

WNE and SNE cases have been compared in the case of a steady nozzle expansion (Fig. 1). They give different results for the freezing level of the vibrational temperature. This is not surprising because in the SNE case the vibrational nonequilibrium is calculated through relaxation equations and in the WNE case the vibrational energy (or temperature) is calculated like a transport term. Of course, in concrete cases, the flow may be treatable successively by both methods and it is therefore necessary to use a universal method (presented below).

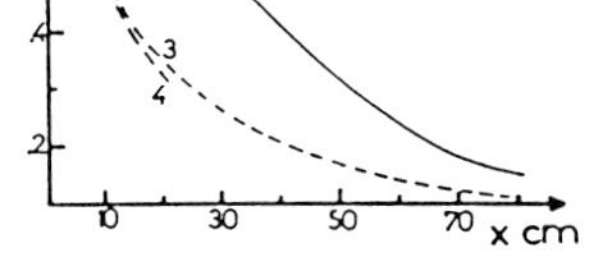

Figure 1. Temperature evolution along a nozzle, N_2, 2500 K; 10^7 Pa; 1: T_V/T_0 (SNE), 2: T_V/T_0 (WNE) 3: T/T_0 (WNE), 4: T/T_0 (SNE). Divergent angle: 4°.

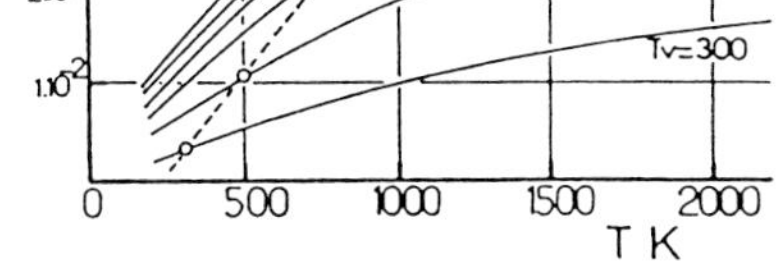

Figure 2. Vibrational conductivity for carbon dioxide; —o— is equilibrium curve.

2.3. Generalized CE method

The matching of WNE and SNE cases is realized by simply adding the term J_{II}^1 in the linearized equation (4), so that the system becomes (Kogan et al. 1979,

Table 1. Comparison of WNE and SNE methods for gases in vibrational nonequilibrium.

Method	WNE	SNE
Zeroth order	• Rot. and vib. modes in equilibrium • No relaxation equations	• Rot. mode in equilibrium vib. mode in nonequilibrium • Vib. relaxation equations
First order	• Navier-Stokes with one temperature T	• Navier-Stokes + vib. relaxation equations
	• Bulk viscosity due to rot. and vib. nonequilibrium	• Bulk viscosity due to rot. nonequilibrium
	• One global conductivity coefficient	• Relaxation pressure
	• Transport terms functions of collision integrals I and II	• Cross terms for each conductivity coefficient
	• T_{trans}, T_{rot}, T_{vib} computed as transport terms	• Transport terms functions of collision integrals I
		• T_{trans}, T_{rot}, computed as transport terms T_{vib} with relaxation eq.

Brun et al. 1984):

$$J_{\text{II}}^1 = 0 \qquad \frac{d f_{\text{ip}}^0}{dt} = J_{\text{I}}^1 + J_{\text{II}}^0 + J_{\text{II}}^1 \tag{5}$$

Of course, solving (5) becomes more complex because of the structure of the operator J_{II}^1. In the same way, approximations of the collision integrals are often necessary. The results concerning the transport terms are formally identical to the SNE case but collisions I and II are present, which allows the matching with the WNE case. The nonequilibrium explicitly appears in these terms but they are not all affected in the same way, the most sensitive being the vibrational conductivity (Fig. 2). Another problem lies in the calculation of inelastic collision integrals; this represents a formidable task but an alternative solution consists in expressing the most affected terms as functions of those which are not too much affected (viscosity, diffusion, etc.) and of global phenomenological relaxation times τ_{R}, τ_{V}, τ_{VV}, τ_{TV}, which are, in principle, easy to determine.

2.4. Vibration-dissociation coupling

The CE method may also be successfully used for the description of nonequilibrium flows behind shocks involving vibrational relaxation and dissociation kinetics simultaneously.

In a wide range of temperatures, the WNE method may be applied because of the different orders of magnitude of the vibration and dissociation times. Thus, at the zeroth order, the dissociative molecules remain in an equilibrium distribution, but the computation of the first-order distribution function leads to an explicit expression of the dissociation rate constant K_f depending on both temperature translation-rotation T and vibration T_v (Méolans et al. 1995). In the same way, the vibrational nonequilibrium depends on K_f. An example of the computation of K_f is presented in Fig. 3 for nitrogen and compared with well-known correlations of experimental results (Park 1993).

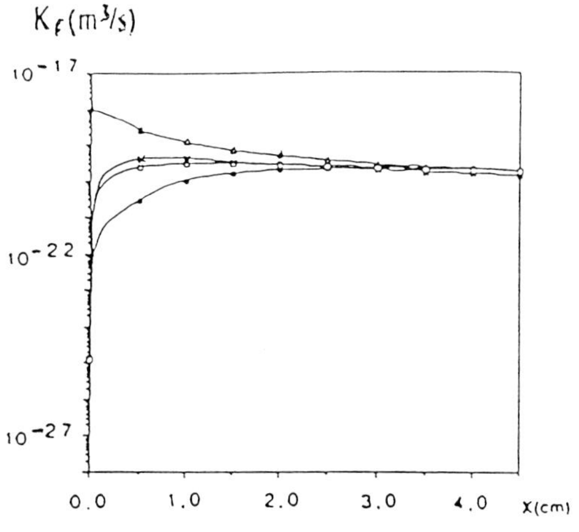

Figure 3. Nitrogen dissociation rate constant evolution behind a strong shock wave: $M = 19$, 235 K, 10 Pa. $\triangle$: K_f (zeroth order); $\times$: K_f (first order); $\bullet$: Park (first correlation); $\square$: Park (second correlation).

3. Experimental data on relaxation and kinetics

3.1. Vibrational populations

The development of laser diagnostic techniques, particularly in the IR domain, allows one to carry out absorption measurements of individual RV lines and therefore to determine the evolution of individual vibrational populations for polar gases (NO,CO...), (Hanson et al. 1976). In spite of difficulties in the interpretation of results, the example shown in Fig. 4 gives a proof of the non-

Boltzmann behavior of these populations in the relaxation zone (Méolans & Brun 1986): the importance of the VV transitions reducing the TV transfers and the existence of an incubation time specific to high vibrational levels are also evidenced. These results contribute to questioning the validity of LT-type equations, as was already the case for the nozzle expansions and the low-temperature regimes. The extension to higher temperatures for which multiple quantum transitions are possible contributes also to the elaboration of new models (Adamovich et al. 1995).

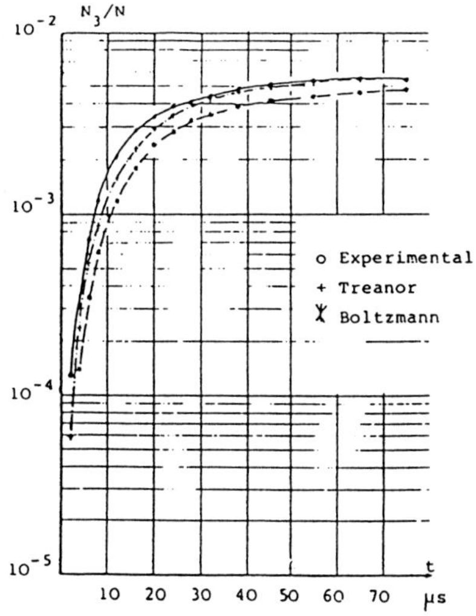

Figure 4. Relative population of the 3rd vibrational level: evolution behind a shock in CO. $M_S = 5.6$. Initial pressure 2.0×10^3 Pa. Initial temperature 291 K.

In the same way, the influence of the dissociation being important at high temperatures, the vibration-dissociation coupling influences the vibrational populations themselves. Thus, a complete analysis of nonequilibrium phenomena is presented with O_2 as a test gas. First, the relaxation is described by the master equations of each level coupled with conservation equations. The transition probabilities appearing in the relaxation equations are computed taking into account the anharmonicity and the potential angular dependence (Méolans et al. 1995). The bound-free transition probabilities of Marrone & Treanor (1963) (MT model) are used in the master equations. The population evolution is not monotonic and a time lag appears, as shown experimentally, and for high levels, the influence of the anharmonicity and VV processes is significant and strongly related (Fig. 5); these VV processes reduce the effects of the TV transfer. Finally, it is shown that the value of the "characteristic probability" U of the MT model has no great importance on the evolution of the level populations.

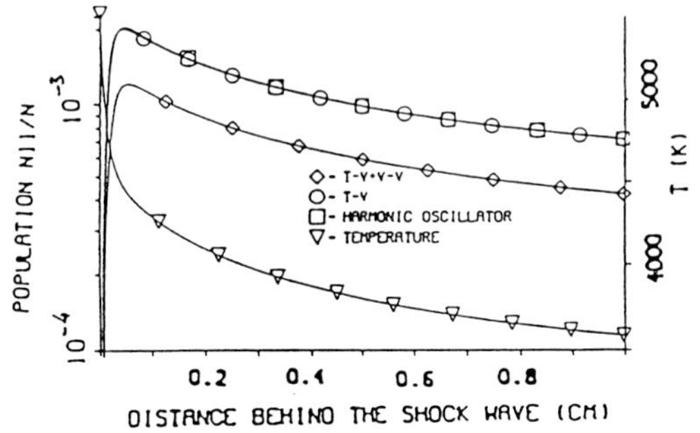

Figure 5. Relative population of the 11th level as a function of the distance from a shock in O_2; 4.5×10^4 Pa, 5525 K, 550 m/s.

However it is important to know that a "universal" curve $U(T)$ may be obtained from several runs when excluding absolutely the strong nonequilibrium part of each run and keeping only the common part of each curve. In Fig. 6, one sees that U may considerably vary around the value T_{diss} for the considered temperature range. Experiments with other gases than O_2 are in progress.

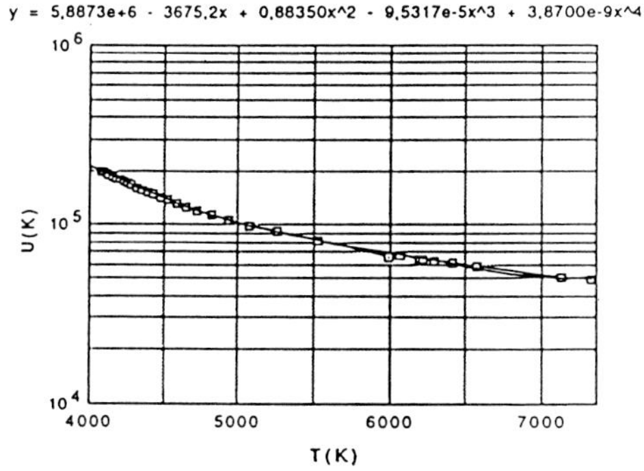

Figure 6. Intrinsic curve $U_{O_2 \to O_2}$ versus the temperature T.

3.2. Chemical kinetics: an example

Large varieties of physico-chemical processes behind shock waves have been experimentally examined, but just one single example will be given here. It concerns one of the last processes studied in the Marseille TCM2 shock tube at large Mach numbers (Dumitrescu 1995).

CN spontaneous emission is recorded behind a shock wave propagating along the shock tube with a velocity of about 6 km/s, in a reconstituted Titan atmosphere $N_2/CH_4/Ar$. At each run, time-resolved emission intensity measurements of three lines belonging to the (0,0) and (1,1) band system of CN violet spectrum (electronic transition) are carried out with three photomultiplier tubes. The results are compared to theoretical calculations and to previous measurements (Park & Bershader 1992), including a systematic analysis of the influence of argon concentration. From emission recordings, the evolution of rotational and vibrational temperatures may also be obtained (Fig. 7). Presently, these measurements are extended to CO_2/N_2 mixtures by recording CN and C_2 emission lines.

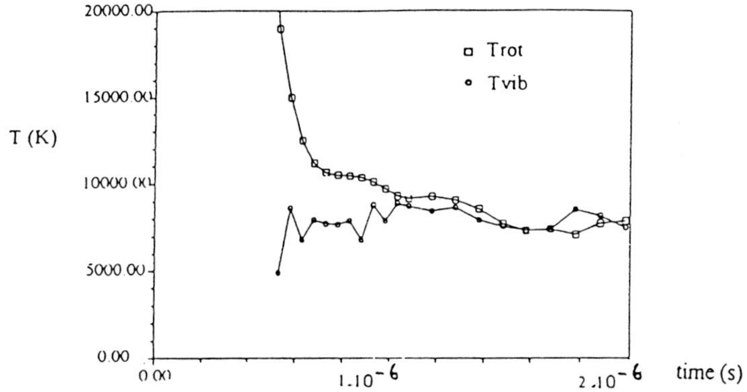

Figure 7. Rotational and vibrational temperatures determined from emission intensities behind a shock in the TCM2 shock tube, $N_2/CH_4/Ar$, 2×10^{-3} bar, $M_S = 18$.

4. Summary and conclusions

An analysis of different possible models of nonequilibrium flows behind shock waves is presented which leads to Navier-Stokes systems differing by their transport and relaxation terms or equations. However, a general model is proposed which avoids misunderstandings and covers all nonequilibrium situations. In the same way, vibrational relaxation models and corresponding experimental results are examined concerning particularly the evolution of individual vibrational populations: the influence of different types of transitions is analyzed and new interpretations and models for the vibration-dissociation coupling are also presented.

For the future, the development of computational techniques will certainly provide quantitative data on the inelastic collision frequencies necessary for precise computation of transport coefficients. The planned use of sophisticated experimental techniques — CCD cameras are a good example — will also provide considerable extension and convenience for recording time-dependent spectra.

Acknowledgement. The author is greatly indebted to his coworkers, particularly to Dr. J. G. Méolans.

References

Adamovich IV, Macheret SO, Rich JW, Treanor CE (1995) Vibrational relaxation and dissociation behind shock waves. AIAA J 33(6):1064.

Brun R (1988) Transport Properties in Reactive Gas Flows, AIAA Paper 88-2655.

Brun R, Villa MP, Méolans SG (1984) Generalized transport terms in vibrationally relaxing flows. In: 14th RGD Symp, H Oguchi (ed), Tokyo.

Dumitrescu MP (1995) Operation and development of the TCM2 free piston hypersonic facility in Marseille. In: 2nd Europ Symp Aerothermod, ESTEC, Noordwijk ESA S P 367:341.

Hanson RK, Monat JP, Kruger CH (1976) Absorption of CO laser radiation by NO. Jour Quant Spectros Rad Transf 16:705.

Kogan MN, Galkin VS, Makashev NK (1979) Generalized Chapman-Enskog method: derivation of the nonequilibrium gasdynamic equations. In: 11th RGD Symp, R Campargue (ed), Paris.

Marrone PV, Treanor CE (1963) Chemical relaxation with preferential dissociation from excited vibrational levels. Phys Fluids 6:1215.

Mason EA, Monchick L (1962) Heat conductivity of polyatomic and polar gases. J Chem Phys 36(6):1622.

Méolans JG, Brun R (1986) Nonequilibrium vibrational population behind shock waves: comparison theory-experiments. In: 15th RGD Symp, V Boffi, C Cercignani (eds), Teubner, Stuttgart.

Méolans JG, Brun R, Mouti M, Llorca M, Chauvin A (1995) Vibration-dissociation coupling in high temperature nonequilibrium flows In: 2nd Europ Symp Aerothermod, ESTEC, Noordwijk ESA SP 367:293.

Park C (1993) Review of chemical-kinetic problems of future NASA mission I: Earth entries. J Therm and Heat Transf 7(3):385.

Park CS, Bershader D (1992) Studies of radiative emission from the simulated shock layer of the Huygens probe. In: Shock Waves, K Takayama (ed), Sendai, p 671.

Pascal S, Brun R (1993) Transport properties of nonequilibrium gas mixtures. Phys Review 47(5):3251.

Tuned operation of a free-piston shock tunnel

K. Itoh
National Aerospace Laboratory
Kakuda Research Center
1 Koganesawa, Kimigaya, Kakuda, Miyagi 981-15 Japan

Abstract: The National Aerospace Laboratory (NAL) has been planning to build a large free-piston shock tunnel, named HIEST, for the reentry aerothermodynamic tests of HOPE. Recently, the HIEST project proceeded to the construction phase, which will be completed in 1997. To support this project NAL has done various experimental and theoretical studies and applied the results to the basic design. In this paper, an experimental study with the medium-sized free-piston shock tunnel HEK in terms of tuned operation is presented. The detailed piston-landing motion was measured with a simple method and it was verified that tuned operation was successfully achieved. The effects of high-efficiency driving on facility performance were also studied. The results suggest that the use of a light piston is expected to be effective to achieve tuned operation when aiming for a more powerful shock-tube driver.

Key words: Free-piston shock tunnel, Tuned operation, Free-piston driver

1. Introduction

A free-piston shock tunnel has great capability to produce high-enthalpy flow, by applying adiabatic compression with the free piston to increase driver-gas temperature and pressure. However, since the driver gas is highly compressed to a very small volume, it is difficult to maintain a useful duration time for a test.

To overcome this essential problem, the concept of high-speed piston operation was given by Stalker (1967) and Hornung & Belanger (1990). According to the concept, a piston is operated to travel at sufficiently high speed even at diaphragm rupture so that the driver-gas pressure can be held nearly constant for a certain duration. Thus, the size of the facility can be drastically reduced. Here, the driver based on this concept is referred to as a high-efficiency driver.

However, because the required piston speed is very high, it is essential to achieve the soft landing operation of the driver, or tuned operation, to avoid damaging the facility. In order to find an effective way to achieve tuned operation, a tuning principle was formulated by Itoh (1994) based on the eigen-value theory. In the present study, an attempt was made at achieving tuned operation with the medium-sized free-piston shock tunnel HEK by following this tuned-operation formula. The detailed piston landing motion was measured with a simple method and it was verified that tuned operation was successfully achieved. The effect of high-efficiency driving on facility performance was also studied. The results suggest that the use of a light piston is expected to be effective to achieve tuned operation when aiming for a more powerful shock-tube driver.

2. Tuned operation of free-piston shock tunnel

During a piston trajectory after diaphragm rupture, there exists a point where the pressures ahead of and behind the piston are balanced and thus minimum piston velocity occurs. This interesting piston motion can be characterized by $\omega \equiv \sqrt{c(\gamma)P_0V_0/W_p(A^*/A)^2a_0^2}$, which acts as a deceleration factor. Here, W_p is piston mass, P_0, V_0, a_0 are initial pressure, volume, and sound speed of the driver gas at piston launch. The quantity A^*/A is the area ratio of the exit to the main part of the compression tube.

When ω is too large, the piston is too much decelerated and rebounds before impact. In contrast, when ω is too small, the piston can not be decelerated enough and impacts directly. Between them, there is good value for which the piston decelerates and comes to rest at the point where the pressures ahead of and behind the piston are just balanced. Then, a soft landing is possible by adjusting the piston stopper height to this point. Therefore, the tuned-operation problem can be recognized as the eigen-value problem to find ω which satisfies $u = 0$ at $du/dt = 0$. In the previous study (Itoh 1994), the formula which represents the relation of ω and β satisfying this condition was derived; β is defined as the ratio of piston speed at diaphragm rupture to reference speed needed to hold the driver-gas pressure constant, thus the high-efficiency driver is characterized by $\beta > 1$.

The relation of β and ω representing tuned operation is shown in Fig. 1. It should be remarked that a sufficiently large ω is needed to achieve tuned operation, and β should be greater than unity.

For given facility and major test conditions, there are various possibilities to achieve a sufficiently large ω, i.e., use of a lower sound-speed driver gas or a smaller area ratio. However, both of these methods cause a reduction in performance. Therefore, we proposed use of a light piston to achieve a more powerful shock-tube driver; and this concept was incorporated into the basic design of HIEST. However, a light piston has difficulty maintaining enough strength against high pressures. Thus, the mechanical design of the piston becomes critical.

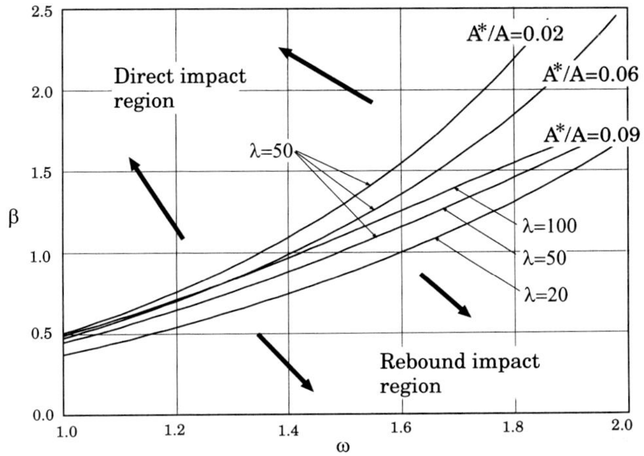

Figure 1. Eigen values for tuned operation.

3. HEK experiments

The above-mentioned theoretical prediction for achieving tuned operation was examined by experiments with the medium-sized free-piston shock tunnel HEK, of which the compression tube is 16 m long and 210 mm in diameter; the shock tube is 6.5 m long and 72 mm in diameter; the orifice between the compression tube and the shock tube is 50 mm to 64 mm in diameter; and the piston mass is 30 kg. The purposes of the HEK experiments were to attempt to carry out tuned operation and to investigate the effect of tuned operation on facility performance.

3.1. Piston-motion measurement system

The unique feature of the HEK is the simple measurement system for detailed piston motion. On the piston surface, black stripes are made with fine pitch using an anode oxidation method. When black stripe comes in contact with a laser beam, the reflection signal of the laser beam disappears. By using this method, detailed piston motion can be measured with a few piston sensors placed in the high-pressure segment of the compression tube. The piston is made of Duralmin and consists of an 240-mm long main body and an 180-mm long skirt which acts as the valve for reservoir gas. The width of the stripes and intervals are 6 mm on the main body and 10 mm on the skirt. The o-rings on the skirt were also detected by the photo sensors.

3.2. Operating conditions

The operating conditions in shots 10–20 (excluding shot 14 in which the data sampling failed) are summarized in Table 1. In all shots, the diaphragm-rupture

Table 1. Operating conditions.

Shot #	Orifice	P_0 (kPa)	P_{R_0} (MPa)
10	$\phi 50$	108.0	5.216
11	$\phi 50$	108.9	5.100
12	$\phi 50$	87.8	4.262
13	$\phi 50$	98.5	4.706
15	$\phi 54$	132.5	6.240
16	$\phi 52$	113.0	5.460
17	$\phi 52$	94.5	4.420
18,19	$\phi 52$	113.0	4.888
20	$\phi 52$	94.5	3.954

pressures were uniformly 52 MPa, and the same piston was used. Therefore, ω and the compression ratio are no longer independent and commonly changed by changing the initial pressure of the driver gas. According to the tuned-operation formula, the β values were decided with respect to ω. Then, the reservoir pressures were adjusted to obtain the intended β values.

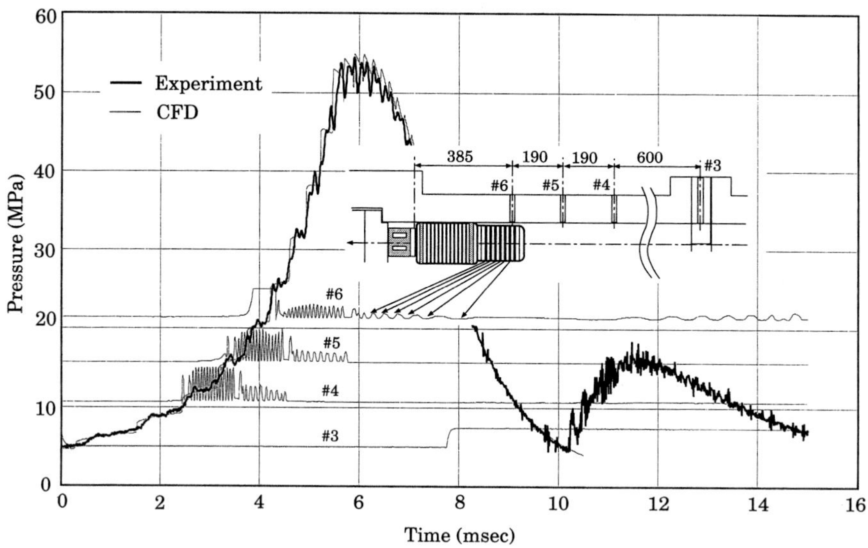

Figure 2. Compression tube pressure and piston motion in shot 16.

3.3. Characteristics of compression-tube pressure

Figure 2 shows an example of the compression-tube pressure as predicted by CFD and as signals detected by the piston sensors obtained in shot 16. As seen in the figure, the detailed piston motion could be successfully detected. The signal from sensor No. 3 was used as a starting point in this measurement. The signal from No. 6 showed that the piston was rapidly decelerated just before landing and almost came to rest at landing. When the reflected shock arrived, the piston was pushed back. In this case, the measured β value was 1.23 so that the driver gas pressure was held nearly constant for about 0.8 msec. Figure 3

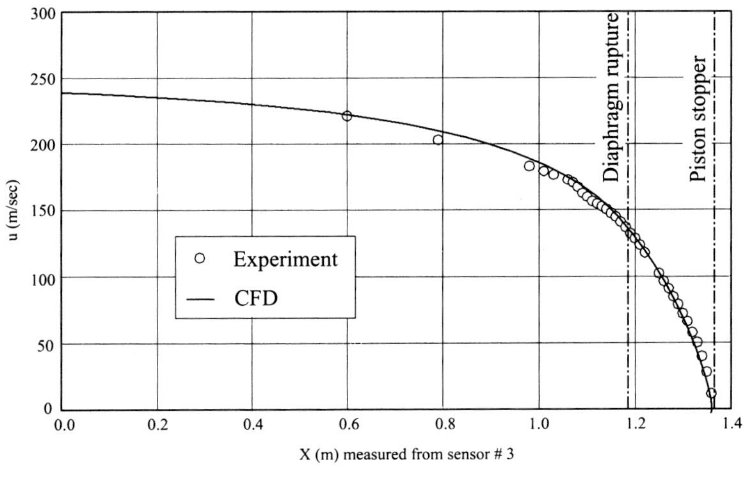

Figure 3. Piston velocity in shot 16.

shows the piston velocity calculated from piston trajectory versus the distance from sensor No. 3. It is seen that the operation was very well tuned so that piston came to rest just at landing.

The characteristics of the compression tube were estimated by the piston-motion measurements and compression-tube pressures and they are summarized in Table 2 and also represented in the $\omega - \beta$ operating map in Fig. 4.

According to the tuned-operation formula, the goal was to carry out tuned operation in all shots except 13, 18, and 20. In shots 13, 18, and 20, the intention was to try to slightly rebound the piston. The results are quite consistent with the theoretical prediction, and it was verified that tuned operation could be successfuly achieved by using the tuned-operation formula.

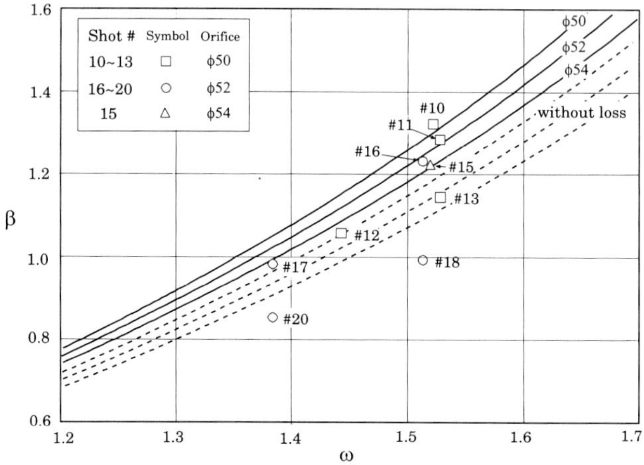

Figure 4. The $\omega - \beta$ operating map.

Table 2. Characteristics of compression tube.

Shot #	Orifice	λ	u_r (m/sec)	ω	β	τ (msec)
10	ϕ50	46.2	135.6	1.522	1.324	1.30
11	ϕ50	46.0	131.4	1.528	1.285	1.20
12	ϕ50	52.2	113.1	1.443	1.058	0.27
13	ϕ50	50.2	117.9	1.528	1.146	0.54
15	ϕ54	40.8	140.6	1.520	1.225	0.84
16	ϕ52	45.6	134.3	1.513	1.232	0.76
17	ϕ52	51.1	110.6	1.384	0.982	–
18	ϕ52	45.8	108.0	1.513	0.993	–
20	ϕ52	54.4	93.0	1.384	0.853	–

3.4. Effect of high-efficiency driving on facility performance

Next, the effect of high-efficiency driving on facility performance was studied. As indicated in Table 2, the useful duration in which the pressure is held nearly constant can be obtained when β is greater than unity; and, when β is less than unity there is no useful duration. This difference in driver-gas condition clearly appeared in shock-tube performance. As shown in Fig. 5, decreasing β led to the pressure level going down; moreover, the pressure profile deviated from the plateau one. The effect of changing orifice size was also investigated with the aim of obtaining a more powerful shock-tube driver. As shown in Fig. 6, the pressure level is clearly increased when orifice size is increased. Both β and orifice-size effects are summarized in terms of the pressure recovery factor in Fig. 7. In order to increase β and orifice size, ω had to be increased by decreasing

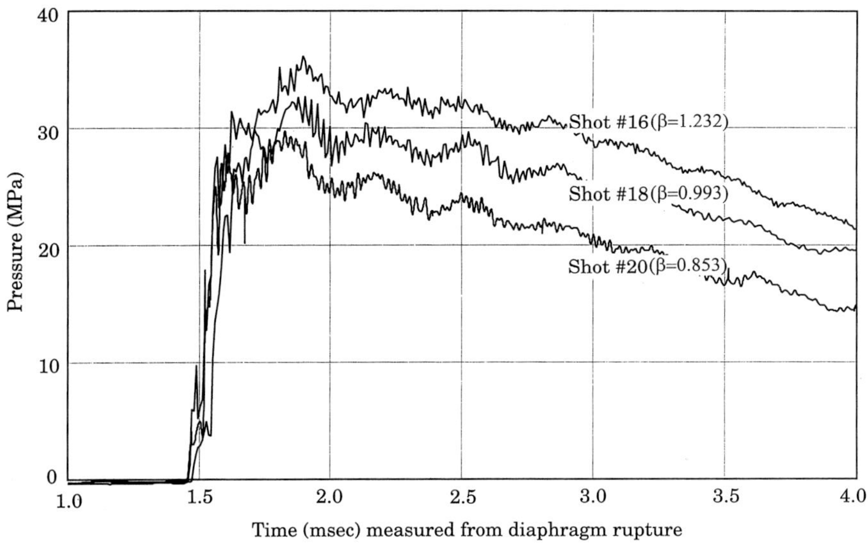

Figure 5. Effect of β on shock-tube pressure (end wall).

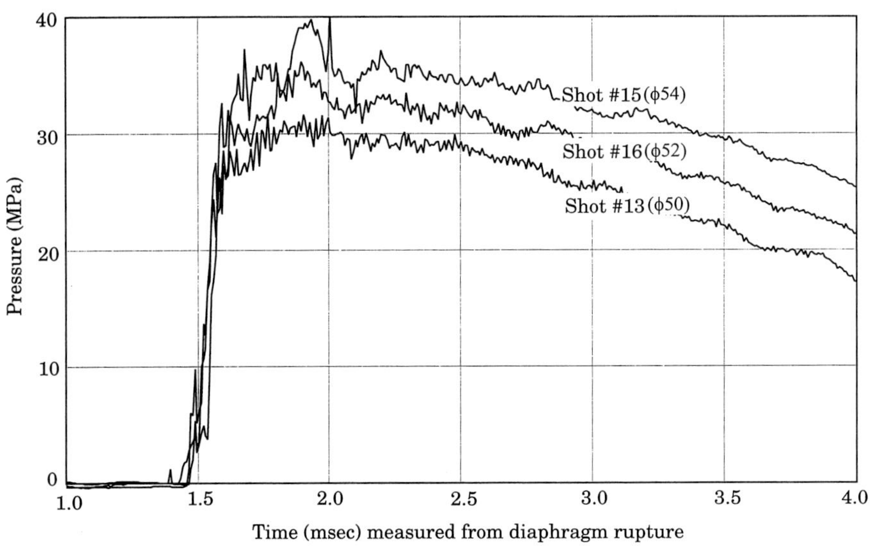

Figure 6. Effect of orifice size on shock-tube pressure (end wall).

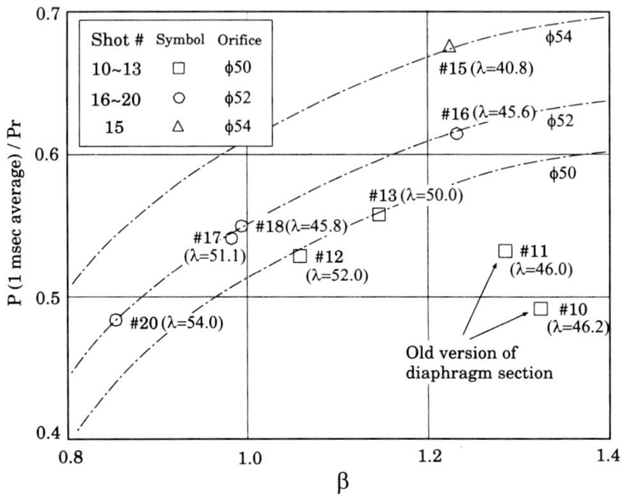

Figure 7. Effect of β and orifice size on pressure recovery factor.

the compression ratio, as indicated by the values in Fig. 7, because the piston mass and diaphragm-rupture pressure were fixed in this series. Therefore, if in practice the piston mass is reduced, even greater effect on ω is expected to be obtained.

4. Conclusions

In the present study, an attempt was made at achieving tuned operation with the medium-sized free-piston shock tunnel HEK by following the tuned-operation formula. Conclusions are as follows:

- Tuned operation was successfully achieved by following the tuned operation formula.
- Effects of high-efficiency driving on facility performance were verified.
- Use of a light piston is expected to be a more effective way to get a more powerful shock-tube driver.

References

Eitelberg G (1994) First results of calibration and use of the HEG, AIAA Paper 94-2525.

Hornung HG (1988) The piston motion in a free-piston driver for shock tubes and tunnels, GALCIT FM88-1, California Institute of Technology.
Hornung HG, Belanger J (1990) Role and techniques of ground testing for simulation of flows up to orbital speed, AIAA Paper 90-1377.
Itoh K (1994) Tuned operation theory for free-piston driver of high-enthalpy shock tunnel. In: Symposium on Shock Waves Japan '94, pp 29–32.
Stalker RJ (1967) A study of the free-piston shock tunnel. AIAA J 5:2160–2165.

High-resolution imaging of hypervelocity metal jets using advanced high-speed photographic techniques

L. L. Shaw, S. A. Muelder
Lawrence Livermore National Laboratory
Box 808, L-281, Livermore, California 94550 USA

Abstract: It is now possible to obtain high-resolution sequential photographs of the initial formation and evolution of hypervelocity metal jets formed by shaped-charge devices fired in air. Researchers have been frustrated by the high velocity of the jet material and the luminous sheath of hot gases cloaking the jet that made detailed observation of the jet body extremely difficult. The camera system that provides the photographs is a large-format multi-frame electro-optic camera, referred to as an IC camera (IC stands for image converter), that utilizes electro-optic shuttering, monochromatic pulsed-laser illumination, and band-pass filtering to provide sequential pictures (in 3D if desired) with minimal degradation due to luminous air shocks or motion blur. The large format (75-mm image plane), short exposure (15 ns minimum), ruby-laser illumination, and band-pass filtering (monochromatic illumination while excluding extraneous light) produces clear, sharp, images of the detailed surface structure of a metal shaped-charge jet during early jet formation, elongation of the jet body, jet-tip evolution, and subsequent particulation (breakup) of the jet body. By utilizing the new camera system in conjunction with the more traditional rotating-mirror high-speed cameras, pulsed radiography, and electrical sensors, a maximum amount of often unique data can be extracted from a single experiment. This paper was intended primarily as an oral presentation. For purposes of continuity and simplicity in these proceedings, we have chosen to concentrate on the development of the IC-camera system and its impact on the photography of high-speed shaped-charge jets.

Key words: Image-converter camera, Laser-illuminated photography, Electro-optic shutter, Multi-pulsed ruby laser, High-speed photography, Shaped-charge jet

1. Introduction

High-speed photography is a major diagnostic tool in the study of fast phenomena such as the acceleration and deformation of high-explosive driven metal surfaces. However, multiple-frame high-speed photography has been of limited benefit in the study of the jets formed by shaped-charge devices fired in air. One problem is poor spatial resolution due to the motion blur caused by the movement of the jet material during the exposure time of the camera frame. A second problem is the shock-induced luminosity cloaking the tip and body of the jet that blocks direct observation of the jet surface. In order to minimize the effect of object motion on picture resolution, exposure times as short as 10^{-8} sec must be possible. To photograph the surface of the jet, the camera must "see through" the shock-induced luminosity and exclude it from the camera while at the same time photographing an adequately illuminated jet surface. The use of the recently developed multi-frame electro-optic shuttered image-converter camera (IC camera) with pulsed-ruby laser illumination meets the performance parameters just stated and provides the means to obtain high-resolution sequential photographs of these jets. A nominal 20-nanosecond exposure time greatly reduces the effect of material motion, even for velocities in the range of 10 kilometers per second. The monochromatic 694-nanometer pulsed-ruby laser illumination with 1.5-nanometer band-pass filtering allows the self-luminosity and other extraneous light to be excluded from the photographs. With this system, sequential pictures of the surface of a metal shaped charge have been obtained that show the detailed metallurgy as the jet proceeds from initial formation through particulation.

2. IC-camera system

The IC-camera system described in this paper is a direct outgrowth of the development work in proximity focused image-converter tubes and in pulsed solid-state lasers that has occurred over the past 40 years. Experimenters and diagnosticians realized that the ultra-short exposure time possible using the image-converter tube as a shutter permitted the photographing of fast phenomena minimizing motion blur, and a number of framing cameras utilizing these tubes were built (Chancel and Pelte 1972, Owren 1959). During the same time period, experimenters realized that the exposure time of mechanical cameras could be dramatically reduced by using pulses from Q-switched or cavity-dumped lasers as light sources. There were also two other desirable attributes of using lasers: extraneous light could be excluded from the camera by the use of band-pass filtering that transmits only the wavelength of the laser light, and the technique proved very adaptable for stereo photography (Carpluk 1970, Paisley 1976). Over the past fifteen years, we have combined the best attributes of the electro-optic framing camera and pulsed-laser illumination into a high-speed framing-photography system (Shaw et al. 1984).

2.1. IC camera

The IC camera has a modular design as shown in Fig. 1. Each module is a complete two-frame camera with its own objective optics and mechanical shutter. An internal 50% mirror creates two image planes from a single entrance optic. Individually triggered 75-mm diameter proximity focused image-converter tubes are positioned at each image plane; these have adjustable exposure times with a minimum of 12 to 15 ns. A detailed description of the camera timing

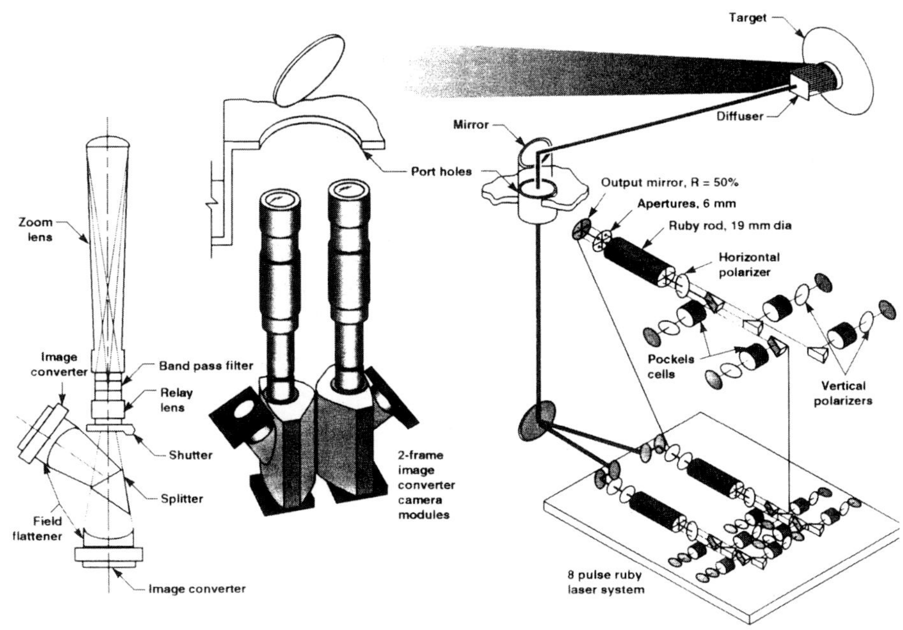

Figure 1. Diagram of the eight-frame IC (image-converter) camera and the eight-pulse ruby-laser illuminator. Note the modular camera construction with two frames per objective optic. The laser is configured to provide four independent pulses from a single rod. Both the camera and the laser are located in a concrete reinforced underground bunker below and to the side of the experimental firing area.

and control system is provided in Shaw et al. (1984, 1993). The optical system for each module consists of a 200-mm to 1140-mm zoom objective lens with a 35-mm image format. A five-to-one magnifying relay lens transmits and enlarges a portion of the image onto the two 75-mm image-converter tubes. The original configuration of eight frames is made up of four modules designed to be clustered as a unit on a single hydraulic stand, but it is often more convenient to use just two modules per stand as shown in Fig. 2. The modular construction and this type of mounting introduce between one and two degrees of parallax between modules. This is acceptable for monocular photography, and proves very useful and quite adequate for stereo photography.

The film used in this camera is Kodak T-MAX 3200 4×5-inch sheet film. The speed, resolution, and low fog level of this film have been major factors in the high quality of the photographs taken with this system. The film is routinely developed at ISO 6400 and quite often pushed to ISO 12800. This allows significant leeway in the choice of illumination light levels and in the adjustment of the aperture to manipulate film exposure and the camera depth of field. Extraneous light from air shocks, explosive products, and other sources is prevented from entering the camera by a 1.5-nm band-pass filter centered at 694 nm (ruby-laser wavelength) and positioned at the output of the objective optics.

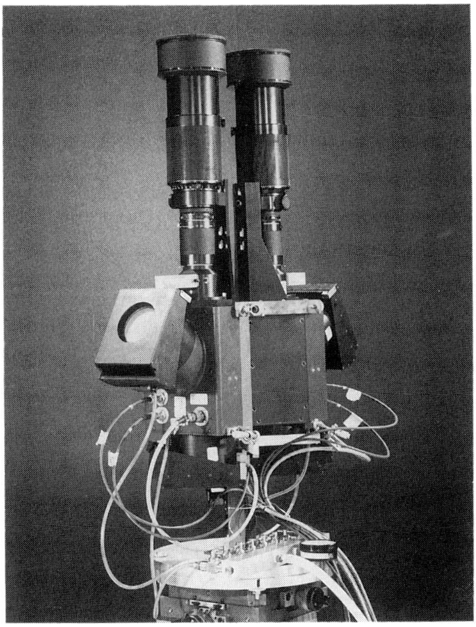

Figure 2. A pair of IC-camera modules mounted on a vertical hoist in the firing bunker camera room; the modules view the experimental firing table through a 30-cm glass porthole in the camera-room ceiling. The 200–1140-mm focal length zoom lens on the camera and the magnification of five by the internal relay lens yield an image on film up to nearly full size at an object distance of 10 meters. Object details as small as 100 microns can be resolved in the resultant photographs.

2.2. Eight-pulse ruby-laser illuminator

The eight-pulse Q-switched ruby laser has a novel design that permits the extraction of four separate pulses from a single ruby rod during a single pumping of the rod. This is illustrated in cartoon fashion in Fig. 1; and the equipment is shown in Fig. 3. A mask with four 6-mm diameter holes allows each quadrant

Figure 3. Shown is the eight-pulse ruby-laser system using two ruby rods located in the flashlamp heads seen at the right center of the photograph. Also clearly visible at the lower left are the four white Pockels cells used to Q-switch the four cavities of the nearer ruby rod. The beams from the two ruby rods are brought together and guided to the firing table through the vertical tube just visible at the top of the photograph.

of a 19-mm diameter rod to be used in a separate laser cavity. The nominal energy of each pulse is 0.5 joules and is approximately Gaussian with a full width at half maximum of 50 ns. Although this design was developed independently, we quickly discovered the approach had previously been developed by others (Jones 1974). Figure 4 is a graphic example of the IC camera's ability to produce detailed photographs in the presence of intense ambient light by using monochromatic laser illumination coupled with a band-pass filter.

The timing and control system, not addressed in this paper, consists of programmed delays and triggers that initiate the individual image-converter tube drivers and the laser Pockels-cell drivers. The jitter of this system is less than 5 ns allowing the 15–20-ns exposure time of each frame to be well centered in the 50-ns laser-illumination pulse reflected from the experiment. Transient digitizers measure laser-to-camera simultaneity and reference timing to the main experimental timing.

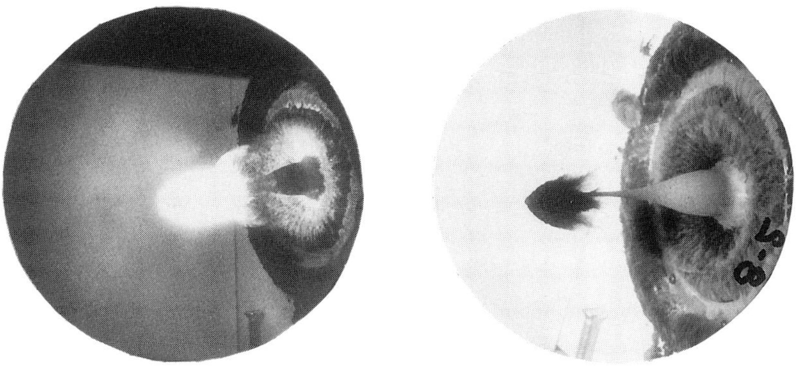

(a) (b)

Figure 4. IC-camera photographs of a jet from a shaped charge with a hemispherical liner. The two pictures were taken 200 ns apart on the same experiment using two different camera modules — one module without the 1.5-nm band-pass filter, and one with it. The exposure time for both photographs was 15 ns. In (a), the only illumination is from the luminous air shock surrounding the jet. In (b), the illumination is from the pulsed ruby laser where the band-pass filter excludes the luminous air shock allowing observation of the underlying metal structure.

3. Experiments

The illustrations used in this paper are of experiments evaluating the performance of explosive shaped charges. Figure 5a is a cartoon of a representative experimental layout. Figure 5b is a cross-sectional view of a typical shaped charge. In diagnosing experiments, we place many diverse diagnostics on the same experiment. These include low-energy pulsed radiography, electrical signal pins, and rotating-mirror framing and streak cameras. These additional diagnostics have been excluded from this paper but are discussed in some detail in Shaw et al. (1979, 1993). The IC camera and laser are located in an underground bunker adjacent to and below the level of the gravel test area known as the firing table. Front surface mirrors direct the incident laser beam to the experiment and the reflected laser light to the camera. The laser pulses strike a ground-glass diffuser placed close to the experiment that scatters the laser light over the experiment. Matte white poster board placed around the experiment provides a light background and helps to "splash" the laser light onto the target.

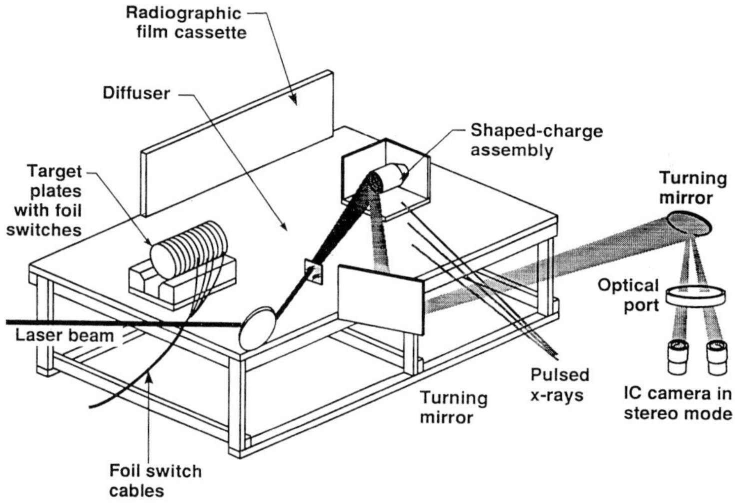

(a)

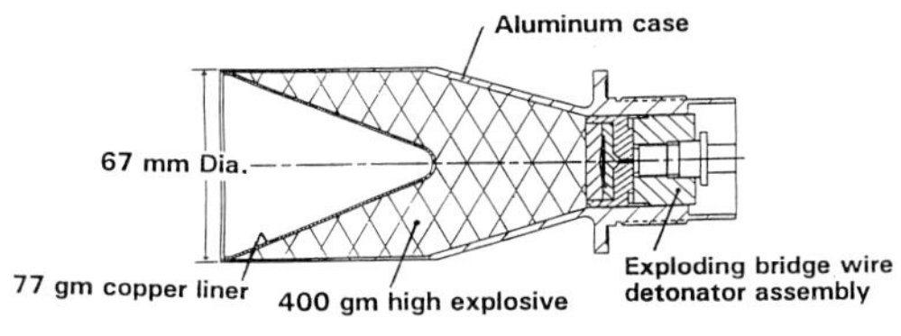

(b)

Figure 5. (a) Diagram of a typical test setup used for shaped-charge experiments. For simplicity, only one IC-camera station is shown. Not shown are some of the other diagnostics often fielded in these experiments such as rotating-mirror framing cameras. (b) Section view of the "Viper" shaped charge used in some of these experiments. The electrically initiated exploding bridge wire on the right end launches a detonation wave through the high explosive. The detonation causes the copper liner to collapse, turn inside out, and "squirt" to the left at velocities approaching 10 km/sec.

4. Results

In the oral presentation, we discussed photographing the entire evolution of the shaped-charge hypervelocity jet. In this paper, we have concentrated on the

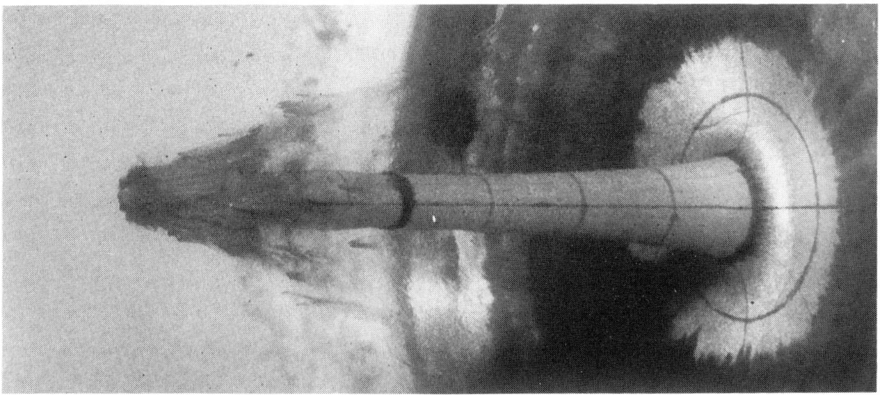

(a)

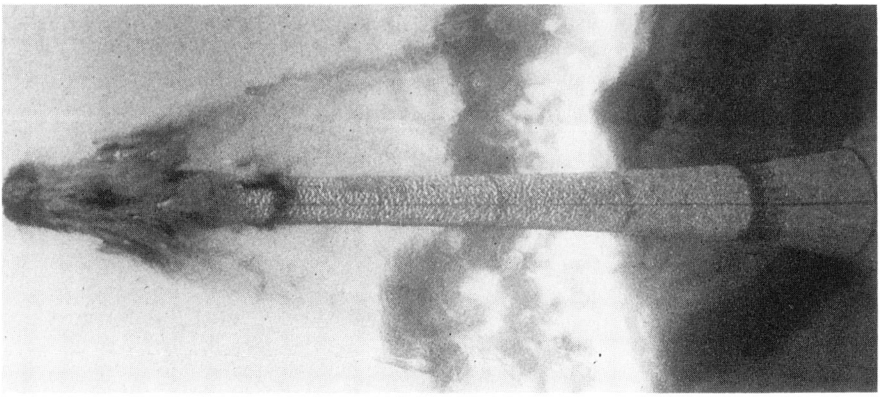

(b)

Figure 6. Laser-illuminated photographs of a shaped charge with a conical liner. Grid lines are made with a fine felt-tipped permanent marker on bare metal. The lines thicken and stretch with the liner deformation providing valuable quantitative strain and timing data. (a) 15-ns exposure taken 40.8 μs after shaped-charge detonator initiation. The jet diameter behind the tip is approximately 10 mm. (b) Same exposure taken 6.7 μs later. Note the ablated copper and the turbulent shock waves in the air behind the tip, and the texture change of the jet surface as it begins to stretch. The jet diameter is now about 6 mm.

high-resolution detail the IC camera is able to record as the metal surface of the jet stretches and cracks during initial elongation and then pulls apart into what are eventually discrete pieces as the jet particulates. In Figs. 6, 7, 10, and

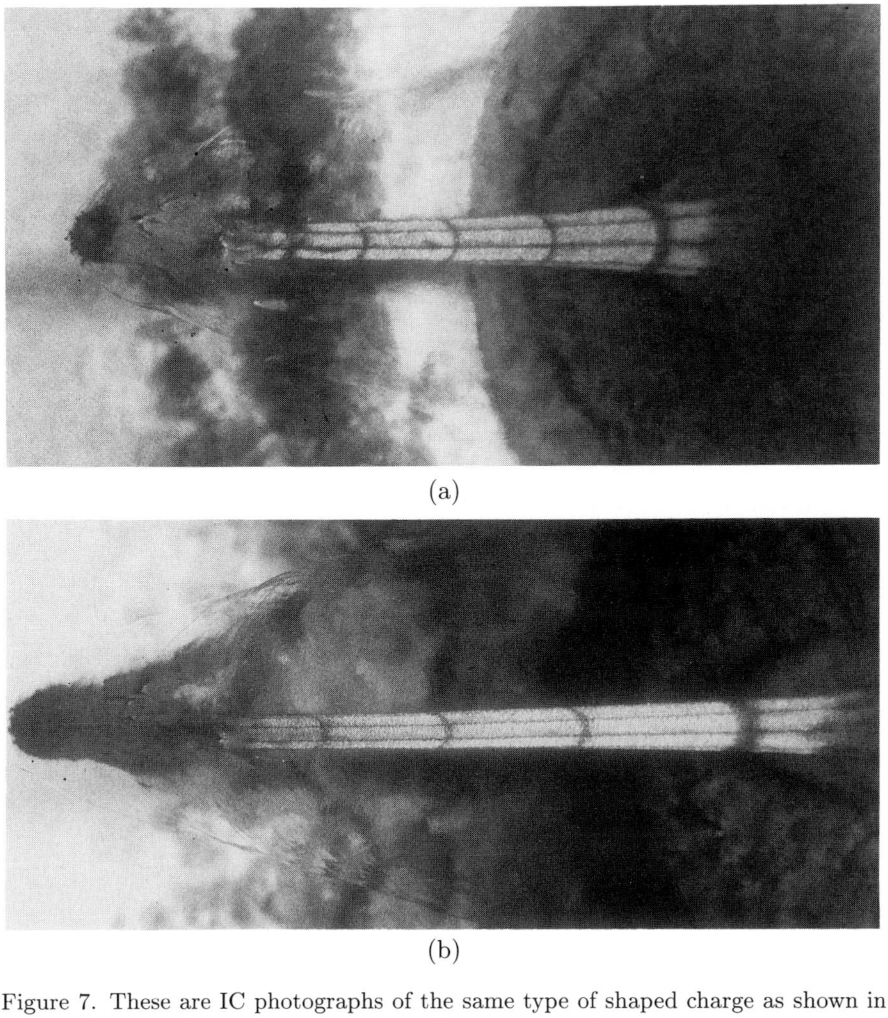

(a)

(b)

Figure 7. These are IC photographs of the same type of shaped charge as shown in Fig. 6. The two photographs were taken 10.0 μs apart. The jets tips extend out from the base of the conical liner by approximately 150 mm (top) and 235 mm (bottom). A total of eight photographs were taken of this experiment, and were used to verify that the fabrication technique used for this liner imparts a spin to the resultant jet. An examination of the ink lines along the jet axis show the rotation as the liner collapses and the jet forms.

11 we have chosen to show IC-camera pictures of jets which, while illustrative of the ability to record overall jet characteristics with unique tip structure and air-shock formations, were chosen primarily to illustrate the surface characteristics of the elongating jet.

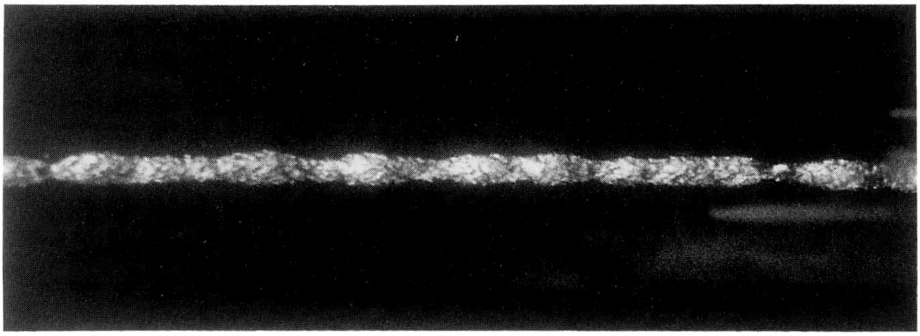

Figure 8. Main body of a jet 320 mm behind the tip. The jet has traveled 870 mm from the exit orifice of the shaped charge. The jet tip has a higher velocity than the tail causing the jet to stretch and thin during flight. This jet had a nominal diameter of 1.4 mm at the time of this photograph, and is beginning to show signs of breaking into separate particles. Note the fine structure on the jet surface including sections that appear cross-hatched.

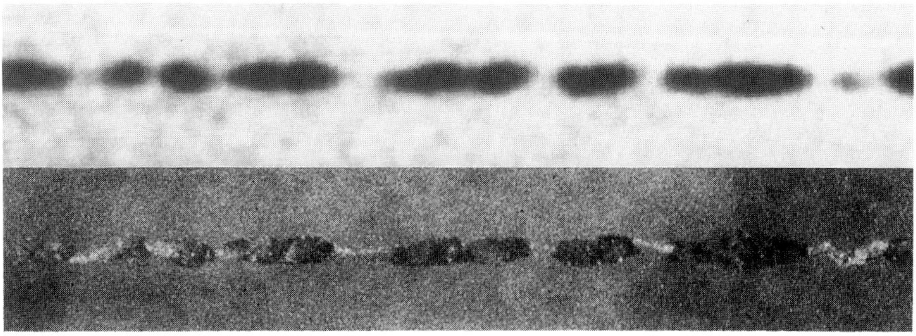

Figure 9. Body of shaped-charge jet 200 mm behind jet tip and approximately 1100 mm from the exit orifice of the shaped charge. The jet has apparently broken into 2-mm diameter particles. Above is a section of a 450-KeV X-ray taken of this jet at the same time as the IC photograph below. Note that the IC photograph clearly shows reflective material between the larger dark particles whereas the radiograph shows little or none. One explanation currently being explored is that there is a thread of liquid copper in the center of the jet that becomes exposed when the jet breaks apart, and that the thread is reflective to the laser illuminator, but too transparent to show in the radiograph.

The particulating jets shown in Figs. 8 and 9 are of unique and recent interest. Particulated jet fragments had previously been observed only in silhouette with low-energy pulsed radiography and occasional rotating-mirror framing photography. The IC-camera photographs show surface crazing and stretching not previously observable. Noted, and not unexpected, are slip dislocation lines running at 45 degrees with respect to the jet axis (Fig. 8).

The similarity to solid metal rods undergoing static tension leads to the conclusion that the Viper jet surface is solid rather than liquid. Totally surprising, however, is the fact that the IC-camera pictures show the particulated segments to be connected by highly reflective threads where the radiographs show clear gaps (Fig. 9). The threads seem to be liquid copper that are radiographically transparent. Since the physical state of the metal is critical to the understanding and computer modeling of these jets, current efforts are being directed towards dynamic measurements of the temperature profiles of the jets in flight.

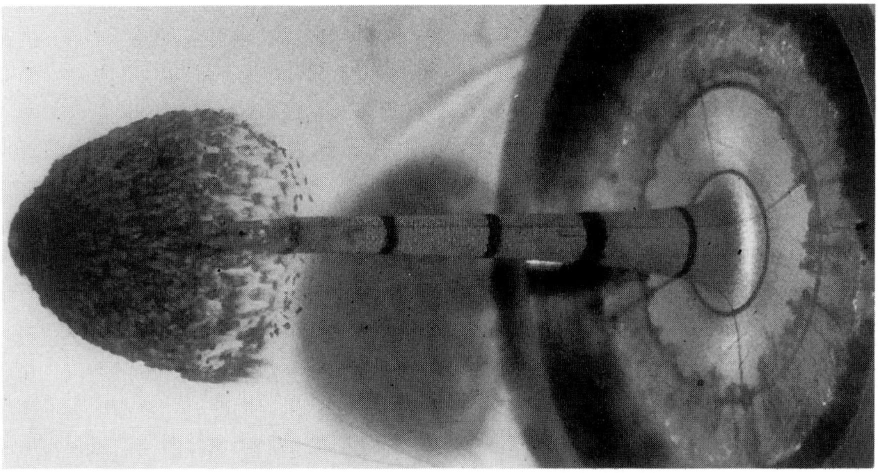

Figure 10. Example of a Viper shaped charge (see Fig. 4) taken when the tip has traveled 80 mm from the exit orifice (31.5 μs after detonator initiation). Tip velocity is 9.2 km/s creating a clearly visible shock structure in the surrounding air and a shadowgraph on the background surface. The "umbrella"-shaped cloud of debris surrounding the tip is a result of ablation and of the rounded apex of the liner cone. The jets shown in Figs. 6 and 7 were from larger shaped charges with conical liners having sharp apexes which produce different debris clouds.

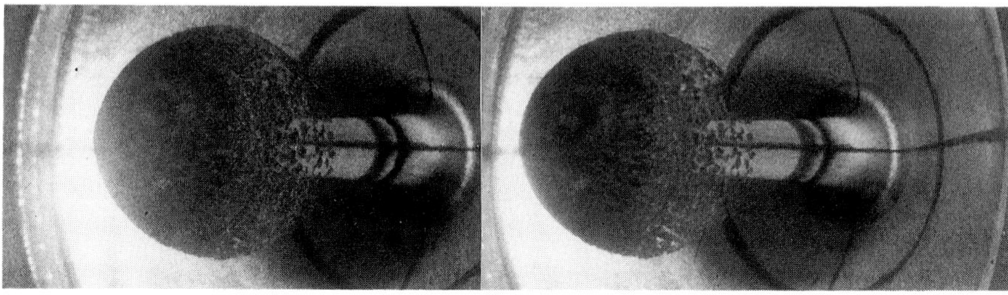

Figure 11. Representative stereo pair of pictures of a Viper taken with the IC camera. To view these images in three dimensions, place the photographs at normal reading distance. Relax your eyes so that you see double images and allow the center two (of the four) to overlap. If you experience difficulty in this free-focus technique, try placing a card between the two images from the page towards your nose so that each eye can see only one image. Relax your eyes until the images overlap to see the three-dimensional effect.

5. Conclusions

The IC-camera system provides a unique capability to photograph very fast phenomena such as the metal jets from shaped-charge devices. This system provides the resolution to observe the minute surface and metallurgical details of the liner deformation and subsequent jet formation and evolution. It has proven invaluable in providing a detailed picture of the breakup process and surface characteristics during the process of jet particulation. Data from the IC-camera pictures have proven indispensable for accurate computer simulations. As was shown in the oral presentation, such data have also has enhanced the quality of the information obtained from more conventional rotating-mirror high-speed cameras viewing the same event, by providing a highly detailed reference when evaluating the data contained in the color frames from these cameras. We continue to improve this system and are currently building the third iteration of the camera that will incorporate a new higher-resolution image-converter tube, greatly improved electronics for pulsing the new tubes, and improved objective optics. Because of our desire to emphasize the IC-camera system's pictures, we have kept the text to a minimum in this paper. We encourage the reader to review the list of references or contact the authors for more detailed descriptions of the camera system and its applications.

Acknowledgement. The authors acknowledge the programmatic support of LLNL's B-Division, the technical expertise on the experiments of R. Don Breithaupt, the leadership in shaped-charge technology of Dennis W. Baum, Leonard C. Haselman Jr., Christian Simonson III, and Kris Winer of the Advanced Conventional Weapons Program in B-Division, and the technical staff of bunkers 851 and 801 at LLNL Site 300. The authors extend a special acknowledgment to Anthony T. Rivera, James T. Wade, and James L. Dunmire for their outstanding contributions to the IC-camera-system fabrication, maintenance, and use.

This work was performed under the auspices of the U.S. Department of Energy by the University of California, Lawrence Livermore National Laboratory under contract W-7405-Eng-48; and was supported in part by the Office of Munitions, U.S. Department of Defense, Washington, D.C.

References

Baum DW, Shaw LL, Honodel CA (1992) Application of IC camera with laser illumination to shaped charge metal jet formation. In: Proc. 13th Int. Symposium on Ballistics, Vol. 2, ADPA, Arlington, VA, USA, pp 575–583.

Baum DW, Shaw LL, Simonson SC, Winer KA (1993) Shaped charge liner collapse and early jet formation. In: Proc. 14th Int. Symposium on Ballistics, Vol. 2, ADPA, Arlington, VA, USA, pp 13–22.

Carpluk GT (1970) Laser stereo photography. In: Proc. 9th Int. Congress on High Speed Photography, SPIE, Bellingham, WA, USA, p 64.

Chancel C, Pelte C (1972) Multi-frame image tube camera. In: Proc. 10th Int. Congress on High Speed Photography, SPIE, Bellingham, WA, USA.

Chavailler, C (1990) High speed photographic systems. In: Proc. 19th Int. Congress on High Speed Photography, Videography, and Photonics, SPIE 1358, Bellingham, WA, USA, pp 210–226.

Jones PA (1974) Four pulse Q-switched ruby laser. J Phys E 7:704.

Muelder SA, Huen T, Firpo CS, Hein DF, Steinmetz LL, Shaw LL (1987) Microchannel plate intensifier version of the IC camera. In: High Speed Photography, Videography, and Photonics V, SPIE 832, Bellingham, WA, USA, pp 57–62.

Owren HM (1959) Product bulletin for single frame image converter camera. Abtronics Inc., Livermore, CA, USA.

Paisley DL (1976) Laser stereo photography. In: Proc. 12th Int. Congress on High Speed Photography, SPIE 97, Bellingham, WA, USA, pp 184–188.

Parker VC, Crews JL (1987) Laser shadowgraphs with a rotating mirror camera. In: High Speed Photography, Videography, and Photonics V, SPIE 832, Bellingham, WA, USA, pp 88–95.

Shaw LL, Sonderman JB, Seppala LG (1979) 70 mm format rotating mirror framing camera. In: Proc. 13th Int. Congress on High Speed Photography, Videography and Photonics, SPIE 189, Bellingham, WA, USA, pp 218–221.

Shaw LL, Steinmetz LL, Behrendt WC, Sonderman JB, Mikkelson K, Beer GK, Seppala LG, Romero E (1984) Eight frame electro-optic camera with laser illumination. In: Proc. 16th Int. Congress on High Speed Photography, SPIE

491, Bellingham, WA, USA, pp 267–273.

Shaw LL, Muelder SA, Rivera AT, Dunmire JL, Breithaupt R (1993) Continuing evaluation of shaped charge metal jet characteristics using multi-frame electro-optic camera with laser illumination. In: Proc. 20th Int. Congress on High Speed Photography, SPIE 1801, Bellingham, WA, USA, pp 92–105.

Winer KA, Shaw L, Muelder S, Breithaupt D, Baum D (1993) Comparison of dynamic behavior of shear-formed shaped charge liners to computer simulations. In: Propellants, Explosives, Pyrotechnics, 18, Weinheim, Germany, pp 345–351.

Winer KA, Breithaupt D, Shaw L, Muelder S, Baum D (1995) High speed, high resolution observations of shaped charge metal jets undergoing particulation. In: Proc. 15th Int. Symposium on Ballistics, ADPA, Arlington, VA, USA, pp 89–96.

Numerical simulation of high-enthalpy impulse facilities

G. J. Wilson*
Thermosciences Institute
Moffett Field, CA 94035 USA

Abstract: Axisymmetric flows inside shock tubes and expansion tubes including the effects of finite-rate chemistry and both laminar and turbulent boundary layers are computed. A simulation of a low-pressure shock tube filled with nitrogen is carried out until the limiting flow regime is reached. Grid-refinement studies are made and the results are compared to Mirels' correlations. Simulations in the acceleration-tube section of the HYPULSE expansion tube are presented for two operating conditions, and the computations are compared with experimental data. One of these operating conditions is characterized by a laminar boundary layer and the other by a turbulent boundary layer. The simulations compare well with experiment and reveal that the nonuniformity in pressure observed during the test time is related to variations in the boundary-layer displacement thickness.

Key words: Shock tube, Expansion tube, Unsteady flow, Boundary layer

1. Introduction

One-dimensional analysis is a valuable tool for understanding and predicting the flow within shock tubes, shock tunnels, and expansion tubes. However, there are many non-ideal multi-dimensional phenomena which occur within these facilities which cause performance to deviate from one-dimensional theory. The result of these non-ideal processes is often temporally and spatially nonuniform test conditions and significantly reduced test times. Examples of these processes include the time-dependent growth of the boundary layer behind the incident shock, diaphragm-rupture mechanics, gas-interface instabilities, the propagation of transverse waves created by an area change or the diaphragm rupture, and the interaction of the reflected shock with the boundary layer. Understanding and quantifying these phenomena are important for interpreting experimental data and form a prerequisite for improving the design and operation of impulse facilities. Improvements in numerical methods and computer power provide opportunities for computational fluid dynamics (CFD) to add to the understanding of impulse-facility flows.

*Mailing address: MS 230-2, NASA Ames Research Center, Moffett Field, CA 94035.

Unsteady simulations of the viscous flow in shock tubes and expansion tubes are computationally expensive because they must include a large physical domain (i.e., the facility length) and simultaneously resolve small length scales associated with features such as the boundary layer or a gas interface. The modeling of a real gas adds to the computational expense by requiring multiple species and energy equations, and chemical kinetics. Recently, researchers have begun to explore the modeling requirements for detailed multi-dimensional simulations of shock tubes and expansion tubes. A few examples of these works include Weber et al. (1994), Daru and Damion (1993), Wilson et al. (1993), and Jacobs (1991). The literature suggests the research community is primarily investigating separate multi-dimensional phenomena in isolation in order to reduce the problem complexity. This approach will probably continue until more experience in simulating these flows is gained and computers become faster.

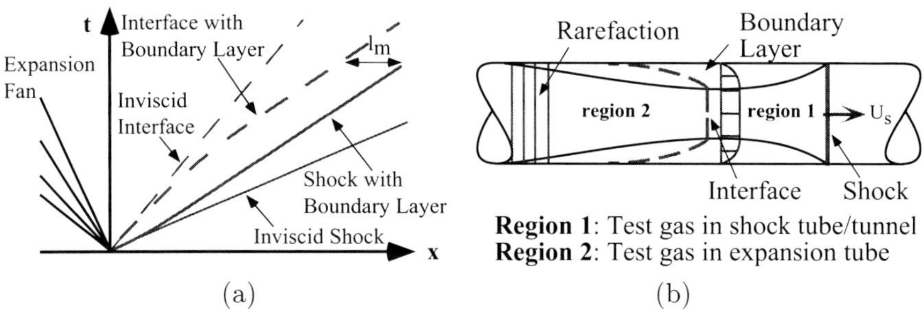

Figure 1. Shock-tube flow including the boundary layer (a) x-t diagram, (b) schematic diagram in laboratory coordinates.

The simulations presented herein focus on the influence of the boundary layer on shock-tube and expansion-tube flow. The x-t diagram of shock-tube flow in Fig. 1a illustrates the effect of the boundary layer on the shock and interface speeds. Shock speed decreases (attenuates) compared to the inviscid shock speed and the interface speed increases. After traveling a sufficient distance, the shock and interface reach the same speed and the separation distance remains constant at a limiting value, l_m. Figure 1b contains a schematic diagram of the major features of shock-tube flow with a boundary layer in laboratory coordinates. There is a large body of analysis, mostly due to Mirels (1963, 1966, 1971), which quantifies the nonuniformity of the gas between the shock and the interface (region 1) due to the boundary layer and which also quantifies the effects of the shock attenuation. However, this theory offers little detailed information about the effect of the boundary-layer growth on the expanded driver gas (region 2). In the acceleration section of an expansion tube, region 2 (expanded intermediate tube gas) is the test gas and the primary region of interest. Since CFD can provide detailed information about both regions, it can be an important tool for studying expansion-tube flow. In addition, CFD simulations can incorporate the effects of thermochemical nonequilibrium and turbulent transition which are not included in Mirels' analysis.

2. Numerical method

The axisymmetric flow through a shock tube or expansion tube is modeled using the thin-layer Navier-Stokes equations for a chemically reacting ideal-gas mixture. The present gas model includes three gas species (He, N_2, and N). A vibrational energy equation is included so that vibrational nonequilibrium can be modeled. Species diffusion is not included. The gas-dynamic equations are solved using a semi-implicit finite-volume form of the Harten-Yee upwind TVD scheme (Yee 1989). Grid points are clustered at the shock and interface and translated with these features to minimize numerical errors. The viscous terms are treated implicitly so that the allowable time-step size is dictated by the explicit stability criteria of the inviscid terms. The source terms representing the finite-rate chemical kinetics and vibrational relaxation are also implicit. The moving grid and single-step implicit algorithm reduce the formal time accuracy to first order. A single-equation algebraic model proposed by Cebeci and Smith (1974) is employed for turbulent-flow cases.

3. Shock-tube flow

Flow conditions within a shock tube are chosen so that the assumptions of Mirels' laminar analysis are valid (Sharma and Wilson 1995). The shock-tube diameter is chosen to be 0.05 m with a driver length of 0.76 m and a driven length of 12.0 m. A helium driver is used and the driven tube is filled with nitrogen at 70 Pa. The incident shock speed is approximately 2.9 km/sec so that there is little dissociation. The simulations are started by instantly removing a planar diaphragm and are run until the distance between the shock and interface has become nearly constant. The axial direction is divided into 700 grid points and the radius into 63 grid points. To assess the accuracy of the simulations, a grid-refinement study is performed with the number of grid points kept constant and minimum spacing at the wall reduced by half. Figure 2 is a plot of the computed distance between the shock and the interface versus time on four grids. The predicted separation distance at the end of the driven tube decreases with each reduction in wall-grid spacing and changes by less than 3% percent on the two finest grids. A computation with 126 points normal to the wall shows little difference compared to a solution with the same wall spacing and 63 points. For these conditions, both Mirels' (1963) analysis and the finest grid solution closely agree and predict a separation distance of 0.151 m.

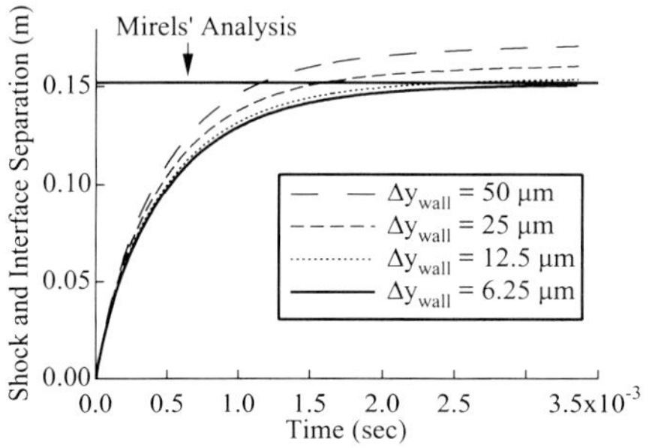

Figure 2. Time history of the separation distance between shock wave and interface for grid-refinement study.

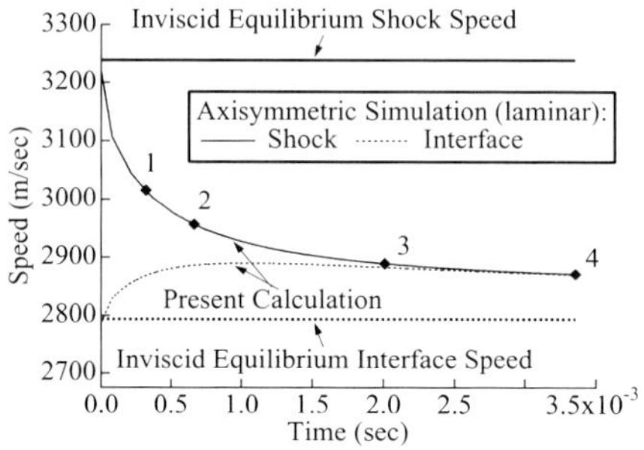

Figure 3. Time history of shock-wave and interface speed starting at diaphragm rupture.

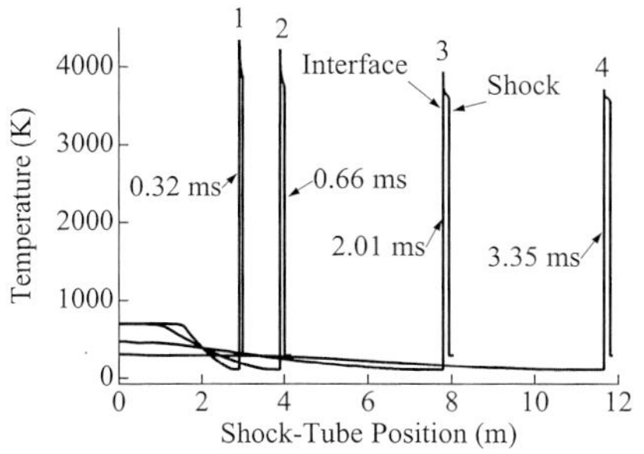

Figure 4. Shock-tube temperature profiles at four times corresponding to the markers on Fig. 3.

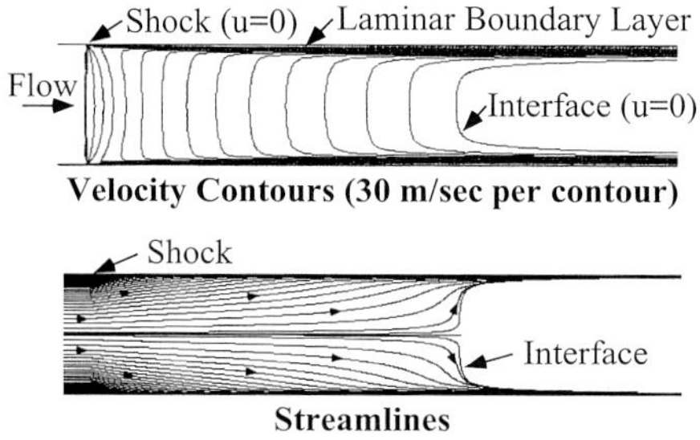

Figure 5. Limiting flow velocity contours and streamlines in shock-fixed coordinates.

Figure 3 shows the variation of the computed shock and interface speed in the driven tube compared with the speeds predicted from inviscid theory. The nearly identical shock and interface speed at the end of the shock tube shows that the limiting flow regime is reached. Figure 4 contains temperature profiles in the shock tube at the four times corresponding to the markers in Fig. 3. The temperature profiles show significant nonuniformity between the shock and interface due to shock attenuation, especially at early times when the rate of change of the shock speed is highest. The highest temperature gas occurs near the interface because this gas is processed by the highest speed shock. The temperature profile of the gas near the end of the tube is the most uniform because the shock speed is nearly uniform and the gas initially processed by the shock has been entrained into the boundary layer. At this point the variations are primarily due to the quasi-steady shape of the boundary layer.

Figure 5 shows computed velocity contours and streamlines in shock-fixed coordinates at the limiting flow regime. The velocity contours show the size and shape of the boundary layer and illustrate the nonuniformity in velocity due to the effective area change created by the boundary layer. The velocity decelerates from the post-shock velocity to zero velocity at the interface so that the shock and interface have the same speed. The streamlines show the gas crossing the shock entering the boundary layer before the interface. At the limiting flow regime, the mass flow across the shock is equal to amount of gas being entrained by the boundary layer. Figure 6 shows a comparison of

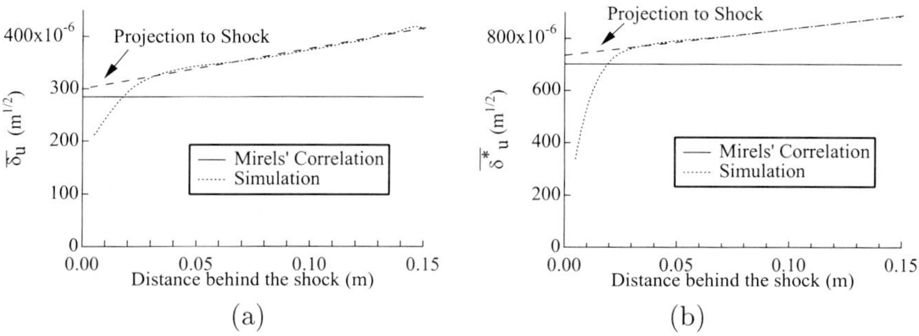

Figure 6. Comparison between Mirels' correlations and simulation (a) boundary-layer-thickness parameter, (b) boundary-layer-displacement-thickness parameter.

the computed boundary-layer-thickness and displacement-thickness parameters with values predicted with Mirels' (1966) correlations. Mirels' values for these quantities are constant because constant post-shock conditions are assumed. Agreement between the correlations and the simulations is expected just behind the shock. One reason for the discrepancy at this location is the curved incident shock captured by the simulations (see Fig. 5). Another possible reason is the lack of grid resolution in the very thin boundary layer immediately behind the shock. If values for the parameters at the shock front are predicted based on the trends observed farther behind the shock, as illustrated in Fig. 6, then the computed values are found to be in good agreement with Mirels' analysis. Differences between the two methods farther behind the shock presumably give an indication of the error introduced by the assumption of constant post-shock conditions made in Mirels' analysis.

4. Acceleration-tube flow

Simulations of the HYPULSE expansion tube are presented for two operating conditions and the computations are compared with experimental data (Wilson et al. 1995). The operating gas for both cases is nitrogen. The first test condition is at a total enthalpy of 15.2 MJ/kg and a pressure of 2 kPa and is referred to as the Mach 17 condition. This case is characterized by a laminar boundary layer and significant chemical nonequilibrium in the acceleration gas. The second test condition is at a total enthalpy of 10.2 MJ/kg and a pressure of 38 kPa and is referred to as the Mach 14 HP (high pressure) condition. This case is characterized by a turbulent boundary layer. These simulations begin with a shock wave in the intermediate tube which travels to the location of the secondary diaphragm. The secondary diaphragm is planar and is broken instantly.

4.1. Laminar boundary layer

A 14.62-m acceleration tube is used for the Mach 17 test condition giving a length-to-diameter (L/D) ratio of approximately 96. A 400×63 grid (63 points normal to the wall) is used and the grid spacing at the wall is varied to assess the required grid resolution in the boundary layer. For this case, the high post-shock temperature and the cold wall create an extremely small thermal boundary layer which is difficult to resolve. The finest grid-wall spacing attempted was 12.5 μm, which gave a value of l_m 5% smaller than the value obtained with a grid-wall spacing of 25 μm. A solution of the flow through the facility on the finest grid requires approximately 700,000 iterations and 70 hours of single-processor CPU time on a Cray C90. The value of l_m predicted by Mirels' (1963) analysis is calculated to be approximately 0.098 m, which compares to the value of 0.103 m from the simulation. Close agreement between the computation and analysis is not necessarily expected, as it was in the shock-tube case, because Mirels' assumption of chemical equilibrium behind the shock is not satisfied for this case.

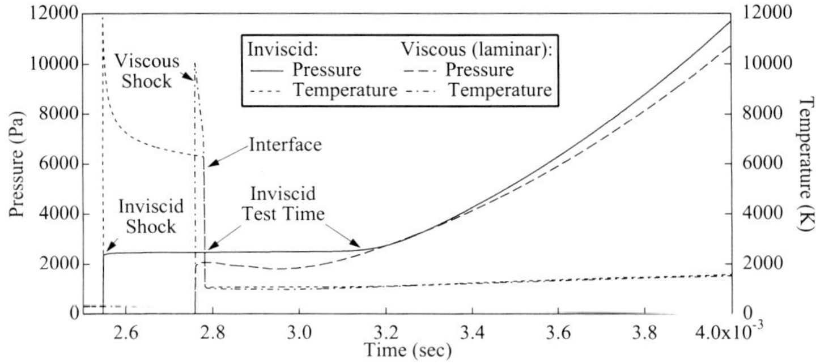

Figure 7. Computed pressure and temperature traces at the end of the acceleration for the Mach 17 condition (inviscid and laminar).

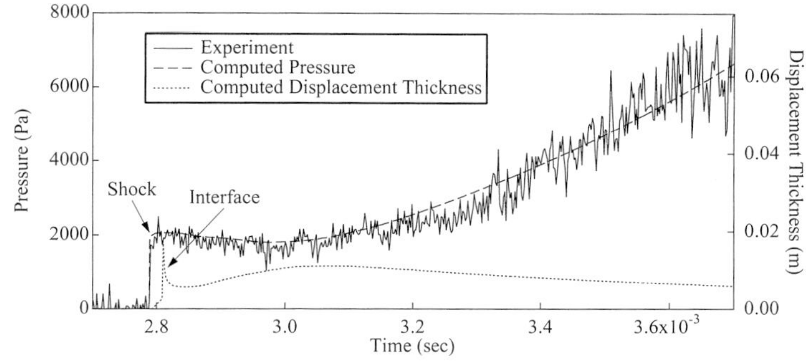

Figure 8. Computed and experimental pressure traces and computed displacement thickness for the Mach 17 condition.

Figure 7 presents pressure and temperature traces at the end of the acceleration tube for the inviscid and viscous solutions. The time of arrival of the interface for the two solutions has been aligned. The large temperature variation behind the acceleration-tube shock is primarily due to chemical nonequilibrium. The pressures and temperatures behind the shock for the viscous solution are lower than the values predicted by the inviscid solution because of shock attenuation. The laminar solution has a much smaller separation distance between the shock and the interface.

A comparison of experimental and computed pressure traces at the end of the acceleration tube is presented in Fig. 8, which shows good agreement for the post-shock pressure level and the variation of pressure during the test time. The computed displacement thickness at the exit of the tube is also presented. The height of the right axis is set equal to the radius of the acceleration tube to allow an easy comparison of the displacement thickness to the radius of the tube. The sharp peak in the displacement thickness at the interface results from the integration path crossing the curved interface. The minimum pressure during the test time is associated with a maximum in the displacement thickness.

For this condition, the agreement between experiment and computation for the temperature and pitot pressure is not as good as it is for the static pressure. Pitot pressure is overpredicted by 10–20% and temperature is similarly underpredicted. Simulations with varied initial conditions and including vibrational nonequilibrium do not account for the observed differences. A possible cause of the discrepancy may be the partial reflection of the intermediate-tube shock off of the secondary diaphragm which is observed in the experimental data. Simulations which delay (10 μsec) the opening of the secondary diaphragm support this conjecture. The processing of the test gas upstream of the secondary diaphragm by the reflected shock decreases the pitot pressure during the test time and increases the static temperature bringing the simulations in better agreement with the data. The static pressure, which is in good agreement with experiment, is found to be relatively insensitive to small delays in the opening of the diaphragm.

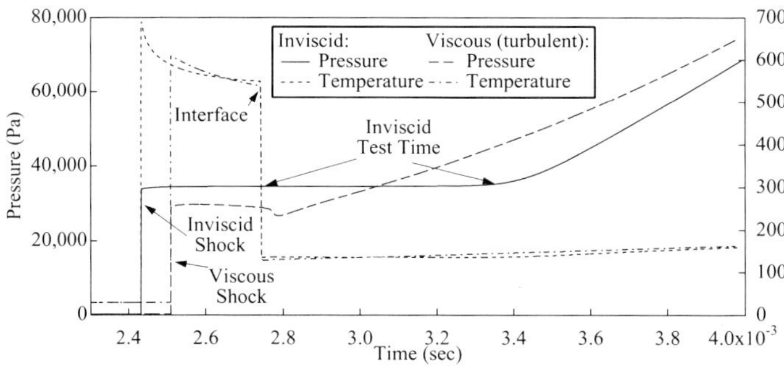

Figure 9. Computed pressure and temperature traces at the end of the acceleration for the Mach 14 HP condition (inviscid and turbulent).

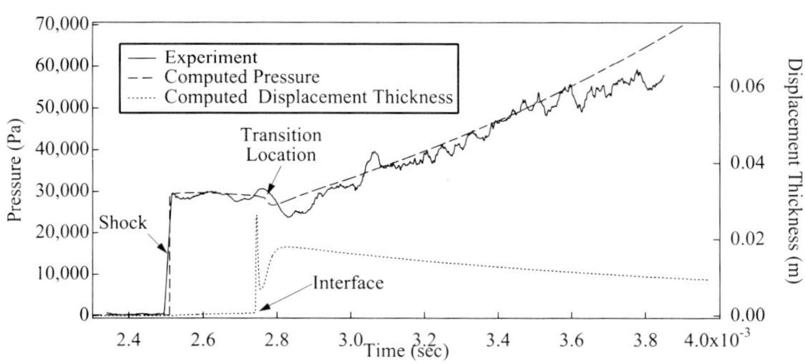

Figure 10. Computed and experimental pressure traces and computed displacement thickness for the Mach 14 HP condition.

4.2. Turbulent boundary layer

The Mach 14 HP condition uses a 10-m acceleration tube resulting in an acceleration-tube L/D ratio of approximately 66. The acceleration-tube fill pressure for the Mach 14 HP case is much higher than it is for the Mach 17 condition (193 Pa versus 7.2 Pa) and the acceleration-tube shock speed is lower (approximately 4 km/sec versus 5 km/sec). As a result, this case has a significantly higher post-shock pressure and a lower post-shock temperature with less dissociation, and the gas is much closer to chemical equilibrium. The flow does not reach the limiting flow regime by the end of the acceleration tube. Experimental heat-transfer measurements (Erdos et al. 1994) suggest that the boundary layer in the region between the shock and interface is laminar, and the transition to turbulence occurs at the start of the test time just behind the interface. For this reason, the simulations specify a turbulent boundary layer starting 2.54 cm behind the interface. A 450×53 grid is used with the wall grid spacing on the finest grid of 25 μm. It is known from the grid-resolution study that this grid accurately captures the time evolution of the shock and interface but slightly underpredicts the boundary-layer thickness in the test gas.

Figure 9 shows the pressure and temperature traces from inviscid and viscous solutions. A comparison of the experimental and computed pressure traces at the end of the acceleration tube is presented in Fig. 10. Both the experiment and simulation show similar characteristics and are in good agreement. There is a drop in pressure near the transition point and a steady rise in pressure during the test time. Figure 10 also shows the computed value of the displacement thickness at the exit of the tube with the plot axis scaled to the tube diameter. As with the Mach 17 simulation, the minimum pressure over the test time is associated with a maximum in the displacement thickness. The decrease in the displacement thickness during the test time seems to be the cause of the rise in pressure during the test time; however, it is difficult to distinguish between this effect and the arrival of the rarefaction. Additional studies of the effect of Reynolds number and acceleration-tube length on the turbulent boundary layer for these conditions are found in Wilson et al. (1995).

5. Conclusions and future work

The present work makes a detailed comparison between axisymmetric numerical simulations and Mirels' correlations for shock-tube flow with a boundary layer. It also represents an initial attempt to characterize laminar and turbulent flow in an expansion tube. These simulations are shown to provide valuable physical understanding about impulse-facility flow. The present numerical approach is computationally expensive and is impractical at higher Reynolds numbers where finer grid spacing at the walls is required to resolve the boundary layer. One goal of future numerical work should be to apply more advanced numerical methods such as higher-order schemes and fully-implicit time-accurate algorithms. There also is a need to identify or develop appropriate turbulence models for these hypersonic internal flows.

In addition to investigations of the boundary layer, numerical studies of the diaphragm-opening processes and pressure disturbances (noise) within shock tubes would be valuable. These topics are related because the diaphragm-rupture process is the likely cause of much of the noise observed in these facilities. Daru and Damion (1993) investigated flow perturbations in a shock tube due to diaphragm curvature. These computations also seem to show a mechanism of driver-gas contamination caused by the diaphragm curvature which has also been observed at higher pressure ratios by the author. Further investigations of these topics may lead to a better understanding of test-gas contamination, as well as facility noise.

Acknowledgement. Support to the author was provided by NASA contract NAS2-14031 to Eloret. The author thanks Dr. M. A. Sussman, Dr. S. P. Sharma, and Dr. R. J. Bakos for their significant contributions. Computer time was provided by the National Aerodynamic Simulation facility at NASA Ames Research Center.

References

Cebeci T, Smith AMO (1974) Analysis of Turbulent Boundary Layers, Academic Press, New York.

Daru V, Damion JP (1993) Analysis of the flow perturbations in a shock tube due to the curvature of the diaphragm. In: Shock Waves @ Marseille, Brun R and Dumitrescu LZ (eds), Springer, Vol 1, pp 451–456.

Erdos J, Calleja J, Tamagno J (1994) Increase in the Hypervelocity Test Envelope of the HYPULSE Shock-Expansion Tube, AIAA Paper 94-2524.

Jacobs PA (1991) Numerical Simulation of Transient Hypervelocity Flow in an Expansion Tube, NASA CR 189615.

Mirels H (1963) Test time in low pressure shock tubes. Phys Fluids 6:1201.

Mirels H (1966) Correlation formulas for laminar shock tube boundary layer. Physics of Fluids 9:1265–1272.

Mirels H (1971) Boundary growth effects in shock tubes. In: 8th Symp (Intl) on Shock Waves, London, pp 6/1–6/30.

Sharma SP, Wilson GJ (1995) Test Times in Hypersonic Shock Tubes, AIAA Paper 95-0713.

Weber YS, Anderson JD, Oran ES, Boris JP (1994) The Numerical Simulation of the Shock Bifurcation Near the End Wall in a Shock Tube, AIAA Paper 94-2307.

Wilson GJ, Sharma SP, Gillespie WD (1993) Time-dependent simulation of the reflected-shock/boundary layer interaction in a shock tube. In: Shock Waves @ Marseille, Brun R and Dumitrescu LZ (eds), Springer, Vol 1, pp 439–444.

Wilson GJ, Sussman MA, Bakos RJ (1995) Numerical Simulations of the Flow in the HYPULSE Expansion Tube, NASA TM 110357.

Yee HC (1989) A Class of High-Resolution Explicit and Implicit Shock-Capturing Methods, NASA TM 101088.

Blast effects of confined and unconfined explosions

S. B. Dorofeev
Russian Research Centre 'Kurchatov Institute'
Moscow, 123182 Russia

Abstract: A brief review of current and past results on the blast effect of confined and unconfined explosions is presented. The processes considered include unconfined gaseous and heterogeneous detonations, deflagrations, and confined explosions. Scaling of unconfined deflagration, detonation, and numerical simulations of confined explosions are discussed.

Key words: Blast waves, Detonation, Deflagration, Explosions, Scaling

1. Introduction

Confined and unconfined fuel-air explosions represent one of the most hazardous events in industrial accidents. A characteristic feature of these events is a distinguishable blast effect. At the same time, these explosions can vary widely in the physical processes involved and in initial and boundary conditions. The physical processes involved include a variety of explosion modes from rather slow deflagrations to detonations. Gaseous and heterogeneous mixtures can form an explosive. The mixtures can differ in stoichiometry, specific energy content, and chemical reaction rate. Shapes of explosive clouds can be different also. For confined conditions, these clouds can occupy different parts of an enclosure. These types of accidental explosions are the subject of much continuing research. The aim of this presentation is a brief review of results and current activity in this field at Russian Research Centre 'Kurchatov Institute.'

2. Unconfined fuel-air explosions

2.1. Gaseous detonations

The effects of unconfined gaseous detonations have been studied in many experiments and calculations (Kogarko et al. 1965; Fishburn 1976; Baker et al. 1983; Zhdan 1983; Brossard et al. 1984; Lannoy 1984; Borisov et al. 1985; Gelfand et al. 1985; Dorofeev et al. 1991, Dorofeev et al. 1993). The main objective of these studies was to obtain relationships for description of air blast, similar to those for high explosives (HE). The parameters of interest in these studies are maximum overpressure P^+, impulse I^+ of the positive phase of the wave, corresponding parameters of the negative phase P^-, I^-, and also duration of these phases. Sachs dimensionless variables are usually used for scaling, similar to those for HE explosions. Corresponding functions for overpressure and impulse of the positive phase are presented in Fig. 1.

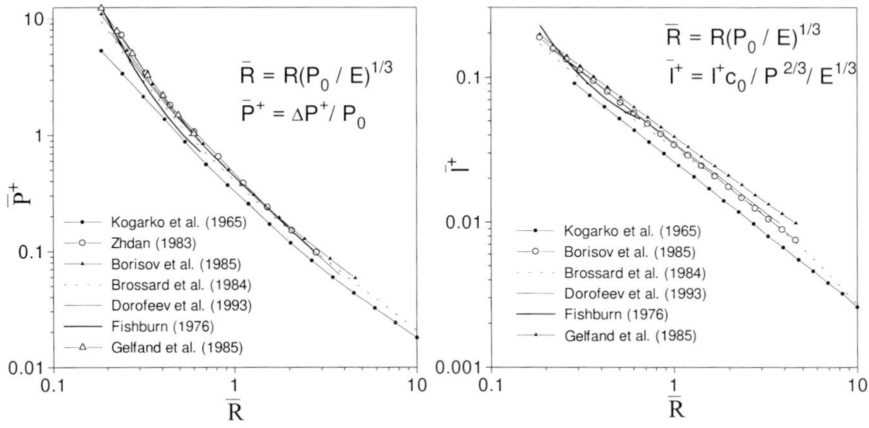

Figure 1. Positive overpressure and impulse in air blast wave versus distance (Sachs' variables) for gaseous detonation.

After about 30 years of experimentation and analysis, we have come to the conclusion that the description of the air blast from gaseous detonation is at the same level as that for HE explosions. The description of the negative phase, however, needs to be improved. In this situation, there is no reason to use TNT equivalents estimating the blast effect from gaseous detonation (Dorofeev et al. 1996a). The dimensionless overpressure and impulse for the positive phase can be calculated from the following expressions (Dorofeev et al. 1993):

$$\bar{P}^{+} = 0.34/\bar{R}^{4/3} + 0.062/\bar{R}^{2} + 0.033/\bar{R}^{3}, \tag{1}$$

$$\bar{I}^{+} = 0.0353/\bar{R}^{0.968}, \qquad 0.21 < \bar{R} < 3.77. \tag{2}$$

Here and in the following discussion all parameters correspond to spherical explosion in air.

2.2. Heterogeneous detonations

The description of the air blast from detonation of heterogeneous clouds can be found in a similar form as that for gaseous explosives. Generally, the efficiency of these explosions is not equal to gaseous detonation because of incomplete combustion of fuel droplets in the detonation wave. The parameters of the blast wave depend on the fuel type, droplet size, vapor fraction, etc. Nevertheless, in a variety of applications, there is a need to have some approximate description. Such a description has been suggested (Dorofeev et al. 1991) on the basis of numerous experiments with air sprays of motor fuel. Corresponding approximations for near-stoichiometric mixtures are given by:

$$\bar{P}^{+} = 0.125/\bar{R}^{1} + 0.137/\bar{R}^{2} + 0.023/\bar{R}^{3}, \tag{3}$$

$$\bar{I}^{+} = 0.022/\bar{R}, \qquad 0.3 < \bar{R} < 2.0. \tag{4}$$

Some of the experimental results are presented in Fig. 2. The data presented show that the impulse depends on the vapor fraction, which is very different in gasoline and kerosene sprays.

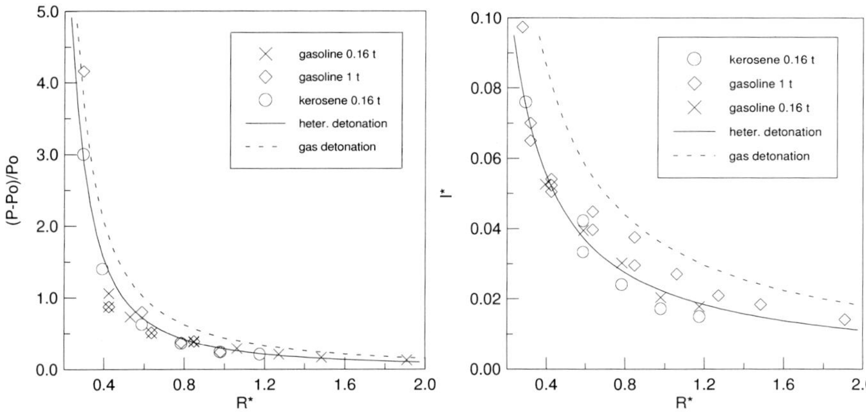

Figure 2. Positive overpressure and impulse in air blast wave versus distance (Sachs' variables) for detonation of heterogeneous clouds.

2.3. Fuel-rich clouds

The previous discussion addressed near-stoichiometric mixtures, while fuel-rich clouds are very typical in industrial accidents. In the case of a small additional amount of fuel (compared with the stoichiometric concentration), it is possible to find the blast parameters using the description for stoichiometric mixtures. It is necessary to calculate only the amount of energy release according to oxygen content or cloud volume. Systematic experiments on the effect of fuel concentration (Dorofeev et al. 1993) have shown that this is the case for all detonable concentrations. Corresponding experimental data for gaseous and heterogeneous mixtures are presented in Figs. 3 and 4. Variables scaled according to the mixture volume V are used in Fig. 4. This type of scaling is used because the clouds were not hemispherical in these experiments. Some data spread for the heterogeneous mixture is defined by difference in cloud shapes and by different vapor fractions.

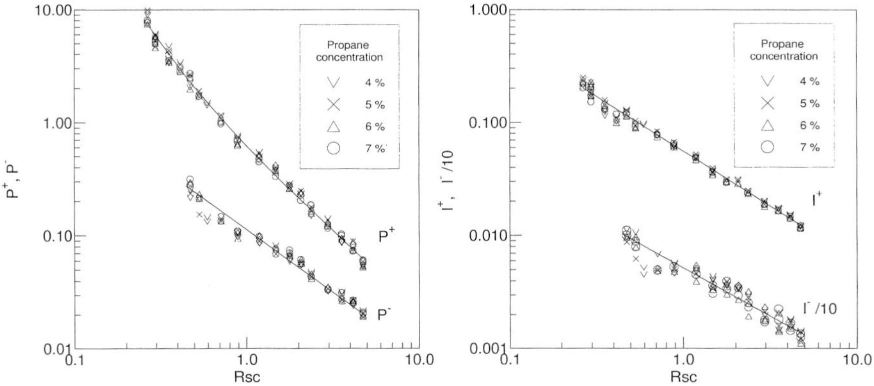

Figure 3. Positive and negative overpressures and impulses in air blast wave versus distance (Sachs' variables) for stoichiometric and fuel-rich gaseous mixtures.

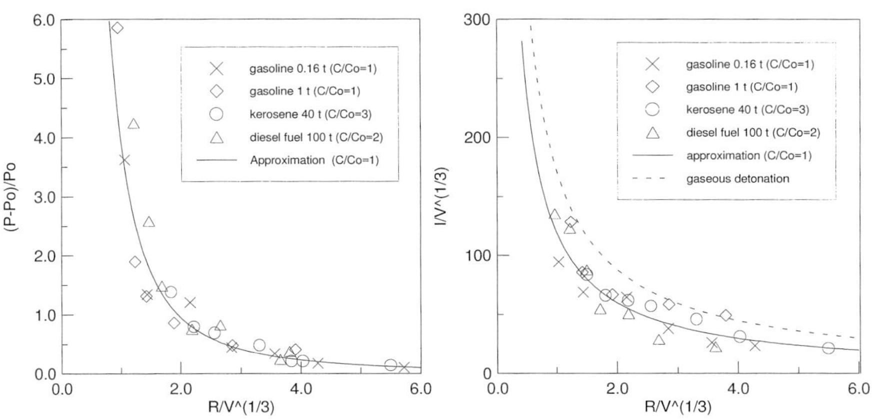

Figure 4. Positive overpressure and impulse in air blast wave versus distance for detonation of stoichiometric (C_o) and fuel-rich ($C/C_0 > 1$) heterogeneous clouds.

2.4. Unconfined deflagrations

While an unconfined detonation defines an upper bound of the explosion effect, this is not a very probable event in industrial accidents. An appropriate description for deflagration should necessarily include the flame speed. In this case, energetic Sachs' scaling can not be used without some modifications. Nevertheless, an estimate for the maximum overpressure and impulse at the cloud boundary as a function of the flame speed can be obtained by quasi-acoustic approximation. Corresponding analysis of the data for different flame speeds available in the literature (Kuhl et al. 1973; Strehlow RA 1975; Desbordes et al. 1978; Geisbrecht et al. 1981; and other studies) has given the following description of the blast parameters (Dorofeev et al. 1991):

$$\bar{P}^{+} = u^{2}\frac{\sigma - 1}{\sigma}(0.83/\bar{R} - 0.14/\bar{R}^{2}), \tag{5}$$

$$\bar{I}^{+} = \frac{u}{c_0}\frac{\sigma - 1}{\sigma}(1 - 0.4\frac{u}{c_0}\frac{\sigma - 1}{\sigma})(0.06/\bar{R} + 0.01/\bar{R}^{2} - 0.0025/\bar{R}^{3}), \tag{6}$$

where u denotes the apparent flame speed, c_0 the sound speed, and σ the expansion ratio. The value of the expansion ratio σ is about 7 for gaseous mixtures and decreases down to 4 for heterogeneous mixtures. The functions (5) and (6) were received for apparent flame speeds $u < 300$ m/s. The experiments conducted with higher flame speeds (Dorofeev et al. 1991) have shown that these functions are applicable also up to about 500 m/s. In this case, the blast parameters are given by:

$$\bar{P}^{+} = \min(\bar{P}_1, \bar{P}_2), \tag{7}$$

$$\bar{I}^{+} = \min(\bar{I}_1, \bar{I}_2), \tag{8}$$

where the values of $\bar{P}_1$, $\bar{P}_2$, $\bar{I}_1$, $\bar{I}_2$ are defined by (1), (5), (2), and (6) for gaseous, and (3), (5), (4), and (6) for heterogeneous mixtures, correspondingly. Figure 5 shows examples of experimental and calculated blast parameters for high-speed deflagrations. Although the prediction of turbulent flame speeds for some particular initial and boundary conditions is a rather complicated problem, an approximate description of the air blast for known flame speed can be suggested.

3. Confined explosions

A description of the blast effect of confined explosions should be different in form from that for unconfined ones. First, a simple engineering approach to find functions of blast parameters on distance is inappropriate, since 3D effects are of prime importance. Second, it becomes more important to predict the explosion mode, because both the effect and the mode appear to be strongly interconnected. Third, details of boundary conditions have a strong influence, and some universal description similar to dimensionless functions for air-blast parameters seems hardly possible. The only useful tool to describe confined conditions is multi-dimensional simulation of explosion processes. A simulation of all possible explosion modes in the frame of one model is not really practical at

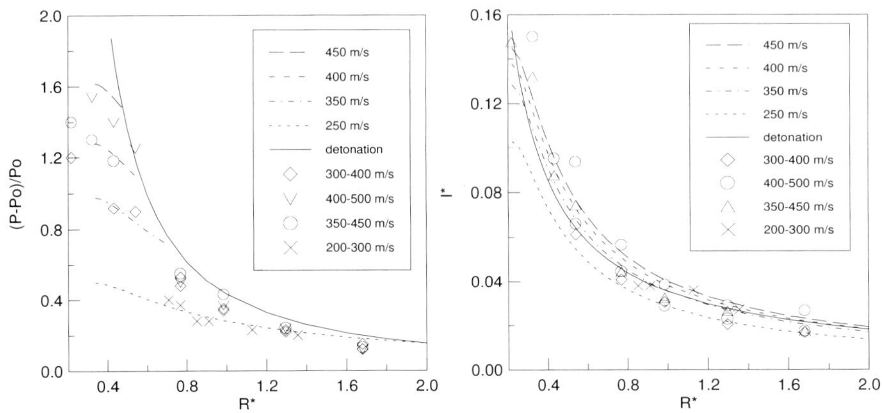

Figure 5. Positive overpressure and impulse in air blast wave versus distance (Sachs' variables) for deflagrations with different flame speeds.

present, since it requires computation of unsteady, turbulent, multi-dimensional compressible reacting flows. Simplified models are necessary, but they should be validated against appropriate experimental data.

3.1. Confined detonations

Confined detonations were studied in a series of large-scale experiments and 3D numerical simulations (Breitung et al. 1994; Breitung et al. 1996). Experiments have been carried out with hydrogen-air mixtures in a large confined volume of 263 m^3 (Fig. 6). The data obtained include all important features of confined 3D detonations, such as curved and oblique fronts, reflections, cellular structure, detonation decay, and reinitiation. Results of 3D simulation show good prediction of detailed load parameters for the case of stable developed detonation waves (Breitung et al. 1996).

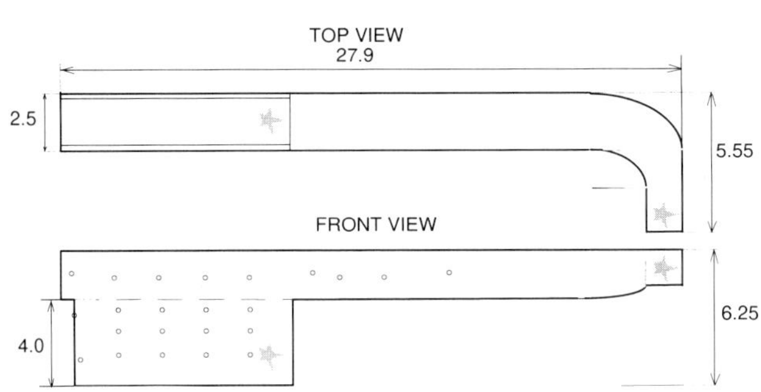

Figure 6. Scheme of the RUT facility. Stars show two initiation positions.

3.2. Transient explosion processes

The case of unstable detonations near the limit of propagation requires much more sophisticated models. First of all, an appropriate description of chemical reaction is necessary. Direct application of detailed kinetic schemes for detonation modeling has many problems (Efimenko et al. 1994). Turbulent effects seem to play a role in the complex 3D structure of detonations. Some simplified models of the rate of energy release for near-limit mixtures show rather reasonable agreement with experimental results (Breitung et al. 1996). An example of such a calculation is shown in Fig. 7. The pressure field for confined turbulent

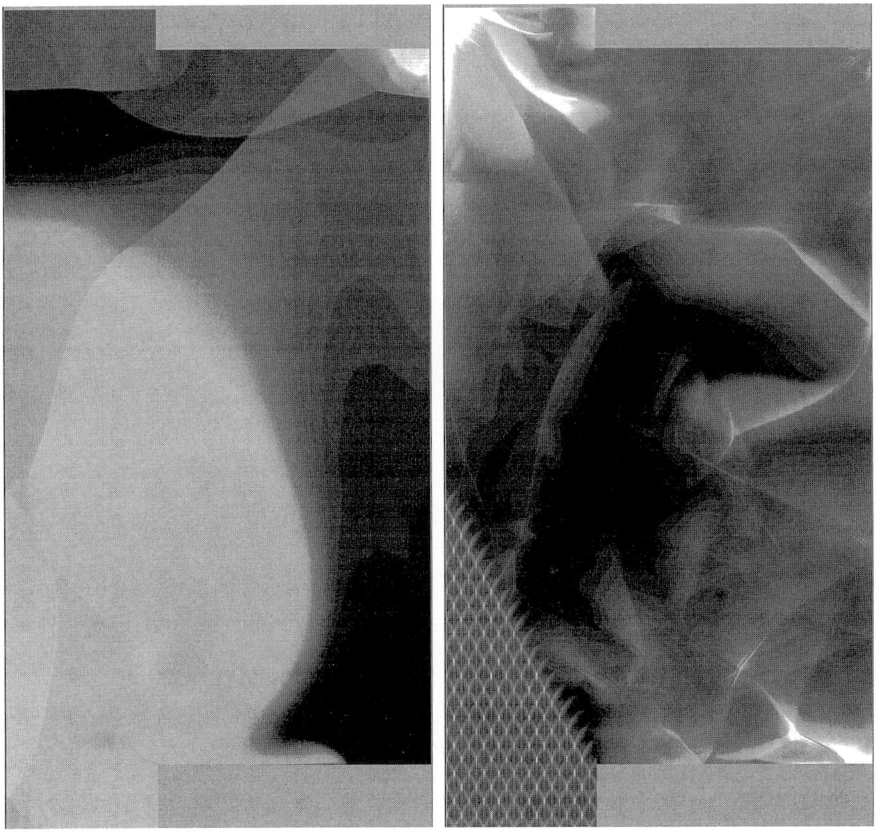

Figure 7. Two-dimensional simulation of marginal detonation in RUT facility. Detonation propagates upward from bottom of the figure. Reinitiation is observed in upper right corner. Temperature field (left) at 18.9 ms ranged from 388 to 2248 K. Maximum pressures (right) ranged from 1.2 to 9 MPa.

deflagration varies greatly in the range of processes from slow deflagration to DDT events (Fig. 8). A predictive modeling of turbulent deflagrations is not possible at present. Nevertheless, corresponding computer models are being developed. Some useful information for the estimation of an explosion effect can be obtained experimentally. An example is that impulse appeared to be almost independent of explosion mode, including fast turbulent deflagrations, detonations, and DDT events (Fig. 9). Another example is an estimation of conditions for spontaneous onset of detonations (Dorofeev et al. 1996b). An approximate scale limitation representing necessary condition for the onset of detonation can be suggested.

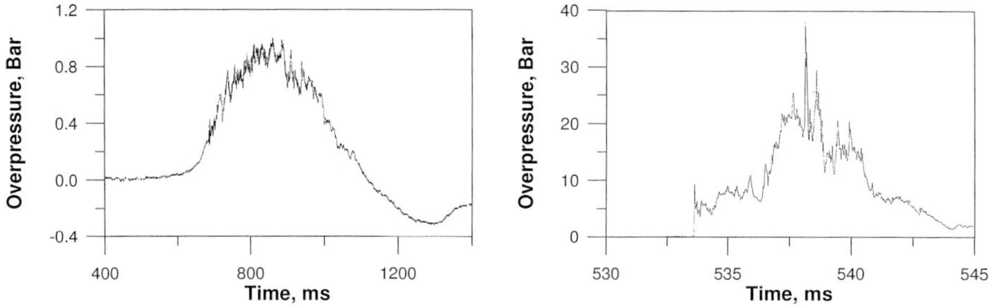

Figure 8. Overpressures versus time for slow deflagration (left) and DDT (right) in large confined RUT volume.

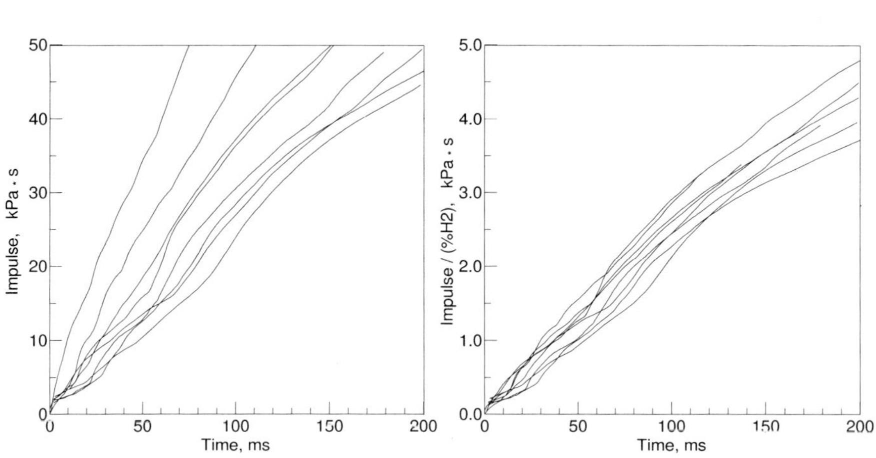

Figure 9. Positive impulse (left) and impulse normalized to hydrogen concentration (right) versus time for different explosion processes in RUT facility. Data range from slow deflagration to fully developed detonations.

4. Concluding remarks

In conclusion, the blast effect of confined and unconfined explosions can be described, with different degrees of confidence, for different types of processes. Confined and unconfined detonations seem to be the most-studied cases. Appropriate numerical tools and simple engineering descriptions of blast parameters are available. More problems are still to be solved for confined and unconfined turbulent deflagrations and for transient explosion processes.

Acknowledgement. The author is thankful to the staff of the Laboratory of Induced Chemical Reactions RRC-KI, and especially to V. Alekseev, A. Dvoinishnikov, A. Efimenko, A. Kochurko, M. Kuznetsov, and V. Sidorov for their technical excellency and dedicated work. This work was sponsored by Russian State Research Program 'Safety,' Project of Nuclear Safety Research (PSF) Forschungszentrum Karlsruhe, Germany, and the U.S. Nuclear Regulatory Commission.

References

Baker WE, Cox PA, Westine PS, Kulesz JJ, Strehlow RA (1983) Explosion Hazards and Evaluation, Elsivier, Amsterdam, Oxford, New York.

Borisov AA, Gelfand BE, Tsyganov SA (1985) On modelling of pressure waves formed by detonation and combustion of gas mixtures. J Phys Combust and Explos (Russian Acad of Sci) 2:90.

Breitung W, Dorofeev SB, Efimenko AA, Kochurko AS, Redlinger R, Sidorov VP (1994) Large scale experiments on hydrogen-air detonation loads and their numerical simulation. In: Proceedings of ARS'94 Int Topical Meeting on Advanced Reactor Safety, Pittsburg, PA, Vol 2, pp 733–745.

Breitung W, Dorofeev SB, Efimenko AA, Kochurko AS, Redlinger R, Sidorov VP (1996) Large-scale hydrogen-air detonation experiments and their numerical simulation. In: Shock Waves, Proc of the 20th ISSW, Sturtevant B, Shepherd JE, Hornung HG (eds), World Scientific, Vol 2, pp 1101–1106.

Brossard J, Leyer JC, Desbordes D, Saint Clouds JP, Hendrickx S, Garnier JL, Lannoy A, Perrot J (1984) Air blast unconfined gaseous detonations. Prog Astron Aeron 94:556.

Desbordes D, Manson N, Brossard J (1978) Explosion dans l'air de charges spheriques non confenees de melanges reactifs gazeux. Acta Astronautica 5:1009.

Dorofeev SB, Sidorov VP, Alekseev VI (1991) Air-Blast Parameters from Detonation of Gaseous, Heterogeneous and Fuel Rich Fuel-Air Explosives, State Research Program 'Safety' Report SRDO IRIS-91/1.6.08, Moscow.

Dorofeev SB, Sidorov VP, Dvoinishnikov AE, Alekseev VI, Kuznetsov MS (1993) Experimental study of air blast parameters from fuel rich mixtures detonation. In: 14th Int Colloquium on Dynamics of Explosions and Reactive Systems (ICDERS), 2:E1.4.1.

Dorofeev SB, Sidorov VP, Dvoinishnikov AE (1996a) Blast parameters from unconfined gaseous detonations. In: Shock Waves, Proc of the 20th ISSW, Sturtevant B, Shepherd JE, Hornung HG (eds), World Scientific, Vol 1, pp 673–678.

Dorofeev SB, Sidorov VP, Dvoinishnikov AE, Breitung W (1996b) Deflagration to detonation transition in large confined volume of hydrogen-air mixture. Comb and Flame (to appear).

Efimenko AA, Dorofeev SB, Shepherd J (1994) Chemical reaction kinetics and detonation modeling. Presented at 25th International Symposium on Combustion, W-I-P Abstracts The University of California, Irvine, p 251.

Fishburn BD (1976) Some aspects of blast from fuel-air explosives. Acta Astronautica. 3:1049.

Gelfand BE, Gubin CA, Mikhalkin VN, Shargatov VA (1985) Computation of shock wave parameters by detonation of combustible gaseous mixtures of a variable composition. J Phys Combust Explos (Russian Acad of Sci) 3:92.

Geisbrecht H, Hemmer G, Hess K, Leuckel W, Stoeckel A (1981) Analysis of explosion hazards on spontaneous release of inflammable gases into the atmosphere. Ger Chem Eng 4:315.

Kogarko SM, Adushkin VV, Liamin AG (1965) Investigation of spherical detonation in gas mixtures. Combust Explos Shock Waves, USSR 1-2:22.

Kuhl AL, Kamel MM, Oppenheim AK (1973) On flame generated self-similar blast waves. In: 14th Symposium International on Combustion, The Combustion Institute, pp 1201–1214.

Lannoy A (1984) Analyse des explosions air-hydrocarbures en milieu libre: Etudes deterministe et probabiliste du scenario d'accident. Prevision des effets de surpression. Bulletin Direct, Etudes et Recherches EDF, A4.

Strehlow RA (1975) Blast waves generated by constant velocity flames. Comb and Flame 24:257.

Zhdan SA (1983) Calculation of gas mixture explosion with regard to shift of chemical equilibrium products. J Phys Combust Explos (Russian Acad of Sci) 1:131.

Anomalous physical effects from artificial numerical length scales

R. Menikoff, K. S. Lackner
Theoretical Division
Los Alamos National Laboratory
Los Alamos, NM 87545 USA

Abstract: Shock-capturing algorithms are widely used for simulations of compressible fluid flow. Though these algorithms resolve a shock wave within a couple of grid points, the artificial length scale from the numerical shock profile can have side effects. The side effects are similar to physical effects that occur when a relaxation process gives rise to fully or partly dispersed shock waves. Two anomalies due to a non-zero shock width are discussed: (i) in one dimension, a non-decaying entropy spike results from a transient when a shock profile is formed or changed; (ii) in higher dimensions, front curvature affects the propagation of a shock wave. We show that both the entropy anomaly and the curvature effect are natural consequences of the conservation laws. The same analysis applies both to the continuum equations and to their finite-difference approximations in conservation form. Consequently, the artificial length scale inherent in a shock-capturing algorithm can mimic real physical effects that are associated with partly dispersed shock waves.

Key words: Shock-capturing, Wave propagation, Shock interaction, Curvature effect, Hugoniot jump conditions

1. Introduction

Fluid flow is described by the Euler equations. When transport effects are neglected, the Euler equations are a system of hyperbolic partial differential equations (PDEs) modeling the conservation of mass, momentum, and energy. An understanding of the wave structure of solutions to the PDEs plays a crucial role in developing numerical algorithms for fluid flow.

In the early days of computer simulations it was found that straightforward finite-difference algorithms of the Euler equations were adequate for smooth solutions but were unstable for shock waves. There are two reasons for the numerical instability: (i) the truncation error associated with the finite-difference discretization is large at a shock discontinuity; (ii) the hyperbolic PDEs have no dissipative terms yet entropy must increase across a physical shock. To cure the numerical difficulty, von Neumann & Richtmyer (1950) introduced the concept of artificial viscosity. The idea is to add dissipation to the equations in such a way that (i) smooth flows are minimally affected, (ii) the state behind a shock is correct, and (iii) the time step for an explicit algorithm is not unduly restricted. This is achieved by smoothing out the discontinuity and giving a profile to a shock wave. The profile is analogous to what occurs for real viscosity. However, the numerical profile must be a few cells wide, and hence is artificially large compared to the real physical shock width.

This numerical procedure works because (i) an artificial profile is a localized error that typically does not affect the global solution, and (ii) the conservation laws give rise to the Hugoniot jump conditions which determine the state behind a shock independent of the form of dissipation generating the entropy. This led to the development of "shock-capturing" algorithms. By modifying the equations such that a shock has a narrow profile, a simple uniform algorithm can be used for the entire flow rather than employing a specialized treatment to detect and track a shock discontinuity. This was especially important in the days when computers were limited in both speed and memory.

With current computers, much more elaborate algorithms and much higher resolution are possible. Shock-capturing algorithms have evolved to use Riemann solvers or approximate Riemann solvers to generate the dissipation necessary for a shock and to account properly for the flow of information along characteristics. They generally give accurate solutions and spread a shock wave over only a couple of grid points. However, calculated solutions with shock-capturing algorithms sometimes display "glitches" which usually manifest themselves as localized errors; see e.g., Quirk (1994). We show that some of these numerical anomalies mimic real physical effects and are a consequence of the artificial shock width.

Two types of physical effects are described. In Section 2, we analyze in one dimension an entropy anomaly that arises from the interaction of fully or partly dispersed shock waves. The entropy anomaly explains a wide class of common numerical errors such as excessive wall heating when a shock reflects from a rigid wall, startup errors when a shock is initialized as a discontinuity, errors from an impedance mismatch when a shock impacts a material interface, and errors from a shock passing through an abrupt change in mesh spacing. In Section 3, we show in higher dimensions that front curvature leads to a modification of the Hugoniot jump conditions. A consequence of the modified jump conditions is the diameter effect in a rate stick of explosive. The curvature effect also explains the artificially large numerical diameter effect that occurs when the reaction zone of a detonation wave is captured and not resolved. Our results are summarized in Section 4.

2. Transient effects in one dimension

The Euler equations for fluid flow in one dimension can be expressed in conservation form as

$$\frac{\partial}{\partial t}\begin{pmatrix}\rho \\ \rho u \\ \rho E\end{pmatrix} + \frac{\partial}{\partial x}\begin{pmatrix}\rho u \\ \rho u^2 + P \\ \rho u(E + PV)\end{pmatrix} = \begin{pmatrix}0 \\ 0 \\ 0\end{pmatrix}, \tag{1}$$

where ρ, e, and u are the density, specific internal energy, and particle velocity. $E = \frac{1}{2}u^2 + e$ is the total specific energy, $V = 1/\rho$ is the specific volume, and the pressure $P(V, e)$ characterizes the gas or fluid of interest. For a steady-state shock wave, the conservation form of the equations leads to the Hugoniot jump conditions. The shock speed and the state behind a shock are independent of the dissipative mechanism that generates the necessary shock entropy.

The jump conditions do not apply to transients when a shock profile is formed or changed. We show that the amount of entropy generated in a transient depends on the dissipative mechanism. When heat conduction is neglected, outside a shock profile entropy is convected along particle trajectories. Consequently, the entropy generated in a transient is frozen into the particle trajectories and does not diffuse. When shock waves interact, a localized entropy anomaly originates in the region in which the shock profiles of the incoming waves overlap.

As an example, we consider the simple case of a partly dispersed shock wave reflecting from a rigid wall. Because this example is equivalent to the symmetric collision of two shock waves, it also serves to illustrate the entropy anomaly caused by the interaction of shock waves. Moreover, this example has been studied in shock-tube experiments. The gas and incoming shock strength can be selected such that a relaxation process causes the shock to have a measurable width. Experiments and numerical simulations with relaxation processes included show that the reflection of a partly dispersed shock wave results in an entropy layer at the wall. This is nicely summarized in a review article by Becker (1972), see in particular his Fig. 9.

Sometime later an idealized problem was formulated by Noh (1987) to test the effect of artificial viscosity in calculations using the Euler equations. In the Noh problem, an ideal γ-law gas is initialized to have a uniform state consisting of a cold gas and a velocity directed towards a rigid wall. The problem has a simple analytic solution consisting of a strong shock wave propagating away from the wall. His calculations showed that a numerical error occurred near the wall. The particle velocity and pressure were correct but the specific energy was too large and was compensated for by a low density. Noh called this numerical anomaly "excess wall heating." It is qualitatively similar to the entropy layer at the wall that had previously been measured in experiments with partly dispersed shock waves.

Both the physical effect and the numerical effect can be explained based on the conservation laws. We outline the analysis given by Menikoff (1994). The solution with dispersed shock waves is compared to the corresponding solution of the Euler equations with discontinuous shock waves. We start by defining the front position for a dispersed shock wave. Let the subscripts a and b denote the states on the shock Hugoniot ahead and behind the shock front. The shock position x_s is defined such that the mass of the dispersed wave matches that of the idealized discontinuous shock. For a right-facing wave,

$$\int_{x<x_s} [\rho_b - \rho(x)]dx = \int_{x>x_s} [\rho(x) - \rho_a]dx \ . \tag{2}$$

Matching the momentum would give the same position. But the position based on matching the total energy is frame dependent. However, the relative energy between the dispersed wave and the discontinuous shock

$$\delta\mathcal{E} = \int_{x<x_s} [\rho(x)E(x) - \rho_b E_b]dx + \int_{x>x_s} [\rho(x)E(x) - \rho_a E_a]dx \tag{3}$$

is Galilean invariant. The key observation is that the relative energy depends on shock strength.

The asymptotic outgoing waves from a shock interaction are the solution to the Riemann problem with initial data corresponding to the states behind the incoming waves. Typically, a shock profile forms and equilibrates on a very fast time scale. The asymptotics set in after the outgoing waves are separated by a distance on the order of the sum of the widths of the equilibrium shock profiles. Otherwise, the dissipative mechanism would affect smooth flows as well as generating the entropy needed for a shock wave.

We consider the case of a symmetric shock collision. This is equivalent to the reflection from a rigid wall. Let the position of the wall be $x = 0$ and let a left-facing shock wave impact the wall at $t = 0$. Suppose at time $t > 0$, the outgoing shock wave has reached its asymptotic steady-state profile. The flux of mass and energy into a fixed region that contains the outgoing wave is given by

$$\hat{m} = -\rho_a u_a \qquad \text{and} \qquad \hat{E} = -\rho_a u_a(E_a + P_a V_a) \ .$$

A uniform state behind the outgoing shock would lead to the equations for conservation of mass and total energy

$$\begin{aligned} (\rho_b - \rho_a) \cdot x_s &= \hat{m}_L t \ , \\ \Big[(\rho E)_b - (\rho E)_a\Big] \cdot x_s &= \hat{E}t + \delta E_{\text{in}} - \delta E_{\text{out}} \ . \end{aligned}$$

Since we are considering a symmetric collision, the total momentum is always zero and momentum conservation is identically satisfied.

The conservation laws reduce to two equations for the one unknown shock position. The constraints from the Hugoniot jump condition imply that there is a solution for the hyperbolic problem when $x_s = \sigma t$, where σ is the outgoing shock speed. The difference in the relative energies of the incoming and outgoing shock waves, $\delta E_{\text{in}} - \delta E_{\text{out}}$, is not zero. This upsets the degeneracy of the equations. Therefore, the states behind the outgoing shock wave can not be uniform. Consequently, an anomaly must occur when the incoming and outgoing wave profiles overlap.

Without heat conduction, the entropy anomaly is frozen into the particle trajectories. Because of the wall, the velocity behind the outgoing shock must be zero. Therefore, the entropy anomaly appears as a boundary layer at the wall. For weak waves the entropy anomaly is small because the entropy change across a shock is third order in the shock strength. The anomaly is large for the Noh problem because the outgoing wave is a strong shock. Thus, by isolating and exaggerating a single numerical effect, the Noh problem is a good test problem for shock-capturing algorithms.

The argument based on the conservation laws can be extended to a non-symmetric wave interaction. In general there are three degrees of freedom corresponding to the positions of the three outgoing waves, and three equations for conservation of mass, momentum, and total energy. Typically, the contact is smeared out and there are small offsets or phase shifts in the asymptotic wave positions compared to the hyperbolic problem. Entropy is an important variable for the transient when wave profiles overlap because of its role as an adiabatic invariant.

A pronounced entropy anomaly at the contact occurs in some cases. A shock wave impacting a material interface is an important example. In this case, there is a separate conservation law for the mass of each material. This leads to four equations for the three unknown wave positions. Again a constant state behind the outgoing waves is not possible. The entropy anomaly can be large when the equations of state of the two materials are very different. When the equations of state are similar, for an Eulerian algorithm, mass diffusion from advection through a mixed cell can cause the entropy anomaly at the contact to artificially diffuse away. A pronounced entropy anomaly also occurs when a shock wave impacts an abrupt change in mesh, and when a shock wave is initialized as a discontinuity. In the last case, in addition to an entropy error which convects along the particle trajectories, there can be a weak wave of the opposite family. The initial data are not compatible with an additional shock. Instead there is an N-wave or discrete diffusion wave, see Chern (1991).

A numerical shock width may have additional side effects. The mass, momentum, and energy within the discrete profile oscillate as the shock front propagates through a cell. The oscillation acts as a source of acoustic noise. For upwind schemes the noise radiates behind the shock front. Numerical stability requires sufficient dissipation to damp short acoustic wavelengths which the mesh can not resolve. However, for a nearly stationary shock in an Eulerian mesh, the acoustic noise has a long wavelength. This is indistinguishable from acoustics waves that the grid is supposed to resolve. Consequently, it is difficult to damp the noise without affecting real signals. This example shows how the discretization can break the Galilean invariance of the continuum equations.

3. Wave propagation with curved front

In higher dimensions, a discontinuous shock wave obeys the Hugoniot jump conditions in the direction normal to the front. However, when the wave has a width, there is a correction to the jump conditions from the front curvature. This gives rise to a curvature effect on a propagating partly dispersed shock wave. A good example of the combined effect of a non-zero wave width and front curvature is the diameter effect in an explosive as measured in rate-stick experiments.

A rate stick is a cylinder of explosive. When detonated at one end, the flow rapidly approaches a steady-state detonation wave. The front is curved due to a rarefaction from the side boundary. Moreover, the asymptotic detonation velocity decreases as the diameter of the rate stick decreases. The diameter effect of the rate-stick experiments is a consequence of the curvature effect; see e.g., Menikoff (1989) and references contained therein.

In the model of Zel'dovich, von Neumann, and Doering (ZND), a detonation wave is a shock followed by a thin reaction zone. It is similar to a partly dispersed shock wave. Because of the chemical energy released, there exists a self-sustaining detonation wave; the lead shock heats the explosive and triggers a chemical reaction which releases energy and drives the lead shock. Moreover, an important effect of the energy release on the Hugoniot locus is that the self-sustaining detonation has the minimum possible wave speed. The curvature effect is an inescapable conclusion of the fact that the detonation speed measured in a rate-stick experiment is less than the minimum value for a planar detonation wave obtained from the Hugoniot jump conditions.

Rate-stick experiments can be simulated using the reactive Euler equations. A systematic numerical study by Donguy & Legrand (1981) used a two-dimensional Eulerian code with a uniform grid and varied both the charge diameter and the cell size. With a fine grid, the calculated diameter effect agreed with experiments. However, on a coarse grid, the reaction zone was not resolved and the numerical diameter effect exceeded the physical effect. Furthermore, the numerical effect increased with cell size. On a coarse grid, the numerical reaction dynamics serves as an artificial process to give the detonation wave a profile, and plays an analogous role to artificial viscosity for a shock wave. In effect, there is a numerical curvature effect that increases with the artificial reaction-zone width. The numerical variation in the curvature effect with cell size is qualitatively similar to experimental results on heterogeneous explosives that used the distribution of grain sizes to vary the reaction zone width; see Aleksandrov et al. (1967) and Moulard (1989).

The analysis by Menikoff et al. (1995) explains both the physical and numerical curvature effects in terms of the conservation laws. We assume that the detonation front is smooth and the reaction zone width w is sufficiently small. Then within the reaction zone the derivatives of the flow variables tangent to the front are small compared to those normal to the front, and the transit time for a particle in the reaction zone is small compared to the time scale for variations in the flow outside the reaction zone. These assumptions imply a long-wavelength, low-frequency approximation that describes the average behavior of a detonation wave and also is consistent with the typical mesh resolution used in numerical simulations. Locally the reaction zone can be approximated as one dimensional and quasi-steady. However, for a curved front, the normal directions are not parallel to each other. This effect can not be neglected. We show that the leading-order effect in higher dimensions can be accounted for by a geometric source term.

Consider the local flow tube generated by the streak lines from a small patch on the detonation front. Let A be the cross-sectional area of the flow tube. The reaction zone is modeled by the reactive flow equations in a duct of variable cross-sectional area

$$\frac{\partial}{\partial t}\begin{pmatrix}\rho A\\ \rho A u\\ \rho A E\\ \rho A\lambda\end{pmatrix}+\frac{\partial}{\partial x}\begin{pmatrix}\rho A u\\ \rho A[u^2+PV]\\ \rho A u(E+PV)\\ \rho A u\lambda\end{pmatrix}=\begin{pmatrix}0\\ P\partial_x A\\ -P\partial_t A\\ \rho A\mathcal{R}\end{pmatrix}, \tag{4}$$

where λ is the mass fraction of the reaction products and $\mathcal{R}$ is the specific reaction rate.

The flow tube is not Galilean invariant. There are two special frames in which to leading order $\partial_t A = 0$ and the equations simplify: (i) the shock-attached frame, and (ii) the normal frame. In these frames the only geometric term that enters the duct equations is the local divergence of the streak lines, $\kappa = \lim_{A\to 0}(\partial_x A/A)$. The shock-attached frame is unsuitable because $\kappa(x)$ is not *a priori* known. The normal frame is defined by the condition that the velocity ahead of the front is on average normal and has magnitude given by

$$\kappa_f u_{\text{ahead}} = \nabla_T \cdot \vec{u} \ , \tag{5}$$

where κ_f is the front curvature and $\nabla_T \cdot \vec{u}$ is the velocity divergence in the plane tangent to the front. In two dimensions the normal coordinates correspond to the streak lines. In three dimensions the streak lines are neither normal to the front nor do they lead to a stationary flow tube. Nevertheless, in this distinguished frame $\kappa = \kappa_f$ is constant to $O(\kappa_f w)$ (Menikoff et al. 1995).

To leading order the reaction zone is quasi-steady and adjusts adiabatically to its local front curvature. The PDEs then reduce to ODEs for the reaction-zone profile. The ODEs extend the ZND model of a planar detonation wave to the case of a curved detonation wave. Integrating the ODEs across the reaction zone leads to the modified jump conditions

$$\Delta[\rho(D-u)] \;=\; \kappa_f w \,\langle \rho u \rangle \ , \tag{6}$$

$$\Delta\left[(\rho(D-u))^2 V + P\right] \;=\; \kappa_f w \,\langle \rho(D-u)u \rangle \ , \tag{7}$$

$$\Delta\left[E + PV + \frac{1}{2}(D-u)^2\right] \;=\; 0 \ , \tag{8}$$

where D is the detonation velocity, $\Delta[f] = f(x_a) - f(x_b)$ is the change of variable f across the detonation wave, and

$$\langle f \rangle = \frac{1}{w}\int_{x_b}^{x_a} dx\, f$$

is the average value of f in the reaction zone. The left-hand sides correspond to the standard jump conditions, and the correction terms on the right-hand sides are proportional to the dimensionless product $\kappa_f w$. Because these relations are derived from the quasi-steady approximation, they are only valid in the normal frame and are only accurate to $O\big((\kappa_f w)^2\big)$.

In contrast to a planar detonation in which the jump conditions depend solely on the equation of state of the detonation products, for a curved detonation the jump conditions are also affected by the reaction-zone dynamics, in particular by the competition between the source term for the heat release and the geometric source term due to the front curvature. The averages are slowly varying functions. The dominant factor in the correction term is the reaction-zone width. Consequently, a large numerical reaction-zone width leads to an artificially large curvature effect. Similarly, changing the grain size of an explosive varies the reaction-zone width and has a corresponding effect on the physical curvature effect.

The one-dimensional approximation breaks down in regions of high front curvature when the tangential flow gradients are comparable to the normal gradients. This also occurs in the interaction region in which wave profiles overlap. When the interaction of discontinuous waves does not have a unique solution based on a shock polar analysis then the dynamics within the wave profiles would resolve the non-uniqueness. For example, the transition between regular and Mach reflection is a threshold phenomenon and could be affected by the dissipative mechanism that gives shock waves a profile. The experiment by Lambourn & Wright (1965) shows that the von Neumann interaction identified by Colella & Henderson (1990) occurs when Mach reflection of detonation waves is expected.

Since the physical dissipative mechanism is difficult to vary experimentally, numerical calculations in which the wave profile is resolved would be very useful for developing an understanding of how the bifurcation of two-dimensional wave patterns can be affected by dissipative mechanisms that take place on short length scales and that for a shock-capturing algorithm represent sub-grid physics. Transitions can depend on dimensionless ratios of length scales associated with different dissipative mechanisms even in the limit as the shock width vanishes. In assessing the accuracy of numerical solutions, it is important to know when localized errors can trigger phenomena that change the global flow.

4. Summary

The conservation form of the fluid equations leads to strong constraints on the wave structure of fluid flow. In one dimension, when a shock wave has a non-zero width, transients can cause an entropy anomaly. This is because of the mass, momentum, and energy contained within a shock profile that must be accounted for when a profile is formed or changed. An entropy anomaly commonly results from interactions involving shock waves. The effect has been measured in experiments with partly dispersed shock waves. In this case, the shock width is due to a physical relaxation process.

In higher dimensions, the leading-order effect on a shock profile is an extra geometric source term due to front curvature. The competition between source terms results in a modification of the Hugoniot jump conditions proportional to the dimensionless product of the shock width and the front curvature. This leads to a curvature effect on the propagation of a partly dispersed shock wave. The curvature effect is most noticeable for a detonation because of the minimum speed at which a steady-state detonation wave can propagate. The effect has been measured in rate-stick experiments.

In order to simulate correctly the propagation of a steady-state shock wave, finite-difference shock-capturing algorithms are written in conservation form.

Numerical solutions are subject to the same analysis as those for the continuum PDEs. A numerical shock width has qualitatively the same effect as the width of a partly dispersed shock wave from a relaxation phenomenon or a chemical reaction. Hence, an artificial numerical length scale can mimic real physical effects. This analogy provides a unifying explanation for a large class of diverse and seemingly unrelated numerical errors.

Acknowledgement. This research supported by the U.S. Department of Energy, Los Alamos National Laboratory.

References

Aleksandrov EN, Veretennikov VA, Dremin AN, Shvedov KK (1967) The detonation mechanism in porous explosives. Fizika Goreniya i Vzryva 3:471–484. (English translation in Combustion Explosion and Shock Wave 3:285–293).

Becker E (1972) Chemically reacting flow. Ann Rev of Fluid Mech 4:155–194.

Chern I-L (1991) Large-time behavior of solutions of Lax-Friedrichs finite difference equations for hyperbolic systems of conservation laws. Mathematics of Computation 56:107–118.

Colella P, Henderson LF (1990) The von Neumann paradox for the diffraction of weak shock waves. J Fluid Mech 213:71–94.

Donguy P, Legrand N (1981) Numerical simulations of non-ideal detonations of a heterogeneous explosive with the two dimensional Eulerian code C.E.E. In: Seventh Symposium (Int) on Detonation, pp 695–702.

Lambourn BD, Wright PW (1965) Mach interaction of two plane detonation waves. In: Fourth Symposium (Int) on Detonation, pp 142–152.

Menikoff R (1989) Determining curvature effect on detonation velocity from rate stick experiments. Impact of Computing in Science and Engineering 1:168–179.

Menikoff R (1994) Errors when shock waves interact due to numerical shock width. SIAM J Sci Comp 15:1227–1242.

Menikoff R, Lackner KS, Bukiet BG (1995) Modeling flows with curved detonation waves. Combustion and Flame (to appear).

Moulard H (1989) Particular aspect of the explosive particle size effect on shock sensitivity of cast PBX formulations. In: Ninth Symposium (Int) on Detonation, pp 18–24.

Noh WF (1987) Errors for calculations of strong shocks using artificial viscosity and an artificial heat flux. J Comp Phys 72:78–120.

Quirk JJ (1994) A contribution to the great Riemann solver debate. Int J for Numerical Methods in Fluids 18:555–574.

von Neumann J, Richtmyer RD (1950) A method for the numerical calculation of hydrodynamic shocks. J Appl Phys 21:232–237.

Curved detonation fronts in solid explosives: collisions and boundary interactions

J. B. Bdzil
Los Alamos National Laboratory
Los Alamos, NM 87545 USA

T. D. Aslam, D. S. Stewart
TAM Department, University of Illinois
Urbana, IL 61801 USA

Abstract: Detonation Shock Dynamics (DSD) can be used to model the effects that shock curvature κ has on detonation speed $D_n(\kappa)$. At the edges of the explosive, $D_n(\kappa)$ is supplemented with boundary conditions. By direct numerical simulation (DNS), we study how the reaction zone interacts with the edge. DSD theory has been integrated with the level-set method of Osher & Sethian and the Los Alamos DNS code Mesa to create a powerful tool for simulating complex explosive-containing systems.

Key words: Detonation, Curvature effect, Edge interactions, Mach reflection

1. Introduction

To accurately predict the propagation of detonation through an explosive, one needs to model the physics that occurs on the chemical reaction-zone scale η_r. In sharp contrast to the structured shocks observed for gaseous detonation, those for heterogeneous solid explosives are broadly curved on the η_r scale. The speed of the detonation is strongly influenced by the curvature of the shock κ; with reductions of speed of 40% in strongly divergent flows.

Safety concerns have led to the use of explosives that satisfy $\mathcal{L}/\eta_r = O(10^3)$, where $\mathcal{L}$ is a representative dimension of the explosive. Curvature effects have a strong influence on detonation in these systems, and highly resolved multidimensional simulations are "expensive." A body of theory and supporting experiments, called Detonation Shock Dynamics (DSD) (Bdzil & Stewart 1989), has been developed that treats these curvature effects (Aslam et al. 1996). The DSD front theory derives a speed function $D_n(\kappa)$ based on a weakly divergent, quasi-one-dimensional (1D) model of the detonation reaction zone. This function can also be determined directly from experiments. The regions of strongest flow divergence are found near the explosives' boundaries. Boundary conditions (BC) must be supplied in addition to $D_n(\kappa)$ to treat these interactions. The dynamics of broadly curved fronts, described by DSD, interacts with the edge through a narrow boundary layer a few η_r thick, where the flow is both reactive and fully two dimensional (2D).

In Section 2 we give a brief review of DSD. Front theories require a front propagation algorithm. In Section 3 we describe an engineering implementation of DSD that uses a level-set (LS) algorithm (Osher & Sethian 1988) to propagate the DSD front (Aslam et al. 1996). To integrate DSD front theory with hydrodynamic simulations, an accurate method is needed to quickly burn the explosive and capture the detonation state consistent with $D_n(\kappa)$. In Section 4 we describe our new burn model and show a full DSD-based simulation. In Section 5 we discuss results obtained from high-resolution simulations of two edge problems: (1) the sudden loss of confinement, and (2) oblique interaction of detonation with a rigid wall. Owing to reaction-zone effects, we find that detonations exhibit von Neumann-like reflection (Colella & Henderson 1990).

2. DSD theory

DSD is the name given to the body of multidimensional detonation theory and experiments that is used to describe the dynamics of detonation with broadly curved shocks on the reaction-zone scale η_r. The model equations used to describe this limit derive from the 2D, reactive Euler equations transformed to shock-attached, intrinsic coordinates. Shown in Fig. 1 is our coordinate net of straight lines normal and curves locally parallel to the shock, all moving with the shock normal speed D_n.

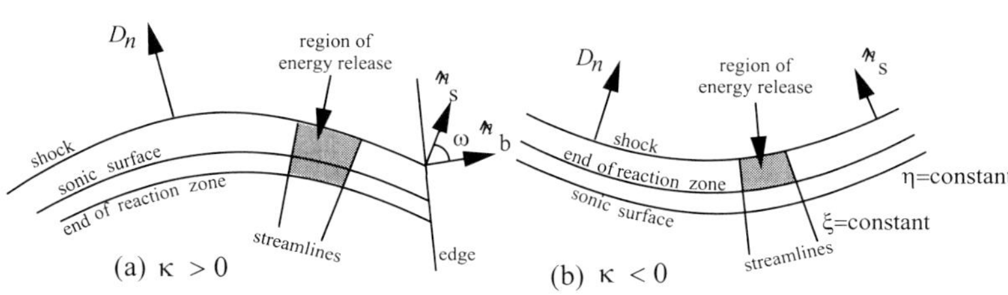

Figure 1. A snapshot showing the intrinsic coordinates and the definition of the boundary-edge angle ω; in (a) for a diverging detonation and in (b) for a converging detonation. The speed of the wave is influenced by two factors, the convergence/divergence and the location of the sonic surface.

2.1. Interior flow

The weak shock-curvature limit defined by $\kappa = O(\epsilon)$, where $\epsilon \equiv (reaction - zone\ scale)/(shock\ radius\ of\ curvature)$, is the basis of most theoretical analysis of 2D detonation. This limit envisions that $O(1)$ changes in the field variables occur over $O(\epsilon^{-1})$ distances in the ξ-direction, and the flow velocity in the ξ-direction is no greater than $O(\epsilon)$. Under these assumptions, the DSD-related time derivatives are $O(\epsilon)$, the flow is principally in the η-direction, and is "nozzle"-like (see Fig. 1), and 2D effects enters the flow only parametrically, via $\kappa(\xi, t)$. This limit allows for time dependence on the $O(1)$ reaction scale, provided no $O(1)$ velocities are generated in the ξ-direction. The various streamtubes then communicate with one another only through the shock-surface compatibility condition (i.e., through the LS equation or whichever propagation method we use). The DSD reaction zone equations are:

$$(\dot{\rho}) + \rho(D_n - U)_\eta = -\rho U \kappa , \tag{1}$$

$$\rho(\dot{U}) - P_\eta = 0 , \tag{2}$$

$$\rho^2(\dot{e}) - P(\dot{\rho}) = 0 , \tag{3}$$

where $(\dot{}) \equiv (\partial/\partial t)_{\eta,\xi} + (D_n - U)(\partial/\partial\eta)_{t,\xi}$ and ρ, U, P, $e(P, \rho, \lambda q)$, λ, and q are the density, laboratory particle velocity in the η- direction, pressure, specific internal energy, degree of reaction, and heat of detonation, respectively. Henceforth, when t, η, ξ, etc. are used as subscripts they denote partial derivatives. The *master equation*, derived from Eqs. (1) – (3) helps us analyze the 2D reaction zone

$$P_t + \rho(D_n - U)U_t = -\rho[(D_n - U)^2 - C^2]U_\eta + \sigma , \tag{4}$$

where $\sigma \equiv -(e_\lambda/e_P)R - \rho C^2 U \kappa$ and C, $-(e_\lambda/e_P)R \geq 0$, and $(\dot{\lambda}) \equiv R \geq 0$ are the sound speed, heat-release rate, and chemical rate law, respectively.

In its simplest realization, DSD theory assumes that $D_n/D_{CJ} = 1 + O(\epsilon)$, that the departures from the Zeldovich-von Neumann-Doring (ZND) limit are small, and neglects the inertia of the reaction zone (setting t derivatives in Eqs. 1 – 4 to zero). The reaction-zone structure is then obtained by solving the steady form of Eqs. (1) – (3), subject to the shock conditions and generalized Chapman-Jouguet (CJ) condition obtained from Eq. (4) (i.e., $(D_n - U)^2 - C^2 = 0$ when $\sigma = 0$, provided that $\kappa \geq 0$). This defines an eigenvalue problem that obtains $D_n(\kappa)$. Figure 1 shows that two effects contribute to $D_n \leq D_{CJ}$: (1) the divergence and (2) the movement of the sonic surface. For weakly convergent systems, effect (1) dominates and $D_n(\kappa) \geq D_{CJ}$ is derived by requiring $(D_n - U)^2 - C^2 = 0$ at the end of the reaction zone (i.e., $\lambda = 1$). Details can be found in (Bdzil 1981), (Bdzil & Stewart 1986), (Stewart et al. 1995), (Aslam et al. 1996) and references therein. Importantly for $D_n(\kappa)$, the shock evolves by parabolic dynamics and so is smooth. For real heterogeneous solid explosives, R is not well known. Then $D_n(\kappa)$ is determined directly from experiment. The $D_n(\kappa)$ for the explosive PBX 9502 is shown in Fig. 2a. Note the $O(1)$ variations of $(D_{CJ} - D_n)/D_{CJ}$.

2.2. Boundary conditions

The slow variations in the ξ-direction that are a part of DSD can break down at explosive boundaries. There the flow can be fully 2D and time-dependent on the $O(1)$ reaction-zone scale (Bdzil & Stewart 1986). On the slow-time DSD scale, this region can appear to be steady. Then the DSD and boundary regions are coupled as follows: (1) the DSD region drives the boundary with information on the shock slope and D_n, and in turn (2) the boundary region uses this data to return a (possibly) modified shock slope, etc. (Bdzil 1981). Using the boundary angle ω defined in Fig. 1, DSD supplies ω_{in} to the boundary flow, which then returns ω_{out} to the DSD region. The boundary is itself characterized by two angles that depend on the explosive/inert pair being considered: a critical angle ω_s and a confinement angle ω_c.

We've distilled these interactions into the following recipe. If $\omega_{in} < \omega_s$, then $\omega_{out} = \omega_{in}$; otherwise, $\omega_{out} = \omega_c$. The boundary dynamics are considered in more detail later in this paper.

3. Front propagators

Together, the $D_n(\kappa)$ function and the boundary conditions provide a complete dynamical description of the detonation front at the DSD level. A second element is needed to propagate the front: a shock compatibility condition that relates how changes in the speed of the front affect its shape. One of the most widely used forms of this condition is Whitham's ray method (Whitham 1973). Every section of the detonation front advances along a system of rays that resembles the bicharacteristic rays of geometrical optics. Although physically appealing, such methods can be logically complex. The front-attached rays converge and diverge with the front, which can lead to numerical problems. This is one member of the family of marker-particle or Lagrangian methods. Such methods are not well suited for engineering applications where the problem geometries are complex.

3.1. Level-set method

Osher and Sethian (1988) devised a powerful algorithm for propagating fronts with curvature-dependent speed. Their method obviates the need for the complex logic to treat collisions and avoids the problem of marker-particle methods. They consider the shock as a level curve $C(x, y, t) = 0$ embedded in a higher-dimensional LS function $\psi(x, y, t)$ that's defined on an Eulerian grid. The evolution equation for the LS function is derived by using the property $\psi(x, y, t) = constant$ along a level curve to obtain

$$\psi_t + \vec{D_n}(\kappa) \cdot \vec{\nabla}\psi = 0 \; . \tag{5}$$

Equation (5) is the LS-method surface compatibility condition.

The precise form selected for $\psi(x,y,0)$ is unimportant, however, $\psi(x,y,0)$ must be single valued. This insensitivity requires that; (1) D_n depend only on data from a single curve and (2) that the level curves not cross with time. The first property is automatic, since $D_n(\kappa)$. The second property follows from the fact that the distance between level curves, as measured by $d(x,y,t) = |\vec{\nabla}\psi(x,y,0)|/|\vec{\nabla}\psi|$, does not go to zero. By way of example, for $D_n = D_{CJ} - \alpha\kappa$ with $\alpha \geq 0$, the d-equation (which is derived from Eq. 5) is

$$d_t + D_{CJ}\vec{n} \cdot \vec{\nabla}(d) = \alpha\kappa^2 d \ , \tag{6}$$

for a cylindrically symmetric system. Since $\alpha\kappa^2 \geq 0$ and $d(x,y,0) = 1$, d increases with time. The result $d(x,y,t) > 0$ continues to hold for fully 2D systems with strong divergence at the boundaries (Aslam et al. 1996). In practice,

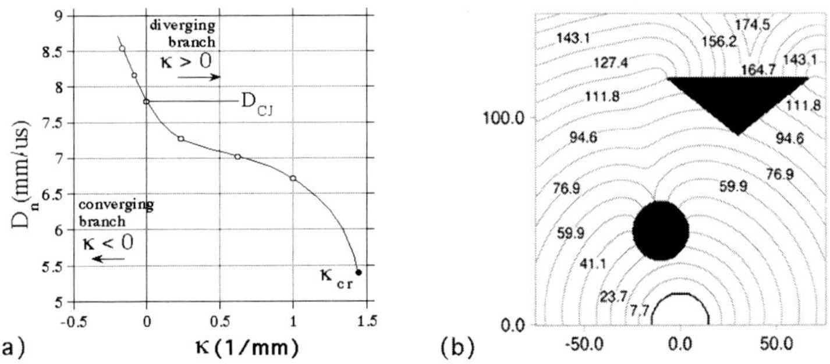

Figure 2. The $D_n(\kappa)$ for a typical condensed phase explosive, PBX 9502 is shown in Fig. 2a. Burn time contours for an engineering-style problem obtained with the LS-method are shown in Fig. 2b. The dark areas are obstacles.

we assign $\psi = 0$ to the shock, with $\psi > 0$ in unburnt material and $\psi < 0$ in burnt material. The burn time $t_b(x,y)$ is taken as the first time that $\psi < 0$ at a point. Figure 2b shows the result of a LS calculation of the burn times for an engineering-style problem. Initially, the detonation is a semi-circle with origin (0,0) and obeys $D_n = 8 \text{ mm}/\mu\text{s} - \kappa \times 66 \text{ mm}^2/\mu\text{s}$ everywhere. Bifurcation, merging, convergence, and divergence of the wave are all well captured. All of these ideas easily carryover to 3D.

4. Hydrodynamic simulations using DSD

The work that has gone into DSD was motivated by the need to capture reaction-zone effects in numerical simulations of engineering systems. Two elements are required to accomplish this goal: (1) accurate front evolution and (2) an ability to deposit the proper detonation state (i.e., $P(D_n)$, $\rho(D_n)$, $U(D_n)$) at the front.

To address element (2), we've developed a numerical reaction-zone model that uses a "pseudo" divergence Ω in place of κ in simulations, so that a detonation state consistent with D_n is captured. To eliminate precursors to the $D_n(\kappa)$ wave, we built a model around the reaction zone equation of state (EOS), $e(P/\lambda, \rho, q)$, which together with the trigger provided by the DSD burn times $t_b(x, y)$, causes the numerical "reaction zone" to be on the weak detonation branch. Starting with $P = 0$ and $\lambda = 0$, at $t_b(x, y)$ the explosive is rapidly burned (i.e., $\lambda \to 1$ in about the time it takes the wave to traverse one computational zone). From Eq. (4) it follows that the ratio κ/R determines D_n. Since this nonphysical numerical reaction rate $\tilde{R}$ is associated with the size of the computational zones, we replace $U\kappa$ in Eqs. (3) – (4), by

$$U\kappa \Rightarrow \Omega\tilde{R}\ , \qquad R \Rightarrow \tilde{R}\ , \tag{7}$$

and then solve for $D_n(\Omega)$. Elimination then yields $\Omega(\kappa)$, the "pseudo" divergence required to get the numerical detonation to be compatible with $D_n(\kappa)$. This 2-part model is implemented in a second-order, 2D, multi-material, Eule-

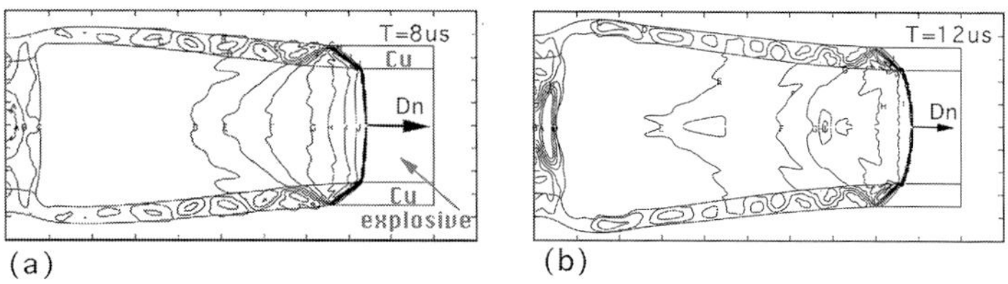

Figure 3. Pressure contours for Mesa calculations of the detonation cylinder test. The detonation wave is moving to the right. Frame (a) shows the results at $t = 8\ \mu$s from the $D_n = D_{CJ}$ model, (b) shows the results for the $D_n(\kappa)$ model at $t = 12\ \mu$s.

rian grid hydrodynamics code at Los Alamos called Mesa (Holian et al. 1989). The results of two calculations of the detonation Cu-cylinder test (one using the standard $D_n = D_{CJ}$ model and the other using DSD) are displayed in Fig. 3. The difference in the timing and wave shapes is striking.

5. Boundary-condition study

A complete theoretical analysis of fully 2D, reactive edge flows is out of reach. Here we present some results obtained using high-resolution numerical simulations. A limited amount of supporting theory is offered to help with the interpretation of these results.

We study a simplified Euler fluid model, and use a polytropic EOS with a large γ to mimic condensed phase explosive, where

$$e = \frac{P}{(\gamma - 1)\rho} - q\lambda \ , \tag{8}$$

with $\gamma = 3$, an initial density $\rho_0 = 2000 \ \text{kg/m}^3$, and a heat of detonation $q = 4 \times 10^6 \ \text{m}^2/\text{s}^2$, so that $P_{CJ} = 3.2 \times 10^{10} \ \text{N/m}^2$ is large. A simplified state-independent rate law is used ($k = 2.51 \times 10^6 \ \text{s}^{-1}$)

$$(\dot{\lambda}) = R = k\sqrt{1 - \lambda} \ , \tag{9}$$

for which the 1D steady-state reaction-zone length is 4×10^{-3} m and the particle reaction time is 0.8 μs. The numerical simulations were done using a second-order Godunov code called Caveat (Addessio et al. 1990). The grid size used in the calculations was 2×10^{-4} m in the streamwise direction. Two types of boundary interactions are studied here. For both, the detonation is initially a plane ZND wave whose direction of propagation is colinear with a *flat rigid* wall (i.e., $\omega = 90°$). Problem (1) considers the response of the detonation to a sudden loss of confinement, while the interaction of the detonation with a converging, rigid wedge is studied in problem (2).

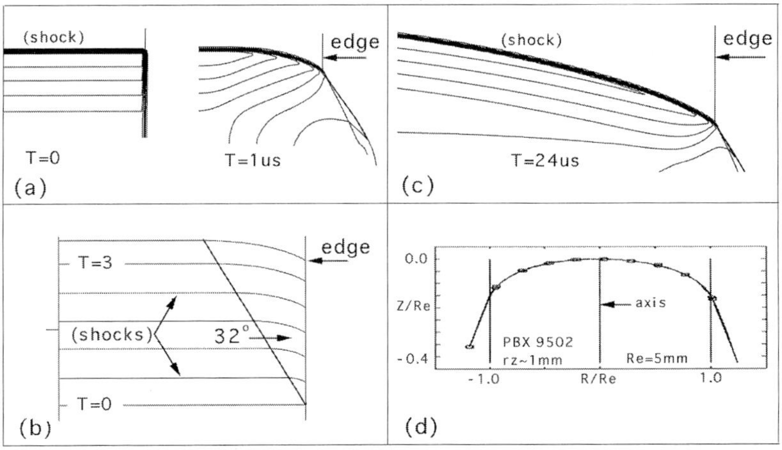

Figure 4. Response of a detonation reaction zone to a sudden loss of confinement. Frames (a), (b), (c), and (d) show the reaction-zone pressure contours, the evolving shock-front shape, late-time pressure contours, and an experimentally measured shock-front shape, respectively.

5.1. Loss of confinement

A collection of the results from problem (1) is shown in Fig. 4. Propagating upwards initially, the detonation loses confinement along its right boundary at $t > 0$. Figure 4a shows strong 2D flow at the edge as evidenced by $P_\xi = O(1)$ and the rapid change in the shock slope (ω changes from 90° to 55°). The Prandtl-Meyer expansion that develops limits the decrease in pressure at the shock which results in a fixed shock-edge angle after a very short transient, as shown in Fig. 4b. The rarefaction moves along the shock at $(\sqrt{C^2 - (D_n - U)^2})_{\eta=0} = 0.7 D_{CJ}$. Figure 4c shows a broadly curved DSD wave at $t = 24$ μs. The value $\omega_c = 55^\circ$, corresponds to a flow that is exactly sonic as measured with respect to the shock-edge intersection point. A discussion of the effect that increasing confinement and changes in ω_{in} have on these results is given elsewhere (Bdzil & Stewart 1986, Aslam et al. 1996). The experimentally measured shock-arrival trace for a $Re = 5$ mm radius detonating cylinder of PBX 9502 is shown in Fig. 4d. The $\omega_c = 45^\circ$, corresponds to the sonic angle for this material. For this explosive, $\eta_r = 1$ mm. Measured on this scale, the shock for this 10-mm diameter explosive is *broadly curved.*

5.2. Converging wedge

A collection of results from problem (2) is shown in Fig. 5. A ZND detonation propagating to the left meets a rigid, converging wedge at $x = 190$ mm ($x = 0$ is at the left). Standard three-shock theory for a CJ detonation predicts Mach reflection for $\omega_{in} \geq 45^\circ$. A simulation of the 50° wedge problem shows regular reflection. The evolving shock and D_n for both a 40° and 20° wedge shown in Fig. 5a display irregular reflection. The pressure contours at $t = 24$ μs shown in Fig. 5b reveal that the irregular reflection grows slowly for the 40° case. Although the structure looks somewhat classical, the reflected wave is clearly dispersed.

The pressure contours at $t = 24$ μs for the 20° wedge reveal a totally non-classical von Neumann-like reflection (Colella & Henderson 1990). The leading shock is broadly curved with no evidence of a reflected shock. This is a consequence of the diffraction of the reflected wave by the reaction-zone flow gradient. By contrast, the Mach reflection of an inert shock of comparable strength, shown in Fig. 5d, is classical. Figure 5e shows "Mach" reflection during the collision of two PBX 9502 detonations. These measurements were made at Los Alamos by Larry Hull (Hull 1995). With the progress of time (time advances to the left), the wave-interaction region (the left-most part of each trace) becomes wider and more "rounded."

We present a qualitative theoretical argument to help understand these observations. Equation (4) is exact along the shock:

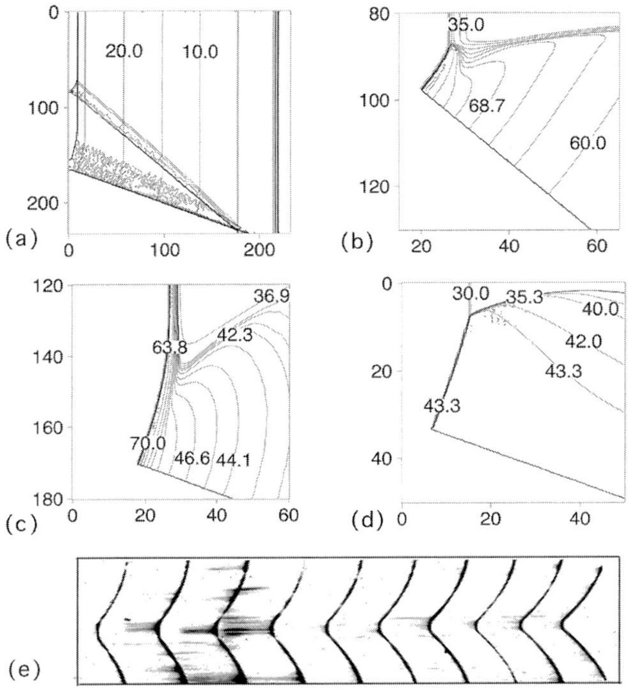

Figure 5. The oblique interaction of a resolved reaction zone detonation with a rigid wedge. Frames (a), (b), (c), (d), and (e) show the evolving lead shocks for a 40° and 20° wedge, pressure profiles at $t = 24$ μs for the 40° and 20° wedge, a strong inert shock over a 20° wedge and the curved "Mach" interaction for PBX 9502, respectively.

$$(D_n^2)_t = \frac{(\gamma+1)^2}{3}\left(\frac{(\gamma-1)}{(\gamma+1)}D_n^2\right)(U_\eta)_{\eta=0} + \frac{(\gamma+1)}{(\gamma-1)}\frac{D_{CJ}^2}{6}(R)_{\eta=0} - \frac{4\gamma}{3(\gamma+1)}D_n^3\kappa\ . \tag{10}$$

Differentiating Eq. (10) with respect to ξ, using the shock-compatibility condition, and then transforming the resulting equation to a reference frame moving at acoustic speed along the shock ($\nu = -(\sqrt{C^2-(D_n-U)^2})_{\eta=0}$ with $d\omega \equiv d\xi - \nu dt$), allows us to derive an equation for the amplitude of the leading edge of a weak 2D disturbance moving into a 1D ZND detonation

$$4\tilde{K}_\tau = -\frac{2(\gamma+9)}{3(\gamma+1)}\tilde{K}^2 - \frac{(\gamma+1)}{6(\gamma-1)}(1-\sigma)\tilde{K} + \frac{2(\gamma-3)}{3\sqrt{\gamma^2-1}}\tilde{K}_{\tilde{\omega}} + O((U_{\eta\omega})_{\eta=0})\ . \tag{11}$$

The variable $\tilde{K} \equiv (\kappa D_{CJ}/k)_{wavehead}$ is the scaled shock curvature, $\tau = kt$

is the scaled time, and σ measures the state dependence of the rate (e.g., $R = k\sqrt{1-\lambda}\exp(\sigma D_n/D_{CJ})$). The strength of the 2D disturbance is $\tilde{K}$, with $|\tilde{K}| \to \infty$ denoting triple-point formation on the shock. Neglecting the $O((U_{\eta\omega})_{\eta=0})$ term in Eq. (11), setting $\gamma = 3$, and solving, one finds that for $\sigma < 1$, $\tilde{K} \to 0$ when $\tilde{K}_{t=0} > -(1-\sigma)/6$ and $\tilde{K} \to -\infty$ otherwise. Thus, sufficiently weak convergence does not lead to triple points. This is what we see in our simulations. Suppressing the reaction-related terms in Eq. (11), one finds that all levels of convergence lead to triple points for inert flows. The DSD boundary conditions derived from this example are: (1) when $\omega_{in} < \omega_s = 35°$, then $\omega_{out} = \omega_{in}$ and (2) when $\omega_{in} > \omega_s = 35°$, then $\omega_{out} = \omega_c = 90°$.

Acknowledgement. We thank Rudy Henninger for Mesa-DSD. John Bdzil was supported by the U.S. Department of Energy. Tariq Aslam and Scott Stewart were supported by the U.S. Air Force, Wright Laboratory, F08630-92-K0057.

References

Addessio FL, Baumgardner JR, Dukowicz JK, Johnson NL, Kashiwa BA, Rauenzahn RM (1990) Caveat: A Computer Code for Fluid Dynamics with Large Distortions and Internal Slip, Los Alamos Report, LA-10613.

Aslam TD, Bdzil JB, Stewart DS (1996) Level set methods applied to modeling detonation shock dynamics. J Computational Physics (accepted).

Bdzil JB (1981) Steady-state two-dimensional detonation. J Fluid Mech 108:195–226.

Bdzil JB, Stewart DS (1986) Time-dependent two-dimensional detonation: the interaction of edge rarefactions with finite-length reaction zones. J Fluid Mech 171:1–26.

Bdzil JB, Stewart DS (1989) Modeling two-dimensional detonation with detonation shock dynamics. Phys Fluids A 1:1261–1267.

Colella P, Henderson LF (1990) The von Neumann paradox for the diffraction of weak shock waves. J Fluid Mech 213:71–94.

Holian KS, Mandell DA, Adams TF, Addessio FL, Baumgardner JR, Mosso SJ (1989) Mesa: A 3-D computer code for armor/anti-armor applications. In: Supercomputing World Conference, Oct 17–20, 1989, San Francisco, CA.

Hull LM (1995) Los Alamos National Laboratory, private communication.

Osher S, Sethian JA (1988) Fronts propagating with curvature-dependent speed: algorithms based on Hamilton-Jacobi formulations. J Computational Physics 79:12–49.

Stewart DS, Aslam TD, Yao J, Bdzil JB (1995) Level-set techniques applied to unsteady detonation propagation. In: Modeling in Combustion Science, Buckmaster J and Takeno T (eds), Springer-Verlag, Berlin.

Whitham GB (1973) Linear and Nonlinear Waves, Wiley, New York.

Bubbly cavitation in intense rarefaction waves and its effects

V. K. Kedrinskii
Lavrentyev Institute of Hydrodynamics
Novosibirsk 630090 Russia

Abstract: The paper presents a review of experimental results, mathematical models, and numerical investigations of wave processes and the dynamics of liquid media with microinhomogeneities under impulse loading. The presence of microinhomogeneities can substantially change the state of the liquid and the structure of the applied wave field as well as the mechanisms of physical processes that take place in liquid media.

Key words: Microinhomogeneities, Cavitation structure, Rarefaction focusing, ESWL, Shock loading, Detonation initiation mechanism, Vapor explosion

1. Introduction

Microinhomogeneities are microbubbles of free gas, solid microparticles, and their combinations (Besov et al. 1992), which are called the cavitation nuclei.

According to modern data for distilled water, based on the experimental studies of the dynamics of scattering indicatrix of a laser beam on microinhomogeneities and on the analysis of the tracks of diffraction spots of nuclei illuminated by a laser beam, the spectrum of nuclei distribution can be described by a simple dependence (Kedrinskii 1989) where $N_o \simeq 10^5 - 10^6$ cm^{-3} is the total density of nuclei (see Fig. 1). When pulsed tensile stresses appear in the fluid, the nuclei begin to grow, forming a cluster whose form and dynamics are determined by a spectrum of initial nuclei sizes, by the character of the stress applied, and by the boundary conditions.

The above data enable us to make conclusions regarding the two-phase structure of a real liquid and to use a two-phase mathematical model for average characteristics to describe the wave processes in a liquid cavitating behind a rarefaction wave front (Kedrinskii 1975). This model can be supplemented by the conditions of heat and mass exchange (for instance, this can be the energy equation with phase transition, Nakoryakov et al. 1990) and by the equation of chemical kinetics for bimolecular reactions (Todes 1933). When numerically analyzing wave processes in chemically active media one usually considers the

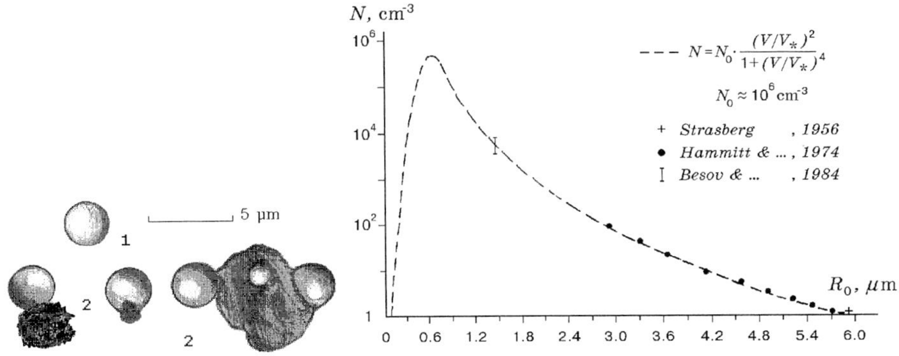

Figure 1. Structure of microinhomogeneities (left): 1 – free gas bubble, 2 – combination structure; spectrum of cavitation nuclei (right).

model of the so-called adiabatic explosion (Kedrinskii and Zamaraev 1989). The study of different stages of a chemical reaction occurring under explosion was carried out by Hasegava and Fujiwara (1982).

As is obvious from practice, the physics of cavitation has direct relation to: the problem of detonation initiation in liquid High Explosive (HE); the problem of the mechanics of fragmentation of kidney stones in lithotriptors; the fundamental problem of liquid strength under explosive loading; as well as to well-known but not completely understood effects of large-scale explosions of fuel containers during transport. Some approaches to the problems mentioned for which cavitation effects can be of principal importance will be considered.

2. Disintegration mechanism of target at shock-wave focusing in ESWL systems

Experimental studies carried out by many researches confirm that the role of the cavitation zone developing around the target when a shock wave focuses on it can appear essential for the process of the target disintegration in lithotriptor systems. However, in laboratory studies of wave focusing in a liquid, as a rule, the rarefaction phase receives little attention, although it naturally arises when using open segments of ellipsoids and spheres as reflectors or sources.

Grönig (1989) in his survey gives a short review of ESWL (External Shock Wave Lithotripsy), distinguishing the works of Jutkin, Goldberg, Forssman, Chaussy et al., Reichenberger and Naser, Ziegler and Wurster, Muller, Delius and Brendel, Takayama et al., and Coleman and Saunders. As a rule, in these papers cavitation effects are simulated primarily as the results of the interaction of shock waves with artificially produced gas bubbles and have no connection to cavitation, as a new medium state, which considerably varies the wave-field parameters (Kedrinskii 1976) and the conditions at which bubble collapse with cumulative jet formation is possible.

Physical aspects of wave focusing in the context of ESWL were analyzed by Sturtevant (1989). He reports that the role of rarefaction waves in the process is to cause the sharply peaked wave form and to limit the amplitude of the shock wave.

Kuwahara (1989) and Kitayama et al. (1987) noted that after the shock wave hits the stone, cavitation phenomenon is seen around the stone; thereafter, the stone deforms and disintegrates. Delius (1990) related the damage effect to the formation of cumulative microjets. Prat (1993) concluded that a shock-induced cavitation has a potential for selective and noninvasive destruction of deep tumors through an extra-corporeal approach. Nagoya et al. (1993) discovered a hysteresis effects in studying human blood. This effect was also described and explained by Besov et al. (1992). Its nature appeared to be determined by the two-phase state of the medium containing microinhomogeneities which played the role of cavitation nuclei and changed their structure under the effect of the wave field.

Sato et al. (1993) suggested the use of a soft-flash radiography system to visualize shock-wave-induced cavitation bubbles in living tissue. Earlier, the technique of pulse X-ray was used for the same purpose in studying distilled water (Bayikov et al. 1984, Berngardt et al. 1984). Church and Crum (1990) suggested the idealized pressure-wave forms produced by commercially available lithotriptors.

The references could be continued, but the explanations of the reasons for fracture would remain the same as mentioned above: cumulative jets arising under collapse of cavitation bubbles and shear stresses inside the target resulting in formation of microcracks. On the basis of the above statements and known experimental facts, we shall try to analyze other possible mechanisms of the evolution of the process of target disintegration.

Experimental simulation. First of all we will note the statements of Delius and Philipp (Philipp 1991). They have carried out experiments on the process of formation of a bubble cluster on the surface of a gall stone as a result of focusing a shock wave generated by Dornier Lithotriptor XLI with amplitude in the front of 65 MPa and a rarefaction phase of maximum amplitude 6 MPa. In these experiments a gall stone was positioned onto foil in the geometrical focus of the Lithotriptor. High-speed photo registration has shown that during the process of cavitation development a gall stone (target) was completely surrounded by a dense bubble cluster which then was transformed probably into a gas-vapor layer on the surface. Analysis and prediction of further behavior of the cluster and the effect of its dynamics on the target are impossible to state directly.

However, principal moments of the process can be simulated by the example of cavitation effects arising at the bottom of a tube filled with liquid, the tube being accelerated vertically downward by the impact (as in the experiments of Hansson et al. 1982). The amplitude of acceleration was about $2 \cdot 10^4$ ms^{-2} with the width of the pulse near the base of the acceleration peak about 300 μs. The high-speed photos show that a dense cavitation zone transformed over 4 ms into a gas-vapor layer.

One can consider that the pressure in this cluster is of the order of the pressure of saturated vapors (for water 2.34 kPa), and the pressure at the outer boundary of the cluster is equal to atmospheric pressure ($p_\infty = 10^2$ kPa). Then under the effect of pressure difference, the layer starts collapsing and over the time 7 – 8 ms disappears. In this time the gauges placed on the tube bottom record a sharp jump in acceleration (Fig. 2, $t = 7.735$ ms) with the amplitude exceeding the initial value and pressure pulse (Fig. 3b); with the amplitude of about 1.5 MPa and duration 1.5 ms (unpublished data received by the author together with I. Hansson and K. Morch). It is clear that the effect was caused by hydraulic impact. Under the effect of the second acceleration pulse a new cavitation cluster develops near the bottom and the process repeats. The oscillograms of the acceleration (Fig. 2) record repeated intense hydraulic impacts ($t = 13.69$ and 18.425 ms) and thus the hydraulic impacts can be considered as the principal mechanism of failure. The pressure oscillograms (Fig. 3c) confirm this conclusion. One should not exclude the possibility of the generation of a series of intermediate weaker hydraulic impacts (Fig. 3b) that are probably connected with the collapse of individual large bubbles or their clusters near the bottom. The digits (Fig. 3) show the parameters of $p(t)$-curve in the location points of the marks.

3. Some problems of pulse fracture of liquid

The process of fracturing a free-boundary liquid under explosive loading is called cavitation fracture, since all features of its development are completely determined by the phenomena of bubble cavitation. This is a complex of phenomena whose sequence can be presented as the stages of bubble cluster formation, its transformation into a foam-like structure, and then into a droplet system. The first two stages were studied in detail using the following methods and measuring techniques (Kedrinskii 1993): (1) two-diaphragm hydrodynamic and electromagnetic shock tubes, (2) flash radiography, (3) scattering indicatrix (statics

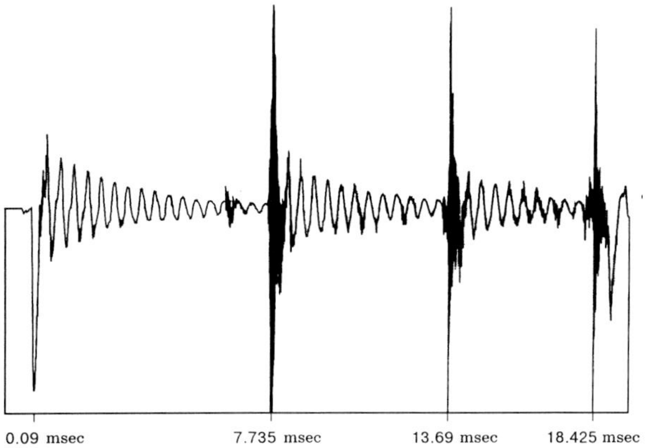

Figure 2. Oscillogram of acceleration: repeated hydraulic impacts.

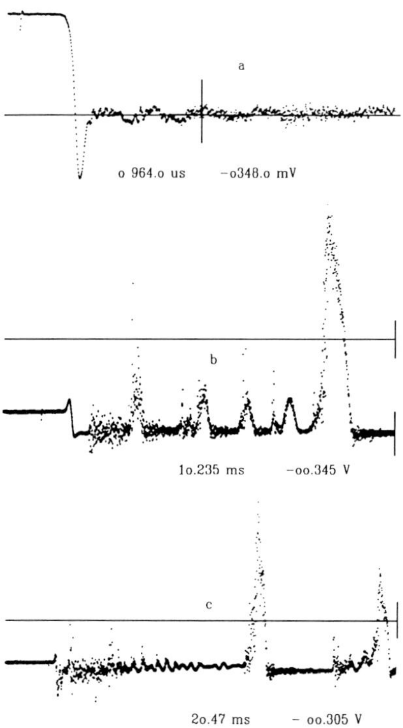

Figure 3. Pressure oscillograms: the tube bottom accelerated by an impact.

and dynamics), (4) electromagnetic capacitance and pressure gauges, the dynamic thrust and electric potential methods, and (5) computer processing.

The mechanism of development of cavitation zones has been formulated (Kedrinskii 1986): in weak rarefaction waves a cluster of visible bubbles is gradually formed depending on the initial position of the nuclei in the spectrum; during the high-intensity rarefaction phase all nuclei attain visible size simultaneously; the liquid becomes a monodisperse medium that is very essential for the mathematical model.

Application of the two-phase model made it possible to calculate the dynamics of a visible cavitation zone in the problems of underwater explosion and rarefaction wave profiles, as well as to show that the steepness of the rarefaction wave front plays a major role in the problem of extreme tensile stresses admitted by a real liquid (Kedrinskii 1976).

A number of criteria of the development process have been obtained (Kedrinskii 1968, 1989; Berngardt et al. 1995): (1) characteristic distance of wave attenuation, (2) relaxation time of tensile stresses and ultimate tensile stresses, and (3) relaxation time of cavitating medium to the state of "bulk density." For example, for $k_o = 10^{-10}$, $R_o = 10^{-4}$ cm, and $p = -30$ MPa in the rarefaction wave, the relaxation time equals $t_{cav} \simeq 8.3 \cdot 10^{-8}$ s, and the liquid withstands only $p_* \simeq -3.8$ MPa at $\tau = 1$ μs.

The third stage poses the principal problems in the investigation of the process of cavitation fracture and consists in revealing a physical transition mechanism of a cavitating liquid state from foam to droplet phase. The problem was solved using the process of the disintegration of a liquid drop as a result of ultra-short impact. A drop of distilled water (a few millimeters in size) was placed on the diaphragm of the electromagnetic shock tube which under the action of a pulse magnetic field impacted on the drop. As a result a shock wave of amplitude 5–6 MPa and a duration of the order of 3–4 μs was formed in the drop.

We could observe a circular cumulative jet as a thin sheet and the formation of a cavitation zone. By 70 μs the whole drop "boils," acquiring a clear cellular structure. The base of this structure consists of large clusters formed by junction of cavitation nuclei growing in the tensile stress field.

Figure 4 shows a frame (left) and detail (right) of a high-speed photo of the last stage of the fracture (original drop diameter was 5 mm) when the characteristic size of the structure reached a few centimeters at $t \simeq 1.5$ ms. The drop transforms into a peculiar foam phase with clear boundaries, as a mesh of liquid bundles in which the main mass of the original drop is likely to be concentrated. The elements of the liquid mesh gradually decompose into

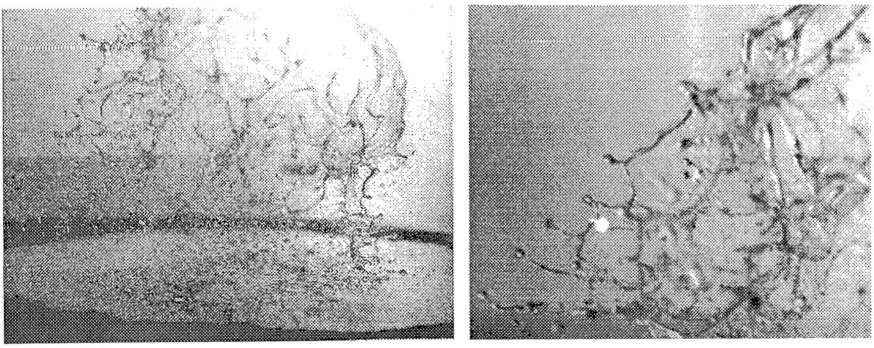

Figure 4. Cellular structure of fracturing drop at pulse impact.

separate fragments under tension, transforming into jets that later disintegrate into separate drops. Notice that even in the drop state the flow conserves the cellular structure pattern for a rather long time.

4. Cavitation effects in chemically active liquid media

In this survey we select from a large number of papers only those dealing with the problem of the effect of microinhomogeneities. This survey does not claim to span the information completely, but it is intended to analyze possible mechanisms of bubbly initiation proposed in the literature.

Bowden and Yoffe (1958) pointed to the thermal nature of explosion when initiating liquid HE by an impact. Small gas bubbles that are usually available in a liquid were recognized as one of the possible seats of heating up. To initiate the explosion the temperature in the seat should be of the order of 800 K with heating duration in the range of 1–100 μs.

Andreev (1957) and Johansson (1957) formulated a model of initiation seat according to which the possible availability of small particles or gaseous phase of HE in bubbles is more important. They concluded that the fast increase of pressure in gaseous combustion products of suspended particles or vapors of HE inside the bubble is responsible for the initiation of detonation.

Campbell et al. (1961) experimentally proved that the mechanism of shock initiation has its origin in the interaction of the wave with microinhomogeneities resulting in the formation of hot spots, and its successive amplification. Dubovik and Bobolev (1978) consider that, in spite of high adiabatic temperatures, a heat exchange in the system "gas bubble-liquid" is unlikely to explain the mechanism of liquid HE initiation. They analyzed possible mechanisms of the occurrence of the smallest droplets of HE in a bubble and estimated their critical value when the droplets could be quickly heated at the time of bubble collapse as well as the value of adiabatic period of induction of mixture explosion inside the bubble.

The experimental facts of increasing the mechanical sensitivity of HE as a result of a successive series of loading pulses (a so-called fractional impact) are known. The first shock wave in a specimen of liquid HE interacts with its free surface and transforms into a rarefaction wave. The subsequent interaction of successive shock waves with cavitation bubbles results in HE initiation (Dubovik and Bobolev 1978).

According to the results of many experiments the shock sensitivity of liquid HE really depends on the quantitative characteristics and the character of inclusions of different nature (Winning 1956, Campbell et al. 1961, Dremin et al. 1970) and the detonation processes in liquid HE develop only on the condition that the liquid becomes nonuniform owing to the above-mentioned loadings (Dubovik et al. 1967, Amster et al. 1967).

Field et al. (1992) and Bourne and Field (1991) considered the effects of stronger shocks of 3 GPa passing over single cavities and cavity arrays, as well as the collapse of cavities in reactive emulsion systems. Roberts and Field (1993) showed that voids of diameter 1.5 mm introduced into sample of octol and cyclotol in the vicinity of the point of projectile impact were the main source of ignition.

This question has not been adequately investigated, in particular, because the numerical models of wave phenomena in cavitating liquid HE have not been seriously studied. Only in recent years the wave processes in a chemically active bubble medium and a model of bubbly detonation have attracted consideration (Hasegawa, Fujiwara, Sychev, Pinaev, Kedrinskii, Mader, Beylich, Gulhan, Lee, Frost, etc.).

From the viewpoint of hot-spot mechanisms, the gas-phase heating realized in a bubbly medium can be a dominant mechanism, according to Field et al. (1992), as large bubbles will be collapsed relatively slowly by an incident shock wave. Therefore a model of an initiation of a cavitating liquid HE by single or fractional impacts is rather close to the approaches developed at the analysis of the wave processes in a liquid medium with the natural inhomogeneities considered above. Unfortunately, there is practically no information on the initial parameters of microinhomogeneities in liquid HE or combustible liquids. But we can assume that the possible range of their volumetric concentrations is from fractions to several percent, and radii are of the order of tens and hundreds of microns.

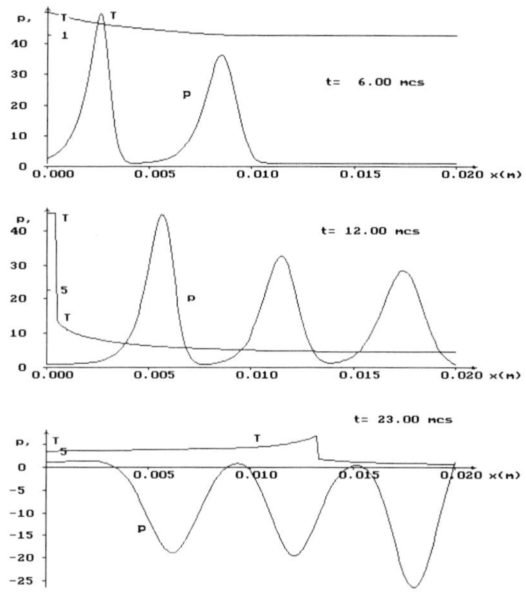

Figure 5. Simulation of a fractional impact. Exponential shock wave sequence propagation, $k_o = 10^{-4}$, $R_o = 0.05$ cm, $l = 2$ cm, $p_{1,2,3} = 300$ bar, $\tau = 0.5$ μs.

Two problems of initiation (a fractional impact and the influence of a rarefaction phase) were considered within the framework of a two-phase model of bubbly detonation (the system of equations is analogous to the system used by Kedrinskii and Zamaraev 1989).

Figure 5 shows the process of bubbly detonation development due to the propagation of a sequence (in 4 μs) of three exponential shock waves with amplitudes 30 MPa and $\tau = 0.5$ μs ($k_o = 10^{-4}$, $R_o = 0.05$ cm) along the sample with a 2-cm length.

The main reason for initiation turned out to be not related to the reflection of the first shock wave. The energy of the incident first and second shock waves was not enough to initiate the process (Fig. 5, $t = 6$ μ s) and the explosion of the gas mixture appeared only after the third wave ($t = 12$ μs, the temperature T of the gas mixture increased tenfold in the vicinity of left boundary of the sample). The influence of the right free boundary of the sample turned out to be contrary: the reflection of the shock-wave sequence interrupted a process of bubbly detonation propagation ($t = 23$ μs) at the point with coordinate $x \simeq 13$ mm. At least, in 10 μs a front of a temperature jump remained at the same place.

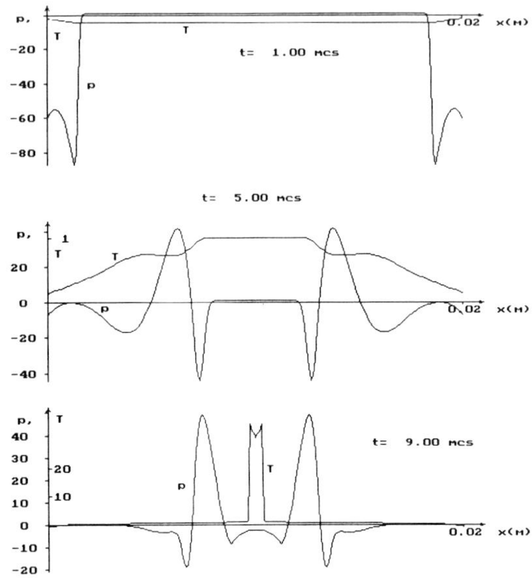

Figure 6. Ignition of a chemical reaction in bubbles at rarefaction wave focusing, $k_o = 10^{-3}$, $R_o = 0.01$ cm, $d = 2$ cm, $p_{l,r} = 100$ bar, $\tau = 3$ μs.

An unexpected effect was discovered upon analysis of the second statement regarding the free-boundary influence. Clearly, as a result of the impact on the plane of a thin sample (one-dimensional case) two exponential rarefaction waves will propagate from its free boundaries to the center (Fig. 6, $t = 1$, 5, and 9 μs). Under the collision of these waves the bubbly detonation process turned out to be excited at the center of the sample and then propagate to the periphery (Fig. 6, $t = 9$ μs, the temperature jump T). The length of the sample is 2 cm, $k_o = 10^{-3}$, $R_o = 0.01$ cm, the amplitudes of rarefaction waves are -10 MPa, and $\tau = 3$ μs. The reason for this effect is the transformation of an exponential profile of a rarefaction wave in the oscillating structure ($t = 5$ μs, Fig. 6) with an intense positive phase behind the front. In both cases the microbubbles in liquid samples contained the acetylene-oxygen mixture. The calculations were carried out using a program developed by Dr. V. A. Vshivkov.

5. Conclusion

The above experimental data, statements, calculation results, and survey have shown that many physical phenomena appearing under impulsive loading of a liquid and the mechanisms of their development can be considered from the viewpoint of wave processes in cavitating media.

Application of the two-phase mathematical model and model experiments have shown that the mechanism of the destruction of kidney stones in lithotriptor systems with high probability is determined by hydraulic impact of the liquid under collapse of dense cavitation clusters. The calculations have shown that the same effect can be achieved when focusing rarefaction waves without initial shock positive phase.

It was found that at the last stage of liquid fracture under pulse loading, the initial liquid volume is transformed into the structure of the liquid mesh type. The decomposition of the mesh cells into separate jets and further into drops forms the basis for the process of inversion of the two-phase state of the liquid.

In the problem of initiating the detonation of liquid HE the mechanisms concerned with the effect of microinhomogeneities of the type of bubbles of inert or chemically active gas play a special role. Application of two-phase models to describe and to estimate the wave-field parameters in such systems makes it possible to extend the spectrum of mechanisms governing the development of the phenomena.

Acknowledgement. This work was supported by the Russian Foundation for Fundamental Studies, Grant 93-01-16383, and by ISF, Grant J7N100.

References

Amster AB, Woolfolk RW (1967) Anomalies in detonation initiation of liquid explosives. Nature 215(5108):1370–1371.

Andreev KK (1958) Some considerations on the mechanism of initiation of detonation in explosives. Proc R Soc London A 246(1245):257–267.

Bayikov I, Berngardt A, Kedrinskii V, Pal'chikov E (1984) Experimental methods of study of cavitation cluster dynamics. J Appl Mech and Tech Phys (translated from Russian by Plenum Publ Corp) 5:30–34.

Berngardt A, Bichenkov E, Kedrinskii V, Pal'chikov E (1984) Optic and x-ray investigation of water fracture in rarefaction wave at later stages. In: Proc of IUTAM Symposium on Optical Methods in the Dynamics of Fluids and Solids, Prague, Czechoslovakia, M Pichal (ed), Springer-Verlag, pp 137–142.

Berngardt AR, Kedrinskii VK, Pal'chikov EI (1995) Evolution of inner structure of a liquid fracture zone under impulsive loading. J Appl Mech and Tech Phys (translated from Russian by Plenum Publ Corp) 36(2):99–105.

Besov A, Kedrinskii VK, Pal'chikov E, Matsumoto Y, Ohashi H (1992) Microinhomogeneity structures and hysteresis effects in cavitating liquid. In: Proc of 14th Int Congress on Acoustics, 3–10 September, Beijing, China.

Bourne NK, Field JE (1991) Bubble collapse and the initiation of explosion. Proc R Soc London A 435:423–435.

Bowden FP, Yoffe AD (1958) Fast Reaction in Solids, Butterworths Scientific Publications, London.

Campbell A, Davis W, Ramsay J, Travis J (1961) Shock initiation of solid explosives. Phys of Fluids 4:511.

Campbell AW, Davis WC, Travis JR (1961) Shock initiation of detonation in liquid explosives. Phys of Fluids 4(4):498.

Church C, Crum L (1990) A theoretical study of cavitation generated by four commercially available ESWL. In: Frontiers of Nonlinear Acoustics, Proc of 12th ISNA, M Hamilton, D Blackstock (eds), ESP Ltd, London, pp 433–438.

Delius M (1990) Effect of lithotripter shock waves on tissues and materials. In: Frontiers of Nonlinear Acoustics, Proc of 12th ISNA, M Hamilton, D Blackstock (eds), ESP Ltd, London, pp 31–46.

Dubovik AV, Bobolev VK (1978) Sensitivity of Liquid Systems to an Impact, Moscow, Nauka Publishers.

Dremin AN, Savrov SD, Trophimov VS, Shvedov KK (1970) Detonation Waves in Condensed Media, Moscow, Nauka Publishers.

Dubovic AV, Bobolev VK (1967) Exciting and propagation of detonation processes at weak initiation of liquid explosives. Fizika Goreniya i Vzryva 3(4):493–501.

Field JE, Bourne NK, Palmer SJ, Walley SM (1992) Hot-spot ignition mechanisms for explosives and propellants. Phil Trans R Soc London 339:269–283.

Grönig H (1989) Past, present and future of shock focusing research. In: Proc of the Int Workshop on Shock Wave Focusing, Inst of High Speed Mechanics, Sendai, Japan, March 22–23, pp 1–38.

Hasegawa T, Fujiwara T (1982) Detonation in oxyhydrogen bubbled liquids. In: Proc of 19th Int Symp on Combustion, Haifa, pp 675–683.

Hansson I, Kedrinskii V, Morch K (1982) On the dynamics of cavity clusters. J Physics D: Appl Phys 15:1725–1734.

Johansson CH (1958) The initiation of liquid explosives by shock and the importance of liquid break-up. Proc R Soc London A 246(1245):160–166.

Kedrinskii VK (1989) On relaxation of tensile stresses in cavitating liquid. In: Proc of 13th Int Congress on Acoustics, Belgrade, Yugoslavia, Vol 1, pp 327–330.

Kedrinskii VK (1968) Wave propagation in a liquid with gas bubbles. J of Appl Mechanics and Technical Physics (in Russian) 4:29–34.

Kedrinskii VK (1975) Dynamics of cavitation zone at underwater explosion near free surface. J of Appl Mechanics and Technical Physics (translated from Russian by Plenum Publ Corp) 5:68–78.

Kedrinskii V (1976) Negative pressure profile in cavitation zone at underwater explosion near free surface. Acta Astronautica, July-August 3(7-8):623–632.

Kedrinskii VK, Zamaraev FN (1989) Wave amplification in chemically active bubble medium. In: AIP Conf Proc, YW Kim (ed), Current Topics in Shock Waves, 17th ISSWST, Bethlehem, pp 51–62.

Kedrinskii VK (1986) On multiplication mechanism of cavitation nuclei. In: Proc of 12th Int Congress on Acoustics, Toronto, Canada.

Kedrinskii VK (1993) Nonlinear problems of cavitation fracture at explosive loading. J of Appl Mechanics and Technical Physics (translated from Russian by Plenum Publ Corp) 3:74–91.

Kitayama O, Ise H, Sato T, Takayama K (1987) Non-invasive gallstone disintegration by underwater shock focusing. In: Proc of the 16th Int Symposium on Shock Tubes and Waves, H. Grönig (ed), Aachen, pp 897–904.

Kuwahara M (1989) Extracorporeal shock wave lithotripsy. In: Proc of the Int Workshop on Shock Wave Focusing, Inst of High Speed Mechanics, Sendai, Japan, March 22–23, pp 65–89.

Nakoryakov VE, Pokusaev BG, Shreiber IR (1990) Wave Dynamics of Gas- and Vapour-Liquid Media, Moscow, Energoatomizdat.

Nagoya H, Obara T, Takayama K (1993) Underwater shock wave propagation and focusing in inhomogeneous media. In: Shock Waves @ Marseille, Brun and Dumitrescu (eds), Springer, Vol III, pp 439–444.

Philipp A (1991) Private communication.

Prat F, Arefiev A (1993) Cellural and tissular effects of shock waves-induced cavitation: potential application to digestive cancers. In: Shock Waves @ Marseille, Brun and Dumitrescu (eds), Springer, Vol III, pp 21–28.

Roberts P, Field JE (1993) Simulated fragment attack on cased munition. In: Proc of 10th Int Det Symp, Boston, MA, 12–16 July.

Sato E, Takahashi K, Sagae M, Shikoda A, Oizumi T, Ojima H, Takayama K, Sasaki K, Tamakawa Y, Yanagisawa T (1993) Soft flash x-ray system for shock wave research. In: Shock Waves @ Marseille, Brun and Dumitrescu (eds), Springer, Vol III, pp 449–454.

Sturtevant B (1989) The physics of shock focusing in the context of ESWL. In: Proc of the Int Workshop on Shock Wave Focusing, Inst of High Speed Mechanics, Sendai, Japan, March 22–23, pp 39–64.

Todes OM (1933) Adiabatic term explosion. In: J of Phys Chem 4(1):71.

Winning CH (1956) Enhanced initiating effect of multiple shock intersection. Nature 177(4497):33–34.

Wave propagation in vapor explosions

D. L. Frost
McGill University
Mechanical Engineering Department
817 Sherbrooke St. W., Montreal
Quebec, Canada H3A 2K6

Abstract: If a hot liquid such as molten metal contacts a cooler volatile liquid, rapid phase transformation of the volatile liquid may lead to a violent vapor explosion. Propagating interactions in melt-coolant mixtures are typically inefficient in converting the available thermal energy to mechanical energy and highly nonideal in comparison with combustion processes in premixed gases. The propagation of the interaction is closely tied to the rates of melt fragmentation and energy transfer to the coolant. The characteristics of vapor explosion waves that have been observed in melt-water mixtures are discussed with reference to recent experimental results. The relevance of wave solutions calculated using equilibrium thermodynamics to the experimental results is described together with a summary of current outstanding issues that need to be resolved to provide a more realistic description of vapor explosion waves.

Key words: Vapor explosion, Fuel-coolant interaction, Rapid phase transformation

1. Introduction

If a hot liquid contacts a cooler volatile liquid, the rapid phase transition of the volatile liquid can have an explosive character. Such explosions are called vapor explosions (or thermal or physical, in contrast with chemical explosions) or steam explosions if the cooler liquid is water, and are a hazard in a wide variety of technological situations. Explosions involving spills of molten metal into water are a concern in the metallurgical, paper, and nuclear industries and often are referred to as fuel- or melt-coolant interactions in this context. Reviews of the various aspects of vapor explosions can be found in the papers by Corradini et al. (1988) and Fletcher & Anderson (1990) as well as the proceedings of two recent specialist conferences (e.g., Park et al. 1993, Hohmann et al. 1994).

In a vapor explosion, the rapid conversion of thermal to mechanical energy may occur across a propagating wave front. Phase-transition waves have been observed in *homogeneous* liquids that have been superheated by heat transfer from a hot liquid (e.g., Shepherd & Sturtevant 1982) or by a sudden drop in the ambient pressure (e.g., Hill 1990). The focus of the present paper is

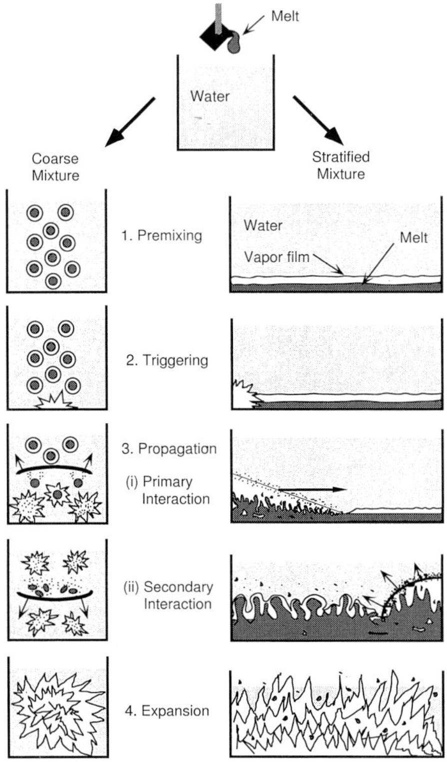

Figure 1. Schematic of the stages of a vapor explosion for two different contact modes following a spill of melt into water.

on the characteristics of propagating interactions in *heterogeneous* melt-coolant mixtures. This is analogous to the combustion-wave propagation in heterogeneous combustible mixtures such as liquid-fuel droplets or a dust suspension in air in contrast to homogeneous premixed gaseous mixtures. Figure 1 shows a schematic of a common accident scenario in which a heterogeneous mixture is formed when a quantity of melt is spilled into water. As the melt jet breaks up, a coarse mixture will be formed consisting of hot-liquid fragments dispersed in the cold liquid. A stratified system will be produced if the melt settles to the bottom of the vessel. Propagating interactions can occur in both of the contact modes shown in Fig. 1. In the initial premixing stage, stable vapor films separate the two liquids physically and thermally. The interaction is triggered by some disturbance which initiates local collapse of the vapor film, resulting in rapid heat transfer and in vaporization of the cold liquid forming a high-pressure vapor bubble. Subsequent expansion of the bubble causes pressurization of the surrounding liquid causing adjacent vapor films to collapse and leading to spatial propagation of the interaction. Multiple interactions may occur if the melt is not completely quenched during the initial interaction.

The dynamics of the propagating interactions shown in Fig. 1 are closely tied to the rate of thermal energy transfer to the coolant which is governed by the rate at which the melt is fragmented and dispersed in the coolant. The fine fragmentation process is the most important process in a melt-coolant interaction and determines if a local interaction will escalate to form a coherent propagating wave which transports momentum and energy in the direction of propagation. For the case in which melt drops are dispersed in a coolant, a global wave phenomenon will only exist if the local explosions of individual drops can couple to form a macroscopic front with associated global pressure and velocity fields. If the drops are too far apart, or too few in number to provide sufficient confinement, sequential explosions of the drops may still occur, but a macroscopic wave will not be formed.

In this paper, the dynamics of propagating vapor explosions in melt-coolant mixtures will be discussed with reference to recent experimental results (using molten tin-water or molten alumina-water mixtures) and compared with ideal wave solutions calculated using equilibrium thermodynamics. To characterize the propagating interactions observed experimentally, it is necessary to examine the characteristic length- and time-scales associated with the local explosion of single drops within the reaction zone and compare them with the appropriate scales associated with the macroscopic mixture.

2. Experimental results

2.1. Explosion of a single molten drop

If a hot melt drop is immersed in water, an interaction may be initiated by disrupting the vapor film around the drop with a pressure or flow disturbance. If the vapor film is collapsed with a weak underwater shock wave, the fragmentation of the drop will be dominated by so-called "thermal-fragmentation" mechanisms in which the energy dissipated in the melt breakup is derived from the internal energy of the melt itself. Conversely, if the drop is subjected to a large relative velocity between the drop and the surrounding coolant, direct "hydrodynamic fragmentation" of the drop will occur due to the acceleration of the drop.

Figure 2, taken from the work of Ciccarelli & Frost (1994), shows the characteristics of the explosion of 0.5-g drops of molten tin at 700°C triggered by a shock wave. The drop breakup is driven by thermal energy transfer to the

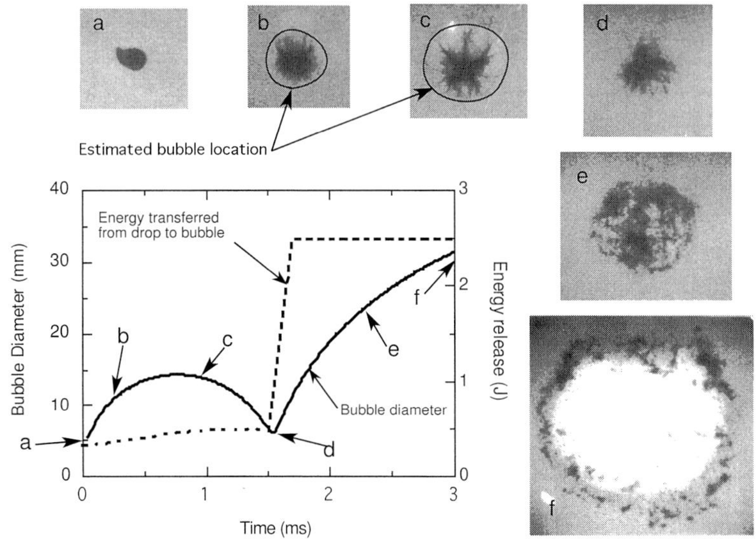

Figure 2. Dynamics of vapor explosion of single 0.5-g drop of tin at 700°C immersed in water at 65°C. X-ray radiographs show fragmentation of the melt drop at various stages of the interaction. Each radiograph corresponds to a different experiment.

water and the most significant feature of the interaction is that the thermal energy is released in several bubble-growth and collapse cycles. Following the initial vapor-film collapse, a steam bubble is formed and the drop surface becomes highly convoluted as shown in the X-ray radiographs. A simple model for the bubble dynamics indicates that about 3% of the thermal energy of the drop is transferred to the thermal energy of the vapor bubble, with the majority transferred at the time of bubble collapse (at $t = 1.5$ ms) when the drop is completely fragmented (Ciccarelli and Frost 1994).

Figure 3 contains X-ray radiographs showing a comparison between the fragmentation of hot and cold drops subjected to an ambient water flow (moving from bottom to top) of 40 m/s. The drops at the top of the figure are cold liquid metal drops (a low melting-point alloy consisting of 45% bismuth, 23% lead, 19% indium, 8% tin and 5% cadmium) and the drop fragmentation is strictly due to the relative flow. For the hot drops shown at the bottom of the figure, a vapor bubble is still generated, but the fragmentation within the bubble is enhanced by the internal turbulent vapor flow. At later times, the bubble is displaced and condensed, again bringing the melt in direct contact with the water. It is conventional to define a nondimensional time $T = t/t^*$,

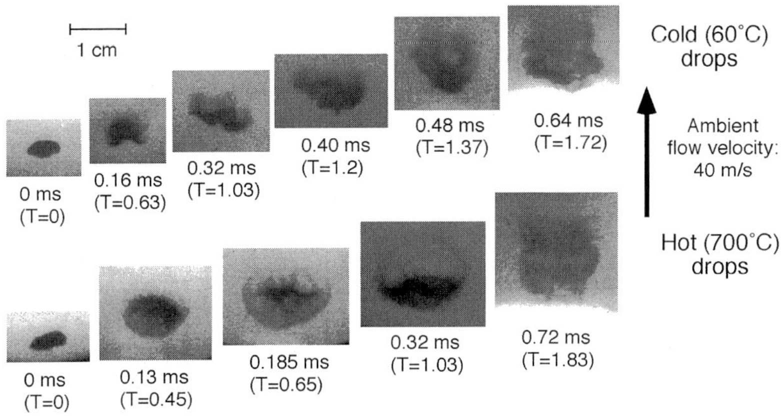

Figure 3. X-ray radiographs showing a comparison of the effect of ambient flow velocity on the fragmentation of *cold* 0.5-g liquid metal drops (low melting-point alloy) at 60°C shown on top and *hot* 0.5-g tin drops at 700°C shown on the bottom. Each picture corresponds to a different experiment. The vertical position of the drops shown is approximate. T corresponds to the nondimensional time as defined in the text.

where the characteristic time $t^* = (d/v_{rel})\sqrt{\rho_d/\rho_c}$, where d is the initial drop diameter, v_{rel} is the initial relative velocity between the drop and coolant, ρ is the density, and the subscripts d and c refer to the drop and coolant, respectively. Physically t^* corresponds to the time for the drop to move a distance approximately equal to its initial diameter due to the acceleration by the flow. Note that for both the hot and cold drops, complete fragmentation of the drop occurs at a time of about $T \approx 2$.

2.2. Propagating interactions in melt-coolant mixtures

The characteristic length scale for the fragmentation of a tin drop in a propagating interaction will depend on the pressure and particle velocity fields within the reaction zone following the leading pressure front. If the flow velocities are small (e.g., < 5 m/s), then the characteristic dimension will correspond to the size of the steam bubble generated, which is of the order of 3–4 cm for the explosion of 0.5-g molten tin drops. If the propagating interaction escalates to generate high particle velocities behind the wavefront, the characteristic reaction-zone thickness $\delta \sim D\tau$, where D is the propagation velocity and τ is the time for complete fragmentation of a drop. If we take a typical propagation velocity of 200 m/s for well-confined tin-water mixtures (e.g., Bürger et al. 1991), a relative flow velocity of 40 m/s and a nondimensional breakup time of 2 based on

Fig. 3, then $\delta \approx 15$ cm. For both of these cases, propagating interactions will be strongly influenced by the boundary conditions unless the scale of the mixture is much greater than the reaction-zone thickness (otherwise expansion waves due to the lateral expansion of the flow will attenuate the pressure within the reaction zone). This implies that mixture scales of greater than 1 m are required for the propagating interaction to be relatively unaffected by the confinement of the mixture.

For an unconfined linear drop array, an incoherent propagating interaction can occur, although little kinetic energy is transferred in the direction of propagation. This is shown by the series of photographs in Fig. 4a which illustrates the propagation of an interaction through an array of 20 5-g drops of tin at 800°C dropped into a water tank. The interaction propagates from one drop to the next due to the blast wave (duration ≈ 10 μs) generated during bubble collapse. The time for the shock to propagate from one drop to another (~ 10 μs) is negligible compared with the characteristic time for bubble growth and collapse, t_{bubble}, so the explosion propagates in a pulsating manner through the drops with an average propagation speed of 8 m/s with the speed proportional to Δ/t_{bubble}, where Δ is the characteristic drop spacing for which the explosion of one drop is sufficient to trigger the explosion of the adjacent drop.

The experiment described above was repeated, with the drops confined laterally by two glass plates forming a channel with a spacing of 1.25 cm, and the results are shown in Fig. 4b. Since the channel width is smaller than the size of the steam bubbles generated, spherical bubble growth is prevented and the flow velocity generated by an expanding bubble is enhanced. This has the effect of reducing the induction time for the explosion of one drop to trigger the next and the interaction sweeps through the drops to the left with a velocity of 50–60 m/s, generating a pressure pulse (with a rise time of ≈ 1 ms and peak pressure of ≈ 0.5 MPa measured at the tank base). Although the expansion is constrained in one dimension, the vapor generated is free to expand vertically which limits the overpressure within the reaction zone. Due to the relatively low pressure amplitudes attained and the high compressibility of the upstream mixture, the velocity of the propagating interaction is of the same magnitude as the homogeneous two-phase sound speed for the upstream mixture (the sound speed for homogeneous water-steam mixtures varies from 39 to 54 m/s for void fractions from 10 to 5%, respectively). A similar propagation velocity is obtained if the molten tin is allowed to fall to the base of the channel, forming a stratified mixture (see Frost et al. 1995). The energy yield (i.e., the conversion of thermal to mechanical energy) per surface area of the tin as well as the overpressures generated are also of the same order for both the dispersed- and stratified-geometry experiments.

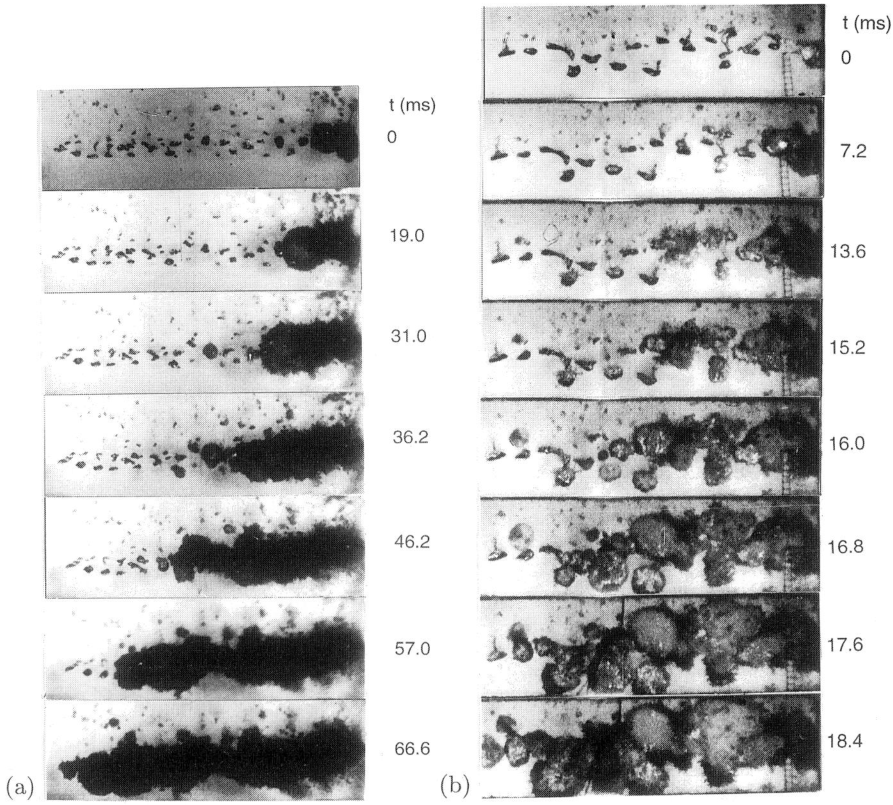

Figure 4. Effect of confinement on the propagation of an interaction in 20 5-g drops of tin at 800°C in water at 60°C. Drops on the left (a) are within a tank with a width of 20 cm. Drops on the right (b) are confined laterally within glass plates 1.25 cm apart. Explosion is initiated on the right by either the contact with a local region of cold water (unconfined drops) or by the impact with a wire mesh (confined drops).

A molten tin-water mixture may be more fully confined by placing it within a narrow tube. If the interaction is initiated at the closed end, the expansion of the high-pressure vapor in the reaction zone will act like a moving piston, driving a pressure wave to supersonic speeds into the upstream melt-coolant mixture. Experiments with this geometry have been carried out in the KROTOS (Bürger et al. 1991) and WFCI (Park et al. 1993) facilities. In both cases, a quantity of tin (between 5 and 7.5 kg) was heated to between 1000°C and 1100°C and poured into a vertical tube filled with a quantity (between 7.5 and 8.7 liters) of water. Propagation velocities and peak pressures recorded were typically on the order of 200 m/s and 5 MPa, respectively. Since the tube diameter (in both cases about 9 cm) is of the same order of magnitude as the reaction-zone thickness estimated earlier, these propagating interactions are strongly influenced by the boundary conditions (i.e., the confinement provided by the tube walls). The reaction-zone thickness is likely even longer than the earlier estimate since the interference between neighboring drops will reduce the relative flow velocity between the drop and coolant and hence the rate of fragmentation.

The KROTOS facility has also been used to study the propagation of vapor explosions in high-temperature molten alumina-water mixtures that represent a better simulant of the molten fuel produced in a hypothetical loss-of-coolant scenario in a nuclear reactor. For example, in the KROTOS-28 test (Hohmann et al. 1994) 1.45 kg of aluminum oxide melt at 2400°C was released into 7.5 liters of water at 87°C. An interaction propagated in the mixture with a velocity between 650 and 1000 m/s. The supercritical pressures attained were poorly characterized due to saturation of the transducers, but were greater than 50 MPa. Due to the high pressures, the corresponding high density of the fluid in the reaction zone will enhance the hydrodynamic fragmentation rate and rate of energy transfer to the coolant. An order-of-magnitude estimate of the reaction-zone thickness can be made by taking a propagation velocity of 1000 m/s, an average relative velocity between the flow and a 1-cm drop of 100 m/s, and a nondimensional total fragmentation time of 2, giving a value for the reaction-zone length of about 10 cm. Since this value is of the same order as the tube diameter, the dynamic parameters of the propagating interaction will again be strongly dependent on boundary conditions.

3. Wave propagation models

3.1. Steady models

The nature of possible steady wave solutions supported by a melt-coolant mixture can be investigated without regard to the complex dynamical processes involved. The analysis involves solving the one-dimensional conservation laws of mass, momentum, and energy with appropriate equations of state and mixture rules across a steady control volume that contains the wave. The downstream locus-of-equilibrium states calculated in this manner define the Hugoniot curve as in a chemically reacting medium. A unique Chapman-Jouguet (CJ) detonation solution can be determined by using the classical construction of the Rayleigh line tangent to the Hugoniot curve. This procedure was first carried out by Board et al. (1975) for mixtures of molten tin and water. For a low-temperature melt-coolant mixture, the Hugoniot curve has two distinct sections separated by a discontinuity or kink at the saturation boundary. As a result, classical CJ detonation solutions can only be obtained for initial void fractions below a certain value. The analysis was refined by McCahan & Shepherd (1993) by incorporating a compressible equation of state for the tin.

Two assumptions in the ideal Hugoniot analysis are not satisfied, in general, for propagating interactions observed in melt-coolant mixtures. First, although the reaction-zone length does not come into the Hugoniot analysis, since boundary conditions do not influence the solution, the dimensions of the mixture must be much greater than the reaction-zone thickness, a condition clearly not satisfied for the propagation experiments. Secondly, the attainment of thermal equilibrium at the downstream state is hindered by the low thermal conductivity of the coolant and the presence of vapor which isolates the coolant from hot melt fragments.

Despite the limited scale of current propagation experiments, it is of interest to determine the properties of the ideal wave solutions for a high-temperature melt-coolant mixture, assuming that the mixture dimension is large enough to support a thermal detonation wave. If a molten alumina-water mixture is considered, with an initial melt temperature (2400°C) and melt, coolant, and vapor volume fractions (0.08, 0.88, and 0.04, respectively) corresponding to the KROTOS-28 test (Hohmann et al. 1994), and a compressible equation of state for molten alumina is used (following McCahan & Shepherd 1994), the equilibrium Hugoniot may be computed and is shown in Fig. 5a. The CJ detonation pressure and velocity (400 MPa and 2600 m/s, respectively) are much higher than the values observed experimentally. A second ideal wave solution, first proposed by McCahan & Shepherd (1993) for wave propagation in a closed-end tube, may be computed which consists of a leading shock wave that precompresses the mixture followed by a deflagration wave that brings the flow to rest downstream. To obtain a unique solution we can use the condition that the deflagration satisfies the CJ criterion. The results for the same initial conditions as for the detonation solution are shown in Fig. 5b. State 0 is the upstream state, state 1 is downstream of the shock wave, and the final state 2 is downstream of the CJ deflagration wave. The structure of this wave solution differs from that of the detonation solution in that it is explicitly assumed that a shock wave is present. To determine the state downstream of the shock, the shock is assumed to be isentropic with the melt and coolant downstream of the shock in mechanical but not thermal equilibrium. This ad-hoc assumption leads to a deficit in the conservation of energy across the shock by about 17% in this case. In this solution, the shock and deflagration waves are not coupled as in a detonation wave and move at different velocities (1690 m/s and 1208 m/s, respectively). Following a high pressure of 474 MPa across the shock, the CJ deflagration pressure is 77 MPa. Although the choice of the CJ criterion cannot be justified on a fundamental basis, the predicted velocities and pressure at the CJ plane are closer to the experimental values suggesting that the shock/deflagration wave complex may be more representative of the observed propagating interaction than a fully coupled detonation wave. However, in this model, uncertainties remain concerning the validity of the assumptions used in calculating the jump conditions across the shock in the multiphase mixture.

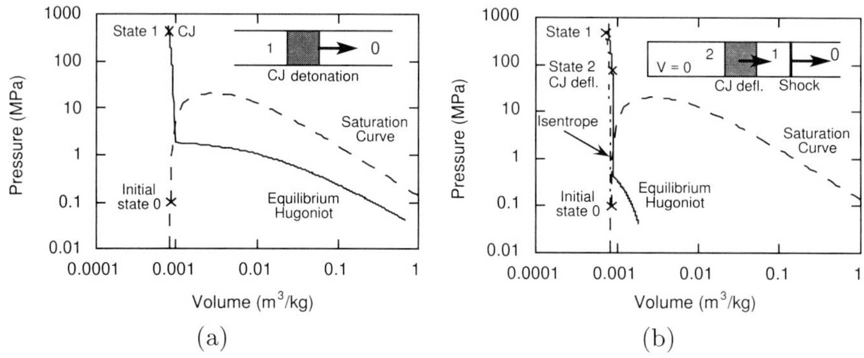

Figure 5. Hugoniot curves showing (a) ideal CJ-detonation and (b) isentropic shock/CJ-deflagration solutions for a mixture of molten alumina at 2400°C and water and steam.

The Hugoniot analysis can be modified to include the effect of momentum losses due to friction at tube walls. For example, Gelfand et al. (1992) included frictional effects by using the traditional method based on the ZND model for detonations in combustible gas mixtures to find the singularity or eigenvalue conditions at the CJ plane that represent valid wave solutions. However, this adds an additional complication because if downstream equilibrium is not assumed, a model for the rate of energy release (involving the rates of fragmentation of the melt and heat transfer to the coolant) must be included in the model. However, even if a realistic fragmentation model is incorporated into the analysis, there is no guarantee that such a wave exists in reality or in fact can be attained following the transient development of the explosion.

3.2. Transient models

A number of multiphase models have been proposed to address the transient development of a propagating interaction by considering the complex processes within the reaction zone. The current generation of models (e.g., CULDESAC, Fletcher & Anderson 1990; ESPROSE, Yuen et al. 1992; IDEMO, Bürger et al. 1991; TEXAS-III, Tang & Corradini 1994) is useful for studying the parametric dependence of various properties on the propagation dynamics. However, uncertainties in the constitutive physical models prevent quantitative predictions from being made with any confidence. The lack of well-characterized experimental data is the limiting factor in the validation of the transient propagation models. This is due largely to two factors: first, it is difficult experimentally, particularly at large scale, to produce a well-defined stable homogeneous mixture of melt and coolant (other than a stratified mixture) and secondly, it is very difficult to obtain time- and space-resolved measurements of temperature, pressure, and the distribution of phases within the reaction zone of a propagating interaction sufficient to resolve the important fine fragmentation process.

4. Outstanding issues

A detailed description of the fine fragmentation process within a propagating melt-coolant interaction remains the most important outstanding factor that must be addressed to improve our ability to estimate the dynamic parameters of vapor explosion waves. In particular, the relative importance of thermal- and hydrodynamic-fragmentation mechanisms and the effect of interference between adjacent fragmenting drops must be determined as a function of the transient pressure and velocity fields within the reaction zone. Realistic models for other processes at the micro-level, such as for the transient vaporization of the nonuniformly heated coolant following melt-coolant contact, are not available. A second major challenge is the theoretical description of how the explosion of discrete melt drops interact and couple to form a coherent macroscopic wave. The critical energy to directly initiate the propagation of a coherent wave and the propagation limits (analagous to the corresponding dynamic parameters for combustible mixtures) are poorly understood for melt-coolant mixtures. The scale of current propagation experiments is typically of the same order as the fragmentation zone length and therefore the interaction is strongly dependent on the boundary conditions, unlike the ideal wave solutions that may be determined using equilibrium thermodynamics. Although progress is being made on the description of the constitutive physical processes within a vapor explosion, complete models for propagating vapor explosion waves with a truly predictive capability will likely remain elusive for some time to come.

Acknowledgement. The author would like to acknowledge the work of his former graduate students Gaby Ciccarelli, Phil Watts, and Barbara Bruckert as well as the assistance of Walter Lei and Patton Chan in the Hugoniot calculations. Valuable comments were also provided by John Lee. This work has been supported by the Natural Sciences and Engineering Research Council of Canada, Atomic Energy Control Board of Canada, and Sandia National Labs.

References

Board SJ, Hall RW, Hall RS (1975) Detonation of fuel coolant explosions. Nature 254:319.

Bürger M, Müller K, Buck M, Cho SH, Schatz A, Schins H, Zeyen R, Hohmann H (1991) Examination of thermal detonation codes and included fragmentation models by means of triggered propagation experiments in a tin/water mixture. Nucl Eng Design 131:61.

Ciccarelli G, Frost DL (1994) Fragmentation mechanisms based on single drop steam explosion experiments using flash X-ray radiography. Nucl Eng Des 146:109.

Corradini ML, Kim BJ, Oh MD (1988) Vapor explosions in light water reactors: a review of theory and modeling. Prog Nucl Energy 22(1):1.

Fletcher DF, Anderson RP (1990) A review of pressure-induced propagation models of the vapour explosion process. Prog in Nucl Energy 23:137.

Frost DL, Bruckert B, Ciccarelli G (1995) Effect of boundary conditions on the propagation of a vapor explosion in stratified molten tin/water systems. Nucl Eng Des 155:311.

Gelfand BE, Bartenev AM, Frolov SM, Tsyganov SA (1992) Thermal detonation in molten Sn-water suspension. Archivum Combustionis 12(1):5.

Hill LG (1990) An Experimental Study of Evaporation Waves in a Superheated Liquid, Ph.D. Thesis, California Institute of Technology.

Hohmann H, Magallon D, Schins H, Yerkes A (1994) FCI experiments in the aluminumoxide/water system. In: Proc CSNI Specialist Meeting on Fuel Coolant Interactions, NUREG/CP-0127, Washington, p. 193.

McCahan S, Shepherd JE (1993) Models of rapid evaporation in nonequilibrium mixtures of tin and water. AIAA Prog in Astr and Aero 154:432.

McCahan S, Shepherd JE (1994) A thermodynamic model for aluminum-water interactions. In: Proc CSNI Specialist Meeting on Fuel Coolant Interactions, NUREG/CP-0127, Washington, p. 134.

Park HS, Yoon C, Corradini ML, Bang KH (1993) Vapor explosion escalation/propagation experiments and possible fragmentation mechanisms. In: Int Sem on the Physics of Vapor Explosions, Tomakomai, Japan.

Shepherd JE, Sturtevant B (1982) Rapid evaporation at the superheat limit. J Fluid Mech 121:379.

Tang J, Corradini ML (1994) Modelling of the complete process of one-dimensional vapour explosions. In: Proc CSNI Specialist Meeting on Fuel Coolant Interactions, NUREG/CP-0127, Washington, p. 204.

Yuen WW, Chen X, Theofanous TG (1992) On the fundamental microinteractions that support the propagation of steam explosions. In: Proc NURETH-5, Vol. 2, p. 627.

Irving Glass Memorial Session

Wave machinery for chemical processing and high-efficiency heat engines

W. H. Christiansen, A. Hertzberg
Department of Aeronautics and Astronautics
P.O. Box 352400
University of Washington
Seattle, WA 98195-2400 USA

Abstract: Wave machines using nonsteady gasdynamic processes, i.e., mechanizations of shock tubes to produce steady gas flows, have been used in energy and chemical processes. The potential of this class of machinery is investigated here using on-rotor combustion as a means of extending the usefulness of this type of machinery. On-rotor combustion adds yet another variable in the complex gasdynamic process on the rotor, but permits the possibility of using the device as a stand-alone heat engine.

Key words: Wave machinery, Heat engine

1. Introduction

Wave machinery involving direct pressure exchange between hot and cold gases has always been an appealing concept, since the average temperature of the pressure-exchange tubes can be maintained at a relatively low level, allowing extremely high temperatures to be explored. However, it was not until after World War II that Brown-Boveri (now Asea Brown-Boveri) under the direction of Claude Seippel (Seippel 1946) built wave machinery in a topping unit using a comprex (trade name, but meaning compression and expansion in the same device, used by Brown-Boveri). Kantrowitz at Cornell University in the late 1940s also studied the comprex concept as a heat engine component (Kantrowitz et al. 1948). In 1950 at Cornell Aero Lab (now Calspan Corporation), A. Hertzberg laid out the design of a device which could, in principle, produce NO efficiently (Glick et al. 1953). Also the design was capable of very efficient pyrolysis and production of acetylene. Al Flax of Cornell Lab adapted these ideas and pushed forward to the construction of a machine capable of heating air to about $3,500°$K for periods of seconds, to be used as a hypersonic nose-cone testing device (Hertzberg et al. 1959). The machine was successful and proved the high temperature capability of wave machinery. In 1960 using the knowledge gained from these efforts in wave machinery, Hertzberg and Weatherston (Hertzberg et al. 1959) developed a concept for a high-temperature gas-turbine device which promised significant increases in efficiency over existing gas turbines of that period. During this period Brown-Boveri solved the mechanical problems of the comprex and developed the concept into an effective supercharger for diesel trucks. Wave machinery is now currently being studied by Asea Brown-Boveri as a method of increasing the efficiency of gas turbines. NASA is supporting efforts at Cornell University under Dr. Ed Resler (Resler et al. 1994) and Dr. Jakob Keller (private communication) at the University of Washington for cycles that would be suitable for aircraft gas turbines. These machines are attractive because they can operate at high temperatures and higher efficiency.

In this paper we would like to discuss a study on alternate cycles which promises relative simplicity of operation, for either high-temperature chemical pyrolysis, or gas-turbine power cycles. A "fly-wheel"-type of driver system was suggested (Hertzberg and Christiansen 1976) in which the compression energy inside the cycle is returned to keep the cycle operating, permitting a stand-alone wave device without auxiliary equipment. Losses in the circulating system were originally anticipated to be made up by an overall pressure drop through the machine with careful adjustment of the wave system, when used as a pyrolysis-inducing device for ethylene, acetylene, or NO production. The term "fly-wheel" is used representing the motion of a massless imaginary piston, whose operating motion can be adapted at will to the cycle. Thus, there is basically one inlet duct (and one exit duct) to let in the fresh unprocessed gas, and to exhaust the spent processed gas from the machine.

2. Details of the fly-wheel driver concept

Figure 1 is the wave diagram (x-t diagram) Hertzberg and Christiansen (1976) used to illustrate the fly-wheel driver concept. In this diagram the reactant and driver gas are idealized, for illustration only, to be perfect air, thus ensuring tailored operation and minimum interface mixing. The temperatures are given in degrees Kelvin, and the pressures in atm — in this case chosen to be in the appropriate range for NO production. At the bottom of the figure high-pressure reactant air is taken into the device from the left at a Mach number of 0.68 and static pressure of 3.48 atm. To start this cycle, the reactant gas is brought to rest by a weak hammer shock. Additional isentropic pressurization is carried out by driver gas coming in from the right through wave action until the reaction conditions are met — in this case 50 atm and 3300°K. The dashed line in this drawing represents the interface between the reactant and driver gas as this process is carried out. After a suitable dwell time, the pressurization process is reversed and the driver gas again leaves to the right, but at pressures slightly higher than those at which it came in. This is carefully carried out until the original driver gas has left the rotor or until the interface reaches the right-hand edge of the rotor. The remaining reactant gas is further removed with the help of new reactant gas from the left, thus arriving at the same gas conditions in order to repeat the cycle.

Further close scrutiny of the total pressures, total enthalpies, and mass flows indicated that it was possible to exhaust the driver gas with excess total pressure so that it could be collected via suitable manifolds and used again without the need of pumps on the driver side of the wave machine. Of course the energy gain in the driver gas was made up by a corresponding energy loss in the reactant gas loop. This illustrates the concept of the "fly-wheel driver," a driver-gas system not requiring any external pumping device. But observation of this drawing shows that the driver over pressures are very small O (3% max) and the excess total enthalpy is also O (1% max). This is a marginal pumping situation, particularly in view of any calculational errors present and external mixing processes, and one which may not be changed easily. Thus, a method to improve on this notion is needed.

3. Modification of concept via on-rotor combustion

This cycle has been modified recently to use the addition of either a layer of combustible gas at the interface, or by modifying the inlet mixture, so that a certain amount of combustion on the rotor in the manner of a reciprocating combustion engine can be employed to provide significant driving energy for the system consistent with the needs of the system. The combustible gas mixture undergoes compression on the rotor. At the appropriate point in the cycle, say at maximum compression, it is ignited. If combustion takes place on the rotor, there are two limiting cases. In one, the combustion time is slow compared to the acoustic transit time, L/a, so that pressure continuity is maintained at the interfaces. While the overall volume, indicated by L, is constant, the individual volumes of the driver gas and the combustible mixture are not constant during combustion. The second possibility is that the combustion time is short compared to L/a, which results in pressure discontinuities at the interfaces because the individual gas volumes remain constant during the process. This situation will produce a shock-wave relaxation process in addition to the wave processes already present in the wave machine. For the rest of the study, it is assumed that only slow combustion takes place.

Following the details in Fig. 1, we imagine that the reactant gas is burnable. At maximum compression, i.e., $p = 50$, there is no flow and the wave system is at rest. One can add significant dwell time, or in this case, combustion time here without affecting the subsequent wave processes necessary to close the cycle. The schematic of the wave diagram under this scenario is shown in Fig. 2, wherein after combustion the interface has moved to the right. Since there is no outflow during this part of the cycle, the pressure of the system

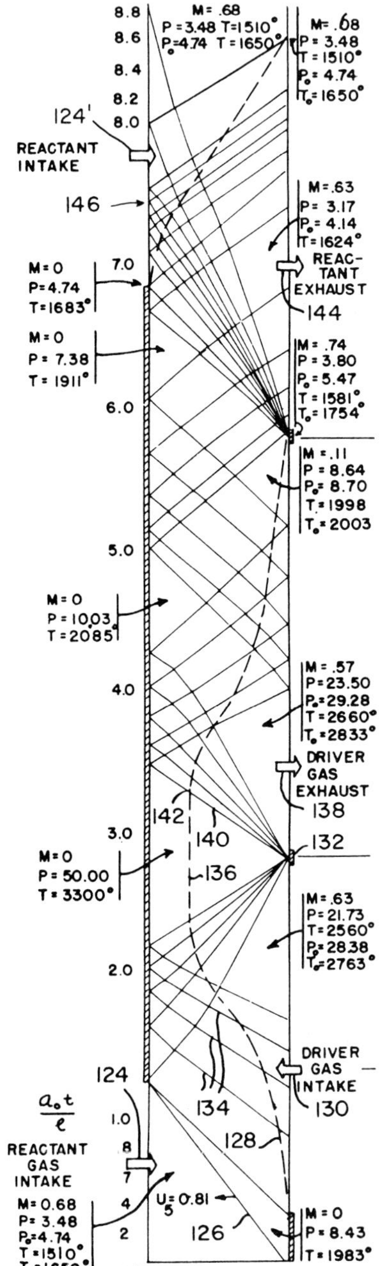

Figure 1. Wave diagram.

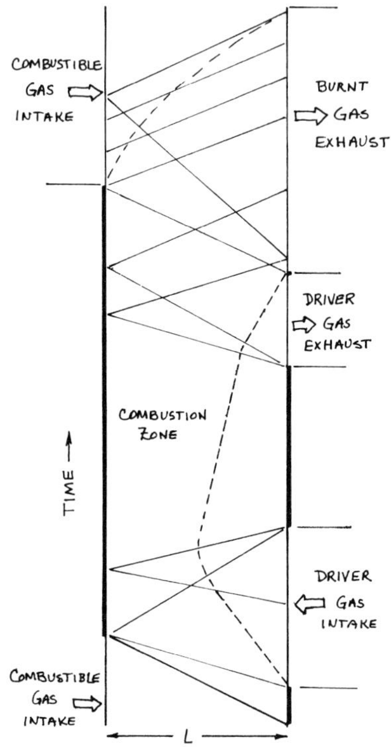

Figure 2. Wave diagram with combustion.

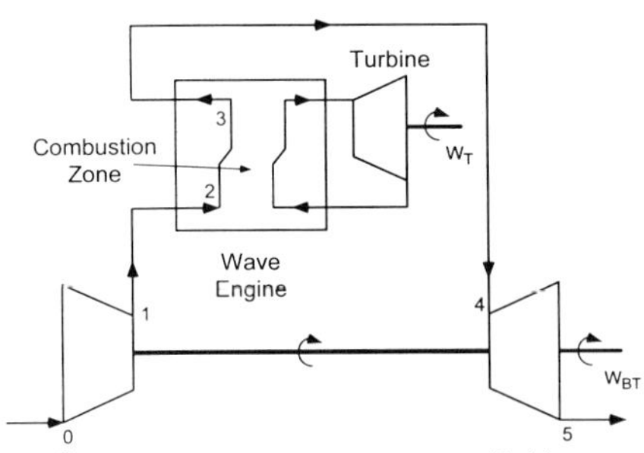

Figure 3. Block diagram of topping unit.

increases, including the driver gas which is being compressed by the movement of the interface. After combustion there is pressure continuity and a temperature discontinuity at the interface, which results in a slight off-tailoring condition. But the point to notice is that significant compression occurs in the inert driver gas, depending on the degree of combustion on-rotor. Examining this cycle more carefully, we can see that by using a combustible mixture as the process gas, the machine becomes in itself a self-driving hot-gas generator with considerable pre-compression. More than enough energy can easily be drained off to continue the fly-wheel action, and in some cases shaft energy could be withdrawn from the driver outflow.

4. Example use with a topping unit

Figure 3 shows a sketch of a wave machine used as a high-temperature combustor in a normal Brayton cycle. In this schematic, the combustor is imagined to run along the lines of the wave diagrams depicted earlier. First, reactant gas is compressed by an external compressor which is then led to the wave engine. It is further compressed in the wave engine and burned there. The driver gas is tailored, i.e., by using a lower molecular-weight gas, to operate at lower driver-cycle temperatures so that current turbine technology can be used. The driver gas is also sufficiently compressed by the combustion process so that additional work can be extracted by a topping turbine. The normal high temperatures involved in combustion are mitigated in the rotor because of the work exchanged in the process. The resulting reactant gas is drawn off the rotor, still at modest temperature and relatively high pressure, to be used in the bottoming turbine. The resulting thermodynamic cycle is more efficient than a stand-alone Brayton cycle as shown in Fig. 3. Of course, other versions including additional external burning off the rotor are also possible.

References

Glick H, Squire W, Hertzberg A (1953) A New Shock Tube Technique for the Study of High-Temperature Gas-Phase Reactions, Project SQUID Report.

Hertzberg A, Glick HS, Squire W, Wetherston RC (1959) Methods and Apparatus for Carrying Out Gas-Phase Reactions Which Require a High Temperature to Promote the Reaction and Rapid Cooling to Preserve the Reaction Product, U.S. Patent Numbers 2,832,665 and 2,832,666 (1958) and 2,902,337.

Hertzberg A, Christiansen W (1976) Method for Continuously Carrying Out a Gas Phase Reaction and Apparatus Therefore, U.S. Patent Number 3,998,711.

Kantrowitz A, Hertzberg A, McDonald E, Resler E (1948) Heat Engines Based on Wave Processes, Cornell University Graduate School of Aeronautical Engineering Report.

Resler EL Jr, Mocsari JC, Nalim MR (1994) Analytic design methods for wave rotor cycles. J Propulsion and Power 10(5): 683–689.

Seippel C (1946) Pressure Exchanger, U.S. Patent Number 2,399,394.

Holographic interferometric animation of shock diffraction and reflection using a highly repeatable shock tube

J. M. Yang[*] H. Babinsky[†] K. Takayama
Shock Wave Research Center, Institute of Fluid Science, Tohoku University
Katahira 2-1-1, Aoba, Sendai 980, Japan

T. Saito
Cray Research Japan Ltd.
2-5-5 Shinyokohama, Minatokita-ku, Yokohama, Japan

J. P. Sislian
Institute for Aerospace Studies, University of Toronto, Canada

Abstract: An animated holographic observation was made of shock diffraction and reflection in a 60 mm × 150 mm diaphragmless shock tube which has a high degree of repeatability. Two animations were made. One is a shock diffraction over a 90° sharp corner; the other examines the variation of the shock-wave reflection configuration with changing inclination of a wedge. It was found that as the inclination of the wedge increases, the reflection pattern continuously changes from von Neumann reflection through simple Mach reflection to regular reflection. The animation of shock-wave diffraction showed some non-self-similarity near the corner.

Some new results of 3-dimensional shock-wave reflection over an oblique cylinder are also reported. This experiment was conducted in the 100 mm × 180 mm shock tube (Glass 1991) which was recently moved from the Institute for Aerospace Studies, University of Toronto to the Shock Wave Research Center, Institute of Fluid Science, Tohoku University.

Key words: Reflection, Diffraction, Animation, Holographic interferometry, Diaphragmless shock tube

1. Introduction

The shock-wave reflection over a straight wedge or shock-wave diffraction past a 90° corner are two of the benchmark tests of shock dynamics (Skews 1967; Ben-Dor 1992; Han and Yin 1993). Motion of such shock waves are usually vi-

---

*Present address: Department of Modern Mechanics, University of Science and Technology of China, Hefei, Anhui, 230026, China.

†Present address: Department of Mechanical Engineering, University of Cambridge, England.

sualized by means of high-speed cameras. Unfortunately, the spatial resolution of existing high-speed cameras is generally at most 20 lines/mm. So far, very precise sequential observations have been recognized to be impossible. Double-exposure holographic interferometry has a very high spatial resolution of more than 100 lines/mm, however, this method can only supply a single image at each test. By using the combination of holographic interferometry and a shock tube which has very high repeatability, it is possible to record a large number of interferograms under nearly identical initial conditions. Images obtained in such a fashion are equivalent to individual frames in a high-speed movie. Through image processing, these frames can be combined into an animation sequence with little degradation of the spatial resolution (Babinsky et al. 1995). Two configurations have been investigated with this technique. One is shock-wave diffraction over a 90° sharp corner, the other examines shock-reflection patterns on a wedge. The wedge inclination was increased by a small amount between each test, while the shock position was kept constant. The thus obtained animation shows the change of the shock-wave reflection configuration with increasing wedge inclination.

2. Experiment

The experiments were conducted in a diaphragmless shock tube at the Shock Wave Research Center, Tohoku University. The shock tube consisted of a low-pressure channel of 60 mm × 150 mm, and a high-pressure chamber whose diameter is 290 mm. Upon the quick movement of a rubber membrane which seals the high-pressure and low-pressure sections, a shock wave can be generated with high repeatability. The scatter of the generated shock Mach number was found to be within ±0.25 % for more than 500 runs (Yang et al. 1994).

Double-exposure holographic interferometry was used to visualize the flow field. The basic optical arrangement is similar to a conventional shadowgraph system (Takayama 1983). A double-pulse ruby laser (Apollo Laser Inc. 22DH, 25-ns pulse duration and 2 J/pulse) was used as the light source. In the experiments of shock-wave reflection over wedges, a 60-mm wide wedge whose inclination angle can be adjusted continuously was installed in the 60 mm × 150 mm test section, and the inclination angles of the wedge varied from 3° to 55° in steps of about 0.7°. Nitrogen was used as the test gas in the reflection experiments. The shock Mach number was 1.20 and its scatter was found to be less than ±0.15% over a total of 156 runs. For the experiments of shock-wave diffraction over a 90° corner, a sudden backward-facing step from 60 mm × 150 mm to 60 mm × 360 mm was used as the test section. Such a large test section allowed an extension of the observation time of up to 800 μs after the shock wave reached the corner. Two shock Mach numbers, $M_s = 1.20$ and 1.50, were tested for this case and the scatter in both cases was below ±0.2 %.

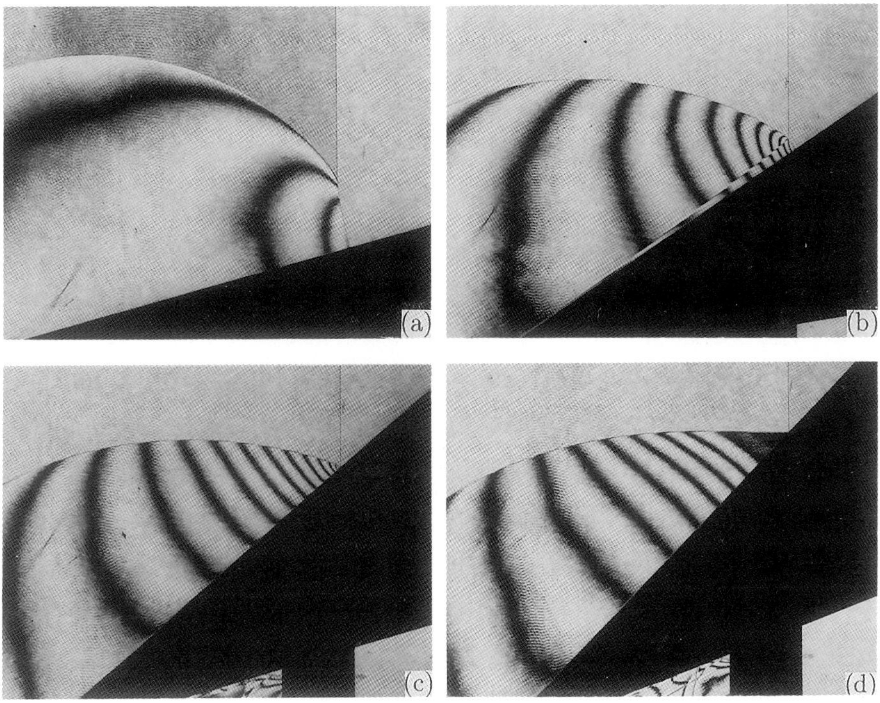

Figure 1. Holographic interferograms of shock wave reflection over wedges of different inclination, $M_s = 1.20$: (a) $\theta = 15°$; (b) $\theta = 34°$; (c) $\theta = 41°$; (d) $\theta = 45°$.

3. Results and discussion

Figure 1 shows some holographic interferograms of the animation result. For the the case of shock-wave reflection over a wedge, as shown in Fig. 1a, when the inclination angle of the wedge is small, the pattern of shock reflection appears to be what is known as von Neumann reflection (Sasoh and Takayama 1994), in which the incident shock wave and Mach stem appear to intersect with zero angle and no slip line can be recognized near the triple point. With increasing inclination angle, a discontinuity at the triple point appears and the slip line becomes clearly observable; this configuration is known as simple Mach reflection and is shown in Fig. 1b. With further increase of the inclination angle, the Mach stem reduces in length and finally the Mach reflection terminates and regular reflection occurs. Since the shock tube is highly repeatable, it is easy to observe the shock-wave reflection pattern near the transition between Mach reflection and regular reflection. Figure 1c shows the configuration near the transition point and the transition angle was found to be about 42° for shock Mach number 1.20. Figure 1d shows regular reflection. The animation sequence clearly shows that as the inclination angle of the wedge increases, the reflection pattern almost continuously changes from von Neumann reflection through simple Mach reflection to regular reflection, with no sudden jump observable. It should be pointed out that this animation cannot be obtained with a high-speed camera because the motion animated here never exists in the real world. It was found to be helpful for the understanding of shock-wave reflection phenomena.

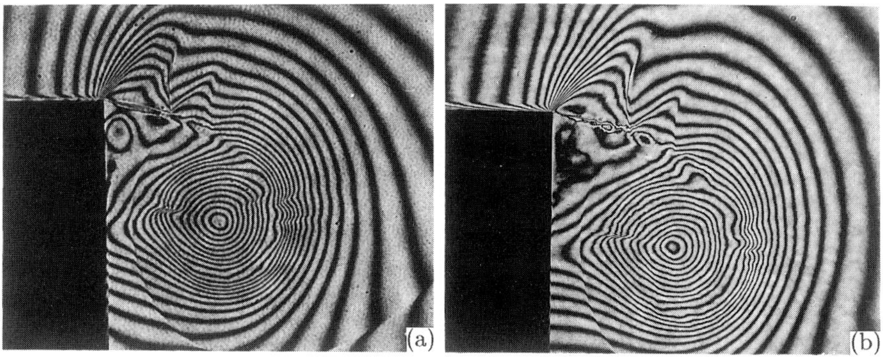

Figure 2. Holographic interferograms of shock wave diffraction over a 90° corner, $M_s = 1.50$: (a) $t = 300\mu s$, (b) $t = 865\mu s$.

The animation of shock diffraction shows the dynamic development of the shock wave diffracting around the corner. Not only the shock front but also the flow field behind the shock wave can be observed. Especially in the case of shock Mach number 1.50, the secondary shock-wave system, the vortex, the slip stream, and the contact surface, nearly self-similarly changing with time, can be clearly seen. With the aid of the high resolution of the hologram, a careful study can be made of the flow field near the corner. It is not difficult to see that there is a secondary vortex near the corner and that this vortex becomes weaker and disappears with time. Figure 2a shows the shock-wave configuration at 300 μs after the shock wave reaches the corner, while Fig. 2b is the visualization result at 865 μs. The secondary vortex can be easily observed in Fig. 2a but is almost invisible in Fig. 2b.

4. Shock reflection over an oblique cylinder

For observation of three-dimensional shock-wave reflection, the situation becomes much more complicated. A shock reflection over an oblique cylinder was visualized by using diffuse holography. This experiment was carried out in a

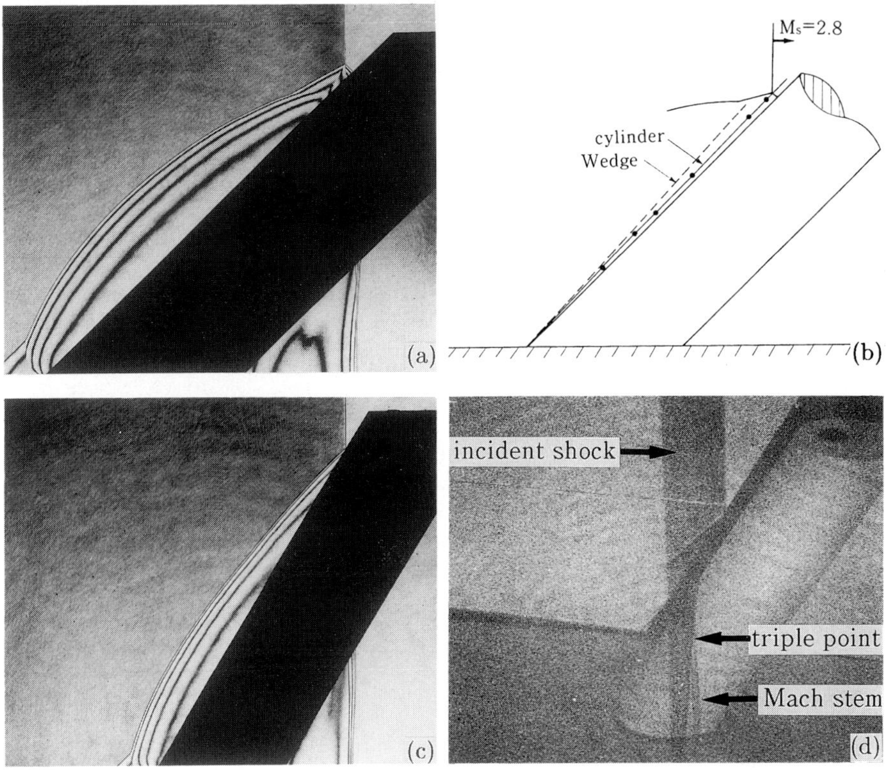

Figure 3. Visualization results of shock-wave reflection over an oblique cylinder, $M_s = 2.8$: (a) $\theta = 45°$, interferogram; (b) $\theta = 45°$, triple-point trajectory; (c) $\theta = 60°$, 2D interferogram; (d) $\theta = 60°$, diffuse (3D) interferogram.

100 mm × 180 mm shock tube which has been used for about three decades at the Institute for Aerospace Studies, University of Toronto, and was recently moved to the Shock Wave Research Center, Tohoku University.

A preliminary result is shown in Fig. 3. The incident-shock Mach number was 2.8 in air. At an inclination angle of the oblique cylinder of 45°, double Mach reflection was observed, as shown in Fig. 3a, and this reflection pattern is different from the two-dimensional one. The triple-point trajectory, unlike the straight line in two-dimensional cases, bends towards the cylinder surface and becomes parallel to the wall (Fig. 3b). When the inclination angle of the cylinder is 60°, regular reflection occurs at the front surface (Fig. 3c), whereas the diffuse interferogram shows that Mach reflection starts on the side wall (Fig. 3d).

Further investigations will be carried out in the near future and experimental results will be used to verify the three-dimensional numerical scheme.

5. Conclusions

By using the combination of holographic interferometry and a shock tube which has very high repeatability, an animated observation was successfully made for two-dimensional shock-wave reflection and diffraction. It was found that for $M_s = 1.20$, the reflection pattern continuously changes from von Neumann reflection through simple Mach reflection to regular reflection. In the case of shock diffraction over a 90° corner, dynamic motion of the complicated flow field behind the diffracted shock is clearly visible and some non-self-similarity was also observed. A three-dimensional shock-wave reflection over an oblique cylinder has been visualized at $M_s = 2.8$.

Acknowledgement. The authors would like to express their thanks to Professor I. I. Glass for his encouragement of shock-wave research at the Institute of Fluid Science, Tohoku University. The authors also express thanks to Messrs. O. Onodera, H. Ojima, and T. Ogawa of the Shock Wave Research Center, Institute of Fluid Science, Tohoku University, for their help in conducting the experiments.

References

Babinsky H, Yang JM, Takayama K (1995) Animated visualization of shock wave flowfields for dynamic comparison between experiment and computational prediction. In: Proc. SPIE 2410:101.

Ben-Dor G (1992) Shock Wave Reflection Phenomena, Springer-Verlag, New York.

Glass II (1991) Over forty years of continuous research at UTIAS on nonstationary flows and shock waves. Shock Waves 1:75.

Han ZY, Yin XZ (1993) Shock Dynamics, Kluwer Acadamic Publishers/Science Press, Dordrecht, The Netherlands.

Sasoh A, Takayama K (1994) Characterization of disturbance propagation in weak shock-wave reflections. J Fluid Mech 277:331.

Skews GW (1967) The perturbed region behind a diffracting shock wave. J Fluid Mech 29:705.

Takayama K (1983) Application of holographic interferometry into shock wave research. In: Proc. SPIE 398:174.

Yang JM, Onodera O, Takayama K (1994) Design and performance of a quick opening shock tube using a rubber membrane for weak shock wave generation. J JSME (in Japanese) 60-570:473.

Universal attenuation law of planar shock waves propagating in dust-gas suspensions

F. Aizik, G. Ben-Dor, T. Elperin, O. Igra, M. Mond
Pearlstone Center for Aeronautical Engineering Studies
Department of Mechanical Engineering
Ben-Gurion University of the Negev, Beer Sheva, Israel

H. Grönig
Stosswellenlabor, RWTH Aachen, Aachen, Germany

Abstract: In a previous study, Olim et al. (1990) proposed an attenuation law for weak to moderate planar shock waves propagating through dust-gas suspensions with relatively high loading ratios of the solid particles. In the present study, a more general attenuation law has been developed. It is valid for shock waves having an initial Mach number in the range $1 < M_0 < 5$ and dust-loading ratios in the range $0.25 < \eta < 10$.

Key words: Attenuation law, Dust-gas suspensions

1. Introduction

A few years ago, Olim et al. (1990) proposed a general law for describing the instantaneous shock-wave Mach number as it attenuates while propagating through dust-gas suspensions. Their proposed attenuation law was

$$M_s = (M_0 - 1)\exp(-x/\chi) + 1 \tag{1}$$

where M_0 is the initial shock-wave Mach number, i.e., the shock-wave Mach number when it first encounters the dust-gas suspension whose edge is at $x = 0$. The attenuation coefficient χ was found to be linearly related to the diameter of the dust particles, D, and inversely to their loading ratio, η, through the expression:

$$\chi = \frac{1}{\eta}\left[\alpha(M_0) + \beta(M_0)D\right] \tag{2}$$

In this expression α and β are constants for a given initial shock-wave Mach number. Olim et al. (1990) also calculated the appropriate values of these two constants for $M_0 = 1.5$. The appropriate values as obtained by them were $\alpha = 2.275$ m and $\beta = 3.07e4$.

The correlation proposed by Olim et al. (1990) implies that when $x \to \infty$ the instantaneous shock-wave Mach number degenerates to a sound wave, i.e.,

$M \to 1$. This was found experimentally by Sommerfeld (1984, 1985) to be the case only if the initial shock wave is weak to moderate and the dust-loading ratio is relatively high. If, however, the initial shock wave is strong and the dust-loading ratio is low to moderate, the shock wave does not attenuate to a sound wave, but to a finite-strength shock wave whose Mach number, M_e, is larger than unity.

Owing to this observation, it was suggested by Olim et al. (1990) that Eq. (1) be replaced by the following more general expression

$$M_s = (M_0 - M_e) \exp(-x/\chi) + M_e \tag{3}$$

Finding the functional dependence of both χ and M_e on the initial shock-wave Mach number, M_0, the diameter of the solid particles, D, and the dust-loading ratio, η, was the purpose of the study presented in the following sections.

2. Present study

In order to find the functional dependence of both χ and M_e on M_0, D, and η, the governing equations describing the propagation of a planar shock wave through a dust-gas suspension, which were formulated by Olim et al. (1990), were solved numerically for various combinations of M_0, D, and η.

Unlike Olim et al. (1990), whose numerical simulation was based on the flux-corrected-transport technique (FCT), the numerical simulation in the present study was based on the random-choice method (RCM).

The computer code was validated by comparing its predictions to all the experimental results which were reported by Sommerfeld (1984). Good to very good agreement was evident. Details of the comparison can be found in Aizik et al. (1995), where a detailed derivation of the governing equations and their final form are also given. In the following only the assumptions upon which the governing equations were based are presented.

The assumptions

1. The flow field is one dimensional and unsteady.

2. The gaseous phase behaves as a perfect gas.

3. The solid particles are rigid, spherical, and inert. They are identical in all their physical properties including their diameter.

4. The solid particles are uniformly distributed in the suspension.

5. The number density of the solid particles is high enough so that the solid phase could be considered as a pseudo-gas.

6. The solid particles do not interact with each other. As a result their partial pressure in the suspension is negligibly small.

7. The volume occupied by the solid particles is negligibly small.

8. Beside the momentum and energy exchange between the solid and the gaseous phases, the gaseous phase is assumed to be an ideal fluid, i.e., $\mu_g = 0$ and $k_g = 0$, where μ_g and k_g are the dynamic viscosity and the thermal conductivity, respectively.

9. The weight of the solid particles and the buoyancy force are negligibly small compared to the drag force acting on them.

10. The solid particles are too large to experience Brownian motion.

11. The temperature within the solid particles is uniform.

12. The dynamic viscosity, the thermal conductivity, and the specific-heat capacity at constant pressure of the gaseous phase depend solely on its temperature.

13. The heat capacity of the solid particles is constant.

14. The diameter of the solid particles is much larger than the shock-wave thickness.

The governing equations were solved for the following ranges of parameters: $1 < M_0 < 5$, $25 < D < 500\ \mu$m, and $0.25 < \eta < 10$. The upper limit of the initial shock-wave Mach-number value, M_0, is due to assumption 2, namely, the gas behaves as a perfect gas. The diameter, D, is limited from above by assumptions 6, 7, 9, and 11, while its lower limit is governed by assumptions 10 and 14. The highest loading-ratio value, η, is limited by assumption 7. In order to quantify the latter, the volume of the solid particles was limited to maximum 1% of the suspension volume. All the one hundred $(4 \times 5 \times 5)$ possible combinations of the following values of the above-mentioned three parameters were used as initial conditions for the numerical simulation undertaken during the course of this study.

$$D = 25, 50, 100, 500\ \mu\text{m}; \quad \eta = 0.25, 0.75, 2, 6, 10; \quad M_0 = 1.1, 1.5, 2, 4, 5$$

As mentioned earlier, the purpose of the present study was to find the functional dependence of both χ and M_e, which are the free parameters in Eq. (3) on M_0, D, and η. The process of evaluating this functional dependence is given in detail in Aizik et al. (1995). In the following only the final results are provided.

3. The general attenuation law

For the reader's convenience the general attenuation law, which was developed during the course of this study, is presented in the following

$$M_s(x) = (M_0 - M_e)\exp(-x/\chi) + M_e \tag{4}$$

where the attenuation coefficient, χ, as a function of the parameters affecting the shock-wave attenuation, is given by

$$\chi(M_0, D, \eta) = (b_1 D/M_0)\,\eta^{-\left(b_2 / M_0^{b_3} + b_4\right)} \tag{5}$$

and the equilibrium shock-wave Mach number, M_e, which is obtained at $x \to \infty$ is expressed by

$$M_e = A(M_0) + [M_0 - A(M_0)]\exp[-B(M_0)\,\eta] \tag{6}$$

where

$$A(M_0) = b_5 + b_6 M_0 + b_7 M_0^2 \tag{7}$$

and

$$B(M_0) = b_8/\ln(b_9 M_0) \tag{8}$$

The values of $b_1, b_2, \ldots, b_9$ as obtained in the course of the present study and which ensure a 99% correlation that the curve fits with the numerical data are:

$b_1 = 0.02728$ $b_2 = 1.19194$ $b_3 = 6.63244$ $b_4 = 0.36737$ $b_5 = 1.15934$
$b_6 = -0.23607$ $b_7 = 0.10080$ $b_8 = 0.52311$ and $b_9 = 0.92796$

It could be seen that the general attenuation law given by Eq. (4) contains all the major flow parameters and dust physical properties which are known to affect the shock-wave attenuation, namely, the initial shock-wave Mach number, M_0, the dust-loading ratio, η, and its diameter, D.

The general attenuation law enables one to calculate the instantaneous shock-wave Mach number within a 5% accuracy. As a result, the jump conditions across the attenuating shock wave could be readily obtained without the need to numerically solve the entire flow field.

4. Validation of the general attenuation law

The shock-wave attenuation, as predicted by the above-developed general attenuation law, were compared to the experimental results of Sommerfeld (1984, 1985). A typical comparison is shown in Fig. 1 for $M_0 = 1.7$, $D = 27$ μm, and $\eta = 1.4$. As can be seen, the agreement between the attenuation predicted by the general attenuation law (Eq. 4) and the actual attenuation is very good.

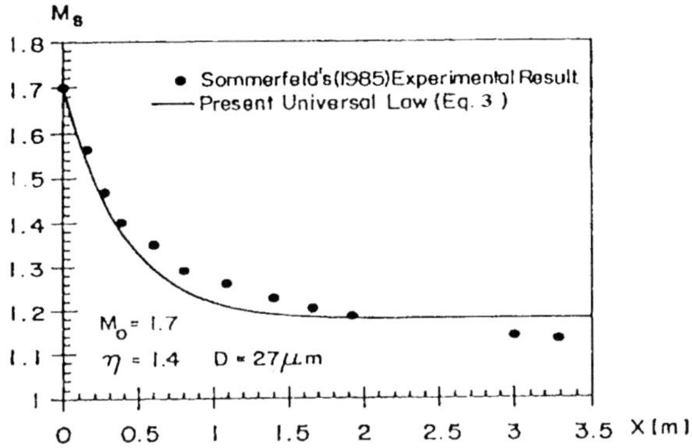

Figure 1. Comparison between the attenuation predicted by the "General Attenuation Law" developed in the course of the present study and the experimental results of Sommerfeld (1984) for $M_0 = 1.7$, $\eta = 1.4$, and $D = 27\mu$m.

A comparison between the general attenuation law (Eq. 4) and the attenuation law developed by Olim et al. (1990) (Eq. 1), which as mentioned earlier is applicable only to weak to moderate planar shock waves and relatively high loading ratios, is shown in Fig. 2 for $M_0 = 1.5$, $D = 200$ mm, and $\eta = 4.0$. It is evident that the general attenuation law predicts an attenuation similar to that predicted by Olim et al.'s (1990) limited law. The difference of about 3 to 4% which is developed at large values of x is due to the fact that while the present general attenuation law dictates a realistic equilibrium shock-wave Mach number, $M_e > 1$, Olim et al.'s (1990) specific attenuation law inherently dictates $M_e = 1$ when $x \to \infty$.

5. Conclusions

The governing equations describing the flow field which develops when a planar shock wave propagates inside a dust-gas suspension were solved numerically using the random-choice method.

The shock-wave attenuation was investigated numerically. Based on the numerical results, a general law for describing the shock-wave attenuation as it propagates inside the dust-gas suspension was developed. It is an improvement of a simplified specific atenuation law which was developed a few years ago by Olim et al. (1990).

The general attenuation law was validated by comparing its predictions to actual experimental results. Very good agreement was evident.

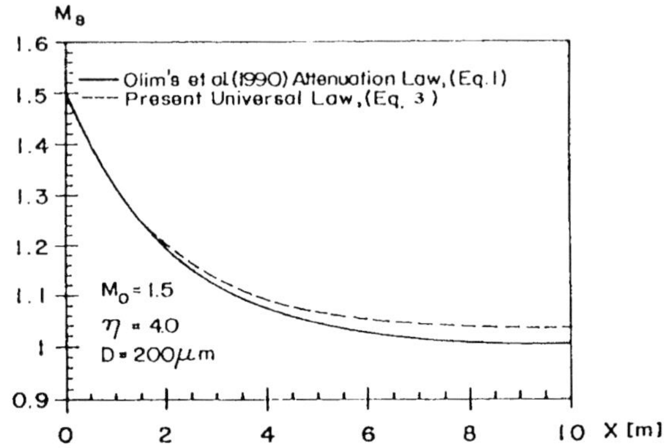

Figure 2. Comparison between the predictions of the general attenuation laws developed during the course of the present study (dashed line) and the limited attenuation law developed by Olim et al. (1990).

The general attenuation law enables one to simply and cheaply calculate the instantaneous shock-wave Mach number, and as a result to immediately obtain the jump conditions across it without the need to conduct a tedious numerical simulation.

Acknowledgement. This research was supported by a grant from the German-Israeli Foundation for Scientific Research and Development. Their support is acknowledged with thanks.

References

Aizik F, Ben-Dor G, Elperin T, Igra O, Mond M, Grönig H (1995) Attenuation law of planar shock waves propagating through dusty-gases suspensions. AIAA J 33(5):953–955.

Olim M, Ben-Dor G, Mond M, Igra O (1990) A general attenuation law of moderate planar shock waves propagating into dusty gases with relatively high loading ratios of solid particles. Fluid Dynamics Research 6:185–200.

Sommerfeld M (1984) Instationare Stosswellenausbreitung in Gas-teilchengemiscen, Ph.D. Thesis, Faculty of Mechanical Engineering, RWTH Aachen, Aachen, Germany.

Sommerfeld M (1985) The unsteadiness of shock waves propagating through gas-particle mixtures. Experiments in Fluids 3:197–206.

Development of a detonation driver for a shock tunnel

M. Lenartz, B. Wang*, H. Grönig
Shock Wave Laboratory
Templergraben 55, 52056 Aachen, Germany

Abstract: The detonation driver represents a new type of high-pressure section for a conventional shock tunnel to increase its performance range. The detonation driver consists of a detonation section which is comparable to the conventional high-pressure section and an additional damping tube. The high driver energy is provided by an oxyhydrogen detonation initiated at the main diaphragm. Simulation calculations show that it is theoretically possible to obtain a high stagnation enthalpy in front of the nozzle and to adjust tailored interface conditions by the proper choice of the pressure ratio between the driver and the driven section and of the composition of the oxyhydrogen. A detonation-driven shock tube was built to investigate the real operation performance (total length about 19 m, inner diameter 140 mm).

Key words: Detonation driver, Shock tunnel, Detonation

1. Introduction

A detonation-driven shock tunnel consists of the "normal" low-pressure side (i.e., driven section, nozzle, test section) of a conventional shock tunnel, a detonation section including an initiation tube at the main diaphragm, and an additional damping section (see Fig. 1). The damping and detonation section,

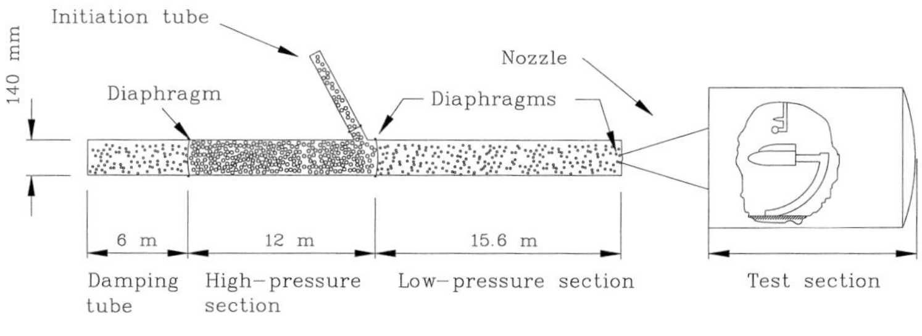

Figure 1. Set-up of a detonation-driven shock tunnel.

*On leave of absence from College of Chemical Engineering, Nanjing University of Science and Technology, Nanjing, China.

and the low-pressure section and the nozzle are separated by diaphragms. The detonation section is filled with oxyhydrogen, the damping section is evacuated, and the medium in the low-pressure section acts as test gas.

The experiment starts with the ignition of the oxyhydrogen at the end of the initiation tube. After a certain induction distance a fully developed detonation wave is established which propagates through the initiation tube into the main tube. There, the detonation front moves towards the damping section, breaks the diaphragm nearly without being reflected, and expands into the damping section which acts to reduce the load on the structure and to increase the duration of the constant stagnation state in front of the nozzle. The velocity of the detonation front is stable and depends mainly on the initial oxyhydrogen composition. The detonation velocity and the state of the burned gas can be predicted by the classical one-dimensional Chapman-Jouguet theory. The planar detonation front is followed by a Taylor expansion fan (Nettleton 1987) which reduces the flow velocity to zero. This state is comparable to the initial state of the high-pressure section of a conventional shock tube. Its pressure level is about 5–8 times higher than the initial pressure and its temperature lies between 3500–4200 K depending on the initial oxyhydrogen composition and its pressure. The wave phenomena inside the low-pressure section are comparable to those of a conventional shock tunnel.

2. Numerical simulation

Numerical simulations of the detonation-driven shock tube have been performed using a program (KASIMIR) which was developed at the Shock Wave Laboratory (Esser 1992, Kanira 1990) and has been extended by a routine to calculate the Chapman-Jouguet detonation for a given oxyhydrogen mixture. The computer code is based on a one-dimensional exact Riemann solver, takes into account high-temperature effects (e.g., dissociation, vibrational excitation), and uses the method of characteristics to calculate the wave diagram of a shock tube with given initial conditions; but it does not consider any losses (e.g., shock attenuation, diaphragm rupture). Figure 2 shows the wave diagram of a detonation-driven shock tube with the same lengths as given in Fig. 1. Two results obtained by the numerical simulation are particularly important (Yu et al. 1992): it is possible to obtain a high stagnation enthalpy in front of the nozzle (see Fig. 3) and to adjust the tailored interface condition for a wide range of pressure ratios p_4/p_1 by matching the oxyhydrogen composition. The dashed line in Fig. 3 determines the initial conditions in the high- and low-pressure sections for a maximum test time under tailored interface conditions. Thus, for a stagnation enthalpy of 10 MJ/kg, a pressure ratio p_4/p_1 of 170 is required with an initial ratio H_2/O_2 of about 1.

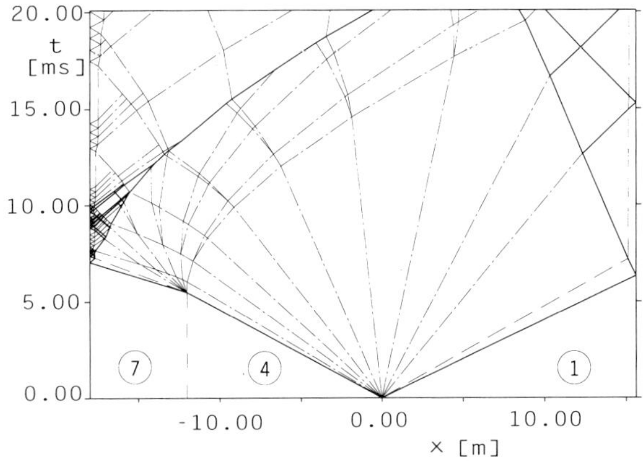

Figure 2. Wave diagram of a detonation-driven shock tube for the following initial conditions: $p_4 = 4$ MPa, $p_1 = 4$ kPa, $p_7 = 5$ kPa; composition of the oxyhydrogen: 60 vol% H_2 and 40 vol% O_2; the temperature of all sections is 293 K; the damping gas is N_2 and the test gas is air.

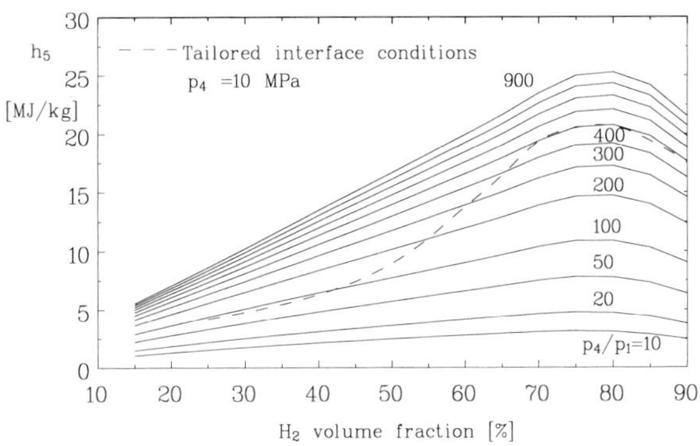

Figure 3. Specific enthalpy h_5 of a detonation-driven shock tube depending on the oxyhydrogen composition and the pressure ratio p_4/p_1; $p_4 = 10$ MPa.

3. Set-up of a detonation-driven shock tube

The detonation-driven shock tube which was built to check the numerical results and to study its performance is sketched in Fig. 4. Its total length is about 19 m; the lengths of the damping, the detonation, and the driven sections are about 6 m each; the inner diameter of all sections is 140 mm, separated by hydraulically clamped diaphragms. The length of the initiation tube is about 900 mm with an inner diameter of 30 mm. Mass-flow controllers adjust the

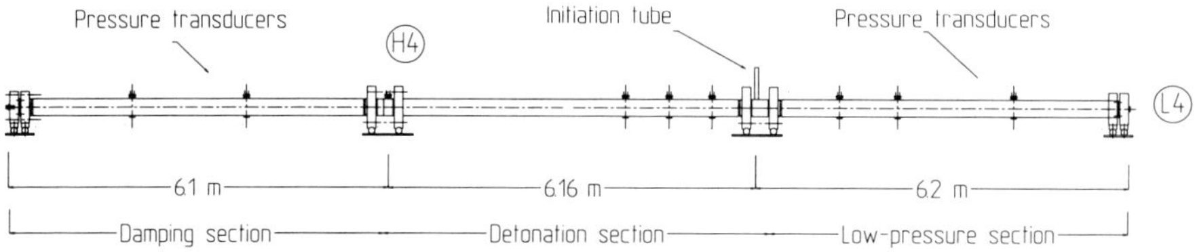

Figure 4. Set-up of the detonation-driven shock tube.

oxyhydrogen composition up to a maximum filling pressure of about 7.5 MPa. There are several locations for piezoelectric pressure transducers in all three sections (detonation section: KISTLER 601H, 6205, 6215; damping and driven section: KISTLER 603B, 601H). The signals of the pressure transducers were recorded by a transient memory system with a maximum sampling rate of 10 MHz.

4. Shock-tube experiments

In order to answer the question if any quenching and reignition processes take place, the first experiments in the shock tube dealt with the transition of the detonation from the initiation tube into the main tube.

These experiments were performed with the detonation and damping sections only. The main diaphragm was replaced by an end plate where a pressure transducer was located in the center. The measurements yielded the direct initiation of an oxyhydrogen detonation in the main tube and proved that the ignition equipment (exploding wire, initiation-tube design) provided a well-working system.

The following experiments concern the complete detonation-driven shock tube to investigate its operational performance. Figure 5 shows the wave diagram based on the following initial conditions providing a nearly tailored interface: $p_4 = 1.5$ MPa, $H_2/O_2 = 1$, $p_1 = 8$ kPa, $p_7 = 5$ kPa; the initial temperature of all sections is 288 K; the damping gas is nitrogen and the driven gas air. Measured and calculated pressure curves at two positions along the tube are presented in Figs. 6 and 7. The agreement between the simulation and the experiment is very good for the detonation itself (see Fig. 6 between 3 and 10 ms) because the state of the burned gas is determined by the CJ condition and is not very sensitive to disturbing influences (Bartlmä 1975). The shift between the measured and simulated pressure curves increases with time because all losses are neglected in the calculation. Figure 7 shows the signal of the pressure transducer mounted on the end plate of the low-pressure section and confirms the tailored operation mode: the pressure level behind the reflected shock has

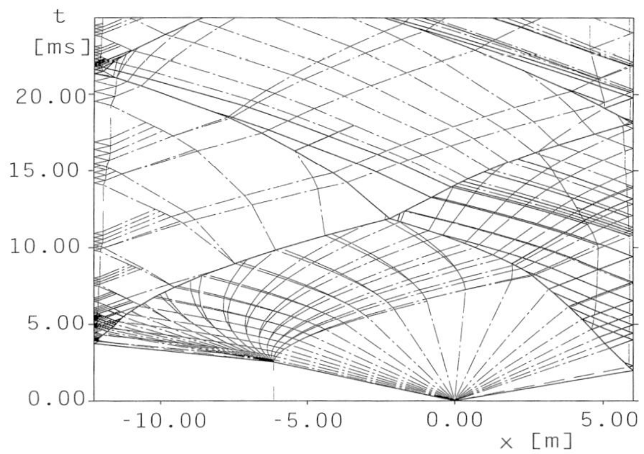

Figure 5. Wave diagram: $p_4 = 1.5$ MPa, $H_2/O_2 = 1$, $p_1 = 8$ kPa, $p_7 = 5$ kPa; the initial temperature of all sections is 288 K; the damping gas is nitrogen and the driven gas air.

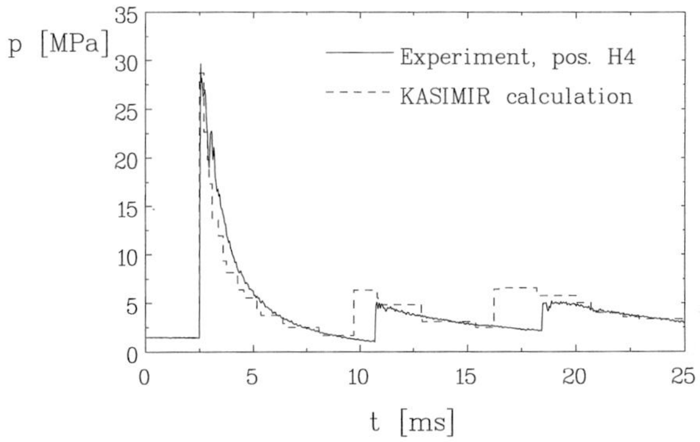

Figure 6. Measured and calculated pressure in the detonation section 6 m from the main diaphragm. Same initial conditions as given in Fig. 5.

a constant value of about 6 MPa during 2 ms; the corresponding stagnation enthalpy is about 8 MJ/kg.

5. Conclusion

The simulation calculations and the experiments performed show that using a detonation driver increases the performance range of a conventional shock tunnel. It is possible to increase its specific enthalpy and to obtain longer test times by operating the tube in the tailored interface mode. The initiation system presented (initiation tube, exploding wire) provides a direct initiation

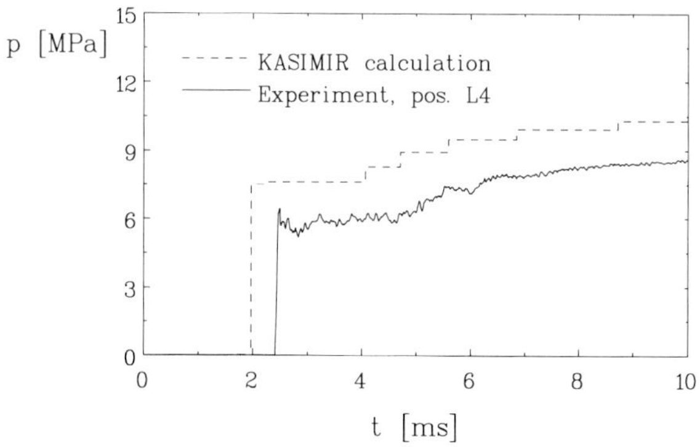

Figure 7. Measured and calculated pressure at the end plate of the low-pressure section. Same initial conditions as given in Fig. 5.

of the oxyhydrogen detonation inside the main tube. The construction of the whole tube is comparably simple, only an additional damping tube has to be added. Compressors to fill the high-pressure section are not needed. But the requirements regarding the safety equipment (e.g., design of the tube structure, leakage detection, gas-mixing system) are much higher. The water generated by the detonation is also a problem (corrosion of the materials and the operation of the facility itself) which has to be treated very carefully.

Acknowledgement. This research project was partially sponsored by the Federal Ministry of Science and Technology, Bonn, by the Ministry of Science of the State of North Rhine-Westfalia, Düsseldorf, and by ESA/ESTEC, Noordwijk.

References

Bartlmä F (1975) Gasdynamik der Verbrennung, Springer, Wien, New York, pp 97–101.

Esser B (1992) Die Zustandsgrößen im Stoßwellenkanal als Ergebnis eines exakten Riemannlösers, Doctoral Thesis, RWTH Aachen, Shock Wave Laboratory.

Kanira I (1990) Simulation der Wellenausbreitung im Stoßwellenrohr, Master Thesis, RWTH Aachen, Shock Wave Laboratory.

Nettleton MA (1987) Gaseous Detonations, Chapman and Hall, London, Chapter 2.

Yu H-R, Esser B, Lenartz M, Grönig H (1992) Gaseous detonation driver for a shock tunnel. Shock Waves 2:245–254.

Behavior of laser-induced bubbles and pressure waves in liquid nitrogen

H. Sato, M. Odagawa, X. W. Sun, K. Maeno, H. Honma
Dept of Mechanical Engineering, Faculty of Engineering, Chiba University
1-33 Yayoi, Inage-ku, Chiba 263 Japan

Abstract: This paper deals with the behavior of laser-induced vapor bubbles in liquid nitrogen. The pulsed irradiation of a ruby laser beam is focused in a special cryostat to generate bubbles. The laser-induced bubbles are visualized by high-speed photography. The pressure signal from bubble motion in liquid nitrogen is measured under near-equilibrium and subcooled conditions. Furthermore, numerical simulation of the dynamics of a single spherical bubble is also performed under the conditions corresponding to our experiments.

Key words: Cryogenic liquid, Cavitation, Bubble dynamics, Pulse laser

1. Introduction

When the pressure in a liquid decreases locally below the saturated value of its temperature due to high flow velocity, bubbles are generated, which expand and collapse, and pressure waves from the bubbles are generated and propagate into the liquid. These phenomena taken together are called cavitation. Cavitation phenomena, mainly in water, have been investigated and widely reported. Recently, in various fields such as aerospace engineering, superconductivity technology, and liquefaction plants for LPG and LNG, cryogenic liquids have been very important objectives of research especially regarding cavitation phenomena in relation to their transportation and preservation. There have been, however, only a few experimental reports which treat cavitation phenomena and two-phase flow in cryogenic liquids, and in particular laser-induced cavitation bubbles (Maeno et al. 1990, Sato et al. 1995).

This paper deals with an investigation of the behavior of laser-induced vapor bubbles in liquid nitrogen. The bubbles produced by pulsed irradiation of a ruby laser beam and pressure waves combined with the bubble behavior are visualized by using both the framing and streak modes of an image-converter camera. The pressure variation is measured by a piezoelectric transducer under equilibrium and pressurized subcool conditions. Furthermore, a numerical simulation of the dynamics of a single spherical bubble including phase-change phenomena at the bubble wall is performed. The results obtained from both experiments and numerical simulations are compared.

2. Experimental setup

In the present experiment, a special cryostat (shown in Fig. 1) has been developed for a cavitation experiment in cryogenic liquids. It is possible to use liquid nitrogen and argon as the test fluid, and by using liquid helium for the coolant, we can also do the experiment with liquid hydrogen. In the present report, liquid nitrogen is employed as the test fluid. The inner vessel to be filled with the test liquid is 150 mm in height and 38 mm in diameter, and two pairs of quartz windows of 20-mm diameter and 8-mm thickness are equipped for visualization and laser-beam introduction. It is also possible to pressurize the test liquid by the gas for an experiment under nonequilibrium conditions. The pressure variation in liquid nitrogen is measured by a transducer (PCB-102A11B) and the temperature in liquid nitrogen is monitored by a silicon diode thermometer (Scientific Instruments, Inc., Si-400).

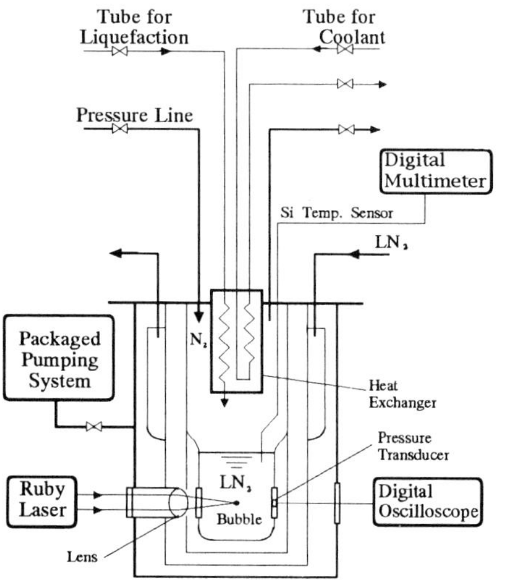

Figure 1. Experimental apparatus (cryostat).

Figure 2 shows a schematic diagram for bubble generation and visualization. A pulsed ruby laser (Nihon Kagaku Engng. Ltd., NAL-707P1) is introduced for bubble formation through the lens system. Iris and stop are used to adjust the light intensity and aberration of the focusing lens. The behavior of the laser-induced bubble and pressure waves are visualized by using the framing or streak mode of an image-converter camera (Hadland Photonics, Imacon-790).

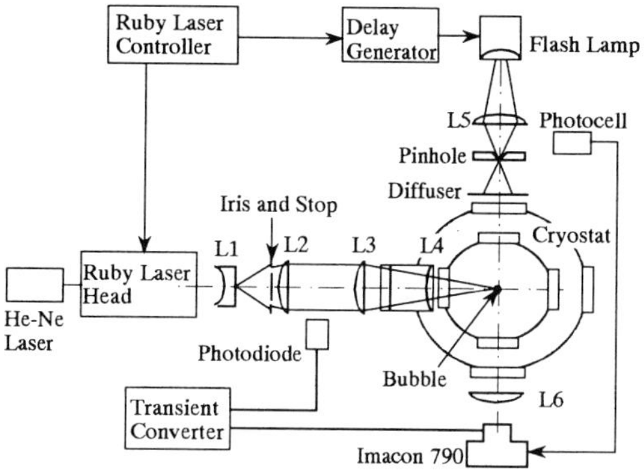

Figure 2. Laser focus and visualization.

3. Numerical analysis

A numerical analysis of a single spherical vapor bubble in liquid nitrogen is carried out to compare with our experimental results. Let us consider a single spherical bubble containing saturated vapor with initial radius R_0 in an infinite volume of liquid nitrogen. The surrounding pressure increases instantaneously to p_∞ at time zero, and the bubble starts to collapse with mass transfer at the interface. The following equations are employed to determine the behavior of the bubble

$$R\ddot{R} + \frac{3}{2}\dot{R}^2 = \frac{1}{\rho_\infty}\left(p_v - p_\infty - \frac{2\sigma}{R} - \frac{4\mu\dot{R}}{R}\right), \tag{1}$$

$$\frac{d}{dt}\left(\frac{4}{3}\pi R^3 \rho_v \hat{U}_v\right) = -p_v \frac{d}{dt}\left(\frac{4}{3}\pi R^3\right) - 4\pi R^2 \Delta U_{v\to L}, \tag{2}$$

$$\frac{d}{dt}\left(\frac{4}{3}\pi R^3 \rho_v\right) = 4\pi R^2 \dot{m}, \tag{3}$$

where $\mu, \sigma, \hat{U}_v$, and $\Delta U_{v\to L}$ denote viscosity, surface tension, specific internal energy, and net rate of the flux of internal energy from the vapor to liquid, respectively. Subscripts L and v denote liquid and vapor, respectively. It is assumed that the vapor within the bubble is described by the equation of state for a perfect gas. In Eq. (3), $\dot{m}$ denotes mass transfer per unit time and area, and is given as follows

$$\dot{m} = \frac{\alpha_M}{\sqrt{2\pi R_v}}\left\{\frac{p_L^*(T_L)}{\sqrt{T_L}} - \frac{p_v}{\sqrt{T_v}}\right\}, \tag{4}$$

where α_M indicates the coefficient of condensation or evaporation. The following energy equation for the liquid

$$\frac{\partial T}{\partial t} + \frac{R^2\dot{R}}{r^2}\frac{\partial T}{\partial r} = D\left(\frac{\partial^2 T}{\partial r^2} + \frac{2}{r}\frac{\partial T}{\partial r}\right), \tag{5}$$

where D denotes thermal diffusivity of the liquid, is solved by means of an integral method (Tomita and Sima 1979) to estimate the liquid temperature at the interface. In this method we assume that the temperature boundary layer in the liquid has a parabolic temperature profile. For easier calculation, a simulation is also performed with the assumption that the liquid temperature at the interface has the same value as at infinity and both results are compared with each other.

4. Results and discussion

4.1. Measurement of the pressure signal and behavior of the bubble

The time history of the pressure variation in liquid nitrogen under atmospheric conditions is shown in Fig. 3. In the figure, compression waves generated by laser irradiation and bubble formation appear on the positive side, and rarefaction waves generated by reflection at the liquid surface appear on the negative side. These pressure waves reflect randomly at the liquid surface and at the wall of the test section. The pressure signal from rebound is not detected clearly, because the distance between the pressure transduser and focal point of the laser is 26 mm in length.

Figure 4 shows a typical series of diffusive shadowgraphs of a laser-induced bubble in liquid nitrogen taken under atmospheric conditions. The frame speed is 5×10^4 frames/s and first frame is taken 5 μs before the ruby-laser pulse. The induced bubble is not spherical but ellipsoidal in its initial stage of bubble expansion because of the small angle for laser focusing. In these shadowgraphs, the bubble expands to maximum radius in the seventh frame. Then it shrinks by the twelveth frame and rebounds. The bubble oscillation period measured from the shadowgraphs is about 200 μs. In the case of pressurized conditions, the bubble radii are smaller and the period of bubble oscillation is shorter than for the case of atmospheric conditions (Sato et al. 1995). The instability of the bubble surface is clearly observed in Fig. 5, this is a characteristic phenomenon of bubble motion in cryogenic liquid. In the present experiment the generated bubble is shaped as an ellipsoidal sphere in the initial stage of bubble motion, as discussed before. So the generated bubble originates with a second-order mode of shape oscillation. But the bubble and its surface in cryogenic liquid are strongly influenced by thermal effects, and higher-order modes of shape oscillation can be excited.

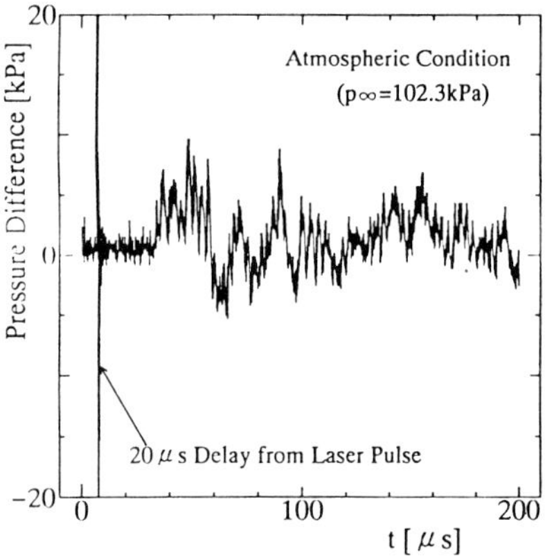

Figure 3. Signal of pressure transducer.

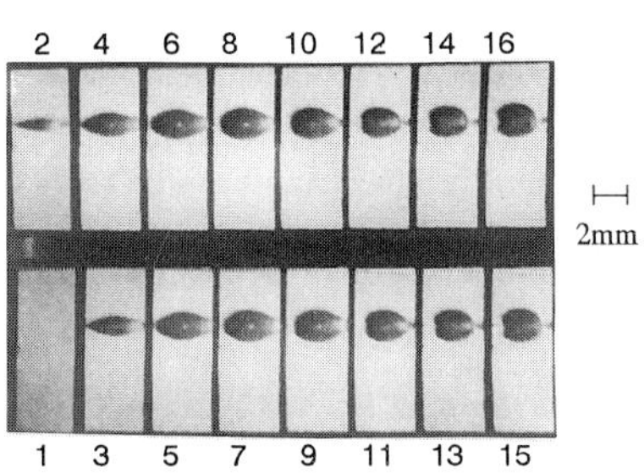

Figure 4. Shadowgraph of induced bubble.

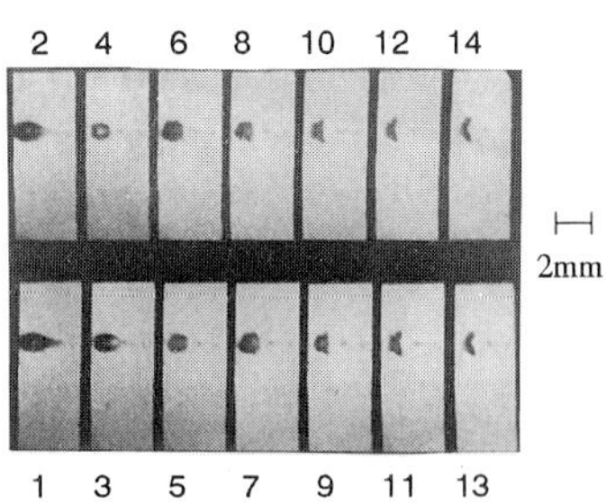

Figure 5. The instability of bubble surface.

Taking notice of the bubble formation, the initial deposit of energy results in the optical breakdown of the cryogenic liquid nitrogen under high-power laser irradiation. This breakdown is characterized by an ionization of the liquid nitrogen by absorption of the incident laser pulse leading to the formation of luminous plasma. Consequently, a shock wave is emitted and a small volume of liquid nitrogen is heated up to the bulk vaporization.

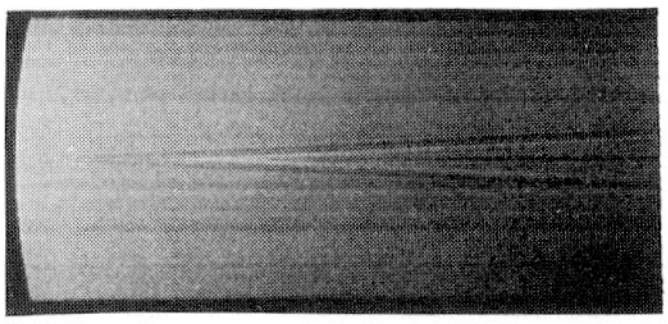

Figure 6. Streak shadowgraph of bubble initiation.

A streak shadowgraph taken 0.39 μs before the laser irradiation under atmospheric conditions is shown in Fig. 6. Parallel light of a Xe flash lamp is employed for visualization of pressure waves. The streak speed is 50 ns/mm. The black line in the middle of the frame indicates the bubble, and a shock wave caused by the laser irradiation is emitted outward from the bubble center. The wave velocity and shock Mach number measured from the shadowgraph are 1187.5 m/s and $M_s = 1.26$, respectively.

4.2. Results of numerical simulation

Figure 7 shows numerical results with nondimensionalized radius for a single spherical bubble in liquid nitrogen under pressurized conditions with $\alpha_M = 0.01$. The solid line and broken line indicate the result when $T_L = T_\infty$ and the result obtained by means of the integral method, respectively. Nondimensional radii of the bubble obtained by the integral method rebound faster than when $T_L = T_\infty$. It is possible to know the bubble behavior by the integral method up to its rebound point assuming the temperature distribution in the thermal boundary layer is not valid after the rebound point.

In the present experiments we consider the shapes of the laser-induced bubbles, such as the series of shadowgraphs in Fig. 4 shows, to be ellipsoids of revolution, and then we estimate an equivalent radius that the bubble would have if it were a sphere of the same volume. The experimental results for temporal motion of a laser-induced bubble in cryogenic liquid nitrogen are compared with numerical results, as shown in Fig. 8. The Rayleigh solution for an empty cavity is also plotted in Fig. 8, where from $t = 0$ to the maximum radius the Rayleigh curve is folded at the maximum point. This solution is for only the inertia-dominant case in liquid nitrogen. In this figure, note that the dynamics in the growth phase of bubble are in good agreement with the folded Rayleigh solution. In the collapse stage, however, a significant retardation appears in the bubble motion due to thermal effects and phase change.

The initial inflating phase of the bubble is assumed as a modified adiabatic change, i.e., the temperature within bubble is obtained when the bubble radius is 1/3 of the maximum radius from the adiabatic relation, and pressure is adjusted. The numerical results with these assumption are shown in Fig. 8 with initial conditions. Present numerical results are in good agreement with our experimental results from the bubble inflating phase to the first rebound of the bubble. Therefore, the effects of phase change explain well the behavior of a laser-induced bubble in cryogenic liquid nitrogen.

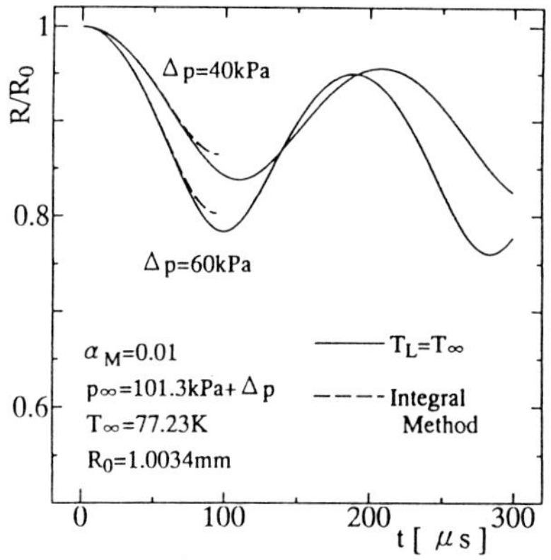

Figure 7. Calculated bubble radius.

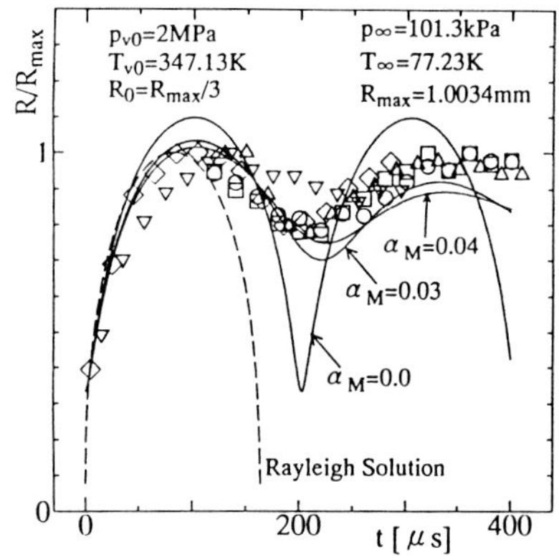

Figure 8. Measured and calculated bubble radius.

5. Conclusion

The behavior of a laser-induced vapor bubble in liquid nitrogen is studied both experimentally and numerically. The induced bubble and initial shock wave combined with pulsed laser irradiation are visualized by high-speed photography. Numerical results are in good agreement with experiments from the bubble inflating phase to the first rebound, and phase-change phenomena influence the bubble motion significantly in liquid nitrogen.

References

Maeno K, Yokoyama S, Hanaoka Y (1990) Study on laser-induced bubble in cryogenic liquids. In: Proc 17th Int Symp on Shock Waves and Shock Tubes, AIP-208:383–388.

Sato H, Sun XW, Odagawa M, Maeno K, Honma H (1995) An investigation on the behavior of laser induced bubble in cryogenic liquid nitrogen. In: Proc Int Symp on Cavitation, CAV'95, Deauville, France, pp 391–397.

Tomita Y, Shima A (1979) The effects of heat transfer on the behavior of a bubble and the impulse pressure in a viscous compressible liquid. ZAMM 59:297–306.

The interaction of a two-dimensional strong compressible vortex with a shock wave

C. T. Kao, K. von Ellenrieder, D. Bershader*
Department of Aeronautics and Astronautics
Stanford University
Stanford, CA 94305 USA

Abstract: The two-dimensional compressible vortex-shock interaction is studied both experimentally and numerically. Shadow and schlieren pictures indicate that the flow field of the strong compressible vortex has a significant effect on the shock configuration in the interaction. A numerical simulation utilizing a first-order shock-capturing scheme shows good qualitative agreement with the present experimental results and previous research. The generation and propagation of diffracted and/or reflected waves due to the interaction are also revealed in the numerical simulation.

Key words: Shock-vortex interaction, Flow visualization, Shock tube

1. Introduction

This investigation is aimed at obtaining a fuller physical insight into the familiar but complex problem of the head-on collision of a shock wave and a two-dimensional strong compressible freely moving vortex of essentially a circular cross section. The obvious questions are: what happens to the shock, what happens to the vortex, and what secondary phenomena are generated, if any? The existence of continuous density, pressure, and velocity variations across the vortex means that the immediate upstream conditions also vary continuously from point to point along the shock, and they vary with time as the inter-penetration proceeds. Also, it is evident that there is a point-wise angular variation of the flow incident on the shock. Clearly, it is expected that an initially plane uniform shock will undergo distortion in both strength and direction. Corresponding variations take place in the downstream flow, including changes in vorticity and entropy.

The density and temperature variations within the vortex imply that the shock is traversing a medium of variable acoustic impedance, a feature which results in the modification of wave properties. The latter has been observed in experimental studies (Dosanjh and Weeks 1965, Naumann and Hermanns 1973)

*In memoriam, deceased on May 30, 1995.

and in numerical simulations (Pao and Salas 1981, Meadows et al. 1991, Ellzey et al. 1995). Also, Ribner (1985) provided a theoretical analysis of the problem. It is evident that the nature of the distorted wave configuration and alteration of the vortex structure will depend on shock Mach number, vortex strength, density level, as well as the structure of the external flow field. However, proper scaling of the overall problem needs further examination. An improved understanding of the shock-vortex interaction is warranted both in connection with the interest in reducing wing-tip noise generation in the transonic flow regime of rotary wing aircraft, and the enhancement of sound radiation from supersonic jets.

2. The compressible freely moving vortex

The method of vortex generation in this study has been described by Bershader (1988), in which the detailed physical configuration of the free vortex prior to shock interaction is also illustrated. Thus, for example, the density as a function of radius and time is expressed by

$$\frac{\rho}{\rho_\infty} = 1 - \frac{\rho^* e^{-\frac{t}{\tau}}}{1 + \frac{r^2}{(r_o+mt)^2}}, \tag{1}$$

where ρ^*, r_o, m, τ are empirical constants equal to 0.898, 0.143 cm, 57.2 cm/s, 1325 μs respectively, while ρ_∞ is the free-stream density. The pressure distribution is measured by a fast-response pressure transducer. The tangential velocity profile, which is especially significant for the shock-vortex interaction, was obtained by insertion of the density and pressure data into the simplified radial momentum equation:

$$\rho\frac{{V_\theta}^2}{r} = \frac{\partial P}{\partial r}. \tag{2}$$

Note that the terms involving radial velocities have been ignored since the radial velocity is at least two orders of magnitude smaller than the tangential velocity. The tangential-velocity profile shown in Fig. 1a is quite similar to that of the classical Rankine vortex, normally linear with radius, r, (solid body rotation) in the core and decaying as $1/r$ outside the core, characteristic of a potential vortex. It is seen that the maximum tangential velocity of the vortex relative to its center is $\sim$ 230 m/s near the edge of the core. Figure 1b gives the corresponding vorticity distribution indicating that the vorticity is concentrated within the vortex core, and the overall circulation is close to 2.5 m^2/s. The vortex is considered strong in the sense that the radial density and pressure distributions exhibit in excess of a 60% drop from free stream to the vortex center.

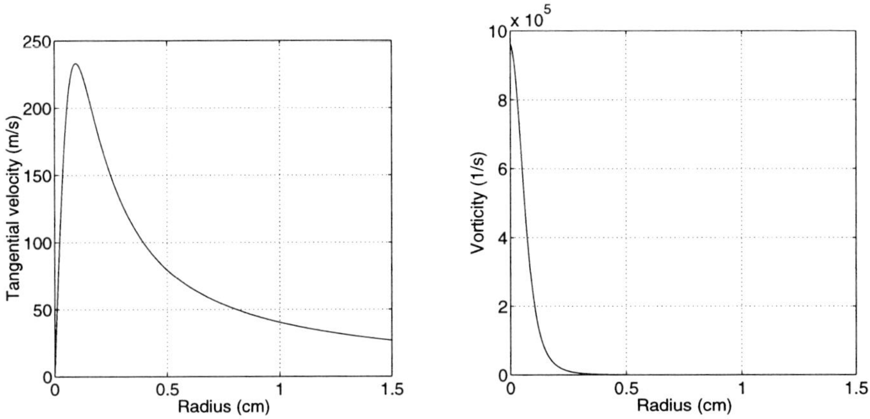

Figure 1. (a) Tangential velocity and (b) vorticity profiles of the free vortex.

3. Experimental observation

The shock-vortex interaction experiments reported here are two dimensional, taking place in a test section bounded by two lucite windows but open to the laboratory atmosphere in the other dimensions. This is in contrast to earlier studies where the interaction occurred in a closed shock tube or channel. The interacting shock originates in the incident shock reflection from a downstream shock-tube model; the portion interacting with the vortex shows only slight curvature. Figure 2 shows both shadow and schlieren pictures of one stage of the interaction. In these photos, the vortex moves to the right approximately with the free-stream velocity of $\sim$ 180 m/s, while the shock moves to the left with a Mach number of $\sim$ 1.2 relative to the incoming free stream. (Other features visible on both photos represent contact surfaces resulting from triple points occurring during earlier shock progress.) The obvious asymmetry of the shock structure occurs because the normal components of the velocity vectors face toward the shock on the lower half of the counter-clockwise rotating vortex, while the opposite is true along the upper half. It is seen that the portion of the shock emerging from the upper half of the vortex is relatively weak; the part diffracting into the lower region of the picture undergoes substantial further curvature and additional weakening. On the other hand, the lower shock portion is stronger, and its progress with respect to the laboratory is diminished due to the opposing velocity of the counter-clockwise vortex fluid rotation. The increased strength corresponds to an increased effective Mach number relative to the opposing rotational vortex velocity. The latter reaches a maximum with respect to the vortex center at the edge of the core, a radius of about 1 mm. It is seen from the shading on both the shadow and schlieren pictures that the density derivatives become especially large within the core, a feature that is also true for the pressure and tangential velocity profiles (see Bershader 1988). Figure 2 also shows an apparent termination of the edge of the lower shock

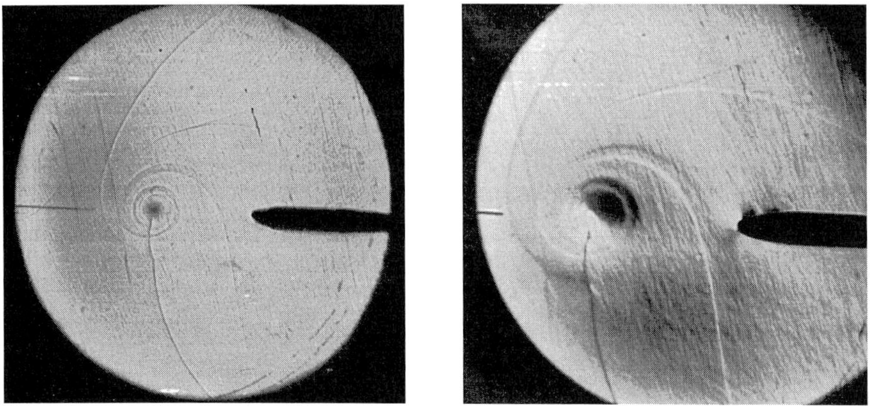

(a) (b)

Figure 2. (a) Shadow and (b) schlieren pictures of the shock-vortex interaction.

portion within the vortex core. That indicates the possible existence of an instability due to the presence of high gradients of physical properties parallel to the shock front. Finally, one may say qualitatively that in the present shock-vortex interaction, the shock "gets the worst of it" in the sense that it is more drastically affected than the vortex. Presumably, the torque generated by the shock during the interaction is not very successful in modifying the angular momentum of the vortex.

4. Numerical simulation

A preliminary numerical study has been performed utilizing the finite-volume formulation to simulate the physical phenomena observed in the experiment. The unsteady two-dimensional Euler equations in conservative form are solved explicitly using the modified Steger-Warming flux-splitting technique. This method is essentially a first-order upwind scheme, in which the fluxes between two adjacent cells are evaluated by averaging the flow properties of the two cells. In the present numerical simulation, the experimentally obtained vortex structure is used as initial conditions, and the shock is assumed initially to be planar. Pressure distributions at two time stages of the interaction are shown in Figs. 3 and 4. The distorted shock structure shown in Fig. 2 is clearly resolved in Fig. 3 at a slightly earlier time stage. Figure 4 illustrates the flow field at a later stage, in which the so-called "acoustic waves" can be identified especially in the 3D plot. The somewhat "thick" shock waves shown in Figs. 3 and 4 are due to the numerical dissipation of the first-order discretization in space. The calculated flow-field features agree with those of the present experiments and previous numerical calculations (for example, Ellzey et al. 1995).

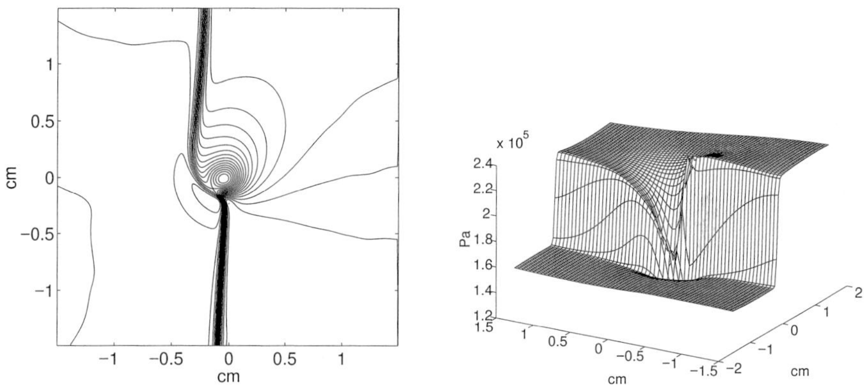

(a) (b)

Figure 3. (a) Contour and (b) 3D plot of pressure distributions.

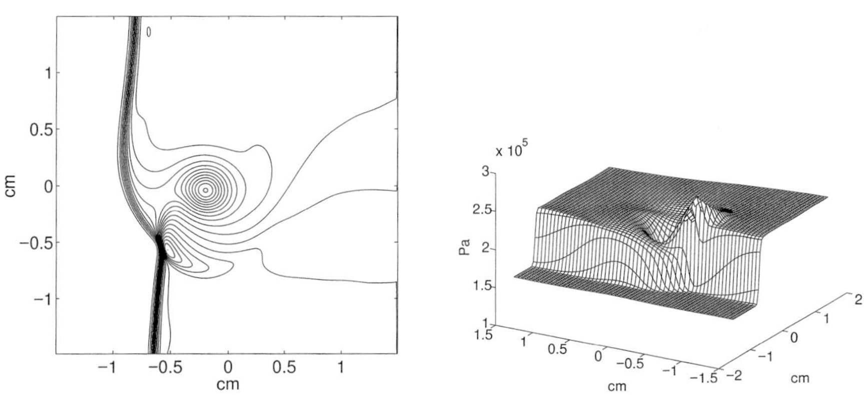

(a) (b)

Figure 4. (a) Contour and (b) 3D plot of pressure distributions (approximately 15 μs later than those in Fig. 3).

5. Discussion and conclusions

Diffracted and/or reflected waves due to shock-vortex interaction have been described by some authors, e.g., Dosanjh and Weeks (1965), as *acoustic* waves. Of course, the acoustic wave is a limiting case of a shock of near-zero strength. However, the term "acoustic" can be misleading in that it implies a train of waves of either narrow or broad frequency spread. The strength of the pressure pulse shown in Fig. 4 is approximately one third of the pressure increase across the shock wave, corresponding to a sound pressure level over 180 dB. In other words, the pressure pulse is about one hundred times stronger than the pressure waves one would receive near a jet engine, typically 140 dB.

The present investigation has increased understanding of the physical basis for most of the admittedly complex phenomena associated with the shock-vortex interaction. The vortex flow field has a drastic effect on the configuration of the shock, including point-wise shock-strength variations. Also, the apparent termination of the edge of the lower shock portion in the vortex core raises a basic question of the nature of non-linear wave behavior in the presence of strong physical gradients parallel to the wave surface. A higher order shock-capturing scheme, e.g., a TVD (total variation diminishing) scheme, including viscous effects is a logical extension of the present numerical work to yield quantitative data for further specification of the phenomenon.

Acknowledgement. The authors are grateful for partial support of this work by NASA Consortium agreement number NCC2-5002 and the NASA Graduate Student Researcher's program. Valuable suggestions from Professor Robert W. MacCormack at Stanford University are also highly appreciated.

References

Bershader D (1988) Shock tube studies of vortex structure and behavior. In: Proc 16th Int Symp on Shock Tubes and Waves, Aachen, Germany, pp 5–18.

Dosanjh DS, Weeks TM (1965) Interaction of a starting vortex as well as a vortex street with a traveling shock wave. AIAA J 3:216.

Ellzey JL, Henneke MR, Picone JM, Oran ES (1995) The interaction of a shock with a vortex: shock distortion and the production of acoustic wave. Phys Fluids A 7:172.

Meadows KR, Kumar A, Hussaini MY (1991) Computational study on the interaction between a vortex and a shock wave. AIAA J 29:174.

Naumann A, Hermanns E (1973) On the Interaction Between a Shock Wave and a Vortex Field, AGARD-CP-131.

Pao SP, Salas MD (1981) A Numerical Study of Two-Dimensional Shock Vortex Interaction, AIAA Paper 81-1205.

Ribner HS (1985) Cylindrical sound wave generated by a shock-vortex interaction. AIAA J 23:1708.

Hypersonic Flow

Interference heating effects caused by a 3D transverse jet in hypersonic flow

N. R. Mudford
Department of Aerospace and Mechanical Engineering
Australian Defence Force Academy
Canberra ACT 2604 Australia

G. T. Roberts, P. H. Schuricht, G. J. Ball, R. A. East
Department of Aeronautics and Astronautics
University of Southampton
Hants SO17 1BJ UK

Abstract: Experiments have been performed in which the heating distribution near a single, circular, sonic jet issuing normally from a flat plate into a hypersonic freestream flow has been quantitatively measured using thermochromic liquid crystals. The results indicate that the heating distribution is sensitive to both injectant gas composition and jet stagnation pressure.

Key words: Control jets, Interference heating, Hypersonic flow

1. Introduction

When a jet is directed transversely into a hypersonic flow, the jet interaction with the freestream creates an enhanced surface pressure field near the jet, resulting in a reaction force which may be several times greater than that of the jet itself. Therefore, such jets are particularly effective for controlling vehicle attitude and stability or creating large side forces for rapid maneuvering.

Powrie et al. (1992, 1993) have recently mapped the surface-pressure distributions on a zero-incidence flat plate with 2D and axisymmetric jets issuing normally into a hypersonic (Mach 6.7) freestream. Qualitative surface visualization using thermochromic (i.e. temperature-sensitive) liquid crystals also showed that large variations in local surface heating were associated with the jet interaction. Reported here are the results of further experiments with a single, circular, sonic transverse jet on a flat plate in which the surface-heating distributions have been quantitatively measured using a recent development of the liquid-crystal thermographic technique (Roberts and East 1995). As with the former work, the influence of injectant-gas composition and jet stagnation pressure was explored under conditions where the upstream (undisturbed) boundary-layer flow on the plate was laminar.

2. Experimental details

2.1. Experimental facility and model design

The experiments were conducted in the Southampton free-piston hypersonic wind tunnel. This facility provides a uniform, steady low-enthalpy flow for a period of $\sim$ 0.5 s; its design and operation have been described elsewhere (Qaswari 1977). For the present work the run conditions were: stagnation pressure $\sim$ 13 bar; stagnation temperature $\sim$ 630 K; freestream Mach number $\sim$ 6.7; unit Reynolds number $\sim 4.3 \times 10^6$ m^{-1}. The freestream gas was nitrogen.

The surface of the 200-mm $\times$ 100-mm flat-plate model, onto which the liquid crystals were sprayed, was made from a black plastic (Acetal). The depth of the plastic (3 mm) was chosen so that it remained effectively semi-infinite (as far as heat conduction is concerned) throughout the duration of the run. The initial temperature of the model (measured by platinum-resistance thermometers) was regulated by an integral heating/cooling system; this also allowed in situ calibrations of the liquid crystals to be performed. A single, 3-mm diameter, sonic jet orifice was located $\sim$ 140 mm downstream of the leading edge of the plate at the mid-span. The injectant gases (Ar, He, N_2, CH_4) were fed to the nozzle via a large-volume plenum chamber; the plenum pressures (p_j) used here were approximately 0.6, 1.0, and 1.6 bar (corresponding to jet-to-freestream stagnation pressure ratios of 0.048, 0.083, and 0.127) and the jet orifice remained choked (and under-expanded) under all conditions.

2.2. Heat-flux measurement technique

The liquid-crystal response to the suddenly applied aerodynamic heating was captured by a video recorder operating at 50 frames/second. Each of the 20–25 frames recorded was analyzed digitally (via a PC and frame grabber) to produce a map of the surface-temperature distribution. To do this, the grabbed red, green, and blue (RGB) values were converted to hue, saturation, and intensity. By comparing the measured hue values (which contain all the color information) with those obtained from an in situ static calibration of hue vs. temperature, the surface-temperature distribution on the model in each frame was obtained. From this, the convective heat-transfer coefficient was determined at all locations on the plate by employing the 1D non-steady heat-conduction equation (see Roberts and East 1995 for details). The above procedure was carried out for both jet-on and jet-off (i.e. undisturbed flat plate) flows. For the latter the local heat-transfer coefficients deduced (h_u) were shown to be in good agreement with laminar boundary-layer predictions using the reference enthalpy correlation of Eckert (1955) (Fig. 1). This correlation was then used to normalize the data to obtain the local ratio of jet-on to jet-off heating, h/h_u. The heat fluxes measured were $O(10^4)$ Wm^{-2}.

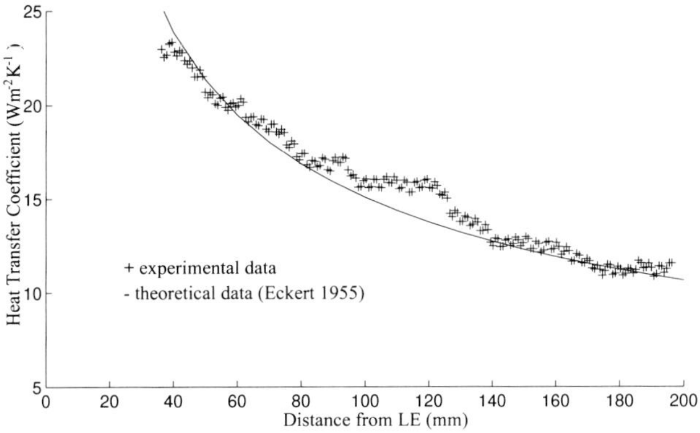

Figure 1. Variation of heat-transfer coefficient h_u with distance along plate.

3. Results and discussion

Figure 2 is an isometric plot of normalized heat-transfer coefficient in the region surrounding an argon jet with $p_j = 1.0$ bar. Data from only one side of the centerline are presented and the region covered extends just upstream of the zone of influence of the jet and to the side and rear of the plate. The spatial resolution of the data (corresponding to the pixel size in the original video image) is approximately 1 mm, although the pixels have been 3×3 neighborhood-averaged to reduce noise. Evidently, the liquid-crystal mapping technique is well able to resolve and measure the large-amplitude features present in the flow field.

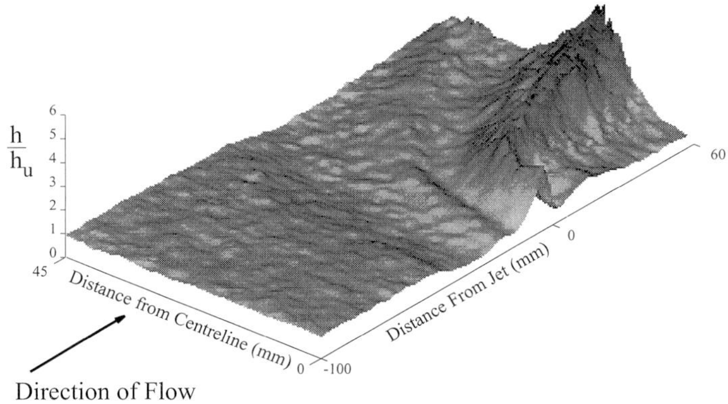

Figure 2. Map of heat-transfer coefficient near jet: argon injectant, $p_j = 1$ bar.

Figure 3 shows the normalized heat-transfer variation along the centerline for the same conditions. From Figs. 2 and 3, and the oil-flow experiments

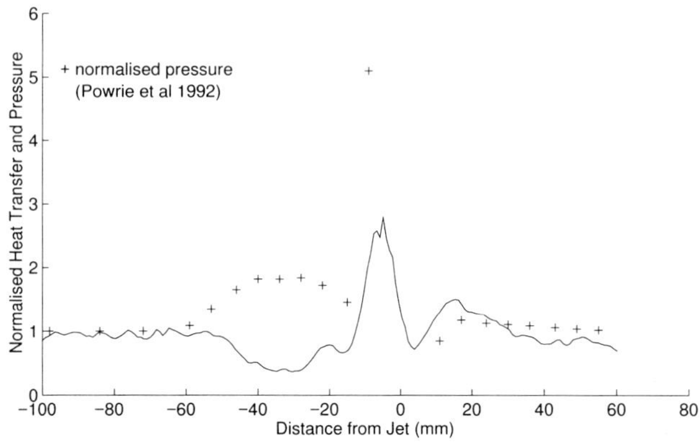

Figure 3. Variation of normalized heat-transfer and pressure along centerline.

reported in Powrie et al. (1993), several features can be identified: the pressure rise associated with the jet interaction (as noted by the centerline pressure measurements, also illustrated in Fig. 3) causes the oncoming boundary layer to separate and roll up into a series of horseshoe vortices. The initial reduction in heat transfer to a minimum value and rise in pressure to a plateau level upstream of the jet are characteristic of a laminar separation (Holden 1978), whereas the peaks in heat transfer correspond to the positions of reattachment of the vortical flows on the plate surface. Evidently, there are at least two of these vortices within the separation zone upstream of the jet and it is probable that the local trough and peak in heating just downstream of the jet are also caused by a relatively small separation zone which is confined close to the jet itself.

As, on the centerline (Fig. 3), the peak heating associated with the reattached flow of the inner vortex approximately coincides with the measured peak pressure, it is tempting to invoke the pressure-interaction theory in which the peak heat transfer is related to the measured peak pressure. Although this theory has been applied reasonably successfully to 2D laminar shock-boundary layer interactions (e.g., Hung 1973), we consider it inappropriate for this type of reattaching flow field as it does not take into account the inherent three dimensionality of the flow nor the differing thermodynamic properties (e.g., stagnation temperatures and heat capacities) of the two gases.

Unlike the blunt-fin problem (Haq et al. 1991), with which the flow is otherwise qualitatively similar, the heat flux along the reattachment line increases as the vortices are swept round the jet; indeed, it is probable that the normalized heat flux would have continued to increase had the plate been longer. Table 1 shows the maximum normalized heating values (measured at the trailing edge of the plate) along the reattachment lines of the outer and inner horseshoe vortices ($(h_p/h_u)_1$ and $(h_p/h_u)_2$, respectively) for argon jets with $p_j = 0.6$, 1.0, and 1.6 bar. These indicate that the amplitude of heating associated with the outer vortex is relatively insensitive to jet pressure whereas that associated with the inner vortex increases with injectant pressure. It has also been observed that the size of the separation zone (evident by the onset of reduction in heat flux from the undisturbed flat-plate value) increases with jet pressure. This agrees with the oil-flow data of Powrie et al. (1993) who showed a strong dependence of the separation length on the jet-to-freestream (stagnation) pressure ratio and a rather weaker dependence on the specific-heat ratio of the jet gas: both influence the effective size of obstruction the jet presents to the oncoming flow.

Table 1. Variation in peak heating (h_p/h_u) with p_j and injectant gas.

Gas	Ar			He	N_2	CH_4
p_j (bar)	0.6	1.0	1.6	1.0	1.0	1.0
$(h_p/h_u)_1$	2.45	2.32	2.53	2.17	1.67	1.37
$(h_p/h_u)_2$	4.33	5.34	7.25	3.22	3.73	2.64

Similar heating features are also observed with different injectant gases. Table 1 also shows the corresponding maximum heating values measured with He, N_2, and CH_4 injectants at $p_j = 1$ bar. Evidently, for a given jet pressure, the heating associated with the reattaching vortical flows is sensitive to the injectant gas composition. Although force data for reaction control jets are often correlated in terms of the jet-to-freestream momentum ratio (e.g., Spaid and Cassel 1973) the data do not incorporate the relevant thermodynamic properties of the injectant gas and cannot, therefore, correlate the present heat-transfer data. Further work is required to identify the appropriate heat-flux correlation parameter.

With helium injectant there are regions where the jet interaction has actually cooled the flat-plate surface. The mechanism for this is as yet unknown, but presumably results from the relatively cool injectant gas (the jet-to-freestream stagnation temperature ratio was ~ 0.48 in these experiments) being brought into contact with the plate surface before it has had an opportunity to mix and exchange energy with the freestream gas. This could be related to the relatively high heat capacity of helium compared with the other injectant gases. Again this warrants further study.

4. Conclusions

The heating distribution on a flat plate near a circular, sonic, transverse jet in a hypersonic freestream flow has been quantitatively measured using thermochromic liquid crystals. The results show that the jet interaction causes the heating to be locally amplified by an amount which is dependent on both jet-injectant pressure and composition. However, further work is necessary in order to identify the relevant correlation parameters for this type of interaction.

Acknowledgement. The work has been carried out with the financial support of the UK Defence Research Agency. P. H. Schuricht is a CASE student supported by the EPSRC and British Aerospace (Dynamics) Ltd. The authors would like to acknowledge the help of Mr. R. P. Clarke in carrying out the experiments and Ms. H. Powrie for her helpful suggestions and advice.

References

Eckert ERG (1955) Engineering relations for skin friction and heat transfer to surfaces in high velocity flows. J Aerospace Sciences 22:585.

Haq ZU, Roberts GT, East RA (1991) Interference heating near fin/body junctions on hypersonic vehicles. In: Proc 1st European Symposium on Aerothermodynamics of Space Vehicles, ESTEC, Netherlands, pp 171–176.

Holden MS (1978) A Study of Flow Separation in Regions of Shock Wave - Boundary Layer Interaction in Hypersonic Flow, AIAA Paper 78-1169.

Hung FT (1973) Interference Heating Due to Shock Wave Impingement on Laminar Boundary Layers, AIAA Paper 73-678.

Powrie HEG, Ball GJ, East RA (1992) Experimental study of the interaction of a two dimensional reaction control jet with a hypersonic flow. In: Proc IUTAM Symposium on Aerothermochemistry of Spacecraft and Associated Hypersonic Flows, Marseille, pp 362–367.

Powrie HEG, Ball GJ, East RA (1993) Comparison of the interaction of two and three dimensional transverse jets with a hypersonic freestream. In: Proc Agard Conf on Computational and Experimental Assessment of Jets in Cross Flow, Winchester, UK, Agard CP534, pp 20.1–20.7.

Qaswari AMS (1977) Measurement of Hypersonic Dynamic Stability of Pitching Blunt Nosed Bodies in a Short Duration Facility, Ph.D. Thesis, University of Southampton.

Roberts GT, East RA (1995) The Use of Thermochromic Liquid Crystals for Heat Transfer Measurement in Short Duration Hypersonic Wind Tunnel Facilities, AIAA Paper 95-6115.

Spaid FW, Cassel LA (1973) Aerodynamic Interference Induced by Reaction Controls, AGARD-AG-173.

Comparison between shock-tunnel measurements on a planetary-probe configuration and DSMC predictions

M. S. Holden, K. M. Chadwick
Calspan-University of Buffalo
New York, New York 14225 USA

M. A. Gallis, J. K. Harvey
Imperial College
London SW7 2BY UK

Abstract: This paper describes an experimental and numerical study of the flow field of a sting-mounted planetary-probe configuration. Two issues are addressed: the flow establishment time of the base flow and the effects of transition to turbulence in the shear layer and reattachment flows. The experimental studies have been conducted at enthalpies up to 10 MJ/kg in the Calspan LENS facility where heat-transfer and pressure measurements have been made for the complete body. Transient Direct Simulation Monte Carlo (DSMC) computations have been made which confirm the wind-tunnel findings.

Key words: 70-deg cone, Hypersonic, Base flow, Transition, High-enthalpy effects, Flow establishment times

1. Introduction

The experimental studies conducted in the LENS facility (Holden 1994) and HEG high-enthalpy shock tunnels with the 70-deg cone planetary-probe configuration coupled with the numerical simulations conducted with the DSMC and Navier-Stokes solvers have resulted in a number of issues that must be resolved to provide an understanding of the key fluid dynamic and chemical modeling of these flows, particularly in the wake. The flow establishment time (FET) for steady attached and separated flows in short-duration facilities is an important factor in the design of an experimental investigation. During the past 30 years, a number of studies have been conducted to determine suitable criteria to judge FET for simple and complex configurations. Typically, from 3 to 4 flow lengths (L_{fl}) based on body diameter are required to establish a steady attached flow, while from 20 to 50 are required for complex shock-induced separated regions and base flows. However, most of these studies were conducted in relatively low-temperature flows where real-gas effects were unimportant. The measurements in the base flow of the planetary-probe configuration in continuum and non-continuum flows for energy levels of 5 MJ/kg and above may be significantly influenced by non-equilibrium flow effects. Thus, the recent high-enthalpy experiments made in the ground tests on the planetary-probe are of great interest.

Another question of central importance, which has arisen during our analysis, is whether transition to turbulent flow occurs in the shear layer or just downstream of the reattachment point in the base-flow region. It is interesting to note in these hypervelocity base flows that there is a relatively narrow laminar-flow regime between the transition associated with non-continuum to continuum flow and the transition to turbulence. It is important that the experimental studies be conducted over a range of Reynolds numbers sufficient to ensure that the transition to turbulence occurs well downstream of the rear stagnation point to provide confidence in the comparisons between experimental and numerical laminar-flow simulations.

2. Time to establish steady flow in the base region

The FET for the base region of blunt bodies has been examined in earlier shock-tunnel studies by Holden (1971) for a sphere supported by fine wires. In these studies, heat-transfer and pressure measurements were made in continuum low-Re flow which in the base flow indicated that this region remained laminar. The pressure measurements showed that the base-flow structure was set-up within $25L_{fl}$ while the heat-transfer measurements suggested that the flow took close to $50L_{fl}$ to become steady.

The determination of FET is complicated by the start time of the tunnel which is controlled principally by establishing the steady flow in the nozzle. In these latter studies, nozzle-flow starting times were less than one ms; nearly an order of magnitude smaller than to establish a steady base flow. In the present studies, covering a wide range of Re and total enthalpies, uniform flow in the test section was established within the first ms of the typically 6-ms flow duration. The steady test duration gave from 15 to 25 m of flow or from 100 to $160L_{fl}$ in which to establish a steady flow over the base region. A typical set of time histories obtained under the 5-MJ/kg conditions was presented in Holden (1994). These measurements suggested that the region of flow close to the shear-layer reattachment point on the sting is the most sensitive indication of FET, and both the heat-transfer and pressure measurements suggest that close to $50L_{fl}$ are required for the position of the shear layer to stabilize in this region. Typically, the recirculation flow in the base region, which is established during tunnel start-up, changes in size causing the pressure and heat transfer at the base of the planetary probe to rise slightly and along the sting to increase or decrease depending on the position in the reattachment region. This adjustment requires about 2 ms to occur. Figure 1 shows the time-history traces for pressure and heat transfer obtained at the 10-MJ/kg condition. We again observe that the tunnel flow is established in one ms, and that while there is relatively minor adjustment to the pressure and heat-transfer levels close to the base of the body, the heat transfer and pressure in the recompression region are sensitive to small movements of the reattaching shear layer as it stabilizes. A steady base flow is reached in 2 ms (or $50L_{fl}$) after the flow stabilizes over the forebody leaving approximately 3 ms over which the flow in the base region is steady.

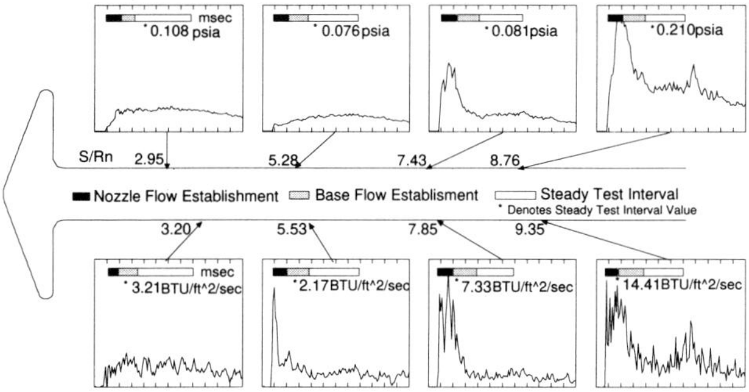

Figure 1. LENS Case E, base flow heat-transfer and pressure time establishment.

3. High-enthalpy non-continuum and continuum flows

Measurements of heat transfer and pressure were made over a range of unit Re spanning the continuum to non-continuum flow regime. Experiments were conducted at stagnation enthalpies of 5 and 10 MJ/kg to investigate the effects of oxygen dissociation. The initial series of studies was designed to provide information for fully laminar base flows over the sting-mounted planetary-probe configuration for direct comparison with solutions from DSMC and Navier-Stokes codes, and was selected so there could be some overlapping in the range of the applicability of these two prediction techniques. Measurements were then made at higher Re to examine when transition to turbulence occurred in the base region. Figure 2 (left) shows the heat-transfer and pressure measurements on the forebody and in the base region compared with DSMC solutions obtained independently by Moss et al. (1995)and Gallis & Harvey (1995) for a rarefied flow condition.

Both theory and experiment indicate that the pressures and heat-transfer rates in the base-flow region are close to 1% of the cone value, rising to 8% at the end of the reattachment compression process. Increasing the Reynolds number to obtain close to continuum conditions in the base-flow region allows us to compare the measurements with solutions by Gallis & Harvey (1995) and Moss et al. (1995) (see Fig. 2, right) employing the DSMC and Navier-Stokes codes, respectively. Again, we see the heat-transfer and pressure levels in the recirculation region at the level of 1% of the cone value, with the heating values downstream of reattachment approaching 10% of the cone value. The DSMC solution is only an instant in the evolution of the flow corresponding to 0.3 ms and shows an overprediction of the heat transfer near the base of the body.

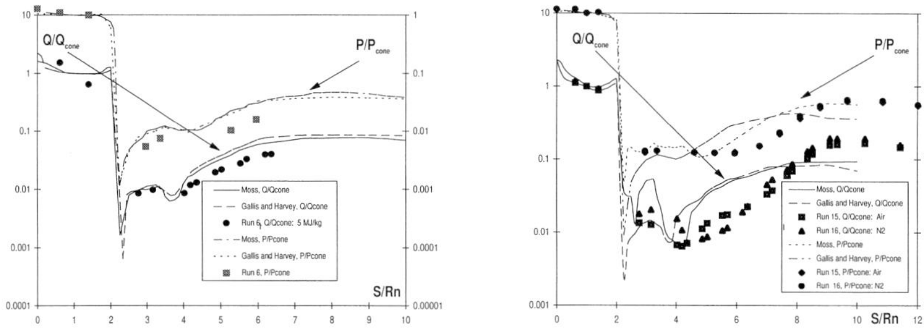

Figure 2. Comparisons between heat-transfer and pressure measurements and computations with DSMC and Navier-Stokes codes.

Transient computations have been performed for the 5-MJ/kg flow to determine the FET using the Imperial College Maximum Entropy DSMC code (Gallis & Harvey 1995). Density contours, non-dimensionalized by the free-stream value, for 0.5, 1.1, 3.0, and 5.0 ms are given in Figs. 3 and 4. From these figures we can identify from the 0.3 contour the changes in the approximate extent of the recirculating region immediately behind the body which has not reached steady state by 3.0 ms. The reattachment shock can be identified by the cluster of 0.4 to 0.6 contours emanating from the sting. It is diffuse because of the low densities in the wake and it plays a dominant role in the formation of the sting boundary and hence on the heat transfer at and downstream of the

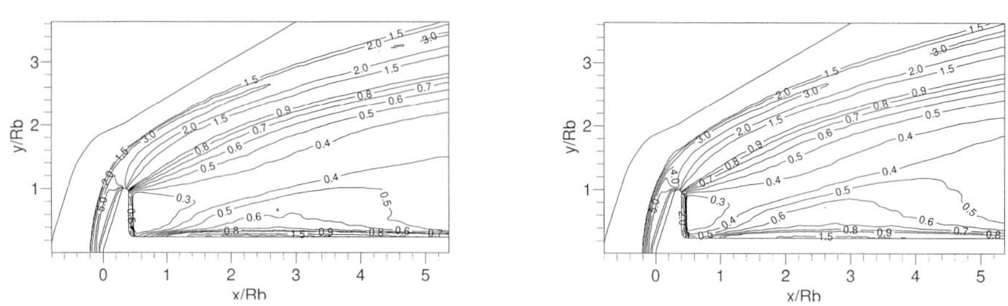

Figure 3. Density contours at: 0.5 ms (left) and 1.1 ms (right).

reattachment point. The wake is a low-density region and flow entering it from the forebody is far from equilibrium due to the rapid expansion around the rim. The FET is thus determined principally by the (a) convection of particles as

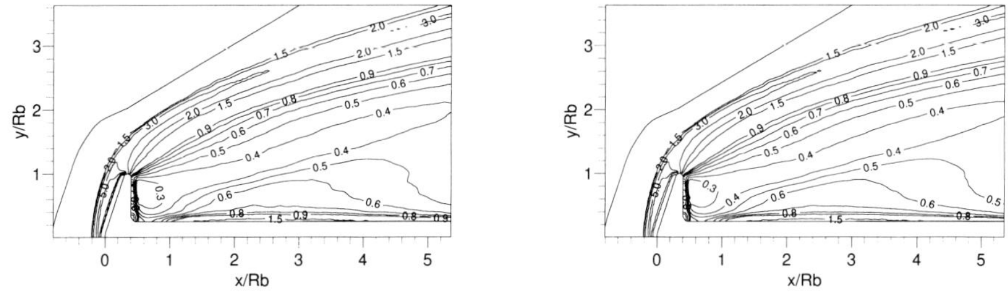

Figure 4. Density contours at: 3.0 ms (left) and (b) 5.0 ms (right).

they recirculate and (b) by the relatively slow real-gas relaxation processes. The evolution of the heat transfer on the sting is presented in Fig. 5 (left). While C_p converges after 1 ms we note that C_h does not reach steady state until 2 ms. This observation is supported by the predicted evolution of the wake.

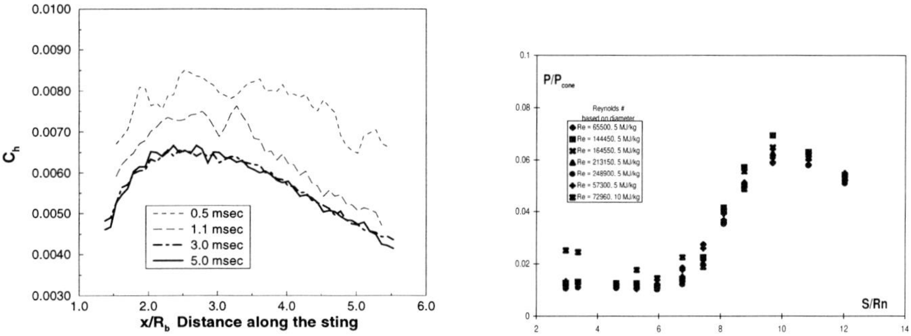

Figure 5. Left: C_h evolution along the sting. Stagnation point at $x/R_b = 0$. Right: LENS planetary-probe data, Reynolds number effect on pressure.

To investigate whether transition to turbulence influenced the pressure and heat transfer in and downstream of the base recirculation region, measurements were conducted at a series of higher Re conditions. These indicated that while the shape and magnitude of the pressure distribution remains roughly independent of Re, above 10×10^4, there is an increase in the maximum heat transfer in the reattachment region accompanied by a forward movement of the peak value (see Figs. 5, right and 6, left). Both these effects suggest that transition has occurred in the recompression region which moves forward with increasing Re, but has yet failed to influence significantly the properties in the recirculation region. For measurements made at similar Re, doubling the total enthalpy of the flow results in a significant decrease in the heating levels in the recompres-

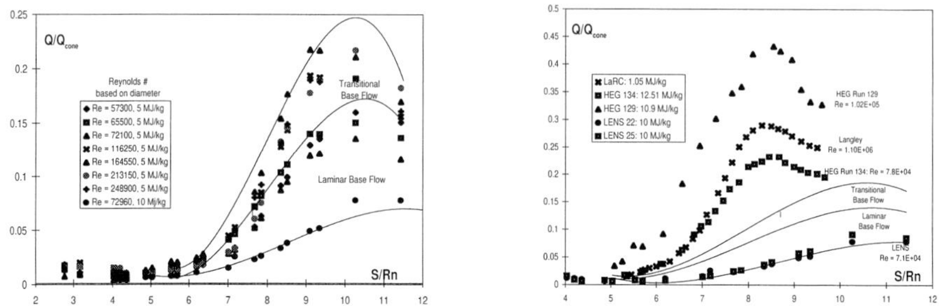

Figure 6. Left: Reynolds number and total enthalpy effects on planetary probe sting pressure and heat-transfer distribution. Right: Comparisons between heat-transfer distribution from LENS, HEG, and NASA LaRC experimental studies.

sion process coupled with an enlarging of the base-flow interaction region. Such a trend was observed in the HEG studies (Kastell et al. 1994), however, in these studies, heating levels in the reattachment region were found to be significantly larger, as shown in Fig. 6, right. Also shown are measurements made in the NASA-Langley Mach 10 tunnel for very large Re conditions ($Re_D = 10^6$) where the base flow is surely turbulent. Plotting our measurements for fully laminar and transition base flows (Fig. 6, right), it could be speculated that the differences between the measurements at the 10-MJ/kg condition could result from differences in the turbulent structure of the shear layer. However, at these high-enthalpy conditions where non-equilibrium flow effects may play an important role in controlling flow-field structure and surface heating, a more detailed examination of these effects and the measurement techniques employed is required to provide a definitive explanation of the differences in heating values observed.

References

Holden MS (1994) Development and Code Evaluation Studies in Hypervelocity Flows in the LENS Facility, AIAA Paper 95–0291.

Holden MS (1971) The establishment time of laminar-separated flows. AIAA J 9(11):2296–2298.

Gallis MA, Harvey JK (1995) Comparison of the Maximum Entropy DSMC Code with Flowfield Measurements, AIAA Paper 95–0413.

Kastell D, Horvath T, Eitelberg G (1994) Nonequilibrium flow expansion experiment around a blunted cone. In: 2nd European Symp on Aerothermodynamics for Space Vehicles, ESTEC, The Netherlands.

Moss JN, Price JM, Dogra VK, Hash DB (1995) Comparison of DSMC and Experimental Results for Hypersonic External Flows, AIAA Paper 95–2028.

Hyperboloid flare experiments in the HEG

R. M. Krek, G. Eitelberg, W. H. Beck
DLR – Institute for Fluid Mechanics, Bunsenstraße 10
37073 Göttingen, Germany

Abstract: Tests on a hyperboloid flare model have been undertaken in the HEG facility. Pressure and heat-flux profiles were measured and the flow field was visualized by a schlieren system. Transitional behavior of the separation shear layer was exhibited by the two high Reynolds number conditions. The experimental heat-flux profile for a low Reynolds number, high total enthalpy condition compares well with Navier-Stokes calculations, assuming a non-catalytic wall. The model forebody pressure and heat-flux profiles for the six conditions tested are self-similar until separation occurs. It is argued, that at constant viscous interaction parameter, which is also at constant relaxation length, the separation bubble size decreases with increasing enthalpy.

Key words: HEG shock tunnel, Hyperboloid flare, Separation, Hypervelocity

1. Introduction

Control surfaces such as body flaps, elevons, and rudders on hypersonic vehicles can produce separation regions that reduce the efficiency of these surfaces. To investigate the flow about deflected surfaces, an axisymmetric body with a flare extension, provided by ESA as a standard model, was utilized. The hyperboloid flare has been tested in many cold wind-tunnel facilities (Krogmann 1994, Koppenewallner et al. 1994, Niezgodka & Henckels 1994, Sagnier 1994a) and one moderate enthalpy facility F4, where the flow is chemically frozen (Sagnier 1994b). The hyperboloid flare model was tested in the High Enthalpy shock tunnel located in Göttingen, HEG, that produces flows with a total enthalpy range of $h_0 \approx 10 - 24$ MJ/kg, with highly nonequilibrium chemistry. Pressure and heat-flux measurements were made as well as flow visualization using a schlieren arrangement.

2. Experimental

The hyperboloid flare is an axisymmetric model whose profile describes the windward centerline of the Hermes re-entry vehicle, at an incidence angle $\alpha = 30°$ and a flap deflection $\delta \approx 20°$. The length of the model is $L_{ref} = 111.4$ mm and the base diameter is $d = 150$ mm. The model was instrumented with 20

Kulite pressure transducers, type XCS-093-50D, and 17 ONERA-made thermocouples. The HEG is a free-piston shock tunnel that can produce flows to simulate the binary scaling parameter of re-entry to Earth's atmosphere (Eitelberg et al. 1992). The model was tested at the HEG operating conditions given in Table 1.

Table 1. Summary of flow properties of the six HEG standard operating conditions.

Condition	p_0 MPa	T_0 K	h_0 MJ/kg	p_∞ kPa	T_∞ K	ρ_∞ g/m^3	u_∞ m/s	M_∞	Re_∞ /m ($\times 10^{-6}$)
I	39	9050	21.1	0.43	790	1.64	5920	9.7	0.27
II	91	9730	22.3	1.21	1040	3.59	6160	9.0	0.52
III	45	7280	13.2	0.47	550	2.83	4810	10.0	0.48
IV	110	8110	14.8	1.32	720	6.15	5150	9.5	0.93
V	51	6340	10.6	0.55	480	3.91	4370	9.8	0.65
VI	94	6870	11.5	1.02	540	6.57	4570	9.8	1.07

Here p is pressure, T is temperature, h is enthalpy, ρ is density, u is velocity, M is Mach number, and Re is Reynolds number.

3. Flow establishment and transition

The flow about the hyperboloid flare is complicated by the shock/shock and shock/boundary layer interactions that occur in the region of the flare, Fig. 1. Flow establishment was monitored by sensors located near the flare hinge line. For all conditions, the model forebody boundary layer is expected to be laminar. For the two higher Reynolds number conditions, the separation bubble displays an oscillatory behavior which was not detected in the other conditions. Figure 2 displays the heat-flux signals taken from sensors located upstream, on the edge of, and inside the separation bubble. The signal produced by the middle sensor displays in an alternating manner the separated and the attached boundary layer heat flux. This is possibly due to shear-layer transition. At the shear layer Reynolds numbers $\approx 1.2 \times 10^5$, determined for the higher pressure conditions, according to the criteria by Holden & Kolly (1995), the shear layer becomes transitional. The transition of the separation shear layer also causes an oscillation in the reattachment point, but to a greater spatial extent than the separation. Figure 4 shows plots of the normalized pressure and heat-flux profiles along the model. The arrows on the symbols indicate the oscillation of the signals during the test time and the symbols are the average of the oscillations. For both the pressure and heat-flux results, the oscillation is many times larger for condition VI than for condition I. The signals on the forebody are steady. It should also be noted that the oscillation amplitudes in the pressure signals are less than those in the heat-flux signals.

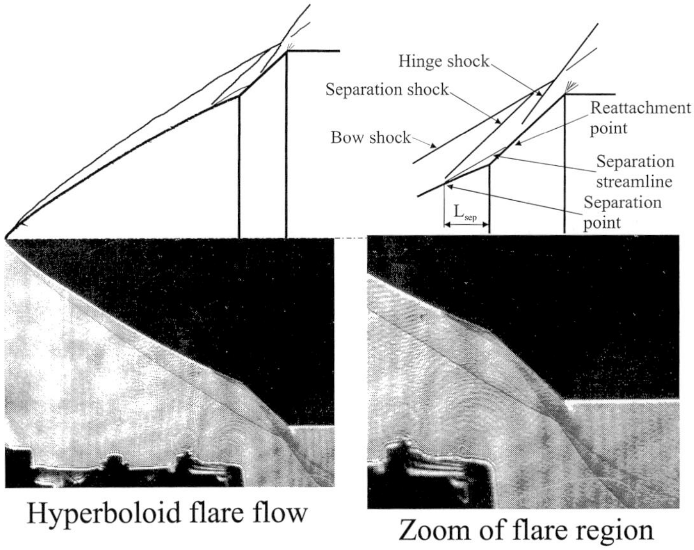

Figure 1. Schlieren photographs and sketch of the flow structure about the hyperboloid flare and close up of the flare region for flow condition III.

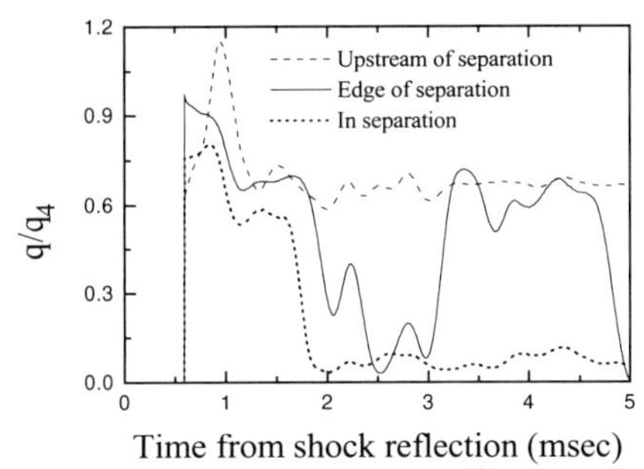

Figure 2. Heat-flux histories for sensors located upstream, on the edge of, and inside the separation bubble for flow condition VI.

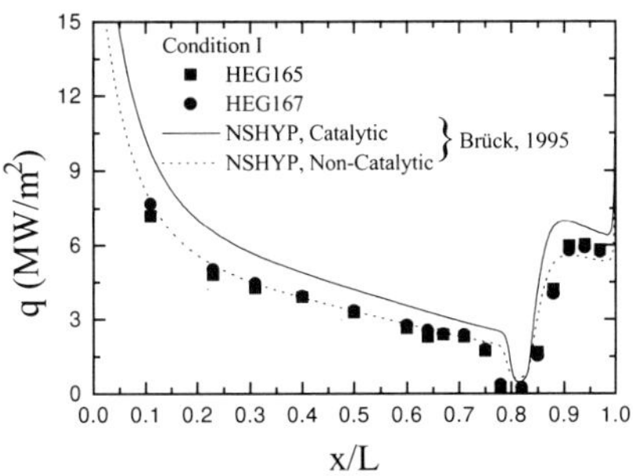

Figure 3. Comparison of the measured and calculated catalytic/non-catalytic heat-flux rates on the hyperboloid flare model for flow condition I.

4. Results

Figure 3 shows the heat-transfer rate distribution measured on the model, compared with Navier-Stokes calculations for catalytic and non-catalytic surfaces. The numerical calculations (Brück 1995) use the NSHYP code (Brenner 1994).

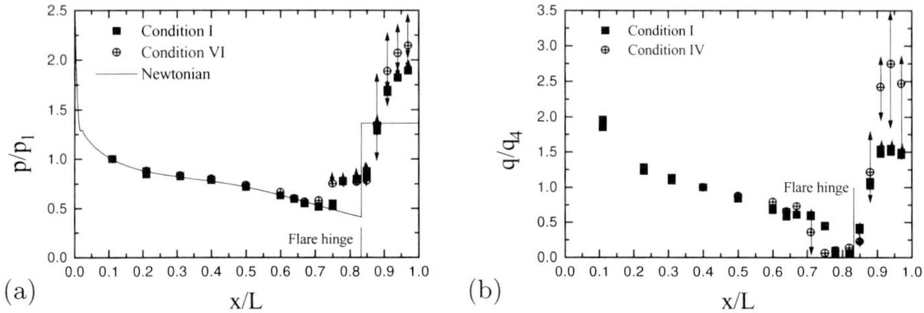

Figure 4. Oscillation of the signals during the test time. Comparison of conditions I and VI: (a) p/p_1 versus x/L_{ref}; (b) q/q_4 versus x/L_{ref}.

From Fig. 3, the measured heat-transfer rates follow the non-catalytic calculations on the forebody. This is in agreement with the prediction by Inger (1994) for large sweepback angles. The experimental results on the flare display slightly higher heat fluxes than the non-catalytic calculations. Due to the complex, unsteady flow in the reattachment region, no conclusions about catalycity effects can be drawn at present. Figure 4 shows examples of pressure and heat-flux distributions for two of the HEG conditions. The results are normalized with respect to measured model quantities to eliminate the influence of non-uniformities in the nozzle flow. The pressure and heat-transfer profiles are self-similar on the forebody of the model until separation of the flow occurs. The Newtonian pressure profile is an excellent approximation for the experimental results.

5. Discussion

Figure 5 shows the separation length versus the viscous interaction parameter for the six HEG conditions and for other hypersonic wind tunnels. The separation length L_{sep} is measured from the separation point, determined from the schlieren images, to the flare hinge line (refer to Fig. 1). The viscous interaction parameter is defined by

$$\bar{V}_\infty = \frac{M_\infty}{\sqrt{Re_\infty}}. \tag{1}$$

All the wind tunnels, except for the HEG and F4, are cold, with the F4 facility being operated from cold conditions to conditions with similar enthalpies as the HEG. Figure 5 shows that all the cold wind-tunnel results, including the cold F4 results, follow the same curve that has a maximum at $\bar{V} \approx 0.015$. The hot F4 and HEG results deviate from this curve, suggesting that the enthalpy is affecting the separation size. The size of the separation bubble is also affected

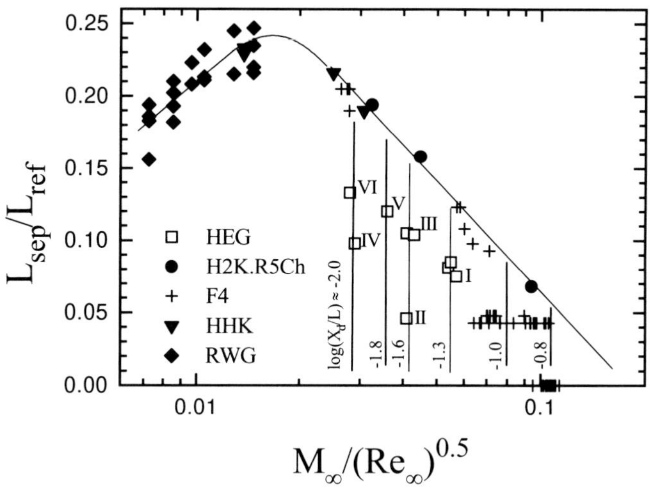

Figure 5. Separation length L_{sep}/L_{ref} versus the viscous interaction parameter $\bar{V}$ for the six HEG conditions and for various other wind tunnels.

by the extent of flow nonequilibrium. The extent of nonequilibrium can be described by the characteristic length for dissociation, calculated assuming a strong shock and ideal dissociating gas, which is given by Eitelberg et al. (1992)

$$X_d = \frac{u_\infty}{C\rho_\infty \Theta_d^\eta}, \tag{2}$$

where Θ_d is the characteristic dissociation temperature, and C and η are constants. The dissociation length X_d is then normalized by a characteristic length scale, in this case, the length of the model L_{ref}. Assuming Sutherland's law for viscosity and frozen freestream, combining Eqs. (1) and (2), the following relationship can be developed:

$$\bar{V}_\infty \propto \sqrt{\frac{X_d}{L_{ref}}} \frac{1}{\sqrt[4]{T_\infty}}. \tag{3}$$

Figure 5 plots lines of constant X_d/L_{ref} calculated for the six HEG conditions and the hot F4 data. The curve representing the cold wind-tunnel data is the frozen limit. For constant $\bar{V}$, and therefore for constant X_d/L_{ref}, as the enthalpy increases, the separation length decreases. Conditions IV and VI are exceptions, but this is possibly due to these conditions being transitional or the uncertainty in determining the freestream conditions. Also for constant $\bar{V}$, e.g., conditions II and III, as the enthalpy increases and the separation point moves downstream, the reattachment point moves upstream.

6. Conclusions

Tests on a hyperboloid flare have been conducted in the HEG at six conditions. The two highest Reynolds number conditions exhibit transitional behavior of the separation shear layer. The experimental results for condition I were compared with numerical calculations and yielded good agreement. All pressure and heat-flux profiles on the model forebody are self-similar for the six HEG conditions. For the HEG range of total enthalpies, it was shown that the viscous interaction parameter is proportional to the square root of the dissociation length scale for the freestream. The separation size was found to be affected by not only the total enthalpy of the flow, but also by the relaxation process.

Acknowledgement. The authors wish to thank ESA for supplying the hyperboloid flare model and for partial funding of the tests. Thanks also go to D. Kastell for the optical visualization, J. Lenz for the HEG operation, and D. Garbe for the installation and interfacing of the model.

References

Brenner G (1994) Numerische Simulation von Wechselwirkungen zwischen Stößen und Grenzschicten in chemisch reagierenden Hyperschallströmungen, DLR FB 94-04.

Brück S (1995) Private communication.

Eitelberg G, McIntyre TJ, Beck WH, Lacey J (1992) The High Enthalpy Shock Tunnel in Göttingen, AIAA Paper 92-3942.

Holden M, Kolly J (1995) Measurements of Heating in Regions of Shock/Shock Interaction in Hypersonic Flow, AIAA Paper 95-0640.

Inger GR (1994) Nonequilibrium-Dissociated Stagnation Boundary Layer on an Arbitrarily Swept Wing, AIAA Paper 94-2072.

Koppenwallner G, Müller-Eigmer R, Johannsmier D, Holze K (1994) Hyperboloid Flare Tests in the HHK, HTG Report November 18 1994.

Krogmann P (1994) Hyperboloid Flare Experiments at Mach 6.8 in RWG, DLR IB 223-94 C 44.

Niezgodka J, Henckels A (1994) Pressure Distribution and Stanton Number Distribution Measurements on a Hyperboloid Flare Model at $M_\infty = 8.7$, DLR IB 39113 94 C 07.

Sagnier P (1994a) Synthesis of Measurement Results Over a Hyperboloid Flare Configuration in ONERA R5Ch Wind Tunnel, ONERA RT No. 64/6121SY.

Sagnier P (1994b) Synthesis of the Hyperboloid Flare Campaign in F4, ONERA RT No. 70/6121SY.

Experiments on the Halis axisymmetric configuration in the T5 shock tunnel

J. Davis, B. Sturtevant
Graduate Aeronautical Laboratories, California Institute of Technology
Pasadena, CA 91125 USA

J. Muylaert
Aerothermodynamics Section YPA, ESA-ESTEC
2200AG, Noordwijk, The Netherlands

Abstract: A model of the Halis Axisymmetric Configuration (HAC) has been built and tested in the T5 hypervelocity shock tunnel, a high-enthalpy impulse facility. Experiments were performed at four test conditions using both air and nitrogen. Static pressures and heat-transfer rates were measured on the model surface. Flow visualization was performed using infinite-fringe differential interferometry. A separated region at the flare root was observed under all test conditions. Pressure and heat-transfer results showed reasonable repeatability, but exhibited unsteadiness in the region downstream of reattachment. Transition to turbulence was found to occur in the separation region. Separation length was seen to increase with viscous interaction parameter $\bar{V}_\infty$, with a possible high-enthalpy effect decreasing the separation length at constant $\bar{V}_\infty$.

Key words: Hypersonic, Hypervelocity, Real gas, Separation, Shock tunnel, Measurements, T5, High enthalpy, Halis, Shock/boundary-layer interaction

1. Introduction

The T5 facility at GALCIT is a free-piston driven shock tunnel capable of reaching very high stagnation temperatures and pressures at relatively high densities, simulating real gas flows around vehicles flying at sub-orbital speeds through the atmosphere. The HAC is an axisymmetric geometry extracted from the windward centerline profile of the STS orbiter. The particular configuration used in the present experiments, shown in Fig. 1a, corresponds to an angle of attack $\alpha = 35^\circ$ and a flap deflection angle $\delta = 20^\circ$. The HAC is a generic axisymmetric form that reproduces flow phenomenon found on 3D flight geometries like the orbiter. While not an exact duplication of the orbiter's geometry (the HAC flare length is much longer than the orbiter flap length), the HAC does incorporate the essential curvature just upstream of the flap, which is expected to decrease the size of the separation region and increase the sensitivity of the separation region to high-enthalpy effects. Separation lengths can be measured from experiments at different enthalpy levels in T5 to look for real-gas effects. A separate paper in this symposium (Muylaert et al. 1996) uses the data obtained from these experiments in conjunction with computational results to study real-gas effects on flap-induced separation.

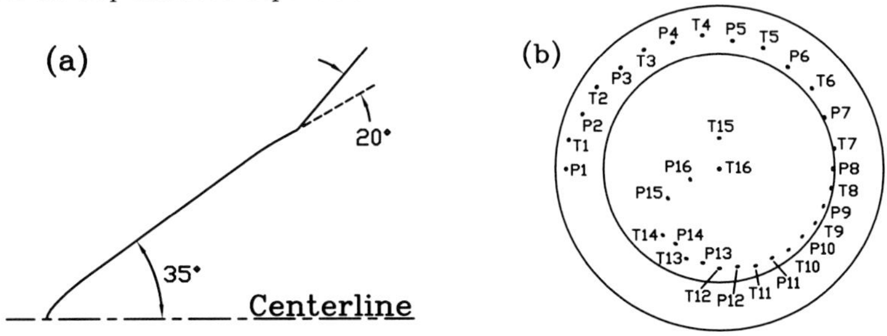

Figure 1. (a) Profile of the HAC geometry; (b) front view of the HAC model.

2. Test model and instrumentation

The stainless steel HAC model of 17-cm base diameter holds 16 pressure sensors and 16 thermocouples, arrayed in a spiral pattern as shown in Fig. 1b. The axial positions of pressure- and temperature-measurement locations are coincident except near the nose, where the pressure sensors and thermocouples are staggered. A thermocouple is used at the stagnation point because the pressure there could be inferred from previous measurements of the free-stream pitot pressure measured for different test conditions.

Surface-temperature measurements were made using type-E coaxial thermocouples with a 1-μs response time. The amplified voltage traces were converted to temperature using a fourth-order polynomial fit for type-E thermocouples. Then time-resolved heat fluxes were calculated using a one-dimensional semi-infinite slab theory for unsteady heat transfer via an indirect method due to Schultz & Jones (1973) and detailed in Germain (1994).

Static-pressure measurements were made using high-frequency, acceleration-compensated, piezoelectric pressure transducers. The recessed mounting of these transducers was designed for fast response time and incorporates several features that protect the sensors from the harsh environment in T5. The possible error in pressure measurement due to the relatively large hole size on the HAC model was estimated to be on the order of 1–2% (Davis 1994); this error was ignored when computing uncertainties for the pressure measurements because it is small and cannot be accurately predicted.

3. Experiments

A detailed description of the T5 facility and its performance envelope can be found in Hornung (1992). The test conditions for the HAC experiments form a matrix of two stagnation pressures and two stagnation enthalpies. Table 1 shows a typical run for each of the four test conditions. Reservoir enthalpy h_0 was computed using ESTC (McIntosh 1970), an equilibrium shock-tube calculation. Free-stream conditions were computed using SURF (Rein 1989), a 2D nonequilibrium, inviscid nozzle-flow code. Re_∞ is based on the model diameter (17 cm) and $\bar{V}_\infty = M_\infty/\sqrt{Re_\infty}$ is the viscous interaction parameter.

Table 1. Typical reservoir and free-stream conditions for HAC tests.

Test Cond.	Run No.	p_0 (MPa)	h_0 (MJ/kg)	ρ_∞ (kg/m^3)	u_∞ (m/s)	T_∞ (K)	M_∞	Re_∞	$\bar{V}_\infty$
1	662	58.5	10.8	0.050	4140	1320	5.7	7.33×10^5	0.0067
2	730	59.3	19.1	0.034	5310	2560	5.0	3.96×10^5	0.0079
3	720	23.5	10.9	0.020	4120	1240	5.8	3.08×10^5	0.0105
4	719	27.2	17.2	0.017	5050	1880	5.4	2.24×10^5	0.0114

Each test condition had four possible types of runs: (1) the original run with test gas of air; (2) a run to show repeatability; (3) a run with the model rotated 180° in roll to show axisymmetric flow uniformity; and (4) a run with N_2 test gas to look for chemical effects. Typical repeatability is shown in Fig. 2 by comparing all four types of runs at test condition 2. Error bars and spline fit are shown only on Run 726. The error bars are based on the variance due to flow unsteadiness during the test time. For pressure measurements they also include uncertainties in sensor calibration and free-stream pitot pressure p_{t2}. Previous measurements of the T5 nozzle flow (Rousset 1995) showed $p_{t2} \propto p_0$ within 5% for air, thus normalization by p_{t2} accounts for variation of p_0 between runs of the same nominal test condition. Figure 2 shows no significant effects due to roll angle or use of N_2 test gas. The unsteadiness and scatter in the data are both larger on the flare than on the forebody.

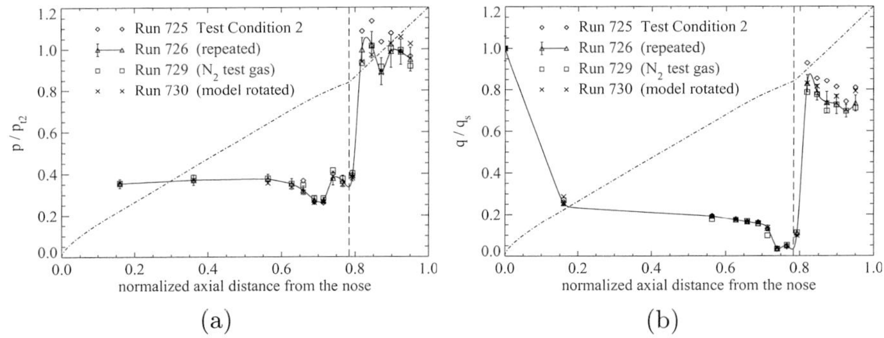

Figure 2. Repeatability of measurements for test condition 2: (a) normalized static pressure, (b) normalized heat flux. Vertical dashed line is hinge-line location, dash-dot line represents model profile.

4. Results

The pressure and heat-flux distributions in Fig. 2 and Fig. 3 have a shape typical of all runs. The pressure is relatively constant over the conical forebody, drops slowly at the curvature upstream of the flare, rises to an intermediate plateau in the separated region, and jumps to stagnation level at reattachment. The drop and subsequent rise in pressure downstream of reattachment appeared in all runs and does not correspond to any flow structure visible in the interferograms; it may be due to the curvature upstream of the flare, which produces a local expansion and causes the reattachment shock to be slightly curved. The heat flux falls to a very low value in the separated region, then overshoots at reattachment and falls off over the flare without a subsequent rise. The experiments were originally intended to encompass fully laminar flow, but preliminary computations (Muylaert et al. 1996) indicate that transition to a turbulent boundary layer at or near reattachment is required to attain the observed heat flux on the flare. This is also consistent with the increased unsteadiness in measurements on the flare.

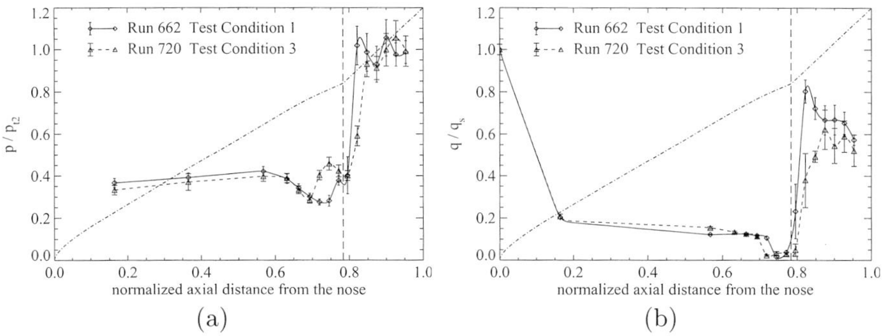

Figure 3. Comparison of data for test conditions 1 and 3: (a) normalized static pressure, (b) normalized heat flux.

The separated region was observed for all runs. Comparison of data between test conditions 1 and 3 in Fig. 3 shows a large difference in separation length due to variation in Re_∞. Figure 4a and Fig. 4b are two differential interferograms, with beam shear aligned at 35° and 135° respectively (measured counterclockwise from the flow direction). The interferograms allow higher resolution measurement of separation length than obtainable with pressure data. The separated shear layer and a corresponding weak compression wave are visible in the interferograms. The separation point is taken where this compression wave intersects the body.

The separation length was measured from interferograms for each of the four runs in Table 1 as the axial distance from the flare root to the separation

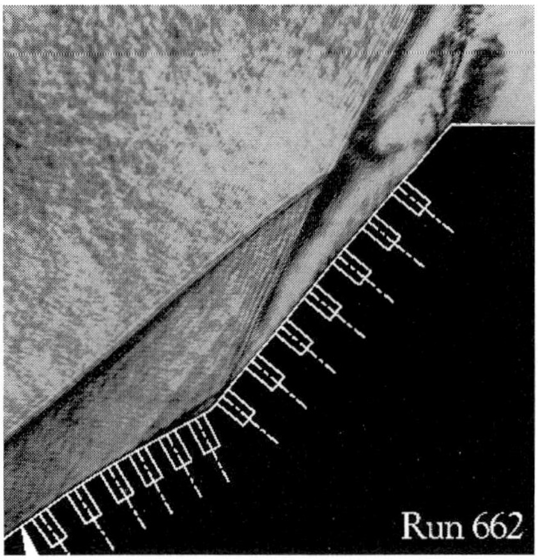

(a)

(b)

Figure 4. Differential interferograms for: (a) test condition 1, and (b) test condition 3. Flow is from left to right. The overlay shows axial locations for measurements, and the arrow indicates start of curvature downstream of conical forebody.

point. These lengths are normalized by the HAC profile axial length (0.1 m) and plotted in Fig. 5 using two different correlations. Figure 5a is an empirical correlation from experiments on a 2D compression corner in cold hypersonic flow (Needham & Stollery 1966), where p_{flare} and p_{sep} are measured surface pressures averaged over the flare and separation regions respectively. M_0 and Re_{x0} at the boundary-layer edge just upstream of separation have been taken from computational results (Muylaert et al. 1996) for laminar flow on the forebody. The decreasing trend with Re_{x0} is consistent with Needham's results for transitional flows (laminar separation and turbulent reattachment) and is opposite the trend for fully laminar or fully turbulent flows. This correlation does not, however, show any clear enthalpy effect with the present data.

The second correlation, Fig. 5b, has been used recently for the hyperboloid flare, an axisymmetric configuration similar to the HAC (Krek et al. 1996). The upward trend with $\bar{V}_\infty$ is the same found for perfect-gas experiments on the hyperboloid flare with $\bar{V}_\infty < 0.15$ and opposite that found for low-density, high-enthalpy experiments with $\bar{V}_\infty > 0.15$, again indicating transitional flows. The high-enthalpy test conditions 2 and 4 show increased separation length compared to the low-enthalpy conditions 1 and 3. This is possibly an enthalpy effect on the separation length, though the evidence is weak considering there are only four data points with significant error bars. Further experiments are planned for T5 using a high-area-ratio conical nozzle to achieve lower density (and fully laminar) test conditions of reasonable free-stream uniformity and to increase the number of data points for correlating separation length.

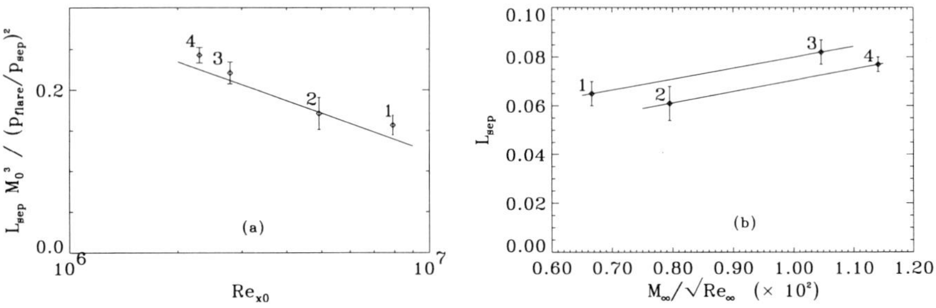

(a) (b)

Figure 5. Two correlations for separation length: (a) empirical correlation due to Needham, and (b) against interaction parameter $\bar{V}_\infty$. Labels indicate test condition.

References

Davis J (1994) Tests on the HALIS Axisymmetric Configuration in the T5 Hypervelocity Shock Tunnel, GALCIT Report FM 94-4, Pasadena, CA.

Germain P (1994) The Boundary Layer on a Sharp Cone in High-Enthalpy Flow, PhD Thesis, California Institute of Technology, Pasadena, CA.

Hornung HG (1992) Performance Data of the New Free-Piston Shock Tunnel at GALCIT, AIAA Paper 92-3943.

Krek RM, Eitelberg G, Beck WH (1996) Hyperboloid flare experiments in the HEG. In: Shock Waves, Proc of the 20th ISSW, Sturtevant B, Shepherd JE, Hornung HG (eds), World Scientific, Vol 1, pp 185–190.

McIntosh MK (1970) A Computer Program for the Numerical Calculation of Equilibrium and Perfect Gas Conditions in Shock Tunnels, Tech Note CPD 169, Australian Defence Scientific Service, Salisbury, South Australia.

Muylaert J, Walpot L, Hironde N, Tumino G, Simeonides G, Koppenwallner G, Davis J (1996) Computational analysis of experiments in T5 and HHK for the study of flap induced separation on an axisymmetric configuration. In: Shock Waves, Proc of the 20th ISSW, Sturtevant B, Shepherd JE, Hornung HG (eds), World Scientific, Vol 1, pp 203–208.

Needham DA, Stollery JL (1966) Boundary Layer Separation in Hypersonic Flow, AIAA Paper 66-455.

Rein M (1989) Surf: A Program for Calculating Inviscid Supersonic Reacting Flows in Nozzles, GALCIT Report FM 89-1.

Rousset B (1995) Calibration and Study of the Contoured Nozzle of the T5 Free-Piston Hypervelocity Shock Tunnel, Engineer Degree Thesis, California Institute of Technology, Pasadena, CA.

Schultz DL, Jones TV (1973) Heat Transfer Measurements in Short Duration Facilities, Technical Report 165, AGARD.

Shock-tube measurements for aero-thermodynamic gas-wall interaction

J. Maul
Institut für Strömungslehre und Strömungsmaschinen
Universität Karlsruhe (TH), Kaiserstrasse 12
D - 76128 Karlsruhe, Germany

Abstract: The motivation for this paper about aerothermodynamic gas-wall interaction comes from the re-entry of a capsule. During the re-entry, the flow is in a thermochemical non-equilibrium state in the higher atmosphere. As a result, the surface catalycity becomes important for the total heat flux into the capsule's heat shield. Due to the surface recombination process an additional heat flux occurs. To determine the influence of the surface catalycity, experiments were performed with oxygen in a shock tube using flow conditions which differed from re-entry conditions. Heat flux and the partial density distribution of atomic and molecular oxygen within the boundary layer of a flat plate coated with aluminium-oxide (Al_2O_3) and silicium-dioxide (SiO_2) were measured. A newly developed two-wavelength laser differential interferometer was applied to measure the partial density distribution. With these investigations, the parameters of a gas-wall interaction model, the atom recombination coefficient γ'_O, the recombination velocity k_{WO}, and the energy accommodation coefficient β_{Oi} were determined.

Key words: Gas-wall interaction, Optical measuring technique, Shock tube, Experiment

1. Introduction

Catalytic recombination is a complex process which includes the absorption of atoms, the chemical reaction at the wall, and the desorption of molecules. Because catalytic recombination cannot be calculated in detail with numerical programs, it is described by a simplified model (Oertel et al. 1994). A common model used in numerical programs is explained by Scott (1990). It consists of three parameters. The first parameter is the atom recombination coefficient γ'_i, which is the ratio of the mass flux of the recombined atoms, $\dot{m}_{i,recomb}$, divided by the mass flux of the atoms hitting the surface, $\dot{m}_i$. The second parameter is the energy accommodation coefficient β_i, which describes the amount of the available energy of formation transformed in an additional heat flux into the structure. The third parameter is the recombination velocity k_{Wi}. The equations for these parameters are:

$$\gamma_i' = \frac{\dot{m}_{recomb}}{\dot{m}_i} \tag{1}$$

$$\beta_i = \frac{\dot{q}_{recomb}}{\dot{m}_{recomb} h_{Di}} \tag{2}$$

$$k_{Wi} = \sqrt{2\pi T_{tr}/kM_i} \tag{3}$$

where $\dot{q}_{recomb}$ is the heat flux due to the catalytic recombination, h_{Di} is the energy of formation, and i specifies species i. T_{tr} is the translational temperature at the wall, k is the Stefan-Boltzmann constant, and M_i is the mass of species i.

To determine these parameters the following terms have to be measured (Scott 1990): the density ρ, the mass fraction c_i, the partial density ρ_i of the species i at the wall, the gradient of the mass fraction of the species i $\partial c_i/\partial z$ at the wall, and the heat flux due to the catalytic recombination $\dot{q}_{recomb}$.

2. Experiments

The experiments were performed in the new shock tube at the Institute for Fluid Mechanics of the Technical University of Braunschweig (Maul and Oertel 1995). This shock tube has a total length of 20 m, an inner diameter of $d_i = 270$ mm, and a test section with a squared cross section of $A = 170 \times 170$ mm^2. In this facility $M_S \approx 11$ can be reached.

A flat-plate boundary layer was investigated. The test gas was oxygen since numerical sensitivity studies have shown that the heat conduction into the structure of a re-entry capsule reacts sensitively to the dissociation of oxygen (Holthoff 1995). Two surface materials were tested, Al_2O_3 and SiO_2. Heat fluxes for different flow conditions from $M_S = 7$ to $M_S = 11$ have been measured. The partial density distribution of atomic and molecular oxygen within the boundary layer of the flat plate has been measured for the following flow conditions: $M_S = 10$, $p_0 = 0.4$ mbar, $M_1 = 3.0$, and $T_1 = 2900$ K (Lieseberg 1994, Maul 1994).

2.1. Two-wavelength laser differential interferometer

To determine the partial density distribution of atomic and molecular oxygen, a two-wavelength laser differential interferometer which works with two different wavelengths simultaneously (Lieseberg 1994) has been developed. The principle is based on laser differential interferometry. It uses the relationship between the density ρ and the refraction index n, the Gladstone-Dale relation, and the assumption that for a gas mixture the Gladstone-Dale constant $G(\lambda)$ consists of the sum of the Gladstone-Dale constant of the single species i weighted with the mass fractions of species i:

$$n - 1 = \sum_i \frac{\rho_i}{\rho} \cdot G_i \cdot \rho \tag{4}$$

Equation 4 can also be regarded as the difference of the refraction index Δn between the light beam 1 and 2 on the basis of a integral density difference $\Delta\rho$:

$$\Delta n_{12} = G(\lambda) \cdot \Delta\rho \tag{5}$$

For two different wavelengths, λ_1 and λ_2, the following systems of equations for the unknown partial densities can be developed:

$$\underbrace{\begin{pmatrix} \Delta n_{12}(\lambda_1) \\ \Delta n_{12}(\lambda_2) \end{pmatrix}}_{\text{measured}} = \underbrace{\begin{pmatrix} G_{O_2}(\lambda_1) & G_O(\lambda_1) \\ G_{O_2}(\lambda_2) & G_O(\lambda_2) \end{pmatrix}}_{\text{given}} \cdot \underbrace{\begin{pmatrix} \Delta\rho_{O_2} \\ \Delta\rho_O \end{pmatrix}}_{\text{unknown}} \tag{6}$$

This system is explicitly solvable when the determinant $||G|| \neq 0$. To guarantee this, one wavelength has to operate in the vicinity of an absorption line of the investigated gas, where $G(\lambda)$ changes drastically. With $\lambda = 760.8$ nm, a specific wavelength for molecular oxygen exists. The choice of the second wavelength is uncritical. For these experiments $\lambda_2 = 632.8$ nm was chosen.

2.2. Thin-film thermometer

Thin-film thermometers (Jessen and Grönig 1992) were used for the determination of the heat flux. They were coated to a thickness of 100 nm with the investigated materials. An estimation shows that the temporal signal is changed only slightly by the coating.

For these flow conditions, the recombination process changes the temperature profile unessentially. Because SiO_2 has non-catalytic behavior, the heat flux due to the catalytic recombination can be obtained. It is the difference between the heat fluxes for the two materials.

3. Experimental results

One necessity for the evaluation of the oxygen partial densities is the knowledge of the Gladstone-Dale constants. Experiments have been performed to determine the Gladstone-Dale constants for the two wavelengths. The results are shown in Fig. 1.

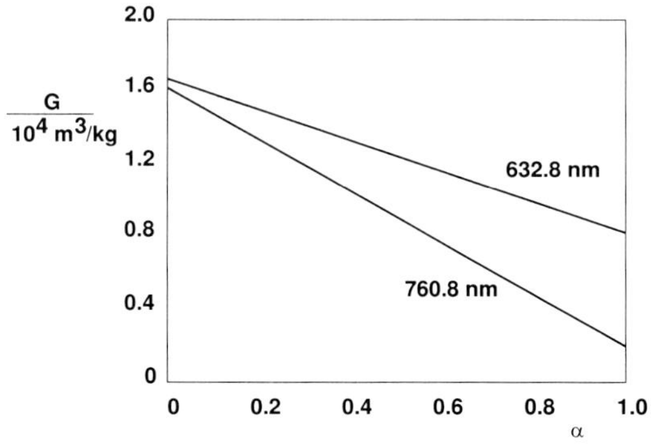

Figure 1. Gladstone-Dale constants for $\lambda_1 = 760.8$ nm and $\lambda_2 = 632.8$ nm.

The diagram shows $G(\lambda)$ plotted against the degree of dissociation α. Obviously the curves have different gradients. Therefore, Eq. (6) has a non-ambigous solution and the partial densities can be determined.

The partial densities for atomic and molecular oxygen have been measured for differences between $z = 0.4$ mm and $z = 2.37$ mm distant from the surface. Figure 2 shows the atomic mass fraction of oxygen c_O for both materials plotted against the dimensionless boundary layer coordinate z/δ.

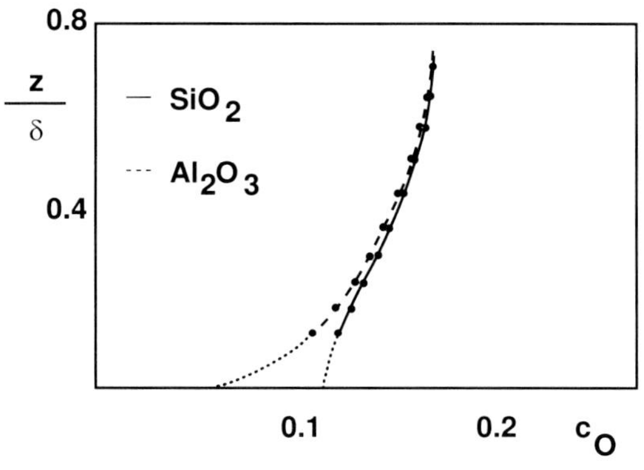

Figure 2. Atomic mass fraction for SiO_2 and Al_2O_3.

For a height above $z/\delta \approx 0.5$ the curves are nearly identical; the surface material has no influence. Below $z/\delta \approx 0.5$, the difference grows. Due to the finite extension of the laser beam, the last measuring point was $z = 0.4$ mm above the surface. To determine c_O and $\partial c_O/\partial z$ at the wall the partial densities have been extrapolated to the wall. The extrapolation is dotted in

Fig. 2. In spite of a detailed inspection, the extrapolation represents the greatest experimental error in the determination of the gas-wall interaction parameters. The error for these investigations is 20%.

Figure 3 shows the heat fluxes for both materials plotted against M_S. For $M_S \leq 7$ the difference between the two curves vanishes because no recombinable oxygen exists in the flow behind the shock. With increasing M_S the difference between the two investigated materials grows due to the increasing degree of dissociation α_O. For $M_S = 10$ the difference is approximately 20%. In Table 1 the experimental results and the evaluated parameters are summarized.

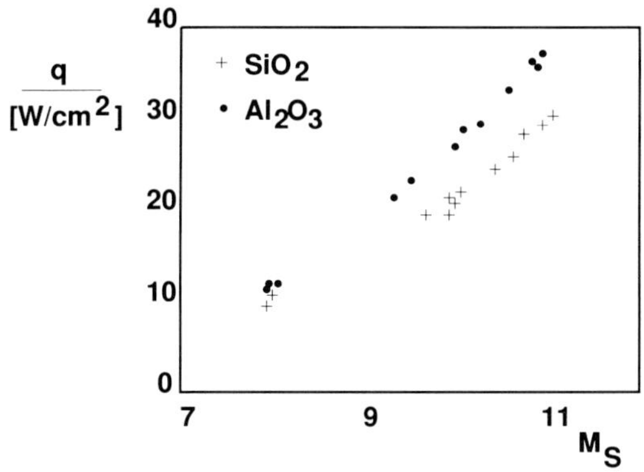

Figure 3. Heat fluxes for SiO_2 and Al_2O_3.

Table 1. Experimental results and evaluated parameters.

	c_O	$\partial c_O/\partial z$ $\left[\frac{1}{mm}\right]$	γ'_O	k_{WO} $\left[\frac{cm}{s}\right]$	β_O
SiO_2	0.117	0.01	0.0003	5.0	—
Al_2O_3	0.067	0.15	0.009	139.0	0.78

The results show that SiO_2 behaves non-catalytically. In comparison with values known from the literature, γ'_O and k_{WO} agree within the measuring error. Al_2O_3 behaves partly catalytically, but γ'_O and k_{WO} obtained by these experiments are greater than literature values. But the advantages of these experiments were the exact knowledge of the flow conditions and that no assumptions for the evaluation of the experimental data needed to be made. Contrary to previous assumptions, the experimental results show that Al_2O_3 does not emit the entire available energy of formation during the recombination. That means the recombined molecules leave the surface either in an excited state or with a higher velocity than that corresponding to the wall temperature. To determine β_O for SiO_2, numerical calculations for the experimental conditions have to be carried out. By a comparison between experiment and numerical results β_O can be obtained.

4. Summary

Experiments have been carried out to determine the parameters of a model describing the interaction of a hot dissociated gas with a catalytic (Al_2O_3) and non-catalytic (SiO_2) surface. A two-wavelength laser differential interferometer has been developed and applied to measure the partial density distribution of atomic and molecular oxygen in a boundary layer. The results show that Al_2O_3 behaves partly catalytically while SiO_2 is non-catalytic. Furthermore, Al_2O_3 does not emit the entire available energy of formation during the recombination. The molecules leave the surface either in an excited state or with a higher velocity than that corresponding to the wall temperature.

Acknowledgement. This research was supported by the Deutsche Forschungsgemeinschaft, Germany. The author wants to thank Dr. A. Lieseberg, who has carried out the optical experiments.

References

Holthoff H (1995) Numerische Simulation des Wärmeübergangs in Hyperschallströmungen, ZLR-Bericht der TU Braunschweig.

Jessen C, Grönig H (1992) Sensoren zur Messung von Wärmestrom und Wandschubspannung innerhalb von Millisekunden, Stosswellenlabor, RWTH Aachen.

Lieseberg A (1994) Interferometrische Messung der aerothermodynamischen Wand-Wechselwirkung an einer längsangeströmten Platte, ZLR-Bericht der TU Braunschweig.

Maul J (1994) Stossrohrmessungen zur aerothermodynamischen Wand-Wechselwirkung, ZLR-Bericht 94-09, TU Braunschweig.

Maul J, Oertel H (1995) Stossrohrmessungen zur aerothermodynamischen Wand-Wechselwirkung (revised manuscript submitted to ZFW).

Oertel H, Böhle M, Delfs J, Hafermann D, Holthoff H (1994) Aerothermodynamik, Springer-Verlag.

Scott CD (1990) Wall Catalytic Recombination and Boundary Conditions in Nonequilibrium Hypersonic Flows — With Application, Third Joint Europe/US Short Course in Hypersonics, RWTH Aachen.

Computational analysis of experiments in T5 and HHK for the study of flap-induced separation on an axisymmetric configuration

J. Muylaert, L. Walpot, N. Hironde, G. Tumino, G. Simeonides
Aerothermodynamic Section YPA, ESA-ESTEC
P.O. Box 299, NL-2200 AG, Noordwijk, The Netherlands

G. Koppenwallner
Hypershall Technology Göttingen
Max Planck Strasse 1, D-37191 Kattlenburg Lindau, Germany

J. Davis
Graduate Aeronautical Laboratories
California Institute of Technology, Pasadena, California, 91125 USA

Abstract: Experiments on an axisymmetric configuration to investigate the influence of real-gas and viscous interaction effects on flap-induced separation and heating were carried out in T5 and the HHK. Nonequilibrium Navier-Stokes computations were performed in conditions including nozzle-flow nonuniformity. For T5, the influence of the pitot-pressure nonuniformity lays within the experimental error. Experiments performed in the perfect-gas Mach 9 facility HHK, with the same viscous interaction parameter as for T5, show a larger separation bubble. For all the T5 experiments, transition to a turbulent boundary layer occurred at the point where the separated shear layer reattaches on the flare. This was confirmed by comparing the experimental heating levels on the flare with Navier-Stokes computations and simpler boundary-layer methods assuming that the flow transitions to turbulence at reattachment.

Key words: Shock tunnel, Shock/boundary-layer interaction, Real-gas effects, Nonequilibrium Navier-Stokes, T5, HHK

1. Introduction

Flap efficiency and associated peak-heating assessment for stability and trimming is a critical design point for space transportation systems. Weilmuenster et al. (1993) showed that the space shuttle STS1 pitching-moment anomaly was due to real-gas effects. However, a detailed systematic analysis of real-gas and viscous interaction effects on flap efficiency and heating has not been done. It is the purpose of the present work to analyze these effects on an "affordable" axisymmetric geometry which incorporates all relevant physical effects associated with the shuttle geometry i.e., the entropy layer from the nose at incidence, the expansion on the windward side upstream of the body-flap hinge, and the compression due to flap deflection. We refer to this configuration as the Halis Axisymmetric Configuration (HAC).

2. Model design and experimental campaigns

The axisymmetric configuration was generated by the windward centerline of the Orbiter at high incidence with body flap deflected. The incidence and flare angles were designed to maximize the laminar separation extent to obtain the largest resolution. Different configurations were tested in the Ludwieg tube HHK in Lindau, Germany. Incidence was varied from 30° to 40° and flap angle from 10° to 20°. The aim was to freeze a configuration which is free from base effects and which produces a steady shock/boundary-layer interaction. The frozen configuration has an equivalent incidence of 35° and a flare angle of 20°. The T5 HAC model was designed and instrumented at Caltech; experiments were performed by Davis et al. (1996). The stainless steel T5 model with a 17-cm base diameter holds 16 pressure sensors and 16 thermocouples. The HHK model contains 20 pressure sensors and is exactly half the size of the T5 model. The T5 and HHK conditions are shown in Table 1 with reference length $L = 0.1$ m for T5 and 0.05 m for HHK, where $\chi_L = M_\infty/\sqrt{Re_L}$.

Table 1. Free-stream conditions.

Test Cond.	Run No.	P_0 (MPa)	H_0 ($\frac{\text{MJ}}{\text{kg}}$)	ρ_∞ ($\frac{\text{kg}}{\text{m}^3}$)	u_∞ ($\frac{\text{m}}{\text{s}}$)	T_∞ (K)	Tv_∞ (K)	M_∞	Re_L ($\times 10^5$)	χ_L
T5 1	662	58.5	10.8	0.049	4216	1185	1769	6.0	4.5	0.009
T5 2	730	59.3	19.1	0.026	5433	2076	2342	5.6	2.2	0.012
T5 3	723	22.7	10.8	0.017	4170	969	2161	6.5	1.8	0.015
T5 4	719	27.2	17.2	0.013	5143	1528	2580	6.1	1.3	0.017
HHK	–	4.26	–*	0.014	1105	37	–	9.06	4.3	0.015

*T_0=645K

3. Computational analysis of experiments

3.1. Codes and free-stream conditions

To have the proper nonequilibrium free-stream data for T5, the LORE 1D code (nonequilibrium multi-temperature, with Edenfield boundary-layer correction) was used (Walpot et al. 1996). The computed free-stream conditions were obtained in such a way that the computed pitot pressures match the measured ones. The matching was done by reducing accordingly the displacement thickness in Edenfield. Table 1 shows the free-stream conditions; they correspond well with those computed by Davis et al. (1996) using the SURF code. The 2D nonequilibrium computations on the HAC geometry were done with the TINA code developed by Netterfield (1992). For the HHK, the perfect-gas code Euranus was used.

3.2. Influence of nozzle nonuniformity

First the influence of the nonuniformity of the pitot-pressure distribution at nozzle exit was assessed. A series of quasi-1D computations were performed along equivalent streamlines, all having the same reservoir pressure and enthalpy; pitot pressures at the exit corresponded to those obtained by measurements at the respective radial positions. These exit conditions were used as 2D inlet boundary conditions for the laminar HAC computations. The influence of the nonuniform pitot was verified for all 4 conditions. Figure 1 shows the extent of pitot-pressure nonuniformity compared to HAC model size. Figure 2 shows the influence of the pitot nonuniformity on pressure distribution. For this configuration the influence of pitot nonuniformity lays within experimental error.

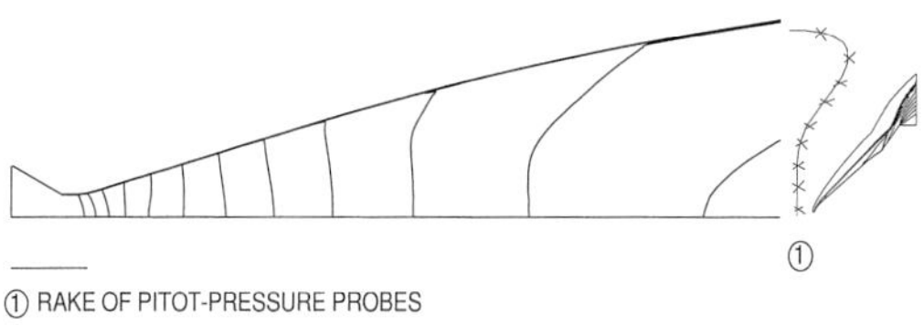

Figure 1. Numerical reconstruction of the HAC model installed in the T5 nozzle.

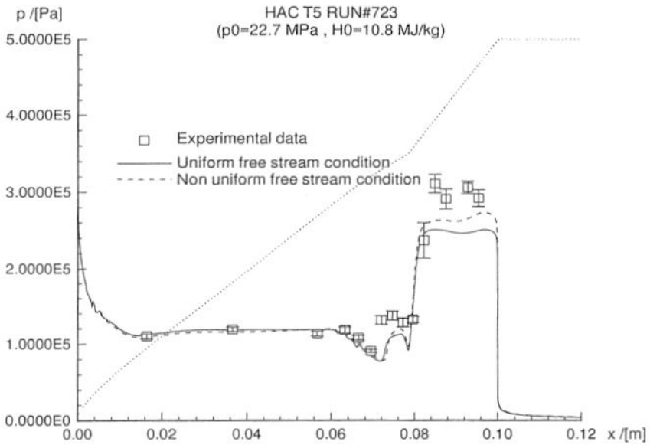

Figure 2. Influence of pitot-pressure nonuniformity.

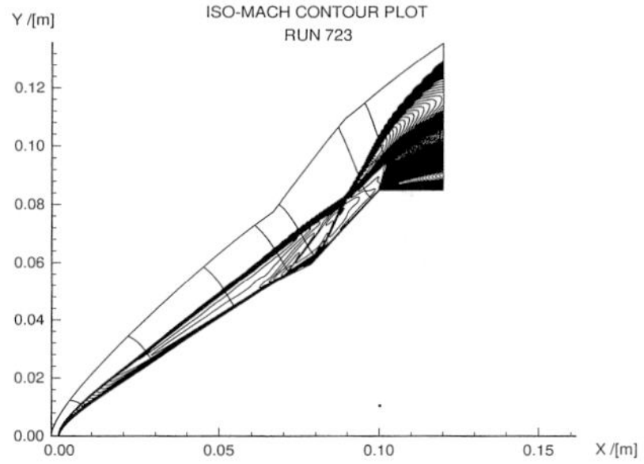

Figure 3. Iso-mach contour for 723.

3.3. Flap-induced separation and reattachment

Figure 3 shows a typical iso-mach plot where all features of the interaction are clearly visible. When comparing the pressure and heat-flux data in Figs. 4 and 5 we notice excellent agreement between experiments and computations on the attached forward part. The computations show only a small influence of catalycity. Large discrepancies occur between experimental and computed heat fluxes on the flare region. It was postulated that the flow at the reattachment point

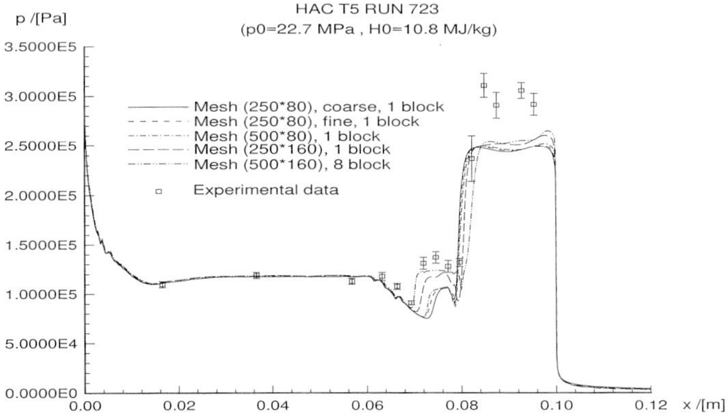

Figure 4. Pressure distribution: TINA results with various computational meshes.

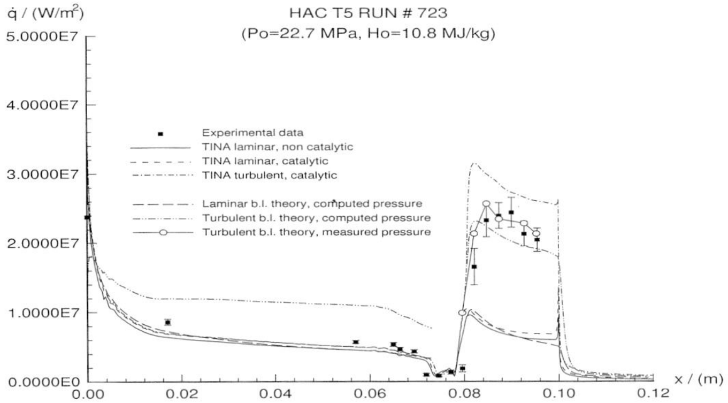

Figure 5. Heat flux distribution: TINA vs. boundary-layer method.

was turbulent; different ways to prove this were considered. First, TINA computations were carried out where turbulence, using the Baldwin-Lomax model, was switched on at reattachment. Second, a boundary-layer method was applied where the computed boundary-layer edge pressure and velocity served as input for the computation of the laminar or turbulent heating. Note that the figure shows an even better match when the measured pressures on the flare are taken instead of the computed ones. And third, the measured and computed peak heating were compared to a correlation derived from earlier work by Simeonides (1995). Figure 6 shows this correlation when accounting for the conicity of the flow on the forebody and on the flare. It shows that for T5, the peak-heating ratio corresponds to a critical Re* in the vicinity of the intersection between laminar and turbulent regimes. The critical Re* is the Re at the location of the peak heating using Eckert's reference enthalpy and free-stream pressure at equilibrium conditions. In Fig. 8 the separation length

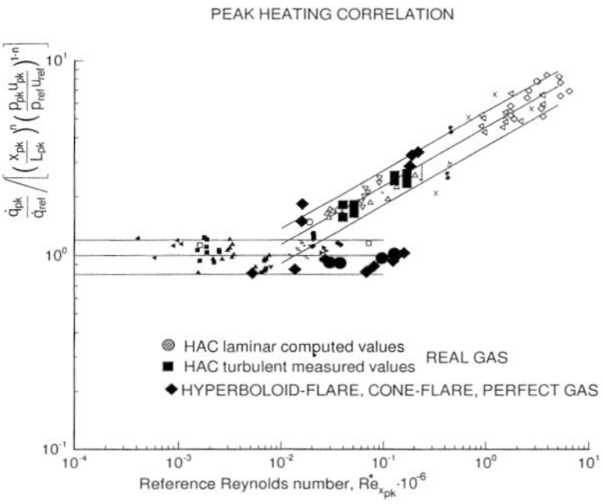

Figure 6. Peak-heating correlation from ramp flows: comparison with transformed HAC data.

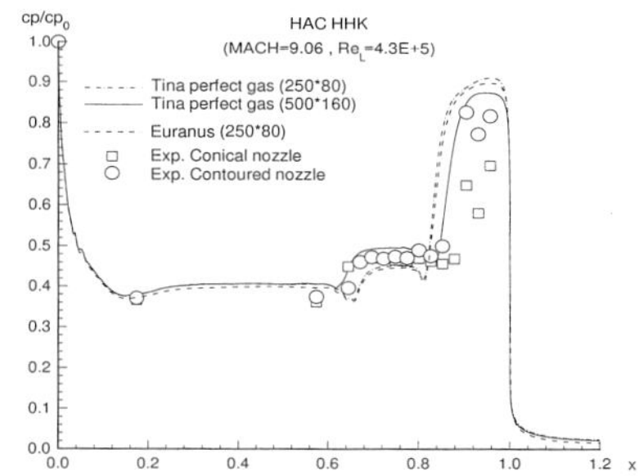

Figure 7. Measured and computed pressure distribution on the HHK HAC model.

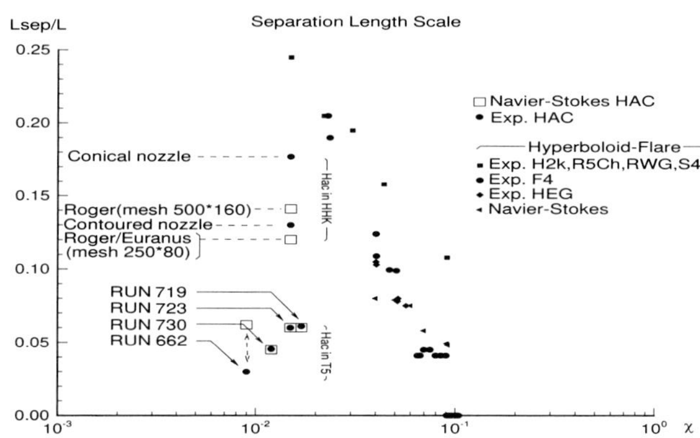

Figure 8. Separation length vs. viscous interaction parameter.

upstream of the hinge-line is plotted as a function of the viscous interaction parameter for the HAC model and compared with hyperboloid-flare data from Sagnier (1994). The trend of the computations with Re follows that of similar hyperboloid-flare laminar data, but the experimental data follow a different trend, confirming again the fact that the flow became turbulent at reattachment. To obtain the influence of real gas at a constant viscous interaction parameter, HHK tests were carried out with reservoir conditions such that for both HHK and T5 the viscous interaction parameter is 0.015. The HAC geometry is much more sensitive to real-gas effects than the hyperboloid flare due to the increased nose bluntness and the expansion upstream of the flare. Figure 7 shows the HHK pressure distribution for conditions as shown in Table 1 for the conical and contoured nozzles and a comparison with the perfect-gas Euranus and TINA results using a 250×80 grid. The perfect-gas computations nearly match the contoured-nozzle data, however, the latter differ considerably from the conical experimental data. T5 HAC experiments show a separation-length reduction with increasing enthalpy. More grid refinement studies are needed for the nonequilibrium laminar computations to detect a similar reduction in separation length.

4. Conclusion and recommendation

An experimental campaign took place in T5 and in the HHK to study the influence of real-gas and viscous interaction effects on flap efficiency and heating. An axisymmetric configuration was designed to enhance the sensitivity to real-gas effects and ease computational analysis. Transition to turbulence occurred at reattachment, making the analysis much more complicated than initially envisioned. It is recommended that the analysis be completed by studying reference laminar interactions at high enthalpy to understand the influence of real-gas effects at constant viscous interaction without the additional complexity of having transition at reattachment.

Acknowledgement. The authors wish to acknowledge H. Wong and M. Netterfield for their support in running TINA and EURANUS.

References

Davis J, Sturtevant B, Muylaert J (1996) Experiments on the Halis axisymmetric configuration in the T5 shock tunnel. In: Shock Waves, Proc of the 20th ISSW, Sturtevant B, Shepherd JE, Hornung HG (eds), World Scientific, Vol 1, pp 191–196.

Netterfield M (1992) Validation of a Navier Stokes code for thermo-chemical non equilibrium flows. In: Proc. 27th Thermo-physics Conf., AIAA 92-2878.

Sagnier P (1994) Synthesis of the Hyperboloid Flare Campaign in F4, Onera RT 70/6121 SY December 1994.

Simeonides G, Haase W (1995) Experimental and computational investigations of hypersonic flow about ramp. J Fluid Mech 283(1):17–47.

Walpot L, Muylaert J, Bakker P, Simeonides G (1996) High-enthalpy nozzle flow sensitivity study and effects on heat transfer. In: Shock Waves, Proc of the 20th ISSW, Sturtevant B, Shepherd JE, Hornung HG (eds), World Scientific, Vol 2, pp 1587–1592.

Weilmuenster KJ, Gnoffo PA, Greene FA (1993) Navier Stokes Simulations of the Shuttle Orbiter Aerodynamic Characteristics with Emphasis on Pitch Trim and Body Flap, AIAA 93-2814.

Shock-interference heating in hypervelocity flow

S. R. Sanderson*, B. Sturtevant
Graduate Aeronautical Laboratories
California Institute of Technology
Pasadena, CA 91125 USA

Abstract: The interaction of a weak oblique shock with the strong bow shock ahead of a blunt body in supersonic flow produces extreme heat-transfer rates and surface pressures. Experiments are conducted in a free-piston shock tunnel to determine the effects of thermochemistry on the problem at high enthalpy. The flow topology is simplified by studying the nominally two-dimensional flow about a cylinder with a coplanar impinging shock wave. Increases in shock-interference heating rates due to real-gas effects are observed to be significantly less than previously expected. A model is discussed that describes the thermochemical processes occurring in the interaction region. The model accurately reproduces the experimental observations.

Key words: Shock interference, Shock impingement, Shock-on-shock

1. Introduction

Although the shock-interference heating problem has been studied extensively in low-enthalpy flows (Edney 1968), the influence of high-enthalpy real-gas effects is poorly understood. Edney (1968) observed and classified six distinct interaction regimes that are known as types I–VI. Severe heating occurs for the type-IV flow where the interaction causes the formation of a supersonic jet that impinges on the surface of the body (Fig. 1). By mapping the shock-wave interactions that occur in the supersonic jet into the pressure-flow deflection angle (p-δ) plane, Edney (1968) studied the variation of the pressure at the jet-impingement point for a perfect gas. It was demonstrated that reductions of the ratio of specific heats, γ, at high enthalpy produce greatly increased surface pressures. The jet-impingement pressure also increases strongly with Mach number. Although the ratio of specific heats is only a crude measure of the influence of equilibrium thermochemistry, the magnitude of the effect predicted by Edney (1968) motivated the current experiments. The objective of the present work was to utilize the capabilities of the GALCIT T5 free-piston

*Present address: GE Corporate Research & Development, P.O. Box 8, Schenectady, NY 12301.

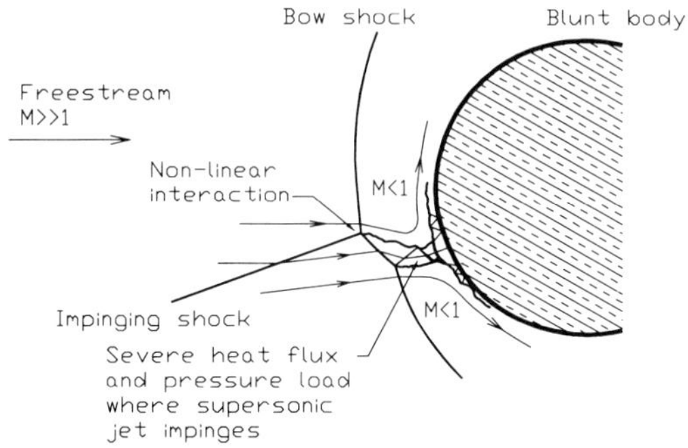

Figure 1. Schematic representation of the type-IV interaction.

shock tunnel to determine the quantitative effects of thermochemistry on the shock-impingement phenomenon.

2. Experimental investigations

The experimental arrangement is shown in Fig. 2. High-temperature and pres-

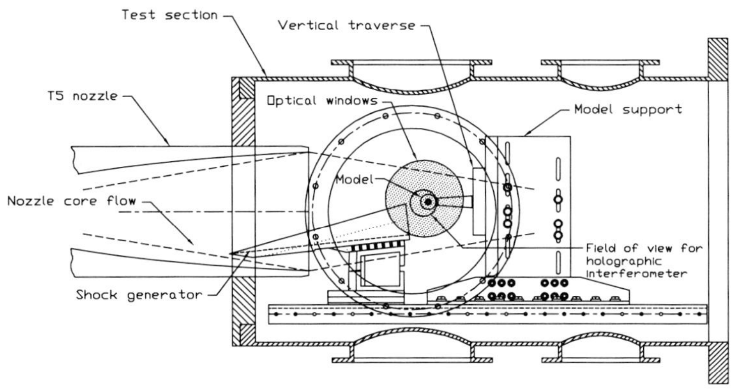

Figure 2. Layout of apparatus in the test section of T5.

sure gas was generated in the nozzle reservoir by the T5 hypervelocity shock tunnel. Nitrogen test gas was used for all experiments (Table 1).

The cylindrical model (ϕ40 mm $\times$ 180 mm) was positioned in the free-jet test section and a large shock generator was partially inserted into the uniform nozzle core flow upstream of the nozzle exit. Holographic interferometry was used to investigate variations in the shock-impingement flow field as the model was vertically translated with respect to the exit of the shock genera-

Table 1. Summary of the free-stream conditions.

Test condition		A	B	C
Test section conditions	Velocity (m/s)	2540	4450	5350
	Density (kg/m^3)	0.0218	0.0155	0.0157
	Pressure (kPa)	1.03	5.48	11.4
	N concentration (mole/kg)	9.9×10^{-5}	6.9×10^{-1}	3.65×10^{0}
	M_∞	9.9	6.3	5.3
	Re per mm	5500	1600	1350

tor. Surface-junction thermocouple gauges provided time-resolved heat-transfer measurements at 24 circumferential locations around the model forebody. Figure 3 shows a holographic interferogram of a type-IV interaction at the high-

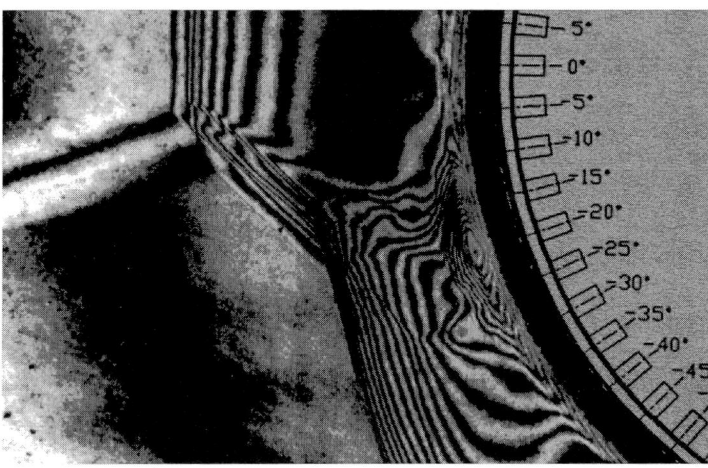

Figure 3. Holographic interferogram of type-IV interaction; condition C.

enthalpy condition, C. Two mutually inverted λ-shocks are observed along with a terminating strong shock and the region of elevated density behind it. The peak heat fluxes recorded during the entire sequence of shots at each test condition are listed in Table 2. The high bandwidth of the heat-flux gauges used for these experiments permitted examination of the temporal characteristics of the heat-transfer signals. Sanderson (1995) discusses observations of coherent unsteady fluctuations of the type-IV jet that are consistent with oscillations observed in recent numerical simulations (Gaitonde 1993, Zhong 1994). Additional data are reported in detail by Sanderson (1995).

Table 2. Measured and predicted peak heat-transfer intensification for type-IV flows.

Test condition		A	B	C
Peak measured $\frac{\dot{q}_{jet}}{\dot{q}_{body}}$	Instantaneous	15.2	10.8	7.9
	100 μs averaged	13.5	7.6	6.0
Predicted $\frac{\dot{q}_{jet}}{\dot{q}_{body}}$	Frozen	14.1	7.2	5.1
	Nonequilibrium	14.1	6.7	4.4
	Equilibrium	14.1	7.4	7.2

The conclusion of the experimental investigation was that the stagnation density and heat-flux intensification decreased at high enthalpy. Care must be exercised in the interpretation of this trend since the Mach number of the flow in T5 decreased as the stagnation enthalpy was increased (Table 1). The net result was that further increases in the large heat fluxes produced by shock impingement, due to real-gas effects, were observed to be much lower than those predicted by existing models (Section 1).

3. Model of the influence of real-gas effects

Modeling of the influence of real-gas effects followed the pattern of the existing perfect-gas model of the type-IV flow (Edney 1968). Normal shock solutions were obtained using the ideal dissociating gas model of Lighthill (1957). The normal shock solutions were extended to oblique waves and p-δ plane methods were used to obtain locally valid solutions for the λ-shock points. These inviscid shock-interaction solutions determine the state of the gas in the type-IV jet. Heating due to impingement of the type-IV jet was modeled using the equilibrium reacting-flow similarity solution of Fay and Riddell (1958). Sanderson (1995) shows that the heat-transfer intensification at the jet-impingement point has the following functional dependence:

$$\frac{\dot{q}_{jet}}{\dot{q}_{body}}\sqrt{\frac{\Delta x}{D}\frac{\zeta_{1_{body}}^{2}\zeta_{2_{body}}}{\zeta_{1_{jet}}^{2}\zeta_{2_{jet}}}} = \text{func}(P_{\infty}, H_{0_{\infty}}, \hat{\rho}_d, \alpha_{\infty}, \beta_1). \tag{1}$$

The parameters are $P_{\infty} = p_{\infty}/\rho_{\infty}u_{\infty}^2 \sim 1/M_{\infty}^2$, $H_{0_{\infty}}$ is the stagnation enthalpy normalized with respect to the dissociation energy of the gas, $\hat{\rho}_d$ is the equilibrium constant, α_{∞} is the dissociated mass fraction of the free stream, and β_1 is the angle of the impinging wave. The geometrical terms on the left-hand side specialize the result for a particular configuration. The similarity coefficients, ζ, are determined from the work of Fay and Riddell (1958) and the Newtonian solution for the velocity gradient at the outer edge of the boundary layer. The jet-width-to-body-diameter ratio, $\Delta x/D$, is empirically scaled with the undisturbed shock standoff distance, χ/D, and the divergence angle of the oblique wave and shear layer at the shock-impingement point, δ;

$$\frac{\Delta x}{D} = \frac{\chi}{D} \sin \delta. \tag{2}$$

All of the quantities appearing in Eqs. (1) and (2) may be predicted a priori.

The behavior of the local solution (1) with stagnation enthalpy, H_{0_∞}, is illustrated in Fig. 4 for typical values of the remaining parameters. Three solutions are obtained using different thermodynamic models to demonstrate the influence of nonequilibrium thermochemistry. The dotted lines represent chemically frozen solutions that apply when the reaction rates are slow with respect to the rate of flow over the body. The dash-dot curves are equilibrium solutions that apply when the reaction rates are fast and are obtained using the Lighthill (1957) model for the shock-jump computations. The dashed curves represent an approximate nonequilibrium solution that is motivated by the disparate shock strengths produced at the λ-points. The weak oblique waves in the supersonic jet are assumed to remain frozen whilst the strong bow shock and terminating jet shock relax to equilibrium. The validity of the nonequilibrium approximation is assessed by computing the reaction rates (Freeman 1958) immediately downstream of the translational discontinuities. Figure 4 shows that equilibrium real-gas effects increase the heat-transfer intensification with respect to the frozen solution. This occurs because of the increased density ratio across the weak oblique jet shocks due to dissociation. Dissociation across the jet shock waves is precluded in the nonequilibrium approximation and the heat-transfer intensification decreases slightly with respect to the frozen solution. Within the scope of the approximation, non-monotonic behavior with the reaction rate is indicated. The significance of real-gas effects declines at higher Mach numbers because the oblique jet shocks become weaker relative to the bow shock and terminating jet shock. This causes reduced dissociation levels and reaction rates in the supersonic jet. Heating rates initially increase with impinging shock strength before reaching a maximum for modest-strength waves. Further analysis (Sanderson 1995) demonstrates that the relative influence of real-gas effects is minimized at this maximum heating condition.

Specific predictions of the model for the test conditions listed in Table 1 are compared with the experimental data in Table 2. The extent of chemical reactions at the low-enthalpy condition A was negligible and so only the frozen solution is shown. Reaction rates for condition C were sufficiently high that departures from equilibrium were expected to be slight. The calculated values accurately reproduce the experimental heat-transfer measurements.

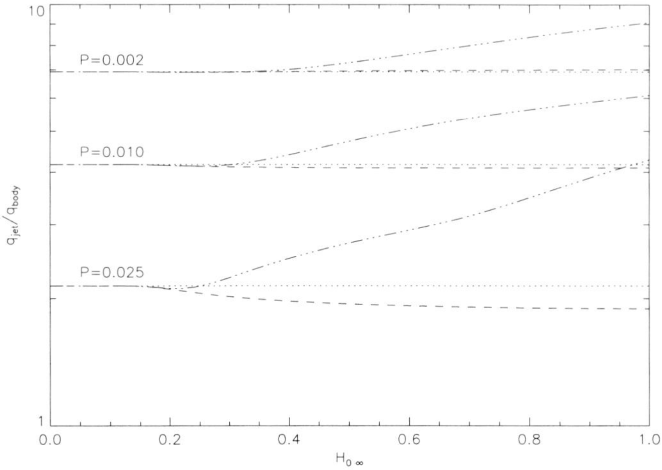

Figure 4. Influence of dimensionless stagnation enthalpy, H_{0_∞}, Mach number, $\sim\ 1/P_\infty^{0.5}$, and reaction rate on type-IV heat-transfer intensification, $\frac{\dot{q}_{jet}}{\dot{q}_{body}}(\Delta x \zeta_{1_{body}}^2 \zeta_{2_{body}}/D\zeta_{1_{jet}}^2\zeta_{2_{jet}})^{0.5}$. The parameters are $\hat{\rho}_d = 1 \times 10^7$, $\alpha_\infty = 0$, and $\beta_1 = 16°$.

Acknowledgement. This work was supported by the Air Force Office of Scientific Research under Grant Nos. F 49620-92-J-0110 and F 49620-93-1-0338.

References

Edney BE (1968) Anomalous Heat Transfer and Pressure Distributions on Blunt Bodies at Hypersonic Speeds in the Presence of an Impinging Shock, FFA Report 115.

Fay JA, Riddell FR (1958) Theory of stagnation point heat transfer in dissociated air. Journal of the Aeronautical Sciences 25:73.

Freeman NC (1958) Non-equilibrium flow of an ideal dissociating gas. J Fl Mech 4:407.

Gaitonde D (1993) Calculations on Unsteady Type IV Interaction at Mach 8, Wright Laboratory, WL-TR-93-3002.

Lighthill MJ (1957) Dynamics of a dissociating gas. Part I. Equilibrium flow. J Fl Mech 2:1.

Sanderson SR (1995) Shock Wave Interaction in Hypervelocity Flow, Ph.D. Thesis, California Institute of Technology.

Zhong X (1994) Application of essentially non-oscillatory schemes to unsteady hypersonic shock-shock interference heating problems. AIAA J 32:1606.

Investigation of shock-shock interactions in hypersonic reentry flows

S. Brück
DLR Institute of Fluid Mechanics
Bunsenstraße 10
D-37073 Göttingen, Germany

Abstract: Numerical simulations of shock-shock interactions in front of a circular cylinder have been performed. The numerical results of a validation test case are compared with experimental data. Models of a perfect gas and a mixture of chemically reacting perfect gases have been applied to reentry flow conditions. Differences of the results of the two models are demonstrated in view of the calculated shock configurations and the peak loads on the surface.

Key words: Shock-shock interactions, Viscous nonequilibrium flows, TVD method, Cylinder flow

1. Introduction

The understanding of the physics of complex flow phenomena in hypersonic flow is important, because the phenomena may result in strong local pressure and temperature and/or heat-flux peaks on the considered body surface. One example of these phenomena is the shock-shock interaction. The physics of such flow phenomena was experimentally investigated in a two-dimensional supersonic model flow by Edney (1968). A weak shock generated by means of a wedge interacts with the bow shock in front of a cylinder. Based on these experiments, six different types of shock-shock interactions were classified for cold hypersonic flows, depending on the position of the interaction location and the strength of the impinging shock wave. In chemically reacting flows, the typical shock configuration of the various interaction types may change due to dissociation effects. Therefore, it is necessary to investigate such flows to predict the influence of shock-shock interactions at free-flight conditions.

2. Numerical method

The numerical code NSHYP (Brenner 1994) used for the investigation of the flow phenomena is based on the solution of the time-dependent thin-layer Navier-Stokes equations. The numerical algorithm employs an implicit finite-difference scheme which is solved by using a line Gauß-Seidel relaxation method. The spatial order of the evaluation of the inviscid fluxes is formally second order and based on a TVD upwind discretization and a non-MUSCL approach. For the simulation of flows in chemical nonequilibrium, a mixture of monatomic and diatomic electrically neutral gases is considered. It is assumed that the density of the gas is sufficiently low. Thus, each component can be regarded as a perfect gas. The model under consideration includes the conservation of the partial densities of the chemical components N_2, O_2, NO, N, and O for air, the momentum in the three spatial coordinates, the total energy per volume, and the vibrational energy per volume of the diatomic species NO excluded. The rates of the chemical reactions used here as well as the modeling of the vibrational relaxation can be found in Park (1989). For the nonequilibrium calculations, a fully catalytic wall is assumed.

3. Validation

At first, numerical investigations should be compared with standard wind-tunnel tests before extrapolating to free-flight conditions. A test case of shock-shock interactions is considered in order to validate the numerical code for defined conditions. The two-dimensional flow field in front of a circular cylinder is calculated where an impinging shock is present at the inflow boundary of the computational domain at a defined position. The interaction with the detached bow shock produces a complex flow field which requires a good local resolution. The test case was experimentally investigated by Wieting & Holden (1987) in a cold wind tunnel ($Ma_\infty = 8.03$, $T_\infty = 122$ K) and is widely used for comparative studies in the literature. For the presented test case where only perfect

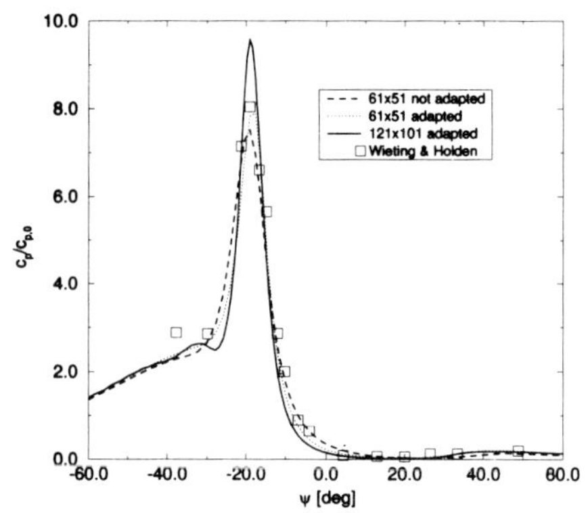

Figure 1. Pressure distribution of several simulations and the experiment, flow conditions of Wieting & Holden (1987).

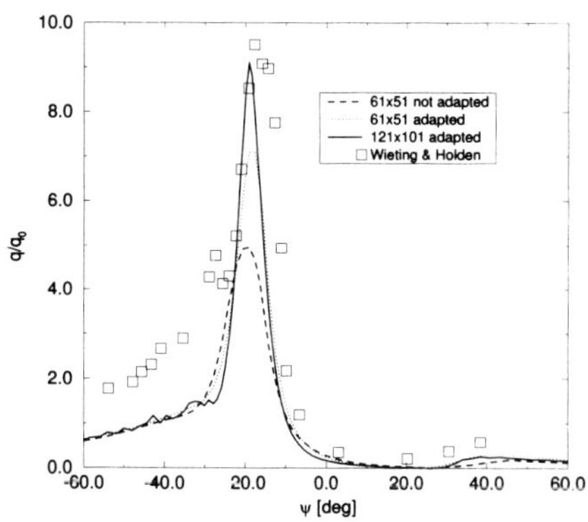

Figure 2. Heat-flux distribution of several simulations and the experiment, flow conditions Wieting & Holden (1987): $q_0 = 681,\ 395\ \mathrm{W/m^2}$.

gas has to be considered, a type-IV interaction is achieved: a supersonic jet impinges on the body surface and produces a high peak in the pressure and heat-flux distribution, as shown in Figs. 1 and 2. The results produced by the finest grid show good agreement with the experimental data. The code has also been applied to similar flow fields including the modeling of chemical and thermal nonequilibrium for pure nitrogen. Results have been presented by Brück (1993) and Hannemann et al. (1993).

4. Reentry-flow simulation

For the flow around a cylinder with the presence of an impinging shock, computations have been performed to investigate the influence of the shock position and of nonequilibrium effects due to dissociation. The free-stream conditions were taken from COESA (1976) and correspond to an altitude of 53.5 km. The free-stream velocity is 4.0 km/s. The incident-shock positions at the inflow boundary of the computational domain are obtained by using the Rankine-Hugoniot relations. For small shock angle, i.e. weak shocks, the composition and the vibrational relaxation can be assumed to be frozen. The shock positions are chosen such that shock configurations of type-III and type-IV interaction occur. Another point of interest is the location of the incident shock for which the jet turns around the body surface, named here type-IVa interaction.

In Fig. 3, Mach-number contours and contours of O_2 concentration (only for chemically reacting flow) are presented for several interaction types and chemistry modeling (perfect gas, thermal and chemical nonequilibrium). For the comparison between perfect gas and a chemically reacting flow field, the vertical position of the incident shock is chosen such that the stagnation point has nearly the same location. Three flow patterns of type-IV interaction are shown to demonstrate the effects of the different flow modeling. The pictures in the first row (Figs. 3a–c) show the flow field of a type-IVa interaction where the supersonic jet turns around the surface and remains supersonic. The dissociation rate within the nonequilibrium flow field indicates that the flow inside the jet is only weakly dissociated, compared with the areas behind the strong bow shock where the oxygen is nearly completely dissociated. This behavior can be observed in all nonequilibrium flow fields. In the next case shown in Figs. 3d–f, a typical type-IV interaction is achieved where the supersonic jets impinge on the body surface. Just in front of the surface, the flow becomes subsonic through a strong normal shock. Downstream of the stagnation point, the flow is accelerated to supersonic. Behind the shock at the end of the jet, the concentration of molecular oxygen decreases due to the higher temperature. Only a small jet is present in the last flow pattern shown in Figs. 3g–i, which can also be classified as a type-IV interaction. In this case the supersonic jet flow also becomes subsonic by a shock just in front of the stagnation point. Downstream

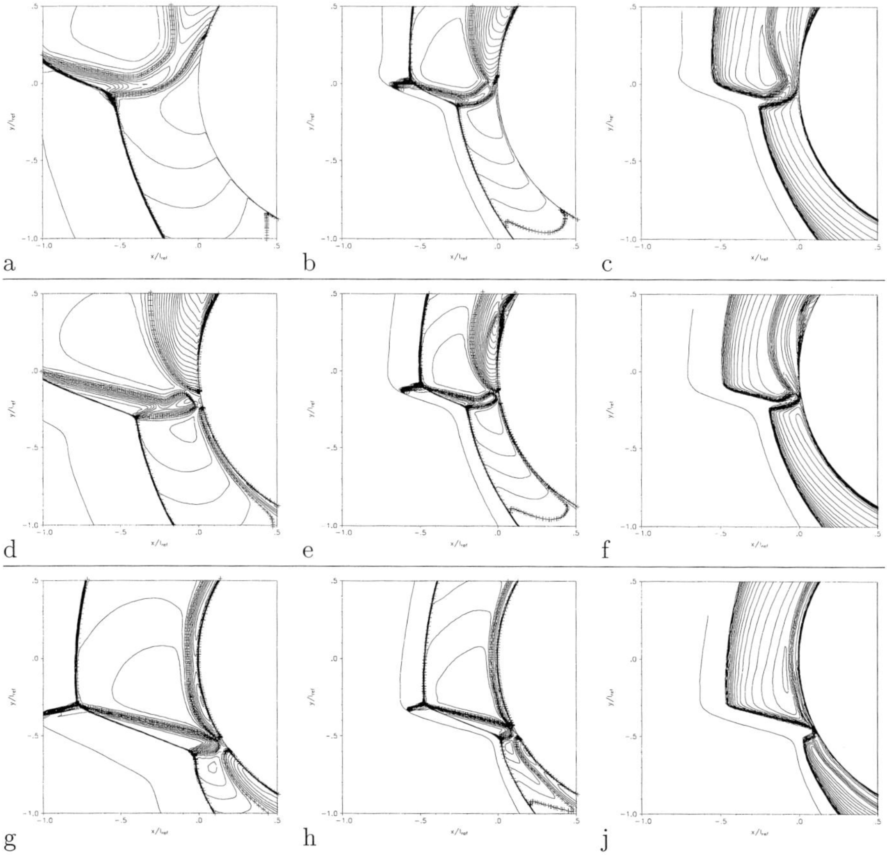

Figure 3. Calculated free-flight flow field for various incident shock positions, $u_\infty = 4$ km/s, H = 53.5 km (left: Mach-number perfect gas; center: Mach-number thermal and chemical nonequilibrium, +++ sonic line; right: concentration of O_2; top: type-IVa interaction; middle and bottom: type-IV interaction).

of the stagnation point, the flow expands and interacts in the upper part with the jet.

Starting from a type-III interaction, the flow field develops towards type IV by shifting up the incident shock, where in the transitional region the highest peak value is observed in the pressure distributions in Fig. 4. The typical peak-pressure levels vanish with the development of a type-IVa interaction where the supersonic jet runs along the surface. The heat-flux distributions in Fig. 6

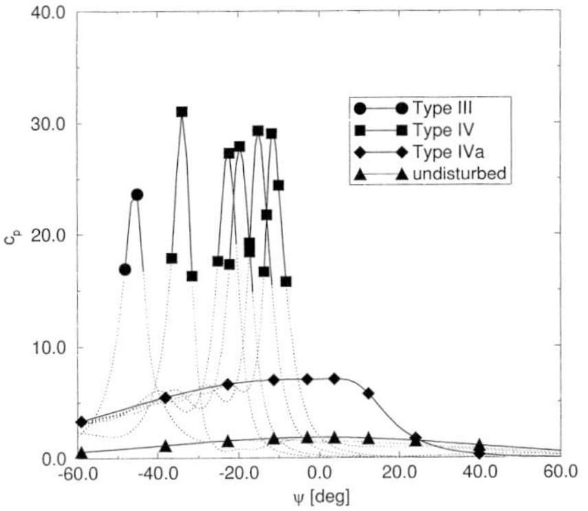

Figure 4. Pressure distributions for different shock positions (perfect gas).

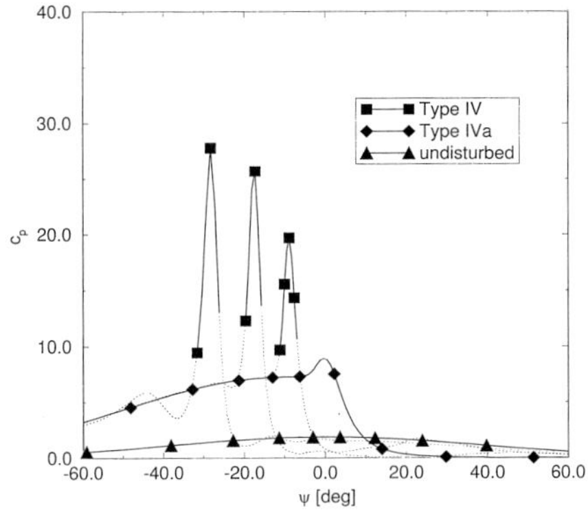

Figure 5. Pressure distributions for different shock positions (nonequilibrium).

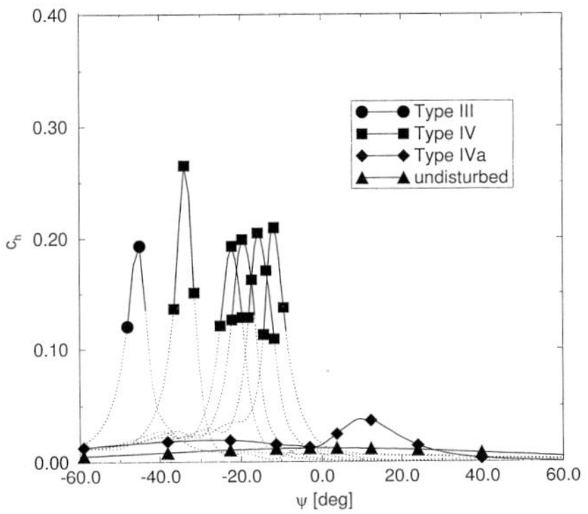

Figure 6. Heat-flux distributions for different shock positions (perfect gas).

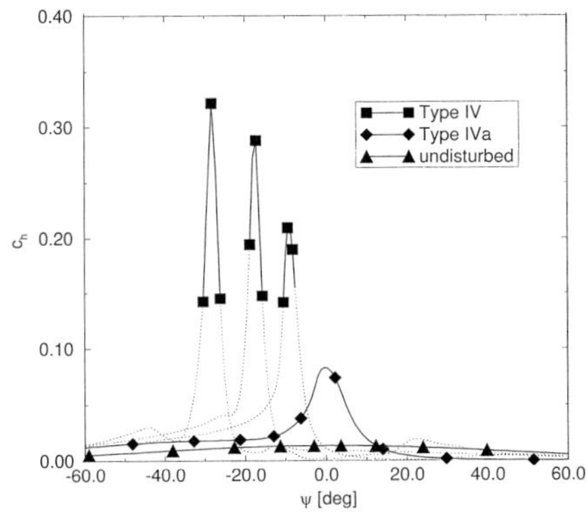

Figure 7. Heat-flux distributions for different shock positions (nonequilibrium).

show the same behavior. Only for type IVa can an increase of the heat-flux be observed where the jet comes close to the wall. The transition between type IV and IVa is sharp for the perfect gas flow, while the nonequilibrium results are showing a continuous transition (Figs. 5 and 7). The maximum loads for the nonequilibrium results are achieved when the impinging shock location for a type-IV interaction is close to the impinging shock location for type III. The peak values in the pressure distribution have nearly the same levels as observed for the perfect-gas solutions, while the heat-flux level is increased due to recombination.

5. Conclusion

The results of the validation test case indicate that the chosen grid resolution is sufficient to capture the most important effects of a typical shock-shock interaction. The predicted surface values agree well with the experimental data. The results of the reentry-flow simulations do not show qualitative changes of the shock configuration due to dissociation, while the shock stand-off distance is decreased. Only small dissociation effects are found in the jet-flow regions. Differences can be seen in the distribution of the surface loads, where the nonequilibrium results show lower peak values in the transitional region from type IV to type IVa. On closer examination we see that for nonequilibrium flows the range of shock positions is decreased when type-IV interactions are observed. The highest peak values are predicted for a jet impingement interacting with an expanding flow.

Acknowledgement. This work is partially funded by the Deutsche Forschungsgemeinschaft and ESA/CNES. Furthermore, I wish to thank K. Hannemann and G. Eitelberg for helpful discussions and their interest.

References

Brenner G (1994) Numerische Simulation von Wechselwirkungen zwischen Stößen und Grenzschichten in chemisch reagierenden Hyperschallströmungen, DLR-FB 94-04, Göttingen.

Brück S (1993) Numerical Simulation of Shock-Shock Interactions in Transitional/Turbulent Flow, DLR IB 221-93 A 23, Göttingen.

Edney B (1968) Anomalous Heat Transfer and Pressure Distributions on Blunt Bodies at Hypersonic Speeds in the Presence of an Impinging Shock, FFA Report 115, Stockholm.

Hannemann K, Brück S, Brenner G (1993) Numerical Simulation of Reacting Flows Related to the HEG. In: Shock Waves @ Marseille, Brun R, Dumitrescu LZ (eds), Springer, pp 251–256.

Park C (1989) Nonequilibrium Hypersonic Aerothermodynamics, John Wiley & Sons, New York.

United States Committee on Extension to the U.S. Standard Atmosphere (COESA) (1976), U.S. Standard Atmosphere, NASA Technical Memorandum X-74335, Washington, D.C.

Wieting AR, Holden MS (1987) Experimental Study of Shock Wave Interference Heating on a Cylindrical Leading Edge at Mach 6 and 8. In: AIAA 22nd Thermophysics Conference, AIAA-87-1511.

Recompression of nonequilibrium flow in the wake of a blunted cone

D. Kastell, K. Hannemann, G. Eitelberg
DLR, Institut für Strömungsmechanik, Bunsenstr. 10, 37073 Göttingen, FRG
T. J. Horvath
NASA Langley RC, GDD AB, Hampton, VA 23681 USA

Abstract: The entry of a planetary probe, which is a 140° total-included-angle blunted cone with a spherical nose, rounded edges, and a cylindrical afterbody is significantly influenced by nonequilibrium flow effects. Therefore its aerodynamic characteristics have been investigated in the continuum flow of the High Enthalpy Shock Tunnel at the DLR in Göttingen (HEG). Calculations with a Thin-Layer Navier-Stokes code which is capable of simulating chemical and thermal nonequilibrium flows have been carried out and compared to measured heat-transfer and pressure distributions. Depending on the operating condition of the tunnel, up to 63% of the stagnation heat-transfer rate was reached on the front-body corner junction and up to 13% was measured on the afterbody sting due to the reattachment of the separated shear layer. The numerical prediction on the forebody compares well with the experimental data, whereas the maximum heat loads on the sting were underestimated by the calculations.

Key words: HEG, Nonequilibrium flow, Hypersonic, Experiment, Numerical simulation, Blunted cone, Wake flow, Shear layer

1. Introduction

Almost any flight mission that involves changes in orbital altitude and/or inclination near an atmosphere-bearing planet is a candidate for the application of aeroassisted transfer vehicles (ASTVs). These vehicles take advantage of the aerodynamic forces generated during an atmospheric pass to change orbit with a much smaller expenditure of energy than is required for an all-propulsive maneuver. Due to the presence of high-temperature effects, which dominate the main flight regime of an ASTV, the influence of chemical and thermal nonequilibrium on the flow field has to be quantified. Experimental data and validated numerical codes are not available for this particular problem at the moment (Deiwert 1993). To overcome this deficiency, experiments have been carried out in the HEG along with numerical simulations with a three-dimensional Thin-Layer Navier-Stokes code including chemical reactions at the DLR in Göttingen, Germany. This investigation was part of the working group 18 activity of the AGARD fluid dynamics panel.

2. Description of the experiments

The HEG is a 60-m total-length free-piston-driven shock tunnel producing a hypervelocity continuum flow with a flow core diameter of 450 mm. At the maximum reservoir enthalpies of 23 MJ/kg it is capable of dissociating oxygen as well as nitrogen behind the bow shock of a model. An extended description of the facility is given by Eitelberg et al. (1992). The model used in the present investigation is a blunted cone with a base radius of $R_B = 76.2$ mm, a nose radius of $R_N = 38.1$ mm, a corner edge radius of $R_C = 3.81$ mm, and a total angle of 140° ($R_N/R_B = 0.5$, $R_c/R_B = 0.05$). The model is supported in the test section with an instrumented cylindrical sting (diameter 38.1 mm) having a length of 457.32 mm. Two models were built, each individually instrumented for three different test campaigns. The first model was instrumented with 15 thermocouples, most of them located on the sting, and two pressure gauges on the forebody. The second model had a forebody of chromel to match the thermal properties of its coaxial sensors and was instrumented with 72 thin-film gauges, located on the sting and the back of the body. For the latest experiments the first model was redesigned with a thin-film sensor insert in the sting. Pressure gauges were added to the sting and to the back of the body. Holographic interferometry and laser schlieren were conducted. A more detailed description of instrumentation, data-reduction procedures, test conditions, and criteria used to determine near-wake flow establishment can be found in Kastell et al. (1994).

3. Numerical analysis

The numerical simulations were performed using the implicit upwind-TVD Euler/Thin-Layer Navier-Stokes code NSHYP (Brenner et al. 1993). The three-dimensional, time-dependent Thin-Layer Navier-Stokes (TLNS) equations in conservation form for a multicomponent mixture of perfect gases are considered. For the simulations discussed in this paper a fully laminar air flow is considered. The isothermal wall ($T_{wall} = 300$ K) of the blunted cone is taken to be fully catalytic. The relevant chemical reactions were taken from Park (1989). The determination of the vibrational energy of the molecular species is based on the assumption of a harmonic oscillator. The transition of energy between the vibrational modes and the translational modes is approximated using the Landau-Teller formulation. The vibrational relaxation times were obtained from correlations of Millikan and White and Park (1989). Table 1 shows the free-stream conditions used for the computations. The data were calculated with NSHYP from nominal reservoir conditions for the axisymmetric contoured nozzle of the HEG (Hannemann et al. 1996). The actual run-to-run conditions vary and have to be calculated individually. Free-stream data presented in Kastell et al. (1994) and Legge (1995) have been calculated with a one-dimensional nozzle code.

Table 1. Nominal flow conditions used for the TLNS calculations.

	u_∞ $[\frac{m}{s}]$	ρ_∞ $[\frac{g}{m^3}]$	Re_∞ $[\frac{1}{m}]$	M_∞	T_∞ [K]	p_o [MPa]	h_o $[\frac{MJ}{kg}]$
Condition I	5700	2.19	$2.7{\cdot}10^4$	9.2	797	38.6	20.35
Condition II	6037	4.45	$4.7{\cdot}10^4$	8.5	1073	90.8	22.27
Condition III	4737	2.99	$3.7{\cdot}10^4$	9.5	574	44.9	13.33

4. Results and comparison

Most experiments and numerical calculations related to ASTVs have only been conducted in continiuum flow at low reservoir enthalpy. They also concentrated only on forebody measurements (see e.g. Stewart & Chen 1994). None of the investigators have focused their attention on the influence of high-temperature effects nor have they considered the flow in the near wake of the aeroshell-shaped vehicle. This region is of interest for the design of aeroassisted flight vehicles because shielding of the payload on its back might have to be considered. The rapid expansion at the shoulder turns the flow towards the centerline of the body where it impinges onto the sting, leading to a local peak in heat transfer and the formation of a recompression shock. The present experiments and numerical investigations were focused on the above-described topics. For three different flow conditions, with varying reservoir enthalpy and free-stream Reynolds number, the heat-flux and pressure distributions on the forebody, back, and sting of

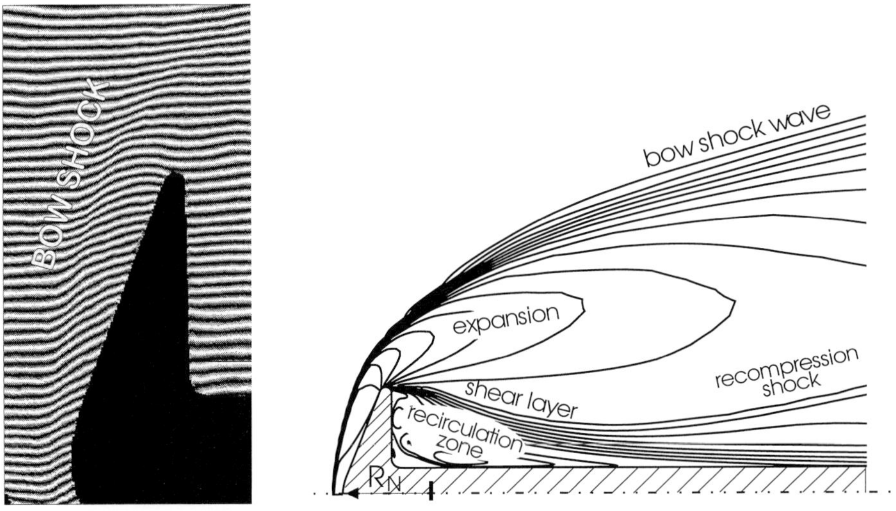

Figure 1. Interferogram (left) and Mach-number distribution for condition II (right).

the model are presented. Figure 1 (right) shows the calculated Mach-number distribution for flow condition II in comparison with an experimental interferogram in Fig. 1 (left). The primary flow features like the bow shock wave, the separated shear layer with the enclosed recirculation zone, and the recompression shock are evident in the calculations. In the interferogram, only the bow shock wave with a stand-off distance of $\Delta/R_N \approx 0.21$ can be identified.

Figure 2 shows the measured pressure on the blunt body together with the numerical calculations as a function of the normalized geometric arc-length S/R_N. To compare the numerical calculations with the experiments, the data were normalized by a value close to the stagnation point ($S/R_N \approx 0.68$), because this point was instrumented with a thermocouple. As can be seen in the plot the calculated pressure distributions on the forebody are equal for all conditions, and only on the sting a small variation is present. These discrepancies are more obvious for the heating distribution shown in Fig. 3. The data were normalized by the values measured at $S/R_N \approx 0.34$. This was done to minimize the influence of stagnation-point heat-transfer measurement and because of possible uncertanties due to grid singularity in the numerical code at this point. On the forebody and on the back of the model surface, the experimental data taken with two different models during three different runs follow the numerical curves well. On the sting, the location of the maximum heat-transfer load is predicted correctly, but the maximum heat-transfer rate of 20% of the reference value is underestimated by the calculations. This discrepancy appears for all conditions, but is more pronounced at lower reservoir enthalpies (cond. III).

Forebody-surface-heating distributions for conditions I, II, and III are shown in Fig. 4. The numerical curves fall on top of each other except at the stagnation point. This indicates that different total enthalpies affect mainly the heat flux on the spherical nose of the model. The experimental data follow the trend of the curves well. The situation on the back and on the sting is different, as shown

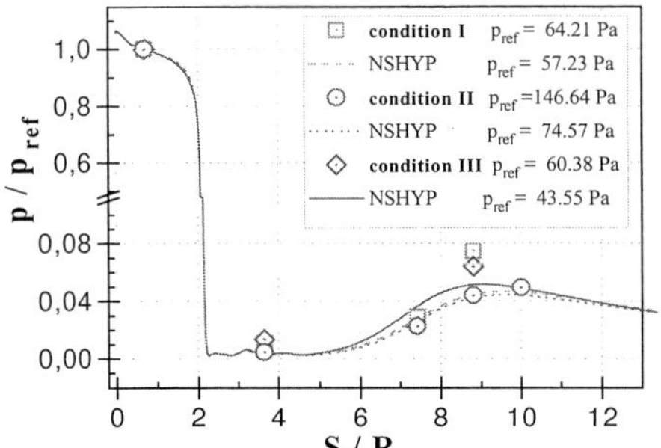

Figure 2. Surface-pressure distribution.

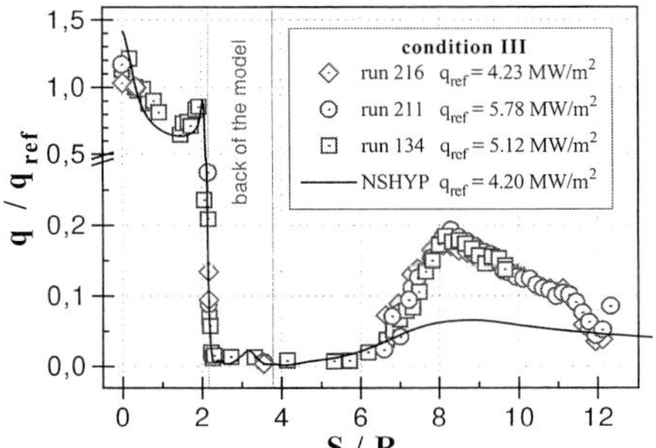

Figure 3. Surface-heating distribution.

in Fig. 5. The experimental data are taken from two different test series. On the back of the model and on the beginning of the sting the experimental data follow the numerical curves. The discrepancy between measured and predicted data becomes evident at the point of maximum heat transfer, where the calculated values are lower than the measured ones. Also, the measured peaks are narrower than the numerical ones. The higher measured heating indicates that shear-layer transition might be a significant feature of the near wake flow field. Tests recently conducted with schlieren visualization and intrusive hot-wire, hot-film, and pitot-probe surveys in a perfect-gas Mach-6 flow at NASA LaRC (with the same model) by Horvath (1995) support this possibility. A similar transition effect has also been noted by Baker & Martin (1966) for a backward-facing step. Significant influence might also cause change of the continuum behavior of the wake flow by the strong shoulder expansion. However the potential effects of high-temperature real-gas behavior on shear-layer transition remain coupled.

Experimentally, the location and magnitude of the peak heating associated with the reattaching shear layer appear to be very sensitive to free-stream conditions, whereas the numerical simulations indicate very little effect. Additional calculations with an adapted grid have shown almost no or little influence. For a noncatalytic-wall-model solution the normalized values are approximately half the magnitude of the presented ones. The calculated and the experimental data show both a shift towards the model base and an increase of the maximum heat-transfer load with increasing Reynolds number and decreasing the total enthalpy respectively (cond. III relative to cond. I). This shift is believed to be a viscous phenomenon whereas the heat-transfer change is influenced by the high temperature. At high-reservoir enthalpy and Reynolds number (cond. II) the high-temperature effects predominate leading to a low maximum heat transfer.

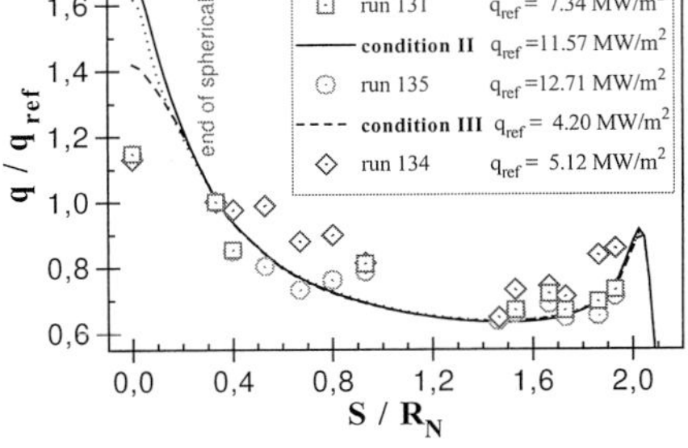

Figure 4. Forebody surface heating.

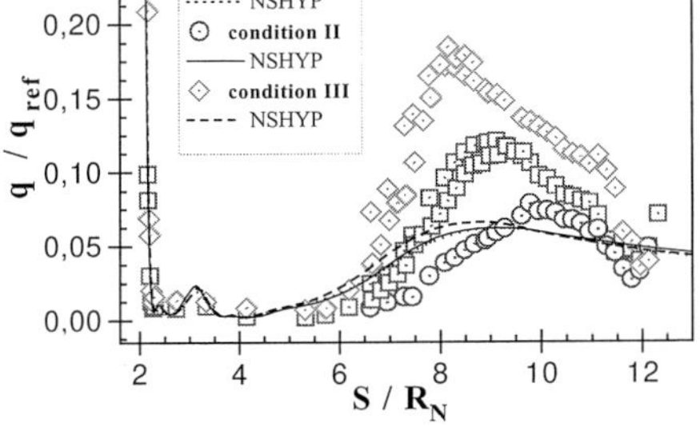

Figure 5. Sting-heating distribution.

5. Summary and conclusions

The heating and pressure distributions on the forebody, back, and afterbody of a 140° total-included-angle blunted cone have been measured in nonequilibrium flow. The reservoir enthalpies were varied between 14 and 23 MJ/kg. The measured data were compared with the results of a Thin-Layer Navier-Stokes code with embedded chemical reactions. If normalized by a value close to the stagnation point, the calculated heat flux on the forebody, except on the spherical nose, is not and on the sting is little affected by altered flow conditions. The same holds for the experimentally measured forebody heating. Nevertheless, the heat-transfer on the sting is strongly influenced by the different free-stream flow conditions. Increasing the Reynolds number moves the maximum heat load on the sting upstream and lowering the total enthalpies shrinks the peak. After all, the maximum heat load on the sting due to the recompression of the flow is underestimated by the numerical code. These discrepancies between the calculated and the experimental data in the wake flow of the blunted cone are not fully understood at this time and are the main issue of continued research.

References

Baker PJ, Martin BW (1966) Heat transfer in supersonic separated flow over a two dimensional backward facing step. JHM Transfer 9:1081–1088.

Brenner G, Gerhold T, Hannemann K, Rues D (1993) Numerical simulation of shock/shock and shock-wave/boundary-layer interactions in hypersonic flows. Computers Fluids. 22(4/5):427–439.

Deiwert GS (1993) Issues and Approach to Develop Validated Analysis Tools for Hypersonic Flows: One Perspective, AGARD CP-514, Ref 24.

Eitelberg G, McIntyre TJ, Beck WH, Lacey J (1992) The High Enthalpy Shock Tunnel in Göttingen, AIAA Paper 92-3942.

Hannemann K, Krek R, Eitelberg G (1995) Latest calibration results of the HEG contoured nozzle. In: Shock Waves, Proc of the 20th ISSW, Sturtevant B, Shepherd JE, Hornung HG (eds), World Scientific, Vol 2, pp 1575–1580.

Horvath TJ (1995) Private communication. NASA Langley RC, VA USA.

Kastell D, Horvath TJ, Eitelberg G (1994) Nonequilibrium flow expansion around a blunted cone. In: 2nd European Symposium on Aerothermodynamic for Space Vehicles, Nordwijk, NL, ESA SP-367 (1995), pp 383–389.

Legge H (1995) Experiments on a 70 Degree Blunted Cone in Rarefied Hypersonic Flow, AIAA Paper 95-2140.

Park C (1989) Nonequilibrium Hypersonic Aerothermodynamics, J Wiley, NY.

Stewart DA, Chen YK (1994) Hypersonic convective heat transfer over 140-deg blunt cones in different gases. J Spacecraft 31(5):735–743.

Flow in the TCM2-Marseille hypersonic facility: some new experimental results

R. Brun, Y. Burtschell, A. Chaix, M.-C. Druguet, L. Z. Dumitrescu, M.-P. Dumitrescu, R. Martin, D. Zeitoun
Lab. I.U.S.T.I. (U.R.A. CNRS 1168) Dept. MHEQ
Université de Provence, Marseille, France

Abstract: Some questions arising during the development of a new free-piston hypersonic facility (TCM2 Marseille) are discussed. It is shown that, to ensure a disturbance-free flow, in addition to having a tuned piston velocity (Hornung 1988), one must also keep this velocity practically constant for a duration of about 1.6 times the desired test duration. A method to ascertain the achievement of these conditions is presented. Then, the use of argon addition for fine adjustment of interface tailoring is advocated. If the above requirements are fulfilled, it is shown that the actual reservoir pressure, behind the reflected shock, coincides with the computed value (from M_S), and no supplementary loss mechanism has to be invoked. Disturbance-free operation of TCM2 for over one msec is demonstrated. Attempts to determine the reservoir enthalpy from the end-wall heat flux yielded much higher values than expected. No explanation is yet available for these aberrant results.

Key words: Hypersonic facility, Free-piston shock tunnel

1. Introduction

The TCM2 facility (Burtschell et al. 1991) is a medium-size free-piston shock tunnel designed for basic studies in hypersonics, especially on the interconnection between fluid-dynamic and physico-chemical phenomena at high enthalpies. The present paper will report on the work carried out since the start of the project: a number of novel developments will be presented, as well as some unexpected findings, for which an explanation is still uncertain.

2. The facility

A schematic of TCM2 is shown in Fig. 1. No attempt has been made by the design to duplicate free-flight conditions, in order to avoid going to exceedingly high pressures: instead, the emphasis was on producing well-controlled flow conditions, at moderate enthalpies (up to 12 MJ/kg), reasonably free of impurities.

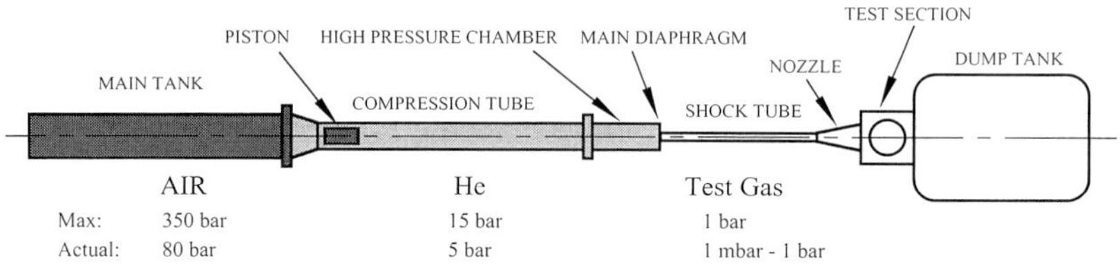

Figure 1. Schematic of the TCM2 free-piston shock tunnel.

Following erection of the facility, a systematic study of the gasdynamic processes taking place in the various compartments has been undertaken, with the aim of a full characterization of the operation of the machine.

3. Gas-dynamic processes in the compression tube and the "tuned-piston" problem

The first problem encountered was the discovery that the pressure build-up in the compression tube does not proceed smoothly, but as a series of successive pressure steps, on which strong oscillations are superimposed (Dumitrescu et al. 1993). Later is was found that in other facilities the pressure displays the same behavior (see, e.g. Fig. 10 in Maus et al. 1993). The cause having been uncovered (acoustic resonance of a cavity near the diaphragm), a remedy has been found: by a special design of the piston (Dumitrescu et al. 1993), the pressure rise in the compression tube of TCM2 is made smooth, while the main diaphragm bursting pressure, p_R, is controlled to within a few percent.

A second problem is that of tuning the piston motion (Hornung 1988). This term designates the setting up of the operating parameters (piston mass, initial pressures, etc.) so that, at diaphragm burst, the piston velocity, V_R, is such as to push the driver gas through the shock-tube entrance with exactly the required mass-flow rate, thus preventing the generation of an expansion wave that would then travel downstream the shock tube and catch up with the primary shock. Upon closer investigation it became clear, however, that the condition on the velocity is not sufficient: it is furthermore required that the piston's inertia be sufficiently great, so that the piston could continue to travel, with almost constant velocity (say, $\pm 10 - 20\%$), during a certain time, t_{**}, necessary to ensure a given test duration, t_* in the facility. To derive a relationship between t_* and t_{**}, one may examine the wave diagram of the shock tube (Fig. 2): it may be seen that, to avoid disturbing the constant regime at the end wall, behind the reflected shock, at least for a duration t_*, the head of the expansion wave coming from the piston should not reach the end wall earlier than t_B (point B). For TCM2, at a nominal M_s of about 10, one gets: $t_{**} \geq 1.6 \times t_*$ (for $t_* \approx 1$ msec). Inspection of Fig. 2 reveals also that the useful test time disappears altogether if

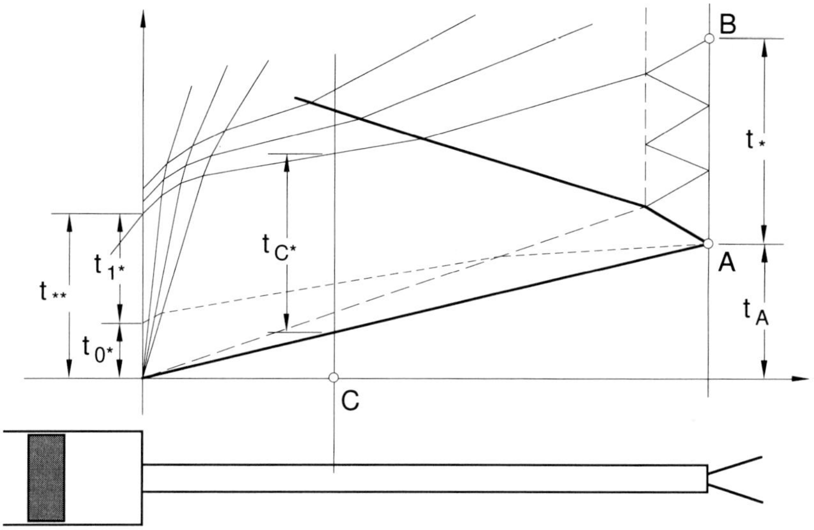

Figure 2. Wave diagram of the shock-tube processes.

the expansion wave generated by the slowing-down motion of the piston arrives at the end wall at the same time as the primary shock (point A); for TCM2, one finds $t_{0*} \approx 0.6$ msec. This means that the relationship between t_{**} and t_* is nonlinear, and that any reduction in the constant-velocity stroke of the piston may cut dramatically the available test duration. The holding time t_{0*} has, in addition, to be increased by about 10 – 20%, and t_{**} has to be supplemented with this quantity, to account for the primary-shock viscous attenuation.

Now, if the required constant-velocity piston stroke is allowed for, the piston will come very close to the end of the compression tube; this raises the question of finding a means to stop the piston without damage over the remaining very short stroke. The problem has been solved at TCM2 by a special piston design (Dumitrescu M-P 1996a).

4. Gas-dynamic processes in the shock tube

While the compression tube acts as a driver for the machine, it is the correct evolution of the gas-dynamic processes in the shock tube that has to be controlled. One important question is then how to verify that the required piston velocity (and its constancy) have been achieved. Recording the pressure in the compression tube provides an indication; but one would like to have a direct check. The evolution of the reflected-shock pressure cannot be used for this purpose, since it is affected by other phenomena (see below). However, an unambiguous way to ascertain the correct piston tuning is simply to inspect the pattern of the pressure trace at a station sufficiently far upstream along the shock tube (point C, Fig. 1).

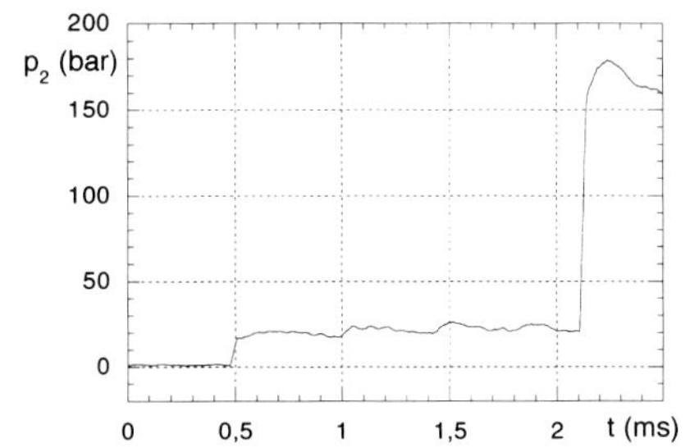

Figure 3. Pressure behind the primary shock at station C.

It is known that, behind a shock wave being attenuated only by viscous effects, the pressure p_2 registers a gradual increase; whereas it would start to fall if an expansion wave came from upstream. Therefore, if the duration of non-decreasing pressure p_2 at station C, t_{C*}, is at least equal to the required t_{**} (for a desired test duration t_*), then the proof is made. Such a recording in TCM2 is presented in Fig. 3, showing at least 1.6 msec of quasi-constant p_2, and thus giving about 1 msec of test time.

Fine tuning the piston motion is nevertheless insufficient to ensure that constant reservoir conditions are obtained: one has also to produce so-called tailored interface operations, that is, to have equal acoustic impedances across the contact surface when it is struck by the reflected shock. Viscous shock attenuation tends to shift conditions towards undertailoring at a given M_S. So, a convenient way to adjust finely the degree of tailoring is to add a small percentage of heavy gas (argon) into the helium driver, in the compression tube, thus increasing its impedance. Tailoring can be verified on the end-wall pressure trace, when a quasi-constant level, lasting the required test duration t_*, is seen. In the undertailored condition, the pressure evolves as a series of plateaus (progressively smeared), whose levels alternate above and below that of the first (just after the arrival of the primary shock) (Fig. 4); whereas overtailoring is signaled by a progression of increasing pressure steps.

Of course, adding argon reduces somewhat the shock Mach number and the enthalpy, but this is the price to pay for having well-defined test conditions. As an example, Fig. 5 shows what can be achieved by careful tuning of the machine, in terms of pressure steadiness.

A final comment is due: if all is set properly, the measured pressure, p_5, should coincide with the value computed (keeping account of real-gas effects) from the *measured* shock Mach number M_S at a station close to the end wall: no supplementary losses need to be invoked to explain any discrepancy. For instance, in the test of Figs. 4 and 5 the actual pressure level is within 3% of the computed value. We maintain, therefore, that, were a significant pressure

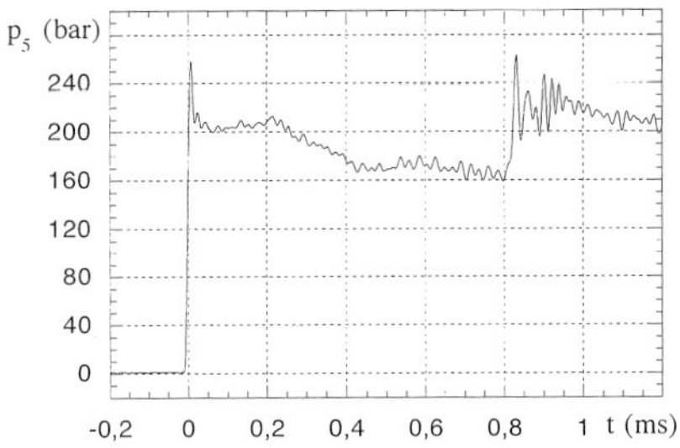

Figure 4. Undertailored end-wall pressure, driver pressure is 170 bar.

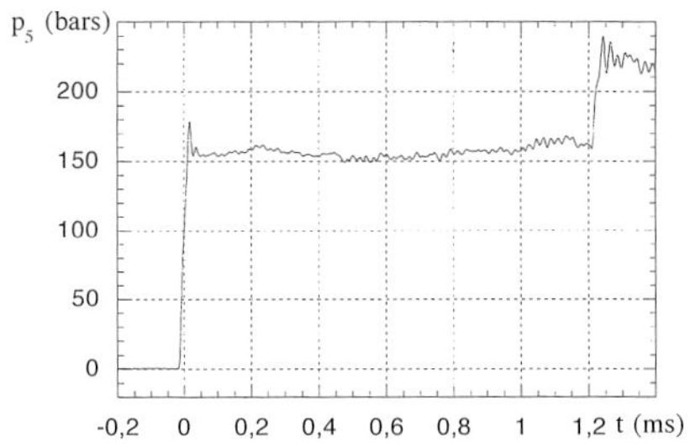

Figure 5. End-wall pressure for tuned and tailored operation, $M_S = 5.0$.

loss to be found, the cause should be pinned down to a poor tuning of the piston velocity and/or decceleration.

5. Determination of the reservoir enthalpy

All data in a hypersonic test are referred to reservoir conditions (p_5, h_5). But, whereas p_5 is readily measured, determination of the reservoir enthalpy remains problematic. Knowing the actual shock Mach number, M_S, real-gas computations give T_5 (and p_5); now, if the measured pressure coincides with the computed value, one can say that the phenomena are really taking place as expected, and infer that the actual temperature should also be equal to that deduced from M_S. Of course, this reasoning fails if significant discrepancies persist between measured and computed p_5; and an independent determination of T_5 (or h_5) would be welcome.

One way to achieve this is by measuring the heat flux from the hot, stagnant gas to the end wall. In the absence of any gas motion, a thermal boundary layer develops near the wall after shock reflection. The relevant equations are readily written and solved, using, as for an ordinary boundary layer, the von Mises transform. Equilibrium real-gas properties are assumed. Theory shows that the wall temperature should display a step-like evolution, consistent with a heat flux varying essentially as the inverse square root of time.

We have placed a number of coaxial thermocouples flush with the end wall of the facility, and recorded the signals: the expected step-like behavior was obtained, but the resulting heat flux (and gas temperature), as computed by the procedure described, was several times higher than expected. In the presence of such aberrant results, all common-sense checks were made, and the tests were repeated, using thin-film platinum gauges and different electronic hardware (but similar results); unfortunately, no definite explanation can yet be advanced, and further tests are now in progress. Anyway, as explained above, provided the machine is properly tuned, the stagnation temperature can confidently be inferred from the shock Mach number, correlated with p_5. The survey of the heat flux at several points across the end-wall surface has produced other surprising results, which are discussed in a companion paper (Dumitrescu M-P 1996b).

6. The nozzle

The TCM2 facility has been equipped with a conical hypersonic nozzle, designed for a free-stream Mach number of about 6. Thorough tests are planned; presently a discussion of the most relevant points concerning the nozzle design is reported (Dumitrescu LZ et al. 1996).

Acknowledgement. This research was carried out within the framework of a project sponsored by the European Space Agency, aimed at a better understanding of the operation and use of ground test facilities for hypersonic flow. The authors wish to acknowledge the continuing support and valuable contributions of G. Durand, J. Muylaert, and D. Vanmol, from ESA.

References

Burtschell Y, Brun R, Zeitoun D (1991) Two-dimensional numerical simulation of the Marseille University free-piston shock tunnel TCM2. In: Shock Waves, Proc 18th Int Symp on Shock Waves, Takayama K (ed), Sendai, Japan, Springer-Verlag, Heidelberg, pp 583–590.

Dumitrescu LZ, Burtschell Y, Zeitoun D (1996) Throat design for hypersonic nozzles. In: Shock Waves, Proc of the 20th ISSW, Sturtevant B, Shepherd JE, Hornung HG (eds), World Scientific, Vol 2, pp 1569–1574.

Dumitrescu M-P, Brun R, Billiotte M, Bertoni JM, Burtschell Y, Canova A, Dumitrescu LZ, Houas L, Labracherie L, Zeitoun D (1993) Dealing with pressure oscillations in Stalker tubes. In: Shock Waves @ Marseille, Proc 19th Int Symp on Shock Waves, Brun R, Dumitrescu LZ (eds), Marseille, France, Springer-Verlag, Heidelberg, pp 247–250.

Dumitrescu M-P (1996a) Discarding the tuned-piston condition for Stalker tubes: possible gains and drawbacks. In: Shock Waves, Proc of the 20th ISSW, Sturtevant B, Shepherd JE, Hornung HG (eds), World Scientific, Vol 2, pp 1487–1490.

Dumitrescu M-P (1996b) Trapping the boundary layer: a method to diminish flow contamination in shock tunnels. In: Shock Waves, Proc of the 20th ISSW, Sturtevant B, Shepherd JE, Hornung HG (eds), World Scientific, Vol 2, pp 1581–1586.

Hornung HG (1988) The Piston Motion in a Free-Piston Driver for Shock Tubes and Tunnels, GALCIT Report FM 88-1.

Maus J, Blanks J, Dewitt J (1993) Calibration and Tests of a New Free-Piston Shock Tunnel, AIAA Paper 93-5003.

Leading edge bluntness effects in hypervelocity compression corner flow

S. G. Mallinson, S. L. Gai, N. R. Mudford
Department of Aerospace and Mechanical Engineering, University College
University of New South Wales, Australian Defence Force Academy
Canberra ACT 2600 Australia

Abstract: Compression corner flow is examined under the combined effects of real gas behavior and leading edge bluntness. Pressure and heat transfer distributions have been obtained using a free-piston shock tunnel at two enthalpies: 2.8 MJ kg^{-1} and 18.0 MJ kg^{-1}. On the flat plate and upstream of the corner, the data for the blunt leading edge configuration were found to lie above the corresponding sharp leading values, whilst on the ramp, there was a considerable reduction in the pressure and heat transfer when the leading edge was blunt. Also, the extent of the interaction was seen to be smaller with the blunt leading edge. These results are similar to those from perfect gas studies. The differences between the sharp and blunt leading edge data seem to be less pronounced at the higher enthalpy and this is thought likely to be due to the reduced shock stand-off as a result of dissociation.

Key words: Compression corner, Hypervelocity, Leading edge bluntness

1. Introduction

The compression corner is a convenient configuration for examining shock wave / boundary layer interactions. It has been widely studied with both sharp and blunt leading edges (see Délery 1989; Kumar & Stollery 1994). The blunt leading edge has been found to increase the pressure and heat transfer upstream of the corner, and to decrease them downstream of the corner, in comparison to the sharp leading edge flows. Also, for small leading edge bluntness, the transition location is moved upstream and separation is delayed, whilst for large leading edge bluntness, the transition location is moved downstream and separation is hastened. Small and large bluntness may be conveniently delineated in terms of β, the combined bluntness-viscous interaction parameter (Holden 1971). The leading edge bluntness is considered small if $0.5 > \beta > 0.1$ and large if $\beta \leq 0.1$. An examination of leading edge bluntness effects using triple-deck theory arrived at similar criteria (Lagrée 1992).

The majority of investigations of the compression corner have considered perfect gas flow with only a few experimental studies (Stalker & Rayner 1985;

Mallinson et al. 1995) and theoretical studies (Ikawa 1979; Grasso & Leone 1992) examining the real gas effects. Certainly, there has been little if any attention paid to the combined effects of leading edge bluntness and real gas behavior on compression corner flow. This paper therefore examines the high enthalpy, hypersonic flow over a compression corner with blunt leading edges. The data are compared with that of a sharp leading edge obtained previously in the same facility (Mallinson et al. 1995).

2. Experimental details

The experiments have been performed using the free-piston shock tunnel T3, located at the Australian National University (Stalker 1972). Table 1 shows typical values of the reservoir and free-stream conditions. The surface pressure and heat transfer were measured using PCB 113M165 pressure transducers and co-axial chromel-alumel (Type K) thermocouples, respectively. The uncertainty in the measurements from these gauges is $\pm 10\%$ and $\pm 15\%$, respectively (Mallinson et al. 1995; Gai & Joe 1992).

Table 1. Reservoir and free-stream values for conditions B and G; h = enthalpy, p = pressure, T = temperature, ρ = density, U = velocity, M = Mach number, γ_{fr} = frozen ratio of specific heats, subscripts 0 and ∞ refer to reservoir and free-stream conditions, respectively.

	h_0 MJ kg^{-1}	p_0 MPa	T_0 K	p_∞ kPa	T_∞ K	ρ_∞ g m^{-3}	U_∞ m s^{-1}	M_∞	γ_{fr_∞}	β_{cyl}	β_{sqr}
B	18.0	20.0	8100	0.85	1080	2.5	5300	7.5	1.45	0.07	0.05
G	2.78	22.0	2300	0.65	150	16.0	2300	9.1	1.40	0.03	0.02

The model employed in the present experiments is shown in Fig. 1. It consisted of a flat plate with a detachable leading edge and a ramp plate. These plates rested upon gauge housings. The corner angle was adjusted by inserting wedges beneath the housings. In the present paper, we will discuss results for corner angles of 0°, 15°, 18° and 24°.

Two blunt leading edges were employed: hemi-cylindrical and square. These have nose drag coefficients of $C_D = 4/3$ and 2, respectively (Anderson 1989). The sharp leading edge, the results of which are used to compare the present data, is described by Mallinson et al. (1995). The three leading edges are shown in Fig. 1.

3. The flat plate flow

The flat plate pressure and heat transfer distributions for conditions B and G with the hemi-cylindrical and square leading edges are compared in Figs. 2 and 3.

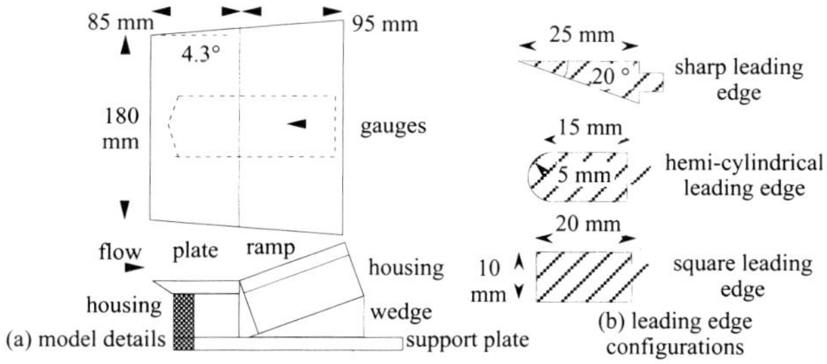

Figure 1. Schematic of the model and leading edges.

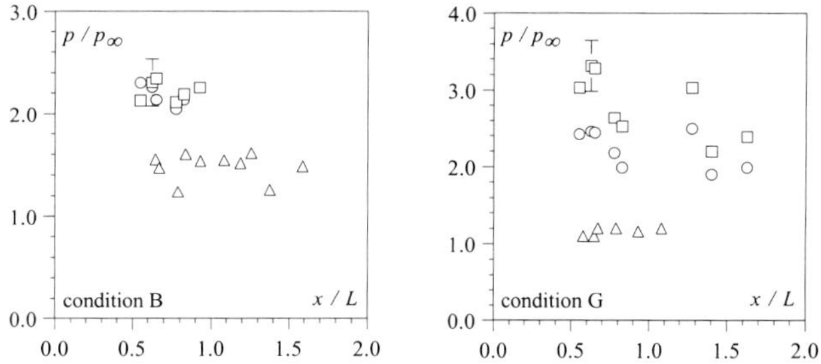

Figure 2. Flat plate pressure distributions. ○ = hemi-cylindrical leading edge; □ = square leading edge; △ = sharp leading edge.

Also shown are the data obtained with a sharp leading edge (Mallinson et al. 1995).

Firstly, consider the pressure distributions. For the high enthalpy condition B, the blunt leading edge data seem to lie closer to the sharp leading edge data than is the case for low enthalpy condition G. Also, the differences between the hemi-cylindrical and square leading edge data are less for condition B than for condition G.

It is seen that the heat transfer for the square leading edge lies above both the hemi-cylindrical and sharp leading edge data. Again, this difference is more pronounced for condition G. In fact, the sharp leading edge data could be said to be in good agreement with the hemi-cylindrical leading edge data for condition B. This may be due to the reduced shock stand-off that is experienced under high enthalpy conditions (Hornung 1972). This would lead to a thinner entropy layer and hence a smaller effect of the blunt leading edge.

4. The compression corner flow

The pressure and heat transfer distributions for conditions B and G with corner angles of 15°, 18° and 24° are shown in Figs. 4 and 5, respectively for both hemi-

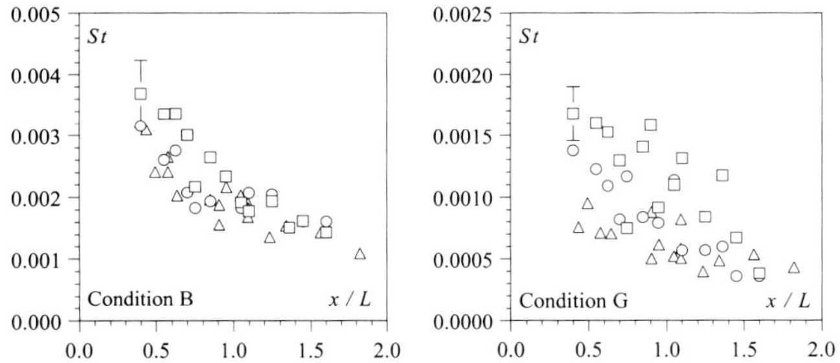

Figure 3. Flat plate heat transfer distributions. Refer to Fig. 2 for explanation of symbols.

cylindrical and square leading edge compression corners. Also shown are the pressure and heat transfer distributions from the sharp leading edge experiments (Mallinson et al. 1995).

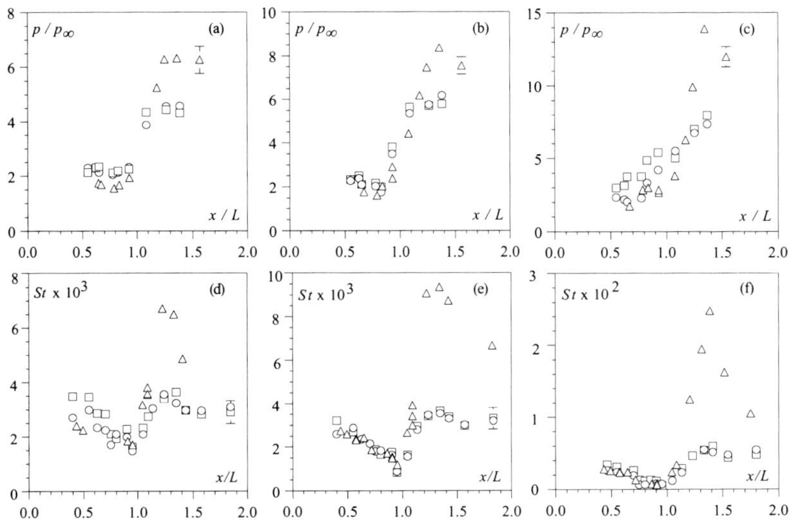

Figure 4. Compression corner pressure (a–c) and heat transfer (d–f) distributions, condition B. (a, d) = 15°, (b, e) = 18°, (c, f) = 24°. Refer to Fig. 2 for explanation of symbols.

We note that on the ramp face, the pressure and heat transfer for the blunt leading edge configuration are lower than that for the sharp leading edge. Overall, there does not seem to be much difference between the hemi-cylindrical and

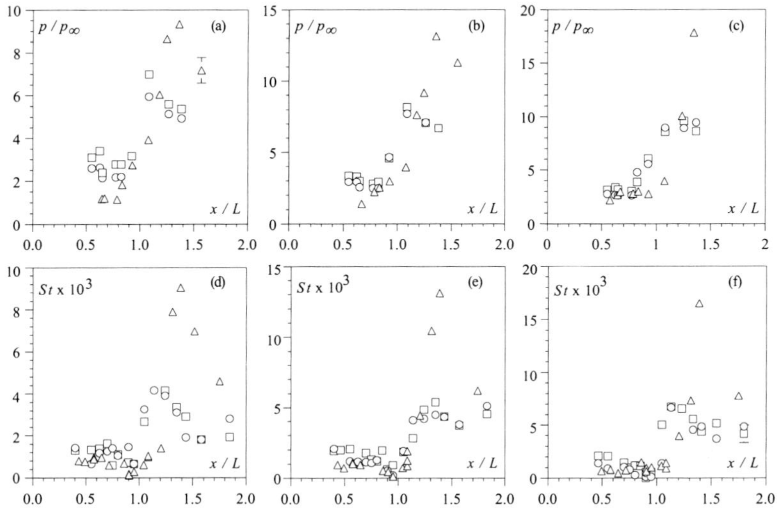

Figure 5. Compression corner pressure (a–c) and heat transfer (d–f) distributions, condition G. (a, d) = 15°, (b, e) = 18°, (c, f) = 24°. Refer to Fig. 2 for explanation of symbols.

square leading edge flows with regards to the levels of pressure and heat transfer on the ramp face.

The values of β for conditions B and G, as given in Table 1, indicate that the leading edges employed here have large bluntness. From the distributions presented, it would seem that the scale of the interaction is reduced for the blunt leading edges which is, therefore, in agreement with previous perfect gas studies (see Kumar & Stollery 1994).

5. Conclusions

The combined effects of leading edge bluntness and real gas behavior on flat plate and compression corner flow have been examined.

On the flat plate, the pressure and heat transfer lie above the sharp leading edge values, which is similar to the previous studies performed under perfect gas flow conditions. The difference between sharp and blunt leading edge flows seems to be more pronounced at low enthalpy. The smaller differences observed at the high enthalpy flow condition are attributed to the smaller bow shock stand-off due to dissociation.

For the compression corner, the pressure and heat transfer on the ramp face and the scale of interaction were reduced, which again is similar to the previous studies performed under perfect gas flow conditions with large leading edge bluntness.

Acknowledgement. The authors would like to acknowledge Mr. P.M. Walsh of the Department of Physics and Theoretical Physics, The Faculties, Australian National University, for assistance with the experiments.

References

Anderson JD Jr (1989) Hypersonic and High Temperature Gas Dynamics, McGraw-Hill.

Délery J (1989) Shock/Shock and Shock-Wave/Boundary-Layer Interactions in Hypersonic Flows, AGARD Rept 761, Part 9.

Gai SL, Joe WS (1992) Laminar heat transfer to blunt cones in high-enthalpy hypervelocity flows. J Thermophysics 6:433.

Grasso F, Leone G (1992) Chemistry effects in shock wave boundary layer interaction problems. In: Aerothermochemistry of Spacecraft and Associated Hypersonic Flows, Brun R, Chikhaoui AA (eds), Proc IUTAM Symp, Marseille, p 220.

Holden MS (1971) Boundary-layer displacement and leading-edge bluntness effects on attached and separated laminar boundary layers in a compression corner. Part II: Experimental study. AIAA J 9:84.

Hornung HG (1972) Non-equilibrium dissociating nitrogen flow over spheres and circular cylinders. J Fluid Mech 53:149.

Ikawa H (1979) Real Gas Laminar Boundary Layer Separation Methodology as Applied to Orbiter Control Surface Effectiveness Prediction, AIAA Paper 79-0212.

Kumar D, Stollery JL (1994) Hypersonic control flap effectiveness. ICAS Paper 94-4.4.3 In: 19th Congress of the International Council of Aeronautical Sciences.

Lagrée PY (1992) Influence of the entropy layer on viscous triple deck hypersonic scales. In: Aerothermochemistry of Spacecraft and Associated Hypersonic Flows, Brun R, Chikhaoui AA (eds), Proc IUTAM Symp, Marseille, France, p 358.

Mallinson SG, Gai SL, Mudford NR (1995) High Enthalpy, Hypersonic Compression Corner Flow, AIAA Paper 95-0155.

Stalker RJ (1972) Development of a hypervelocity wind tunnel. Aeronautical J 76:374.

Stalker RJ, Rayner JP (1985) Shock wave - laminar boundary layer interaction at finite span compression corners. In: 15th Int Symp on Shock Waves and Shock Tubes, Bershader D, Hanson R (eds), Stanford University, p 509.

Hypersonic high-enthalpy flow over a blunt-stepped cone

S. L. Gai, J. F. Milthorpe
Department of Aerospace and Mechanical Engineering, University College
University of New South Wales, Australian Defence Force Academy
Canberra ACT 2600 Australia

Abstract: The paper describes some experimental and computational results of a high-enthalpy flow over a blunt-stepped cone. The enthalpy range covered in the experiments was from 2.6 MJ/kg to 19.8 MJ/kg with equivalent velocities ranging from 2.1 km/s to 5.6 km/s. The results are of particular interest to proposed hypersonic aerospace planes.

Key words: Blunt cone, Backwards step, Hypervelocity

1. Introduction

There has been a resurgence of interest in the study of aero-thermodynamic problems of hypervelocity flight with the proposed aerospace planes and aeroassisted space transfer vehicles (ASTVs). This new generation of spacecraft will be subjected to prolonged radiative and convective heating during the trajectory through low-density upper atmosphere at speeds ranging from 5 km/s to 10 km/s. It is important, therefore, to understand the physical and chemical gas-dynamic processes in the flow over such vehicle configurations.

A simple configuration amenable to both theoretical and experimental studies is a spherically blunted cone. Flow around sphere cones has been studied for a long time (Lees 1956, Griffith and Lewis 1964, Gai and Joe 1992). However, data is sparse regarding configurations which may contain separating and reattaching flows as part of the total flow around the vehicle. These separating and reattaching flow phenomena are important because they may affect the control effectiveness and maneuverability of the vehicle.

In the present investigation the flow over a blunt stepped cone under high-enthalpy hypersonic conditions is studied using a free-piston shock tunnel T3 at the Australian National University in Canberra, Australia (Stalker 1985).

2. Model, facility, and experimental conditions

The configuration studied is shown in Fig. 1. It was a spherically blunted cone of nose radius (RN) 21.7 mm and cone half angle of 12°. The cone had a recessed step at a distance of 101 mm from the nose and was then extended parallel to the fore-cone surface for another 120mm. An axisymmetric rearward-facing step of height either 6 mm (0.30 RN) or 3 mm (0.15 RN) was thus formed where the flow separated at the end of the fore-cone and reattached on the aft-cone.

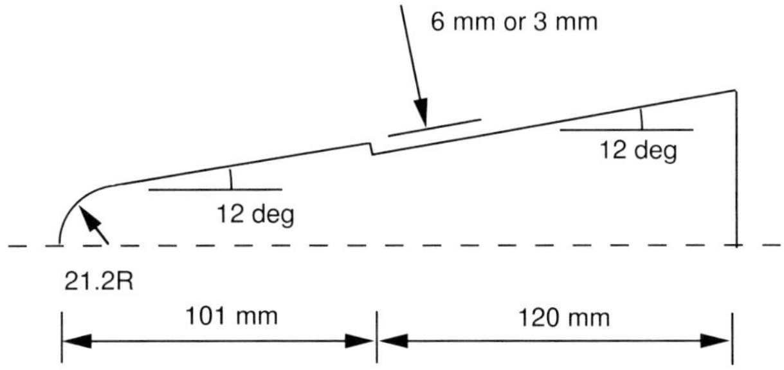

Figure 1. Schematic of the model.

The tests were conducted at the Australian National University free-piston shock tunnel T3 (Stalker 1985) with air as test gas and helium as driver gas. The flow was generated through a conical nozzle of 305 mm exit diameter and a 12.7 mm throat with a 7.5° half angle. This provided a test gas of frozen composition in the Mach number range 7.7 to 10 depending on the stagnation enthalpy and pressure. The tunnel was operated at three different conditions. These conditions, designated as B, D, and G, were typical of high, medium, and low-enthalpy flow, respectively. The flow conditions are summarized in Gai and Joe (1992).

3. Numerical method

The algorithm used to calculate the flow values is based on conservation of mass, momentum, and energy as the simulation proceeds over discrete time-steps. The flow domain is divided into rectangular cells. Each cell contains, at any given time, a certain block of material which possesses a certain momentum and energy. In the course of the next time-step, this block, with its associated mass, momentum, and energy, is convected a small distance. It may also gain or lose mass, momentum or energy by diffusion and the action of pressure gradients. The algorithm evaluates the changes in these quantities for the material block and proportionally redistributes the mass, momentum, and energy to the cells over which the block is located at the end of the time-step.

The mechanism of the algorithm is illustrated in Fig. 2 for convection with positive u and v velocities. The letters A, B, C, D represent the proportion of the material block located in a particular cell at the end of a time-step. These proportions of the mass, the momentum, and the energy in the material

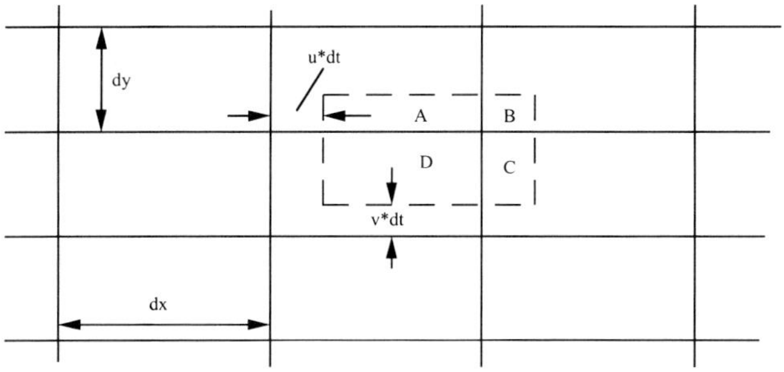

Figure 2. Numerical convection scheme.

block are allocated to their respective cells. A cell is regarded as moving with unchanged dimensions relative to the mesh of cell boundaries, so that it overlaps adjacent cells. The letter A indicates the area of the cell which remains in the original cell at the end of the time-step, while B, C, and D represent the areas of a cell which are located in adjacent cells. The quantities of mass, momentum, and energy in the cell are divided in proportion to the areas A, B, C, and D. Those proportions represented by B, C, and D are subtracted from the quantities in the original cell and added to the appropriate adjacent cells.

When the adjustment of mass, momentum, and energy has been completed for all cells, the flow variables ρ, u, v, p, and t can be calculated for all cells.

4. Stability limits

It is more economical to use the largest possible time-step in the computations, but there are a number of limits imposed by accuracy and stability considerations. The first is that, as convection is only calculated into the adjacent cells, the distance traveled in a time-step must not exceed the cell dimension. The second limit is that due to over-prediction of diffusion, as described by Roache (1976), which restricts the time-step to

$$\Delta t < \frac{\rho \Delta x^2}{2k}, \tag{1}$$

where Δx is the step size, ρ is the density, and k is the diffusion coefficient for any quantity.

5. Results and discussion

Figure 3 shows heat-transfer results for flow condition G normalized with the stagnation-point heat transfer measured on the nose. Also shown on the figure is a numerical result calculated for a similar flow condition. The calculations were made at Reynolds numbers similar to those in the experiments. It is seen that the code is able to predict the fore-cone heat transfer reasonably well.

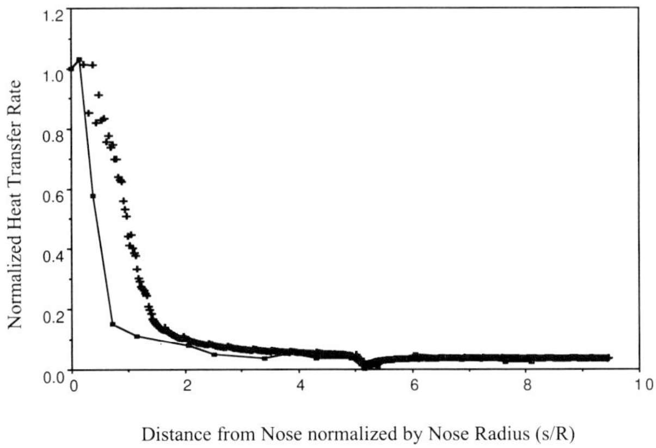

Figure 3. Normalized heat-transfer rate on blunt-stepped cone — measured and computed (condition G; $h = 6$ mm).

Downstream of the step the heat transfer (see Fig. 4) is typical of that in separated flow, viz. a sudden fall in heat transfer very near the step and then a gradual increase. While the experimental data show a decrease in heat transfer after reattachment, the numerical prediction exhibits a plateau for a considerable distance.

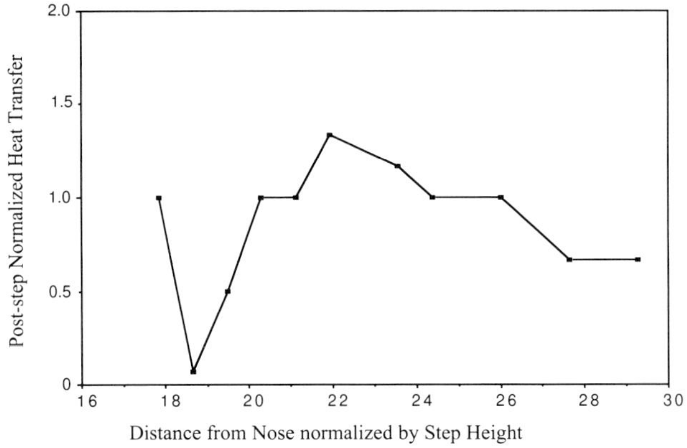

Figure 4. Normalized heat-transfer rate downstream of step.

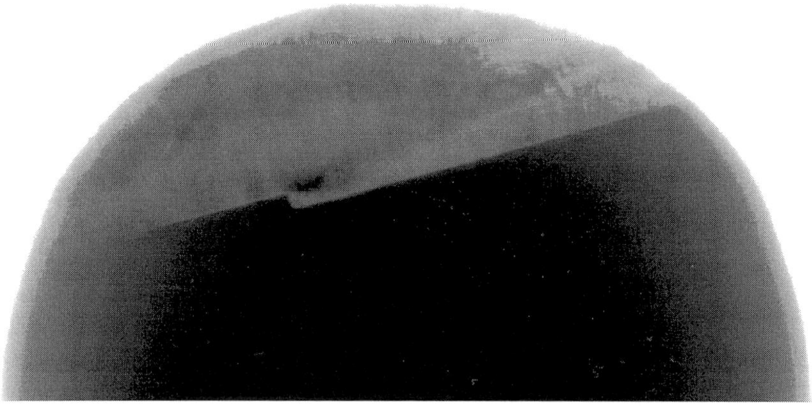

Figure 5. Schlieren photograph of flow behind the step (condition G; $h = 6$ mm).

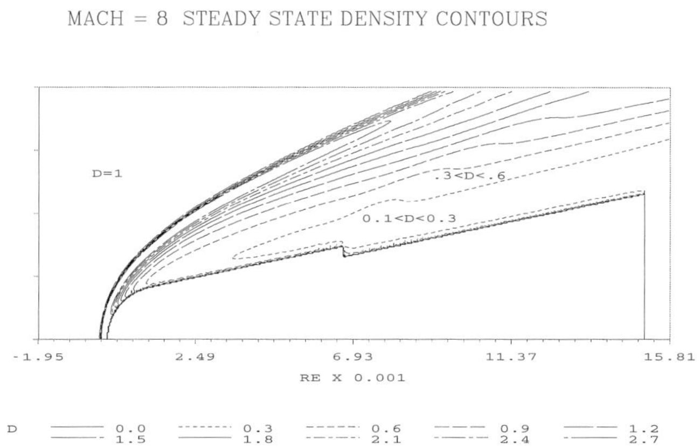

Figure 6. Non-dimensional density — computation.

Figure 5 shows an example of a schlieren photograph. We can clearly see an expansion at the corner, a separated shear layer behind the step, a wake neck, and a clear recompression shock. A notable feature is the strong curvature of the separated shear layer which is in contrast to that observed in a two-dimensional rearward-facing step flow (Gai et al. 1989). This is possibly due to the axisymmetric nature of the flow configuration.

Figure 6 shows the non-dimensional density field which shows a strong expansion, a clearly separated flow behind the step, a wake neck, and a strong recompression shock, which are consistent with the schlieren data.

The results are generally similar to those behind a two-dimensional rearward-facing step (Gai et al. 1989) but the shear layer appears to have a rather strong curvature which is unlike the two-dimensional case. This is attributed to the axisymmetric nature of the flow.

Acknowledgement. The authors would like to acknowledge the Australian Research Council for their generous support. We would also like to thank Dr. S. G. Mallinson for his help in gathering and analyzing the data, and Mr. P. Walsh for his assistance with the experiments.

References

Gai SL, Joe WS (1992) Laminar heat transfer to blunt cones in high enthalpy hypervelocity flows. J Thermo Phys Heat Transfer AIAA 6:433.

Gai SL, Reynolds NT, Ross C, Baird JP (1989) J Fluid Mech 199:541.

Griffith BJ, Lewis CH (1964) AIAA Journal 2:438.

Lees L (1956) Laminar heat transfer over blunt-nosed bodies at hypersonic speeds. Jet Propulsion 26:259–269, 274.

Milthorpe JF (1993) Numerical simulation of two-dimensional compressible flows. Int J Num Methods for Heat & Fluid Flow 3:223.

Roache PJ (1976) Computational Fluid Dynamics, Hermosa Publishers, Albuquerque.

Stalker RJ (1985) Free Piston Shock Tunnel T3 Facility Handbook, University of Queensland, St Lucia.

Surface heat flux to the Viking aeroshell in hypersonic reacting flow

N. R. Mudford, S. L. Gai
Department of Aerospace and Mechanical Engineering, University College
University of New South Wales, Australian Defence Force Academy
Canberra ACT 2600 Australia

Abstract: Measurements are reported of surface heat flux on the windward face of a Viking aeroshell model at zero angle of attack in high-enthalpy, dissociating, hypersonic flow. The reduction of surface heat flux due to dissociation is evaluated and used to infer the extent of recombination in the boundary layer.

Key words: Heat flux, Dissociation, Boundary layer, Hypersonic

1. Introduction

Surface heat flux is an important consideration in the design of atmospheric entry vehicles particularly for Aeroassisted Space Transfer and single stage to orbit vehicles. At Earth orbital speeds and above, molecular dissociation in the body shock layer causes a significant proportion of the total flow enthalpy to be present as chemical enthalpy which may contribute to surface heat flux only if recombination reactions proceed in the gas phase in the boundary layer or via surface catalyzed reaction at the vehicle surface. A thorough understanding of these recombination mechanisms is thus required.

The theory of chemical enthalpy recovery is well developed, e.g. Lees (1956), Fay and Riddell (1958), but experimental tests are rare and have yet to cover the full range of simulated flight conditions. In the present work, surface heat-flux measurements are reported for the windward surface of a Viking aeroshell model at zero angle of attack in hypersonic high-enthalpy, chemically reacting air flows. These are compared with measurements in low-enthalpy flow and with the results of calculations of expected surface heat flux under conditions of equilibrium and frozen boundary-layer chemistry. The effects of molecular dissociation are then assessed.

2. Facility and experimental conditions

The test flows were produced in the Australian National University's (ANU) T3 Free-Piston Shock Tunnel (FPST) (Stalker 1985) operating in the reflected mode with a 12.7-mm diameter throat and a 305-mm diameter exit conical nozzle.

The free-stream flow conditions for the tests are shown in Table 1. They are a subset of the air flows used in the flat plate experiments of East et al. (1980) and Gai et al. (1996). Conditions B and D have high total enthalpies and high levels of dissociation behind the bow shock around the model. Condition G has low total enthalpy and negligible dissociation behind the bow shock.

Table 1. Reservoir and free-stream values for conditions B, D, and G.

	h_0 MJ kg^{-1}	p_0 MPa	T_0 K	T_∞ K	ρ_∞ g m^{-3}	U_∞ m s^{-1}	M_∞	$\alpha_{N\infty}$	$\alpha_{O\infty}$
B	19.0	22.0	8400	1200	2.5	5500	7.5	10^{-7}	0.82
D	13.0	21.0	7100	930	3.5	4600	7.4	10^{-9}	0.43
G	2.78	22.0	2300	150	16.0	2300	9.1	0.00	0.00

Here h = enthalpy, p = pressure, T = temperature, ρ = density, U = velocity, M = Mach number, α_N = nitrogen dissociation fraction, α_O = oxygen dissociation fraction, subcripts 0 and ∞ refer to reservoir and free-stream conditions, respectively.

3. Model and instrumentation

The two mild steel models, A and B, are 70° half-angle sphere cones with base radius r_b = 25.0 mm and nose radius r_n = 12.5 mm. The conical skirt terminates in a sharp corner. In experimental series 1, models A and B were placed in the test section at the same downstream station, with model A on the centerline. Sufficient lateral separation existed to prevent interference. In experimental series 2, only model A was present on the flow centerline. Two shots were fired at each flow condition in each series.

The measurements were made using type-K surface junction coaxial thermocouples manufactured in-house to an ANU design (see Gai and Mudford 1992).

The thermocouple thermal product $(\rho ck)^{1/2}$ was obtained using the 'dipping' technique of Jessen et al. (1993) with distilled water as the fluid. This produced a value of $9500 \pm 500 \mathrm{Ws^{1/2}m^{-2}K^{-1}}$ (Mallinson 1994) which compares well with the $9900 \mathrm{Ws^{1/2}m^{-2}K^{-1}}$ obtained by Lyons and Gai (1988) for type-K thermocouples.

4. Theory

East et al. (1980) investigated the surface heat flux on flat plates at 0° and 12° incidence under the flow conditions of the present study. They argue that the appropriate recovery enthalpy for surface heat flux under conditions of equilibrium chemistry in the boundary layer is given by

$$(h_r)_{eq} = c_{pf}T_e + h_{c,e} + h_{v,e} + \sigma^{1/2}(u_e^2/2) \tag{1}$$

where $h_{c,e}$ is the chemical enthalpy in molecular dissociation and the remaining terms on the RHS are, respectively, the translational and rotational enthalpy, the vibrational enthalpy, and the kinetic energy contributions. The subscript e denotes conditions at the boundary-layer edge. Their corresponding expression for recovery enthalpy for the case of frozen boundary-layer chemistry and a non-catalytic wall is

$$(h_r)_{fr} = (h_r)_{eq} - h_{c,e} \tag{2}$$

Lees (1956) derives an expression (his Eqs. 12 and 12a) for q_E/q_{sE} for a bluff body in hypersonic, high-enthalpy flow. q_E is the surface heat flux with equilibrium boundary-layer chemistry and unity Lewis number, Le, and q_{sE} is the corresponding quantity at the stagnation point of a sphere. The surface heat flux q_{FN} with frozen boundary-layer chemistry and non-catalytic wall, is given by $q_{FN} = q_E(h_r)_{fr}/(h_r)_{eq}$. With $Le = 1$, q_E is the upper limit in a dissociated flow as all of $h_{c,e}$ is recovered and contributes to surface heat flux. Conversely, q_{FN} is the lower limit, as no part of $h_{c,e}$ contributes to surface heat flux.

The boundary-layer edge conditions for the present calculations of q_E/q_{sE} and q_{FN}/q_{sE} were obtained using the BSHOCK program which calculates the full inviscid flow field using the inverse method (Garr and Marrone 1963). It includes nonequilibrium chemistry and equilibrium vibrational excitation using a single temperature model. BSHOCK has successfully predicted the density fields ahead of circular cylinders in reacting flows in T3 (Hornung 1972). The reaction rates employed were those of Park (1990).

5. Results and discussion

The measurements are presented here in the form q/q_{sE} against r/r_b where r is the radius from the model symmetry axis; q_{sE} is for a sphere of radius equal to the model nose radius of curvature r_n, and its value is obtained from the correlation of Fay and Riddell (1958).

The results for condition G are shown in Fig. 1 together with the BSHOCK prediction, shown as a solid line, and the results of Shih and Gay (1985), who measured surface heat flux on a 70° half angle sphere cone in low-enthalpy hypersonic flows. The Shih and Gay data are shown as $*$ on the figures. The r_n/r_b for their model is not known and the trailing edge of the conical skirt is radiused. Also shown are the average values of the present q/q_{sE} at each station, excluding outlying points. The figure shows that the experimental q/q_{sE} is close to 0.8 at the stagnation point. Both the BSHOCK prediction and the Shih and Gay (1985) results are in good agreement with the present results from $r/r_b = 0.2$ to the limit of the instrumented region of the present model. The high q/q_{sE} in the Shih and Gay data for $r/r_b < 0.15$ may be due to $r_n/r_b < 0.5$ for their model. Interestingly, the BSHOCK prediction follows

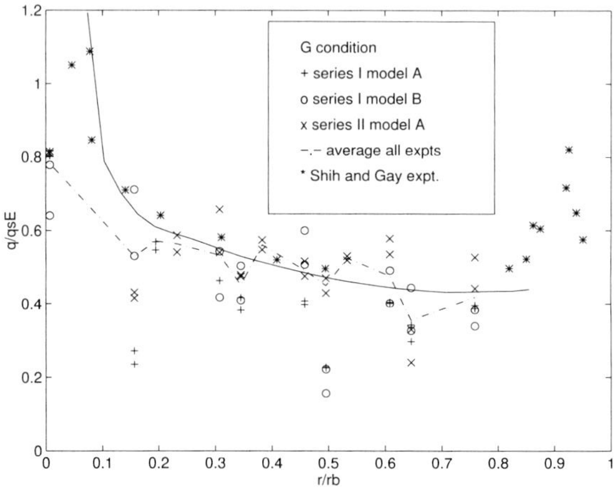

Figure 1. Condition G.

the Shih and Gay data closely in this region. Note that BSHOCK is inaccurate close to the stagnation point.

Figures 2 and 3 show the reacting flow-field results with the BSHOCK predictions for q_E/q_{sE} and q_{FN}/q_{sE} labeled E and FN respectively. Note that for both these conditions $h_{c,e}$, and consequently q_{FN}/q_E, are almost constant throughout the flow field. The negligible difference between the q_E/q_{sE} for the three flow conditions is expected on dimensional grounds (Fay and Riddell 1958).

It can be seen from the figures that q/q_{sE}, at the stagnation point and for $r/r_b > 0.5$, is unaffected by $h_{c,e} \neq 0$ indicating full recovery of $h_{c,e}$ in these regions. On the conical skirt in the region $0.1 < r_n/r_b < 0.5$, the q/q_{sE} for conditions D and B are lower than for condition G and lie between the E and FN limits. Consideration of Figs. 2 and 3 indicates that in this latter region approximately 60% of $h_{c,e}$ is recovered in the D condition flow and 70% in the B condition flow.

6. Conclusions

Measurements in non-reacting flow in the T3 FPST have been shown to agree well with those performed elsewhere. The present theoretical predictions agree well with the non-reacting flow results from all sources. Significant reductions below the value corresponding to full recovery of chemical enthalpy were detected in the inner regions of the conical skirt in the reacting flows. It is not possible from these experiments to determine whether gas-phase or surface-catalyzed reaction is responsible for the observed chemical-enthalpy recovery.

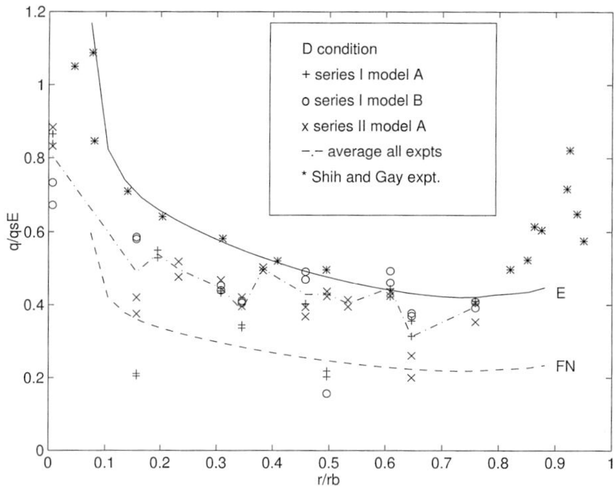

Figure 2. Condition D.

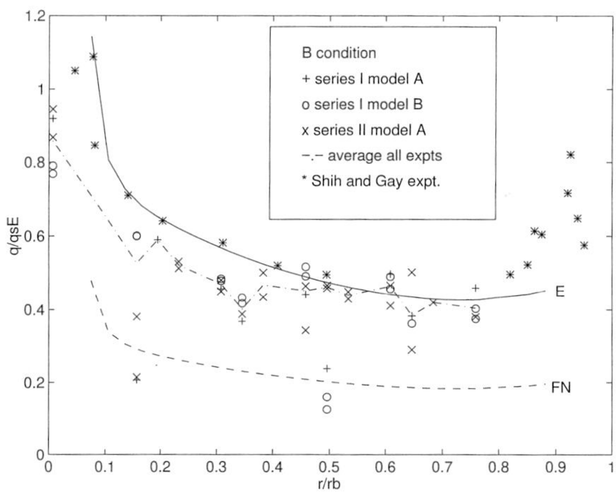

Figure 3. Condition B.

The high levels of chemical-enthalpy recovery observed here contrast with zero recovery found by East et al. (1980) and Gai et al. (1996) on sharp-nosed flat plates at low incidence under the same flow conditions. Clearly, the regime over which the boundary-layer chemistry moves from frozen to equilibrium lies between the two extremes of slender and bluff bodies in the FPST flows. This merits further investigation.

Acknowledgement. The authors would like to thank Mr. G. F. Foppoli of the Department of Aerospace and Mechanical Engineering, ADFA for making the models and the thermocouples to such a high standard. They would also like to thank Mr. P. M. Walsh and Mr. W. S. Joe of the Department of Physics and Theoretical Physics, the Faculties, ANU for their invaluable assistance with the experiments.

References

East RA, Stalker RJ, Baird JP (1980) Measurement of heat transfer to a flat plate in a dissociated high enthalpy laminar air flow. J Fluid Mech 97:673.

Fay JA, Riddell FR (1958) Theory of stagnation point heat transfer in dissociated air. J Aero Sci 25:73–122.

Gai SL, Mudford NR (1992) Stagnation point heat flux in hypersonic high enthalpy flow. Shock Waves 2:43–47.

Gai SL, Mudford NR, Roberts GT, Mallinson SG (1996) Effects of catalycity and boundary layer reactions on surface heat flux in hypersonic high enthalpy flows. In: Shock Waves, Proc of the 20th ISSW, Sturtevant B, Shepherd JE, Hornung HG (eds), World Scientific, Vol 1, pp 317–322.

Garr LJ, Marrone PV (1963) Cornell Aeron Lab Rep QM-1626-A-12(II).

Hornung HG (1972) Nonequilibrium dissociating nitrogen flows over spheres and circular cylinders. J Fluid Mech 53:149–176.

Jessen C, Vetter M, Grönig H (1993) Experimental studies in the Aachen hypersonic shock tunnel. Z Flugwiss Weltraumforsch 17(2):73–81.

Lees L (1956) Laminar heat transfer over blunt nosed bodies at hypersonic flight speeds. Jet Prop 26:259–269, 274.

Lordi JA, Mates RE, Moselle JR (1966) Computer program for the numerical solution of nonequilibrium expansions of reacting gas mixtures. NACA Computer Rept CR-472.

Lyons PRA, Gai SL (1988) A method for the accurate determination of the thermal product $(\rho c k)^{1/2}$ for thin film heat transfer or thermocouple gauges. J Phys E: Sci Instrum 21(5):445–448.

Mallinson SG (1994) Shock wave/boundary layer interaction at a compression corner in hypervelocity flows. PhD Thesis, Mech Eng Dept, University College, UNSW, Australian Defence Force Academy, Canberra, Australia.

Park C (1990) Nonequilibrium Hypersonic Aerothermodynamics, Wiley Interscience, New York.

Shih PK-T, Gay A (1985) Low lift-to-drag aerobrake heat transfer test at Mach 10. In: Thermal Design of Aeroassisted Orbital Transfer Vehicles (Nelson HF, ed), Progress in Astronautics and Aeronautics, AIAA, New York.

Schultz DL, Jones TV (1973) Heat-Transfer Measurements in Short-Duration Hypersonic Facilities, AGARD Rept No 165.

Stalker RJ (1985) Free Piston Shock Tunnel T3, A Facility Handbook, Univ of Qld, Australia.

Heat transfer in a superorbital chemically frozen boundary layer

A. Neely*, R. Morgan
Department of Mechanical Engineering
The University of Queensland
Brisbane, Queensland 4072 Australia

Abstract: The chemical behavior of flat-plate boundary layers and the associated heat-transfer rates have been investigated using 104-$MJkg^{-1}$ and 108-$MJkg^{-1}$ flows produced in the X1 superorbital expansion tube. The Damkohler numbers for these flows indicate that the chemical composition of the boundary layer will remain frozen along a plate at zero incidence, in agreement with measurements. Inclination of the surface to the flow raises the density in the boundary layer and increases the likelihood of the recombination reactions proceeding. This effect has been observed in the measurement of incident heating rates.

Key words: High enthalpy, Heat transfer, Boundary layer, Experiment, Superorbital expansion tube, Recombination, Damkohler number

1. Introduction

The flat-plate heating in high-enthalpy flow depends on the boundary-layer flow chemistry. There is a large component of chemical enthalpy in the form of dissociated species that is available for heating. The heating rate is evaluated for the two extremes of: (i) a frozen non-catalytic boundary layer, in which none of the chemical energy is returned to the flow and (ii) an equilibrium boundary layer or a fully catalytic surface, in both of which the chemical enthalpy is fully returned to the flow. In the frozen case, the wall and reference conditions are evaluated assuming constant composition across the boundary layer. In the equilibrium and catalytic cases, the composition must be determined from the local gas states. Here the chemical enthalpy which may be liberated through recombination must be included in determining the recovery enthalpy.

For this study it has been assumed that the free stream is in equilibrium. This has been shown to be a reasonable approximation for the flows in this facility (Neely & Morgan 1994). Therefore the tendency towards equilibrium from a dissociated non-equilibrium state should not be a driving force for re-

*Present address: Department of Engineering Science, University of Oxford, Parks Road, Oxford, OX1 3PJ, UK.

combination. If the reference temperature is lower than the temperature at the boundary-layer edge there will be a net tendency toward recombination. The gas-phase Damkohler number is evaluated to determine whether recombination will occur. It is the ratio of the diffusion time to the reaction time. For Damkohler numbers less than unity, the likelihood of the reaction occurring is small. The composition is said to be frozen typically for Damkohler numbers less than the order of 10^{-2} (Roberts 1991).

If the boundary layer remains frozen, there is the possibility of the dissociated species diffusing to the surface and recombining there via catalysis. The surface Damkohler number is the ratio of the characteristic surface reaction rate to the characteristic surface diffusion rate. If the surface Damkohler number is small, then the diffusion rate is much greater than the surface reaction rate and the surface reaction will be frozen and all surfaces will be effectively noncatalytic (East et al. 1980). If it is large then virtually all the dissociated species reaching the wall will be recombined, increasing the heat-transfer rate. Metals are far more likely to be catalytic than non metals. Vitreous materials, such as the quartz substrate and the SiO_2 coatings covering the platinum gauges, should not induce recombination, having low rate constants. However, if the SiO_2 layer was to be eroded, uncovering the platinum, surface catalysis may proceed.

2. Experimental measurement of flat-plate heating rates

The geometries and mounting configurations of the flat plates used in the experiments are shown in Fig. 1.

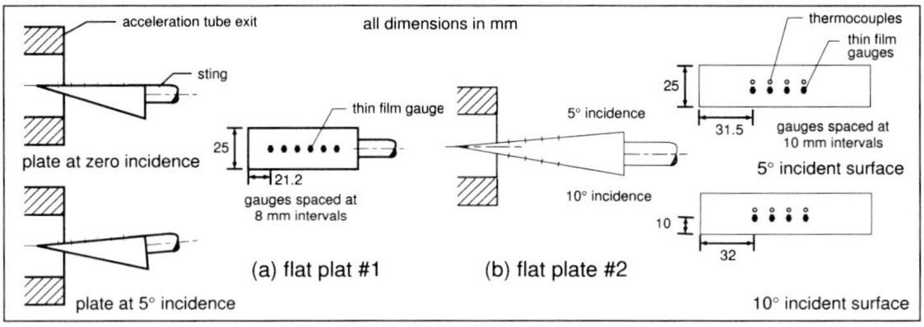

Figure 1. Flat-plate geometries for heat-transfer experiments.

2.1. Discussion of non-incident measurements

The test-flow states in Table 1 were used to calculate conditions at the boundary-layer edge and at the surface using pressure distributions determined by the strong interaction principle for a cold wall. Heating rates calculated from these conditions are compared with measurements for 108 $MJkg^{-1}$ (Fig. 2).

Table 1. Test flows calculated from equilibrium (Neely and Morgan 1994).

Flow condition		104 $MJkg^{-1}$	108 $MJkg^{-1}$
flow velocity	(ms^{-1})	12400	13000
pressure	(kPa)	3.9	5.8
temperature	(K)	6015	5935
density	(10^{-2} kgm^{-3})	0.137	0.215
static enthalpy	($MJkg^{-1}$)	26.5	23.2
total pressure	(MPa)	134	468
dissociation fractions	N, O	0.64, 1.00	0.47, 1.00

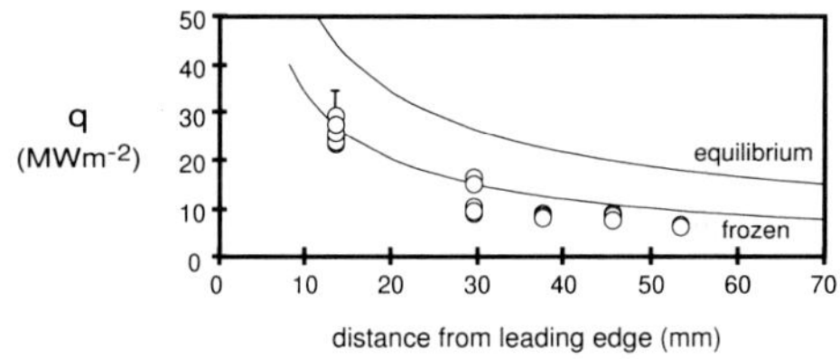

Figure 2. Measured heating rates at zero incidence, 108 $MJkg^{-1}$.

For both recombination reactions the gas-phase Damkohler numbers along the plate are more than two orders of magnitude less than unity, indicating likely chemical freezing in the boundary layer (Fig. 3a). For the length of the

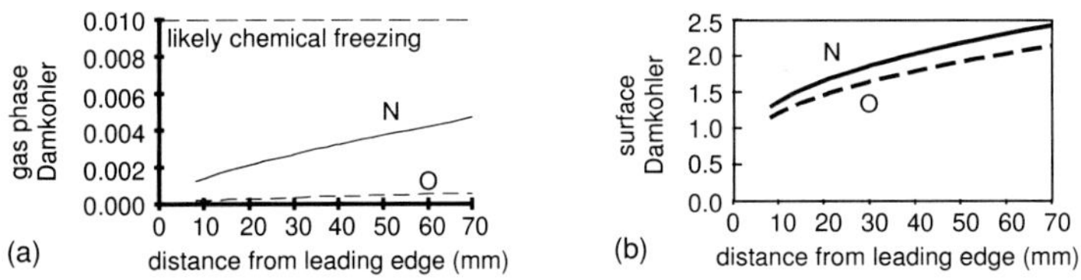

Figure 3. (a) Gas phase and (b) surface Damkohler numbers for recombination along a non-incident surface at 108 $MJkg^{-1}$.

plate, the average temperature within the boundary layer is greater than at the edge despite the cold wall, indicating no net driving potential toward recombination. The net temperature rise within the boundary layer could induce further dissociation of the nitrogen component which, being endothermic, would result in a decrease in the heat-transfer rate, but this is not seen in the equilibrium calculations or the measurements. The surface Damkohler numbers (Fig. 3b) show the diffusion rate and the surface reaction rate to be of similar order, suggesting that recombination will be small.

The measurements agree reasonably with Eckert's (1955) empirical correlation (Fig. 4). Reynolds analogy is found to be less satisfactory. The method of Ott and Anderson (1994), in which Reynolds analogy is corrected to account for nonequilibrium effects, is found to be in better agreement with Eckert's method and the measurements. Van Driest's (1952) numerical calculations are

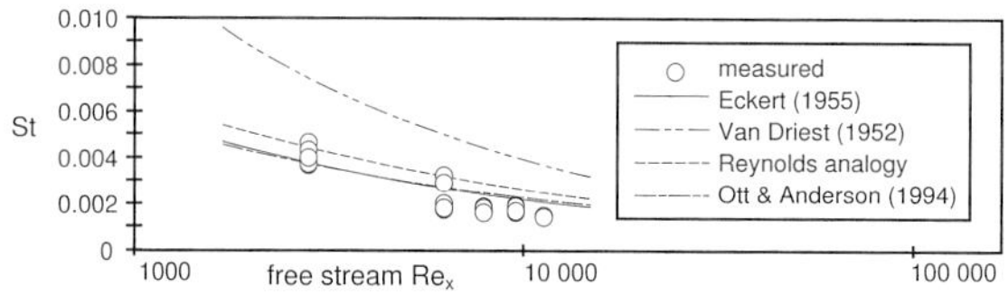

Figure 4. Predictive heating calculations for non-incident 108-MJkg^{-1} flow.

found to significantly over predict the heating rates. This would be expected as these calculations do not account for the possibility of chemical freezing. The experimental results are added to the summary of high-enthalpy heating data presented by Mallinson (1994) (Fig. 5). The measurements at 108 MJkg^{-1} continue the trend observed at lower enthalpies.

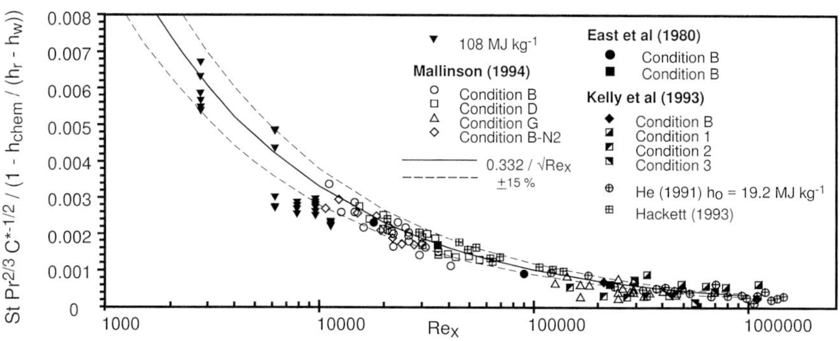

Figure 5. Comparison of non-incident flat-plate heating measurements at high enthalpy (after Fig. 3.5, Mallinson 1994).

2.2. Discussion of incident measurements

The gas-phase Damkohler numbers for the incident flows (Fig. 6a) indicate that nitrogen recombination may no longer be frozen toward the end of the plate at 10° for the 104-MJkg^{-1} flow. At 108 MJkg^{-1} the nitrogen recombination may not be frozen toward the end of the plate at 5°, or along the length of the plate at 10°. The incident data are limited due to deterioration of the gauges via erosion. This may permit recombination catalyzed by the platinum film.

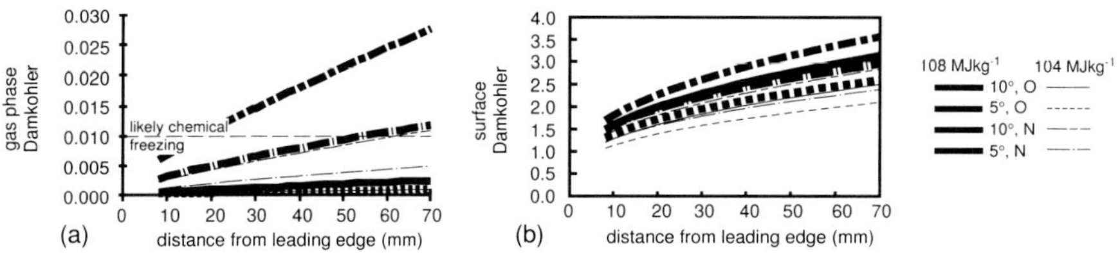

Figure 6. (a) Gas phase and (b) surface Damkohler numbers for recombination along an incident surface.

The surface Damkohler numbers increase with incidence (Fig. 6b), indicating a greater likelihood of recombination. The incident heating rates indicate different behavior (Fig. 7). At the lower incidence, for 108 $MJkg^{-1}$, the measurements follow the frozen predictions as indicated by the gas phase Damkohler numbers. From the limited measurements for 104 $MJkg^{-1}$ at 5°, the agreement is less satisfactory. The leading gauge indicates frozen behaviour as expected but the rearmost gauge indicates recombination.

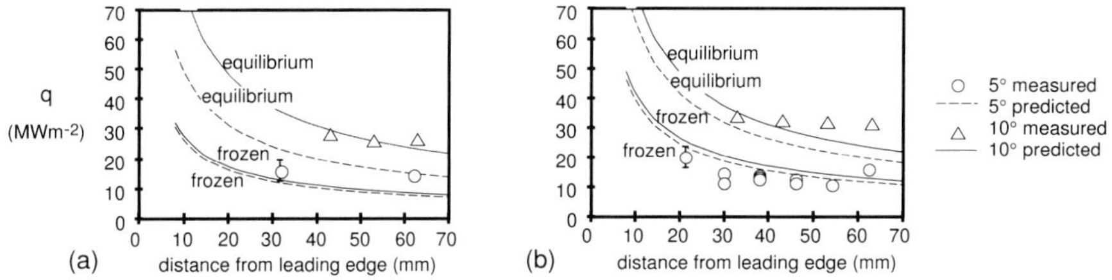

Figure 7. Incident heat-transfer rates for (a) 104-$MJkg^{-1}$ and (b) 108-$MJkg^{-1}$ flows.

At 10°, the measurements for both conditions more closely match the equilibrium predictions, indicating recombination of the dissociated species, in agreement with the Damkohlers. The levels towards the rear of the plate at 10° for 108 $MJkg^{-1}$ are higher than would be expected for even the full contribution of the chemical enthalpy to the heat transfer through recombination. Radiation should not contribute to the heating at the incidences considered.

3. Conclusions

These experiments use the superorbital expansion tube to extend the enthalpy range of heat-transfer measurements in air to beyond 100 $MJkg^{-1}$. At these enthalpies there is total dissociation of the oxygen and significant dissociation of the nitrogen in the free stream. For the geometries considered, the gas-phase and surface Damkohler numbers indicate the likelihood of chemical freezing in the boundary layer on a flat plate at zero incidence in agreement with the measurements. At incidence to the flow, the gas-phase and (to a lesser extent) the surface Damkohler numbers indicate the likelihood of recombination. This is borne out reasonably by the measurements. Hence calculation of the gas-phase and surface Damkohler numbers is found to give a good indication of the chemical behavior in boundary-layer flows at superorbital enthalpies.

Acknowledgement. Financial support for this work was provided by the Australian Research Council.

References

East RA, Stalker RJ, Baird JP (1980) Measurements of heat transfer to a flat plate in dissociated high-enthalpy laminar air flow. J Fluid Mech 97:673.

Eckert ERG (1955) Engineering relations for friction and heat transfer to surfaces in high velocity flow. J Aeronaut Sci 22:585.

Hackett (1993) Aerothermodynamic Heating Due to Shock Wave/Laminar Boundary-Layer Interactions in High-Enthalpy Hypersonic Flow, AIAA 93-3135.

He Y (1991) Transition and Heat Transfer in Compressible Boundary Layer Flow Over a Flat Plate, Ph.D. Thesis, Dept of Mech Eng, The University of Queensland.

Kelly GM (1993) A Study of Reynolds Analogy in a Hypersonic Boundary Layer Using a New Skin Friction Gauge, Ph.D. Thesis, Dept of Mech Eng, The University of Queensland.

Mallinson SG (1994) Shock Wave/Boundary Layer Interaction at a Compression Corner in Hypervelocity Flows, Ph.D. Thesis, Dept of Aerospace and Mech Eng, University College, ADFA, University of New South Wales.

Neely AJ, Morgan RG (1994) The superorbital expansion tube concept, experiment and analysis. Aeronaut J 946:175.

Ott JD, Anderson JD Jr (1994) Effects of nonequilibrium chemistry on the reference temperature method and Reynolds analogy. J Thermophys Heat Trans 8:381.

Roberts GT (1991) Investigation of Surface Recombination Effects on Convective Heat Transfer in Dissociated Air, AASU Memo 91/1, Dept of Aeronautics and Astronautics, University of Southampton.

Van Driest ER (1952) Investigations of Laminar Boundary Layers in Compressible Fluids Using the Crocco Method, NACA TN-2597.

Standard model experiments in high-enthalpy facilities

S. Ueda
National Aerospace Laboratory, Kakuda Research Center
1 Koganesawa, Kimigaya, Kakuda, Miyagi 981-15 Japan
K. Hannemann, G. Eitelberg
DLR SM-SK
Bunsenstrasse 10, 37073 Göttingen, Germany

Abstract: Experiments on the blunt cone model were performed in the HEG high-enthalpy wind-tunnel facility. The model was instrumented for both pressure and heat-flux measurements. It was tested at two diaphragm burst conditions with two respective enthalpies. Holographic interferometry was employed for the flow-field visualization. Test results were compared with numerical calculations. The same model was also tested in the T5 facility at the California Institute of Technology (Caltech). A comparison of the results from these two free-piston driven shock tunnels is also presented.

Key words: Free-piston shock tunnel, Tunnel calibration, Experiment, Heat-flux rate, Wall pressure, Flow visualization

1. Introduction

The development of hypersonic flight vehicles requires accurate knowledge of the flow field around the vehicle. Several facilities have been newly developed for the study of high-enthalpy phenomena. One of these is the HEG in DLR Göttingen Germany (Eitelberg et al. 1992). This facility, a free-piston driven shock tunnel operating in the reflected mode, was established for work in the hypervelocity flow regimen where dissociative effects of air are significant. A standard model experiment was required to evaluate tunnel characteristics. A blunt cone, referred to as ELECTRE, for which flight data exists, was selected as a standard model for reference testing. This model was selected for the following reasons. It exhibits sufficient bluntness to produce dissociation/recombination behind the bow shock in the stagnation region. The expansion from the blunt spherical part to the slender conical section might be rapid enough to display significant non-equilibrium in the expansion fan. It is a simple enough model to be transferred from tunnel to tunnel and to be used to study the tunnel characteristics as they influence the non-equilibrium chemistry as described above. The aim of this study is to provide quantitative data for both numerical code developers as well as facility operators.

2. Experimental

Figure 1 shows a model design 1/6 the size of the ELECTRE flight model. The model came in two halves, the upper one outfitted with pressure transducers and the lower one with thermocouples for heat-flux measurements. A series of 10 shots were conducted on the model in the HEG. Four test conditions, two stagnation enthalpies, and two stagnation pressures were selected based on calibration shots. Table 1 shows test conditions in the HEG. The same model was also tested in the T5 at Caltech (Rousset & Adam 1993). T5 is also a free-piston driven shock tunnel operating in the reflected mode. By comparison, T5 has a smaller nozzle-area ratio than the HEG, and differential interferometry was used instead of holographic interferometry (Kastell et al. 1993). Six test conditions, consisting of two stagnation enthalpies, two stagnation pressures, and three nozzle throat diameters were used.

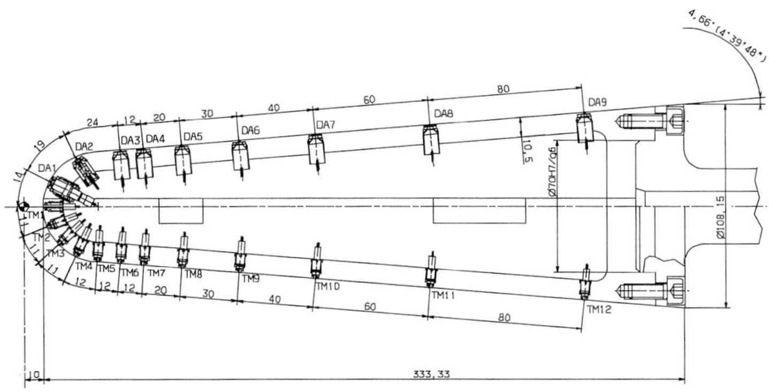

Figure 1. Perspective of the instrumented model.

Table 1. Test conditions in the HEG.

Condition	Reservoir Pressure (MPa)	Enthalpy (MJ/kg)	Nozzle Exit Velocity (km/s)	ρL (g/m^3)
I	38.6	21.1	6.49	0.52
II	90.9	22.3	6.68	1.14
III	45.0	13.2	5.14	0.90
IV	111.0	14.8	5.44	1.95

3. Results and discussion

In free-piston driven shock tunnels like HEG and T5, the mounting of the pressure transducer has to be designed carefully because of the short test time and severe environment. Flush mounting is the best way to improve response time. However, the pressure transducer has to be protected from high temperature and particles in the gas. Pressure-transducer mounting used in the model is shown in Fig. 2a. The pressure is introduced by a hole and goes a long way around a screen to reach the pressure transducer. Several mounting systems were tried to improve response time and prevent damage at the same time. Figure 2b shows the measured pressure history in the conical portion using various transducer mountings. Response time $\tau = \frac{V}{Ac}$, where V is the volume of cavity, A is the area of hole, and c is the speed of sound of the test gas, is required to fill the volume of the mounting system with test gas. Under HEG test conditions, the calculated response times of these mounting are approximately (a) 0.15 ms, (b) 0.07 ms, (c) 0.02 ms at the stagnation region, and (a) 0.4 ms, (b) 0.2 ms, (c) 0.06 ms at the conical portion. These values agree quite well with measured values.

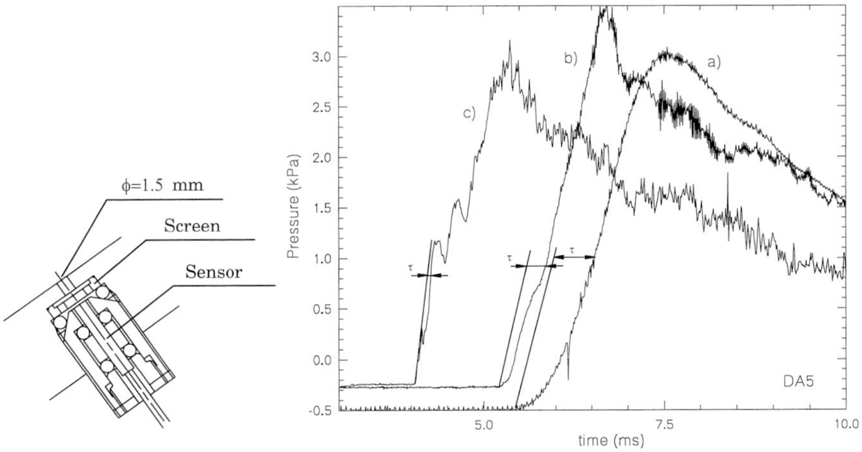

Figure 2. (a) Mounting type b (1.5-mm hole with screen) and (b) response time of various mountings: (a) 1-mm hole with screen, (b) 1.5-mm hole with screen, and (c) 1.5-mm hole without screen.

3.1. Pressure and heat-flux rate

Kulite resistance-type pressure transducers and Medtherm coaxial thermocouples were used for measurement. Measured pressures and heat-flux rates were normalized by calculated free-stream conditions. A non-equilibrium nozzle code was used to calculate these values. Figure 3 shows the pressure coefficient $C_p = \frac{p-p_\infty}{\frac{1}{2}\rho_\infty u_\infty^2}$ and Stanton number $St = \frac{\dot{q}}{\rho_\infty u_\infty (h_s - h_w)}$ distributions along the wall, where ρ, u, p, h, and $\dot{q}$ are respectively density, flow velocity, enthalpy, and heat-flux rate to the wall. Comparison between experimental results and numerical calculations using a non-equilibrium Navier-Stokes code (Brenner et al. 1993) are also presented. Figure 3a shows that slightly higher pressures were measured in the conical portion; this was due to the fact that values of static pressure used to calculate pressure coefficient were insufficiently accurate. Normalization of pressure by pitot pressure is another possible way to compare experimental and numerical results. Figure 3b shows that the Stanton numbers obtained experimentally and numerically are in good agreement. However, the heat-flux rate was not well normalized by Stanton number. The viscous effect has to be considered to obtain better normalization in this regard. Reduced Stanton number by Miller et al. (1983) $St_R = St\frac{(\rho_2/\rho_\infty)^{1/4}}{V^*}$ is one way to consider viscous effect, where V^* and ρ_2 are respectively the parameter of viscous interaction and the density behind a normal shock.

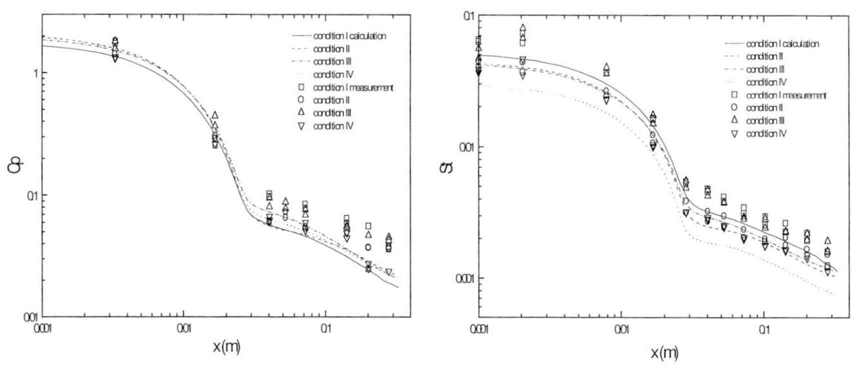

Figure 3. ELECTRE test results obtained in the HEG compared with results obtained by numerical calculation: (a) pressure coefficient and (b) Stanton number.

3.2. Flow visualization

Figure 4a is a finite-fringe interferogram of a HEG shot for the nominal burst pressure condition of 100 MPa at lower enthalpy, condition IV. In the field of view is the spherical part of the model. Figure 4b shows the shock shape obtained by holographic interferometry (HI) and Navier-Stokes calculations. Shock shapes were read from Mach contour graphs and HI photos. The experimentally observed shock shape agrees with the calculated shape downstream of the stagnation region. In the stagnation region, the observed shock stand-off distance lies between that predicted by non-equilibrium Navier-Stokes and perfect gas calculation (Van Dyke & Gordon 1959, Billig 1967).

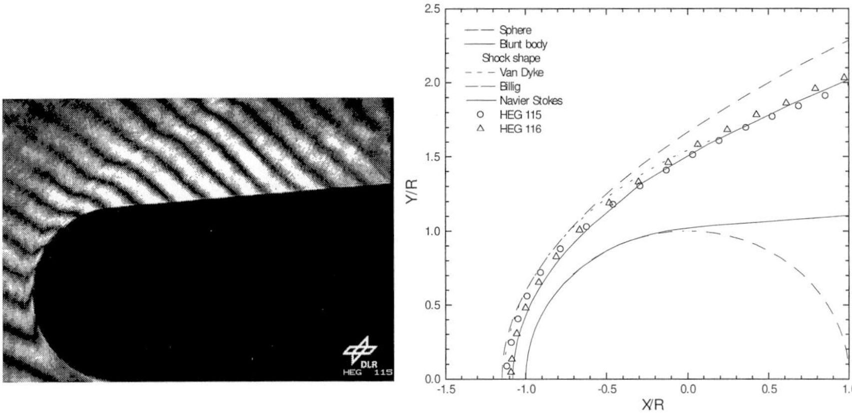

Figure 4. (a) Picture obtained using holographic interferometry. (b) Comparison of shock shape obtained using HI, non-equilibrium Navier-Stokes calculation, numerical calculation according to Van Dyke & Gordon (1959) for an ideal gas around a sphere, and an approximation of Billig (1967).

3.3. Comparison between results obtained the HEG and T5

It is difficult to compare measured results obtained in the HEG and T5 facilities because of the differences in test conditions and nozzle-area ratio. Normalized pressure by pitot pressure and reduced Stanton number by Miller et al. (1983) were used for comparison. Figure 5 shows the results. Test results were averaged for each facility. The vertical line on each set of values shows the standard deviation of test results. The distributions of pressure and the rate of heat flux in the conical portion were flatter at T5 than at HEG. Bellucci et al. (1994) indicated that this is caused by non-uniform density distribution in the T5 nozzle. This is caused by off-design use of contoured nozzle.

4. Conclusions

Standard model experiments were conducted in the HEG and T5 wind-tunnel facilities. Values obtained for the distribution of pressure and heat flux along the wall were normalized and compared. Normalized pressure by pitot pressure and St_R were seen to be good parameters to compare different test conditions and different facilities. Non-uniform density distribution in T5 caused observed differences in the pressure and heat-flux rate distributions. Accurate information on free-stream conditions is required to improve agreement. Shock shape and shock detachment distance were measured and compared with numerical calculations. The observed shock stand-off distance lies between that predicted by non-equilibrium Navier-Stokes and perfect gas calculations.

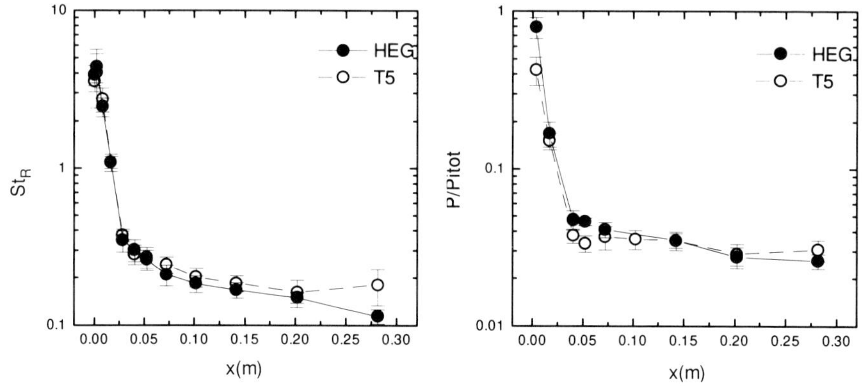

Figure 5. Comparison between HEG and T5 experimental results: (a) pressure normalized by pitot pressure and (b) reduced Stanton number.

Acknowledgement. The authors express their gratitude to J. Muylaert of ESTEC for his support of the model development, D. Kastell for the HI pictures, and the T5 and HEG groups for their active participation in this test series.

References

Bellucci V, Ueda S, Eitelberg G, Muylaert J (1994) Numerical Simulation of Electre in T5 and HEG Hypervelocity Shock Tunnels. EWP-1776.

Billig FS (1967) Shock wave shapes around spherical- and cylindrical-nosed bodies. Journal of Spacecraft and Rockets 4(6):822–823.

Brenner G, Gerhold T, Hannemann K, Rues D (1993) Numerical simulation of shock/shock and shock-wave/boundary-layer interactions in hypersonic flows. Comput Fluids 22(4/5):427–439.

Eitelberg G, McIntyre TJ, Beck WH, Lacey J (1992) The High Enthalpy Shock Tunnel in Göttingen, AIAA Paper 92-3942.

Kastell D, Kortz D, Eitelberg G (1993) In high enthalpy flows studying shock/shock interaction. In: Proceeding of Euromech 309, Optical Flow Diagnostics, Göttingen, Germany.

Miller CG, Gnoffo PA, Micol JR (1983) Heat-transfer distributions for biconics at incidence in hypersonic hypervelocity real-gas flows. In: Proc 14th Int Symp on Shock Tubes and Waves, Sydney, pp 333–341.

Rousset B, Adam P (1993) Electre Experiments in T5. FM 93-2, GALCIT, Caltech, Pasadena, CA.

Van Dyke MD, Gordon HD (1959) Supersonic Flow Past a Family of Blunt Axisymmetric Bodies, NASA TR R-1.

Radiation around a hypersonic ablative flight model

K. Komurasaki, S. Yano, T. Fujiwara
Department of Aerospace Engineering, Nagoya University
Nagoya 464-01 Japan

Abstract: Thermochemical structure around an ablative flight model at hypersonic speed is investigated by using a ballistic range. A polyethylene projectile of 1 cm diameter is launched at 5 km/sec into an observation chamber, and the light emitted from an induced shock layer around the projectile is observed with a spectroscope and a high-response power meter. From the emission spectrograms, ablation products C_2, CN, and CO are found to cause radiation comparable to air molecules N_2 and NO. The emission-power profile along the flight path of the projectile shows a tail layer in which nonequilibriumly excited molecules emit radiation behind the projectile.

Key words: Hypersonic flow, Radiation, Ballistic range, Ablation

1. Introduction

Sample-return missions which return a set of samples from a near-Earth asteroid to Earth have been proposed for scientific interests (Wallace et al. 1985, ISAS 1995). In such missions, the re-entry Mach number is beyond 40 and aerodynamic heating around the capsule will be the order of 10 MW/m^2. Therefore, heat protection becomes a serious problem in developing the capsules (Walberg 1985, Howe 1985, Park 1985). Ablation cooling is an essential technique when the heat flux reaches several MW/m^2 because there exists no material that can endure such an extremely large heat flux.

Reliable data of the heat-transfer rate around the capsule are needed to minimize the ablator weight. However, the prediction is very difficult because at such a hypersonic speed, nonequilibrium radiation plays as important a role as convection. In addition, the interaction between a shock-heated gas and an ablative surface is very complicated (Page and Arnold 1964), and radiation from ablation products, mostly carbon-containing molecules, makes prediction more difficult (Komurasaki et al. 1993). Strong radiation from the carbon-containing molecules enhances radiative heat flux to the vehicle and reduces the effectiveness of ablation cooling.

Available information about chemical reactions and radiative properties of ablation products is very limited. Therefore, measurements on ablative shock layers would be very meaningful for developing a reaction and radiation model of ablation products, and for exactly predicting the radiative-heat-transfer rate over ablation-cooled re-entry vehicles.

The objective of this study is to experimentally examine the thermochemical structure around an ablative projectile to obtain a primary understanding of the phenomena related to ablation, and to develop reaction and radiation models of ablation products.

2. Experimental apparatus

As shown in Fig. 1a, the ballistic range consists of a high-pressure driver, a pump tube, the launch tube, and a test section (vacuum chamber). In the first stage, a high-pressure reservoir and a quick-valve system are utilized, instead of the conventional set of gunpowder and a diaphragm, for the purpose of reproducibility and operational convenience. This gun is capable of launching a projectile of 1 gram weight at the speed of 5 km/sec ($M = 15$) despite its total length of only 6.8 m, which is much shorter than other similar facilities. The projectile flight speed is measured by the laser-cut method. The flight duration of the projectile in the test chamber is enough for the model surface to ablate, due to the extremely large heat flux over the projectile (Yano et al. 1993).

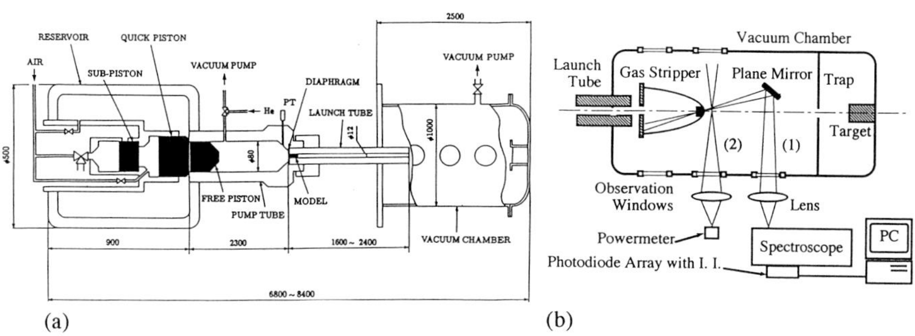

Figure 1. Schematic diagrams of (a) a ballistic range and (b) an optic collection arrangement.

The light emitted from a shock layer is collected from the front of the projectile in an optic system (shown in Fig. 1b). Spectral intensity distributions are obtained using a spectroscope (JAS Co. CT50N). Wide-range spectrograms were taken on instant film (ASA3000), and high-resolution ones are obtained with a multi-channel photo-diode array detector (HAMAMATSU PMA–50). In order to avoid collecting the light induced by the collision between the projectile and a target, a big trap room is assembled around the target. In addition, a gas stripper is set up at the end of the launch tube to prevent the driver gas (helium) from interacting with the shock and tail around the projectile.

The emission-power profile around a projectile is obtained by setting the optical collection path perpendicular to the flight path as shown in Fig. 1b. The temporal variation of the emission power is measured using a Si-PIN photodiode powermeter (YOKOGAWA 3296). The powermeter response time is less than 5 ns, corresponding to a spatial resolution of 0.025 mm. The field of view at the focal point is 7.5 mm high and 1.7 mm wide.

3. Spectral intensity distribution

The spectral intensity distribution of emission for a polyethylene projectile is shown in Fig. 2a. Several band spectra from air molecules like N_2 and NO are identified, while no atomic line spectra are identified.

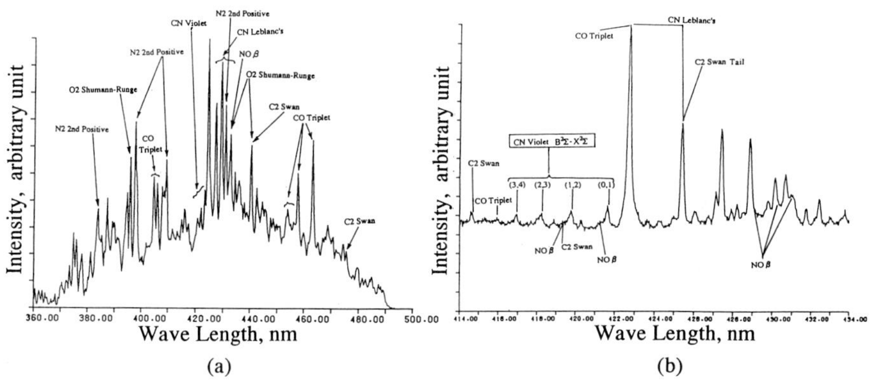

Figure 2. Spectral intensity distributions of radiation: (a) wide-range spectrogram; (b) high-resolution spectrogram, $V = 4.8$ km/s, $p_\infty = 10$ Pa.

In addition to air spectra, those from several carbon-containing molecules, C_2, CO, and CN, are detected. C_2 is thought to come from polyethylene ablation and subsequent dissociation reactions, while the other molecules, CO and CN, are thought to be produced though the exchange or recombination reactions between dissociated carbon atoms and air molecules.

Figure 2b shows the high-resolution spectral intensity distribution. Several CN violet transition spectra appear on the continuum spectrum. The vibrational temperature of the CN molecules can be estimated from those transitions.

The Boltzmann plot is shown in Fig. 3. Although the CN vibrational temperature is determined to be 3500 ± 700 K from the intensities of the (0-1) and (1-2) transitions, the populations are largely separate from Boltzmann equilibrium. The molecules are populated more in the upper vibrational levels than equilibrium ones. This indicates that CN molecules are in a rising nonequilibrium situation where CN number density increases from zero. The molecules in highly excited vibrational states tend to be in equilibrium with those in free states through radiative de-excitation when collisional de-excitation is not predominant.

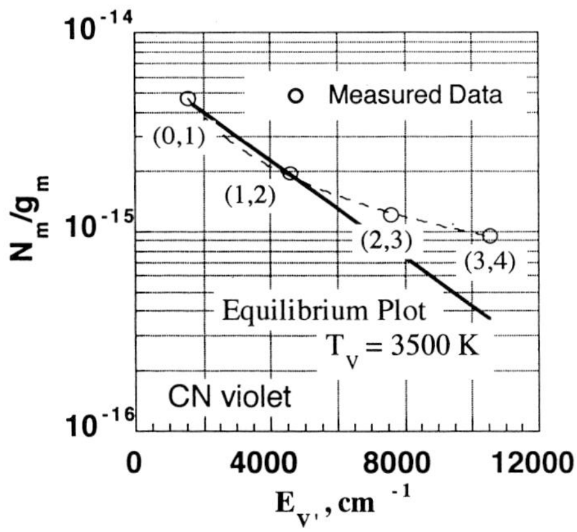

Figure 3. Boltzmann plot for a CN violet band system.

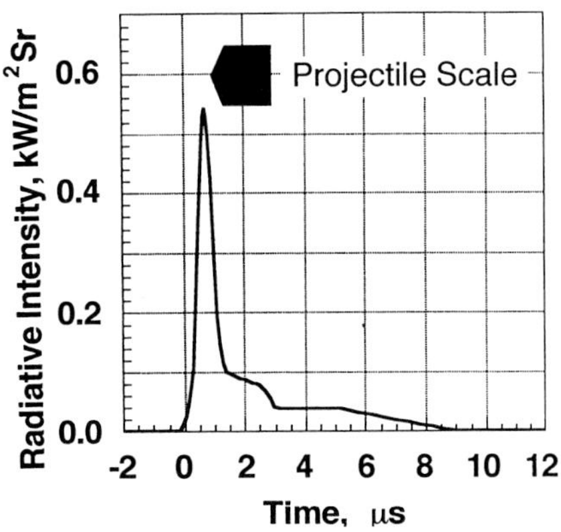

Figure 4. Temporal variation of radiation, $V = 4.5$ km/s, $p_\infty = 1000$ Pa.

4. Emission power distribution

Figure 4 shows the emission-power profile obtained by setting the optical collection path perpendicular to the flight path. Since the flight speed is constant, temporal variation is identical to the spatial distribution in the flight direction. The power increases steeply in the shock layer ahead of the projectile and then decreases. There is still emission in the side region of the projectile, and a weakly illuminated tail follows several centimeters behind the projectile.

The calculated emission power and temperature distributions simulating the experiment are illustrated in Fig. 5. The calculation is based on a two-temperature reaction and nonequilibrium radiation model, not including ablation and carbon reactions (Komurasaki et al. 1994). The calculated emitting region is limited to the portion ahead of the projectile. The discrepancy between measured and calculated profiles might come from carbon reactions. That is, in the side and base regions, the generated carbon-containing molecules (rather than the air molecules) emit radiation.

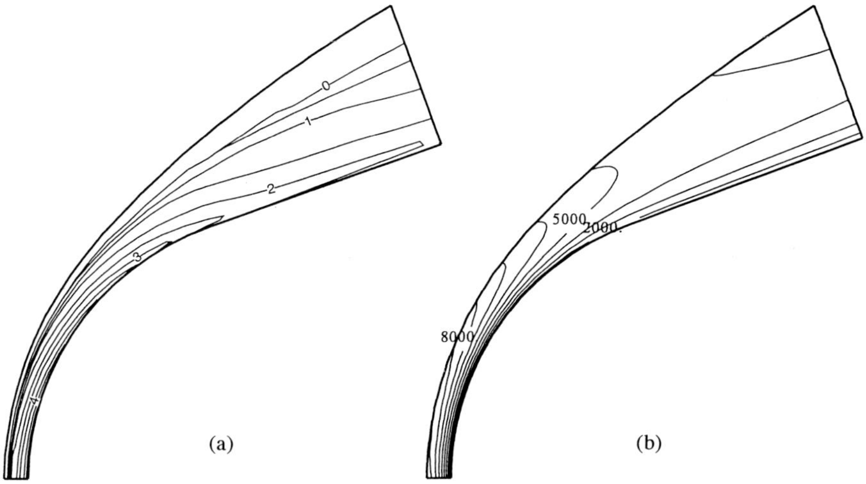

Figure 5. Calculated contour maps of (a) global emission power $\log(J)$, (J, W/m^3) and of (b) translational temperature, T (K); $V = 5$ km/s, $p_\infty = 500$ Pa, $T_\infty = 300$ K.

5. Reactions of ablation products

The ablated carbons are decomposed in the shock layer through the reactions

$$C_3 + M \to C_2 + C - 174 \text{ kcal/mol} \tag{1}$$

$$C_2 + M \to C + C - 142 \text{ kcal/mol} \tag{2}$$

Owing to these reactions, heat flux to the projectile surface is considerably reduced. Dissociated carbon can react with oxygen and nitrogen. Among the carbon associated reactions, CN formation reactions are listed as follows:

$$C + N_2 \to CN + N - 48 \text{ kcal/mol} \tag{3}$$

$$CO + N \to CN + O - 78 \text{ kcal/mol} \tag{4}$$

$$C + NO \to CN + O + 28 \text{ kcal/mol} \tag{5}$$

$$C_2 + N \to CN + C + 36 \text{ kcal/mol} \tag{6}$$

$$C + N + M \to CN + M + 178 \text{ kcal/mol} \tag{7}$$

In the high-temperature region inside the shock layer, endothermic carbon-associated reactions also contribute to the cooling. However, the exothermic ones would occur in an expanding flow in a tail where the gas is of relatively low temperature.

6. Conclusion

The carbon vapor ablated from polyethylene is dissociated into C_2 and C in a shock layer, absorbing the heat over the front surface of a projectile. The absorbed heat is carried away with the carbon to the tail behind the projectile, where flow is expanding and the gas translational temperature decreases quickly. Therefore, the carbon atoms react with air molecules in the tail. The highly excited carbon-containing molecules release a large part of their vibrational energy through spontaneous emission. The tail, emanating radiation, spreads several centimeters in length behind the projectile. Since carbon-containing molecules are known to radiate strongly even at low temperatures, in an actual case, carbon radiation may increase the heat flux to the vehicle and lower the effectiveness of ablation cooling. In addition, radiation in the tail brings heats to the vehicle base, so that a rear heat shield would be important in vehicle design.

References

Howe JT (1985) Introductory aerothermodynamics of advanced space transportation systems. J Spacecraft and Rockets 22:19–26.

ISAS (ed) (1995) A proposal for exploration of the asteroids, MUSES-C (in Japanese).

Komurasaki K, Fujiwara T (1994) Radiation and shock layer structure around a hypersonic flight model. In: Proc 19th Int Symp Space Tech and Sci, Yokohama, pp 399–404.

Komurasaki K, Kasahara J, Yano S, Fujiwara T (1993) Radiative heat transfer from a shock layer generated around a projectile launched in a ballistic range. In: Shock Waves @ Marseille, 1, pp 127–132.

Page WA, Arnold JO (1964) Shock Layer Radiation of Blunt Bodies at Reentry Velocities, NASA TR R-193.

Park C (1985) Radiation enhancement by nonequilibrium in Earth's atmosphere. J Spacecraft and Rockets 22:27–36.

Walberg GD (1985) A survey of aeroassisted orbit transfer. J Spacecraft and Rockets 22:3–18.

Wallace RA, Blume WH, Hulkower ND, Yen CL (1985) Interplanetary missions for late twentieth century. J Spacecraft 22:316–324.

Yano S, Komurasaki K, Kasahara J, Takaoka H, Fujiwara T (1993) Radiative heat transfer to a hypersonic projectile in ballistic range experiments. In: Proc 19th Int Symp Space Tech and Sci, Yokohama, pp 405–410.

Laser-induced fluorescence measurements in high-enthalpy facilities: application to the study of air plasma/surface interaction

L. Robin, D. Honore, P. Vervisch
CNRS URA 230 – CORIA
University of Rouen
76821 Mont Saint Aignan Cedex, France

Abstract: This study presents experimental results obtained with Laser-Induced Fluorescence (LIF) in continuous high-enthalpy supersonic air flows produced at the Central Research Institute of Machine Building (Tsniimach, Moscow) in ground experimental facilities designed for the study of aerodynamics and heat transfer of supersonic and hypersonic spacecraft. The LIF technique is used to obtain the measurements of NO density, rotational temperature, and velocity evolutions in these flows. Experiments are performed with a narrow-band ArF excimer laser in free jets and in numerous test sections within the boundary and shock layers lying at the vicinity of a Thermal Protection System (TPS) model designed to simulate a misalignment of tiles.

Key words: Laser-induced fluorescence, NO, High-enthalpy air flow

1. Introduction

At the Tsniimach Institute (Moscow), the TT1 arc-jet facility is devoted to the study of local aerodynamics problems arising in supersonic high-temperature flows, while a high-frequency facility has been devoted to characterize the physico-chemistry phenomena encountered in reactive-flow/surface interaction (Anfimov 1992). The main objective of the complete program was to improve reliable data for validation of chemical nonequilibrium boundary-layer and Navier-Stokes codes to compute a new test case of supersonic air-heated flow impinging on a TPS surface. The experimental work presented here is devoted to the study of the aerodynamic flow properties at the vicinity of a misalignment of tiles over a SiO_2 model immersed in a dissociated air-heated flow. An additional aim of the experimental investigation is to point out the great potential of the LIF technique for the study of continuous high-enthalpy air-flow/body interaction.

Fluorescence of NO was induced in the plasma flow by a narrow-band tunable ArF excimer laser via the $D^2\Sigma^+_{(v'=0)} \leftarrow X^2\Pi_{(v''=1)}$ epsilon band. The rotational temperature and NO number density have been determined. The spatial resolution of the LIF technique permitted accurate characterization of boundary and shock layers, as well as shock thickness. Finally, flow velocity is deduced from Doppler-shift measurements of excited rovibrational NO-ϵ lines.

2. Model design and flow conditions

Flight-condition requirements are fulfilled in the TT1 arc-jet facility. The plasma source is a classical DC arc jet. The air flow is accelerated through a supersonic nozzle before its expansion in a low-pressure test chamber at 10^3 Pa. The divergent-shaped nozzle provides an air plasma jet at Mach 5 with a diameter of 300 mm. The stagnation chamber parameters are $P_0 = 1.6 \cdot 10^5$ Pa, $H_0 = 6.7$ MJ/kg, and $T_0 = 4500$ K, with a 4-MW input power supply and an air-flow rate of 0.5 kg/s. The free-stream conditions are $P_{st} = 2 \cdot 10^4$ Pa, $T_A = 1000$ K, and $R_{eu} = 10^6$ m^{-1}. The enthalpy leads to a supersonic, weakly dissociated air-plasma flow (oxygen: 17%). Aerodynamic studies have been performed in the supersonic nonequilibrium flow and at the vicinity of a SiO_2 model with 200 mm length and 100 mm spanwidth, with a step 3 mm high located 100 mm from the water-cooled blunt leading edge. The model is immersed at 15° incidence on the jet axis.

3. Experimental set-up

A tunable ArF excimer laser (Lambda Physik LPX 150/T) is used to induce fluorescence of NO in high-enthalpy air flow via the rotational lines of the ϵ-band $D^2\Sigma^+_{(v'=0)} \leftarrow X^2\Pi_{(v''=1)}$. In narrow-band mode, the $1.8 \cdot 10^{-4}$-nm HWHM pulsed radiation (15 ns) can be tuned with a step increment of 10^{-5} nm over the full spectral range of the broad-band ArF emission centered at 193.40 nm with a 0.5-nm HWHM. To follow through wavelengths the NO-ϵ excitation spectrum, the excitation spectrum of the NO-β band is simultaneously carried out in a home-made static cell (NO, 50 mbar, 300 K).

A schematic of the set-up is depicted in Fig. 1. The probe laser beam is moved perpendicularly to the surface of the model. Furthermore, the model is translated along the jet axis in steps of 10 mm to carry out experiments within different test sections. Fine spatial resolution down to 200 μm is achieved perpendicularly to the surface where the most significant gradients are expected. The NO-ϵ fluorescence is collected at a right angle from the excitation beam pathway on the top of the test chamber through a monochromator. The probe volume is imaged onto the entrance slit of the monochromator used as a tuning 4-Åbandpass filter. The signal is recorded on a 2-GHz numerical oscilloscope (Tektronix DSA602A) allowing its temporal resolution.

4. LIF measurements

ArF broad-band mode experiments provide the signing of species in the plasma and information on the boundary and shock thicknesses. In narrow-band mode, excitation of single rotational levels is devoted to the measurement of the rota-

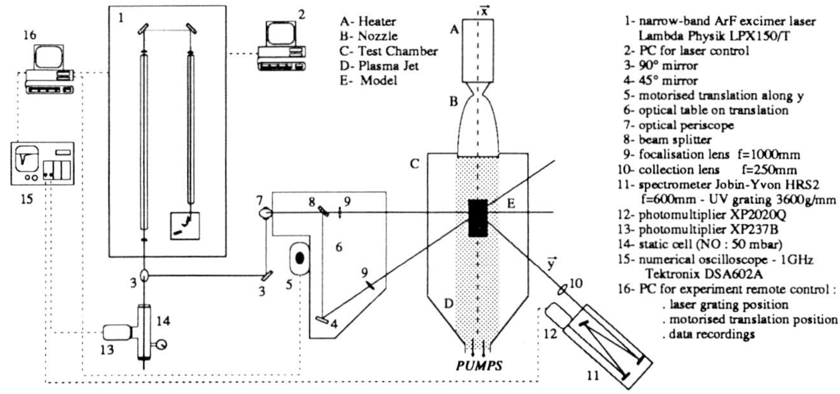

Figure 1. Experimental set-up.

tional temperature and NO density. The highly resolved excitation spectra are used to evaluate flow velocity via the Doppler shift of excitation lines.

The $D^2\Sigma^+_{(v'=0)} \rightarrow X^2\Pi_{(v''=2)}$ NO-ϵ shifted fluorescence is collected after excitation of $D^2\Sigma^+_{(v'=0)} \leftarrow X^2\Pi_{(v''=1)}$ NO-ϵ band via the $R_1(25.5)$, $P_1(38.5)$ and $Q_1(31.5)$ group of lines. These transitions are sufficiently sensitive to temperature. In addition, they are close to the $R_2(30.5)$ NO-β line excited in the NO reference static cell to provide the unshifted reference line for both laser-beam pathways used to measure the Doppler shift. A typical measurement is illustrated Fig. 2 with response of the two incident beam pathways when the excitation wavelength is scanned around the group of ϵ-lines. Temperature is deduced from the intensity ratio of the lines and compared with computed excitation spectra. Density is drawn from the integrated intensity rather than from peak intensity. According to the two excitation pathways separated by a α-degree angle incidence, the shift in frequency $\Delta\lambda$ from the selected transitions in both directions is then proportional to the flow velocity at the beam crossing point ($\frac{\Delta\lambda}{\lambda} = \frac{v}{c}\cos\alpha$). One-step increment of the grating motor fixes the accuracy of the method and corresponds to a shift of $\Delta\lambda = 1.8 \cdot 10^{-4}$ nm and to a flow velocity of 395 m/s for an incidence angle of $\alpha = 45°$.

5. Experimental results

Experiments are performed on both sides of the free jet axis and reveal a sufficient homogeneous region to locate the wide SiO_2 model and assure subsequently a 2D aerodynamic description of the boundary conditions. The unchanged rotational-line distribution and the constant fluorescence intensity point out flat NO number density and temperature (1200 K) profiles in the free-jet section. The Doppler-shifted fluorescence signals provide a flat velocity radial

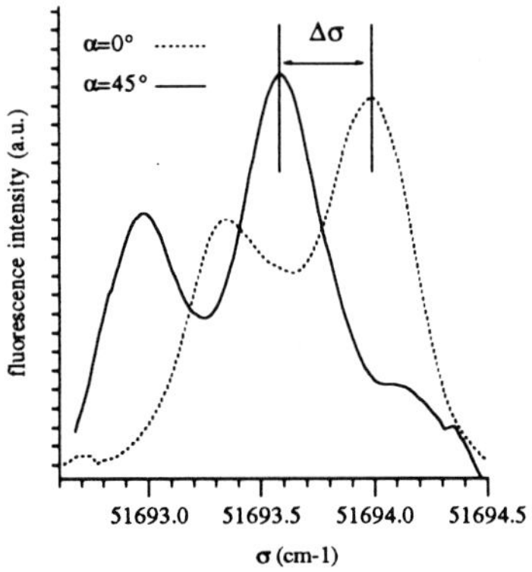

Figure 2. Doppler-shifted NO-ϵ excitation lines observed in the supersonic plasma jet.

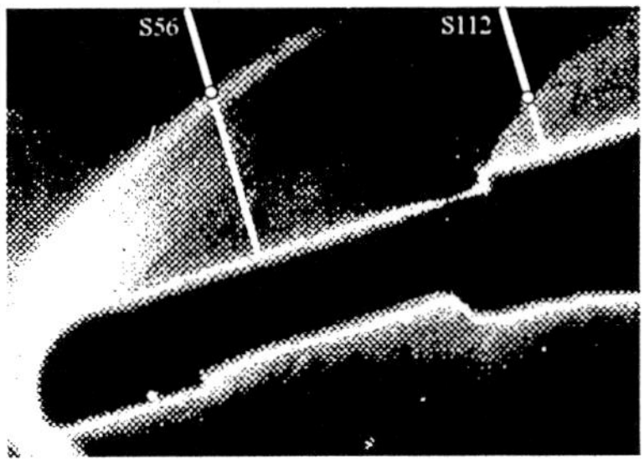

Figure 3. Shadow image of the flow over the model. White lines: sections analyzed by LIF; circle: shock position obtained by LIF.

distribution in the test section with a constant value of 3325 m/s, corresponding to a supersonic flow of Mach 4.9. These results are in good agreement with the calculated parameters: $v = 3300$ m/s and $M = 4.75$, computed with the stagnation-chamber conditions, the nozzle shape, and assumption of perfect-gas flow in equilibrium upstream of the critical convergent throat followed by a frozen expansion downstream in the nozzle divergent region.

The aerodynamic behavior of the supersonic high-enthalpy air flow impinging on the TPS SiO_2 model is shown in Fig. 3. One observes mainly the bow shock which originates from the rounded leading edge and propagates laterally away from the surface, and another less intense shock which rises from the top of the step. Infra-red thermography reveals an important temperature gradient in this region, up to 1500 K at the top of the step and down to 1100 K in the dark region at the foot of the step, where a recirculation zone can be expected. Three sections located at 56, 112, and 156 mm from the leading edge are investigated. The section S_{112} is 12 mm downstream from the step edge.

Figure 4 shows the fluorescence-signal profile from the first section. Measurements are carried out with 500-μm resolution from the free jet ($y > 0$) down to the surface ($y \leq 0$). At the separation region between the free jet and shock layer ($35 \leq y \leq 38$ mm) the fluorescence signal increases exponentially across the shock front because of NO vibrational relaxation, i.e., the higher number density of the NOϵ ($X^2\Pi_{(v''=1)}$)-level and higher vibrational temperature (McMillin et al. 1992). Downstream in the shock layer, the fluorescence intensity evolves slowly down to the model. The number density and rotational temperature do not vary significantly, indicating no significant dissociation of

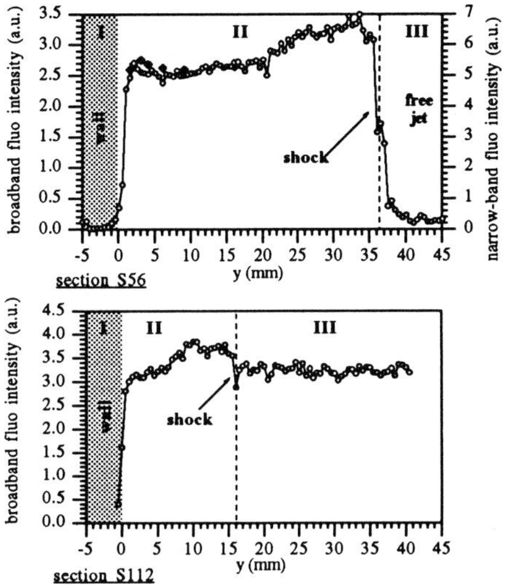

Figure 4. NO-ϵ LIF-signal evolutions. Squares: narrow-band laser excitation; circles: broad-band laser excitation.

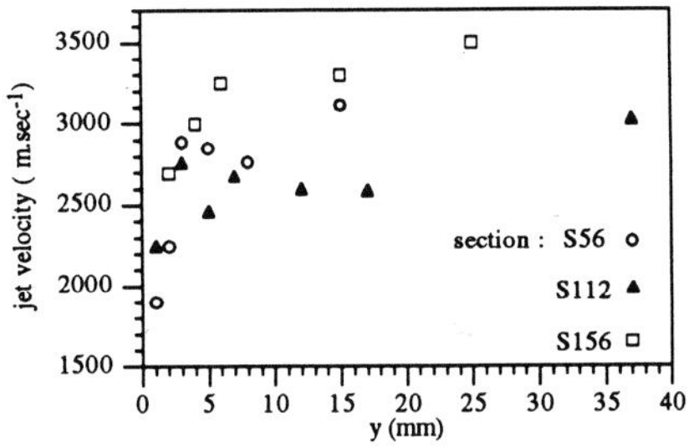

Figure 5. Doppler-shift measurements of the jet velocity over the model.

NO in the shock layer. Near the SiO_2 surface ($0 \leq y \leq 5$ mm), the weak increase of signal does not allow one to locate with accuracy the limit between the boundary and shock layers. In this weakly dissociated jet, the current results do not confirm a predominant effect of wall recombination in comparison to the compression effect of the shock layer near the surface. The equilibrium conditions in the free jet upstream are 1200 K and 10^2 Pa, and the stagnation pressure behind a normal shock is evaluated to $1.9 \cdot 10^3$ Pa, namely, 20-fold greater than the free-jet static pressure. The same order of magnitude is observed on the fluorescence signal jump across the shock propagating laterally away from the model. If one considers first the fluorescence intensity with respect to the distribution of NO $X^2\Pi_{(v''=1)}$ in the absorbent rotational levels and, secondly, the very slight variation of the rotational temperature across the shock, then the NO concentration jump $\simeq$ 22-fold greater in the shock layer than in the free jet, is assumed to be directly proportional to the difference of pressure across the shock. The spatial resolution of 5 points/mm accurately describes the location and intensity of the shock. However, the real thickness of the shock is overestimated because the vibrational relaxation time is lower than the rotational one. The LIF results in this first section are in good agreement with the ombroscopy picture which reveals only the most intense vibrational relaxation zone, without giving information on the intensity on both sides of the shock. The same results are obtained for the location of the second shock ($y \simeq$15 mm) originating from the step (Fig. 4). Its lower intensity indicates no real change in vibrational relaxation. This is confirmed with the insignificant evolution of the rotational temperature and NO number density.

The axial component of the velocity is affected by the presence of the model. The profiles drawn from the three test sections of interest show a velocity gradient mainly located in the last millimeters close to the surface, in the region where a slight decrease of the fluorescence has been previously observed but not explained (Fig. 5). No discrepancy is observed for the axial component of the velocity across the shock front. Its value is about 3300±200 m/s, which corresponds to the initial free-jet velocity. Upstream of the step, the thickness of the dynamic boundary layer is given at $y = 5$ mm in the first section. Under this limit, the velocity decreases continuously down to $y = 1$ mm to reach its lower value of about 1900 m/s. The thickness of the dynamic boundary layer in front of the step can be expected to be of the same order of magnitude as the step height. One of the main features of the velocity distribution in this second section, 12 mm downstream of the step, lies in the continuous increase in velocity up to the limit of the shock generated at the top of the step. The lower values measured near the wall may be due to a deflection of the flow impinging on the step and a normal component of the velocity, non-negligible in comparison to the lower axial one. The most important velocity gradient is located 5 mm above the surface. For $y > 5$ mm the axial value is approximately constant ($\simeq$ 2500 m/s) up to the limit of the secondary shock, then increases far off to reach the maximum value of the free jet.

6. Conclusion

We have demonstrated the potential of the LIF technique for instrumentation of continuous high-enthalpy facilities. NO number density, temperature, and velocity have been measured in the supersonic air-plasma free jet as well as over the surface of a TPS model, allowing characterization of the aerodynamic behavior of the flow, especially in the vicinity of a misalignment of tiles.

Acknowledgement. This work was sponsored by E.S.A., C.N.E.S. under Dassault Aviation contract. The authors wish to thank I. N. Mourzinov, V. B. Knotko, and V. A. Fadeyev for fruitful discussions and their collaborators for their assistance in operating the test campaign at the Tsniimach Institute.

References

Anfimov N (1992) TSNIIMACH Capabilities for Aerogasdynamical and Thermal Testing of Hypersonic Vehicles. AIAA Paper 92-3962.

McMillin BK, Lee MP, Hanson RK (1992) Planar laser-induced fluorescence imaging of shock-tube flows with vibrational nonequilibrium. AIAA J 30:436.

Use of an expansion tube for imaging studies of hypersonic reactive flows around blunt bodies

M. R. Kamel, C. I. Morris, M. C. Thurber,
S. D. Wehe, R. K. Hanson
High Temperature Gasdynamics Laboratory
Department of Mechanical Engineering
Stanford University, Stanford, CA 94305 USA

Abstract: Initial experiments applying an expansion tube to the study of hypersonic reactive flow around blunt bodies are reported. Two fuel mixtures, 16.7%CH_4/33.3%O_2/50%N_2 and 5%C_2H_4/15%O_2/80%N_2, have been successfully accelerated to velocities of 2230 m/s using an expansion tube. The combustion mode observed was that of an adiabatic shock followed by a combustion front, and is consistent with the observations in previous work.

Key words: Expansion tube, Hypersonic flow, Flow imaging, PLIF

1. Introduction

This paper reports on the use of an expansion tube to study hypersonic reactive flows around blunt bodies. Numerous experimental and numerical investigations of such flows have been performed over the past 30 years. Most experimental work has involved firing spheres and blunted cylinders at hypersonic speeds into combustible mixtures, and using schlieren photography to observe the different combustion modes occurring around and behind the body (Lehr 1972). For the cases where the velocity of the body was higher than the Chapman-Jouguet (CJ) velocity of the fuel mixture, a smooth shock-induced combustion front was observed (Chernyi 1968). As the body velocity was decreased, instabilities originating in the induction zone between the bow shock and the combustion region at the front of the body appeared. For body velocities close to the CJ velocities, the instabilities became more pronounced and in some cases led to the formation of cellular detonations.

To demonstrate the potential of the expansion tube for such investigations, a pilot study of hypersonic reactive flow around a spherically blunted cylinder 19 mm in diameter was undertaken in the new Stanford expansion-tube facility. In these initial experiments, methane- and ethylene-based fuel mixtures were accelerated to velocities around 2300 m/s, and OH planar laser-induced fluorescence (PLIF) was used to image the combustion around the body.

A thorough understanding of the characteristics of the hypersonic flows generated by the expansion tube is a prerequisite to such an investigation. Ongoing characterization efforts described by Morris et al. (1995) allowed flow quantities to be determined using the combined information of wall pressure, pitot pressure, IR absorption, and NO PLIF measurements. This foundation in turn allowed selection of the flow condition used in the present study. Following a brief description of the experimental facility, data characterizing this test condition is discussed, with particular attention to flow repeatability and uniformity. Selection of the reactive mixtures is then outlined, with a view towards avoiding premature ignition. Finally, the results of imaging experiments for inert and reactive flows over the blunted cylinder at the chosen test condition are presented and compared to previous studies.

2. Experimental facility

The Stanford expansion-tube facility (Kamel et al. 1995, Morris et al. 1995) consists of four sections: a 180-cm driver with 10-cm inner diameter, followed by a 90-cm buffer, a 275-cm driven section, and a 430-cm expansion section, all with inner diameters of 9 cm. Mounted at 30.5-cm intervals along the downstream end of the driven section and at ports 61.0 cm apart on the expansion section are PCB pressure transducers for shock-speed and wall-pressure measurements. Also located at the second port from the end of the expansion section is an infrared absorption setup used to mark test-gas arrival in characterization runs — a bandpass-filtered InSb IR detector indicates absorption of light from a globar by CO_2 seeded into the test gas.

The expansion section ends in a square viewing chamber in which the test body is mounted, with optical access provided by two opposed pairs of 13 × 13-cm quartz windows. For characterization runs, the test body can be replaced by a pitot rake equipped with four PCB 113A26 pressure transducers mounted at 19.1-mm intervals across the diameter of the tube.

The quartz ports in the viewing section allow horizontal passage of a laser sheet through the test flow and imaging of fluorescence from above. The sheet is formed from the frequency-doubled output of a Lambda Physik FL3002 dye laser pumped by a Lambda Physik EMG 203 excimer. For PLIF work with nitric oxide (NO) transitions near 226 nm, Coumarin 450 dye was used to achieve laser energies of about 0.3 mJ per pulse. OH PLIF used the somewhat higher pulse energies — roughly 0.7 mJ for this work — achievable near 283 nm with Coumarin 540A dye. In both cases, fluorescence from the flow was imaged onto the 578 × 384 pixel array of an appropriately filtered Princeton Instruments intensified CCD camera.

Table 1. Freestream characteristics of selected test condition.

Measured Quantities		Calculated Quantities	
Contact Surface Velocity	2230 m/s	Mach Number	7.1
Static Pressure	0.93 psia	Static Temperature	280 K
Pitot Pressure	58 psia		
Steady Flow Test Time	125 μs		

3. Properties of chosen test condition

Previous characterization efforts assisted in the identification of the test condition described in Table 1. On the basis of further control experiments using gas mixtures of pure N_2 and 10% CH_4 + 90% N_2, both deemed sufficiently close gas-dynamically to the reactive mixtures used later, flow core uniformity and flow repeatability during the test time were determined to be good. Figure 1 shows a cross-section of pitot pressure 2 cm downstream of the tube exit. The figure indicates a uniform flow core 3.8 cm in diameter. To investigate the repeatability of the flow, the control mixture of CH_4/N_2 seeded with 0.15% NO was employed, and NO PLIF images of the bow shock were taken for four successive shots, 75 μs into the test time in each case. The shock shapes and standoff distances showed excellent overlap for all shots. Also, pitot traces for different shots agreed to within 5%, and shock speeds in the driven and expansion sections were repeatable to within 1%.

The observed shock standoff distance and shock shapes agree well with correlations from the literature for $M = 7$ flow. Figure 1b compares the shock shapes observed for the two control mixtures to an experimental correlation given by Billig (1967) for $M = 7.1$.

4. Mixture selection

The primary criterion for the reactive mixture was that it not ignite anywhere inside the tube. The most severe limitation comes from the fact that the test gas is subjected to post-reflected-shock conditions for a few microseconds at the driven-expansion section interface due to the finite diaphragm-opening time. Presently, Mylar diaphragms 25 μm thick are used at that interface, but other diaphragms and diaphragm-opening schemes are under investigation.

Premature ignition of the test gas is characterized by one or all of the following effects. A strong reflected shock is recorded by the pressure transducer in the driven section at a location 15.2 cm upstream of the driven-expansion diaphragm. The shock speeds in the expansion section and the pressure jump upon shock passage are considerably higher than in the corresponding inert case.

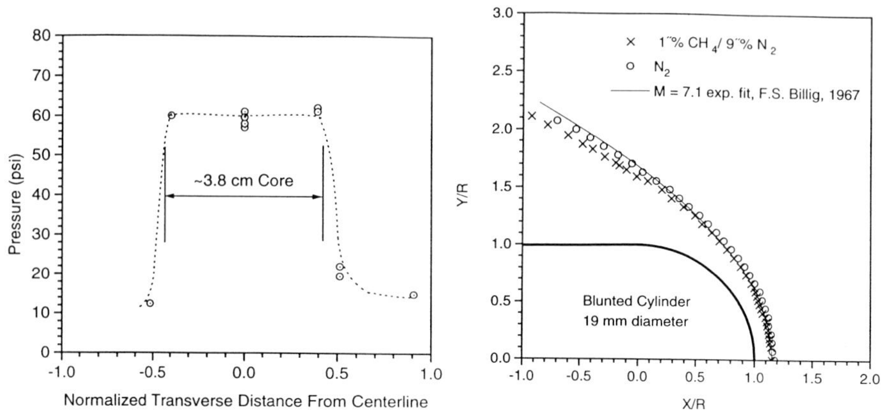

(a) (b)

Figure 1. For control experiments performed with the test conditions of Table 1: (a) the profile of pitot pressure across tube diameter 2 cm downstream of exit plane; (b) comparison of the observed bow-shock shape with correlations obtained from the literature for $M = 7.1$ flow for gases at low temperatures and with $\gamma = 1.4$.

In addition, emission from hot CO_2 combustion products is detected by the IR detector as the test gas flows by its viewing port. Finally, the PLIF images in cases of premature ignition show OH fluorescence signals in the freestream, upstream of the bow shock.

Of the several mixtures tested, the following two exhibited no signs of premature ignition in the tube at the chosen test condition, and were therefore used for these experiments:

$$16.7\%CH_4 + 33.3\%O_2 + 50.0\%N_2 \tag{1}$$

$$5\%C_2H_4 + 15\%O_2 + 80\%N_2 \tag{2}$$

5. Results and discussion of reactive experiments

Previous work has shown combustion structure to depend on the ratio of test-gas velocity to CJ velocity. In these experiments, the CJ velocity is 1960 m/s for the methane-based mixture and 1685 m/s for the ethylene mixture. Given the nominal freestream test-gas velocity of 2230 m/s, the ratios of test-gas velocity to CJ velocity for the mixtures are 1.13 and 1.32, respectively. Hence, based on previous work, the combustion mode expected is that of an adiabatic shock followed after an induction length by a combustion front. Figures 2a and 2b are OH PLIF images of the flow of the methane- and ethylene-based

(a) (b)

Figure 2. OH PLIF images for (a) methane-based and (b) ethylene-based mixtures taken 75 μs into the steady test time at the test conditions specified in Table 1.

mixtures, respectively, both taken 75 μs into the steady test time. Both cases show evidence of shock-induced combustion close to the cylinder. The images differ considerably in the thickness of the combustion region and the standoff distance of the combustion front. That difference is consistent with the fact that the ethylene mixture has a shorter induction time than the methane one.

Another feature of the images, more evident in the methane case, is that the boundaries of the OH PLIF signals around the cylinder become almost parallel to the body. This matches previous observations, and is explained by the fact that the bow shock weakens rapidly in the radial direction away from the body's nose, leading to an increase in induction time of the shocked gas. Eventually the induction time approaches infinity, and the reaction front becomes coincident with the flow-field streamlines around the projectile (Alpert and Toong 1972, Lee 1994).

6. Conclusions

The bow-shock shape and standoff distance observed in the control experiments with NO PLIF agreed well with experimental fits from the literature. For the reactive-flow experiments, the features of combustion observed were in accord with the qualitative analysis and observations of previous researchers at corresponding conditions of V_{flow}/V_{CJ}. Thus we conclude that one of the three modes of combustion known to occur when a blunt body is fired into a reactive mixture can be reproduced using an expansion tube. Future experiments will be directed at reproducing the other two combustion modes, namely, the case of an adiabatic shock followed by an oscillating flame front, and the case of a cellular detonation. This will be done by changing the CJ velocity of the mixture and the dimensions of the body. Such experiments will allow future work in the expansion tube aimed at applying PLIF to simulated ram-accelerator flows to be performed with confidence.

Acknowledgement. The authors gratefully acknowledge the contributions of A.F.P. Houwing, M. Di Rosa, F. Levy, and I. Stouklov to this investigation. The work was supported by the U. S. Army Research Office, with Dr. Dave Mann as technical monitor, and the U. S. Air Force Office of Scientific Research, with Dr. Julian Tishkoff as technical monitor.

References

Alpert RL, Toong TY (1972) Periodicity in exothermic hypersonic flows about blunt projectiles. Astro Acta 17:539–560.

Billig FS (1967) Shock-wave shapes around spherical and cylindrical nosed bodies. J Spacecraft and Rockets 4(6):822–823.

Chernyi GG (1968) Supersonic Flow Past Bodies with Formation of Detonation and Combustion Fronts, Problems of Hydrodynamic and Continuum Mechanics, Society for Ind and App Math, Philadelphia, pp 145–169.

Kamel MR, Morris CI, Thurber MC, Wehe SD, Hanson RK (1995) New Expansion Tube Facility for the Investigation of Hypersonic Reactive Flow, AIAA Paper No. 95-0233.

Lee JHS (1994) On the initiation of detonation by a hypervelocity projectile. Presented at the Zeldovich Memorial on Combustion, Voronovo, Russia, Sept 12–17.

Lehr HF (1972) Experiments in shock-induced combustion. Astro Acta 17:589–597.

Morris CI, Kamel MR, Thurber MC, Wehe SD, Hanson RK (1995) Development of an expansion tube for investigation of combustion in supersonic projectile flowfields. Presented at the 2nd International Workshop on Ram Accelerators, Univ of Washington, Seattle, USA, July 17–20.

Spatially and spectrally resolved laser-induced fluorescence measurements in the HEG

M. Wollenhaupt, M. Rosenhauer, T. Müller, W. H. Beck, M. Scheer*
Institute for Fluid Mechanics, German Aerospace Research Establishment
DLR, Bunsenstrasse 10, 37073 Göttingen, Germany

Abstract: NO excitation and fluorescence spectra, excited by an ArF excimer laser at 193 nm and measured in a hot cell (1200 K) in order to find suitable transitions for use in the high-enthalpy shock tunnel HEG, are discussed. NO laser-induced fluorescence measurements at 193 nm were performed in HEG, with the aim of carrying out 2D/2-line thermometry in the free stream and in gas flows around models. Results in HEG at enthalpies up to 21 MJ/kg are given for the space gliders HERMES and HOPE. The difficulties with contaminant emission at these high enthalpies, and means of overcoming them, are discussed.

Key words: NO spectroscopy, Shock tunnel, Laser-induced fluorescence, LIF, Hypersonic flow

1. Introduction

In order to be able to characterize the free-stream flow and examine the flow around test models in the HEG, laser-induced fluorescence (LIF) measurements of NO at 193 nm were carried out in a hot cell and in the HEG. The unambiguous assignment of spectral lines for the five electronic transitions of NO accessible at 193 nm — $\epsilon(0,1)$, $\gamma(3,0)$, $\gamma(4,1)$, $\gamma(5,2)$, and $\beta(7,0)$ — and a study of the behavior of the spectral fluorescence intensity of the chosen NO transitions as a function of temperature were carried out in a specially constructed hot cell.

The difficult environment found in a shock tunnel at enthalpies up to 21 MJ/kg necessitates spectrally resolved measurements of the laser-excited fluorescence in order to ascertain if only the required NO transition is being excited and that no spurious emission is present; this latter point is still one of the severe limitations and handicaps to spectroscopic diagnostic measurements at high enthalpies, and must be taken into account. How this can be done, with reference to emission measurements carried out in HEG, is described.

---

*Present address: Department of Physics and Astronomy, McMaster University, Hamilton, Ontario, Canada.

Spectrally and 1D spatially resolved LIF images have been recorded in the HEG in the free stream and in the stagnation region of a HERMES model. A quantitative evaluation of these images was partially feasible. Similarly obtained LIF images at the wing leading edge and tip-fin locations of the Japanese HOPE orbital plane could be combined to give 2D temperature field maps (for the present work relative to the free-stream temperature value only).

2. Experimental setup

The experimental setup is described elsewhere (Scheer 1994), and only brief details will be given here. The LIF apparatus consists of two laser sheets counter-propagating through the HEG test section in front of the test model, and two separate image-intensified gated camera systems to capture the fluorescence, either 2D spatially or (by including a monochromator) spectrally/1D spatially resolved. In the latter case, the monochromator slit is placed in the direction of the required spatial resolution. Several checks have to be included in the system to correct for laser spectral and intensity drifts or scatter. The test time in the HEG of ≤ 1 ms necessitates the use of single laser shot measurements. As opposed to single laser measurements, the two-laser/two-camera system, together with an in situ calibration with the heated NO calibration cell carried out in the test section, allows one to combine images so that several apparatus and flow effects then cancel out; this feature showed itself to be very desirable.

3. Results and discussion

3.1. NO spectroscopy

Figure 1 shows NO excitation and fluorescence spectra taken in the heated cell at $p_{NO} = 10$ mbar and $T = 1200$ K. In the 1-nm laser tuning range, over 400 lines could be identified and assigned to their appropriate transitions. Candidate lines suitable for use in the HEG are identified; here the major criteria of single-line excitation and suitably large energy separation of the lower excitation levels ($\Delta E > kT$) are (largely) met. All HEG work was carried out with lines 2 and 8 ($R_{21}(17.5)$ and $R_{22}(27.5)$, respectively) belonging to the NO $\epsilon(0,1)$ band. Tests carried out in the heated cell to determine the behavior of these two lines at different temperatures and pressures revealed that the intensities of the R_{21} satellite branch were stronger compared to the main branches than expected from calculations using published values of Hönl-London factors (Bennett 1970). This is shown in Fig. 2, where it can be seen that the measured satellite branch line intensities relative to their R_{22} neighbors are at least two times stronger than calculated. There is evidence to show that excited electronic-state interactions in NO lead to these unexpected intensity distributions (Amiot and Verges 1982). This topic cannot be further discussed here, but its implications for quantitative LIF measurements must be considered.

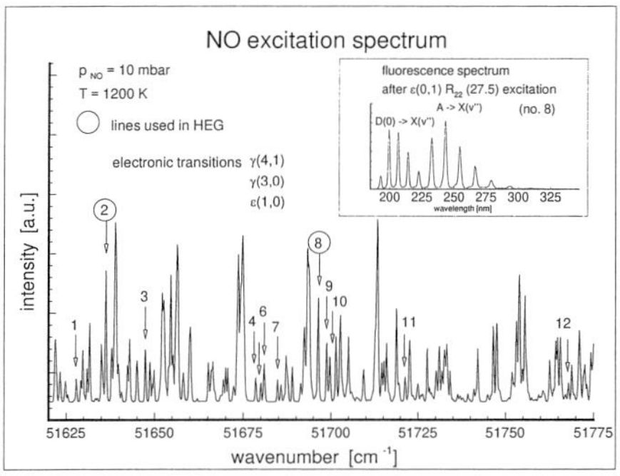

Figure 1. NO excitation spectrum, showing candidate lines (1–12). Fluorescence spectrum for line 8 ($R_{22}(27.5)$) is inset.

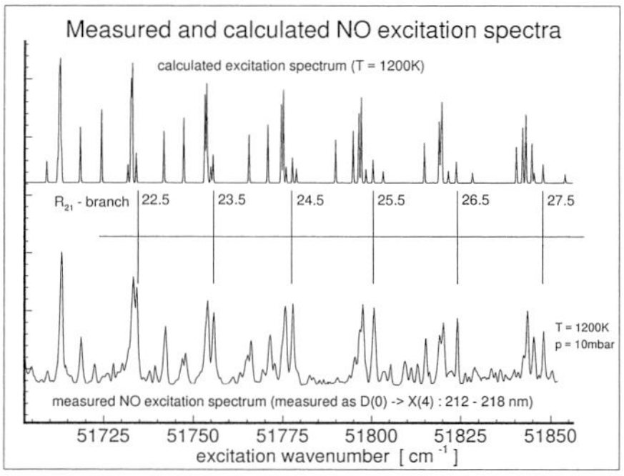

Figure 2. Comparison between measured and theoretical spectra, showing disproportionately large R_{21} branch.

3.2. Spectrally resolved LIF in the HEG

Figure 3 shows one of two images of a spectrally resolved LIF measurement carried out in the stagnation region of a HERMES model placed at an angle of attack of 30° in the HEG at condition I ($p_0 = 41$ MPa and $h_0 = 21$ MJ/kg). A good signal-to-noise ratio for the NO fluorescence in the free stream was obtained, as can be seen in the inset spectrum plot. The luminosity from

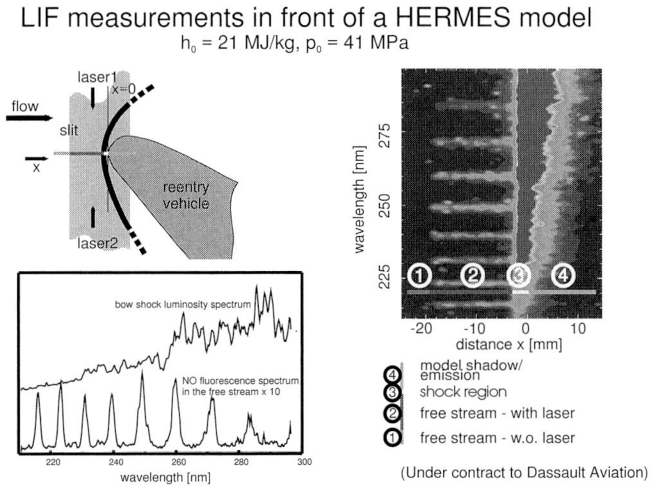

Figure 3. LIF measurement in stagnation region of HERMES. Inset spectrum plot shows characteristic NO lines in the free stream and stray emission behind the shock.

the bow-shock region is considerably stronger than the fluorescence in the free stream, and above about 220 nm completely swamps any fluorescence lines that may be present. However, the two fluorescence lines seen at about 215 and 222 nm in the free stream do not appear behind the shock, even though the luminosity here is not very large. The reasons for this are not yet understood (quenching, absorption, line broadening, etc.).

Evaluation of pairs of such images obtained in the free stream, including corrections for laser shot-to-shot variations, led to values for rotational temperatures which showed scatter that could not be due solely to flow variations. This may be due to the above-mentioned difficulty in carrying out quantitative LIF measurements with the NO lines at 193 nm.

It is clear from the HERMES result that to be able to carry out 2D-resolved LIF measurements under these tunnel conditions, either the spurious emission has to be filtered out, or its nature and source understood and removed. To this end: (i) a special filter was constructed to pass light in the very narrow but tunable region around 200–220 nm where D-X transitions occur; and (ii) emission spectra were recorded in the HEG behind a cylinder bow shock.

3.3. Spurious emission in the HEG

In a series of shots at condition I with test gas N_2, emission spectra were recorded in the stagnation region of a cylinder placed with axis perpendicular to the flow. By suitable normalization, these spectra were combined to give a spectrum over the range 180–850 nm, as shown in Fig. 4. It was possible to identify a large number of the emission lines as originating from Fe, and a simulation of Fe emission spectra, using literature values for line positions and transition probabilities (Fuhr et al. 1981), showed good agreement over major portions of the wavelength range. Nevertheless, several lines were present in the HEG spectrum which were not due to Fe.

Also shown in Fig. 4 are the locations of the NO fluorescence lines (highly magnified). This exemplifies the result shown before — except in a very narrow window around 200–220 nm, the emission completely swamps the LIF signals. Here one can also see why laser excitation at 193 nm has some advantages over the more popular and simpler NO excitation at 226 nm ($\gamma(0,0)$ bands); the latter leads to fluorescence signals in a wavelength range that is overlapped with the spurious emission.

3.4. Spatially resolved LIF in the HEG

Even though the strong emission in the vicinity of the stagnation region makes LIF measurements very difficult, regions further downstream of the model bow (wings, flaps, wakes) are colder and therefore suffer less from this effect. An

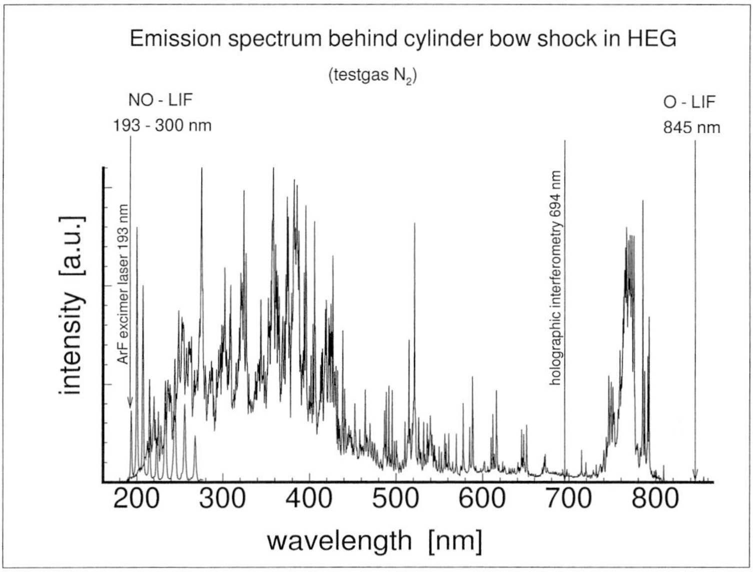

Figure 4. Strong spurious emission behind the HEG bow shock, due largely to Fe lines. A typical NO fluorescence spectrum has been overlaid (not to scale).

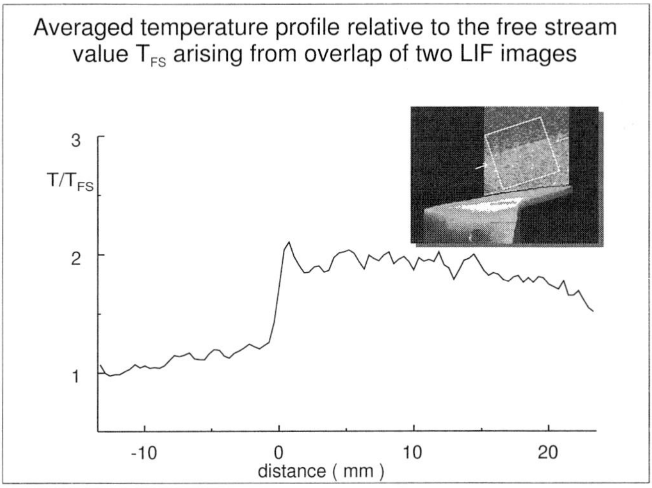

Figure 5. Measured temperature field and plot (relative to calculated free-stream value) at HOPE model wing-tip fin. Model surface is located at $\sim 25 - 29$ mm.

example of this is tests carried out in the HEG on the Japanese space plane HOPE; models with different shapes were placed at an angle of attack of $\alpha = 35°$ and yaw angle $\beta = 0$ and $\pm 3°$ in the HEG such that the two laser beams could probe the regions of interest (wing leading edge and wing-tip fin) on opposite sides of the model. The filters referred to before were used here. The measured LIF images were corrected to provide a visualization of the flow phenomena, as shown in Fig. 5. The LIF image shows clearly the location of the shock relative to the wing. A LIF signal in the free stream can also be seen. By combining the two measured images, it was possible to obtain temperatures (relative to the value in the free stream), as also shown in the figure. The averaged temperature plot shows that temperature varies very little between shock and wing surface.

4. Conclusions

LIF images were obtained in flow studies using two different models in the HEG. As long as one uses laser excitation at 193 nm, the problem of obtaining reliable absolute temperatures (via prior calibrations) still needs to be solved; here further tests are necessary. Steps are underway to suppress the contamination sources in HEG; until then, this still remains a major difficulty. The narrow bandpass filters were essential to carry out 2D LIF work further away from the stagnation region.

Acknowledgement. Our thanks go to J. Lenz and D. Garbe for their support. The financial support of ESA through CNES is gratefully acknowledged. The HERMES and HOPE measurements were carried out under contract to Dassault Aviation (Velizy, France) and Kawasaki Heavy Industries (Gifu, Japan), respectively.

References

Amiot C, Verges J (1982) Fourier transform spectroscopy of the $D^2\Sigma^+ - A^2\Sigma^+$, $E^2\Sigma^+ - D^2\Sigma^+$ and $E^2\Sigma^+ - A^2\Sigma^+$ systems of nitric oxide. Physica Scripta 26:422.

Bennett RJM (1970) Hönl-London factors for doublet transitions in diatomic molecules. Mon Nat R Astr Soc 147:35.

Fuhr JR, Martin GA, Wiese WL, Younger SM (1981) Atomic transition probabilities for iron, cobalt and nickel (a critical data compilation of allowed lines). J Phys Chem Ref Data 10:305.

Scheer M (1994) Temperature and Density Measurements on Hypersonic Wind Tunnels Using Laser Induced Fluorescence of NO, Including a Study of NO Spectroscopy in the Range 193-4 nm, Using a Novel Heated Cell, Diplom thesis, Institute for Fluid Mechanics, DLR, Göttingen.

Bow-shock/disturbance interaction for hypersonic flow over a cylinder

X. Zhong
Mechanical, Aerospace and Nuclear Engineering Department
University of California, Los Angeles
Los Angeles, CA 90095 USA

X. Joubert
UFR de Mécanique
Université Joseph Fourrier, Grenoble, France

Abstract: This paper presents direct numerical simulations of free-stream acoustic and entropy disturbances interacting with the bow shock in front of a cylinder moving at hypersonic speeds. The interaction is an important part of the phenomena of hypersonic boundary-layer receptivity to free-stream disturbances. The back-and-forth reflection and interaction of disturbance waves between the bow shock and body change the disturbance flow field considerably. They create strong entropy and vorticity disturbances behind the shock. It has been found that the smaller the value of the free-stream-disturbance frequency, the closer the entropy and vorticity disturbance spots to the body, which can be important for the boundary-layer receptivity analysis.

Key words: Hypersonic receptivity, Shock-disturbance interaction

1. Introduction

The receptivity of hypersonic boundary layers to free-stream disturbances refers to the processes by which the environmental disturbances initially enter the boundary layers and generate linear instability. Though it is an important aspect of the transition process from laminar to turbulent states, the receptivity phenomena for supersonic and hypersonic boundary layers has not been seriously considered (Reshotko 1990). One difficulty is that the interaction of disturbance waves with the bow shock and other shock waves needs to be considered. Another difficulty is that the real-gas effect becomes important in hypersonic flow.

The presence of the strong bow shock wave in hypersonic flow past a blunt leading edge alters the receptivity phenomena considerably. Figure 1 shows a schematic of a bow-shock/disturbance interaction flow field. When the bow shock is perturbed from its steady configuration, the wave fields behind the bow shock contain three types of disturbance waves: (1) acoustic waves which propagate with the speed of sound relative to the moving fluid and $\delta p \neq 0$, $\delta \rho = \frac{\delta p}{c^2}$,

$\delta u = \delta p/\rho c$, $\delta s = 0$; (2) entropy waves which convect with the moving fluid and $\delta s \neq 0$, $\delta\rho = (\frac{\partial\rho}{\partial S})_p \delta s$, $\delta u = \delta p = 0$; (3) vorticity waves which also convect with fluid velocity and $\delta\omega \neq 0$, $\delta u \neq 0$, $\delta\rho = \delta p = \delta s = 0$. If the disturbances are small, linear analysis can be used to determine the disturbances in flow variables behind the shock and disturbances in shock configurations (Mckenzie and Westphal 1968).

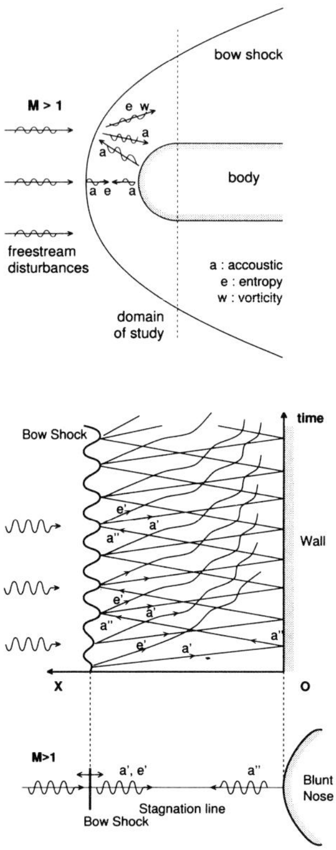

Figure 1. Schematic of bow-shock/disturbance interaction flow field and patterns in the x-t plane.

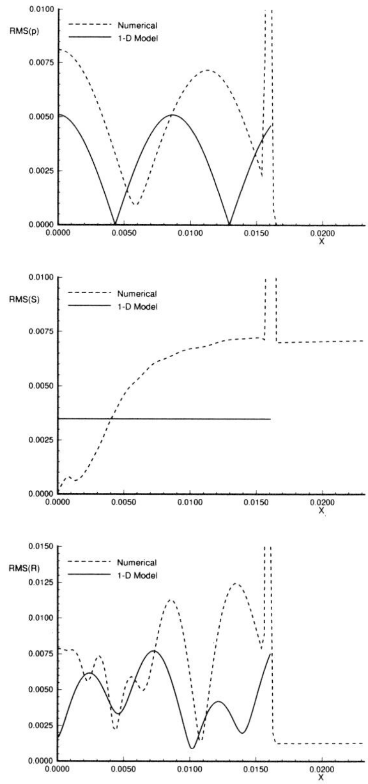

Figure 2. Root-mean-square values of p, s, and ρ perturbations along the stagnation line for Case 2.

The three disturbance waves originating from the bow-shock interaction propagate downstream independently, connected only through the boundary conditions at the shock wave and at the wall. The acoustic waves, which travel along Mach lines, reach the boundary layer and reflect back to the shock upstream. In Morkovin (1960), this reflection and interaction process was assessed using a simplified one-dimensional linear analytical model. A general qualitative knowledge of the process could be obtained. However, this approximate analysis can only be applied to the stagnation line with assumption of uniform flow behind the shock, but the actual flow is nonuniform. It is not clear how the vorticity, a 2D phenomena, is affected by the wave interaction and reflection.

In this paper, we continue our research project (Chiu and Zhong 1995) of studying hypersonic receptivity and stability phenomena involving strong shock waves and real-gas effects. Specifically, we conduct direct numerical simulations of free-stream acoustic and entropy disturbance waves interacting with the bow shock in front of a cylinder moving at hypersonic speeds. The objective is to study the detailed structure of flow disturbances outside the boundary layer which may have strong effects on the stability and transition of the boundary layer.

2. Numerical methods

Time-accurate computations of transient hypersonic flow require high-order spatially and temporally accurate methods with shock-capturing capability. In the present studies, both high-order ENO and TVD schemes solving the Euler equations are used for the numerical simulations. High-order Runge-Kutta schemes are used in temporal discretization. Details of the numerical methods can be found in Chiu amd Zhong (1995).

3. Results and discussions

A Mach 8.03 flow past a circular cylinder is considered as a model of hypersonic flow past a blunt nose. A set of 200 $\times$ 105 body-fitted grids is used. The fluid is assumed to be a perfect gas with $\gamma = 1.4$.

3.1. Case 1: acoustic-wave/bow-shock interaction

We continue our previous studies of free-stream acoustic waves interacting with the bow shock and the blunt body. First, a steady mean flow is obtained numerically. Then, a free-stream acoustic wave is introduced to this numerical mean flow along the free-stream boundary of the computational domain. The amplitude of the acoustic wave is set to be 1% of free-stream unperturbed properties. The following dimensionless wave numbers have been used in the simulations: $\overline{k} = \frac{2\pi}{kd} = 6.260, 3.130, 2.608, 1.956, 1.565$, where k is the wave number in the free stream and d is the standoff distance.

Figure 4 shows the time history of instantaneous pressure, entropy, and density along the stagnation line for the case of free-stream disturbance wave number $\overline{k} = 3.13$. The pressure and entropy contours clearly show the trajectories of independent acoustic and entropy waves behind the bow shock.

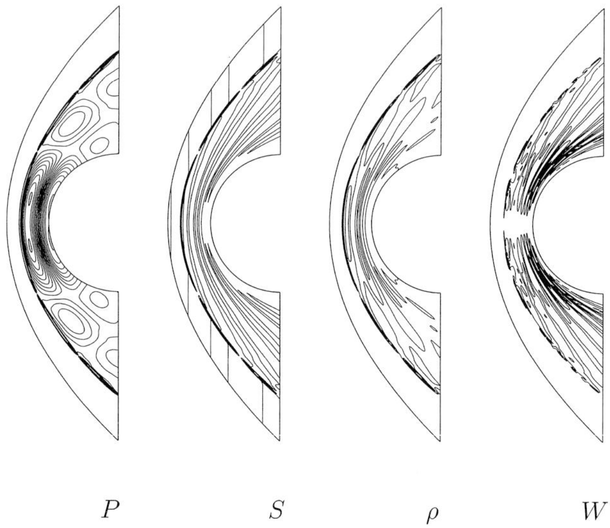

Figure 3. Instantaneous pressure, entropy, density, and vorticity contours, Case 2.

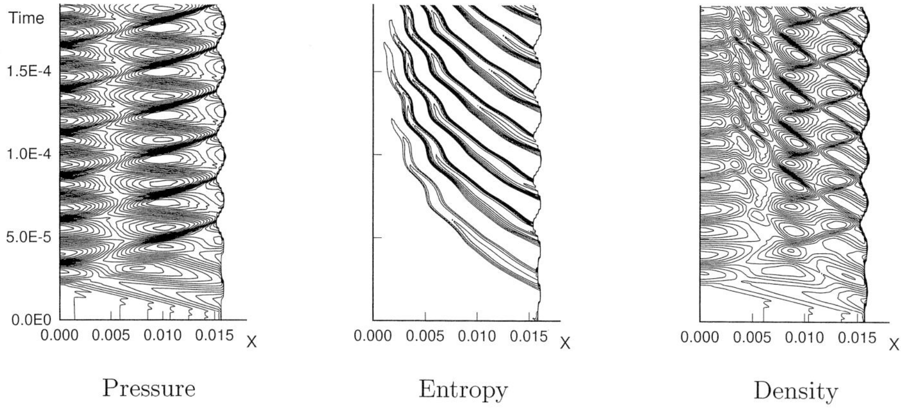

Figure 4. Flow variables along stagnation line as functions of time for Case 1.

The acoustic wave is reflected between the shock and solid body, which results in larger pressure oscillations and bow-shock movements. The oscillation of the bow shock in turn increases the intensity of the entropy and vorticity waves. The figure also shows that the entropy wave is convected independently with the fluid velocity, which decreases to zero at the stagnation point. The nonuniform velocity field behind the bow shock leads to curved entropy-wave lines, which do not reach the wall and oscillate near the wall as velocity does. As a result, the disturbance entropy is very small in a thin layer on the wall. The schematic of the patterns of the acoustic and entropy waves is shown in Fig. 1. On the other hand, the wave pattern of density is more complex than those of pressure and entropy because density disturbances are induced by both acoustic and entropy disturbance waves. The influence of both acoustic and entropy waves can be seen in the density x-t plot. Other results of case 1 can be found in Chiu and Zhong (1995).

3.2. Case 2: entropy-wave/bow-shock interaction

Free-stream entropy waves interacting with the bow shock and the blunt body are computed in this case. The flow conditions are the same as those in Case 1, except 1% entropy disturbance waves are imposed to the free-stream as follows:

$$\delta q = \Delta q e^{ik\{x-u_\infty t\}} .$$

Figure 3 shows the instantaneous contours of flow variables. As explained in the previous case, the density contours have complicated patterns because of the combined effects of the entropy and acoustic waves. The disturbance vorticity intensity has a large magnitude near, but not at, the cylinder surface. It is found that the smaller the free-stream disturbances, the closer are the vorticity spots to the wall. Meanwhile, the induced oscillation of the bow shock also creates strong vorticity disturbances which convect near the body surface along stream lines.

Figure 2 shows the root-mean-square (RMS) values of pressure, entropy, and density disturbances along the stagnation line. The RMS values represent the wave strength in the flow field. In the figure, the results of the one-dimensional model of Morkovin (1960) are compared with the numerical solutions. The RMS pressure of the one-dimensional model agrees reasonably well with the numerical solutions. However, the RMS entropy and density results of the model do not agree with numerical solutions, because the model assumes uniform flow behind the shock. The numerical solutions in this figure show that the reflection of the acoustic wave produces the strongest pressure disturbance at the stagnation point, but the maximum density disturbance is not always located at the wall.

Similar to Case 1, the interaction and reflection of the disturbance waves increase the disturbance pressure and shock movement considerably. However, free-stream entropy-wave/bow-shock interaction produces about one order of magnitude weaker disturbances behind the bow shock than the acoustic-wave/bow-shock interaction of Case 1.

4. Conclusions

The interaction of free-stream acoustic and entropy waves with bow shocks has been studied using numerical simulations as a tool. The results and analyses show that the pressure and density disturbances are amplified considerably by the back-and-forth reflection of disturbance waves between the bow shock and the wall. The interaction and reflection also increase the magnitude of oscillation of the bow-shock configuration. The maximum pressure disturbances are always located at the stagnation point, but not the density disturbances because density waves are the combined effects of the entropy and acoustic waves. The entropy and vorticity disturbances have large magnitudes near, but not at, the cylinder surface. These entropy and vorticity disturbance spots get closer to the wall when the frequency of the free-stream disturbances is smaller.

Work is underway to study bow-shock/free-stream disturbance interaction with real-gas effects. Detailed results of the current research can be found though the world wide web at 'http://cfdlab5.seas.ucla.edu/'.

Acknowledgement. This research was supported by AFOSR grant F49620-94-1-0019 with Dr. L. Sakell as the grant monitor. We are grateful to Theodore Lee for reproducing the linear shock interaction results for comparison with current numerical solutions.

References

Chiu CE, Zhong X (1995) Simulation of Transient Hypersonic Flow Using the ENO Schemes, AIAA Paper 95-0469 (to appeared in AIAA J).

Mckenzie JF, Westphal KO (1968) Interaction of linear waves with oblique shock waves. Physics of Fluids 11(11):2350–2362.

Morkovin MV (1960) Note on the assessment of flow disturbances at a blunt body traveling at supersonic speeds owing to flow disturbance in free stream. J of Applied Mechanics 27:223–229.

Reshotko E (1991) Hypersonic stability and transition. In: Hypersonic Flows for Reentry Problems, Springer-Verlag, Vol 1, pp 18–34.

Monte Carlo study of threshold line models for dissociation in hypersonic shock waves

I. D. Boyd
Cornell University
Ithaca, NY 14853 USA

Abstract: A new dissociation model for the Direct Simulation Monte Carlo (DSMC) method is formulated using the threshold line concept. The model is evaluated in a heat bath under equilibrium and nonequilibrium conditions. The model is also employed to calculate one-dimensional hypersonic shock waves of nitrogen and comparison is made with existing experimental data.

Key words: Dissociation modeling, Monte Carlo method, Hypersonic shock waves

1. Introduction

The present study focuses on the well-known mechanism whereby the rate of dissociation is altered under conditions of thermal nonequilibrium behind shock waves. Specifically, the dissociation rate is reduced when the vibrational temperature is lower than the translational value. Several models have been developed to capture this effect, which is important in hypersonic aerospace applications. Of these, two phenomenological approaches are highlighted here. In continuum computational fluid dynamics (CFD), the two-temperature model of Park (1988) is widely used. In this model, an Arrhenius reaction rate coefficient has its temperature dependence governed by: $T_a = {T_t}^s \times {T_v}^{1-s}$ where T_t is the translational temperature, T_v is the vibrational temperature, and s is a modeling parameter that is usually set to 0.5. In the particle-based Direct Simulation Monte Carlo (DSMC) method, the vibrationally favored dissociation (VFD) model examined by Boyd (1992) is considered. In this model, the probability that a particular collision of two particles leads to dissociation is given by:

$$P_d = f(E_c) \times E_v^{\phi}$$

where E_v is the vibrational energy, $f(E_c)$ is a function of the total collision energy, E_c, and ϕ is a modeling parameter. The parameter ϕ is set to 2 for nitrogen.

These dissociation models are unsatisfactory due to their highly phenomenological nature. The purpose of this study is to consider the use of the more detailed threshold line concept for dissociation proposed by Macheret and Rich (1993). Specifically, a form of that model is introduced for use in the DSMC technique.

2. Threshold line model

The threshold line concept is based on the premise that a certain amount of translational energy in a collision is required above the vibrational energy of the molecule to effect the dissociation reaction. In Macheret and Rich (1993), it is proposed that at high vibrational energy levels the energy needed (the threshold) is simply the difference between the activation energy and the molecular vibrational energy (ΔE). By consideration of the collision dynamics, it is proposed that the threshold energy at low vibrational levels is larger than ΔE.

Using these ideas, continuum models expressed in terms of translational, rotational, and vibrational temperatures have been proposed by Macheret et al. (1994) to model nonequilibrium dissociation processes. The first continuum threshold model (Macheret and Rich 1993) was employed by Boyd et al. (1995) to give the first successful predictions of data from a high-altitude, hypersonic flight experiment.

The purpose of this study is to use the threshold concept to develop a nonequilibrium dissociation model that may be employed directly in the DSMC technique. This requires the formulation of a dissociation probability based on the translational collision energy and the rotational and vibrational energies of the dissociating molecule. This has been performed in the work of Macheret et al. (1994). However, the final results of that study are described in terms of continuum temperatures rather than collision energies.

As described in detail by Macheret et al. (1994) the threshold line probability of dissociation for diatom-atom collisions is:

$$P_d \propto \frac{(E_t - F)^{\frac{3}{2}}}{F[(\sqrt{D^*} - \sqrt{\alpha E_v})(\sqrt{D^*} - (2 - \sqrt{\alpha})\sqrt{E_v})]^{\frac{1}{2}}}$$

where $\alpha = 0.25$ for N_2–N_2 and N_2–N. The effective activation energy is

$$D^* = D - E_r + \frac{2E_r^{\frac{3}{2}}}{3\sqrt{6D}}$$

where D is the dissociation energy. For low vibrational levels ($E_v < \alpha D$), the threshold function is

$$F = \frac{(\sqrt{D^*} - \sqrt{\alpha E_v})^2}{1 - \alpha}$$

and for $E_v > \alpha D$, the function is simply

$$F = D - E_v.$$

In the above expressions, E_t is the translational collision energy, and E_r and E_v are the rotational and vibrational energies of the reacting molecule. The above probability model is employed in the present study for both N_2–N_2 and N_2–N dissociation reactions. This approach is adopted due to the requirement for a relatively simple mathematical expression for use in the numerically intensive DSMC technique. The constant of proportionality for P_d is obtained through calibration of the model against rate data under equilibrium conditions as described in the following section.

3. Results

The DSMC threshold line model is used to calculate dissociation rate constants under conditions where the translational, rotational, and vibrational modes are in thermal equilibrium with Boltzmann distributions. An isothermal heat bath is simulated that consists of 50,000 particles. Energy is exchanged between the various modes during collision, but chemical reactions are not processed. Instead, the average dissociation probability is evaluated over all collisions and then converted into a rate constant. Results obtained in this manner are compared in Fig. 1a with the rate coefficient of Byron (1966) for the N_2–N_2 reaction over a wide temperature range. As discussed by Boyd (1992), the coefficient of Byron is employed as it was measured under conditions of thermal equilibrium. It is quite remarkable that the DSMC model provides a temperature dependence that is so close to the Byron rate. It should be noted that the leading constant in the DSMC model is explicitly chosen to give agreement with the Byron rate at $T = 10,000$ K. In Fig. 1b, the dissociation probabilities for various vibrational energy levels are plotted as a function of total collision energy (expressed as a temperature). It is interesting to note that there is a region for all vibrational levels where the dissociation probability exceeds one. The dissociation rates shown in Fig. 1a limit the dissociation probability to a maximum of one.

Comparison is next made of the DSMC threshold line model with the continuum model of Macheret et al. (1994) under nonequilibrium conditions. For simplicity, the translational and rotational modes are equilibrated. Only the effect of vibrational nonequilibrium is considered. Once again, heat-bath simulations are performed. In this case, the translational and rotational modes are given a temperature of 10,000 K. For each of the vibrational levels of the harmonic oscillator that lie below the dissociation energy, a large number of collisions are calculated to determine the state-dependent dissociation rate. These rates are then summed over the Boltzmann populations at different vibrational

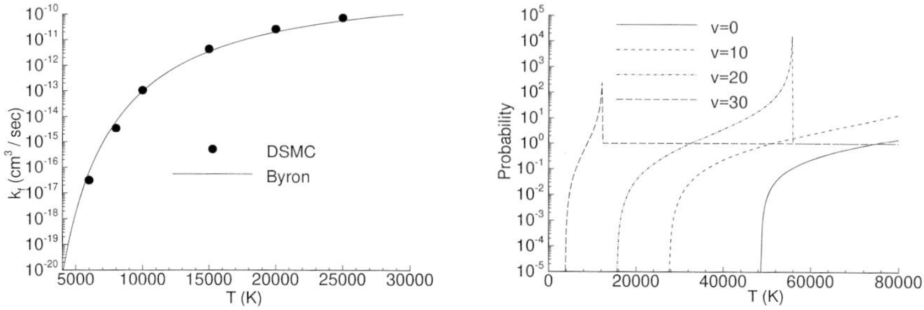

(a) (b)

Figure 1. Properties of the DSMC threshold model: (a) equilibrium dissociation rate constants; and (b) dissociation probabilities as a function of collision energy (temperature) for various vibrational energy levels.

temperatures. The results of this analysis are shown in Fig. 2a for the N_2–N_2 reaction. Comparison is made with the continuum threshold line model. While the two models give good agreement at the lower and upper values of vibrational temperature, the DSMC model predicts a higher dissociation rate

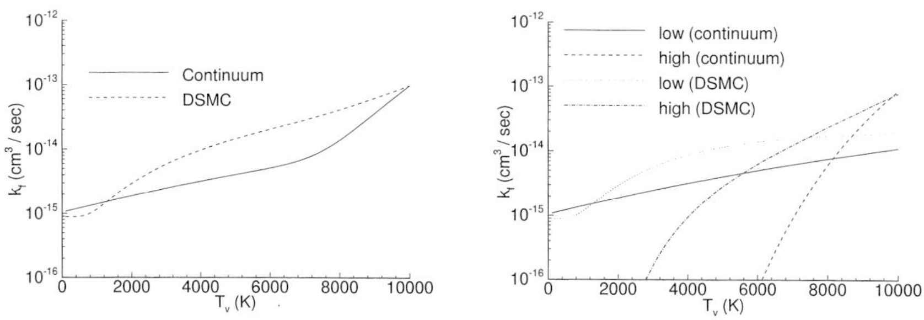

(a) (b)

Figure 2. Comparison of DSMC and continuum threshold models for N_2–N_2 dissociation: (a) nonequilibrium dissociation rate constants; and (b) contribution to k_f of low and high vibrational levels.

for the intermediate values. The reason for this difference is partly explained through consideration of Fig. 2b. This shows the contributions to the overall dissociation rate of the lower ($E_v < \alpha D$) and higher ($E_v > \alpha D$) vibrational levels. The DSMC model generally shows higher dissociation rates from both low and high vibrational levels. It is assumed that these differences arise from the approximations made in obtaining analytical expressions for the continuum model. In particular, integrals are evaluated using the method of steepest descent which can give rise to significant inaccuracies. In addition, the continuum model assumes a continuous distribution of vibrational energy whereas the DSMC approach employs the truncated harmonic oscillator.

Finally, the DSMC threshold model is evaluated through comparison of experimental data in hypersonic shock waves of nitrogen. The data is taken from Kewley and Hornung (1974) who report density profiles obtained from interferograms. Two cases with different upstream velocities are considered. The DSMC simulations of these one-dimensional shocks employ several thousand cells and about 100,000 particles. In the first case, the upstream velocity is 4.8 km/s. This gives a density profile that is dominated by N_2–N_2 dissociation reactions. The DSMC results computed with the threshold line model compare very favorably with the experimental data as indicated in Fig. 3a. The second case considers an upstream velocity of 7.3 km/s. Under these conditions, the density rise across the shock is dominated by N_2–N dissociation reactions. The

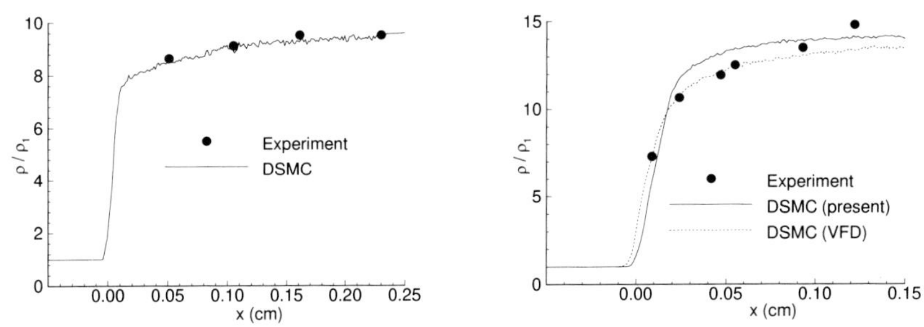

(a) (b)

Figure 3. Profiles of density ratio in nitrogen shock waves: (a) U_1=4.8 km/s, (b) U_1=7.3 km/s.

threshold line model is compared with the experimental data in Fig. 3b. Also shown is the result obtained previously for this case using the VFD dissociation model (Boyd 1992). The threshold line model gives reasonable agreement with the experimental data, although it appears that the dissociation rate is slightly overpredicted. By comparison, the VFD model gives excellent agreement with the data. It should be noted that this level of agreement was only obtained with the VFD model by trying several different values of the parameter ϕ.

The level of agreement obtained with the DSMC threshold model and the data for the second shock wave case could be improved through use of a different rate coefficient for the N_2–N reaction. It appears that the data of Byron may be too high. It should be noted that the temperatures in the immediate post-shock region produced by the 7.3-km/s conditions exceed the range examined in the study of Byron. With this in mind, the performance of the new model is considered satisfactory. A more detailed investigation is reported in Boyd (1996).

4. Concluding remarks

A threshold line dissociation model for the DSMC technique has been formulated that determines the probability of reaction based on the energies of each collision. The model reproduced the temperature dependence of measured rate data under equilibrium conditions. It provided good agreement with measurements of density profiles through hypersonic shock waves. The use of the model is recommended over previous DSMC models for vibration-dissociation coupling. The new model is less phenomenological and only requires the determination of a leading-constant multiplication factor for a particular reaction.

Acknowledgement. This work is funded by the Air Force Office of Scientific Research under University Research Initiative F49620-93-1-0373.

References

Boyd ID (1996) A threshold line dissociation model for the direct simulation Monte Carlo method. Phys Fl (in press).

Boyd ID (1992) Analysis of vibration-dissociation-recombination processes behind strong shock waves of nitrogen. Phys Fluids A 4:178–185.

Boyd ID, Candler GV, Levin DA (1995) Dissociation modeling in low density hypersonic flows of air. Phys Fluids 7:1757–1763.

Byron S (1966) Shock tube measurements of the rate of dissociation of nitrogen. J Chem Phys 44:1378–1388.

Kewley DJ, Hornung HG (1974) Free-piston shock tube study of nitrogen dissociation. Chem Phys Let 25:531–536.

Macheret SO, Rich JW (1993) Nonequilibrium dissociation rates behind strong shock waves: classical model. Chem Phys 174:25–43.

Macheret SO, Fridman AA, Adamovich IV, Rich JW, Treanor CE (1994) Mechanisms of Nonequilibrium Dissociation of Diatomic Molecules, AIAA Paper–94–1984.

Park C (1988) Assessment of a two temperature kinetic model for dissociating and weakly ionizing nitrogen. J Thermophysics 2:8–16.

Analytical hypersonic waverider configurations

A. Merlen*, E. Fabre, S. Cieslak
ONERA-IMFL
5 Bvd Painlevé
59000 Lille, France

Abstract: Waverider configurations are deduced from hypersonic flows around axisymmetrical power-law slender bodies at low angles of attack. These inviscid flows have been calculated by analytical techniques and the stream functions are explicitly tabulated. This allows the generation of many waveriders, the aerodynamic coefficients of which are known.

Key words: Hypersonic flows, Waveriders

1. Theory

1.1. Hypotheses

The gas is assumed to be inviscid and thermodynamically perfect, having the specific heat ratio γ. Except in the vicinity of the blunt nose, the slenderness of the body, the high upstream Mach number M_∞, and the small angle of attack α allow the resolution of the problem in the frame of the hypersonic small-disturbance theory. At the limit $M_\infty \to \infty$ and $\alpha = 0$, which leads to the equivalence principle between explosions and hypersonic flows. Well-known self-similar solutions are then available, particularly for power-law bodies Σ', the equation of which is:

$$R' = \frac{\tau L}{2}\left(\frac{x}{L}\right)^n \quad \text{and for the shock: } R_0 = \frac{\tau L}{2\xi_0'}\left(\frac{x}{L}\right)^n$$

where ξ_0' is a positive constant, less than unity. The theoretical results can be summed up roughly by considering that the equivalence principle yields good results for pressure on the obstacle if the power-law exponent n is given a value between 2/3 and 1. Experiments (Kubota 1957) confirm these values. Nevertheless, this theory provides a poor description of the blunting and of the consequences blunting has on the density on the body surface. However, they do show that if the stream function is known, the entropy really acquired at the

*Present address: Laboratoire de Mécanique de Lille, Cité Scientifique, 59655 Villeneuve d'Ascq Cedex, France.

passage through the shock can be reassigned to each streamline to correct the density (Merlen & Andriamanalina).

1.2. Solutions

The definition $M_\infty^2 = \frac{\rho_\infty V_\infty^2}{\gamma p_\infty}$, shows that under the previous hypotheses, the counterpressure p_∞ only appears by way of a linear disturbance in $\frac{1}{M_\infty^2 R_x^2}$. Moreover, α only appears in the term $\frac{\alpha}{R_{0x}}$, R_{0x} being the first approximation of the slope of the shock. This is a typical result of hypersonic small-disturbance theory. The same result holds for all the unknowns of the problem, apart from the fact that they also depend on r, which introduces a dimensionless variable ξ proportional to $\frac{r}{\tau L}\left(\frac{L}{x}\right)^n$. The conditions on the shock can be used to determine explicitly the θ-dependency of the function of subscript 1 (Merlen & Andriamanalina). Finally, by taking as variables: $\eta = \frac{\alpha x}{R_0} = n\frac{\alpha}{R_{0x}} = \frac{2\xi_0'\alpha}{\tau}\left(\frac{x}{L}\right)^{(1-n)}$, and $\zeta = \frac{n^2}{M_\infty^2 R_x^2}$ we write the solution in the form:

$$R = \frac{\tau L}{2\xi_0'}\left(\frac{x}{L}\right)^n (1+\eta\xi_1\cos\theta+\zeta\xi_2) \qquad R' = \frac{\tau L}{2}\left(\frac{x}{L}\right)^n \qquad \xi = \frac{2r\xi_0'}{\tau L}\left(\frac{L}{x}\right)^n$$

$$\begin{aligned} u &= \frac{2nV_\infty}{\gamma+1}\frac{r}{x}(u_0+\eta u_1\cos\theta+\zeta u_2) & v &= \frac{2nV_\infty}{\gamma+1}\frac{r}{x}(\eta v_1\sin\theta) \\ \rho &= \frac{\gamma+1}{\gamma-1}\rho_\infty(\rho_0+\eta\rho_1\cos\theta+\zeta\rho_2) & p &= \frac{2n^2\rho_\infty V_\infty^2}{\gamma+1}\frac{r^2}{x^2}(p_0+\eta p_1\cos\theta+\zeta p_2). \end{aligned} \tag{1}$$

The terms with subscript 0, 1, and 2 are functions of ξ, except for ξ_0', ξ_1, and ξ_2, which are constants, and ξ_0, which is introduced so that the shock equation may be written $\xi = 1$ in the self-similar solution. Small-disturbance hypotheses are expressed by $\eta \ll 1$ and $\zeta \ll 1$. Despite the low values of α considered here, the effect of the incidence is far from negligible. The solution (1) must verify the general equations for any value of θ, η, and ζ. These equations are identically verified for any θ, which validates the form and reduces the problem to three systems of ordinary differential equations in ξ. The first of them gives the self-similar basic flow. The other systems (subscripts 1 and 2) are given in Merlen & Andriamanalina (1992) and give the effect of the incidence and the counterpressure, respectively. We must emphasize that the solutions are calculated once for each pair (n, γ) and provide universal sets of functions and constants. Hence, a particular physical configuration is directly calculated by analytical formulae using the tabulated universal solutions.

2. Comparison with experiments

As an example, we limit this section to the presentation of the results on the lift coefficient. Starting with the definition, the lift is obtained by $F_z = -\int_0^{2\pi}\int_0^L p'R'\cos\theta\,dx d\theta$ where only the subscript -1 term is nonzero. Letting $S_{ref} = \frac{\pi\tau^2L^2}{4}$, we obtain: $C_z = \frac{F_z}{\frac{1}{2}\rho_\infty V_\infty^2 S_{ref}} = -\frac{2np_1'\xi_0'}{\gamma+1}\alpha$, which has the same sign as α since p_1 is negative. Experiments realized in the ONERA R3Ch wind tunnel are in good agreement with these analytical results showing less than 5% discrepancy (Fig. 1). The wind tunnel was chosen to fulfill a few necessary conditions to make possible the comparison with this inviscid theory, among them: weak viscous interaction, no real-gas effect (cold hypersonic flow), and strong

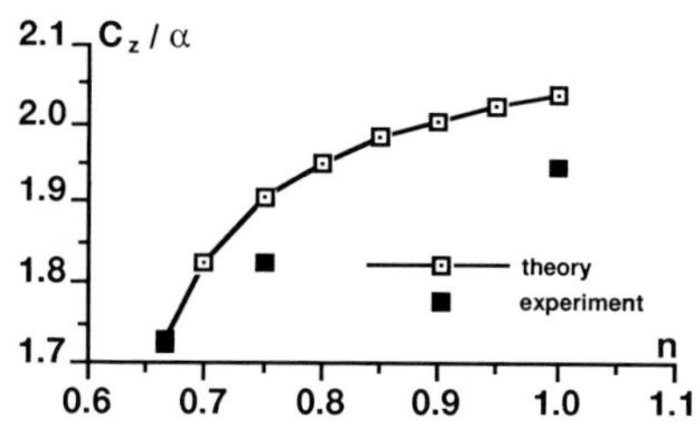

Figure 1. Lift coefficient of the ogives.

shock condition. Of course, this choice led to a compromise on the following physical conditions: $T_i = 1050°K$, $p_i = 120\ 10^5$ Pa, $M_\infty = 9.95$, $\mathcal{R}_\infty = 2.6\ 10^6$. Three ogives have been tested: $n = 1, 3/4, and 2/3$ with $\tau = 0.4$ and $L = 300$ mm. Because of the slenderness of the bodies, the strong-shock condition is not fulfilled on the rear part of the ogives. However, a higher Mach number would have increased the real-gas effect. Actually, this relatively weak Mach number allowed us to emphasize the influence of second-order terms in the development of the solution, particularly on the shock location and wall pressure. The viscous effects have been studied in the case of the $n = 3/4$ body, for which the self-similarity of the external flow can be extended to the coupled viscous-inviscid flow (Fig. 2) for the axisymmetric case (see Fabre 1995 for the small-incidence effect). As the main result of this comparison between theory and experiment, we can say that the first-order theoretical solution is valid for very high Mach numbers and a cold wall. For lower M_∞, the theory needs to be extended to second-order terms to explain the behavior of the shock shift when $\alpha \neq 0$ and to correct the wall pressure. On the contrary, the aerodynamic coefficients are mainly determined by the first-order terms.

3. Stream functions

The agreement between theory and experiment justifies the use of the theoretical solution to build waverider configurations. Many choices are possible but all are

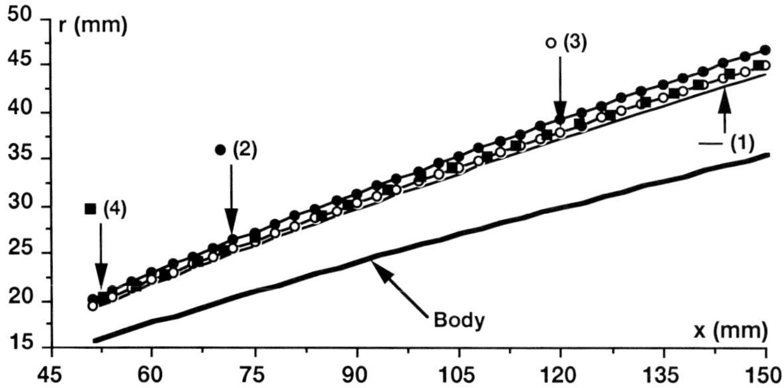

Figure 2. Shock location for $n = 0.75$ and $\alpha = 0.$: (1) shock by inviscid theory; (2) theory for adiabatic wall; (3) theory for cold wall; (4) experiment.

based on a solidification of stream surfaces. In the previous theory, this is made easier by the explicit knowledge of the stream functions which are tabulated in universal dimensionless form. The method for finding the stream function is based on dimensional analysis, but the dimensional considerations are not sufficient if $\alpha \neq 0$ since the flow is not self-similar. Thus, the form of ψ as a function of ζ must be anticipated. From the local equation of the stream functions, it can be shown that, for $\theta \neq 0$, or π:

$$\Psi = \left(\frac{x}{L}\right)^{2n} (|\sin\theta|)^{\frac{2n}{1+n}} \Psi_s \tag{2}$$

where

$$\Psi_s = \left(\frac{x}{L}\right)^{2n} \left\{\rho_0(u_0 - \frac{\gamma+1}{2})\xi^2 + \frac{2n}{1+n}\eta\cos\theta\left[\xi^2\left(\rho_1(u_0 - \frac{\gamma+1}{2}) + \rho_0 u_1\right)\right]\right.$$
$$\left. + n\zeta\left[\xi^2\left(\rho_2(u_0 - \frac{\gamma+1}{2}) + \rho_0 u_2\right)\right]\right\} \tag{3}$$

is the dimensionless form of the stream function in the plane of symmetry. At the zeroth order ($\alpha = 0$), the second stream function is obviously $\Phi = \theta$. We look for the form $\Phi = \theta + \eta\Phi_1(\xi)\sin\theta$. From the local equation, we get a first-order inhomogeneous differential equation for Φ_1. Letting $\Phi_1 = 0$ on the shock, the solution is $\Phi = \theta_i = \theta + \eta\Phi_1 \sin\theta$ with:

$$\Phi_1 = \frac{2}{1-\gamma}\left[\frac{2}{1-\gamma}\rho_0(u_0 - \frac{\gamma+1}{2})\xi^2\right]^{\frac{1-n}{2n}} \int_\xi^1 \frac{\rho_0 v_1 \hat{\xi}}{\left[\frac{2}{1-\gamma}\rho_0(u_0 - \frac{\gamma+1}{2})\hat{\xi}^2\right]^{\frac{1+n}{2n}}} d\hat{\xi}. \quad (4)$$

A more detailed analysis of this stream function, including the Hayes-Probstein entropy correction, is given in Merlen & Andriamanalina (1992).

4. Application to waverider design

As an example, we indicate here a very straightforward principle to build simple hypersonic waveriders. At first we choose a stream surface $\Phi = cst$ as the lower surface (Fig. 3). The upper surface is a cylinder starting from the shock intersection with the lower surface and generated by straight lines parallel to $\overrightarrow{V}_\infty$. Then, the pressure on the upper surface is supposed to be equal to p_∞. Figure 3 shows one of the

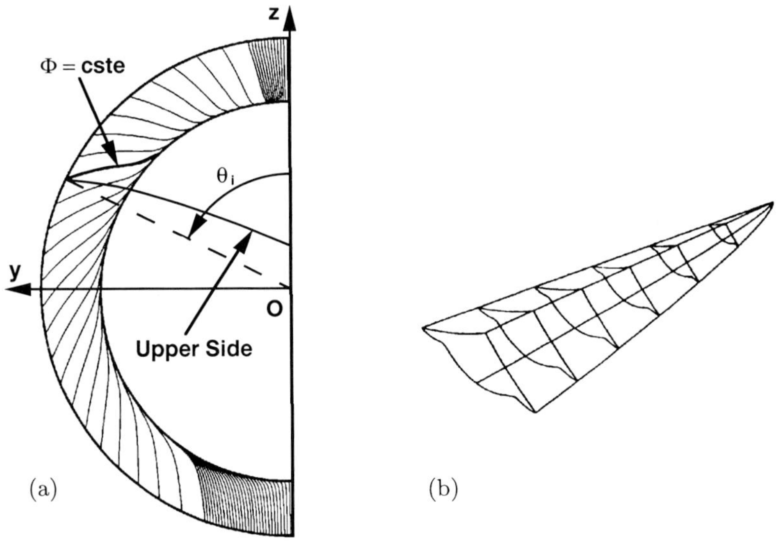

Figure 3. Waverider generation: (a) principle, (b) example.

geometries obtained in this way. Because pressure is known everywhere in the original flow field, we obtain explicitly the aerodynamic coefficients by integration on the surface of the body. It is then possible to evaluate the

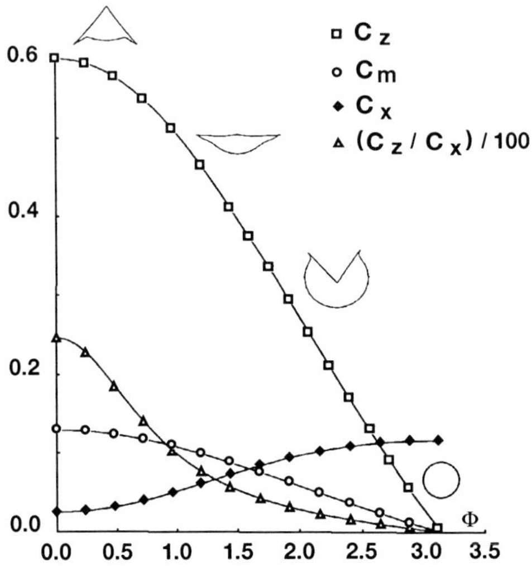

Figure 4. Aerodynamic coefficients versus dihedral angle.

influence of the geometrical parameters on these coefficients. For example, Fig. 4 shows the effect of the dihedral angle on C_z. We must emphasize that the generation of waveriders is reduced to a simple treatment of data files obtained by solving the universal dimensionless problem of Section 1. This allows a very quick evaluation of the characteristics of different configurations. Moreover, optimization is possible taking into account criteria such as a given volume, or a given length, etc.

References

Fabre E (1995) Explosions and Self Similar Hypersonic Viscous and Inviscid flows. Theory-calculation-experiment, Thesis of Université des sciences et technologies de Lille.

Kubota T (1957) Investigation of Flow Around Simple Bodies in Hypersonic Flow, Memorandum No. 40 Guggenheim Aeronautical Laboratory, California Institute of Technology.

Merlen A, Andriamanalina D (1992) Analytical solutions for hypersonic flow past slender power-law bodies at small angle of attack. AIAA Journal 30(11):2683.

A more precise method to determine stagnation-point heat-transfer rates

H. Olivier, H. Grönig
Shock Wave Laboratory, RWTH Aachen
Templergraben 55
52056 Aachen, Germany

Abstract: The theory of Fay and Riddell (1958) shows that the tangential velocity gradient is of great importance for stagnation-point heating. Based on the work of Belotserkovskii (1958) the authors developed an improved method to determine the tangential velocity gradient which, in opposition to the classical Newtonian method, is also valid for high-enthalpy conditions and which, as far as possible, takes into account the real flow properties within the shock layer.

Key words: Hypersonic flow, Stagnation-point heating, Theory of Fay and Riddell, Improved tangential velocity gradient, Integral method

1. Introduction

Most commonly in conjunction with the theory of Fay and Riddell, the Newtonian flow model (Lees 1956) is applied to determine the velocity gradient at the stagnation point of a sphere. It makes use of the one-dimensional momentum equation for a streamline close to the stagnation point. The pressure distribution on the sphere and thus the pressure gradient is computed from Newtonian theory. This finally leads to

$$\left.\frac{du}{dx}\right|_s \frac{D}{u_\infty} = \frac{2}{u_\infty}\sqrt{\frac{2(p_s - p_\infty)}{\rho_s}} \tag{1}$$

where the subscript s indicates the conditions at the boundary-layer edge of the stagnation point and the coordinate x points in the circumferential direction. Since this approach is based on the one-dimensional momentum equation it does not take into account multi-dimensional effects, like the influence of the vorticity produced by the curved bow shock, or the influence of the shock stand-off distance on the shock-layer flow. Especially the shock stand-off distance Δ strongly influences the shock-layer flow and therefore also the tangential velocity gradient. Real-gas effects mainly act on the tangential velocity gradient via the shock stand-off distance.

The aim of the present paper is to derive a method for computing the tangential velocity gradient at a sphere for high-temperature flows, where real-gas effects are important. Vorticity and compressibility effects are also included in the method to be presented. It will be shown that from the conservation equations for the stagnation point an analytical solution can be derived which is valid for perfect as well as real-gas behavior.

2. Improved model of stagnation flow

The conservation equations for mass and and momentum for the radial direction are used in the following form (Hayes and Probstein 1966, Chapter VI, p. 411)

$$\frac{\partial}{\partial \Theta}(\rho u \eta) + \frac{\partial}{\partial r}(\rho v r \eta) = 0 \tag{2}$$

and

$$\frac{u}{r}\frac{\partial v}{\partial \Theta} + v\frac{\partial v}{\partial r} - \frac{u^2}{r} + \frac{1}{\rho}\frac{\partial p}{\partial r} = 0\,, \tag{3}$$

with the velocity component u in the circumferential direction, v in the radial direction, and $\eta = r\sin\Theta$. The conservation equations for energy, for the momentum in Θ direction, and the equation of state are replaced by the integral approaches described below. With the non-dimensional variables

$$\bar{p} = \frac{p}{p_s}, \quad \bar{\rho} = \frac{\rho}{\rho_s}, \quad \bar{u} = \frac{u}{u_{max}}, \quad \bar{r} = \frac{r}{R}, \quad \bar{\Delta} = \frac{\Delta}{R}, \quad u_{max} = \sqrt{2h_0} \tag{4}$$

and

$$Q_v = \bar{\rho}\bar{v}, \quad Q_u = \bar{\rho}\bar{u}, \quad P_v = \bar{\rho}\bar{v}^2 + \frac{p_s}{\rho_s u_{max}^2}\bar{p},$$
$$P_u = \bar{\rho}\bar{u}^2 + \frac{2p_s}{\rho_s u_{max}^2}\bar{p}, \quad P_{uv} = \bar{\rho}\bar{u}\bar{v}, \quad \bar{\eta} = \frac{\eta}{R}$$

and after some algebraic reformulation, similar expressions are achieved as those given by Hayes and Probstein (1966, Chapter VI, p. 412)

$$\frac{\partial}{\partial \bar{r}}(\bar{r}\bar{\eta}Q_v) + \frac{\partial}{\partial \Theta}(\bar{\eta}Q_u) = 0\,, \tag{5}$$

$$\frac{\partial}{\partial \bar{r}}(\bar{r}\bar{\eta}P_v) + \frac{\partial}{\partial \Theta}(\bar{\eta}P_{uv}) - \bar{\eta}P_u = 0\,. \tag{6}$$

The two conservation equations (5) and (6) are integrated along a ray $\Theta = \text{const}$ from the body up to the shock. In order to solve these integral equations it is necessary to make an approximation for the three integrands in these equations. Following the work of Belotserkovskii (1958), the variables $\bar{r}Q_u, \bar{r}P_{uv}$, and $\bar{r}P_u$ are approximated by a linear distribution between the shock and the body. Navier-Stokes computations performed by Wong and Muylaert (1995) show that this approximation is even justified for high-temperature flows with significant real-gas effects (Olivier 1995). The general linear distribution

$$F = F_b + (F_2 - F_b)\frac{\bar{r} - 1}{\bar{\Delta}} \qquad (7)$$

and the following boundary conditions for the stagnation-point streamline are introduced into the integral equations

$$\cos\sigma = 1\,, \;\; \bar{u}_2 = 0\,, \;\; \frac{d\bar{\Delta}}{d\Theta} = 0\,, \;\; \bar{\rho}_b = 1\,, \;\; \bar{p}_b = 1\,, \;\; \bar{v}_b = 0\,, \;\; \sin\Theta \approx \Theta\,. \qquad (8)$$

The subscript 2 characterizes values immediately behind the shock, whereas subscript b indicates the conditions at the body surface. For the stagnation point the tangential velocity is expressed as

$$\bar{u}_b = \Theta \left.\frac{d\bar{u}_b}{d\Theta}\right|_s \,. \qquad (9)$$

With these boundary conditions the following analytical solution is achieved for the tangential velocity gradient (Olivier 1995)

$$\left.\frac{du}{dx}\right|_s \frac{D}{u_\infty} = \frac{2(1+\bar{\Delta})}{\bar{\Delta}} \frac{p_s - p_2}{\rho_\infty u_\infty^2} \frac{\rho_2}{\rho_s}\,. \qquad (10)$$

It is important to note that, for the derivation of (10), no explicit statement is necessary regarding the thermal and chemical properties of the gas. These properties are part of the basic approximation expressed by (7). Therefore, the equation for the velocity gradient is valid as long as the approximation (7) holds. As expected, according to (10) the tangential velocity gradient at the stagnation point depends on the shock stand-off distance, but it also depends on the density ratio ρ_2/ρ_s, which is a measure of the deviation of the real flow from the constant-density flow model for the stagnation region.

3. Results

3.1. Numerical simulation and theoretical results

According to (10) the velocity gradient has been computed for different flow conditions (see Fig. 1). In this example an equilibrium flow was assumed, where the conditions behind the shock and at the stagnation point were determined by a nine-component air model (Esser et al. 1992).

For the data presented in Figs. 1 and 2 the shock stand-off distance was determined by

$$\frac{\Delta}{R} = \frac{\rho_\infty}{\rho_2} \qquad (11)$$

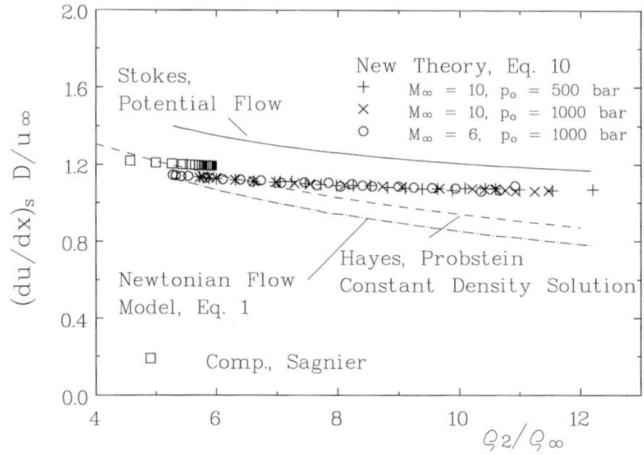

Figure 1. Non-dimensional velocity gradient versus density ratio.

which is a good approximation for flow conditions with high-temperature effects (Olivier 1995). It is worth noting that in Fig. 1, with increasing density ratio the non-dimensional velocity gradient slightly decreases. But, as expected, the dimensional velocity gradient increases, since in this case the shock stand-off distance becomes smaller (for brevity, this graph is not shown in this paper). For the stagnation-point heat flux it was possible to compare the results of non-equilibrium Navier-Stokes calculations with the modified theory of Fay and Riddell which makes use of the new velocity gradient. A collection of numerical results was compiled by Vérant (1994). These computations were carried out at European research institutes, i.e., ONERA-France, ESTEC-Netherlands, and DLR-Germany. Figure 2 shows the collection of these numerical results, where the stagnation-point heat flux is scaled by the square root of the nose radius divided by the stagnation-point pressure. Two curves are also presented according to the theory of Fay and Riddell for the old as well as for the new formulation of the velocity gradient. As expected, for low enthalpies they coincide, but for higher enthalpy flow conditions the formula of Fay and Riddell with the new velocity gradient relation gives higher values which agree very well with the Navier-Stokes computations.

3.2. Experimental results

For high-enthalpy conditions it is very difficult to measure the stagnation-point heat flux with sufficient accuracy in order to prove a theoretical result. Therefore it was decided to determine the derivation of the velocity gradient in the experiment from the Newtonian value (1). With the help of the one-dimensional momentum equation this ratio can be expressed as

$$\frac{du/dx\mid_{s,exp.}}{du/dx\mid_{s,Newt.}} = \frac{D}{2}\sqrt{\frac{-d^2p/dx^2\mid_{s,exp.}}{2(p_s - p_\infty)}}\,. \tag{12}$$

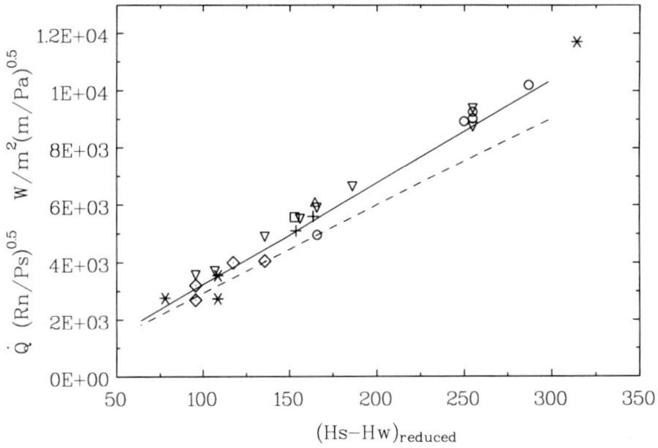

Figure 2. Comparison of Navier-Stokes calculations with the stagnation-point heat flux according to Fay and Riddell and the new velocity gradient: —— Fay and Riddell, Lewis number Le = 1.4, viscosity μ (Sutherland, in Hansen 1959), new velocity gradient; – – – – Fay and Riddell, Lewis number Le = 1.4, viscosity μ (Sutherland, in Hansen 1959), Newtonian velocity gradient; symbols are from Navier-Stokes calculations, Vérant (1994).

The tests were performed in the Aachen shock tunnel TH2 for different flow conditions. The stagnation temperature was varied between 1500 K and 4700 K and the free-stream Mach number between 6 and 12. In the experiments the pressure distribution in the stagnation region was measured with high spatial resolution. The distribution was approximated by a spline of second order. From this function it was possible to determine the value $d^2p/dx^2\mid_{s,exp.}$ in (12). In this way the deviation could be determined between the tangential velocity gradient of the experiment and according to the Newtonian theory. These experimental values are shown in Fig. 3 and compared with those according to the new theory (10). Despite the fact that in the experiments only a few flow conditions could be tested, the experimental values clearly show the tendency of the theoretical ones, and the agreement achieved with the values according to the new theory is quite good. Of course, in particular for high stagnation temperatures, the measurements performed in the stagnation region are very sensitive to several experimental parameters. Therefore, the experiments performed can only support the general behavior of the tangential velocity gradient as determined by the theory.

4. Conclusions

The new expression for the velocity gradient represents a significant improvement and a higher accuracy when it is used to compute the stagnation-point heat

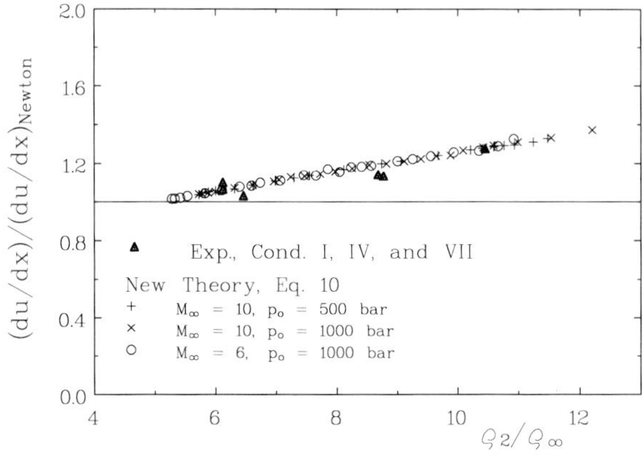

Figure 3. Deviation of the tangential velocity gradient from that according to the Newtonian theory as a function of the density ratio across the normal shock.

flux according to the theory of Fay and Ridell. It is based on a well-developed flow model which is much more realistic than the old one used before. This is confirmed by the good agreement between numerical and experimental results.

References

Belotserkovskii OM (1958) Flow with a detached shock wave about a symmetrical profile. J Math & Mech 22:279–296.

Esser B, Grönig H, Olivier H (1992) High-enthalpy testing in hypersonic shock tunnels. In: Bertin JJ, Periaux J, Ballmann J (eds), Advances in Hypersonics, Defining the Hypersonic Environment, Birkhäuser, Vol. 1, pp 182–258.

Fay JA, Riddell FR (1958) Theory of stagnation point heat transfer in dissociated air. J Aer Sci 25:73–85.

Hansen CF (1959) Approximations for the Thermodynamic and Transport Properties of High Temperature Air, NASA TR R-50.

Hayes WD, Probstein RF (1966) Hypersonic Flow Theory, Inviscid Flow, Academic Press, New York, London.

Lees L (1956) Hypersonic flow. In: Proc. 5th Int. Aeron. Conf., IAS-RAeS, Los Angeles, 1955, pp 241–276.

Olivier H (1995) Influence of the velocity gradient on the stagnation point heating in hypersonic flow. Shock Waves 5:205–216.

Vérant JL (1994) Numerical Enthalpy Rebuilding for Perfect Gas and Nonequilibrium Flows. Application to High Enthalpy Wind Tunnels. Technical report no. 69/6121 SY, ONERA, Chatillon, France.

Wong H, Muylaert J (1995) Private communication, ESTEC, The Netherlands.

Millisecond aerodynamic force measurement using a free-flying side-jet model with built-in gas reservoir

K. W. Naumann, H. Ende, G. Mathieu, A. George
French-German Research Institute Saint-Louis, ISL
5, Rue General-Cassagnou
F-68301 Saint-Louis, France

Abstract: This paper presents the technical realization and first results of a novel development of the ISL millisecond aerodynamic-force measuring technique. It is used to measure the additional force which acts on a model due to interaction between side jets and hypervelocity cross flow. The basic idea is to measure the acceleration of a model, which is set free for the time of test flow and then recaptured again. In addition to prior developments using a free-flying model with built-in pyrotechnical charges, we developed a method to perform such measurements with a free-flying model which carries a gas-filled reservoir. The technical realization includes development of an on-board fast-acting opening device, construction of the model, setup of the gas-supply system, and adaptation of the mounting of the accelerometers. The measurements comprise investigations on the internal flow phenomena, evaluation of the aerodynamic coefficients of the model, and measurements on jet interaction.

Key words: Aerodynamic force measurement, Acceleration measurement, Short-duration flow, Shock tunnel, Jet interaction

1. Introduction

The aim of the work described here is to measure the additional force which acts on a model surface as a result of intense interaction between control side jets and tropospheric hypervelocity cross flow (Naumann et al. 1993a, 1993b).

The experiments to study these effects are carried out in ISL shock tunnel "B" (Fig. 1a), now operating in the reflected mode. It duplicates main sea-level hypervelocity flow conditions up to about 2000 m/s. This significantly facilitates investigations on side-jet interaction, external combustion, or mixing and combustion phenomena in scramjets. The handicap is the short duration of test flow, actually about 3 ms, which complicates aerodynamic force measurements. Our technique uses a fast-acting mechanical support which sets the model free for the time of test flow to measure its acceleration, then recaptures it again. The very short time of free flight allows us to use wires and flexible

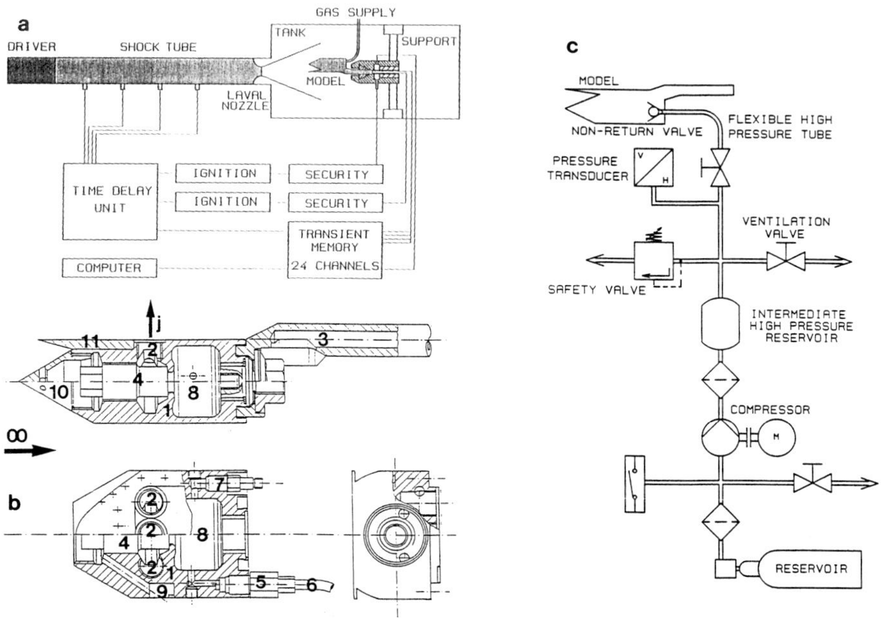

Figure 1. (a) ISL shock tunnel "B" with Laval nozzle and fast-acting mechanical support; (b) model with on-board gas reservoir — (1) body, (2) jet piece, (3) stem, (4) fast-acting opening device, (5) non-return valve, (6) filling tube, (7) pressure gauge, (8) gas reservoir, (9) location for accelerometers, (10) nose cap, (11) wing, (j) jet axis; (c) setup for the gas supply of the model.

pipes without disturbing model motion noticeably. The data-evaluation procedure is straightforward, insensitive to disturbances arising inevitably at the onset of the test flow, and allows us to judge whether the flow field has become quasi-stationary (Naumann et al. 1991a, 1991b). Operation of this technique with free-flying wing-body models equipped with laterally blowing jets supplied by pyrotechnical charges is described in Naumann et al. (1993a, 1993b).

2. Technical realization

The basic problem associated with our measuring task is to supply up to 3 jets of 4-mm throat diameter with at least 10 MPa within about one millisecond, without using a stiff or pressurized tube, in order not to disturb the model's free flight. Of the possible solutions to achieve this (Naumann et al. 1993b), we chose that which uses an on-board reservoir. Figure 1b shows the design of the model; Fig. 1c shows a sketch of the gas-supply system. A compressor fills the reservoir before the test with usually 80.4 MPa. Then the thin and flexible pipe is vented to prevent it from acting upon the model. Immediately before test flow onset the reservoir is opened towards the jets by a fast-acting opening device. It is a development of our own, very compact, and driven by an explosive charge. This accelerates a free piston which weakens the sealing diaphragm of 12-mm diameter so that it ruptures due to the reservoir pressure. Figure 2 shows different aspects of model #3 and its components; Fig. 3 shows the model and the test chamber.

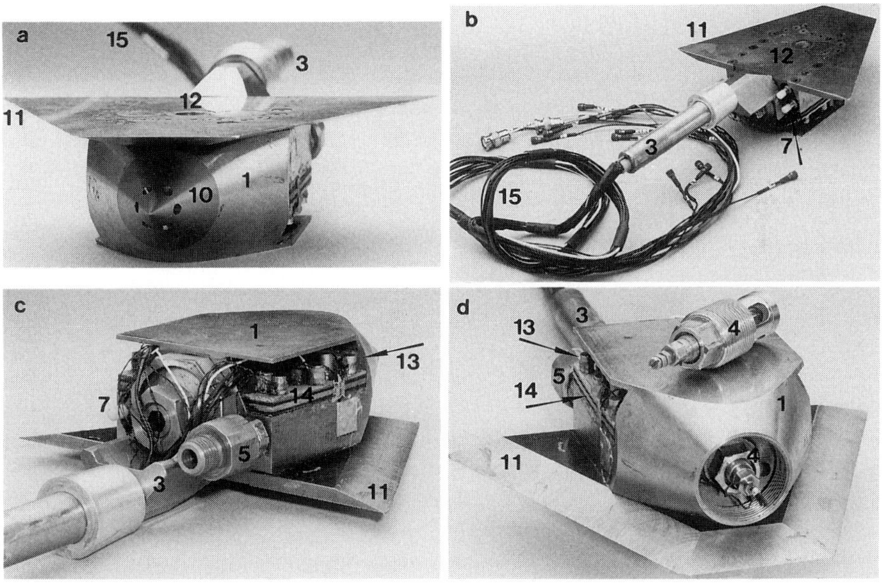

Figure 2. Different aspects of model #3: (1) body, (3) stem, (4) fast-acting opening device, (5) non-return valve, (7) pressure gauges, (10) nose cap, (11) wing, (12) side jet, (13) accelerometers, (14) shock-damping layer, (15) electrical wires.

3. Experiments

For the experiments on internal and nozzle flow we used a nozzle equipped with seven pressure transducers. Figure 4 shows the pressure histories and a sketch of the setup. One millisecond after ignition of the opening device the internal pressure is balanced and provides some milliseconds of quasi-stationary jet flow. The strong pressure drop during the internal equalization of the jet supply gas exhibits significant action of nonideal flow phenomena. These cause a low jet plenum temperature and subsequently condensation effects in the jet flow, which need additional investigation. For our present requirements it is sufficient that the ratio of thrust to plenum pressure is about constant over

Figure 3. Model #3: (a) in the test chamber; (b) retracted and banked for better visibility — (16) model #3, (17) test chamber, (18) mechanical support, (19) Pitot probe with tip aligned to flow direction in the test chamber, (20) flexible gas-supply tube.

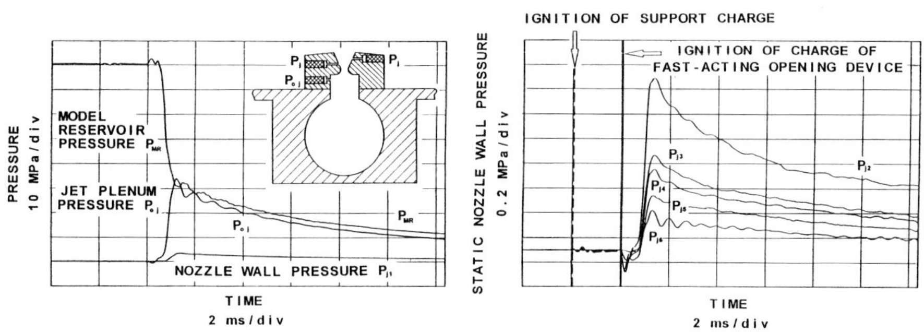

Figure 4. Static pressure histories of model #3 and instrumented nozzle.

time. The aerodynamic coefficients of the model without jet are evaluated as described by Naumann et al. (1993a).

Figure 5a shows the histories of five acceleration components and pitot pressure, Fig. 5b shows lift and drag coefficient histories. The settling times of lift (0.3 ms) and drag coefficient (0.5 ms) are improved compared to prior results (Naumann et al. 1993a) due to improved shock-damping materials for the accelerometers. The evaluation of the additional interaction force acting on the flat wing is carried out as presented in Naumann et al. (1993a) (Fig. 6). The pressure histories yield the relative partial acceleration histories (Fig. 6b). These and the measured total acceleration yield the acceleration component due to the jet/cross-flow interaction (Fig. 6c), and from this and the jet acceleration the interaction coefficient history is calculated (Fig. 6d). In contrast to prior investigations with jets driven by combustion (Naumann et al. 1993a, 1993b), the cold and condensing nitrogen show no quasi-stationary state within 2 ms. Helium

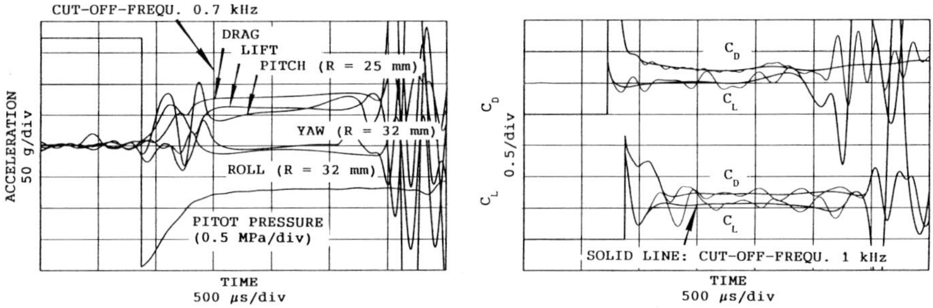

Figure 5. (a) Histories of lift, drag, pitch, roll, and yaw accelerations together with pitot pressure; (b) histories of lift and drag coefficients of both tests with silent jets.

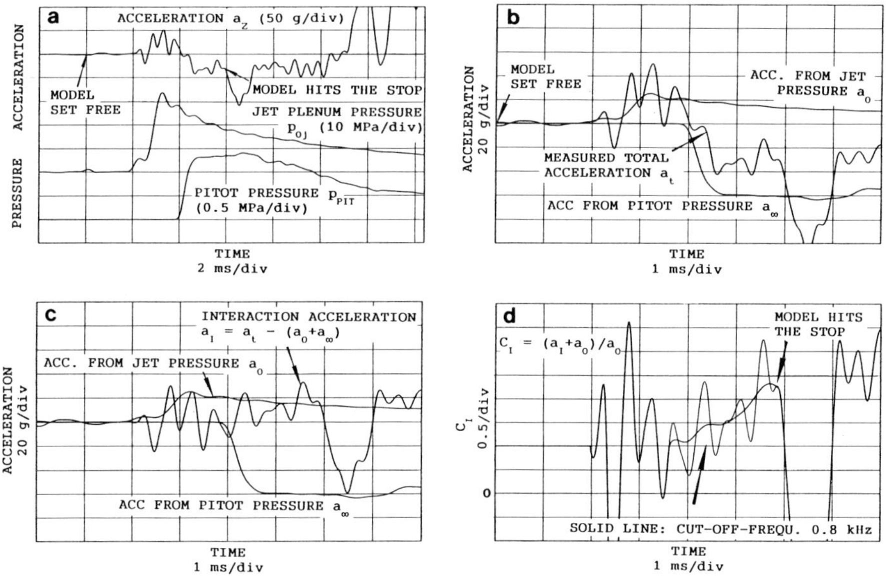

Figure 6. Data histories of an experiment using a nitrogen jet.

(Fig. 7) produced higher model acceleration, less internal pressure loss, and established quasi-stationary flow within 0.5 ms. Hence the nature and state of the gas can significantly influence the time necessary to establish quasi-stationary behavior.

4. Conclusion

The experiments prove that the novel technique to supply jets of free-flying models with stored gas works well. The improved methods to protect the ac-

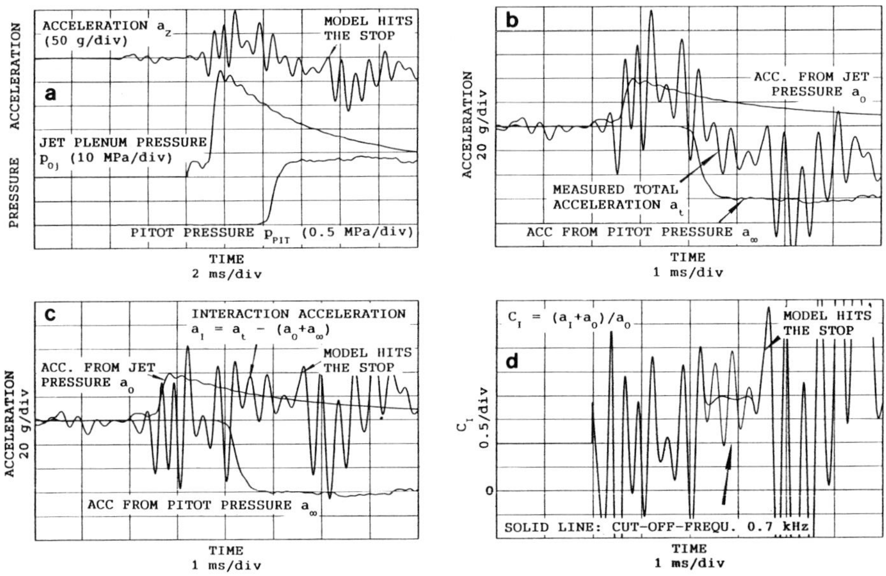

Figure 7. Data histories of an experiment using a helium jet.

celerometers against too heavy shocks originating from the built-in fast-acting opening device brought an improved overall response. Experiments on the internal flow exhibited the existence of strong nonideal and nonisentropic effects. The jet flow shows significant condensation of the nitrogen. The results on interaction force show that with these nitrogen jets, establishing quasi-stationary conditions takes longer than with hot gas from combustibles, or with helium. Further investigations will study these effects.

References

Naumann KW, Ende H, Mathieu G (1991a) Technique for aerodynamic force measurement within milliseconds in shock tunnel. Shock Waves 1:223.

Naumann KW, Ende H, Mathieu G (1991b) Technique for aerodynamic force measurement within milliseconds in a shock tunnel. In: Shock Waves, Proceedings of the 19th ISSW, Sendai, Japan, Vol 2, p 979.

Naumann KW, Ende H, Mathieu G, George A (1993a) Millisecond aerodynamic force measurement with side-jet model in the ISL shock tunnel. AIAA J 31(6):1068.

Naumann KW, Ende H, Mathieu G, George A (1993b) Shock Tunnel Experiments and Approximative Methods on Hypervelocity Side-Jet Control Effectiveness, AIAA Paper 93-1929.

Effects of catalycity and boundary layer reactions on surface heat flux in hypersonic high enthalpy flows

S. L. Gai[†], N. R. Mudford[†], G. T. Roberts[‡], S. G. Mallinson[†]

[†]Department of Aerospace and Mechanical Engineering
University College, University of New South Wales
Australian Defence Force Academy
Canberra ACT 2600, Australia

[‡] Department of Aeronautics and Astronautics
The University, Southampton, United Kingdom

Abstract: The role of chemical reactions in enhancing or inhibiting surface heat flux to bodies exposed to hypersonic, high enthalpy flows is investigated. Attention was directed to both gas phase atomic recombination and catalyzed atomic recombination at the surface. Also examined was the seeming anomaly between the results of an earlier experiment on a flat plate at 0° incidence and the more recent measurements on bluff bodies. In particular, the validity of the hypothesis made in the earlier experiments—that endothermic reactions occurred in the boundary layers, resulting in reduced heat transfer rates beyond what is to be expected when no recombination occurs—has been investigated.

Key words: High enthalpy, Hypervelocity, Catalycity, Surface heat flux

1. Introduction

This study investigates the role of chemical reactions in enhancing or inhibiting surface heat flux to bodies exposed to high enthalpy, hypersonic flows produced in a free-piston shock tunnel. Attention has been given to both gas phase atomic recombination and to surface catalysed atomic recombination at the solid surface. The aim is to test long standing but experimentally unconfirmed theories concerning the extent of the departure of the reactions from chemical equilibrium and the consequent effects on surface heat flux. In particular, attention is directed towards examining the veracity of the hypothesis (see East et al. 1980) that endothermic reactions occur in the boundary layers of sharp leading edge flat plate flows causing reductions in the surface heat flux beyond that which is expected when no recombination occurs via exothermic reactions.

2. Experimental details

The free-piston shock tunnel T3 at the Australian National University in Canberra was used to generate high enthalpy hypersonic flows for the present experiments. The experiments were conducted at two flow conditions, designated B and D, typically representing high and medium enthalpy flows respectively. A conical nozzle of 300 mm exit diameter and two throat sizes 12.7 mm and 25.4 mm diameter were used to generate the free stream. The larger nozzle throat yielded flows with densities about three times higher. Two test gases, air and nitrogen, were used in the experiments. The reservoir and free-stream conditions of these flows are given in Table 1 below.

Table 1. Reservoir and free-stream values for conditions B and D; h = enthalpy, p = pressure, T = temperature, ρ = density, U = velocity, M = Mach number, α_N = nitrogen dissociation fraction, α_O = oxygen dissociation fraction, subcripts 0 and ∞ refer to reservoir and free-stream conditions, respectively.

	Test Gas	Throat mm	h_0 MJ kg^{-1}	p_0 MPa	T_0 K	T_∞ K	ρ_∞ g m^{-3}	U_∞ m s^{-1}	M_∞	α_N	α_O
B	Air	12.7	19.0	22.0	8400	1200	2.5	5500	7.5	10^{-7}	0.82
B	N_2	12.7	19.0	22.0	8800	1250	2.5	5500	7.4	0.08	-
D	Air	12.7	13.0	21.0	7100	930	3.5	4600	7.4	10^{-9}	0.43
D	N_2	12.7	14.0	22.0	8100	1050	3.3	4900	7.4	0.03	-
B	Air	25.4	18.5	21.0	8300	1850	8.8	5200	5.8	10^{-6}	0.72
B	N_2	25.4	18.0	21.0	8800	2100	8.6	5400	5.7	0.07	-
D	Air	25.4	13.5	21.0	7200	1500	11.7	4600	5.8	10^{-7}	0.38

Surface heat flux was measured on a sharp leading edge flat plate at 0° incidence via chromel / alumel surface thermocouples, whose accuracy was ± 15 % (Gai & Joe 1992). In order to assess the effects of varying the degree of catalycity, the model was made such that it was divided lengthwise into two halves, one of which was an aluminium surface and the other was copper plated. The surface thermocouples were located symmetrically with respect to the halves and also staggered such that the fetch of each was unaffected by the other. The effects of a fully non-catalytic surface were investigated in a second series of experiments, by coating the lower half of the model with silicon dioxide SiO_2 which is known to have very low catalycity. A schematic of the model is shown in Fig 1.

3. Theoretical considerations

The Stanton number, S, is defined as

$$S = \dot{q}_w / \rho_e u_e \left(h_r - h_w \right) \tag{1}$$

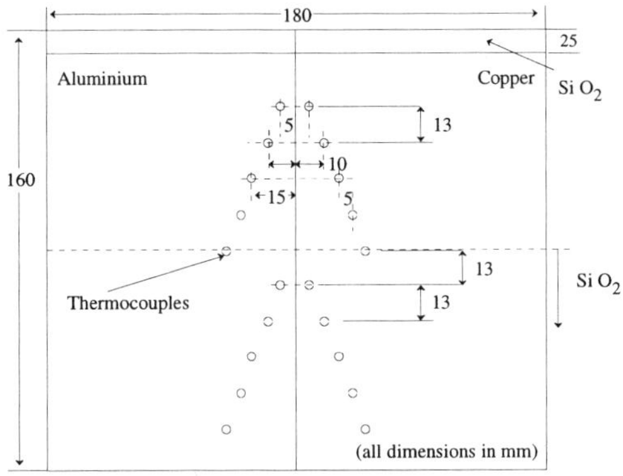

Figure 1. Schematic of the model.

Knowing that $C_f = 0.664\,(C^*/Re_x)^{1/2}$, where $C^* = \rho^*\mu^*/\rho_e\mu_e$, C_f is the skin friction coefficient and Re_x is the Reynolds number based on conditions at the edge of the boundary layer and distance x from the leading edge, we have, by Reynolds analogy,

$$S = 0.5C_f\,(\sigma^*)^{-2/3} \tag{2}$$

where σ^* is the Prandtl number (0.72 for air and 0.67 for nitrogen). Generalizing this to the frozen flow limit, we obtain (see East et al. 1980)

$$S_f = 0.5C_f\,(\sigma^*)^{-2/3}\{1 - h_{chem}/\,(h_r - h_w)\} \tag{3}$$

where h_{chem} is the chemical potential enthalpy of the dissociated species. The value of h_{chem} can either increase due to dissociation and vibrational excitation or decrease due to recombination and vibrational de-excitation. If the flow is in equilibrium, this will correspond to complete recombination either through gas phase boundary layer reactions or an infinitely catalytic wall, so that $h_{chem} = 0$ and we recover the equilibrium expression.

4. Results and discussion

Figure 2 shows results for conditions B and D with the nozzle throat size of 12.7 mm. The heat transfer data are expressed in the terms of the Stanton number against the Reynolds number. Firstly, we note that within the experimental scatter, there are no significant differences between aluminium/copper and SiO_2 coated results. This shows that, for the present experimental conditions, both catalytic and non-catalytic surfaces behave essentially the same indicating negligible surface recombination. Secondly, the results predominantly follow the frozen prediction although over the rear of the plate, there seems to be some tendency to rise towards the equilibrium prediction.

Figure 3 shows results obtained with the larger nozzle throat. We note that, on average, the data seem to lie closer to the equilibrium condition, indicating that with increased density of the flow, greater recombination is taking place, driving the heat transfer towards the equilibrium limit.

Figure 4 shows the air data compared with that of nitrogen at the same flow enthalpies. Using nitrogen eliminates reactions and reaction products involving oxygen. We note that, similar to air, the nitrogen data also fall approximately along the frozen chemistry prediction. In fact, if anything, they are somewhat below the frozen flow prediction on the front part of the plate. Comparing the air and nitrogen results, it is quite evident that the hypothesis of endothermic reactions involving atomic oxygen in the boundary layer, as invoked in East et al. (1980), is not sufficient to explain their lower than frozen boundary layer heat flux results.

Figure 5 shows all the data for condition B in terms of the similitude parameters $M_\infty^3 S$ and $\bar{\chi}$ (Stollery 1970), corrected to take into account the chemical potential enthalpy in a dissociated flow (Mallinson 1994). The data seem to be satisfactorily correlated in terms of these parameters and with the theory (Mallinson, 1994). Also shown in Fig. 5 are the data of Mallinson (1994). Figure 5 summarizes and demonstrates, firstly, that the air and nitrogen data are similar, and, secondly, that the differences between the results for aluminium, copper and glass surfaces are of the same order as the uncertainty in the measurements.

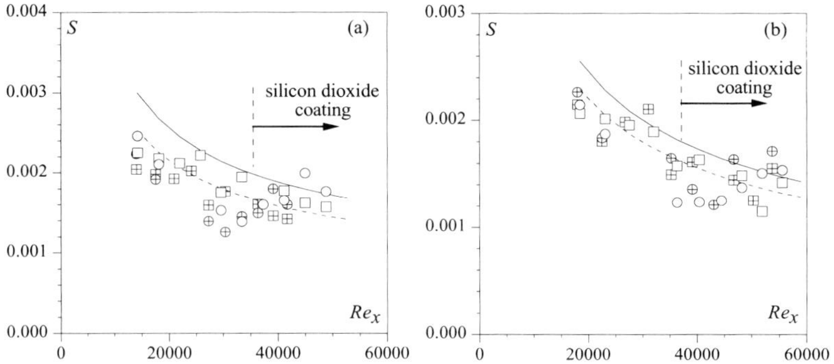

Figure 2. Heat transfer distributions for air with 12.7 mm throat: (a) condition B; (b) condition D. ○ = Al side; □ = Cu side. Open symbols = no SiO_2 coating; crossed symbols = SiO_2 coating on rear of model. Solid line = equilibrium limit; dashed line = frozen limit.

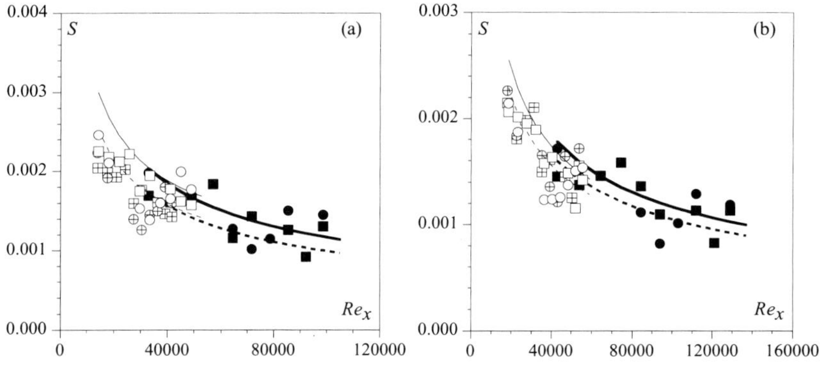

Figure 3. Heat transfer distributions for air with 12.7 mm and 25.4 mm throat: (a) condition B; (b) condition D. ○ = Al side; □ = Cu side. Open symbols = 12.7 mm throat, no SiO_2 coating; crossed symbols = 12.7 mm throat, SiO_2 coating on rear of model; filled symbols = 25.4 mm throat, SiO_2 coating on rear of model. Solid line = equilibrium limit; dashed line = frozen limit; heavier lines pertain to 25.4 mm throat.

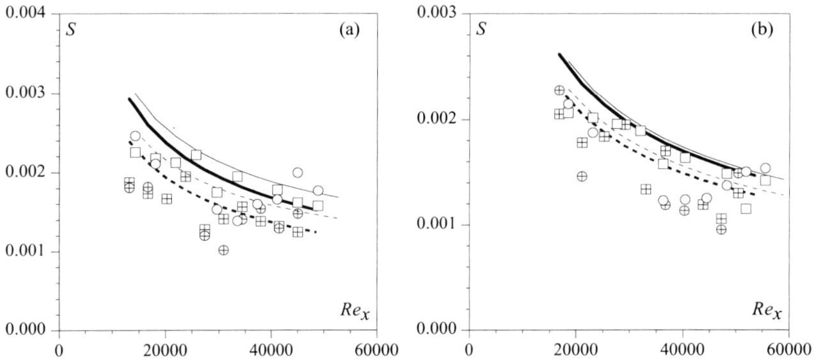

Figure 4. Heat transfer distributions for air and nitrogen with 12.7 mm throat: (a) condition B; (b) condition D. ○ = Al side; □ = Cu side. Open symbols = air , no SiO_2 coating; crossed symbols = nitrogen, no SiO_2 coating. Solid line = equilibrium limit; dashed line = frozen limit; heavier lines pertain to nitrogen.

5. Conclusions

The results show reasonable comparison with the frozen enthalpy theory thus indicating that the boundary layer reactions are frozen and surface reactions are negligible. The nitrogen data are similar but tend to be somewhat below the

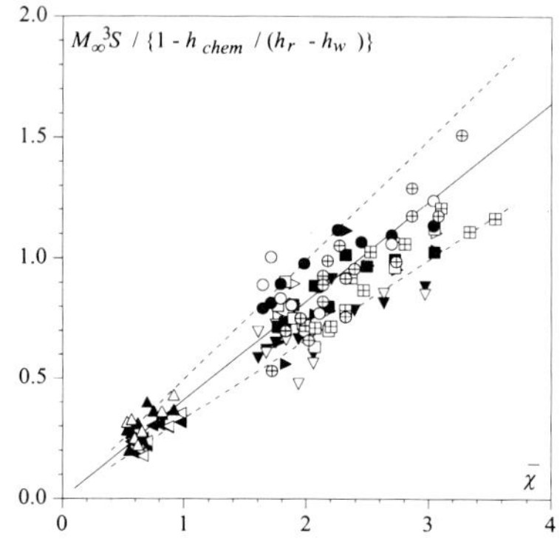

Figure 5. Correlation of flat plate heat transfer for condition B. Open symbols = Al side; filled symbols = Cu side. ○ = condition B-Air, 12.7 mm throat, no SiO_2 coating; □ = condition B-Air, 12.7 mm throat, no SiO_2 coating; △ = condition B-Air, 25.4 mm throat, SiO_2 coating on rear of model; ∇ = condition B-N_2, 12.7 mm throat, no SiO_2 coating; ▷ = condition B-N_2, 12.7 mm throat, SiO_2 coating on rear of model; ◁ = condition B-N_2, 25.4 mm throat, SiO_2 coating on rear of model; crossed circle = condition B-air, Mallinson (1994); crossed square = condition B-N_2, Mallinson (1994). Solid line = frozen theory, Mallinson (1994); dashed lines = ±20%.

frozen value, which cannot be explained by taking recourse to the endothermic reactions.

Acknowledgement. We gratefully acknowledge the generous support of the Australian Research Council (Grant A 89131813). We would also like to thank Mr. P.M. Walsh of the Department of Physics and Theoretical Physics, the Faculties, ANU for his assistance with the experiments and Dr. A. Byrne of the RSPhysSE, ANU, for his help in the preparation of the surface coatings.

References

East RA, Stalker RJ, Baird JP (1980) Measurements of heat transfer to a flat plate in a dissociated high-enthalpy laminar flow. J Fluid Mech 97:673.

Gai SL, Joe WS (1992) Laminar heat transfer to blunt cones in high-enthalpy hypervelocity flows. J Thermophysics 6:433.

Mallinson SG (1994) Shock Wave / Boundary Layer Interaction at a Compression Corner in Hypervelocity Flows, PhD Thesis, University of New South Wales, Australia.

Stollery, JL (1970) Hypersonic viscous interaction on curved surfaces. J Fluid Mech 43:497.

Study of experiments sensitive to vibration-dissociation coupling models

J. Olejniczak, G. V. Candler
Aerospace Engineering & Mechanics and Army HPC Research Center
University of Minnesota
Minneapolis, MN 55455 USA

Abstract: A set of experiments to test vibration-dissociation coupling models has been designed for the T5 shock tunnel at Caltech. Using computational fluid dynamics, numerous experimental configurations were analyzed before choosing double-wedge and double-cone geometries. The predicted flow fields for these geometries strongly depend on the choice of the coupling model. This sensitivity is due to a complex shock-wave/boundary-layer interaction that results in oscillations of flow quantities along the body surface. Subsequent measurements of surface pressure and heat-transfer rate, along with interferometry, will provide new validation data for vibration-dissociation coupling models.

Key words: Hypersonic flow, Vibration-dissociation coupling, Shock interaction

1. Introduction

In many hypersonic flows the characteristic times of the fluid motion, chemical dissociation, and vibrational relaxation are of the same order of magnitude. Therefore, these processes must be modeled in a coupled manner. In compressing flows, the vibrational relaxation reduces the dissociation rate, and the dissociation of high-energy oscillators inhibits the vibrational relaxation. The details of this coupling process affect the pressure distribution on a vehicle, and therefore its trim angle of attack.

There are many vibration-dissociation coupling models, however no model has been adequately validated with experiments. This is because there are no experiments specifically designed to test the coupling models. This paper presents a set of experiments designed for the T5 shock tunnel at Caltech that is sensitive to the choice of the vibration-dissociation coupling model. The experiments were designed using a computational fluid dynamics method (Olejniczak and Candler 1995) and two different coupling models. A combination of flow conditions and geometry was found that produces a measurable difference between the flow fields computed using the two models. We found that certain double-cone and double-wedge geometries produce flows that are particularly sensitive to the choice of the model; this sensitivity is discussed below.

We chose the Park (1987) $\sqrt{TT_v}$ and Marrone and Treanor (1963) CVDV models because previous work showed that under T5 conditions, they reproduce the data at either end of the experimental uncertainty (Olejniczak and Candler 1995). Nitrogen was chosen as the test gas to avoid the uncertainties in modeling reacting air. Interferometry will be used to visualize the flow, while surface pressure and heat-transfer rate measurements will provide quantitative data.

2. Design of experiments

To design the experiments, we used computational fluid dynamics to simulate the flow over many different geometries at T5 freestream conditions. Each case was computed with the two different coupling models. The goal was to identify the combination of geometry and freestream conditions that yields the greatest difference in the interferograms and surface quantities between the two computed flow fields.

We found that certain double-wedge and double-cone geometries are attractive because the two coupling models predict dramatically different flow fields. High-enthalpy and high-density freestreams with certain wedge or cone angles are particularly sensitive to the coupling model. Examples are shown in Figs. 1 and 2, where the computed pressure contours are plotted for the flow over double-wedge configurations. The first wedge produces an attached shock which interacts with the detached shock from the second wedge. This produces a complicated flow field with a separation zone near the corner of the wedges. The size of the separation zone depends on the coupling model. There is a range of wedge angles for which the predicted flow field is particularly sensitive to the choice of coupling model; the reason for this sensitivity is discussed below. Similar results are found for double-cone geometries.

The measurable differences between the predictions of the two coupling models are shown in Figs. 3 and 4. Figure 3 shows the computational Mach-Zehnder interferograms corresponding to Fig. 1. The differences in the shock location, fringe number, and fringe shape are easily seen. We can assess the accuracy of these models by comparing the computed interferograms with the experimental interferogram. Figure 4 plots the surface pressure and heat-transfer rate for this case. Note that the pressure and heat-transfer-rate peaks caused by the shock impinging on the second wedge are shifted and that the separation zones are different sizes. These differences will be greater than the uncertainty in the experimental measurements, provided there is sufficient instrument density on the model. Thus, this set of experiments will serve as validation data for vibration-dissociation coupling models.

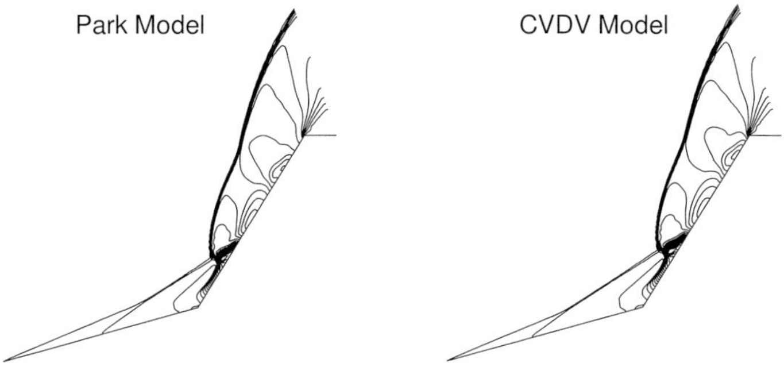

Figure 1. Pressure contours for flow over a double-wedge with a base diameter of 5 cm and angles of 15.5° and 58.5°. The freestream conditions are nitrogen at $u = 6690\,\mathrm{m/s}$, $\rho = 0.01\,\mathrm{kg/m^3}$, and $h_o = 26\,\mathrm{MJ/kg}$.

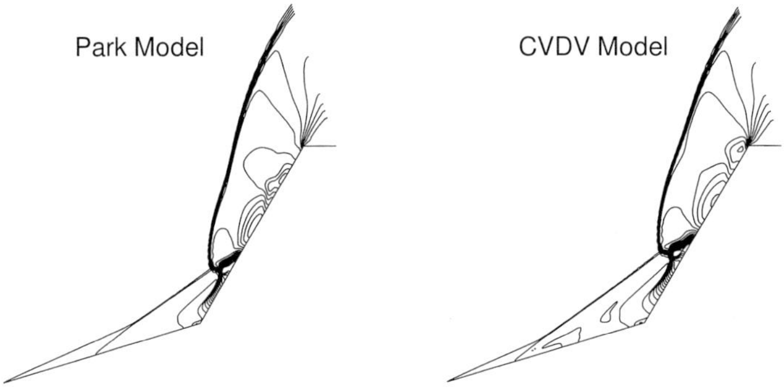

Figure 2. Pressure contours for flow over a double-wedge with a base diameter of 5 cm and angles of 17° and 60°. Freestream conditions as above.

3. Discussion of flow

The flow over the double-wedge and double-cone geometries can be understood by considering a schematic of the flow, shown in Fig. 5. A transmitted shock, formed by the intersection of the oblique shock from the first wedge and the shock from the separation zone, intersects with the bow shock from the second wedge. A transmitted shock from this triple point impinges on the second wedge, causing separation to occur. The gas that is reversed into the separation zone must flow out of the shear layer at the edge of the separation zone. As the second wedge angle is increased the impinging shock becomes more normal,

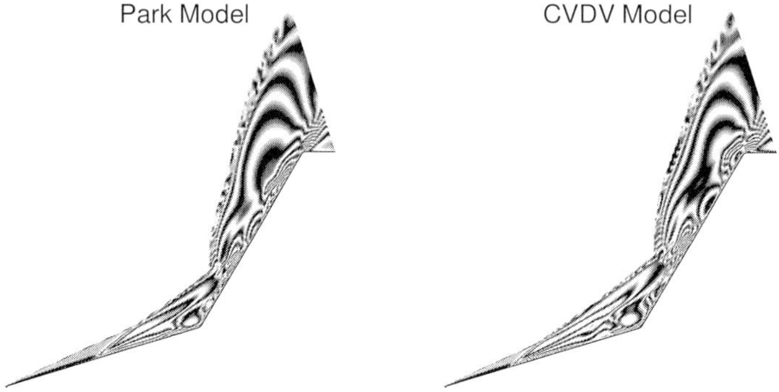

Figure 3. Computed interferograms for the previous figure.

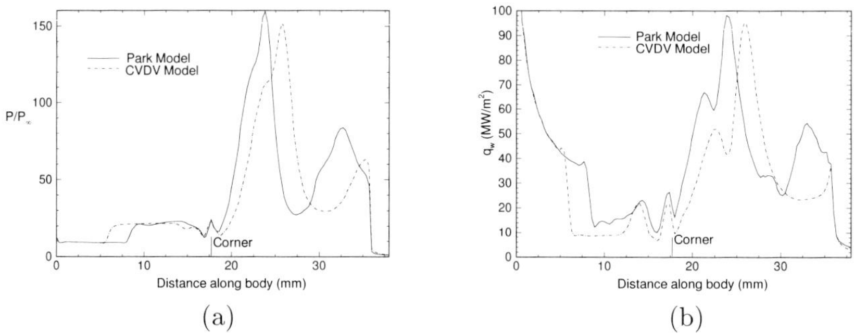

Figure 4. Computed flow over a double-wedge corresponding to Fig. 2: (a) pressure distribution, (b) heat-transfer-rate distribution.

and more gas is reversed into the separation zone. Eventually a steady shear layer cannot eject enough mass, and the flow becomes unsteady and oscillates. This is the oscillation mechanism identified by Maull (1960) for spiked bodies. This process is highly nonlinear near the point of unsteadiness. Small changes in the bow shock, resulting in changes in the location of the triple point, have a large influence on the size of the separation zone. Thus, different models for the nonequilibrium dissociation rate and the vibrational energy source terms predict widely different flow fields.

Another interesting feature of this flow, which provides a quantitative test of the coupling model, is the underexpanded supersonic jet that is formed along the second wedge surface. This jet is bounded by the shear layer emanating from the triple point and is trapped against the body by the subsonic region behind

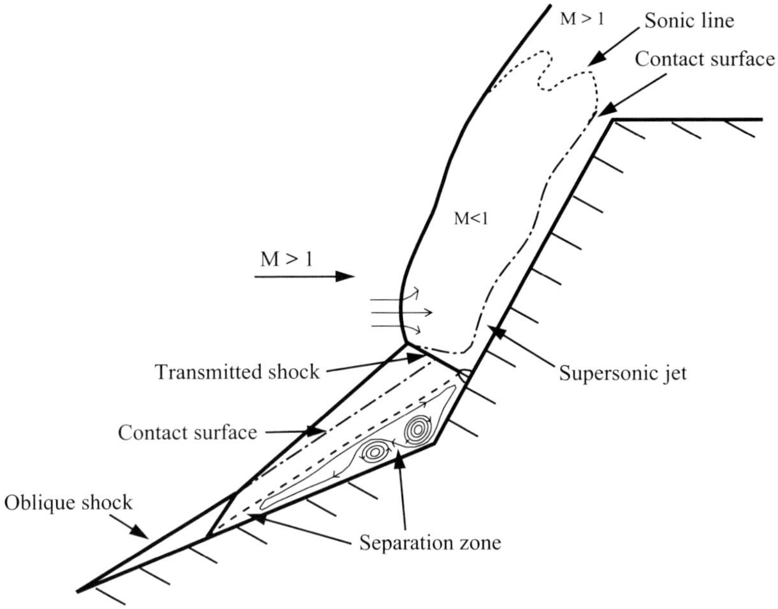

Figure 5. Schematic diagram of the viscous flow over a double-wedge geometry.

the bow shock. This jet produces oscillations in the flow variables along the second wedge surface, as shown in Fig. 6. These oscillations result from the finite rate of transfer between the different energy modes. As in an underexpanded jet, the main mechanism is the reversible transfer between internal and kinetic energy (the $p\nabla\cdot\vec{u}$ term in the kinetic-energy equation). The peak height and wavelength of the oscillations are altered by the rate of vibrational-translational energy transfer and vibration-chemistry energy transfer. This latter term differs among the vibration-dissociation coupling models. Figure 6 clearly shows the differences in the predicted oscillations of pressure and heat-transfer rates between the Park and CVDV models. These oscillations are also present in the density, and therefore are seen in the interferograms.

4. Conclusion

This paper presents the design of a set of experiments which will provide new validation data for vibration-dissociation coupling models. The experiments, planned for the T5 shock tunnel, use double-wedge and double-cone geometries, which we have found to be particularly sensitive to vibration-dissociation coupling physics. The mechanism for this sensitivity is a shock-wave/boundary-layer interaction in which the model of the vibration-dissociation coupling has

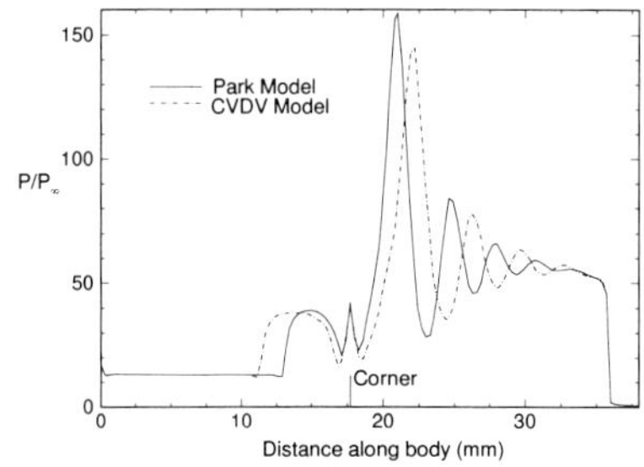

(a)

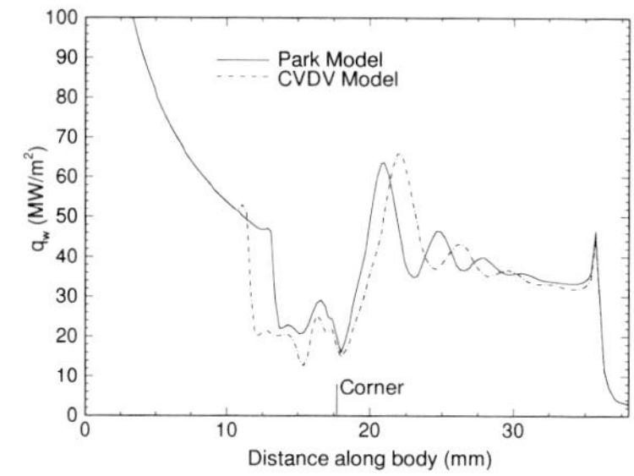

(b)

Figure 6. Computed flow over a double-cone with angles of 25° and 68°. Freestream conditions are the same as the previous figures. (a) Pressure distribution, (b) heat-transfer-rate distribution.

a large influence on the predicted flow field. There are differences in the size of the predicted separation zone and in the oscillation of flow quantities along the body. Therefore, interferometry, along with surface pressure and heat-transfer-rate measurements, will provide data to test vibration-dissociation coupling models.

Acknowledgement. We would like thank Professor H. G. Hornung for providing many useful insights and assisting us in the design of the experiments. This work was supported by the Air Force Office of Scientific Research Grant Number F49620-93-1-0338. Computer time was provided by the University of Minnesota Supercomputer Institute. This work was also sponsored in part by the Army High Performance Computing Research Center under the auspices of the Department of the Army, Army Research Laboratory cooperative agreement number DAAH04-95-2-0003/contract number DAAH04-95-C-0008, the content of which does not necessarily reflect the position or the policy of the government, and no official endorsement should be inferred.

References

Marrone PV, Treanor CE (1963) Chemical relaxation with preferential dissociation from excited vibrational levels. Physics of Fluids 6:1215.

Maull DJ (1960) Hypersonic flow over axially symmetric spiked bodies. J Fluid Mechanics 8:584.

Olejniczak J, Candler GV (1995) Vibrational energy conservation with vibration-dissociation coupling: general theory and numerical studies. Physics of Fluids 7:1764.

Park C (1987) Assessment of Two-Temperature Kinetic Model for Ionizing Air, AIAA Paper No. 87-1574.

Supersonic Flow

Disturbances on the shock front ahead of a concave body due to resonant excitement of vortices

A. O. Serov, L. S. Shtemenko, F. V. Shugaev
Moscow State University, Department of Physics
Chair of Molecular Physics
119899 Moscow, Vorobyovy Gory, Russia

Abstract: A jet is found to arise behind the shock wave reflected from a concave body. The jet gives rise to vortical structures. Experiments were conducted in a shock tube with carbon tetrafluoride used as the test gas. A resonant excitation of the vortical structures arises inside the cavity. Radially unstable cores emit beams of pressure waves which cause disturbances on the shock front.

Key words: Resonant excitement of vortices, Disturbances on shock front

1. Introduction

Sound emission from unstable vortices was investigated by Kopyov et al. (1987) and Vlasov et al. (1978). Milton et al. (1987) showed the formation of vortices behind the shock wave reflected from a cavity. Shock-wave instability was examined in a hypersonic wind tunnel ahead of concave bodies (Johnson 1959). It was suggested that the instability may be the result of the formation of vortices inside the cavity. Disturbances on the shock front were observed in ballistic experiments at high Mach numbers, $M > 8$ (Bedin et al. 1981). However, processes inside the cavity were not investigated.

We studied the formation of vortices and their influence on the detached shock wave in front of concave bodies in supersonic flow.

2. Experimental facility

Experiments were conducted in a single-diaphragm shock tube of a rectangular 34×72 mm^2 cross-sectional area. Carbon tetrafluoride (CF_4) was chosen as the test gas. A model (concave body) was mounted in the test section which had glass windows. The models had the shape of a parallelepiped. The cross-section of some models was a 10-mm square. The flow was three-dimensional. Other models had a width equal to the distance between the glass windows (34 mm), so the flow past the models was close to two-dimensional. The height of the models was $h = 10$ mm. The generatrix of the cavity was either a 10-mm diameter (deep cavity) circle or a 20-mm diameter (shallow cavity) one. Some models had a rectangular cavity. The initial pressure was equal to $p_0 = 1.3 \times 10^3$–2×10^4 Pa. The Mach number of the incident shock wave varied from 2 to 5, the Reynolds number varied from 10^4 to 2.4×10^5 in the flow behind the reflected shock wave, the characteristic length being half the arc of the cavity. A quasi-stationary supersonic flow took place in the test section after the burst of the diaphragm. Its duration was about 150 – 300 μs. We took shadowgraphs and streak-camera records. The pressure on the bottom of the cavity was measured with a Kistler transducer.

3. Results and discussion

The flow pattern inside the cavity was described earlier (Ibrahim et al. 1985; Shtemenko et al. 1993). The characteristic feature of the flow is the formation of the jet. It is laminar at low Reynolds numbers. Vortices are situated at the end of the jet. They also arise near the walls due to the interaction of the transverse shocks with the boundary layer and because of its subsequent separation. In addition, vortices appear as a result of the instability of slipstreams. Figure1 shows successive shadowgraphs. The vortices move toward the plane of

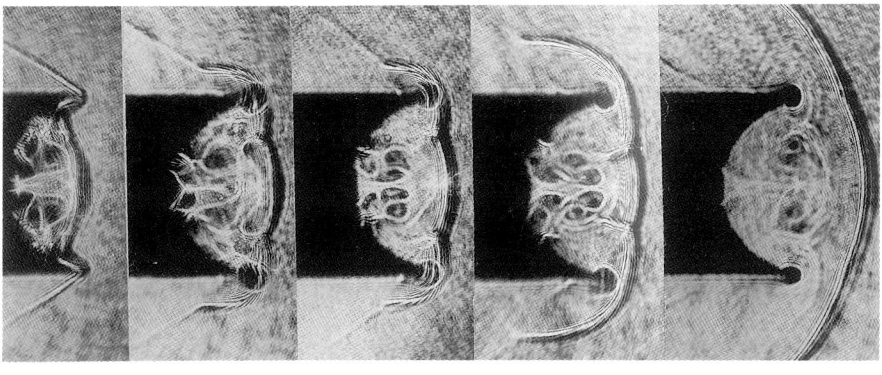

Figure 1. Two-dimensional flow pattern inside the deep cavity: $l/h = 0.5$; $M = 4$; $p_0 = 1.3\times10^3$ Pa; $t = 20$ μs, 30 μs, 35 μs, 50 μs, 80 μs; the time is measured from the instant when the incident shock arrives at the bottom of the cavity (l is its depth).

symmetry and then inside the jet toward the reflected shock. Afterwards they stop and turn into two cylindrical vortices (or into a vortex ring when the flow is a three-dimensional one). If the Reynolds number is large enough then the vortical structures become unstable. Waves propagate along the vortex ring i.e., azimuthal instability arises. Distortion of the vortex ring takes place as one can see from Fig. 2. Radial (axisymmetrical) waves propagate along the

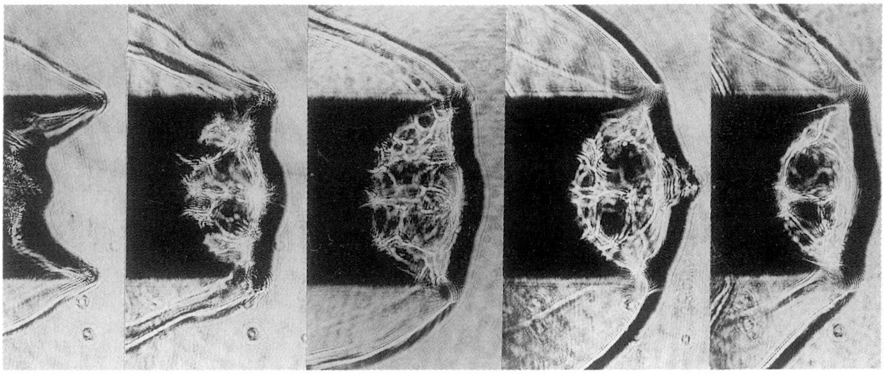

Figure 2. Three-dimensional flow pattern inside the deep cavity: $l/h = 0.5$; $M = 4.5$; $p_0 = 6 \times 10^3$ Pa; $t = 15$ μs; 30 μs, 60 μs, 70 μs, 90 μs.

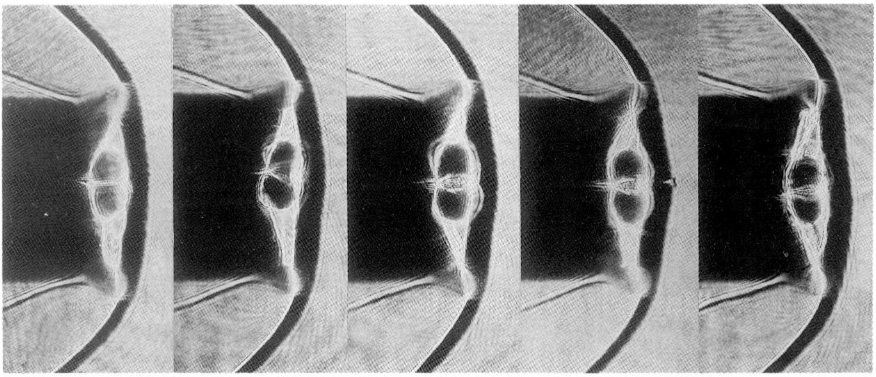

Figure 3. Three-dimensional flow pattern inside the shallow cavity: $l/h = 0.15$; $M = 4.9$; $p_0 = 4 \times 10^3$ Pa; $t = 60$ μs, 80 μs, 90 μs, 100 μs, 130 μs.

vortex ring if the Reynolds number is equal to 2–3$\times10^5$ and the Mach number is equal to 4.5–5. Figure 3 shows the oscillations of the vortex ring.The core of the ring becomes elliptic and disturbances emerge on the shock front. The motion of the shock front is oscillatory. The vortical structures are destroyed periodically, the flow inside the cavity becomes turbulent and the gas flows out of the cavity. Then the vortical structures appear anew and the process repeats. Two oscillations were seen at the streak camera records during 150 μs. Figure 4 shows the flow pattern inside the rectangular cavity. Vortical structures are situated near horizontal walls and not near the plane of symmetry as in the former case. These structures oscillate (the diameter of the vortex core increases and decreases). As a consequence a disturbance appears on the shock front against

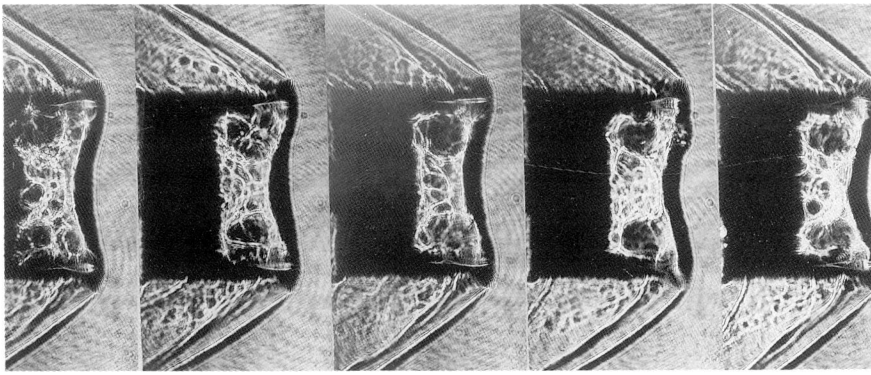

Figure 4. Three-dimensional flow pattern inside the rectangular cavity: $l/h = 0.5$; $M = 4.9$; $p_0 = 4 \times 10^3$ Pa; $t = 50$ μs, 70 μs, 100 μs, 110 μs, 120 μs.

the vortex. The disturbances appear in the two-dimensional flow at the same values of Reynolds number and Mach number.

The amplitude of the pressure oscillations at the bottom of the cavity is equal to 10% of the pressure mean value behind the shock wave in the case of the deep cavity. The period of oscillations coincides with the time during which the vortex moves from the bottom of the cavity to the vortex ring. The pressure waves reflect from the bottom and interact with the vortical structures. At high Reynolds numbers ($Re = 2$–3×10^5, the characteristic length being the diameter of the vortex ring) the vortical structures become unstable and emit acoustic waves that arrive at the shock front.

Disturbances on the shock front appear under appropriate conditions, namely, at $M = 4.5$–5. This fact can be explained as the result of the interaction between acoustic waves and vortices. The interaction may be resonant or not. If there is no resonance then the vortex ring oscillates as a whole, the frequency being equal to that of incident acoustic waves. If the frequency of the acoustic waves is a multiple of the natural frequency of the vortex ring then the ring becomes unstable. Radial waves propagate along its surface. The ring emits acoustic waves and their amplitude may be several times as great as the amplitude of the incident waves.

In our experiments the resonance arises at $M = 4.9$ (shallow cavity). The amplitude of the oscillations grows as the time increases. Radial waves appear at the surface of the core of the vortex ring. One can see this phenomenon from Fig. 3. The frequency of the oscillations of the vortex ring surface is equal to 170–220 kHz as it was found from streak camera records. The frequency of the pressure oscillations at the bottom of the cavity was equal to 150–200 kHz. It is the frequency of the waves that the vortex emits at a resonant excitement.

The frequency of oscillations was equal to 50–70 kHz in the case of deep cavity. In experiments we measured the period of time between two successive disturbances on the shock front. We calculated the amplitude and frequency of the acoustic waves that caused these disturbances. The values coincided with ones recorded with the aid of the pressure transducer. Consequently the disturbances on the shock front are caused by the acoustic waves emitted by unstable vortical structures while being resonantly excited. Figures 2 and 3 show diffraction fringes near unstable vortex rings. We consider them to be a result of the diffraction of light from the grid that is formed by ultrasonic waves in the gas (Shtemenko et al. 1993).

Figure 5 demonstrates the natural frequency of the vortex ring core and the frequency of acoustic waves against Mach number.

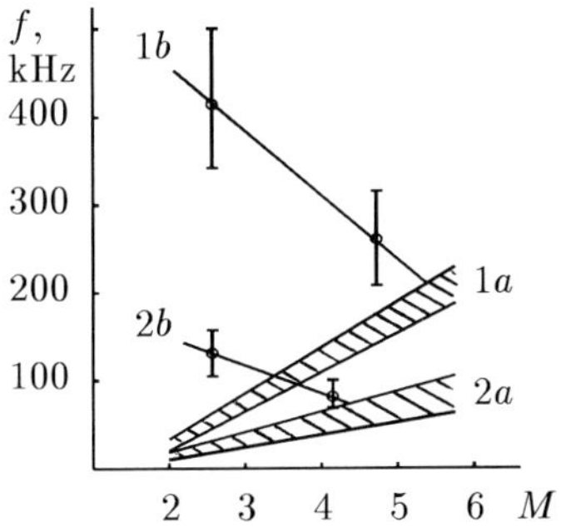

Figure 5. Dependence of the natural frequency of the vortex core and of the frequency of acoustic waves on Mach number: 1 corresponds to shallow cavity, 2 to deep cavity; *a* corresponds to acoustic waves, *b* to vortex ring.

The frequency of acoustic waves may be evaluated if we know the time during which they propagate along the path from the stagnation point of vortices to the bottom of the cavity. The value of the natural frequency of the vortex ring was calculated according to the measured size of the core and the ring velocity. Index 1 relates to the shallow cavity, index 2 relates to the deep cavity, *a* denotes acoustic waves, *b* denotes vortex ring. The size of shaded area denotes the uncertainty in frequency determination. One can see that both values of frequency coincide at M = 4–5, i.e., in the range where disturbances on the shock front exist.

4. Conclusion

An experimental investigation of the flow inside the cavity behind the reflected shock wave has been performed. Disturbances on the shock wave in front of a concave body may appear under appropriate conditions. The origin of the disturbances is a resonant interaction between acoustic waves and vortical structures inside the cavity.

References

Bedin AP, Klimov AI, Koblov AI, Mishin GI (1981) Experimental investigation of the flow of relaxing gas past concave bodies. Zhurn Tekhn Fiz 51(10):2119–2122 (in Russian).

Ibrahim MA, Serov AO, Shtemenko LS, Shugaev FV (1985) The reflection of a plane shock wave from a body with groove. Fluid Dynamics (Izv Akad Nauk SSSR MZhG) 20:822–828.

Johnson RH (1959) Instability in hypersonic flow about blunt bodies. Phys Fluids 2:526–532.

Kopyov VF, Leontyev EA (1987) Radiation and scattering of sound from vortex ring. Fluid Dynamics (Izv Akad Nauk SSSR MZhG) 22:398–409.

Milton EE, Takayama K, Archer RD (1987) An evaluation of two-dimensional and axisymmetric focusing of a plane shock wave. In: Proc 16th Int Symp on Shock Waves, Groenig H (ed), Aachen, pp 567–573.

Shtemenko LS, Shugaev FV, Serov AO (1993) Visualization of acoustic waves by means of diffraction. J Flow Vis and Image Processing 1:113–116.

Vlasov EV, Ginevsky AS, Karavasov RK (1978) Investigation of the flow pattern of a jet at various levels of initial turbulence. Uchenye zapiski TSAGI 1:25–32 (in Russian).

An inverse method for calculating nonequilibrium flow behind curved shock waves

S. Mölder
School of Aerospace Engineering, Ryerson Polytechnic University
Toronto, Ontario, M5B 2K3 Canada

Abstract: The theoretical aerodynamic model described here applies to flow of an ideal dissociating gas between a shock and the body producing that shock. The model allows calculation of flow with either axial or planar symmetry and for concave or convex curved shocks which can be either attached or detached at the leading edge of the body. For non-reactive flows over plane wedges we obtain exact results; for flow over circular cones, extremely good agreement is obtained for both weak and strong shocks; much improved agreement over previous methods and with experimental results is obtained for detached shocks on blunt bodies, particularly at the stagnation streamline. For flows in chemical nonequilibrium we get qualitative agreement with the calculations of previous authors.

Key words: Aerodynamics, Hypervelocity, Nonequilibrium

1. Introduction

For supersonic aerodynamics of blunt bodies the *direct problem* is the one where the body shape is specified and the shock shape and the flow between the shock and the body are calculated. For the *inverse problem* the shock is given, and the body shape and the intervening flow are found. Although the direct problem is of usual practical interest, there are design-oriented approaches, where the desired flow is specified, in which case it is convenient to begin by specifying shock-wave shape. Such situations occur in the design of supersonic and hypersonic air inlets. Van Dyke (1958) solved the inverse problem for an ideal gas and found that the solution is poorly convergent. Maslen (1964) based his inverse calculation on the assumption that the streamline curvature behind the shock is equal to the shock curvature and that this curvature remains constant from the shock to the body. This assumption gives good results for moderately inclined body surfaces in hypersonic flow. However it does not correspond to reality for blunt-body flow where shock and streamline curvature have opposite signs. Also, for the flow over a circular cone Maslen's theory assumes a zero streamline curvature behind the shock and therefore a constant pressure in the region between the shock and the body. This is contrary to the exact Taylor-McColl equation solution. In the present inverse method we calculate an exact flow curvature behind the specified shock. The changing flow field is then calculated between the shock and the body assuming the transverse pressure gradient, at a given station, to be a constant as determined by the streamline curvature at the shock at this station.

2. The aerodynamic model

Consider a shock element located a distance y_1 from a reference axis (Fig. 1). In a plane which contains both the upstream and downstream flow vectors, the radius of curvature of the shock is R_a. In a plane which is perpendicular to this plane and to the shock wave the curvature of the shock wave is R_b. It is easy to show that for axisymmetric flow $R_b = y_1/\cos\theta_s$ where θ_s is the shock angle. The corresponding curvatures of the shock are $S_a = -1/R_a$ and $S_b = -1/R_b$. The curvature of the streamline at the shock is calculated using the curved shock relations (Mölder 1971). This curvature is denoted by $(\partial\delta/\partial s)_1$ where δ is the local flow inclination and s is the distance measured along the streamline. In intrinsic streamline coordinates (s, n), the momentum equation across the streamlines is:

$$\frac{\partial p}{\partial n} = \rho V^2 \frac{\partial \delta}{\partial s} \tag{1}$$

In finite-difference form this can be written as:

$$(\Delta n)_1 = \frac{(\Delta p)_1]}{\rho_1 V_1^2 (\frac{\partial\delta}{\partial s})_1} \tag{2}$$

where Δn is measured along a normal to the streamline such that the pressure change is Δp. At this point we know $\Delta p/\Delta n$ but neither Δp nor Δn individually. Now select a small mass flow $\Delta\dot{m}$. This amounts to selecting the calculation step size. It is the mass flow between the streamline passing through (x_1, y_1) and the streamline immediately below, so that: $\Delta\dot{m} = \rho_0 V_0 (\Delta A)_0 = \rho_1 V_1 (\Delta A)_1$.

Using this we can find the height at which the streamline enters the shock, y_2. This allows us to geometrically locate the point (x_3, y_3), at the intersection of the streamline from (x_2, y_2) and the streamline-normal from (x_1, y_1), and to find the pressure $p_3 = p_1 + (\Delta p)_1$. To calculate the other flow quantities at (x_3, y_3), Maslen (1964) assumed that the flow is isentropic from (x_2, y_2) to (x_3, y_3). Thus knowing p_3 and the conditions behind the shock at (x_2, y_2), p_2, and T_2, we can find T_3 from $T_3/T_2 = (p_3/p_2)^{\gamma/(\gamma-1)}$. The energy equation and the equation of state then allow the determination of all flow variables at (x_3, y_3). This process is then repeated for the next lower streamline. The approximate nature of this method comes about from the assumption that the pressure gradient at any given downstream station remains proportional to the streamline curvature at

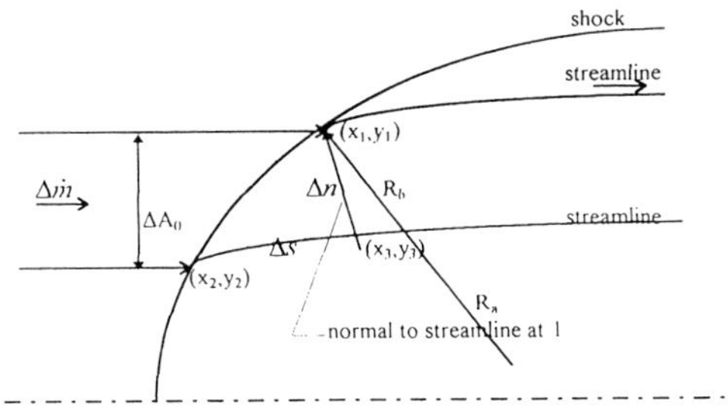

Figure 1. Schematic diagram of curved shock element.

the shock at that station. Thus the method should give results which are exact at the shock and as accurate as the modified Newtonian theory near the body surface.

3. Results

3.1. The aerodynamic model

The present inverse method has been applied to flow behind a plane oblique shock, flow behind a circular cone shock, and flow behind a blunt-body shock consisting of cubic spline segments, in each case for a perfect gas as well as for a chemically reactive gas.

Case A1 — Plane oblique shock. This is the classical oblique shock-on-a-wedge situation for which the streamlines behind the shock are straight and parallel and there is no pressure gradient. Many cases were calculated for Mach numbers ranging from 1.2 to 25 and for shock angles between the Mach angle and 90 degrees. In all cases the calculated wedge angle did not vary from the exact value by more than 0.001 degrees.

Case A2 — Cone flow. For this case the streamline curvature causes the pressure to increase from the shock to the body surface. This flow configuration provides a test of the main feature of this report, namely the assumption of pressure-gradient constancy along the streamline-normals. Figure 2 compares the present results against the classical Taylor-McColl equation solution for flow over a circular cone. For both strong and weak shock waves the difference between the two predictions of cone angle was usually less than one degree, and never larger than 2 degrees.

Case A3 — Blunt body. For blunt-body flow the curved bow shock was specified either as an equation or as a series of points on the shock wave. In the latter case a cubic spline was fitted through the points and the shock

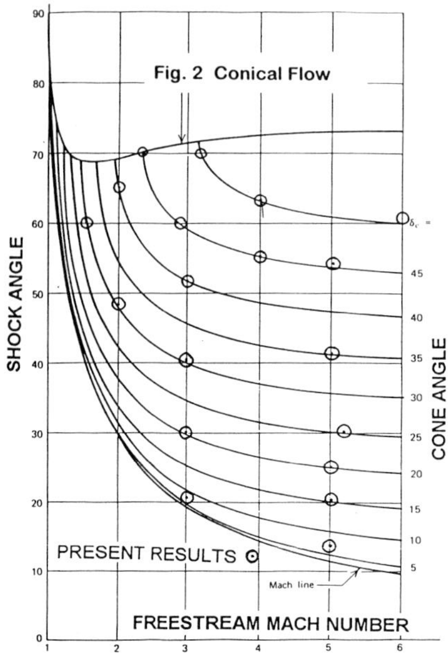

Figure 2. Conical flow.

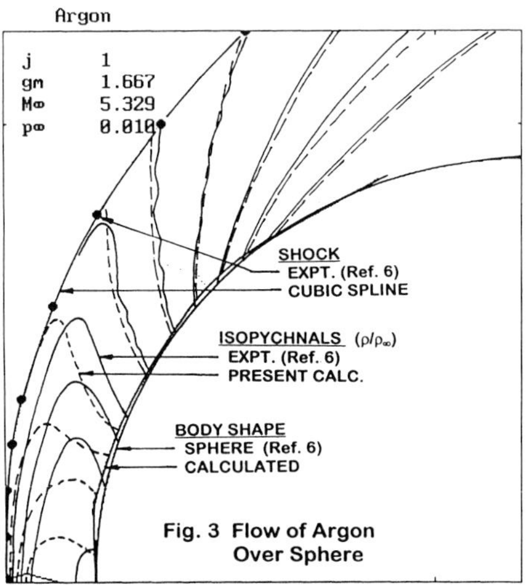

Figure 3. Flow of argon over a sphere.

slope and curvature were then calculated from the spline. In Fig. 3 we compare our calculations with the experimental results of Sedney and Kahl (1961) for Mach number 5.329 flow of argon over a sphere. In this case, of course, no dissociation is present so that we are again testing the aerodynamic aspects of our model. The cubic spline passes through the experimental shock-wave points. Solid contours are lines of measured constant density. Dashed contours are for the corresponding calculations. Good agreement is apparent except in the stagnation region. Lack of agreement is most likely due to the difficulty of specifying the experimental shock shape. Improved agreement in the stagnation region was obtained by perturbing the shock shape within the experimental tolerances. The calculated body is seen to be very close to spherical. Again, small changes in shock shape would lead to even better correspondence. Further refinements were not justified since changes in body shape were of the order of the body boundary-layer thickness, for which no experimental data were available. Aerodynamic features of blunt-body flow are predictable with good accuracy.

3.2. The thermochemical model

The present aerodynamic technique calculates the flow quantities along streamlines, making it particularly suitable for the superposed calculation of chemically reactive flows with time-dependent total differential rate equations. The purpose of this section is to illustrate the application of the simple Lighthill-Freeman nonequilibrium ideal dissociating-gas model (Lighthill 1957, Freeman 1958) where we calculate the dissociation rate (the substantive rate of change of the molecular mass fraction α) along the streamline.

Case C1 — Nonequilibrium flow behind plane shock. Figure 4a shows the nonequilibrium flow of air behind a plane shock at a freestream Mach number of 25. Shown are the lines of constant dissociation rate. As discussed by Stalker (1989), the wedge surface is slightly concave towards the oncoming flow.

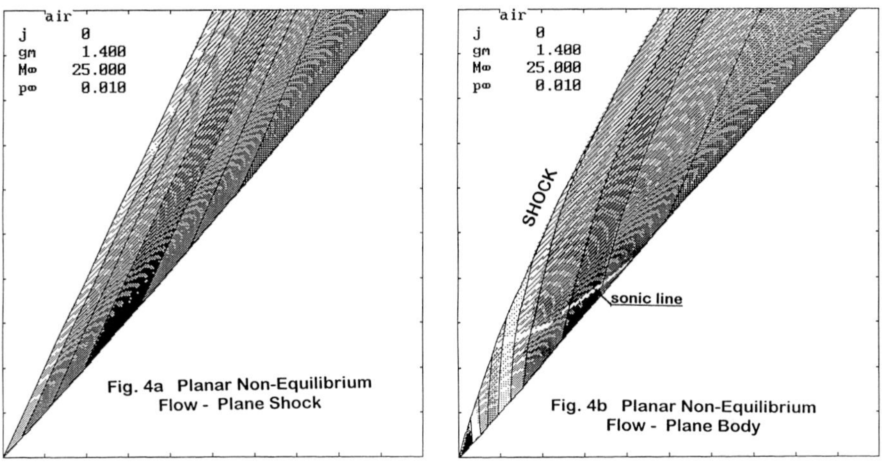

Figure 4. Planar nonequilibrium flow: (a) plane shock; (b) plane body.

Case C2 — Nonequilibrium flow over plane wedge. Figure 4b shows the nonequilibrium flow over a plane wedge. In this case the shock, as well as the lines of constant-dissociation rate, are convex towards the oncoming flow. A sonic line appears, separating the subsonic flow near the leading edge from the supersonic flow further downstream. This clearly illustrates the non-similar nature of nonequilibrium flow over wedge surfaces.

Case C3 — Nonequilibrium flow over bluff body. Figures 5a, b, c are for Mach number 25 flow over a bluff, axi-symmetric, mushroom-shaped body. The three figures are for lines of constant temperature, constant dissociated mass fraction, and constant dissociation rate. The general features of such flows, as described by Hornung (1976), are confirmed.

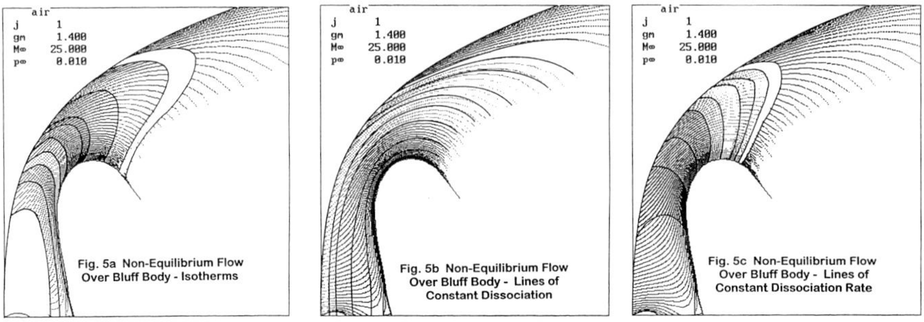

Fig. 5a Non-Equilibrium Flow Over Bluff Body - Isotherms

Fig. 5b Non-Equilibrium Flow Over Bluff Body - Lines of Constant Dissociation

Fig. 5c Non-Equilibrium Flow Over Bluff Body - Lines of Constant Dissociation Rate

Figure 5. Non-equilibrium flow over bluff body: (a) isotherms; (b) lines of constant dissociation; and (c) lines of constant dissociation rate.

4. Conclusions

An inverse method is presented for the calculation of flow of chemically active gas behind a given shock wave. The method gives very good results for non-reactive flows and is in qualitative agreement with previously calculated ideal dissociated-gas flows.

Acknowledgement. This work was begun under USAF Contract No. F33615-87-C-2748 and The Johns Hopkins University Contracts 602235-0 and 603077-S. Completion was made possible through a sabbatical leave opportunity provided by Dr. G. Eitelberg of DLR Göttingen and Dr. S. Matar of Ryerson Polytechnic University.

References

Freeman NC (1958) Non-equilibrium flow of an ideal dissociating gas. J Fluid Mech 4(4):407.

Hornung HG (1976) Non-equilibrium ideal gas dissociation after a curved shock wave. J Fluid Mech 74:(1)143–150.

Lighthill MJ (1957) Dynamics of a dissociating gas. J Fluid Mech 2(1):1.

Maslen SH (1964) Inviscid hypersonic flow past smooth symmetric bodies. AIAA J 2(6):1055–1061.

Mölder S (1971) Reflection of curved shock waves in steady supersonic flow. CASI Trans 25:44–49.

Sedney R, Kahl GD (1961) Interferometric study of the hypersonic blunt body problem. In: Proc. of the 4th AFBMD/STL Symposium, Pergamon Press.

Stalker RJ (1989) Approximations for nonequilibrium hypervelocity aerodynamics. AIAA J 27(12)1761–1769.

Van Dyke MD (1958)The supersonic blunt-body problem — review and extension. J Aerosp Sci 25:485–495.

Shock-cell structure in a jet at the critical pressure ratio

A. Makhsud, E. Brocher, A. Chambarel, P. Cordonnier
Institut de Mécanique des Fluides
13003 Marseille, France

Abstract: For a jet emanating from a nozzle, the one-dimensional gasdynamics equations lead to the well-known relation for the critical pressure ratio $\pi_c = [(\gamma+1)/2]^{\gamma/\gamma-1}$. At this pressure ratio the jet exit Mach number is equal to unity for one-dimensional flow. In many experimental set-up or practical situations (Crouch et al. 1976), the geometrical configurations of the nozzle do not generate a one-dimensional flow. In particular, experiments on jet noise have often been carried out with internal conical shapes (Ponton & Seiner 1992). The free-surface hydraulic analogy of the two-dimensional flow of a compressible gas is a useful experimental tool to study wave and acoustic phenomena. In this analogy the specific heat ratio γ is equal to 2, the pressure is proportional to the square of the water height h, and the Froude number is equivalent to the Mach number. Therefore, the critical pressure ratio of the "hydraulic" gas is equal to $\pi_c = (3/2)^2$ and the critical height ratio h_{tot}/h_a equal to 3/2, where h_{tot} represents the total height of the flow and h_a the water height on the water table. While running experiments with a converging nozzle (5° half angle) on this table, it was observed that a shock-cell structure exists at the critical height ratio. The existence of a shock-cell structure is important since it may lead to the generation of the so-called "screech" noise. It was then decided to carry out a detailed study of this flow configuration, both experimentally and numerically.

Key words: Shock-cell structure, Critical pressure ratio, Numerical computation

1. Introduction

When a convergent nozzle is operated at subcritical ratio, or when a convergent-divergent nozzle is operated at design pressure ratio, the acoustic spectrum is broad band and smooth, and consists of pure turbulent mixing noise. However, if a convergent nozzle is operated at supercritical pressure ratio (underexpanded), and a convergent-divergent nozzle is operated at off-design Mach number (underexpanded or overexpanded), the resulting acoustic spectrum contains an extra noise contribution, due to the presence of shock structure in the jet flow, in addition to basic turbulent mixing noise (Tanna 1977). The shock-cell structure was investigated by Powell (1953) who observed that the shock-cell spacing is not constant along the jet axis because turbulence in the shear layer produces a momentum exchange at the jet periphery. Tam et al. (1985) have used a multiple-scale model to determine the shock-cell structure of imperfectly expanded supersonic jet in detail. Gutmark et al. (1990) have experimentally studied an underexpanded supersonic rectangular jet over a wide range of pressure ratios.

The hydraulic analogy has been shown to be a valuable tool to simulate the complex interaction process between turbulence, sound, and shock waves. It may be demonstrated that, if the vertical acceleration in the shallow-water flow is smaller than the gravitational acceleration at any point in the flow, the conservation equations for this flow have the same form as the equations for a compressible, two-dimensional flow. Comparison of the momentum equation for the two cases shows that the analogy implies the following correspondence between physical quantities:

Compressible-gas flow	Shallow-water flow
pressure p	$gh^2/2$
density ρ	h
temperature T	h
specific heat ratio $\gamma = Cp/Cv$	2
gas velocity u	water velocity u
speed of sound $c = \sqrt{\gamma RT}$	speed of wavelets $c = \sqrt{gh}$
Mach number $M = u/c$	Froude number $F_r = u/\sqrt{gh}$

where h represents the water depth and g the gravitational constant.

In the analogy, sound waves are simulated by water waves. Despite its limitations (ratio of specific heats = 2, some dispersion of wave speed due to surface tension, simulation restricted to two-dimensional flows) the shallow-water analogy offers several interesting experimental advantages: (i) simple experimental set-up; (ii) geometrical configuration easily modified; (iii) possibility of simulating high Mach-number flows (high Froude number) with small energy expenditure; (iv) visualization of shock-wave/sound-wave propagation by shadowgraph optics; (v) measurement of wave spectrum with non-intrusive method (fiber optics or optical displacement sensors); (vi) possibility of flow-field measurement by PIV; and (vii) frequency of phenomena typically several hundred times smaller than in the gas (some of them visible with naked eye and easily recorded with a video-camera). For amplitudes small compared to the mean value, the linearized relation between pressure and water-level fluctuation is $(\Delta p/p) = 2(\Delta h/h)$. In the present work, a convergent nozzle is operated at critical pressure ratio (critical height ratio) and the shock-cell structure is visualized by using shadowgraphy. Flow visualization showed that a shock-cell structure was present even at the critical height ratio, which cannot be the case for a one-dimensional flow.

2. Experiments

The experimental set-up has previously been described (). The nozzle length is 441 mm. It has walls converging at a half angle of 5° and an exit width D equal to 57 mm. The water level h_a on the table is 27.5 mm so that the total height in the water tank is 41.3 mm at the critical height ratio. The velocities are measured with a hot-film probe Dantec gauge 55 R 42 and the water level with a vertical needle mounted on a micrometer. It was found that the Froude number of the flow at the nozzle exit is not equal to unity as would be the case for a one-dimensional flow at the critical height ratio. The measured Froude number is only 0.83 and the flow is therefore *underexpanded.* After leaving the nozzle, the jet expands and reaches a water level which is even lower than h_a. It is then *overexpanded* and the local Froude number is higher than unity. This is only possible if the streamtubes form converging-diverging channels, the converging part having its origin in the particular nozzle shape. Downstream of this overexpansion, successive compression and expansion waves form the shock-cell structure which has been observed (Fig. 1).

3. Numerical computations

With the usual notations, for a compressible fluid, the Navier-Stokes equations can be written:

$$\begin{cases} \dfrac{\partial}{\partial x_j}(\rho u_j) + \dfrac{\partial \rho}{\partial t} = 0 \\ \rho\left(\dfrac{\partial u_i}{\partial t} + u_j \dfrac{\partial u_i}{\partial x_j}\right) = -\dfrac{\partial p}{\partial x_i} + \dfrac{\partial \tau_{ij}}{\partial x_j} \\ \dfrac{\partial}{\partial t}(\rho e) + \dfrac{\partial}{\partial x_j}(u_j(\rho e + p)) = \dfrac{\partial}{\partial x_j}\left(u_j \tau_{ij} + \dfrac{\gamma}{P_r}\dfrac{\partial T}{\partial x_j}\right) \end{cases}$$

$$\text{i.e.:} \quad \frac{\partial}{\partial t}(\rho, u_i, T) = \Phi(\rho, u_i, T) \tag{1}$$

$$\text{and} \quad p = (\gamma - 1)\rho\left[e - \tfrac{1}{2}\left(u_j^2\right)\right] = (\gamma - 1)\rho T\ .$$

The form of variational formulation is:

$$\int_\Omega \delta(\rho, u_i, T)\frac{\partial}{\partial t}(\rho, u_i, T)\, d\Omega = \int_\Omega \delta(\rho, u_i, T)\,\Phi(\rho, u_i, T)\, d\Omega$$

and one obtains a differential system discretized by finite elements for the space and Θ-method for the time:

$$([M] + \Theta \Delta t [K])\,\Delta\{\rho, u_i, T\}^k = \Delta t\{\Phi_n\}^{k+\Theta} \qquad \text{and} \qquad t_{k+1} = t_k + \Delta t\ .$$

An automatic mesh-refinement technique was performed (Sani & Gresho 1992). A parallel computer M.I.M.D. (Ipsc 860) and a parallel algorithm for the resolution are used (Liu 1987). The jet nozzle and the outer space were discretized. A time-dependent flow was calculated before the jet filled a pseudo-cavity.

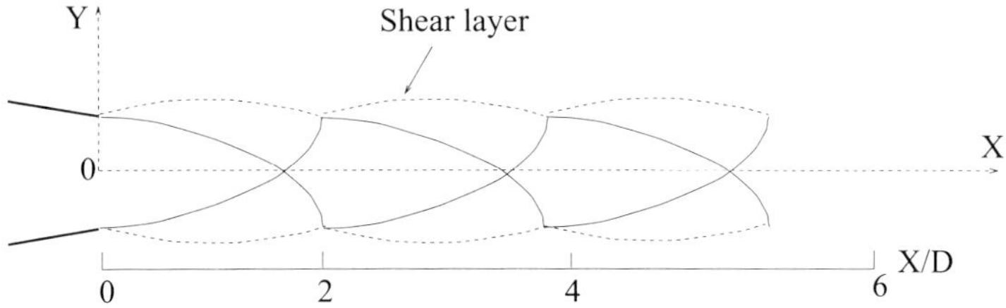

Figure 1. Shock-cell structure.

4. Results and discussion

The Mach number on the jet axis and the velocity profiles have been studied experimentally and computationally. Results for the local Froude-number distribution along the jet axis at the critical height ratio 3/2 are represented in Fig. 2. Experimentally, at the nozzle exit, the flow is subcritical ($Fr = 0.83$), it then accelerates and reaches a supercritical value ($Fr = 1.15$). The jet is then *overexpanded*. This is easily explained by the fact that the flow is conver-

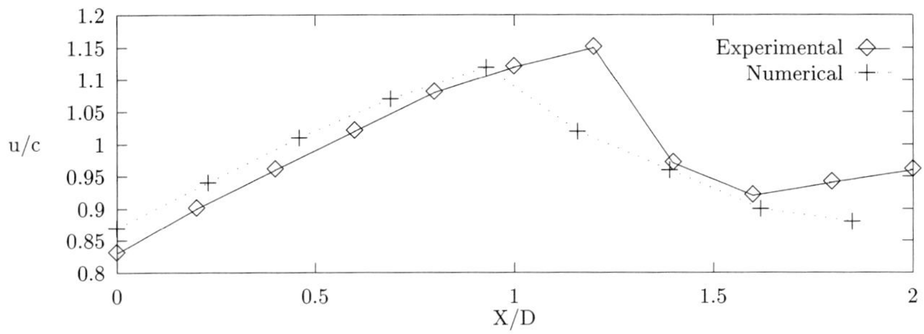

Figure 2. Mach number (Froude number) along the jet axis; X is abscissa, D is nozzle width.

gent at the nozzle exit and has to diverge at some point downstream. At the downstream end of the overexpanded region, a shock wave appears and decelerates the flow to subcritical velocities. The instationary numerical results are in agreement with the experiments on three points:

1. The Mach number increases downstream of the nozzle exit and the maximum Mach number is superior to unity. The difference between the numerical and experimental results at values $X/D \geq 1.75$ is explained by the fact that the numerical domain boundary produces a reflection wave which does not give a stationary flow.

2. Uniformity of the velocity across the jet was tested by measuring the velocity profile at a height $Z = 15$ mm above the surface of the water table. The comparison of measurements and numerical results is shown in Fig. 3. The agreement is seen to be quite satisfactory, except for the shear-layer velocity profile.

3. The velocity vectors obtained numerically clearly show that the flow downstream of the nozzle exit is at first convergent and then divergent, thereby creating a sonic throat (results to be published).

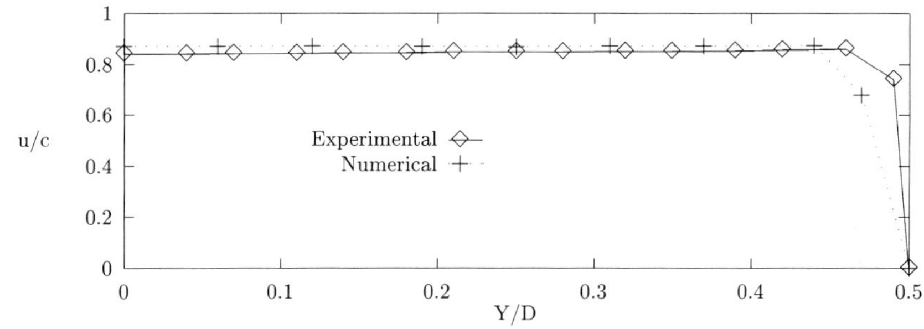

Figure 3. Velocity profiles at the nozzle exit.

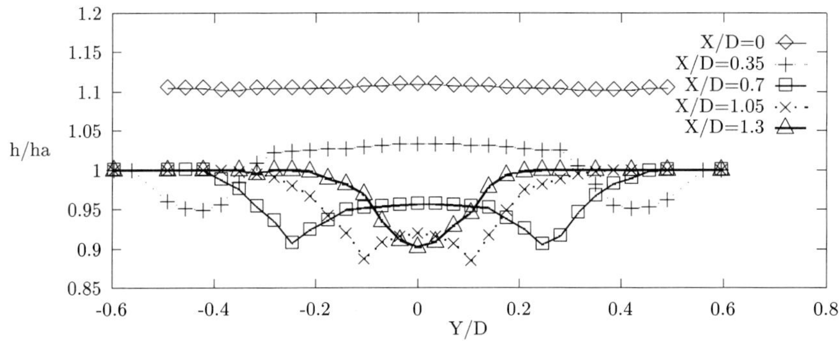

Figure 4. Water-depth profiles downstream of nozzle exit; Y is lateral coordinate.

The local value h of the water depth is interesting for two reasons. First, comparison of this value with the ambient water depth h_a enables one to detect areas of the flow for which either overpressure or underpressure exists, because in the analogy $(h/h_a)^2 = (p/p_a)$. Second, combined with the measurement of the local velocity, it is possible to calculate the local value of the Froude number $Fr = u/\sqrt{gh}$. Figure 4 shows the water-depth profiles at the critical value of $h_{tot}/h_a = 1.5$. It is seen that a fairly strong depression builds up on the jet axis.

5. Conclusions

The hydraulic analogy has shown that a shock-cell structure exists at the critical height ratio (critical pressure ratio) for a convergent nozzle. This process is due to the fact that the streamtubes form converging-diverging channels downstream of the nozzle exit. Good agreement was obtained between the experimental and numerical results, both for the Mach number on the jet axis and for the velocity profiles. Future work with the hydraulic analogy will include detailed measurements of the interaction process of the shear layer with the shock-cell structure by using the PIV technique.

References

Brocher E, Makhsud A (1995) Experimental study on the influence of nozzle inlet geometry on screech noise. In: First Joint CEAS/AIAA Aeroacoustics Conference (16th AIAA Aeroacoustics Conference), CEAS/AIAA Paper 95-067, Munich, Germany.

Crouch RW, Coughlin CL, Paynter GC (1976) Nozzle exit flow profile shaping for jet noise reduction. In: 3rd AIAA Aero-Acoustics Conference, AIAA Paper 76-511, Palo Alto, California.

Gutmark E, Schadow KC, Bicker CJ (1990) Near acoustic field and shock structure of rectangular supersonic jets. AIAA J 28:1163–1170.

Liu WK (1987) Parallel Computation for Mixed-Time Integration. Numerical Method for Transient and Coupled Problems, J Wiley & Sons Ltd.

Ponton MK, Seiner JM (1992) The effects of nozzle exit lip thickness on plume resonance. J of Sound and Vibration 154(2):531–549.

Powell A (1953) On the noise emanating from a two-dimensional jet above the critical pressure. The Aeronautical Quarterly 4:103–122.

Sani RL, Gresho PM (1992) On the finite element modeling of the scalar transport equation. Revue Européenne des Eléments Finis 1(3).

Tam CKW, Jackson JA, Seiner JM (1985) A multiple-scales model of the shock-cell structure of imperfectly expanded supersonic jets. J of Fluid Mechanics 153:123–149.

Tanna HK (1977) An experimental study of jet noise. Part II shock associated noise. J of Sound and Vibration 50(3):429–444.

Mach disc and shock structure in coflowing dual axisymmetric jets

J. Kurian, T. V. Ramana Rao, R. Ramesh Kumar
Department of Aerospace Engineering
Indian Institute of Technology, Madras 600036 India

Abstract: The jet structure of co-axial flows has been investigated experimentally. The effect of variation in the Mach number and semi-divergence angle of the inner nozzle for various combinations of inner and outer chamber pressures were studied. The presence of the outer flow was found to give rise to an exit shock and compress the inner flow, leading to Mach-disc formation. The Mach number and nozzle wall angle of the inner nozzle were found to alter the shock structure of the dual jet. The mass-flow ratio of the outer to inner jet was also seen to modify the flow field.

Key words: Mach disc, Coflowing jets, Shock structure

1. Introduction

Axisymmetric co-axial supersonic jets are found in many high-speed air-breathing propulsion devices. The development and structure of a single freejet has been studied extensively by several investigators. Love et al. (1959) and Crist et al. (1966) studied the factors affecting the jet structure of single underexpanded jets. Love et al. (1959) also studied jets exhausting into supersonic outer streams and observed that an intercepting shock pattern existed at higher pressure ratios before giving way to a Riemann wave, than was the case if the outer stream was subsonic. Buckley (1975), based on his review of earlier studies, concluded that the location of the Mach disc would be unaffected by the coflowing airstream, since the axial Mach-number distribution upstream of the Mach disc is expected to be independent of the external flow conditions. D'Attorre and Harshbarger (1965), however, noted that the distance of the normal shock decreased as the ambient flow velocity was increased. Moreover, Narayanan and Damodaran (1993) observed significant changes in the centerline Mach number distribution upstream of the Mach disc, due to the presence of an outer flow.

The aim of this investigation was to study the influence of the exit Mach number and semi-divergence angle of the inner nozzle, and the chamber pressures of the inner and outer flows on the co-axial jet structure.

2. Experimental facility and test procedure

The experiments were conducted in a two-stream co-axial blowdown freejet facility. Air was the working medium for both streams and was supplied from a common reservoir, but the streams were controlled independently. Suitably designed nozzles could be attached to the inner flow line interchangeably. Nine conical convergent-divergent nozzles with semi-divergence angles (α) of 2.5°, 5°, and 10° designed for exit Mach numbers (M_e) of 1.4, 1.7, and 2.0 were used. The outer line was fitted with co-axial convergent nozzles with semi-convergence angles of 40°. The ratio of the exit area of the outer annular nozzle to the throat area of the inner nozzle (A_O/A_I) was 2. Axial and radial surveys were made using a pitot probe. A cone-static probe (based on the design of Kenworthy and Schetz 1973) was used to make axial surveys. A schlieren system using two concave mirrors in an off-axis arrangement was used for flow visualization. The Mach-disc diameter and location, and the shock cell lengths were measured from the negatives of the schlieren photographs.

3. Results and discussion

The effect of the presence of the outer flow on the jet structure is shown in Fig. 1. Figure 1a is a schlieren photograph of the underexpanded jet ($p_e/p_a = 1.125$) from the $M_e = 1.7$, $\alpha = 2.5°$ nozzle. Alternate expansion (bright) and

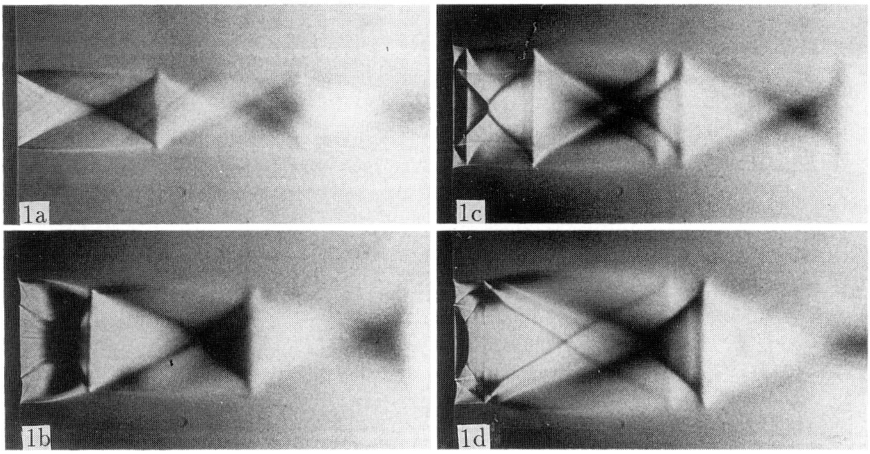

Figure 1. Schlieren photographs: (a) $P_I = 5.56$ atm; (b) $P_0 = 5$ atm; (c) $P_I = 5.56$ atm, $P_0 = 3$ atm; and (d) $P_I = 5.56$ atm, $P_0 = 5$ atm.

compression (dark) regions of the first shock cell are observed and no Mach disc is discernible. Figure 1b is the photograph of the annular jet; the jet is observed to meet the axis and form a shock cell downstream. Figure 1c is the photograph of the dual flow. An annular interaction shock is generated to maintain flow tangency to the inner jet boundary. A similar shock was observed by Love et al. (1959) and called the exit shock — the same terminology will be used here. The intercepting shock formed at the nozzle exit plane meets the jet axis and forms a Mach disc. These shocks proceed further, get reflected at the outer jet boundary to form the second shock cell. As seen in Fig. 1d, increasing the outer-flow chamber pressure moved the first Mach disc closer to the nozzle exit (reduced x_{m_1}) and increased the Mach-disc diameter (d_{m_1}). The greater expanse of the bright region in Fig. 1d relative to Fig. 1b is due to the greater expansion of the outer flow. The variation of d_{m_1} with x_{m_1} for all the nozzles tested is shown in Fig. 2, and is found to be approximately exponential. In contrast, Crist et al. (1966) and Davidor and Penner (1971) observed a linear variation for single sonic freejets. Figures 3a and 3b are the schlieren photographs of the $M_e = 1.7$ and $M_e = 2$, 5° nozzles respectively. The underexpansion ratios of

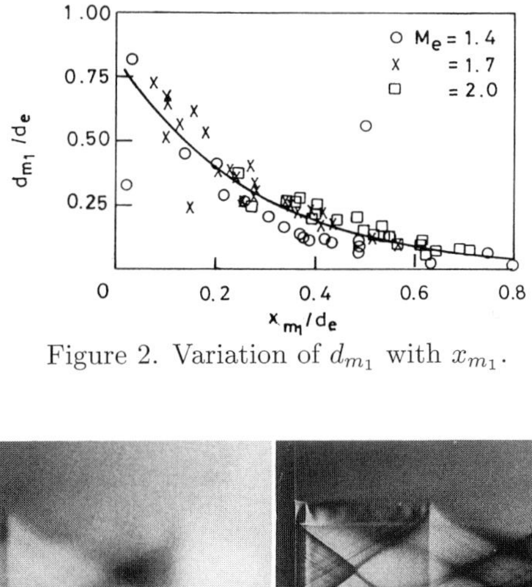

Figure 2. Variation of d_{m_1} with x_{m_1}.

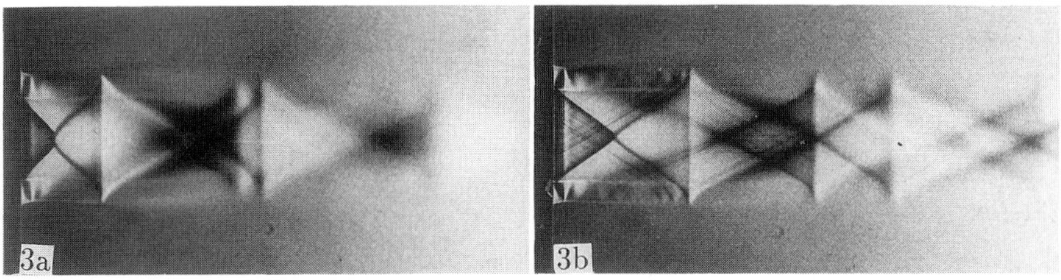

Figure 3. Schlieren photographs: (a) $P_I = 5.56$ atm, $P_0 = 3$ atm; (b) $P_I = 8.82$ atm, $P_0 = 3$ atm.

the inner and outer nozzles are the same; however, considerable difference is observed in the jet structure, indicating a strong Mach-number effect. Increase in the Mach number of the inner jet was found to increase the range of static pressure ratios in which a Mach disc was not formed. A similar trend was

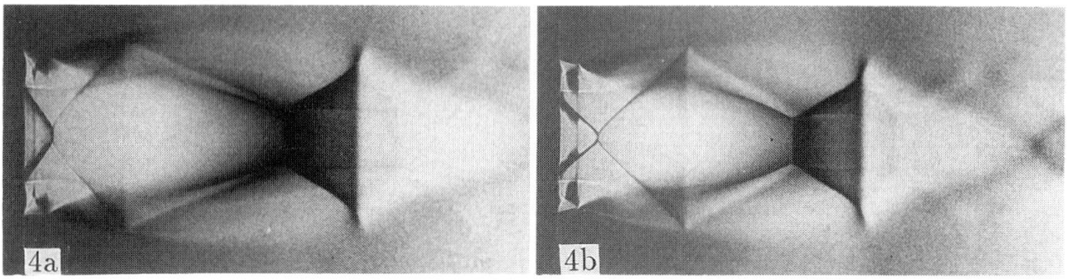

Figure 4. Schlieren photographs, $P_I = 7.8$ atm, $P_0 = 5$ atm, $M_e = 1.4$: (a) $\alpha = 2.5°$; (b) $\alpha = 10°$.

obtained by Love et al. (1959) for freejets exhausting into the ambient. The effect of the semi-divergence angle of the inner nozzle on the jet structure is illustrated in Fig. 4. Increase in the semi-divergence angle is found to increase x_{m_1} and reduce d_{m_1}. However, this trend was not consistently observed for all the nozzles tested. These photographs also show that maximum jet diameter occurs in the second shock cell — a trend followed in almost all the conditions studied. Thus, it is found that the chamber pressures of the inner and outer flows, exit Mach numbers, and semi-divergence angle of the inner nozzle affect the jet structure. Hence, attempts to correlate the jet structure must take all these factors into consideration. The parameter

$$\phi = \frac{(P_I/P_O)(M_e)^{-1/M_e}}{(\alpha/2.5)^{(0.25M_e-0.35)(0.5+1/(M_e)^2)}} \tag{1}$$

was found to reduce the scatter of the data, and reasonably correlate x_{m_1} and the first shock cell length (l_1). The result of the correlation for l_1 is presented in Fig. 5.

Figure 6a gives the variation of x_{m_1} with the chamber pressure of the inner flow, when the chamber pressure of the outer flow is held constant and vice versa. The linear variation observed was partially affected with increase in α. This trend was also not observed for the $M_e = 2$ nozzles. The variation of the axial location of the second Mach disc from the nozzle exit (x_{m_2}) with change in the chamber pressure of the inner and outer flow is shown in Fig. 6b. The linear variation of x_{m_2} with P_I and P_O was not observed to be followed for all the nozzles tested. Plots of d_{m_1} for variation in chamber pressures for the $M_e = 1.7$ nozzles showed that d_{m_1} increased with increase in P_O, and reduced with increase in P_I. Experiments also revealed that the change in mass-flow ratio (due to change in area) of the outer to inner flow affected the dual-jet structure as evident from Figs. 7a and b. Increase in the mass-flow ratio was found to reduce x_{m_1}, increase d_{m_1}, reduce l_1, and increase the second shock

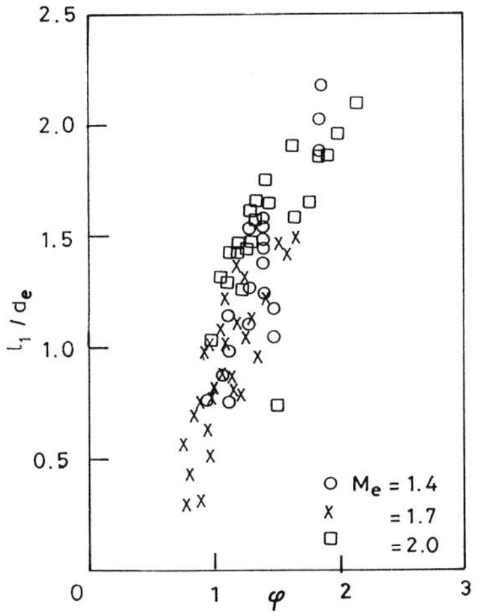

Figure 5. Variation of l_1 with the correlation parameter.

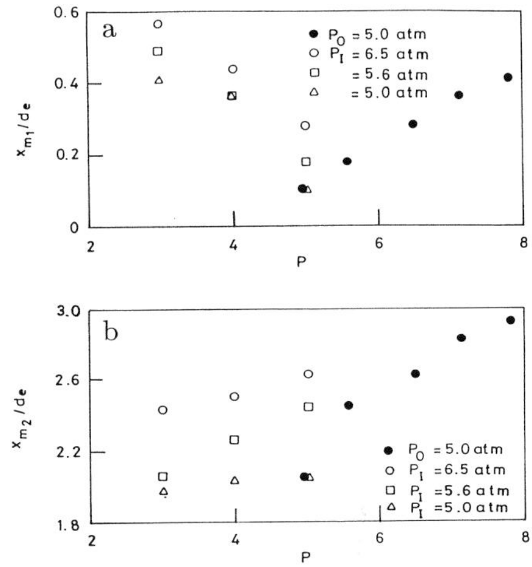

Figure 6. $M_e = 1.7$, $\alpha = 2.5°$: (a) variation of x_{m_1} with chamber pressure; (b) variation of x_{m_2} with chamber pressure.

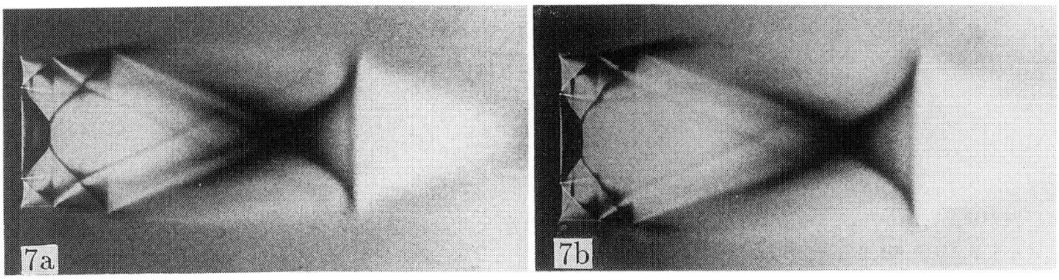

Figure 7. Schlieren photographs, $M_e = 1.7$, $\alpha = 2.5°$, $P_I = 7.12$ atm, $P_O = 5$ atm: (a) $A_O/A_I = 2$; (b) $A_O/A_I = 3$.

cell length. Figure 8 shows the variation in Mach number due to the presence of the outer flow. The Mach-number distribution upstream of the first Mach disc (of the single jet) is altered significantly due to the presence of the outer flow. Lastly, the variation in the shear-layer thickness (b) for the $M_e = 2$ nozzles is shown in Fig. 9. The thickness was defined as the difference in radial locations where the pitot pressure assumed a value of $1 + 0.1 \times (P_{cl} - 1)$ and $1 + 0.9 \times (P_{cl} - 1)$ (cl denotes the centerline value). As seen, the maximum thickness was observed for $\alpha = 5°$ — a similar trend was also observed for the $M_e = 1.7$ nozzles. However, the relative variation at $\alpha = 2.5°$ and $10°$ was not the same at the two Mach numbers.

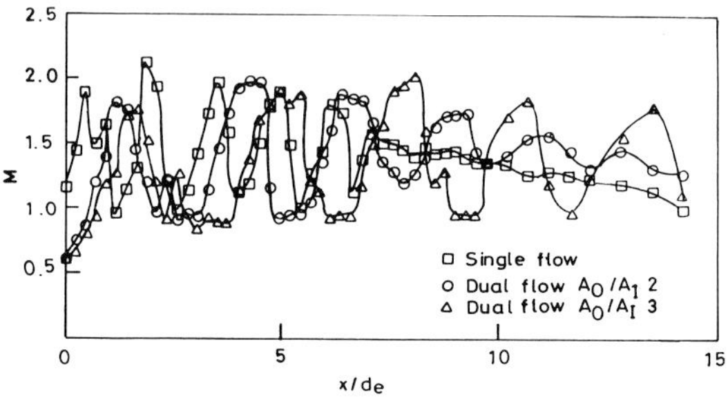

Figure 8. Variation in the centerline Mach-number distribution, $M_e = 1.4$, $\alpha = 10°$.

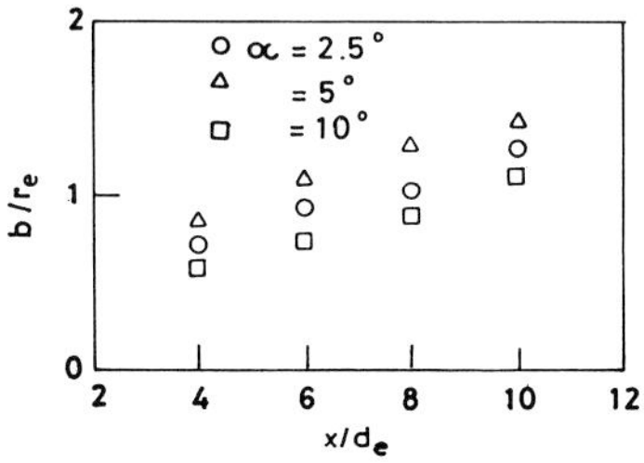

Figure 9. Variation in shear-layer thickness.

4. Conclusions

The presence of the outer flow produced striking changes in the centerline Mach number distribution; and under certain conditions the intercepting shocks formed a Mach disc. The location and size of the Mach disc were found to depend on the chamber pressures of the inner and outer flows. Increase in the Mach number of the inner flow was found to increase the range of the static pressure ratios in which a Mach disc was not formed. While the semi-divergence angle of the inner nozzle was found to modify the shock structure, a discernible trend of variation could not be inferred. A parameter taking into account the variation in the above flow properties was found to reasonably correlate the first shock cell length. The mass-flow ratio of the outer to inner jet was also found to modify the flow field. No discernible trend was observed for the variation in shear-layer thickness with change in the inner-nozzle semi-divergence angle.

References

Buckley FI Jr (1975) Mach disc location in jets in co-flowing airstreams. AIAA J 13:105–106.

Crist S, Sherman PM, Glass DR (1966) Study of the highly underexpanded sonic jet. AIAA J 4:68–71.

D'Attorre L, Harshbarger FC (1965) Parameters affecting the normal shock location in underexpanded gas jets. AIAA J 3:530–531.

Davidor W, Penner SS (1971) Shock standoff distances and Mach disc diameters in underexpanded sonic jets. AIAA J 9:1651–1653.

Kenworthy M, Schetz JA (1973) An Experimental Study of Slot Injection into a Supersonic Stream, NASA CR-2128.

Love ES, Grigsby CE, Lee LP, Woodling MJ (1959) Experimental and Theoretical Studies of Axisymmetric Free Jets, NASA TR R-6.

Narayanan AK, Damodaran KA (1993) Mach disc of dual coaxial axisymmetric jets. AIAA J 31:1343–1345.

A study of three-dimensional supersonic mixing flow phenomena induced by ramp injectors

K. Umii, S. Aso, M. Tannou
Department of Aeronautics and Astronautics, Kyushu University
6-10-1 Hakozaki, Higashi-ku, Fukuoka 812 Japan

Y. Yamane, Y. Ando
Ishikawajima-Harima Heavy Industry Co., Ltd.

M. Fukuda
National Aerospace Laboratory

Abstract: Supersonic mixing flow fields with parallel injection are experimentally investigated. A parallel flow is injected parallel into supersonic flow through a ramp injector. Flow fields are visualized by the schlieren method and oil-flow technique by changing the sweepback angle of the ramp injector. The distributions of surface pressure are also measured in the mixing flow fields. Details of the flow structures of mixing flows due to parallel injection are revealed.

Key words: Stream-wise vortex, Supersonic mixing, SCRAM jet engine

1. Introduction

The development of the SCRAM jet engine has become quite important to the propulsion systems of hypersonic transports and spaceplanes. However, many unsolved problems remain. One of those is the mixing of fuel with supersonic flow in supersonic combustion. For the design of supersonic combustors the mechanism of mixing between free-stream supersonic flow and secondary gaseous flow should be investigated thoroughly. In the case of secondary flow injected normally into supersonic flow, a large total pressure loss is induced. In the present experiments, flow fields with secondary flow injected into supersonic flow with Mach number 3.8 using a ramp injector are investigated.

Flow fields are visualized by the schlieren method and oil-flow technique. Also the distributions of surface pressure are measured. In the experiments the sweepback angle of the ramp injector is varied, and the resulting influence on the geometry of the flow fields is investigated carefully.

2. Experimental apparatus and procedures

A supersonic wind tunnel with nominal Mach number of 4 and with a test section of 150 mm × 150 mm is used. A flat-plate model with a width of 150 mm is installed in the wind tunnel. Secondary gas is injected into supersonic flow through a ramp injector which is attached onto the flat plate. The flow fields interacting between the streamwise vortices generated by the ramp injector, the turbulent boundary layer developed on the plate, and the shock wave induced by the injection of secondary gas are investigated. Nitrogen gas is used as an injectant. Three types of ramp injector are used in order to investigate the differences of the mixing flow fields due to the configuration of ramp injector. A diagram of the ramp injector models is shown in Fig. 1. A circular nozzle is prepared in back of the ramp. The diameter of the circular nozzle is 5 mm. The

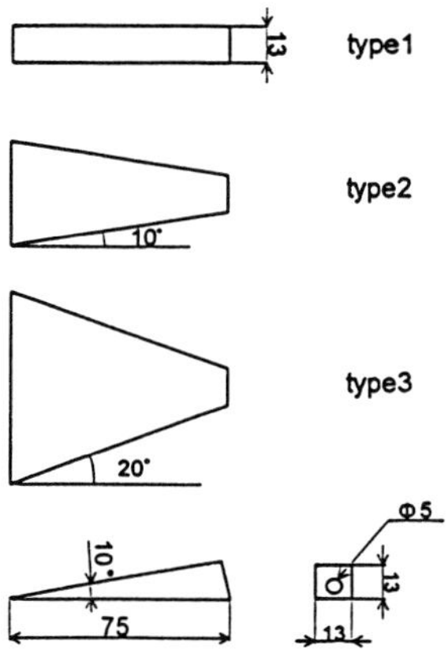

Figure 1. Diagram of ramp injector.

experiments are conducted under the conditions of free-stream Mach number of 3.85–3.90, total pressure of 1.3 MPa, total temperature of 280–290 K and Reynolds number, based on the distance between the leading edge of the flat plate and the forward corner of the ramp, of 2×10^7, with almost adiabatic wall condition.

In the flow fields a primary parameter is the sweepback angle, since the pressure ratio P_c/P_0 (P_c denoting the total pressure of injected secondary flow and P_0 the total pressure of freestream) is fixed as $P_c/P_0 = 0.46$.

3. Experimental results and discussion

Representative schlieren pictures of the flow fields are shown in Fig. 2. All pictures show that a strong shock wave is generated in the forward corner of the ramp, and an expansion wave and shock wave are generated at the apex of the ramp. Downstream of the circular nozzle a barrel shock is generated and a shock wave is seen due to interaction of secondary gas with the flat plate.

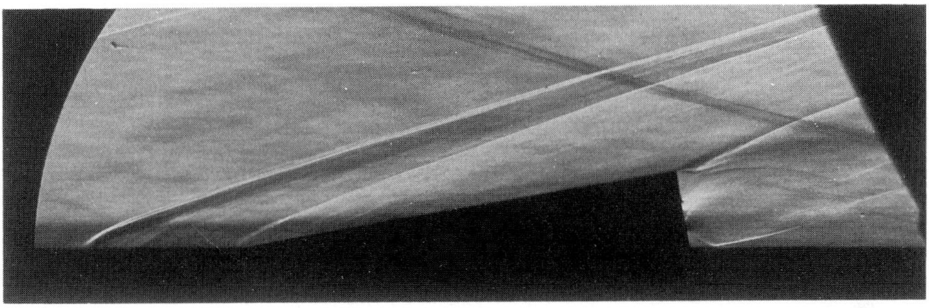

(a)

(b)

Figure 2. Representative schlieren pictures: (a) sweepback angle 0° and (b) sweepback angle 20°.

The flow fields are also visualized by the oil-flow technique. Representative oil-flow pictures are shown in Fig. 3. A primary separation line is observed in front of the ramp. On the ramp a streamline that flows away to side of the ramp is observed and on the plate a streamline that flows from the edge to the center is observed. These results suggest that a streamwise vortex near the corner of the ramp is generated, and the vortex becomes larger with the increase in sweepback angle. As the sweepback angle is increased, the region of influence becomes large. However, in the case of a sweepback angle of 20°, the expansion of the jet on the plate conversely becomes more narrow. This is probably because the streamwise vortices generated in the corner of the ramp decrease the expansion of the jet on the plate.

(a)

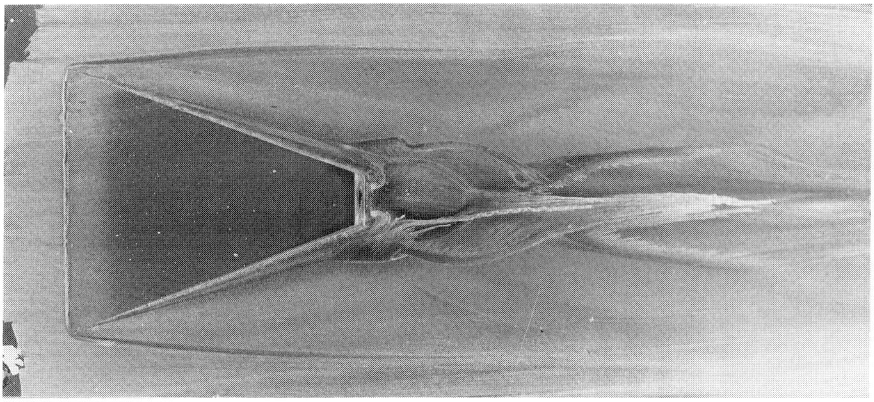

(b)

Figure 3. Representative oil-flow pictures: (a) sweepback angle 0° and (b) sweepback angle 20°.

The distributions of surface pressure are shown in Fig. 4. The pressure is normalized by the static pressure in the undisturbed region. In all pressure distributions a rapid increase of static pressure is observed just behind the ramp, and the higher pressure regions expand parabolically. Further, a rapid drop of static pressure is observed after the increase. This corresponds to the shock wave observed in the schlieren pictures.

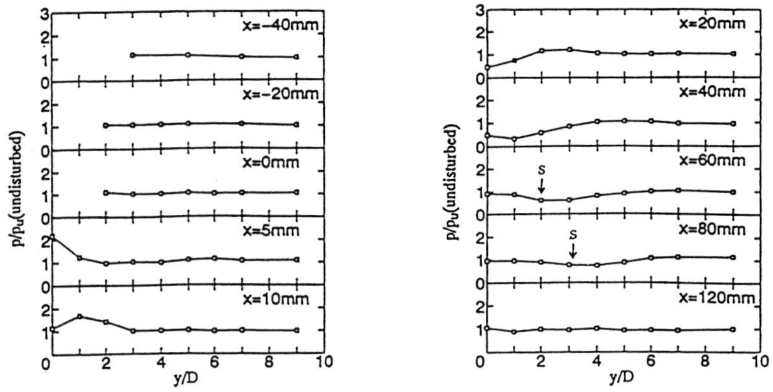

(a)

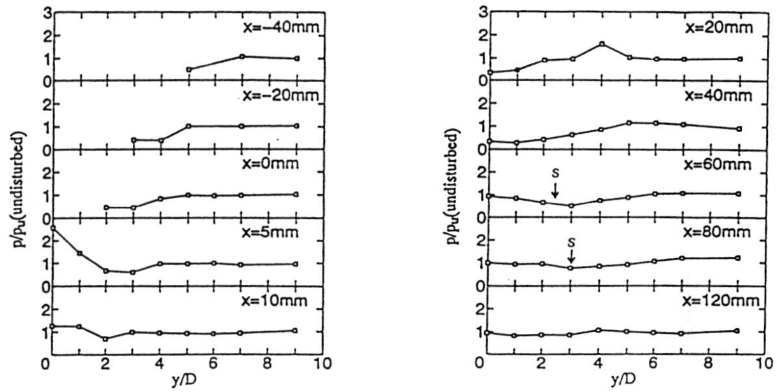

(b)

Figure 4. Representative surface-pressure distributions: (a) sweepback angle 0° and (b) sweepback angle 20°.

On the side of the ramp injector, a lower pressure region is observed. Results point to the existence of streamwise vortices located near the side corner of the ramp; the strength of the vortex increases with the increase of the sweepback angle. In the case of a sweepback angle of 20°, it is observed that these vortices become stronger and induce a region of lower pressure below the vortices, and they decrease the increase of static pressure. These results correspond to those obtained using the oil-flow technique.

4. Conclusions

For all cases, expansion waves and shock waves are generated in the mixing flow fields. It is suggested that streamwise vortices located in the corner of the ramp are generated and these vortices become strong and large with the sweepback angle. Further, it is suggested that the strong streamwise vortices decrease the expansion of the secondary flow on the plate.

References

Anderson GY (1987) An Outlook in Hypersonic Flight, AIAA Paper 87-2074.

Aso S, Tannou M, Okuyama S, Ando Y, Yamane Y, Fukuda M (1994) A Study on Mixing Phenomena in Three-dimensional Supersonic Flow with Circular Injection, AIAA Paper 94-0707.

Drummond JP, Carpenter MH, Riggins DW, Adams MS (1989) Mixing Enhancement in a Supersonic Combustor, AIAA Paper 89-2794.

Waltz IA, Marble FE, Zukoski EE (1991) An Investigation of a Contoured Wall Injector for Hypervelocity Mixing Agmentation, AIAA Paper 91-2265.

Effects of cavity configurations on supersonic internal flow

Y. Watanabe, S. Kudo, F. Higashino
Tokyo University of Agriculture and Technology (TUAT)
2-24-16 Naka-cho, Koganei, Tokyo 184, Japan

Abstract: Effects of cavity configuration and an oblique shock wave impinging over a cavity on supersonic internal flow were investigated experimentally. A shock tunnel with a laval nozzle at the test section was used to generate steady supersonic flows at Mach number $M = 1.83$. The length and the depth of the cavity were varied systematically. The supersonic flow over a cavity was visualized by the schlieren method and by the infinite-fringe method in the form of a Mach-Zehnder interferometer, and the wall static pressure was determined. From the experiment, the effects of the oblique shock wave impinging over a cavity were taken into account. It is found that the incident oblique shock wave does not affect the frequency of the pressure fluctuation inside the cavity, but that the presence of a bow shock wave generated at the rear side of the cavity affects the frequency of the pressure fluctuation downstream of the cavity significantly.

Key words: Supersonic flow, Shock wave, Shear layer, Internal cavity flow

1. Introduction

Supersonic air intake is one of the key components for the development of high-speed propulsion systems, and has been investigated by others (Bogar et al. 1985, Bogar & Sajben 1987). A cavity located at the throat region of the supersonic intake plays a roll in stabilizing the flow. Particularly the normal shock wave, which interacts with the boundary layer and oscillates in the flow direction over the cavity, should be controlled to maintain stable combustion. However, a supersonic flow over a cavity is characterized by a strong oscillation inside the cavity (Zhang & Edwards 1990, Sakamoto et al. 1995), and is very unstable due to the interaction of the shock system with a shear layer or boundary layer. Since the oscillation of the normal shock wave could be affected directly by the pressure fluctuation upstream of the normal shock wave, it is important to investigate the pressure fluctuation downstream of the cavity. However, the mechanism of the pressure propagation downstream from inside the cavity has not been understood. If the flow inside the cavity affects the downstream pressure fluctuation, the flow inside the cavity should be stabilized by some method. In the present study, the effect of cavity configuration on the flow is discussed. The aims of the present study are to observe the change in the vortex structure inside a cavity, and to investigate the frequency and the amplitude of the pressure fluctuation downstream of the cavity.

2. Experimental apparatus and procedures

Figure 1 shows the schematic of the shock tunnel. The high-pressure chamber is 190 mm in inner diameter and 1500 mm in length. The low-pressure section is 77 mm in inner diameter and 3000 mm in length. The test section and the dump tank are located downstream of the low-pressure section. The laval nozzle was installed to obtain a steady supersonic flow at Mach number $M = 1.83$. Figure 2 shows the detail of the cavity configuration. The experiment was performed on two kinds of cavities. One cavity, of length $L = 20$ mm and depth $D_1 = D_2 = 10$ mm, was used to observe the change in the vortex structure in the cavity which corresponds to oscillation of the shear layer. Another cavity, of length $L = 40$ mm, was used to examine the effect of the bow shock wave generated at the rear side of the cavity on the frequency of the pressure fluctuation downstream of the cavity. In this case the depth of the forward side of the cavity D_1 was kept constant at 20 mm, and the depth of the rear side of the cavity D_2 was varied. These changes correspond to a change in cavity dimensions, $D_2/D_1 = 0.5 \sim 1.0$ at $L/D_1 = 2.0$. The location of the incident oblique shock wave was varied from

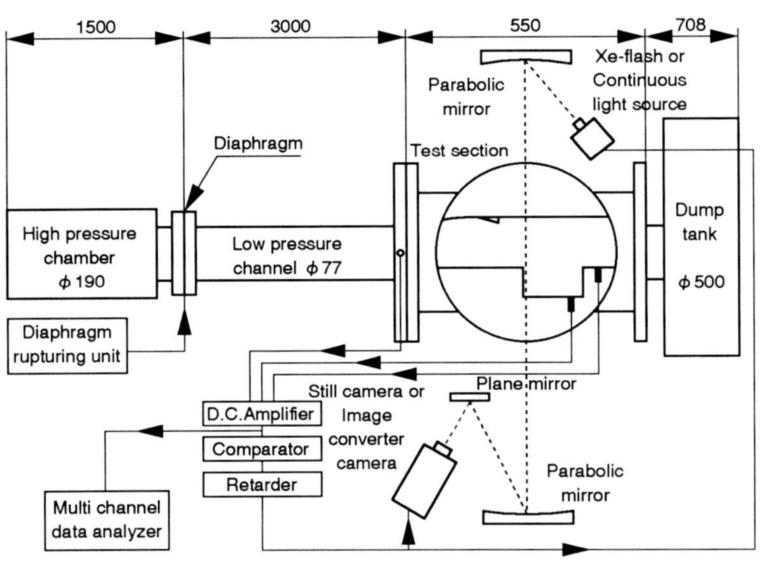

Figure 1. Schematic of experimental apparatus.

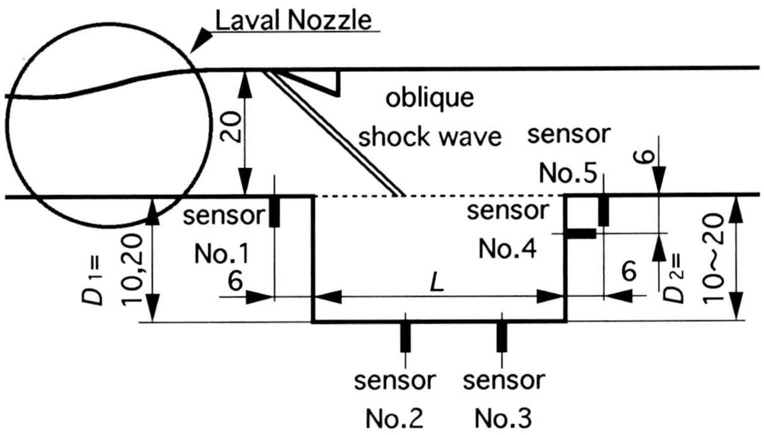

Figure 2. Detail of cavity configuration.

the entrance to the exit of the cavity to examine the effect of impingement of the oblique shock wave on the flow. The flow over the cavity was visualized by the schlieren method and by the infinite-fringe method in the form of a Mach-Zehnder interferometer. An image-converter camera was used to take sequential pictures. The wall static pressure was determined at the points shown in Fig. 2 by piezoelectric transducers.

3. Results and discussion

3.1. The effect of incident oblique shock wave on the flow

Figures 3a and 3b show schlieren photographs which correspond to the case without the incident oblique shock wave and with the oblique shock wave, respectively. The shear layer developed over the cavity and the bow/oblique-shock system were observed by means of these schlieren photographs. The shear layer is bent into the cavity by the incident oblique shock wave. Figure 4a shows the pressure time history determined at the rear side of the cavity (sensor no. 4). Since the shear layer impinges closer to the pressure transducer in the case with the incident oblique shock wave, the amplitude of the pressure fluctuation becomes much larger than in the other case. Figure 4b shows the result of FFT analyses which correspond to the pressure time history shown in Fig. 4a. It is clear from this figure that the frequency of the pressure fluctuation inside the cavity was not affected by the incident oblique shock wave very much. Sequential Mach-Zehnder interferograms were taken by the image-converter camera to clarify the cause of the pressure fluctuation. The frequency of the change in the fringe pattern obtained by sequential interferograms coincides with the

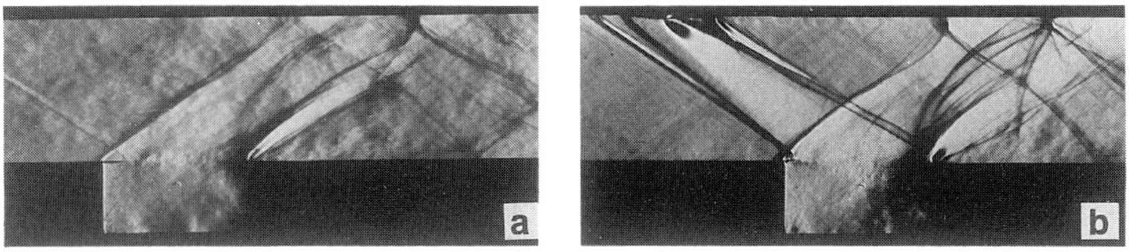

Figure 3. Schlieren photographs: (a) without incident oblique shock wave; (b) with incident oblique shock wave.

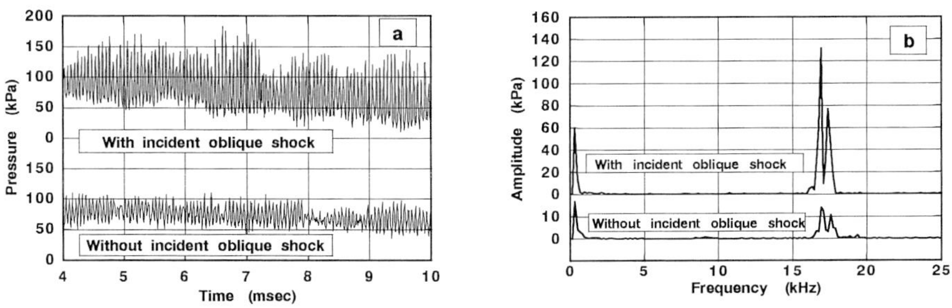

Figure 4. Pressure fluctuation and its frequency inside the cavity: (a) pressure time history; (b) result of FFT analysis.

frequency of the pressure fluctuation. The oscillation frequency of the shear layer also corresponds to the frequency of the change in the fringe pattern.

3.2. The effect of the bow shock wave on the flow

Figures 5a, b, and c show the schlieren photographs. The length-to-depth ratio is $L/D_1 = 2.0$ and the depth of the forward side of the cavity is $D_1 = 20$ mm; these remained the same in all cases. The depth of the rear side wall of the cavity D_2 was changed from 10 mm to 20 mm, which corresponds to the change in cavity dimension D_2/D_1 from 0.5 to 1.0. In the same manner as shown in the schlieren photographs in Fig. 3, the shear layer and the bow/oblique-shock system were observed in Fig. 5a. However, the bow shock wave disappeared in the case of $D_2/D_1 = 0.5$ and 0.75, and instead the attached oblique shock wave was observed. The bow shock wave could be generated by the impingement of the supersonic region in the shear layer on the rear side wall of the cavity. In the case of $D_2/D_1 = 0.5$, the flow passing over the cavity is accelerated abruptly, and deflected downward (with a larger angle compared to the case of $D_2/D_1 = 0.75$) by the expansion fan generated at the corner of the forward side of the cavity. The shear layer impinges on the surface downstream of the cavity with a larger angle in the case of Fig. 5b than in the case of Fig. 5c. This difference of impingement angle causes the difference in the degree of disturbance of the flow in the region downstream of the cavity. Figure 6a shows the pressure time history measured at the point of sensor no. 5 located downstream of the cavity. These measurements correspond to the case shown in Fig. 5. In these waves there are obvious differences in their frequencies and amplitudes. The difference between the case of $D_2/D_1 = 0.75$ and the case of $D_2/D_1 = 0.5$ could be caused by the different angle of impingement of the shear layer. The FFT analysis was

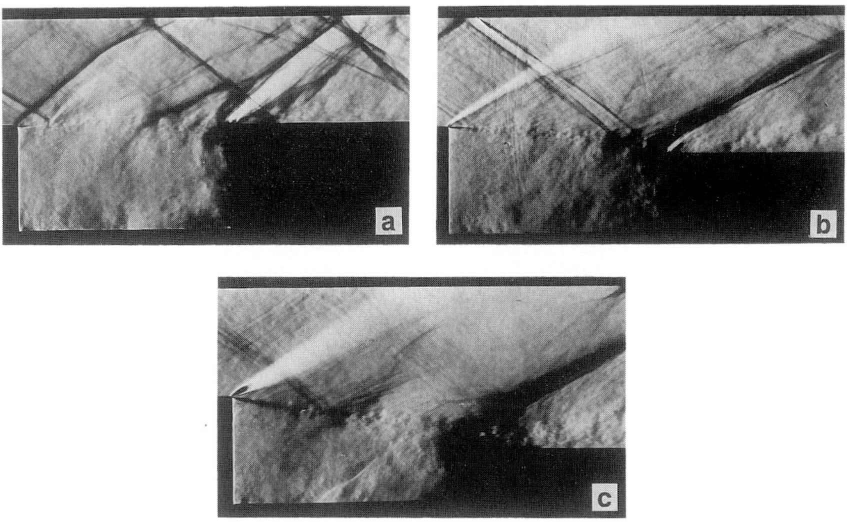

Figure 5. Schlieren photographs: (a) $D_2/D_1 = 1.0$; (b) $D_2/D_1 = 0.75$; (c) $D_2/D_1 = 0.5$.

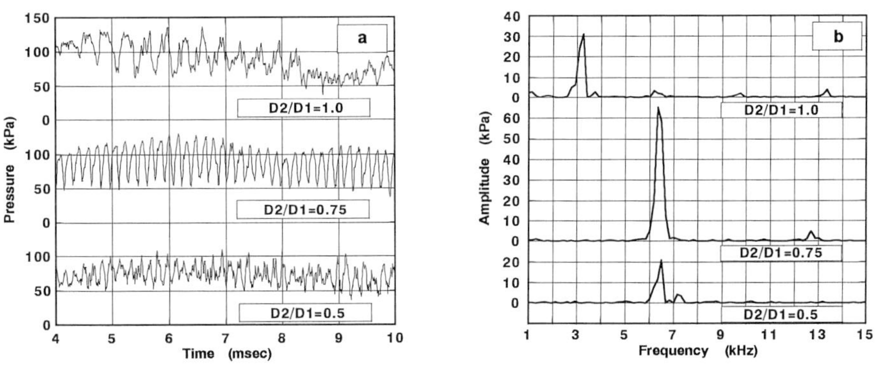

Figure 6. Pressure fluctuation and its frequency downstream of the cavity: (a) pressure time history; (b) result of FFT analysis.

performed on these wave forms to make the difference between the frequency of the pressure fluctuations clear. Figure 6b shows the result of the FFT analysis. The dominant frequency of the cavity with dimension $D_2/D_1 = 0.75$ is almost the same as the value in the case of $D_2/D_1 = 0.5$. Dominant frequencies of the pressure fluctuation inside the cavities are also almost the same value at 6.5 kHz in each case. The only exception is the frequency measured at the point of sensor no. 5 in the case of $D_2/D_1 = 1.0$, which is reduced to a small value by some factor. This might be explained by the fact that as the supersonic region impinges on the rear wall of the cavity, a bow shock wave is generated there only in the case of $D_2/D_1 = 1.0$. The new dominant frequency is thought to be generated by the bow shock wave. These pressure fluctuations are important for the control of the normal shock wave, because the pressure fluctuation could be affected by the oscillation of a normal shock wave significantly. However, it is still unknown how the pressure fluctuation excerts control on the normal shock wave.

4. Conclusions

Effects of the cavity configuration on the supersonic internal flow were investigated experimentally.

The pressure fluctuation inside the cavity corresponds to the change in the fringe structure observed by sequential interferograms, and also corresponds to the oscillation of the shear layer. The dominant frequency of the pressure fluctuation downstream of the cavity was affected significantly by the bow shock wave generated at the corner of the rear side of the cavity.

Acknowledgement. The authors wish to thank to Messrs. H. Kobayashi, T. Osafune, and H. Kakinuma of the Department of Mechanical Systems Engineering, Tokyo University of Agriculture and Technology, for their considerable assistance in conducting the present experiments.

References

Bogar TJ, Sajben M, Kroutil JC (1985) Response of a supersonic inlet to downstream perturbations. J Propulsion 2:118–125.

Bogar TJ, Sajben M (1987) Response of Transonic Diffuser Flows to Abrupt Increases of Back Pressure: Wall Pressure Measurements, AIAA Paper 87–1356.

Sakamoto K, Matsunaga K, Fujii K, Tamura Y (1995) Experimental Investigation of Supersonic Internal Cavity Flows, AIAA Paper 95–2213.

Zhang X, Edwards, JA (1990) An investigation of supersonic oscillatory cavity flows driven by thick shear layers. Aeronautical J December:355–364.

Wave Reflection and Diffraction

Experimental reconsideration of the state of the art of oblique shock-wave reflections in steady flows

A. Chpoun, D. Passerel, J. C. Lengrand
Laboratoire d'Aerothermique du C.N.R.S. Meudon, France
H. Li, G. Ben-Dor
Pearlstone Center for Aeronautical Engineering Studies
Department of Mechanical Engineering
Ben-Gurion University of the Negev, Beer Sheva, Israel

Abstract: The reflection of shock waves over straight surfaces was investigated experimentally. The results for $M_0 = 4.96$ contradict the state of the art regarding the RR $\leftrightarrow$ MR transition. Not only was a hysteresis found to exist, but RR configurations were found to be stable in the dual-solution domain in which theoretically both RR and MR are possible.

Key words: Mach reflection, Transition criterion, Steady flows

1. Introduction

Two shock-wave-reflection configurations are possible in steady flows, namely, regular reflection (RR) and Mach reflection (MR). Two possible transition criteria between these reflections were proposed in the past by von Neumann (1943). They are known as the von Neumann and the detachment criteria. Let the incident shock-wave angles appropriate to these criteria be ω_i^N and ω_i^D, respectively. For $\omega_i > \omega_i^D$, RR is theoretically impossible and for $\omega_i < \omega_i^N$ MR is theoretically impossible. For $\omega_i^N \leq \omega_i \leq \omega_i^D$ both RR and MR are theoretically possible, consequently, the RR $\leftrightarrow$ MR transition could occur at any value ω_i inside this range. Consequently, $\omega_i = \omega_i^N$ and $\omega_i = \omega_i^D$ are the smallest and the largest possible values of ω_i for transition. The dual-solution domain for which both RR and MR are theoretically possible in the (M_0, ω_i)-plane is shown in Fig. 1 for $\gamma = 1.4$. The $\omega_i = \omega_i^D$ and the $\omega_i = \omega_i^N$ criteria divide the (M_0, ω_i)-plane into three domains: a domain in which only RR, a domain in which only MR, and a domain in which both RR and MR are theoretically possible.

The following summarizes the state of the art regarding the RR $\leftrightarrow$ MR transition of planar shock waves in steady flows over straight wedges (see Hornung and Robinson 1982 and Hornung et al. 1979): (1) The RR $\rightarrow$ MR and the MR $\rightarrow$ RR transitions occur at the von Neumann condition. (2) The RR is unstable

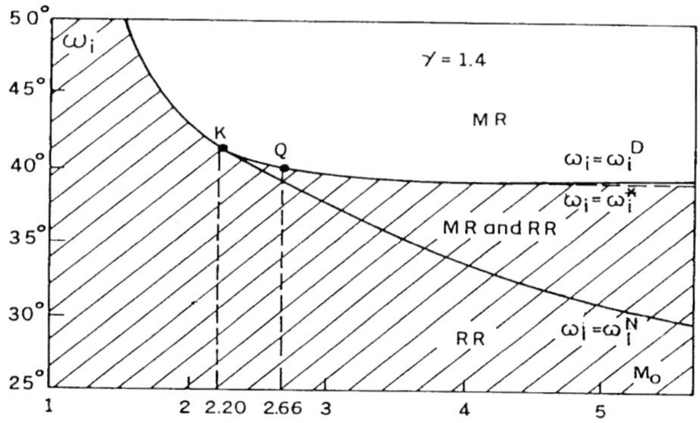

Figure 1. Domains of possible reflection configurations in the (M_0, ω_i)-plane. $\omega_i = \omega_i^D$ and $\omega_i = \omega_i^N$ are the transition lines corresponding to the detachment and von Neumann criteria, respectively. RR is stable for $\omega_i < \omega_i^*$.

in the dual-solution domain. (3) A hysteresis phenomenon does not exist in the RR $\leftrightarrow$ MR transition.

By applying the principle of minimum entropy production, Li and Ben-Dor (1995) showed that the RR is stable in most of the dual-solution domain. The domain $\omega_i \leq \omega_i^*$ in Fig. 1 satisfies the principle of minimum entropy production. This implies that RR wave configurations are physical in that domain. Note that the $\omega_i = \omega_i^*$ line is very close to the $\omega_i = \omega_i^D$ line. Consequently, of the entire $\omega_i^N \leq \omega_i \leq \omega_i^D$ domain for which RR is theoretically possible, only in a very narrow domain, $\omega_i^* < \omega_i \leq \omega_i^D$, is RR aphysical. For this reason it was decided to experimentally investigate the RR $\leftrightarrow$ MR transition with an attempt at establishing a stable RR inside the dual-solution domain.

2. The experimental study

A description of the experimental facility used in the present study is given in Chpoun et al. (1995). Color schlieren photography was employed to record the reflection phenomenon both with still and continuous cameras. When the wind tunnel was started the wedges were placed outside the test section. After a constant flow was established and the nominal conditions were obtained the wedges were brought into the flow one after the other. Then the hysteresis experiments were carried out in two manners: discrete variation of the wedge angle and continuous variation of the wedge angle, with wedge angles changed continuously at a very slow rate of about 0.57 deg/s. A video movie was recorded.

The first set of experiments was aimed at determining the conditions under which the RR $\leftrightarrow$ MR transition occurred. Initially an MR at about $\omega_i \cong 40°$ (in all the experiments $M_0 = 4.96$) was established. Then the angles of attack of the two reflecting wedges were simultaneously and continuously decreased. As a result, the wave angles, ω_i, of the incident shock waves decreased. This resulted in a continuous decrease in the height of the Mach stem until the Mach stem completely vanished. The point where this occurred was defined as the experimental MR $\rightarrow$ RR transition point. Then the direction of the rotation of the wedges was reversed and their angles of attack were simultaneously and continuously increased. While increasing the angles of attack it was observed that the RR was not terminated at the earlier determined MR $\rightarrow$ RR transition point. Instead the RR was maintained until it suddenly changed to an MR. The point where this occurred was defined as the experimental RR $\rightarrow$ MR transition point. The height of the Mach stem of the suddenly formed MR was definitely not zero. The experimental results are shown in Fig. 2 in the (ω_i, ω_r)-plane for $h_{in} = 9$ cm and $w = 6$ cm. MR is marked with triangles and RR is marked with circles. The von Neumann and the detachment transition angles for $M_0 = 4.96$ are $\omega_i^N = 30.88°$ and $\omega_i^D = 39.33°$, respectively. The MR $\rightarrow$ RR transition took place at the von Neumann angle, i.e., $\omega_i^{tr}(\text{MR} \rightarrow \text{RR}) = \omega_i^N = 30.9°$; the RR $\rightarrow$ MR transition occurred at about $\omega_i^{tr}(\text{RR} \rightarrow \text{MR}) = 37.2°$. Thus, it is clear that the hysteresis phenomenon which was hypothesized by Hornung et al. (1979) and up to now has not been confirmed experimentally does exist in the RR $\leftrightarrow$ MR transition in steady flows.

In order to investigate the effect of the geometrical set-up (i.e., length-scale effect) on the transition process, experiments similar to those described above were repeated for three different reflecting wedge surface lengths, w (50, 60, and 70 mm). Each reflecting wedge was tested with three different combinations of inlet cross sections, h_{in} (about 70, 85, and 100 mm). In seven out of the above nine possible combinations of w and h_{in} both the RR $\rightarrow$ MR and the MR $\rightarrow$ RR processes were recorded. The results of these experiments again verified the hysteresis phenomenon in the transition process. All the MR $\rightarrow$ RR experiments showed transition in the vicinity of ω_i^N and all the RR $\rightarrow$ MR experiments indicated transition between ω_i^N and ω_i^D in the vicinity of $\omega_i = 37°$. In addition, the RR $\rightarrow$ MR transition was found to depend on both the inlet cross section and the reflecting-wedge surface length (for details see Chpoun et al. 1995).

The stability of both RR and MR inside the dual solution was also examined. It was found that there is a critical incident shock-wave angle value, say ω_i^{cr}, below which the stable reflection was found to be RR and above which the stable reflection was found to be MR. For the wedge with $h_{in} = 9$ cm and $w = 6$ cm, for which results are shown in Fig. 2, it was found that $\omega_i^{cr} \cong 35.5°$. Consequently, from a stability point of view the dual-solution domain can be divided into two sub-domains. RR is stable in the sub-domain $\omega_i^N \le \omega_i < \omega_i^{cr}$ and MR is stable in the sub-domain $\omega_i^{cr} < \omega_i \le \omega_i^{tr}$ (RR $\to$ MR). The fact that RR was established in spite of the very large disturbance associated with this experimental procedure contradicts the conclusion of Hornung et al. (1979) that inside the dual-solution domain large disturbances should be sufficient to cause the reflection to be MR rather than RR.

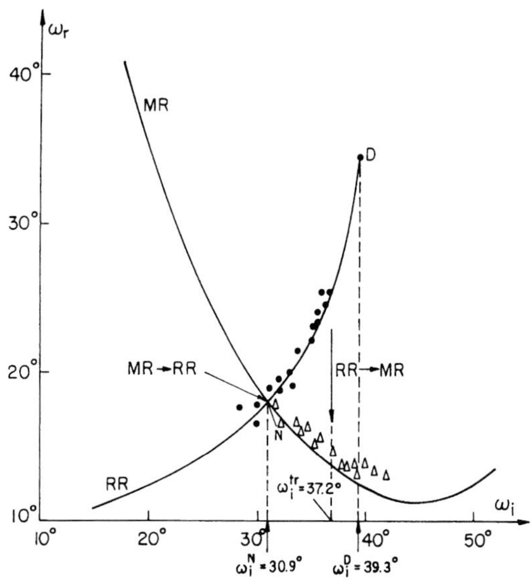

Figure 2. The present experimental results in the (ω_i, ω_r)-plane. △ – MR wave configurations, • – RR wave configurations. The hysteresis phenomenon in the RR $\leftrightarrow$ MR transition is clearly evident.

The experimental results also contradict Henderson and Lozzi's (1975) conclusion that the RR $\to$ MR transition should occur at $\omega_i = \omega_i^N$, "in such a way that mechanical equilibrium of the system is preserved through the process." The sudden transition from RR to MR at $\omega_i > \omega_i^N$ occurred at a point where the mechanical equilibrium requirement was clearly not fulfilled.

3. Conclusions

1. A hysteresis exists in the RR $\leftrightarrow$ MR transition. While the MR $\rightarrow$ RR transition occurs at the von Neumann criterion, i.e., $\omega_i^{tr}(\mathrm{MR} \rightarrow \mathrm{RR}) = \omega_i^N$, the RR $\rightarrow$ MR transition can occur anywhere inside the dual-solution domain $\omega_i^N \leq \omega_i^{tr}(\mathrm{RR} \rightarrow \mathrm{MR}) \leq \omega_i^D$. The exact point at which the RR $\rightarrow$ MR transition takes place depends probably on geometrical parameters associated with the experimental facility and the specific experimental set-up. It is possible that the experimental facilities and set-ups used by Hornung and Robinson (1982) in their experimental investigations were such that the RR $\rightarrow$ MR transition in their facilities occurred at or near the $\omega_i = \omega_i^N$ point. For this reason they failed to obtain a stable RR in the dual-solution domain and to observe the hysteresis associated with the RR $\leftrightarrow$ MR transition.

2. For the geometrical parameters of the wind tunnel and of the wedges and the experimental set-up used in the present study $\omega_i^{tr}(\mathrm{RR} \rightarrow \mathrm{MR}) = 37.2° \pm 1.5°$. This violates Henderson and Lozzi's (1975) argument that the transition should occur at a point where mechanical equilibrium is preserved.

3. There is a critical value, ω_i^{tr}, inside the dual-solution domain below which RR is stable and above which MR is stable.

4. For the geometrical parameters of the wind tunnel, the wedges, and the experimental set-up used in the present study $\omega_i^{cr} = 35.5° \pm 1°$.

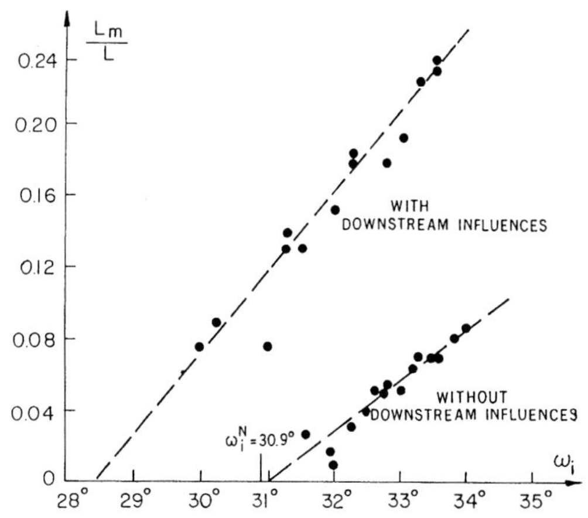

Figure 3. The dependence of the nondimensionalized Mach-stem height l_m/L on the incident shock-wave angle ω_i for two cases with and without downstream influences.

It should be noted here that Azevedo (1989), who obtained Hornung and Robinson's (1982) raw experimental data, showed that the presence of downstream influences at the triple point caused an increase in the height of the Mach stem and, as a result, a shift in the MR $\rightarrow$ RR transition to values of $\omega_i^{tr}(\text{MR} \rightarrow \text{RR}) < \omega_i^N$. This is shown in Fig. 3 in which Hornung and Robinson's unpublished experimental results, as taken from Azevedo's (1989) Ph.D. thesis, for cases with and without downstream influences, are shown. Note that while the results, which were free of downstream influences, indicated that $\omega_i^{tr}(\text{MR} \rightarrow \text{RR}) = 31° \cong \omega_i^N$, the results for which downstream influences were present indicated that $\omega_i^{tr}(\text{MR} \rightarrow \text{RR}) \cong 28.5° < \omega_i^N = 30.9°$. Consequently, it could be concluded that the MR $\rightarrow$ RR transition occurs at the von Neumann point, i.e., $\omega_i^{tr}(\text{MR} \rightarrow \text{RR}) = \omega_i^N$ provided the reflection is free of downstream influences. The fact that Hornung and Robinson (1982) obtained MR at $\omega_i < \omega_i^N$ where MR is theoretically impossible, is yet to be explained.

References

Azevedo DJ (1989) Analytical Prediction of Shock Patterns in a High-Speed Wedge Bounded Duct, Ph.D Thesis, Dept Mech and Aero Eng, State Univ NY Buffalo, NY, USA.

Chpoun A, Passeral D, Li H, Ben-Dor G (1995) Reconsideration of the state-of-the-art of oblique shock wave reflections in steady flows, part I: experimental investigation. J Fluid Mech 301:19-35.

Henderson LF, Lozzi A (1975) Experiments on transition of Mach reflection. J Fluid Mech 68:139–155.

Hornung HG, Oertel H Jr, Sandeman RJ (1979) Transition to Mach reflection of shock waves in steady and pseudo-steady flow with and without relaxation. J Fluid Mech 90:541–560.

Hornung HG, Robinson ML (1982) Transition from regular to Mach reflection of shock waves, part 2: the steady-flow criterion. J Fluid Mech 123:155–164.

Li H, Ben-Dor G (1995) Principle of minimum entropy production in the regular $\leftrightarrow$ Mach reflection transition in steady flow. (Submitted for publication)

Neumann J von, (1943) Oblique reflections of shocks. Explos Res Rep 12, Navy Dept, Bureau of Ordnance, Washington DC, USA.

Experimental and theoretical study of stationary reflection of shock waves

L. G. Gvozdeva, N. B. Sherbak
Gasdynamics Department of High Energy Density Research Center Scientific Association IVTAN Moscow, Russia
A. Chpoun, D. Passerel
Laboratoire d'Aerothermique du C.N.R.S. Meudon, France

Abstract: Results of the joint French-Russian work of a theoretical and experimental investigation of shock-wave reflection phenomenon in steady flows are presented. For experiments, the wind tunnel SH2 in the Laboratoire d'Aerothermique was used. The regular reflection (RR) $\leftrightarrow$ Mach reflection (MR) transition has been studied. It has been found that there exists a phenomenon of hysteresis in the process of transition. The transition from RR to MR occurs near ω_d, the detachment criterion. The transition from MR to RR takes place at $\omega_1 = \omega_N$, the von Neumann criterion. The explanation of the phenomenon is discussed. Numerical simulation was performed. A comparison of numerical-simulation data with experimental data is conducted. The simplest geometry with thin plates is able to provide a qualitative description of hysteresis phenomena.

Key words: Shock wave, Mach reflection, Regular reflection, Hysteresis

1. Introduction

It is well known (see e.g. Bazhenova et al. 1984) that the intersection of two shock waves takes two forms (see Fig. 1). One, which is called regular reflection (RR), takes place at angles of incidence ω_1 close to zero, and the other is Mach reflection (MR), which occurs for angles of incidence close to 90°, near-glancing angles. The transition from one type to the other can occur at the von Neumann criterion, ω_N, or at the detachment criterion, ω_d. In the region between these two boundaries, both systems of reflection are possible. In comparison with the problem of unsteady shock-wave interaction, the steady interaction of shock waves has been studied experimentally and numerically in only a few works. In Pantozopol et al. (1972), it was stated that the transition for air takes place at von Neumann criterion. In Henderson and Lozzi (1979) this observation was confirmed. In the study carried out by Hornung et al. (1979), it was suggested that there should be a hysteresis. In other words, the transition from MR to RR should occur at a different angle than the transition from RR to MR. Nevertheless, the experiments of Hornung and Robinson (1982) showed that there was no hysteresis. The transition occurred at $\omega_1 = \omega_N$.

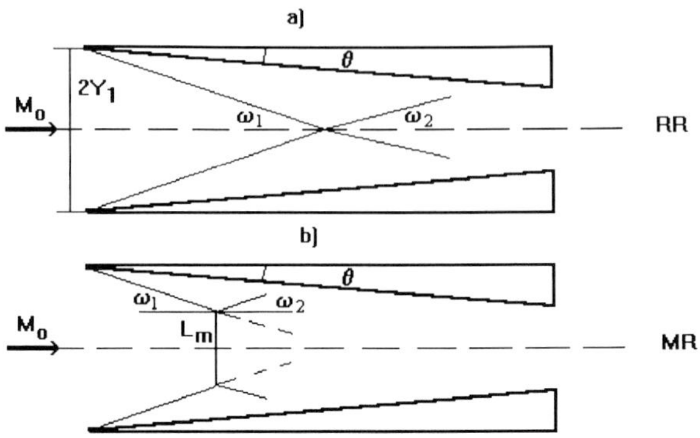

Figure 1. Reflection in stationary case: (a) regular reflection, (b) Mach reflection. M_0 = Mach number of initial flow, θ = wedge angle, ω_1 = angle of incidence, ω_2 = angle of reflection, $2Y_1$ = distance between the plates, L_m = length of Mach stem.

So it was evident that a new series of experiment should be made in order to examine more clearly the conditions of the transition. Gvozdeva suggested that these experiments be undertaken in the Laboratoire d'Aerothermique headed by J. C. Lengrand in Meudon. The work began in 1992; Chpoun and Passerel conducted the experiments and the calculations were done by Sherbak. Passerel's work on this problem will be a major portion of his Doctor's Thesis to be presented in October.

2. Experimental set-up

The wind tunnel SH2 has the following parameters: working gas – air, Mach number of the flow – 4.96, static pressure – 1.683 10^5 N/m^2, static temperature – 76.5 K. A schematic is given in Fig. 2. Two symmetrical wedges allowing a continuous change of the angle between them were placed at the exit of the nozzle. The color schlieren method was used to obtain the photos of the different shock-wave patterns. The stagnation pressure behind the point of intersection was measured. The regular reflection (RR) $\leftrightarrow$ Mach reflection (MR) transition has been studied. The rotation of the plates was around the axis going through the tip of the wedges (see Fig. 2). We believe this experimental design caused the occurrence of the hysteresis phenomenon.

3. Experimental results

3.1. Hysteresis effect

It has been found that a hysteresis can exist through the transition under certain conditions of operation of the special this experimental set-up. The method of

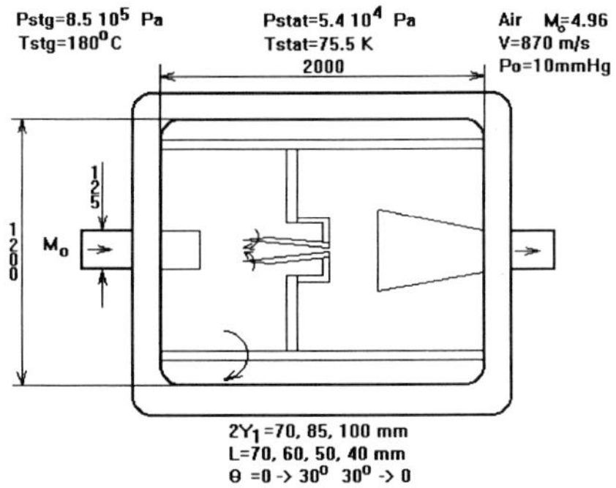

Figure 2. Experimental set-up. Working gas is air, $M_0 = 4.96$, initial pressure in chamber is about 10 torr.

operating is as follows. First, the plates are inserted in the chamber parallel to each other. Then, the flow is started. After reaching stagnation conditions, the plates are rotated smoothly, gradually increasing the angle ω_1 with a velocity $d\omega_1/dt = 0.5°/\text{s}$. The configuration changes from RR to MR. With an opposite rotation, the inverse transition from MR to RR is observed. A number of obtained color schlieren pictures show very clearly that hysteresis takes place. Figure 3 presents the experimental dependence of ω_2 (reflection angle) on ω_1 (angle of incidence). We can clearly see that the transition from MR to RR takes place near the point ω_N and the transition from RR to MR 2° below the ω_d. The length of the Mach stem L_m was measured. The graph of its depen-

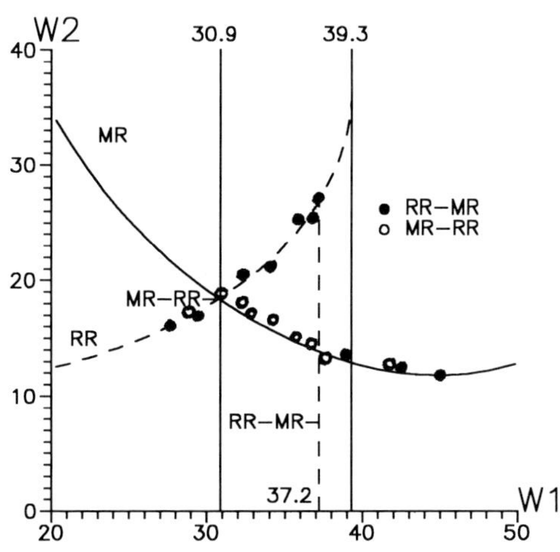

Figure 3. Dependence of reflection angle on incident angle for MR ↔ RR transition.

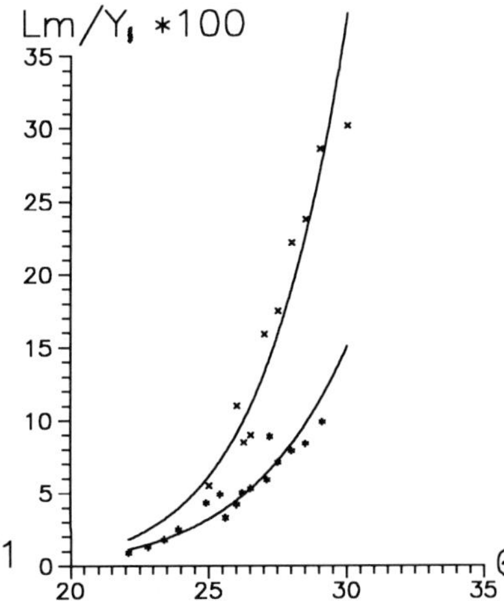

Figure 4. Dependence of Mach-stem height on wedge angle.

dence on θ is given on Fig. 4. It shows the hysteresis effect very clearly. The stagnation pressure was measured with a Pitot tube on the plane of symmetry. The pressure behind the Mach stem was equal to the stagnation pressure of the flow (53.72 KPa). The pressure behind the reflected shock wave in regular reflection was approximately equal to that calculated by the theoretical inviscid values. So it was possible to find what regime was obtained at a given ω_1 not only with schlieren pictures but also with pressure records.

3.2. Dependence of the pattern on geometry

From the theoretical point of view, the transition angle should depend only on local parameters ω, M, γ, P_0, and T_0. It is certainly true for regular reflection for which supersonic flow exists behind the reflected shock wave.

In a Mach reflection, subsonic flow exists behind the Mach stem, so any disturbance can influence the shock-wave pattern. We can expect that the position of the Mach stem depends not only on ω, M, γ, P_0, and T_0, but also on the geometrical parameters of the plates. Furthermore, 3D effects can take place if the width of the plates is too small. Therefore experiments were conducted for different geometrical parameters of the plates. The aim was first to look at the 3D effects and secondly to find the influence of the geometrical parameters. The following set of experiments were made: $2Y_1 = 70$, 80, 100 mm; $L = 70$, 60, 50, 40 mm; plate width = 50, 70 mm.

It was found that there exists a dependence upon geometrical parameters as demonstrated by the fact that the boundary between RR and MR is about 2° less than ω_d, while the boundary from MR going to RR lies at ω_N. For the comparison with computed data, we used the experimental data obtained with the larger plate width (70 mm) and a ratio $Y_1/L = 1$.

3.3. Stability of regular and Mach reflection

Next, an attempt was made to verify the stability of the wave pattern. The plates were arranged so that angle ω_1 was more than 40°. When the flow started, MR occurred. Then the lower plate with its plate holder was removed from the flow and inserted again. This disturbance of the flow was repeated several times. The configuration which appeared at last was determined. With this way of operating it was found that in this region where two types of reflection have been observed to exist, there is in reality no ambiguity. At incident angles greater than 35.5°, MR occurs, and at lesser angles, RR. So for this mode of generating the flow hysteresis does not take place. The transition angle $\omega_{tr} = 35.5^\circ$.

4. Discussion of experimental results

It seems that the suggestion of Hornung et al. (1979) about the possibility of hysteresis in the case when the change of ω_1 takes place very smoothly is correct. It is possible to get hysteresis with a very slow rotation of the plates. The axis of rotation may also be of importance. At any rate the difference between our experimental set-up and Hornung and Robinson's is mainly in the dimension of the surrounding volume and in the disposition of rotating angle. So the effect of hysteresis depends upon all the details of the set-up.

5. Numerical simulation

The shock-wave reflection problem was investigated by solving the two-dimensional unsteady Euler equations for an ideal (inviscid, non-heat-conducting) gas in Cartesian coordinates. TVD schemes constructed on the basis of Godunov and Steger methods were used for integration. Constructed TVD schemes have a second-order accuracy over space on smooth solutions and formally a third-order on uniform nets. The algebraic interpolation technique was used for grid generation.

Two symmetrical infinitely thin plates are placed at an angle θ inside an incident supersonic flow. At the initial time the parameters of an undisturbed incident supersonic flow with a definite Mach number M are set in all computational regions. Then, the stationary regimes of the flow are determined by the solution of unstationary equations.

In this investigation two parameters were fixed: $\gamma = 1.4$ and $M = 4.96$, in correspondence with experimental values. The ratio of Y_1/L was equal to 1. The angle of the plate θ has been varied over the region 20 to 30 degrees. The number of cells in the characteristic length was 75. Free-stream conditions were imposed on the upper and right boundary of the computational domain and the condition of impermeability was taken at the lower boundary (the symmetry plane) and on the surface of the plate. On the upstream boundary, the flow was supersonic.

The calculations showed that the observed incidence angles of the plane shock waves are rather close to the corresponding theoretical values. Observed parameters behind the incident and the reflected shock waves are in good agreement with the theoretical values.

As was done in experiments, the first series of computations was carried out for angles θ changing during the calculations. This was achieved by the following procedure. First, the calculation was performed for some initial angle $\theta_0 = \theta_d$. After the flow became stationary and a MR was observed, the plate was rotated around the leading edge, so that the new plate position generated a shock wave with the angle smaller than θ_d. When the flow became stationary in this position, the plate was rotated again.

Figure 4 shows the dependence of Mach-stem height normalized by the wedge length on ω_1 for calculations (higher curve) and for experiments (lower curve). The boundary for the transition from RR to MR was 26°, the boundary for the transition from MR to RR was smaller than 23°. The simplest geometry with thin plates therefore gives a qualitative description of the hysteresis phenomenon.

A second series of computations was carried out for a set of fixed plate angles generating incident shock waves with a specified angle θ between θ_n and θ_d. This models the case of a starting flow with a big disturbance before the stationary state. In these latter calculations RR was observed up to $\theta = 26°$. When the angle $\theta > 26°$, MR is observed. The agreement of these calculated angles with experiments seems to be accidental. The whole study shows that further investigations in this region are appropriate and will likely provide needed understanding.

References

Bazhenova TV,Gvozdeva LG, Nettleton MA (1984) Unsteady interaction of shock waves. Progress in Aerospace Sciences 21(4):249–332.

Gilding BH (1988) A numerical grid generation technique. Computers and Fluids 16(1):47–58.

Godunov SK (ed) (1976) Numerical Simulation of Many-Dimensional Problems in Gasdynamics, Nauka, Moscow, p 40 (in Russian).

Henderson LF, Lozzi A (1979) Further experiments of transition to Mach reflection. J Fluid Mech 94:541–560.

Hornung HG, Oertel H Jr, Sandeman RJ (1979) Transition to Mach reflection of shock waves in steady and pseudo-steady flow with and without relaxation. J Fluid Mech 90:541–560.

Hornung HG, Robinson ML (1982) Transition from regular to Mach reflection of shock waves. Part 2. The steady flow criterion. J Fluid Mech 123:155–164.

Pantazopol D, Bellet J-C, Soustre J (1972) Sur les conditions d'apparition de l'effet de Mach dans la'reflexion d'une onde de choc oblique on e'qoulement supersonique. C.R. Acad. Sc., Paris, t.275, Serie A-225.

Steger JL, Warming RF (1981) Flux vector splitting of the inviscid gasdynamic equations with application to finite difference methods. J Comp Phys 40(2):263–293.

Sweby PK (1981) High resolution schemes using flux limiters for hyperbolic conservation laws. SIAM J Numer Anal 25(5):995.

Van Leer B (1981) Flux-vector splitting for the Euler equations. Lecture Notes in Physics 170:507–512.

Mach-stem growth in von Neumann reflection

R. J. Virgona,* F. Higashino
Tokyo-Noko University
2-24-16 Naka-cho, Koganei
Tokyo 184 Japan

Abstract: This research reports on a simple method for predicting the growth of the Mach stem as a function of the wedge angle, θ_w, in von Neumann reflection. In experiments in argon for strong, weak, and very weak shocks the behavior of χ against θ_w could be described by a simple quadratic. Curves for χ vs θ_w derived from the simple theory agree well with experiment and numerical computation results. The theory also predicts that another type of reflection is possible in which the incidence angle of the incident shock, i, remains constant and equal to the glancing incidence angle for very small θ_w which is usually less than 7° for a perfect gas and is also dependent on the Mach number of i.

Key words: von Neumann reflection, Anomalous reflection, Shock tube

1. Introduction

For very weak plane shocks transition from regular reflection at the sonic point $\theta_w = \theta_w^*$ results in von Neumann reflection (VNR) as described by Colella and Henderson (1990). For stronger shocks VNR occurs at $\theta_w = \theta_a$ where the von Neumann three-shock theory (von Neumann 1943) predicts that the reflected shock, r, should lean forward so that β_1, the angle between r and the contact discontinuity, *cd*, is required to be $\beta_1 > 90°$ (Virgona et al. 1992, Virgona 1993). In VNR there appears to be no discontinuity in slope between the incident shock, i, and the Mach stem, s, so the 'triple point,' T, is defined as the point where i begins to curve at the confluence with the leading edge of r. The trajectory path angle of T with respect to the surface of the wedge is taken to be χ. In pseudo-stationary co-ordinates the free-stream angle of incidence, ω_0, of i increases as θ_w becomes smaller until it reaches a maximum at ω_b, which is defined by the glancing incidence theory for which $\chi = \chi_{gl} = 90° - \omega_b$. There can be no reflection for $\omega_0 > \omega_b$, so if $\theta_w > 0°$ when $\omega_0 = \omega_b$ a new type of reflection will occur which has been referred to as anomalous reflection (ANR)

*Present address: Thammasat University, Department of Mechanical Engineering, International Institute of Technology, P.O. Box 22, Thammasat Rangit Post Office, Pathumthani 12121, Thailand.

(Virgona 1993, Fig. 1). For ANR, ω_0 remains fixed as θ_w reduces to zero, and $\theta_w + \chi = \chi_{gl}$, which is referred to as the glancing incidence line.

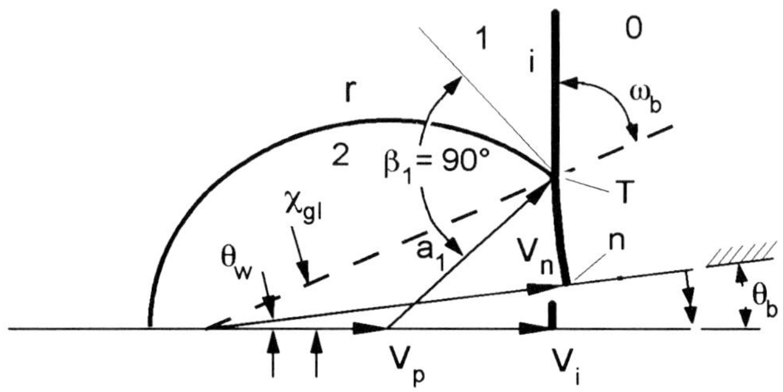

Figure 1. Velocity vector diagram for anomalous reflection, ANR.

2. The theory for χ against θ_w in VNR

In experiments in argon for strong, weak, and very weak shocks the rate of growth of the Mach stem in VNR increased as θ_w was reduced so that the behavior of χ against θ_w could be described by a function of the form

$$\chi = B_2\theta_w^2 + B_1\theta_w + B_0 \tag{1}$$

where the coefficients B_2, B_1, and B_0 are such that (1) is asymptotic to the glancing incidence line at one end (at θ_b, say) and at the other end corresponding to the onset of VNR there is either $d\chi/d\theta_w = 0$ at θ_w^* for very weak reflections (Fig. 2a), or for stronger shocks where transition to VNR takes place from Mach reflection there is a smooth transition in the behavior of χ such that $d\chi/d\theta_w = -Q$ where $-Q$ is taken as $d\chi/d\theta_w$ within the SMR region (Fig. 2b). For very weak shocks the coefficients reduce to

$$\begin{aligned} B_2 &= 1/(4(\theta_w^* - \chi_{gl})), \\ B_1 &= -\theta_w^*/(2(\theta_w^* - \chi_{gl})), \\ B_0 &= \theta_w^{*2}/(4(\theta_w^* - \chi_{gl})), \end{aligned} \tag{2}$$

and $\theta_b = 2\chi_{gl} - \theta_w^*$, while for stronger shocks they become

$$\begin{aligned} B_2 &= \frac{-((\theta_a/2 - Q(\chi_{gl} - Q\theta_a/2 - \chi_a))/(\omega_a - \omega_b) + 1)^2}{4(\chi_{gl} - (Q\theta_a(\chi_{gl} - \chi_a) - (\theta_a/2)^2(Q+1)^2 + \chi_a(\omega_a - \omega_b))/(\omega_a - \omega_b))}, \\ B_1 &= (\theta_a/2 - Q(\chi_{gl} - Q\theta_a/2 - \chi_a))/(\omega_a - \omega_b), \\ B_0 &= (Q\theta_a(\chi_{gl} - \chi_a) - (\theta_a/2)^2(Q+1)^2 + \chi_a(\omega_a - \omega_b))/(\omega_a - \omega_b), \end{aligned} \tag{3}$$

which give $\theta_b = 2(\chi_{gl} - B_0)/(B_1 + 1)$. For $\xi_i \leq \xi_v$ it is necessary to know Q, θ_a, and χ_a in order to evaluate (1), so these must be obtained from experiment or

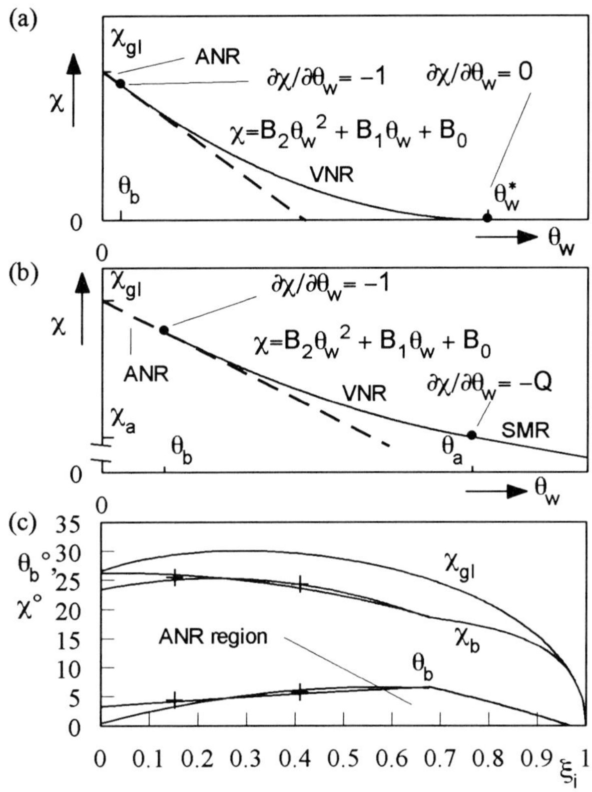

Figure 2. The relationship between χ and θ for VNR and ANR: (a) For $\xi_i \geq \xi_v$; (b) for $\xi_i \leq \xi_v$; (c) existence region for ANR for $\gamma = 5/3$, cross symbols are data derived from Godunov-code numerical computation.

numerical computation. Figure 2c shows the region for θ_b and χ_b derived from the above and using two numerical computation points for the region $\xi_i \leq \xi_v$. There are two curves shown for this region because one curve passes through the $(0, 0)$ point while the other is a smooth fit through the two numerically derived points and the end point of the curve for $\xi_i \geq \xi_v$.

3. Results

In Fig. 3 are shown the curves derived from the above theory against experimental and numerical data for three shock strengths with average $\xi_i = 0.9120$, 0.4106, and 0.1534 for argon. The derived curves for χ vs θ_w agree well with experiment and numerical computation results especially for the case of

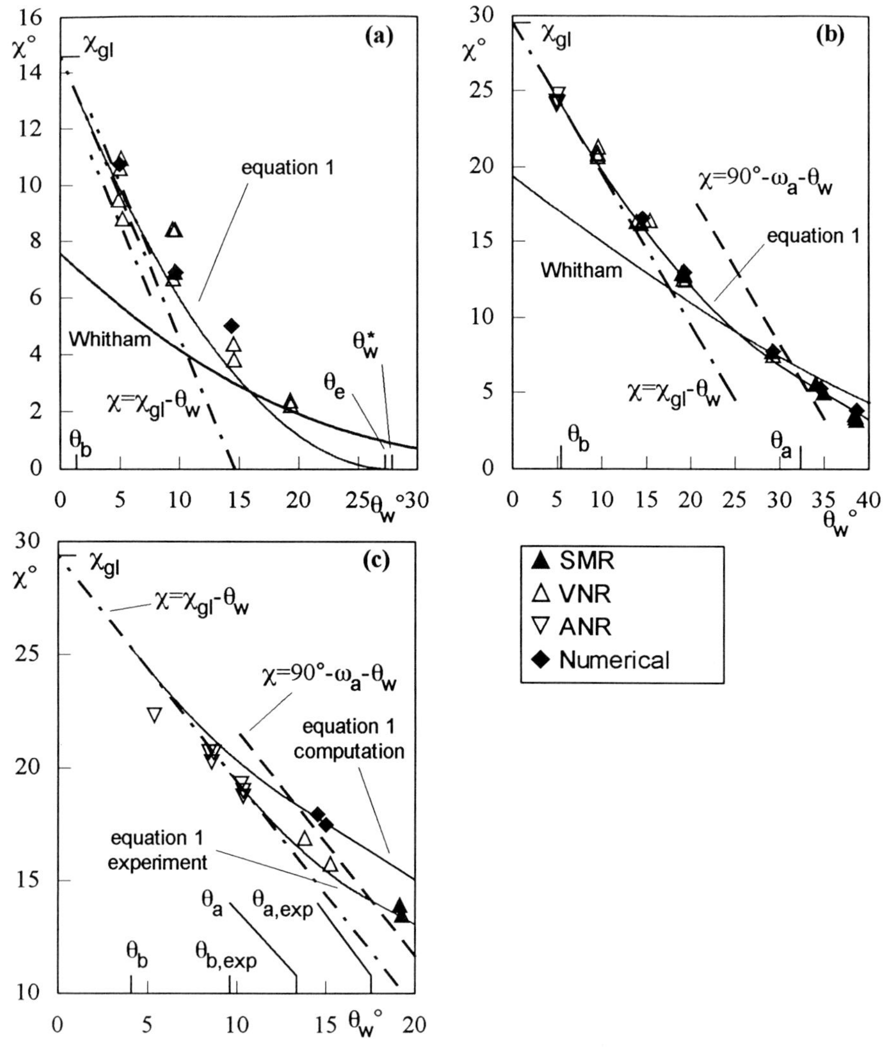

Figure 3. Results for a perfect gas $\gamma = 5/3$: (a) $\xi_i = 0.9120$; (b) $\xi_i = 0.4106$; (c) $\xi_i = 0.1534$. Diamond symbols are numerical computation, other symbols experiment. For Whitham theory, see Henderson (1980).

$\xi_i = 0.4106$. It can be seen that a number of points lie on or close to the ANR line for $\theta_w > 0°$, thus showing that ANR is possible for very small θ_w. Some representative schlieren photographs of the ANR and VNR ranges are shown in Fig. 4, for the values of ξ_i given above.

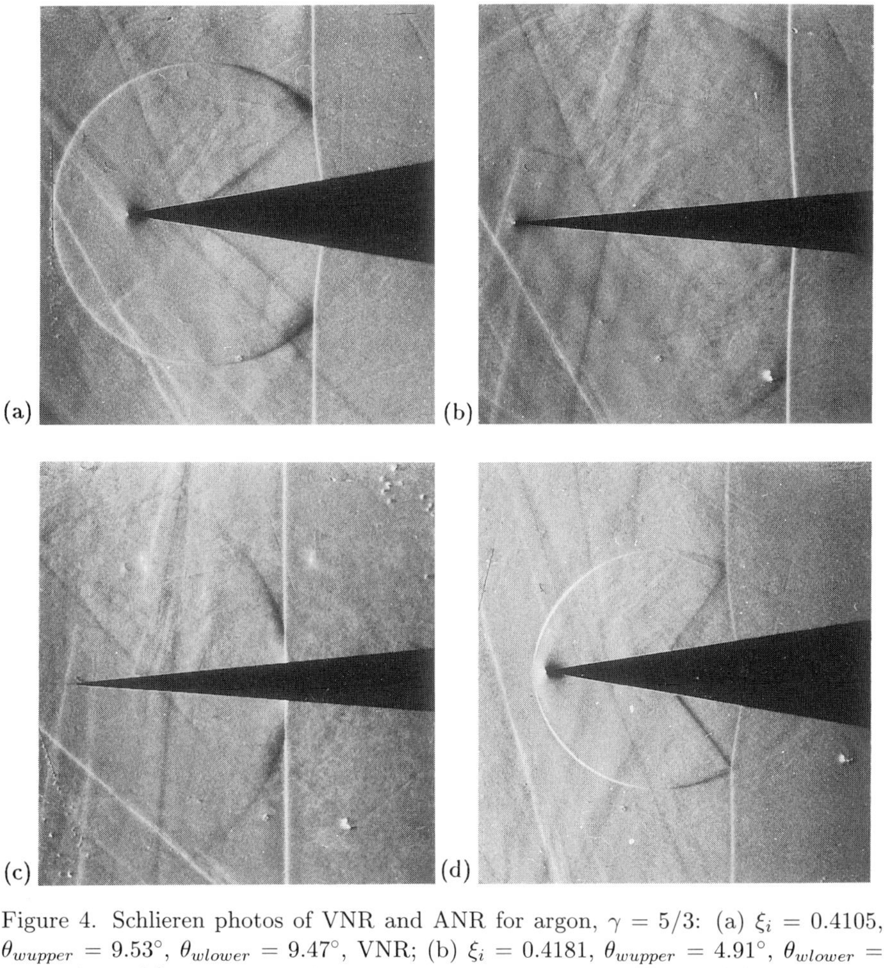

Figure 4. Schlieren photos of VNR and ANR for argon, $\gamma = 5/3$: (a) $\xi_i = 0.4105$, $\theta_{wupper} = 9.53°$, $\theta_{wlower} = 9.47°$, VNR; (b) $\xi_i = 0.4181$, $\theta_{wupper} = 4.91°$, $\theta_{wlower} = 5.09°$, ANR; (c) $\xi_i = 0.9243$, $\theta_{wupper} = 5.16°$, $\theta_{wlower} = 4.84°$, VNR; (d) $\xi_i = 0.1802$, $\theta_{wupper} = 8.08°$, $\theta_{wlower} = 10.88°$, ANR.

References

Colella P, Henderson LF (1990) The von Neumann paradox for the diffraction of weak shock waves. J Fluid Mech 213:71–94.

Henderson LF (1980) On the Whitham theory of shock-wave diffraction at concave corners. J Fluid Mech 99:801–811.
Virgona RJ, Henderson LF, Honma H, Xu DQ (1992) Discrepancies between theory and experiment for shock reflection. In: Takayama K (Ed) Proc 18th Int Symp on Shock Waves, Sendai, Japan, Springer-Verlag, Berlin, pp 189–192.
Virgona RJ (1993) A Study of Shock Wave Diffraction From Regular to Irregular Reflection and the Effects of Gas Viscosity, PhD Thesis, The University of Sydney, Sydney, Australia.
von Neumann J (1943) Oblique reflection of shocks. Explos Res Rep 12, Navy Dept, Bureau of Ordnance, Washington, DC.

Mach-reflection flow at hypersonic speeds

J. Le, H. Ni, G. Hu
China Aerodynamics Research and Development Center
P.O. Box 211, Mianyang, Sichuan, China

Abstract: Experiment and computation on the reflection of a shock wave of $M_s = 7.6$ from a wedge of $\theta_w = 40°$ are investigated. Experiments are carried out in the CARDC shock tube. Computation of both equilibrium and nonequilibrium Mach-reflection flow is based on the laminar full Navier-Stokes (N-S) equation in which the inviscid and source terms are calculated by a second-order upwind TVD scheme and a point-implicit method, respectively. Computational results are in good agreement with those of experiment in all flow fields.

Key words: Mach reflection, Hypersonic flow, TVD scheme

1. Introduction

As judged by comparison with a great number of shock-tube experiments, numerical methods based on the Euler equation with second-order upwind difference schemes such as the Godunov or TVD scheme are sufficiently accurate in the analysis of the perfect inviscid, complex-wave structure of Mach-reflection flow (Glaz et al. 1986). However, for a very strong incident shock wave, the temperature behind the shock is high enough to bring about vibrational excitation and dissociation of the air, resulting in strong effects of the viscosity and nonequilibrium on the experimental flow field. Therefore, further development of the numerical modeling of the viscous nonequlibrium flow field is necessary for the analysis of experimental data and for the validation of computer codes.

To date, there have been a few typical quantitative measurements of density fields widely used for the validation of computer codes for viscous nonequilibrium usteady flow. The purpose of the experiments carried out in CARDC's shock tube at $M_s = 7.6$, $\theta_w = 40°$ is to study the detailed flow structure of the region about the two triple points, including the reflected shock wave, weak slip layer, and Mach stem.

With regard to the numerical simulation, there have been several results only based on the Euler equation (Glaz 1987, Needham 1991, Le and Hu 1992, Clinton and Groth 1993). In the present paper, the laminar full N-S equation is used for numerically simulating both thermal and chemical nonequilibrium Mach-reflection flow. The high-temperature air is assumed to consist of 11 species: O_2, N_2, NO, O, N, O_2^+, N_2^+, NO^+, O^+, N^+ and the reaction-rate coefficients in 32 chemical reactions are taken from the data set compiled by Dunn and Kang (1973). The forward reaction-rate constants K in the chemical source terms are based on Park's two-temperature model of T and Tv: $K = K(\overline{T})$, $\overline{T} = \sqrt{TTv}$. Source terms of three vibrational energy equations with vibration temperature Tv_{O2}, Tv_{N2}, Tv_{ON}, are based on the Landau-Teller method for harmonic oscillators. The relaxation times are obtained from the semi-empirical expressions from Millikan and White. The high-temperature gas-transport coefficients are taken from accurate curve fit (Roop and Gupta 1990).

2. Experiments and results

The experiments were performed in CARDC's shock tube with a test section of 132 mm × 132 mm, which has two pairs of observation windows of 80-mm diameter. The distance between the windows is 400 mm. The model located in the first window is a 2D wedge with a 40° compression corner and height of 88 mm. Double-exposure holographic interferometry was used for quantitatively measuring the density fields and differential interferometry was used for studying the flow structure. Figure 1 is a differential interferogram in which both the first and second slip layers can be distinguished easily. A strong jump in interferomentric fringes against the wall can also be seen. This jump means that values of density begin to increase sharply due to the effect of the cold wall at 298 K. Figure 2 is a holographic interferogram from which the density of the flow field can be evaluated in a very simple way by directly counting the fringes without relying on analytical considerations of three-shock theory near the triple points. The fringe-counting data-reduction technique is obviously inaccurate in chemically reacting flow, as the Gladston-Dale constant K is dependent on the mass concentration of the species. However, in the present

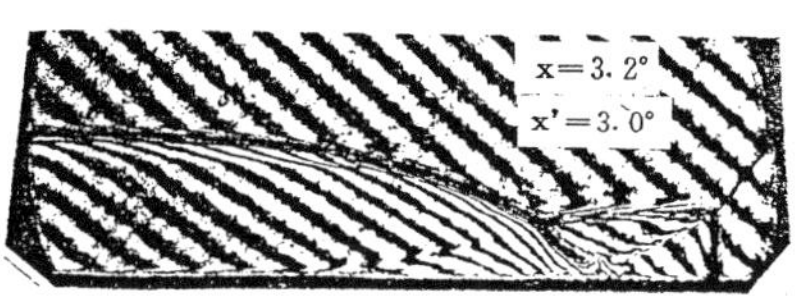

Figure 1. Differential interferogram from CARDC shock-tube experiment. $M_s = 7.6$, $\theta_w = 40°$, $P_0 = 2.63$ KPa, $T_0 = 289$ K.

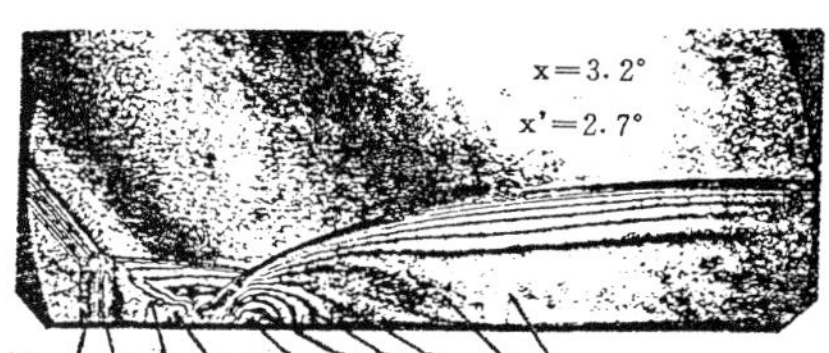

Figure 2. Holographic interferogram from CARDC shock tube. $M_s = 7.6$, $\theta_w = 40°$, $P_0 = 2.63$ KPa, $\rho_0 = 3.21 \times 10^{-5}$ g/cm^3; $\Delta\rho_0 = \lambda/KL$, $\lambda = 6.943 \times 10^{-3}$ cm, $K = 0.225$ cm^3/g, $L = 13.2$ cm; $\rho_N/\rho_0 = 1 + N(\Delta\rho_0/\rho_0) = 1 + N \times 0.729$.

test case, the value of the density some distance behind the incident shock obtained by counting fringes with $K = 0.225 \cdot 10^{-3}$ m^3/kg is very close to the value calculated from the Rankine-Hugoniot conditions for the air equilibrium state. Therefore, the approximate wall density distribution shown in Fig. 3 is evaluated with the same value of K. Strong relaxation effects both behind the incident shock and reflected shock are also indicated in this figure.

3. Numerical calculation and results

Governing equations and difference schemes. The 2D laminar conservative full N-S equations with a gas of N reacting species and M vibrational excited species can be written:

$$\frac{\partial U}{\partial t} + \frac{\partial F}{\partial x} + \frac{\partial G}{\partial y} = \frac{1}{Re}\left(\frac{\partial Q}{\partial x} + \frac{\partial R}{\partial y}\right) + S, \tag{1}$$

where $U = (\rho_s, \rho u, \rho v, \rho_m e_{vm}, E)^T$, $F = (\rho_s u, \rho u^2 + p, \rho uv, \rho_m e_{vm} u, \rho u h_0)^T$, $G = (\rho_s v, \rho uv, \rho v^2 + p, \rho_m e_{vm} v, \rho v h_0)^T$, and

$$\begin{aligned} Q = & \left(\rho D_s \frac{\partial x_s}{\partial x}, \tau_{xx}, \tau_{xy}, \rho D_m \frac{\partial x_m}{\partial x} e_{vm} + k_{vm} \frac{\partial T_{vm}}{\partial x}\right., \\ & \left.\sum_{S=1}^{N} \rho D_s \frac{\partial x_s}{\partial x} h_s + k \frac{\partial T}{\partial x} + \sum_{m=1}^{M} k_{vm} \frac{\partial T_{vm}}{\partial x} + u\tau_{xx} + v\tau_{xy}\right)^T \\ R = & \left(\rho D_s \frac{\partial x_s}{\partial y}, \tau_{xy}, \tau_{yy}, \rho D_m \frac{\partial x_m}{\partial y} e_{vm} + k_{vm} \frac{\partial T_{vm}}{\partial y}\right., \\ & \left.\sum_{S=1}^{N} \rho D_s \frac{\partial x_s}{\partial y} h_s + k \frac{\partial T}{\partial y} + \sum_{m=1}^{M} k_{vm} \frac{\partial T_{vm}}{\partial y} + u\tau_{xy} + v\tau_{yy}\right)^T \end{aligned}$$

and $S = (\dot{\omega}_s, 0, 0, \rho_m \dot{e}_{vm} + \dot{\omega}_m e_{vm}, 0)$. The multispecies state equation is $P = \sum_{s=1}^{N} \frac{\rho_s}{M_s} RT$. Introducing the transformation of variables $\tau = t$, $\xi = \xi(x, y)$, $\eta = \eta(x, y)$, Eq. (1) becomes

$$\frac{\partial \widehat{U}}{\partial \tau} + \frac{\partial \widehat{F}}{\partial \xi} + \frac{\partial \widehat{G}}{\partial \eta} = \frac{1}{Re}\left(\frac{\partial \widehat{Q}}{\partial \xi} + \frac{\partial \widehat{R}}{\partial \eta}\right) + \widehat{S}. \tag{2}$$

The discrete conservation equation (2) can be written in the following form:

$$\begin{aligned} D_{i,j}^n \Delta \widehat{U}_{i,j} = & -\frac{\Delta\tau}{\Delta\xi}\left(\widetilde{F}^n_{i+\frac{1}{2},j} - \widetilde{F}^n_{i-\frac{1}{2},j}\right) - \frac{\Delta\tau}{\Delta\eta}\left(\widetilde{G}^n_{i,j+\frac{1}{2}} - \widetilde{G}^n_{j,j-\frac{1}{2}}\right) \\ & + \frac{1}{Re}\left[\frac{\Delta\tau}{\Delta\xi}\left(\widetilde{Q}^n_{i+\frac{1}{2},j} - \widetilde{Q}^n_{i-\frac{1}{2},j}\right) + \frac{\Delta\tau}{\Delta\eta}\left(\widetilde{R}^n_{i,j+\frac{1}{2}} - \widetilde{R}^n_{j,j-\frac{1}{2}}\right)\right] + \Delta\tau \widehat{S}^n_{i,j} \end{aligned}$$

$$D_{i,j} = \left(I - \Delta\tau \frac{\partial \widehat{S}}{\partial \widehat{U}}\right)_{i,j} \tag{3}$$

The following two-step Range-Kutta method is used so that the conserved variables $\widehat{U}$ can be obtained with second-order accuracy in time and space

$$\overline{U}_{i,j}^{n+1} = \widehat{U}_{i,j}^{n} + \Delta\widehat{U}_{i,j}^{n} \qquad \widehat{U}_{i,j}^{n+1} = \frac{1}{2}(\widehat{U}_{i,j}^{n} + \overline{U}_{i,j}^{n+1}) + \frac{1}{2}\Delta\overline{U}_{i,j}^{n+1}, \quad (4)$$

where numerical flux in ξ-direction is $\widetilde{F}_{i+\frac{1}{2},j} = \frac{1}{2}(\widetilde{F}_{i,j} + \widetilde{F}_{i+1,j} + R_{i+\frac{1}{2},j}\Phi_{i+\frac{1}{2},j})$. The vector $\Phi_{i+\frac{1}{2},j}$ in the upwind TVD scheme can be expressed by $\Phi^{l}_{i+\frac{1}{2},j}$, $\Phi^{l}_{i+\frac{1}{2},j} = \delta(a^{l}_{i+\frac{1}{2},j})(g^{l}_{i+1,j} + g^{l}_{i,j}) - \Psi(a^{l}_{i+\frac{1}{2},j} + \gamma_{i+\frac{1}{2},j})a^{l}_{i+\frac{1}{2}}$; $l = 1, 2, ..., N + M + 3$. The numerical flux $\widetilde{G}$ in the η direction can also be defined similarly.

3.1. Results and discussion

To judge the predictive capabilities of present computer codes, a double Mach-reflection flow of incident $M_s = 8.7$ impinging on a 27° wedge is numerically calculated using N-S and Euler equations. Figure 4 is the schematic diagram

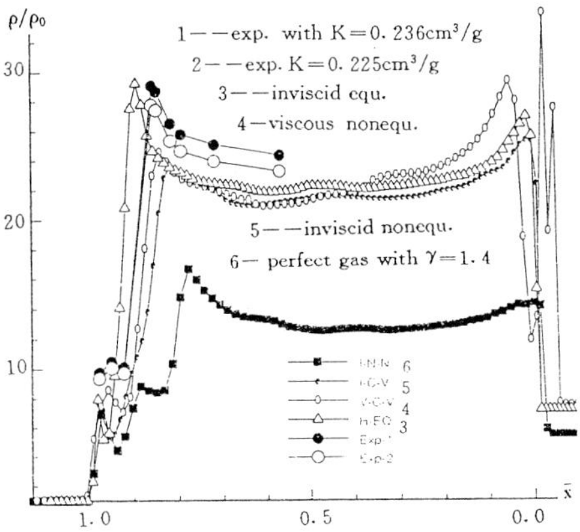

Figure 3. Comparison of density distribution on the wall. $M_s = 7.6$, $P_0 = 2.63$ KPa, $T_0 = 289$ K.

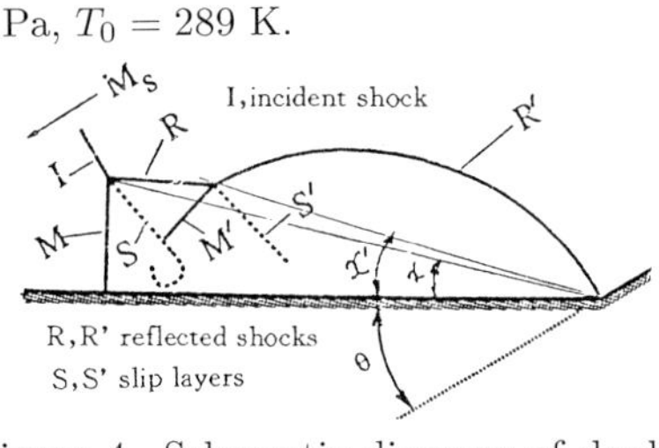

Figure 4. Schematic diagram of double Mach reflection; X, X' first and second triple-point angles.

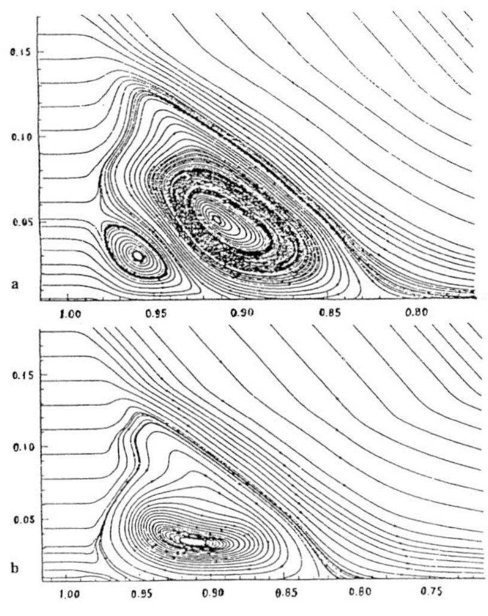

Figure 5. Computed streamlines near Mach-stem region: viscous nonequilibrium (top); inviscid nonequilibrium (bottom). $M_s = 8.7$, $\theta_w = 27°$, $P_0 = 4.1$ KPa, $T_0 = 299.2$ K.

of this flow. Figures 5a and 5b show the computed streamlines of inviscid and viscous nonequilibrium flow near the Mach-stem region. The rollup pattern with two vortices (shown in Fig. 5a) is in substantial agreement with the inter-

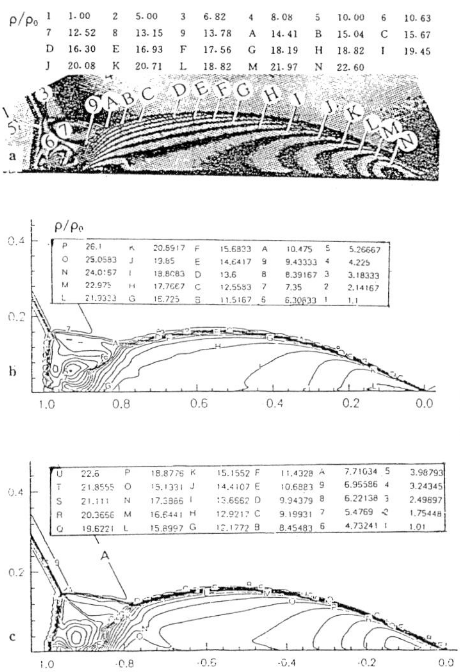

Figure 6. Density contours. $M_s = 8.7$, $\theta_w = 27°$, $P_0 = 4.1$ KPa, $T_0 = 299.2$ K. (a) Exp. from Groth (1993); (b) computed viscous nonequilibrium; (c) computed inviscid nonequilibrium.

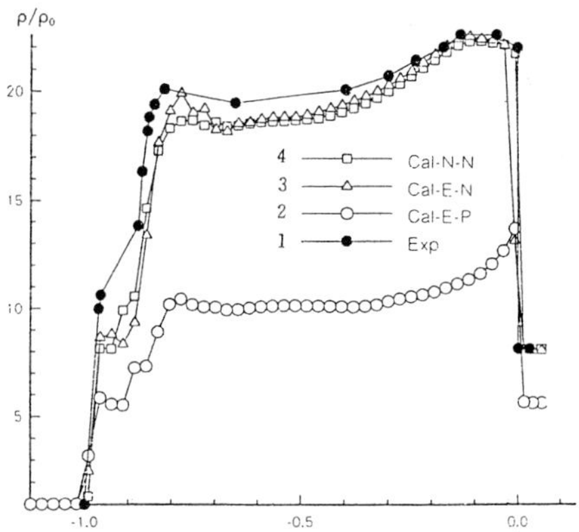

Figure 7. Comparison (cal. and exp.) of density distribution on the wall.

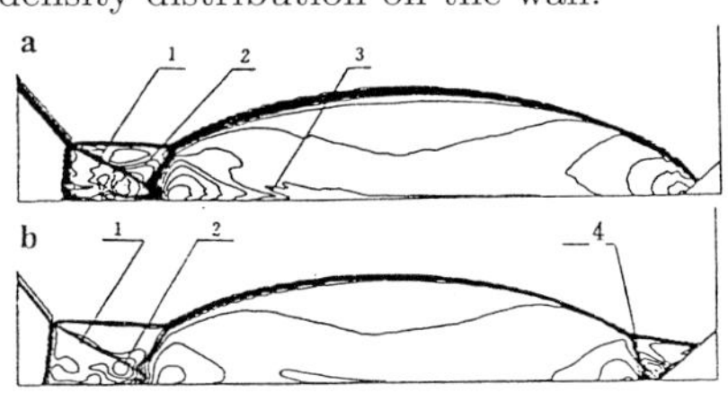

Figure 8. Computed density contours of perfect gas. $M_s = 7.66$, $\theta_w = 40°$, $P_0 = 2.63$ KPa, $T_0 = 289$ K. (a) Inviscid, 1 – first slip layer, 2 – vortex, 3 – second slip layer, $Re = 6.5 \times 10^5$; (b) viscous, 4 – "λ" shock wave.

ferogram from a UTIAS experiment (shown in Fig. 6a). The computed density contours are also displayed in Figs. 6b and 6c. The computed wall density distribution shown in Fig. 7 indicates the good agreement with the measured results of UTIAS tests. A relaxation length of about $0.1L$ can be determined from Fig. 6c, which can be confirmed by the UTIAS interferogram of Fig. 6a. As a second example, the calculation of the double Mach-reflection flow in the present experimental case with $M_s = 7.66$, $\theta_w = 40°$ has also been performed. Figures 8a and 8b are the computed density contours for inviscid and viscous perfect gases, respectively; the first and second slip layer, vortex, and "λ" shock wave in the compression corner are displayed clearly. Comparison of the density distribution on the wall between experiment and computation is illustrated in Fig. 3. The computed results for $\gamma = 1.4$, equilibrium and nonequilibrium gases are reasonable and at the location of $\overline{x} = 0.4$–0.95; densities seem to tend toward equilibrium; the reason for some deviations from experimental results is now being analyzed.

4. Conclusion

The present experiment is successful in attaining quantitative density fields by directly counting fringes without relying on analytical relations of three-shock theory at the triple points. However, in chemically reacting flow, fringe-counting data-reduction techniques provide very rough estimation of the density fields, therefore the reasonable determination of Gladstone-Dale constant K is still a problem to be solved for the accurate comparison of experiment with computation. The present computer codes, validated by experiments, represent a substantial predictive capability for the detailed modeling of the complex nonstationary nonequilibrium and viscous flow with complicated shock structure.

References

Clinton PT, Groth S (1993) TVD Flux-Difference Split Methods for High-Speed Thermochemical Nonequilibrium Flows With Strong Shocks, UTIAS Report No. 350

Dunn MG, Kang SW (1973) Theoretical and Experimental Studies of Reentry Plasmas, NASA CR 2232.

Glaz HM (1987) High Resolution Calculation of Unsteady Two Dimensional Nonequilibrium Gas Dynamics With Experimental Comparisons, AIAA Paper 87-1293.

Glaz HM, Collella P, Glass II (1986) A Detailed Numerical, Graphical, and Experimental Study of Oblique Shock Wave Reflections, UTIAS Report No. 285.

Gupta R, Yos J, Thompson R, Lee K-P (1990) A Review of Reaction Rates and Thermodynamic and Transport Properties for an 11 Species Air Model for Chemical and Thermal Nonequilibrium Calculation to 30000 K, NASA RP 1232.

Le J, Hu G (1992) A high resolution calculation of reflection of oblique shock wave from a compression corner in chemical nonequilibrium gas flow. In: Proceeding of Int Conf on the Methods of Aerophysical Research, Novosibirsk, p 84.

Needham CE (1991) Chemical nonequilibrium effects of the shock reflection. In: 18th Int Symp on Shock Tube and Waves, pp 897–902.

Wave configuration of Mach reflection in steady flows: analytical solution and dependence on downstream influences

H. Li, M. Schotz, G. Ben-Dor
Pearlstone Center for Aeronautical Engineering Studies
Department of Mechanical Engineering
Ben-Gurion University of the Negev, Beer Sheva, Israel

Abstract: Azevedo and Liu's (1993) analytical model for predicting the size of the Mach-reflection wave configuration in steady flows has been improved. Predictions based on the modified analytical model were compared to available experimental results. The agreement was found to be better than that obtained by Azevedo and Liu's (1993) original model. The reason for the better agreement is due to the fact that downstream effects were not neglected in our modified model.

Key words: Mach reflection, Wave configuration, Steady flow

1. Introduction

It was pointed out, both in Ben-Dor (1991) and Ben-Dor and Takayama (1992), that one of the yet unsolved shock-wave reflection problems in steady flows is the mechanism by which the size of a Mach reflection is determined. In such flows, the height of the Mach stem of the Mach-reflection pattern is not uniquely determined. Consider Fig. 1 where the solid lines describe the four discontinuities of a steady Mach-reflection wave configuration, namely: the incident shock wave (i), the reflected shock wave (r), the Mach stem (m), and the slipstream (s). These four discontinuities meet at a single point, the triple point, T. If one selects a new point along the incident shock wave (i) and draws there three lines parallel to the reflected shock (r*), to the Mach stem (m*) and to the slipstream (s*), then one has a new triple point, T*, with the appropriate four discontinuities. The two triple points, T and T*, as well as other triple points which could have been obtained by choosing a different location for T* along the incident shock wave (i), completely satisfy the governing equations of the three-shock theory. However, when experiments with identical initial conditions are repeated in a given facility, only one of the above-mentioned infinite theoretically possible Mach-reflection wave configurations is obtained. Thus, the three-shock theory is incapable of predicting the actual size of the Mach reflection since it is inherently independent of any physical length scale.

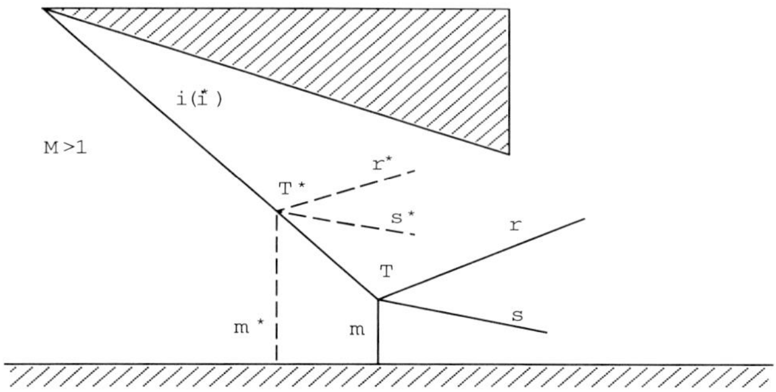

Figure 1. Two theoretically possible Mach-reflection wave configurations for identical initial conditions.

Azevedo and Liu (1993) suggested a physical model for predicting the height of the Mach stem. Consider Fig. 2 where a schematic drawing of a reflecting wedge which generates a Mach reflection is shown. In their model they assumed that the Mach stem, the slipstream, and the bottom surface form an effective one-dimensional converging nozzle, and that the flow in region (3) is isentropic and reaches sonic conditions at the minimum cross-section part of the effective conduit. In addition Azevedo and Liu (1993) assumed that the downstream flow has no influence on the location of the throat and that the throat of this converging nozzle is located at the point where the leading characteristic of the expansion fan, generated at the trailing edge of the reflecting wedge, intersects the slipstream (point E in Fig. 2).

By applying the conservation laws of mass and momentum to the control volume given by ABCDEFA, Azevedo and Liu (1993) developed an analytical model by which the Mach-stem height could be calculated. The predictions of their model are shown in Fig. 3. Although their analytical predictions showed a trend similar to the experimental results obtained for the Mach-stem height dependence on the angle of incidence, the predicted values were far from being satisfactory. This lack of satisfactory agreement is attributed to the above-mentioned oversimplifying assumptions incorporated into their model.

The purpose of the present study is to propose a more realistic model for predicting the Mach-stem height in a Mach reflection observed in steady flows. While adopting Azevedo and Liu's (1993) first two assumptions, the model proposed in the present study takes into account the influence of the downstream flow on the Mach-stem location. Therefore the flow downstream of the Mach stem is analyzed and its effect is incorporated into the analytical model in order to obtain a better model for predicting the Mach-stem location and height.

2. Present study

A physical model describing the interaction, shown in Fig. 2, between the expansion fan (generated at the trailing edge of the reflecting wedge) and the reflected shock wave of the Mach reflection was developed. Unlike Azevedo and Liu (1993) who assumed that the throat of the converging-diverging nozzle is located at the point where the head of the expansion fan intersects the slipstream, in our model the shape of the slipstream was calculated and the throat was located at the minimum cross section. For details see Schotz et al. (1995).

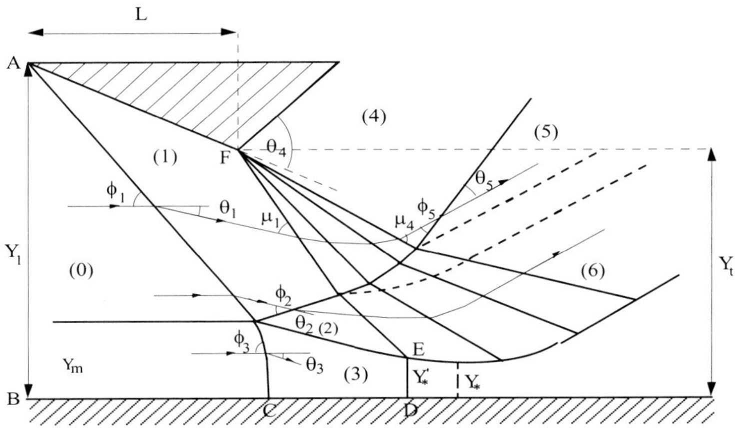

Figure 2. Schematic illustration of Azevedo and Liu's (1993) simplified model. Region (3) is an isentropic converging nozzle.

3. Results and discussion

The normalized Mach stem height y_m/L as a function of the incident shock-wave angle Φ, along with Hornung and Robinson's (1982) experimental results (obtained for $M_0 = 2.84$, 3.49, 3.98, and 4.96) are shown in Figs. 3a–3d, respectively. In addition, Azevedo and Liu's (1993) analytical predictions are also shown in these figures by dashed lines. It is apparent that their predictions consistently underestimate the actual Mach-stem height. Predictions based on the present modified model, for $y_t/L = 0.37$ and a variety of trailing-edge expansion angles θ_4, are shown in Figs. 3a–3d by solid lines. It is evident from these figures that the trailing-edge expansion angle significantly affects the Mach-stem height. Consequently, ignoring its effect, as was done in Azevedo and Liu's (1993) model, is wrong. The results shown in Figs. 3a–3d indicate that the normalized Mach-stem height, y_m/L, increases with increase in the trailing-edge expansion angle. Since Hornung and Robinson (1982) and Azevedo and Liu

(1993) did not specify completely the geometrical shape of the reflecting wedges used in the course of the experiments (no information about θ_4 was given) it is not known whether one or more reflecting wedges were used. However, the present analytical model suggests that:

$$45^\circ > \theta_4 > 20^\circ \text{ for } M_0 = 2.84, \quad 50^\circ > \theta_4 > 38^\circ \text{ for } M_0 = 3.49$$

$$35^\circ > \theta_4 > 25^\circ \text{ for } M_0 = 3.98, \quad 32^\circ > \theta_4 > 25^\circ \text{ for } M_0 = 4.96.$$

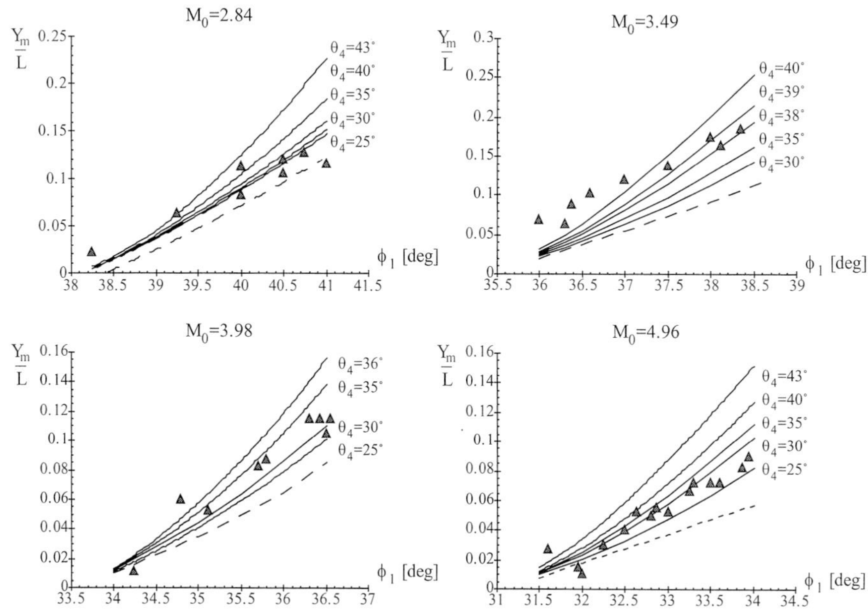

Figure 3. Comparison between experimental results and analytical predictions of the normalized Mach-stem height versus the shock-wave angle of incidence for (a) $M_0 = 2.84$, (b) $M_0 = 3.49$, (c) $M_0 = 3.98$, and (d) $M_0 = 4.96$. Solid triangles – experimental results (Hornung and Robinson 1982), dashed line – Azevedo and Liu's (1993) model, solid line – present model.

Note also that based on Azevedo and Liu's (1993) arguments, the results of Hornung and Robinson (1982) for $M_0 = 3.49$ were most likely influenced by the downstream conditions which caused the Mach stem to be larger in its length than had the reflection been free of downstream influence. Consequently, if the results for $M_0 = 3.49$ are excluded then it is evident that the actual wedge trailing-edge angles in Hornung and Robinson's (1982) experiments were probably within the range $45^\circ > \theta_4 > 25^\circ$.

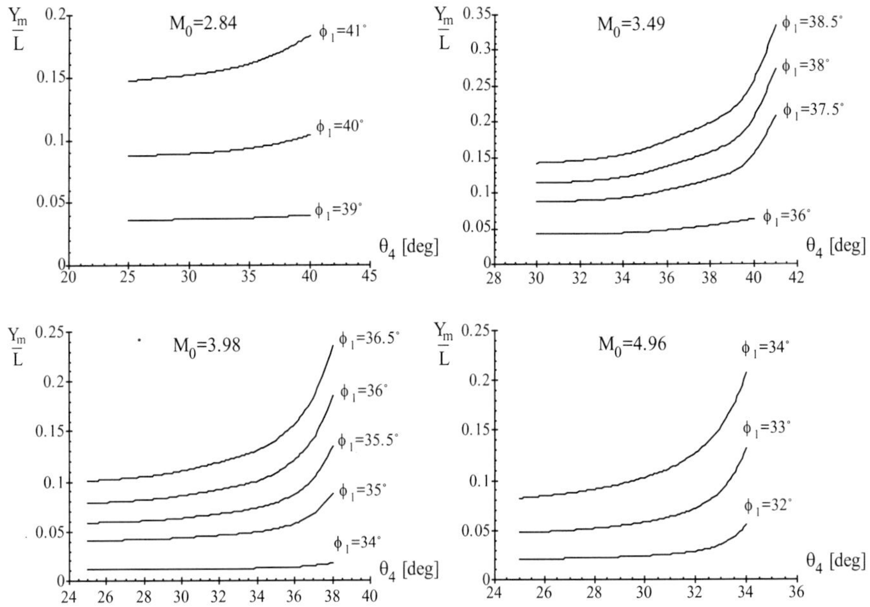

Figure 4. The dependence of the normalized Mach-stem height on the trailing edge angle θ_4 for different shock-wave incidence angles: (a) $M_0 = 2.84$, (b) $M_0 = 3.49$, (c) $M_0 = 3.98$, and (d) $M_0 = 4.96$.

The normalized Mach-stem height, y_m/L, is shown in Figs. 4a–4d as a function of the trailing-edge expansion angle θ_4 (obtained for $M_0 = 2.84$, 3.49, 3.98, and 4.96). The results indicate that the normalized Mach-stem height, y_m/L, decreases as the trailing-edge expansion angle, θ_4, decreases, until it reaches asymptotically a constant value. It is also evident from these figures that as θ_4 increases the rate of increase of the Mach-stem height increases. Figs. 4a–4d also indicate that higher shock-wave angles of incidence result in larger Mach stems for identical wedge trailing-edge angles.

4. Conclusions

The analytical model for predicting the Mach-stem height in steady Mach reflections which was suggested by Azevedo and Liu (1993) has been modified by replacing one of its oversimplified assumptions by a more realistic one. Unlike Azevedo and Liu's (1993) model, which ignored the trailing-edge angle of the reflecting wedge, the present model accounts for this parameter and hence takes into consideration the entire expansion fan and its interaction with both the reflected shock wave and the slipstream whose shapes inside the interaction zone are assumed to be second-order polynomials.

Predictions of the improved model together with those of Azevedo and Liu's (1993) model were compared with the experimental results of Hornung and Robinson (1982). The predictions based on the present model were found to be in better agreement with experimental results than those of Azevedo and Liu (1993).

References

Azevedo DJ, Liu CS (1993) Engineering approach to the prediction of shock patterns in bounded high-speed flows. AIAA J 31(1):83–90.

Ben-Dor G (1991) Shock Wave Reflection Phenomena, Springer, New York.

Ben-Dor G, Takayama K (1992) The phenomena of shock wave reflection—a review of unsolved problems and future research needs. Shock Waves 2:211–223.

Hornung HG, Robinson ML (1982) Transition from regular to Mach reflection of shock waves, part 2: the steady-flow criterion. J Fluid Mech 123:155–164.

Schotz M, Levy A, Ben-Dor G, Igra O (1995) Analytical prediction of the wave configuration size in steady flow Mach reflections. AIAA (to be published).

Non-self-similarity of the initial stages of Mach reflection

J. M. Dewey, A. A. van Netten
University of Victoria
Victoria, B.C., V8W 3P6 Canada

Abstract: Multiple shadowgraph and schlieren photography were used to obtain sequences of photographs showing the initial stages of a Mach reflection progressing along a wedge surface. The wedges used were about 30 cm long. The experiments confirmed the delayed initiation of Mach reflection, and included several cases in which the initiation of Mach reflection occurred more than 20 cm from the tip of the wedge.

The delay in initiation of a Mach reflection was strongly influenced by the ambient gas density and the wedge roughness. After the delayed initiation of Mach reflection, the shock configuration first expanded in a non-self-similar fashion, and in the case of double Mach reflection, the two triple-point trajectories did not extrapolate back to the same point on the wedge. In general, it was observed that the shock configurations eventually became self-similar relative to the tip of the wedge.

Key words: Self-similarity, Shock wave, Mach reflection, Experiment

1. Introduction

Several authors (e.g. Walenta 1980, 1983, 1987; Henderson & Grey 1981; Hornung 1985; Schmidt 1989; and Fuchs & Schmidt 1991), have shown that at low ambient densities Mach reflection does not begin at the tip of the wedge but at a distance of about 100 mean-free-paths from the leading edge. These authors studied the phenomenon using techniques in which a single record was obtained from each experiment. Because of the difficulty of exactly reproducing all the physical conditions from one shock-tube experiment to another, it was difficult to define accurately the triple-point trajectory for a specific incident-shock Mach number, wedge angle, and ambient gas condition. In addition, most of these studies were done using relatively short wedges, less than 10 cm long. These problems have been overcome in the present work by studying the initial stages of Mach reflections using multiple shadowgraph or schlieren photography so that a sequence of photographs was obtained of the development and progression of a Mach reflection in a single experiment on a wedge about 30 cm long.

2. Experimental procedures

2.1. Photographic methods

The experiments were carried out in a 7.5-cm × 25-cm rectangular shock tube, and the wedges were placed in a 25-cm × 25-cm window section which allowed the use of wedges up to 30 cm long. The light source was a ruby laser that could be Q-switched at frequencies up to 25 kHz, and with a pulse duration of about 50 ns. For shadow photography a 6.35-cm × 20.3-cm strip of film was placed directly against the second window, parallel to the surface of the wedge, so as to obtain direct-contact shadowgraphs of the reflection process. For schlieren photography, a system similar to that described by Dewey & Walker (1975) was used. The individual images were recorded with a rotating prism camera, and up to 10 images could be obtained as the shock reflection progressed along the wedge in a single experiment.

2.2. Experimental conditions

Experiments were carried out using plane incident shocks with Mach numbers of approximately 1.3, 2.0, or 2.7. Air at room temperature was the working gas. The reflecting wedges were made of steel or aluminum, and were rigidly mounted with the leading edge about 3 cm above the floor of the shock tube so that the boundary layer from the floor would not be swept up the wedge. The leading edge of the wedge was symmetrical relative to the flow direction, i.e., the angle between the two surfaces was twice the nominal wedge angle, θ_w. To investigate the effect of surface roughness, three types of wedge surface were used: polished (0.05 μm); smooth (1.6 μm); and rough (200 grit sandpaper). The smoothness was measured with a roughness comparator. A range of wedge angles was used so as to form single (SMR), complex (CMR), and double (DMR) Mach reflections.

The experiments in which the incident-shock Mach number was close to 2.7 were carried out at an initial pressure of approximately 11 Torr. Those at Mach 2.0 were done at ambient pressures of 20 or 130 Torr, and those at Mach 1.3 were carried out at either atmospheric pressure or approximately 13 Torr. This was done to investigate the effect of gas density on the initiation of Mach reflection.

The Mach number of the incident shock was measured in each experiment with two piezo-electric time-of-arrival gauges 20 cm apart in the roof of the shock-tube window section. This measurement was confirmed from the photographic record by measuring the shock displacement in the measured time interval between the laser pulses.

2.3. Method of analysis

The shadowgraph or schlieren images were projected at a magnification of 1:1 or 20:1, respectively, on to the flat bed of an x-y digitizer. The surface of the wedge and the positions of all the shocks and other density discontinuities were digitized manually, and the coordinates stored in electronic format. Fiducial markers in the fields of view were also digitized to provide scaling factors and to ensure that there was no significant optical distortion. Figure 1 shows a sequence of digitized shock configurations.

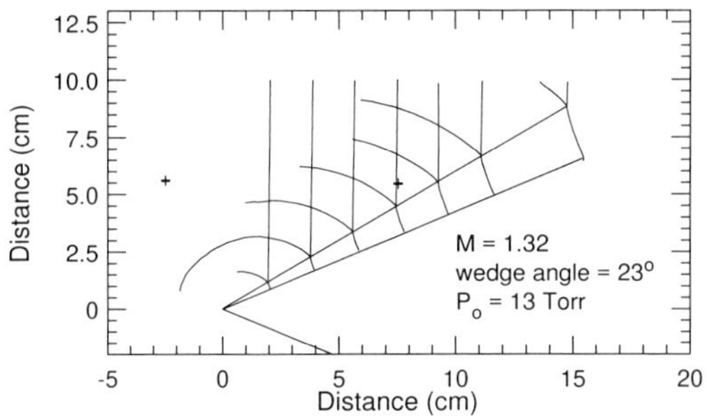

Figure 1. Digitized images of a series of schlieren photographs of a Mach reflection obtained in a single experiment.

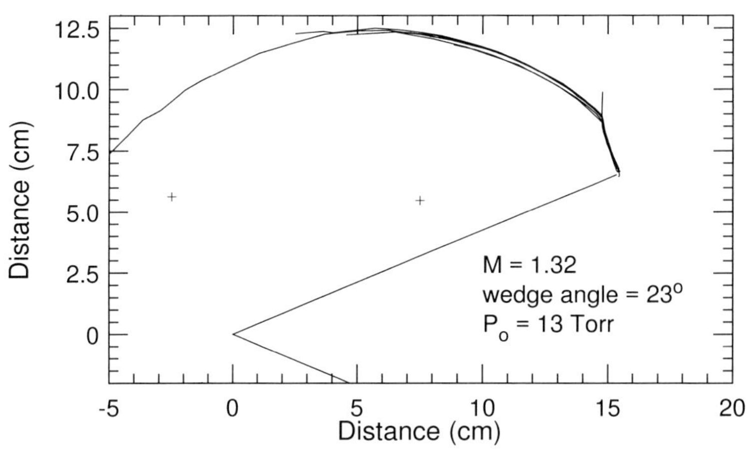

Figure 2. The shock configurations from Fig. 1 replotted after scaling relative to the distance of the incident shock past the leading edge of the wedge. This shows a high order of self-similarity.

The digitized shock configurations from a single experiment were then replotted after being scaled relative to the distance of the incident shock past the leading edge of the wedge. Using this technique, if two or more shock configurations were self-similar, when replotted in this way they would exactly overlay one another. For example, Fig. 2 shows the same series of digitized images as Fig. 1, but replotted after scaling, and showing a high order of self-similarity.

In contrast, Fig. 3 shows four nominal DMRs from a single experiment, replotted after scaling. Here there is a high degree of non-self-similarity.

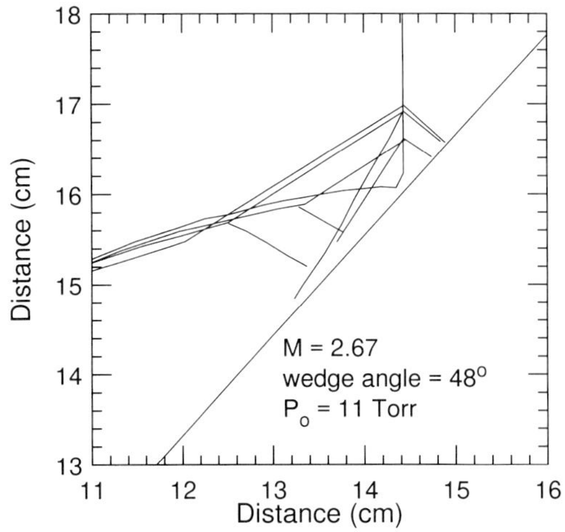

Figure 3. The development of a double Mach reflection after scaling relative to the distance of the incident shock past the leading edge of the wedge. This reflection was clearly non-self-similar, initially appearing as a regular reflection, but gradually developing into a self-similar form.

3. Summary of results

All the experiments with a smooth ($\epsilon = 1.6\mu$m) 23° wedge and incident-shock Mach numbers of 1.3 and 2.7 produced self-similar SMRs at 11 Torr and 761 Torr within 1 cm of the tip of the wedge. The reflections at both ambient pressures were identical, and the scaled shock configurations at the two pressures were exactly similar.

Most of the other experiments were done with wedge angles in the range from 46° to 49°, i.e., close to the angle for transition from regular to Mach reflection. Incident-shock strengths were approximately 2.0 or 2.7 in air at ambient pressures of approximately 11, 20, or 130 Torr. In all cases there was a lack of self-similarity in the initial development of the CMR or DMR. Even on a polished ($\epsilon = 0.05\mu$m) 48° wedge, DMR could not be identified until the incident shock had travelled up to 9 cm along the wedge. Before that, the incident shock appeared to extend to the surface of the wedge. The reflected shock also appeared to extend to the wedge, but to a point a short distance behind the incident shock. There was then a transition period in which the reflection at first appeared to be a CMR, before developing into a DMR. During this phase the two triple-point trajectories did not extrapolate back to the same point on the wedge. Eventually, at a point beyond 20 cm from the tip of the wedge, a self-similar DMR configuration was established. It should be noted that this self-similarity was relative to the tip of the wedge, i.e. the triple-point trajectories extrapolated back to the leading edge, and not to the points on the wedge where they had first appeared.

This pattern of development for Mach reflection extended further up the wedge surface as the ambient-gas density decreased and/or rougher surfaces were used. In the case of very rough surfaces, e.g. 600 grit sandpaper, an apparently self-similar type of regular reflection was maintained for a distance of 30 cm from the tip of the wedge, with wedge angles and shock Mach numbers for which DMR developed on smooth surfaces. Again, in these situations there appeared to be a gap between the the incident and reflected shocks at the wedge surface.

4. Discussion and conclusions

The experiments reported here confirm that there is a delay in the initiation of Mach reflection on a smooth wedge. Where others have suggested that the delay is on the order of 100 mean-free-paths, the present work shows delays many orders of magnitude greater than this. The distance from the leading edge of the wedge at which Mach reflection is initiated increases with decreasing ambient-gas density and increasing surface roughness.

The typical development of DMR is as follows. The initial reflection is similar to a regular reflection, albeit with a gap between the incident and reflected shocks at the wedge surface. At distances which may be more than 20 cm from the tip of a smooth wedge, the Mach-reflection pattern begins to emerge and grows in a non-self-similar manner. The two triple-point trajectories do not extrapolate back to the same point on the wedge. Eventually, the shock configuration becomes self-similar with respect to the tip of the wedge, i.e., the triple-point trajectories are straight and can be extrapolated back to the leading edge.

These findings raise doubts about the validity of the bounds between the various reflection regions determined by experimental observations and reported in such sources as Ben-Dor (1992). It appears that these observations were made using wedges which were not longer than 10 cm and which were machined smooth but not polished, and with ambient pressures of the order of 10 Torr. Under such conditions, the present work indicates that the final reflection state would not have been reached in those experiments; and had longer, smoother wedges been used, Mach rather than regular reflection may have been observed. Such results would have the effect of raising the experimental boundaries between the different reflection regions in the (M_s, θ_w)-plane, thus bringing them closer to the theoretical transition bounds.

Acknowledgement. The authors gratefully acknowledge the assistance of Susan Watkins in the preparation of this paper.

References

Ben-Dor G (1992) Shock Wave Reflection Phenomena, Springer-Verlag, New York.

Dewey JM, Walker DK (1975) A multiply pulsed double-pass laser schlieren system for recording the movement of shocks and particle tracers within a shock tube. J Appl Phys 46(8):3454–3458.

Fuchs J, Schmidt B (1991) Formation of the shock reflection on a wedge. In: Shock Waves, Takayama K (ed), 1, pp 293–298, Springer-Verlag, Berlin/New York.

Henderson LF, Grey PM (1981) Experiments on the diffraction of strong blast waves. Proc Roy Soc Lond A 377:363–378.

Hornung H (1985) The effect of viscosity on the Mach stem length in unsteady strong shock reflection. In: Lecture Notes in Physics, 235, Flow of Real Fluids, pp 82–91, Springer-Verlag, Berlin/New York.

Schmidt B (1989) Structure of incipient triple point at the transition from regular reflection to Mach reflection. In: Rarefied Gas Dynamics: Theoretical and Computational Techniques, Muntz EP et al. (eds), AIAA Inc., Washington DC, 118:597–607.

Walenta ZA (1980) Microscopic structure of the Mach-type reflexion of the shock wave. Arch Mech Warszawa 32(5):819–825.

Walenta ZA (1983) Formation of the Mach type reflection of shock waves. Arch Mech Warszawa 35(2):187–196.

Walenta ZA (1987) Mach reflection of a moving, plane shock wave under rarefied flow conditions. In: Shock Tubes and Waves, Grönig H (ed), VCH Verlagsgesellschaft, Weinheim, pp 535–541.

Numerical confirmation of the hysteresis phenomenon in the regular to Mach reflection transition in steady flows

A. Chpoun
Laboratoire d'Aérothermique, CNRS
4 ter Route des Gardes, 92190 Meudon, France
G. Ben-Dor
Pearlstone Center for Aeronautical Engineering Studies
Department of Mechanical Engineering
Ben-Gurion University of the Negev, Beer Sheva, Israel

Abstract: Numerical calculations based on the Navier-Stokes equations were carried out to investigate the reflection of shock waves over straight reflecting surfaces in steady flows. The results for a flow Mach number of $M_0 = 4.96$ confirmed the recent experimental findings of Chpoun et al. (1995) concerning the transition from regular to Mach reflection. Numerical calculations as well as experimental results showed a hysteresis phenomenon during this transition and the regular reflection was found to be stable in the dual-solution domain in which theoretically both regular- and Mach-reflection wave configurations are possible.

Key words: Hysteresis, Regular reflection, Mach reflection

1. Introduction

In steady flows the reflection of a shock wave over a straight surface leads to two possible wave configurations, namely, regular reflection (RR) and Mach reflection (MR). These wave configurations and the definition of some flow parameters are illustrated in Figs. 1a and 1b, respectively. The regular reflection (RR) consists of two shock waves, namely, the incident shock wave (i) and the reflected shock wave (r). They meet at the reflection point (R), located on the reflecting surface. The flow states are (0) ahead of (i), (1) behind it, and (2) behind (r). The Mach reflection (MR) consists of three shock waves, namely, the incident shock wave, (i), the reflected shock wave (r), and the Mach stem (m). They all meet at a single point known as the triple point (T) which is complemented with a slipstream (s). The Mach stem (m) is usually a curved shock wave which is perpendicular to the reflecting surface at the reflecting point (R).

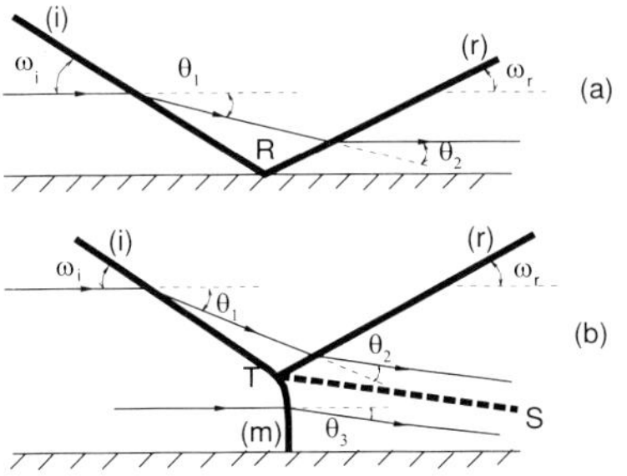

Figure 1. (a) Regular-reflection and (b) Mach-reflection wave configurations.

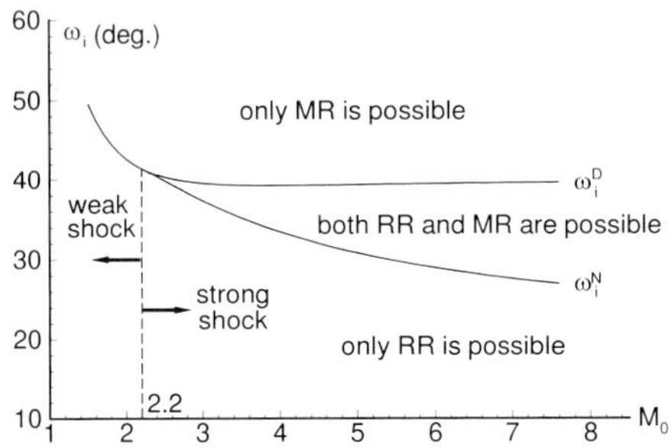

Figure 2. RR $\Longleftrightarrow$ MR transition criteria vs. flow Mach number.

Mainly from theoretical considerations, two criteria were suggested for the RR $\Longleftrightarrow$ MR transition. The first one called the von Neumann (or "mechanical equilibrium") criterion was first suggested by von Neumann in the early 1940s. This criterion corresponds to the situation where the pressure jump across the incident and reflected shock waves (i and r) becomes equal to the pressure jump across the Mach stem (m). The appropriate incident shock angle is noted ω_i^N. The second one, called the detachment criterion, was also suggested by von Neumann and corresponds to the limit where an attached oblique shock can be obtained from the flow at state (1) impinging on the reflecting surface. The corresponding incident shock angle is noted ω_i^D. It should be noted that $\omega_i^N < \omega_i^D$ for strong incident shock waves (Fig. 2). From the analysis of graphical solutions (i and r shock polars), it can be shown (Ben-Dor 1991) easily that for $\omega_i < \omega_i^N$ a Mach-reflection wave configuration is theoretically impossible. On the other hand for $\omega_i > \omega_i^D$, the regular-reflection wave configuration is theoretically impossible. The situation where $\omega_i^N < \omega_i < \omega_i^D$ corresponds to the dual-solution domain where both (RR) and (MR) wave configurations are theoretically possible (Fig. 2).

Since the pioneering work of von Neumann, have been only few expèrimental studies related to the RR $\Longleftrightarrow$ MR transition in steady flows. The experiments of Hornung and Robinson (1982) are the more recent ones. Based on their own experiments, as well as those of Henderson and Lozzi (1975, 1979), Hornung and Robinson concluded that in steady flows, the transition from regular to Mach reflection of strong shock waves occurs at the von Neumann condition ($\omega_i = \omega_i^N$) and not at the detachment condition ($\omega_i = \omega_i^D$) as mentioned in previous studies. Based on their length scale concept, Hornung et al. (1979)

hypothesized that a hysteresis should exist in the RR $\Longleftrightarrow$ MR transition. If one starts with an MR at $\omega_i > \omega_i^D$ and then slowly decreases ω_i while M_0 is kept constant, there is a decrease in the Mach-stem height until it vanishes at $\omega_i = \omega_i^N$ where a smooth transition from MR to RR takes place. If the decrease in ω_i continues, then the RR is maintained. If now ω_i is slowly increased, since the flow field is free of significant disturbances, the RR should be maintained until $\omega_i = \omega_i^D$ where a sudden transition from RR to MR should take place. Hornung and Robinson's (1982) experimental attempts to confirm this hypothesis regarding the hysteresis in the RR $\Longleftrightarrow$ MR transition failed and they concluded that "the hysteresis effect predicted by Hornung et al. (1979) could not be confirmed." For this reason they refer to the RR in the dual-solution domain $\omega_i^N < \omega_i < \omega_i^D$ as an unstable regular reflection. Meanwhile, in recent experimental work (Chpoun et al. 1995) which was conducted in the SH2 hypersonic wind tunnel of CNRS Meudon, France, the hysteresis loop hypothesized by Hornung et al. (1979) was observed. The experiments were conducted in a continuous, free jet facility. A few geometrical parameters (plate length, plates cross inlet section, etc.) were varied and the hysteresis was reproduced in all experimental dispositions. Furthermore, unlike previous reports, regular-reflection wave configurations were found to be stable in the dual-solution domain. The hysteresis phenomenon which was observed by Chpoun et al. (1995) is shown in Fig. 3 in the (ω_i, ω_r)-plane. While the MR $\Rightarrow$ RR transition occured at $\omega_i = 30.9°$, which is equal to the von Neumann condition ω_i^N, the RR $\Rightarrow$ MR transition occured at $\omega_i = 37.2°$, which is about 2° smaller than the corresponding detachment condition, $\omega_i^D = 39.3°$. Based on the foregoing

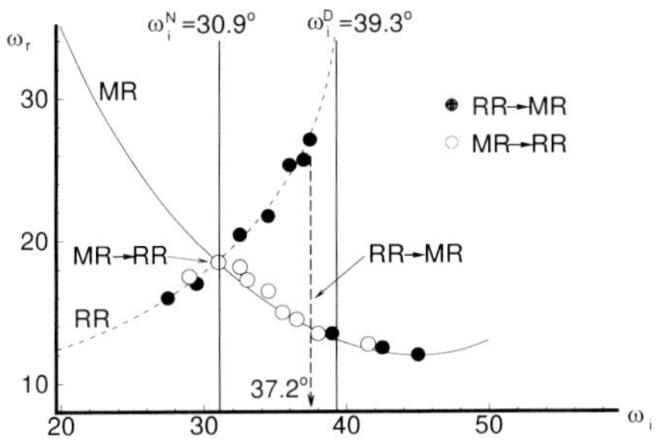

Figure 3. The experimental hysteresis loop in (ω_i, ω_r)-plane.

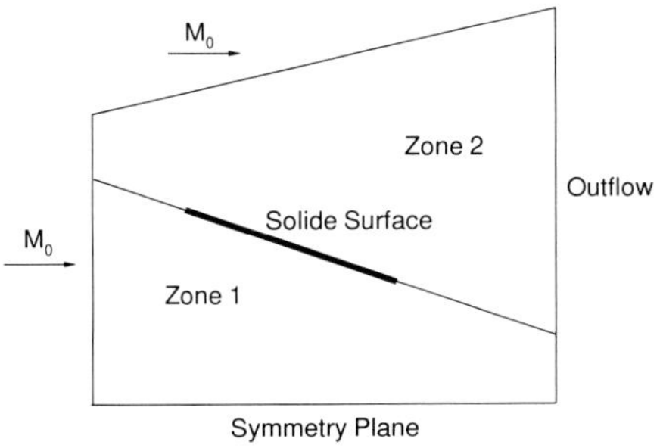

Figure 4. Computational domain and boundary conditions.

introduction and particularly on the recent experimental findings of Chpoun et al. (1995), the following summarizes the state of the art regarding the RR $\Longleftrightarrow$ MR transition in steady flows: (1) the RR $\Rightarrow$ MR transition takes place at nearly the detachment criterion and the MR $\Rightarrow$ RR transition occurs at the von Neumann criterion, therefore, a hysteresis exists in the RR $\Longleftrightarrow$ MR transition; (2) regular-reflection wave configurations are stable in the dual-solution domain. The following sections present numerical work related to shock-wave reflections in steady flows over straight surfaces. The results confirmed the principal experimental findings of Chpoun et al. (1995).

2. Computational work

The numerical results were obtained using a finite-volume zonal code. The code solves the complete Navier-Stokes equations by means of a time-marching method. The evaluation of the inviscid terms are based on the flux-splitting method of Steger and Warming (1981) in combination with upwind-based differencing. The diffusion terms are evaluated using standard central differences. The LU-SGS algorithm is used to solve the system of point Gauss-Seidel relaxation. Viscosities are evaluated using Sutherland's law and thermal conductivities are defined by assuming a constant Prandtl number. The computational domain is divided into two zones shown in Fig. 4. It contains $(2 \times 40 \times 60)$ cells. An infinitely thin solid surface was considered to separate the two zones. The computational domain was large enough for the upstream and upperside boundaries to be specified as freestream conditions, and downstream outflow quantities were extrapolated. To avoid viscous effects, the reflecting straight surface was replaced by a symmetry plane. Although the numerical model deals with an infinitely thin plate model, it reproduces the essential features of the flow which are obtained in the experiments, e.g., the flow compression and expansion at the windward side and the leeward side of the plate, respectively. The sensitivity of numerical results to mesh refinement was analyzed. The study showed that 4800 cells having a size of 0.2 mm near the wall were sufficient to ensure the grid independency of the results. In the same manner, the downstream boundary was moved at different locations and non-significant change was observed in the numerical results. The upstream flow conditions were those of Chpoun et al.'s (1995) experiments. The upstream Mach number was 4.96. The stagnation temperature and pressure were 453 K and 8.5 bar, respectively, leading to a high unit Reynolds number of 1.27×10^7/m which minimized the thickness of the viscous layer along the plate. The plate angles were set to a succession of prescribed values ranging from small values (RR) to large values (MR), then from large to small values with a step of 0.5°. For each angle the numerical results were carried out with initial flow conditions which were obtained from the previous angle's calculations. For each case of calculation, a steady-state flow was achieved. The first calculation was initiated with uniform flow conditions.

3. Results and discussion

Figures 5a–5l (from left to right and from top to bottom) show a sequence of the calculated flow-field density isolines obtained for plate angles ranging from small values (RR) to large values (MR) and then from large values to small values. Figure 5a shows a regular reflection; the incident and reflected shocks as well as

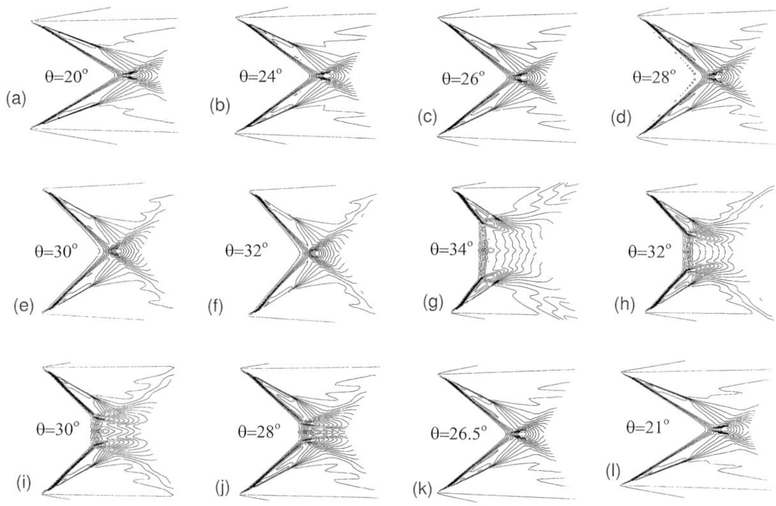

Figure 5. Flow-field numerical density isolines.

leading-edge expansion waves on the leeward side of the plate are evident. When the plate angle is increased step by step the RR is maintained until about 32°, then a sudden transition to Mach reflection with a big Mach stem occurs. The corresponding transition incident shock angle is approximately 44°. Inversely, as the plate angle decreases, a smooth transition to regular reflection is obtained for the plate angle of approximately 27°. The appropriate shock angle is about 36°. In cases with small Mach stems the clear evidence of slipstreams from the density isolines avoids confusion with a regular reflection. It should be noted that, as shown in Fig. 2, the incident shock angles corresponding to the von Neumann and the detachment criteria are, respectively, 30.9° and 39.3° in these flow conditions. The 5° differences between the theoretical and numerical transition angles originate probably from viscosity and the simplified numerical model, i.e., infinitely thin plate rather than a finite-size reflecting wedge.

On the other hand the regular-reflection wave configuration was found to be stable in the dual-solution domain, i.e., the calculations converged systematically to RR wave configurations as the initial flow was uniform or arbitrary. In the same manner, in the experimental part of their work, Chpoun et al. (1995) had shown that in the dual-solution domain an external disturbance changes MR wave configuration to RR. Therefore, the discrete variation of the plate angle shunts the theoretical MR $\Longrightarrow$ RR transition inside the dual-solution domain and the numerical transition angle was found to be larger than the von Neumann criterion. A small angle step probably will lead to a better agreement with the theroretical prediction, i.e., ω_i^N. Nevertheless, a hysteresis loop is clearly evident in the numerical results. In the dual-solution domain, for the same plate angle (28°) both RR and MR solutions were obtained (compare Figs. 5d and 5j).

4. Conclusion

A numerical study has been carried out to investigate the RR $\Longleftrightarrow$ MR transition in steady flows over straight surfaces. The numerical results confirmed the principal experimental findings of Chpoun et al. (1995) and contradicted previous reports on the subject: (1) a hysteresis exists in the RR $\Longleftrightarrow$ MR transition; (2) regular-reflection wave configurations are stable in the dual-solution domain. Furthermore, numerical solutions of flows which contain shock-wave interaction and reflection depend on the initial flow conditions. Therefore, the dependence of numerical solutions on initial flow conditions should be analyzed.

References

Ben-Dor G (1991) Shock Wave Reflection Phenomena, Springer-Verlag, New York.

Chpoun A, Passerel D, Li H, Ben-Dor G (1995) Reconsideration of the state-of-the-art of oblique shock wave reflection in steady flows, part I: experimental investigation. J Fluid Mech (to be published).

Henderson LF, Lozzi (1975) Experiments on transition of Mach reflection. J Fluid Mech 68(1):139–155.

Henderson LF, Lozzi (1979) Further experiments on transition to Mach reflection. J Fluid Mech 94(3):541–559.

Hornung HG, Oertel H, Sandeman RJ (1979) Transition to Mach reflection of shock waves in steady and pseudo-steady flow with and without relaxation. J Fluid Mech 90:541–560.

Hornung HG, Robinson ML (1982) Transition from regular to Mach reflection of shock waves, part 2: the steady-flow criterion. J Fluid Mech 123:155–164.

Hornung HG (1986) Regular and Mach reflection of shock waves. Ann Rev Fluid Mech 18:33–58.

Steger JL, Warming RF (1981) Flux vector splitting of the inviscid gasdynamic equations with application to finite-difference methods. J Comp Phys 40:263–293.

The existence of reflected finite-compression waves in weak pseudo-steady Mach reflections over not-"small" wedge angles

J. J. Liu
Department of Engineering Science, National Cheng-Kung University
Tainan, Taiwan ROC

Abstract: It is well known that the classical three-shock theory fails badly to describe the wave configuration of weak pseudo-steady Mach reflections (MR). The fundamental triple-point trajectory angle (χ) cannot be predicted by the three-shock theory, and the angles defined by the shocks and contact surface near the triple point cannot be determined by the theory even when an experimentally observed χ is used in the three-shock theory. Using existing experimental data for not-"small" reflecting wedge angles, it is shown analytically that the reason for this failure is owing to the existence of reflected finite-compression waves, instead of reflected shock waves, in weak pseudo-steady MR. The predicted χ using the linear pressure-deflection weak oblique shock approximation for the reflected backward-facing finite-compression wave in the otherwise same three-shock theory is found to be in very good agreement with Smith's data (1945) and in reasonably good agreement with Olim and Dewey's data (1992) for incident shock Mach number (M_s) up to 1.51. The "smallness" of a reflecting wedge angle is analytically defined as a function of M_s and the specific heat ratio (γ). Boundaries separating regimes of "small" from not-"small" reflecting wedge angles are determined for M_s up to 2 and $\gamma = 1.4$.

Key words: Mach reflections, Finite-compression waves

1. Introduction

It is a well-known unresolved issue that the von Neumann three-shock theory succeeds in describing regular reflection (RR) and the Mach reflection (MR) of strong incident shocks, but fails badly to describe the MR of weak incident shocks. For weak MR, it fails to predict correctly the angles defined by the shocks and contact surface where they meet at the triple point. This occurs even when experimentally observed triple-point trajectory angles are used in the three-shock theory (Olim and Dewey 1992). In seeking an explanation for the incapability of the three-shock theory to adequately describe weak MR, fundamental assumptions of the theory, such as viscosity, finite curvature of reflected waves, non-self-similarity, and flow non-uniformity, must be reviewed as has been reported by Sternberg (1959), Henderson and Siegenthaler (1980), and Sakurai et al. (1989); but most to not much avail. Recently, Colella and Henderson (1990) numerically simulated weak MR and showed that there is one kind of weak irregular wave reflection which cannot be explained by the conventional three-shock theory, called a von Neumann reflection (vNR). The reflected wave of vNR is not a shock but a smoothly distributed compression wave of finite thickness, and the incident and Mach shocks appear to form a single wave with a continuously turning tangent. Their analysis therefore explains why the conventional theory cannot be applied to vNR. On the other hand, Olim and Dewey (1992) confirmed the observations of other workers that the three-shock theory is unable to describe MR of incident shocks with strengths between Mach numbers 1.1 and 1.5. The resulting three-shock configuration is such that the reflected wave is supersonic and there is a measureable discontinuity in the slopes of the incident and Mach-stem shocks. In this work, we present experimental and analytical evidence to show that the reason for the inapplicability of the three-shock theory to describing weak pseudo-steady MR over not-"small" reflecting wedge angles is owing to the existence of the reflected finite-compression wave, instead of the reflected shock wave. The analysis is then carried out by replacing the reflected shock wave with a linear or a second-order small deflection weak oblique shock approximation for the reflected finite compression wave in an otherwise same three-shock theoretical formulation. The "smallness" of the reflecting wedge angle is analytically defined. For "small" reflecting wedge angles, Liu (1993) showed that the reflected wave in weak MR is a normal Mach wave, when the corresponding three-shock theory requires forward-facing reflected shock solutions.

2. Analysis

A typical pseudo-steady single MR is shown schematically in Fig. 1: there are four discontinuities — the incident shock i, the reflected shock r, the Mach stem m, and the slipstream s — which coincide at the triple point T, making a triple-point trajectory angle χ with the wedge surface having an inclination angle θ_w. Since the analytical formulation of the pseudo-steady von Neumann three-shock theory for a perfect gas is well known, the theory is not discussed here. Subscripts 0, 1, 2, and 3 refer to the flow states as defined in Fig. 2. Following Colella and Henderson's (1990) numerical finding of a band of reflected smoothly distributed compression waves of finite thickness in weak MR, the weak oblique shock approximation linearly relating finite changes between the pressure and flow deflection across a weak oblique shock, as discussed by Liepmann and Roshko (1957) valid for small deflection, is used to describe the reflected finite-compression wave replacing the reflected shock wave in von Neumann type MR.

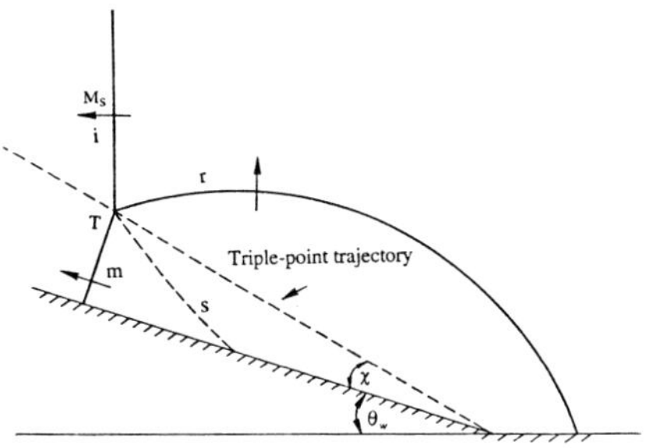

Figure 1. Schematic illustration of a propagating MR over an oblique wall.

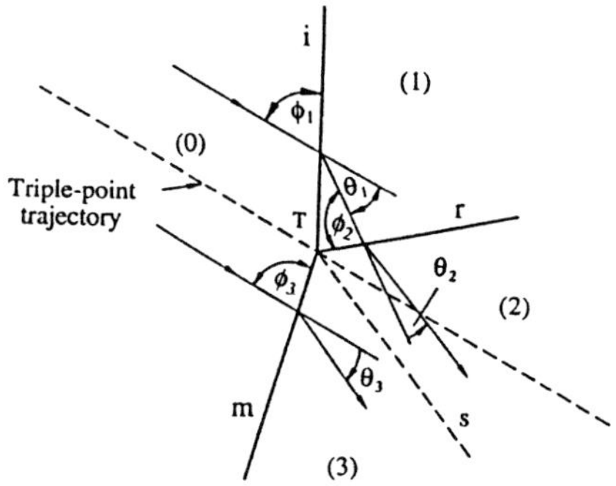

Figure 2. The wave configuration of a MR viewed from the triple point.

The thickness of this reflected compression wave is characterized by a small but finite $\Delta\theta$, so that the weak oblique shock approximation is applicable. Ben-Dor (1992) suggested this simple gas dynamic consideration for reflected compression waves in vNR. However, the pressure-deflection (P-θ) polar solution discussed by him is forward-facing of the triple point. The polar solution should be obtained by drawing a line with the above-calculated gradient, in the direction of decreasing deflection while increasing pressure, on the incident shock polar. This is because Liu (1993) has shown that sound waves downstream of incident shock waves of pseudo-steady MR cannot coalesce into Mach waves which propagate upstream from the triple point. A schematic drawing of a shock polar presentation of a vNR, using MR data reported by Olim and Dewey (1992), is given in Fig. 3. The reflected shock polar is also shown in the figure. It can be seen that the tiny reflected shock polar does not intersect the incident polar except for the Mach-line degeneracy point. The computed χ using the linear weak oblique shock approximation is in close agreement with the experimentally measured value of 12°. The reported uncertainty in the measurement of χ is $\pm 1°$. The γ used is 1.4. The use of a second-order small deflection weak oblique shock approximation for the reflected compression wave gives $\chi = 13.20°$. Figure 4 gives the shock polar solution from the three-shock theory for the same vNR, and one can see the reflected shock wave is forward-facing. The predicted χ (20.47°) differs significantly from the experiment. A graphed (P-θ) solution of vNR over a large wedge angle, where the above-mentioned second-order weak oblique shock approximation is applied, is shown in Fig. 5. The predicted χ (3.95°) agrees closely with the measured value (4°); while the linear weak oblique shock approximation gives $\chi = 2.42°$.

The necessity of classifying reflecting wedge angles of pseudo-steady MR as "small" arises from the existence of physically unrealistic forward-facing

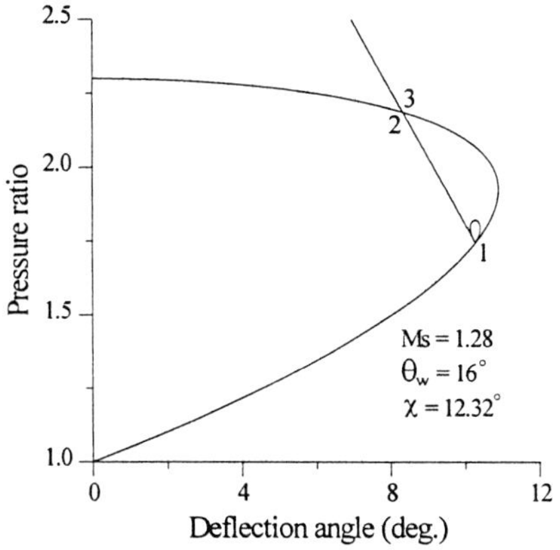

Figure 3. The (P-θ) shock polar solution of a vNR using the linear weak oblique shock approximation.

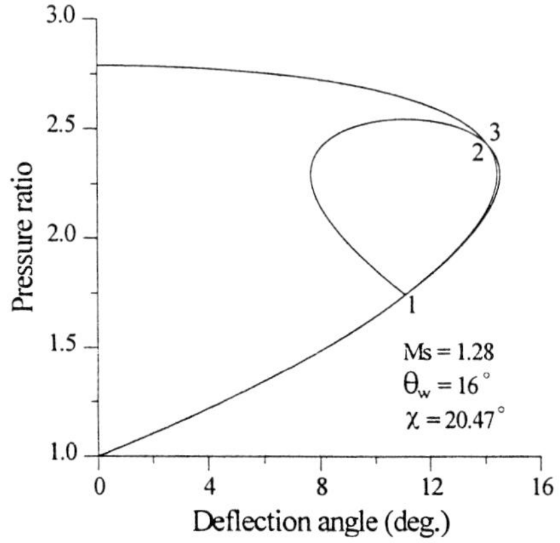

Figure 4. The three-shock theoretical forward-facing reflected shock solution for the same case as shown in Fig. 3.

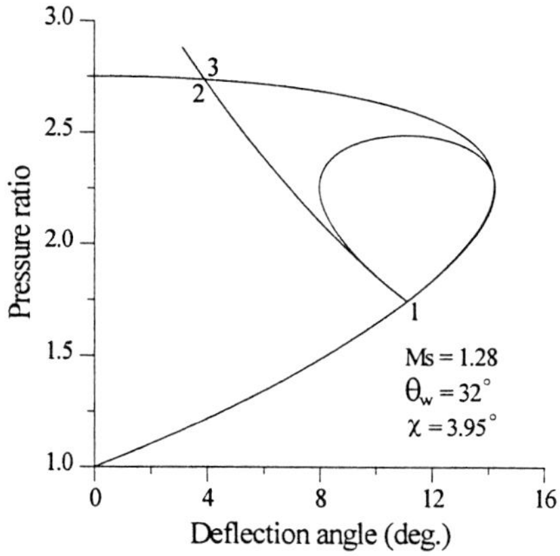

Figure 5. The (P-θ) shock polar solution of a vNR using the second-order weak oblique shock approximation.

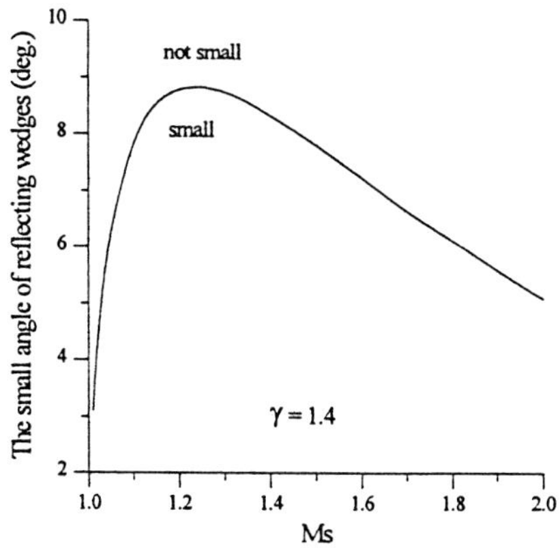

Figure 6. Theoretical limiting small reflecting wedge angles of pseudo-steady MR.

reflected finite-compression solutions when they become sufficiently small for given M_s and γ. A "small" reflecting wedge angle of pseudo-steady MR is thus defined as, for given M_s and γ, a wedge angle below which a backward-facing reflected finite-compression wave solution satisfying the boundary condition assuming a straight Mach stem normal to the wedge surface is theoretically unobtainable. Figure 6 shows the calculated boundary separating "small" from not-"small" angles for $\gamma = 7/5$ and $1.0 < M_s < 2.0$.

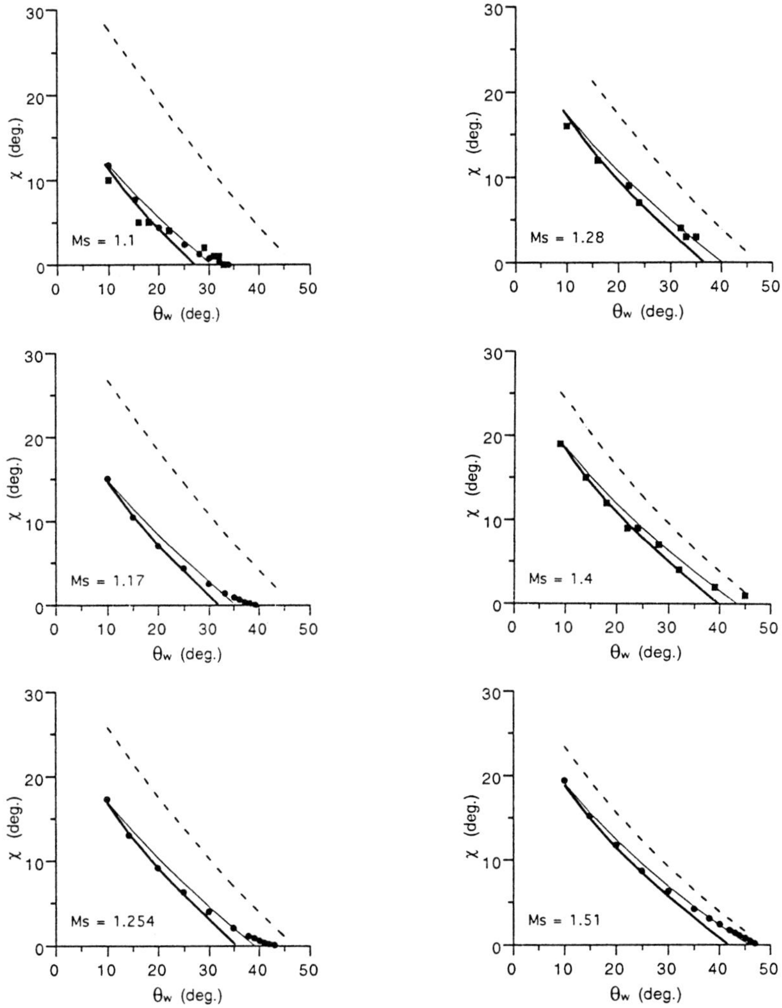

Figure 7. Comparisons between predicted variations of the triple-point trajectory angle with wedge angle for given incident-shock Mach numbers from the three-shock theory (dashed line) and from the first-order (heavy solid line) and second-order (light solid line) weak, reflected oblique shock approximations. Experimental data taken from Smith (1945) (circle) and Olim & Dewey (1992) (square).

3. Discussions and conclusions

In the following, we apply the above-discussed modified three-shock theory to analyze the experimental data of Smith (1945) and Olim and Dewey (1992). Figure 7 shows that predicted triple-point trajectory angles using a linear pressure-deflection weak oblique shock approximation for the reflected backward-facing finite-compression wave in the otherwise same three-shock theory are in very good agreement with Smith's data and in reasonably good agreement with Olim and Dewey's data for M_s up to 1.51. The good agreement holds for flow deflection across reflected finite-compression waves ranging from $\leq 3.3°$ for $M_s = 1.10$ to $\leq 14.8°$ for $M_s = 1.51$, giving the upper limit of the reflecting wedge angle, below which the present theory applies, ranging from $25°$ for $M_s = 1.10$ to $38°$ for $M_s = 1.51$. The computed flow Mach numbers downstream of incident shock waves (responsible for the formation of a reflected shock) at these upper limiting wedge angles for $M_s = 1.10$, 1.17, 1.254, 1.28, 1.4, and 1.51 are 1.07, 1.12, 1.20, 1.22, 1.86, and 2.00, respectively. The corresponding reflecting wedge angles are 25°, 29.9°, 34.9°, 35°, 39°, and 38°. One finds that the angles of Mach waves downstream of these incident shock waves are all less than 60°. It can then be argued that the subsequent nonlinear coalescence of downstream Mach waves into a reflected shock wave may take place to render the first-order weak oblique shock approximation for the reflected wave inaccurate. On the other hand, application of the second-order small deflection weak oblique shock approximation gives considerably better agreement with experiment only for wedges within 10° near their transition angles to RR. Figure 7 also shows for a given wedge angle the discrepancy between experiment and three-shock theory, for χ increases rapidly as M_s decreases from 1.51 to 1.1. Therefore, the reflected waves in pseudo-steady weak MR over not-"small" reflecting wedge angles, with inclinations not greater than about 10° less than their transition wedge angles to RR, are backward-facing compression waves of small, finite thickness.

References

Ben-Dor G (1992) Shock Wave Reflection Phenomena, Springer-Verlag.

Colella P, Henderson LF (1990) The von Neumann paradox for the diffraction of weak shock waves. J Fluid Mech 213:71–94.

Henderson LF, Siegenthaler A (1980) Experiments on the diffraction of weak blast waves: the von Neumann paradox. Proc Roy Soc Lond A 369:537–555.

Liepmann HW, Roshko A (1957) Elements of Gasdynamics, Wiley.

Liu JJ (1993) Sound wave structures downstream of incident propagating oblique shock waves. In: Shock Waves @ Marseille, Springer.

Olim M, Dewey JM (1992) A revised three-shock solution for the Mach reflection of weak shocks ($1.1 < M_i < 1.5$). Shock Waves 2:167–176.

Sakurai A, Henderson LF, Takayama K, Walenta Z, Colella P (1989) On the von Neumann paradox of weak Mach reflection. Fl Dyn Res 4:333–345.

Sternberg J (1959) Triple-shock-wave interactions. Phys Fluids 2:179–206.

Smith LG (1945) Photographic Investigation of the Reflection of Plane Shocks in Air, OSRD Rep. 6271, Office Sci. Res. Dev., Washington, DC.

Shock-wave reflections from a cone with attack angles in a symmetric plane

T. Suzuki, Y. Sakamura
Toyama Prefectural University
Kosugi, Toyama 939-03 Japan
X-Z. Yin
University of Science and Technology of China
Heifei, Anhui 230026 China

Abstract: Results from preliminary experiments for three-dimensional shock-wave reflections from a cone are presented. The Mach number of the incident planar shock wave was 1.34. The configurations of shock waves reflected from a cone at a given angle of attack are compared with those from a wedge, which is set at the same angle of attack, and those from other cones with the same angle between the shock-propagating direction and the cone ray.

Key words: Shock-wave reflections, Three-dimensional flows, Cone

1. Introduction

The reflections of oblique shock waves from wedges have been investigated experimentally, analytically, and numerically by many researchers. However, almost all of the previously reported studies belong to two-dimensional (Ben-Dor 1991) or axi-symmetrical (Takayama & Sekiguchi 1977, Milton & Archer 1995) reflections. Few works regarding three-dimensional shock-wave reflections have been reported (Hookham et al. 1995). In three-dimensional shock-wave reflections, interesting and technically important phenomena occur. Shock-wave reflection from a cone positioned at an angle of attack is an example of reflections with a three-dimensional nature. In shock-wave reflection from a cone at a finite angle of attack, the local angle between the incident shock wave and the cone surface varies continuously in the circumferential direction. Under certain conditions, we can simultaneously observe two types of reflections (i.e. a regular reflection (RR) and a Mach reflection (MR) on the cone's lower and upper surfaces, respectively) in a symmetric plane (the vertical meridian plane) as shown in Fig. 1. Hence the *spatial* transition from MR to RR, which cannot be observed in reflections from a wedge, may occur, and there may exist a crescent-shaped Mach shock wave. The phenomena of three-dimensional shock-wave reflections are too complicated to be analyzed theoretically and there is

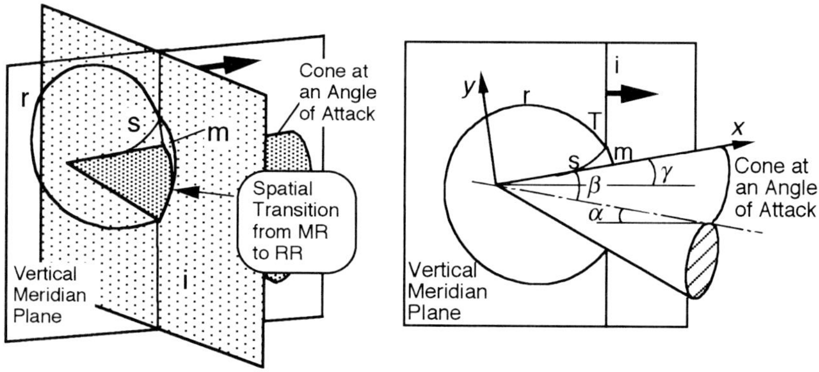

Figure 1. Sketches of shock-wave reflection from a cone at an angle of attack α: i: incident shock; r: reflected shock; m: Mach stem; s: slip surface; T: triple-point; β: semi-apex angle.

little understanding of them. The experimental work reported in this paper is the first step in our plan to study three-dimensional shock-wave reflections from a cone. The configurations of shock waves reflected from a cone at a given angle of attack are compared with those from a wedge, which is set at the same angle of attack, and those from other cones with the same angle between the shock-propagating direction and the cone ray, γ ($\beta - \alpha$ for the upper side of the cone, and $\beta + \alpha$ for the lower, as shown in Fig. 1).

2. Experimental method

The experiments were performed in a shock tube of Toyama Prefectural University. The lengths of the driver and driven sections are 1.45 m and 4.2 m, respectively. The test section of the shock tube has a rectangular cross section of 30 mm $\times$ 100 mm and viewing windows of 90 mm $\times$ 200 mm. Both the driver and driven gases were air and the Mach number of the produced planar shock wave, M_s, was 1.34. A shadowgraph optical system was used to photograph the wave configuration of the resulting shock-wave reflection.

The cone and wedge models used in the present study are shown in Fig. 2. We prepared cone models with $\beta = 20^\circ$, 30°, and 40°, and a wedge model with $\beta = 30^\circ$. These models can be set in the shock-tube test section at arbitrary α ranging from 0° to 20° as shown in Fig. 2.

In the present experiments, shadowgraphs were taken at four different shock locations for each case in order to visualize the shock-wave configurations. Note that only reflections in the vertical meridian plane from the upper and lower sides of the models at an angle of attack can be visualized by the present

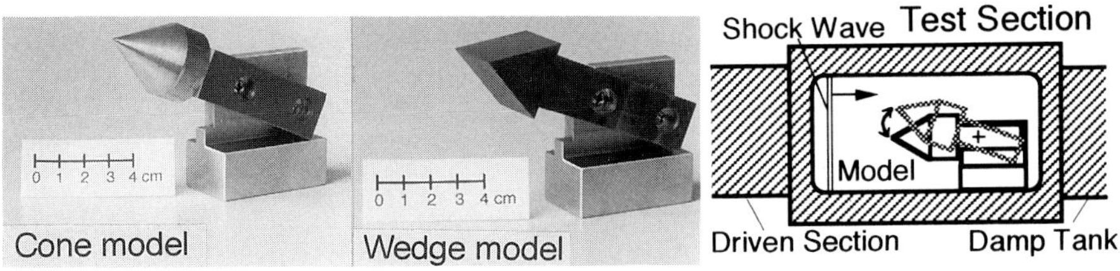

Figure 2. Cone and wedge models.

experimental system. In the case where an MR occurs, the triple-point locations were also measured and the obtained triple-point trajectories were compared. The measurements from photographic negatives (35mm film) were made with the aid of a profile projector of 50 magnifications (Nikon, V-12A).

3. Results and discussion

Figures 3a to 3d show shadowgraphs of shock-wave reflection from a cone model with $\beta = 30°$ at $\alpha = 0°$, $5°$, $10°$, and $20°$, respectively. In the cases of $\alpha = 0°$ and $5°$, both shock-wave reflections from the upper and the lower sides are of the MR type. In all pictures of Fig. 3, the reflected shock waves and slip surfaces are hardly observed because the shadowgraphs only show the information of density change, integrated along the light path. Especially, shock waves reflected from the upper surface of the cone model with high angles of attack ($\alpha > 10°$) were so weak that an accurate determination of the triple point from the shadowgraph pictures was hardly possible. It was found from many additional shadowgraphs taken at smaller intervals of α that the transition from MR to RR, on the lower side, occurred at $\alpha = 10°$ for the present conditions. Thus, the *spatial* transition from a Mach to a regular reflection may occur on the surface of the cone ($\beta = 30°$) at larger attack angles than $10°$. It is smaller than the observed effective wedge angle for a transition in the two-dimensional case, which was $13°$ under the present conditions.

Figure 4 shows the measured locations of the triple points for a cone case and a wedge case, both at zero angle of attack. The x is the distance measured from the apex along the cone ray, and the y is the normal distance measured from the cone surface (see Fig. 1). The symbols show the experimental results. In the case of the Mach reflection on a cone or a wedge placed at $\alpha = 0°$, the triple-point trajectory can be approximately predicted by the shock dynamics theory (Han & Yin 1992). The trajectories calculated using this theory are also plotted for both cone and wedge cases on Fig. 4. In this figure it is seen that

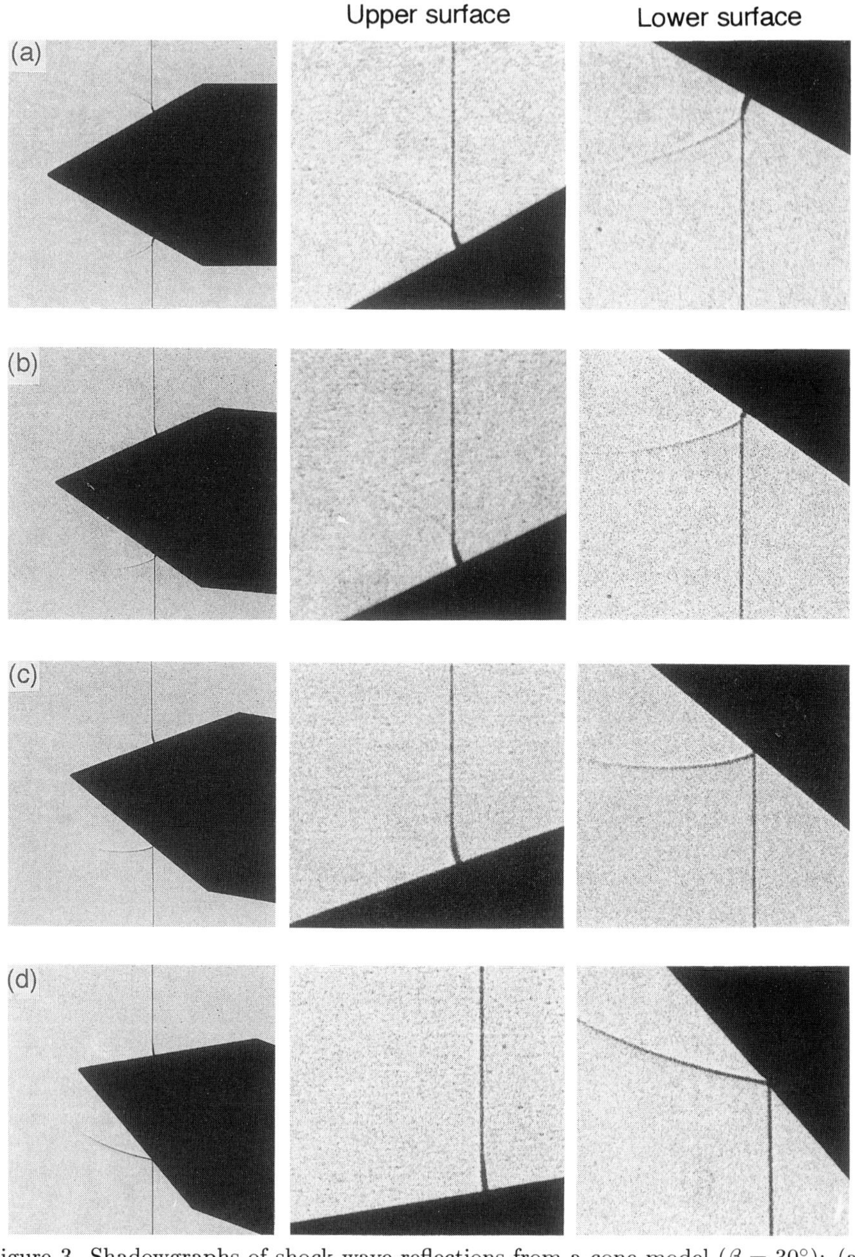

Figure 3. Shadowgraphs of shock-wave reflections from a cone model ($\beta = 30°$): (a) α=0°, (b) α=5°, (c) α=10°, (d) α=20°.

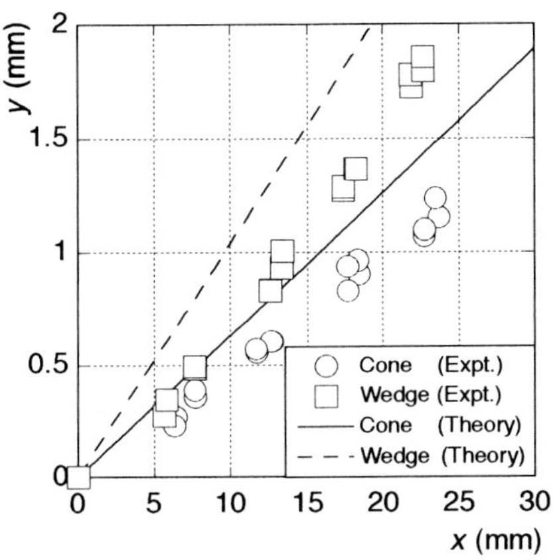

Figure 4. Comparison of the triple-point trajectory for the cone case with that for the wedge case (M_s=1.34, $\alpha = 0°$, $\beta = 30°$).

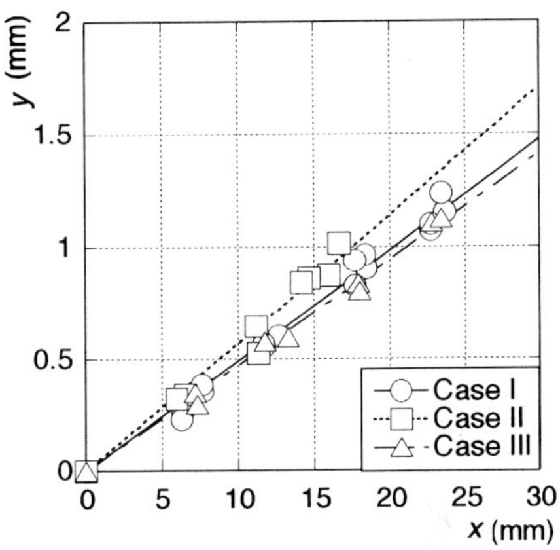

Figure 5. Comparison of the triple-point trajectories for Case I ($\alpha = 0°$, $\beta = 30°$), Case II ($\alpha = 10°$, $\beta = 40°$), Case III ($\alpha = 10°$, $\beta = 20°$) (M_s=1.34).

the triple-point trajectory angle for the cone case is smaller than that for the wedge case although the difference between the two is small. Furthermore it can be said that the trajectories predicted by the shock dynamics theory do not agree well with the experimental ones for both the cone and wedge cases under the present conditions.

In Suzuki & Sakamura (1995), it was shown that the shock-wave reflections from wedge models at a given angle of attack were determined by the effective wedge angle, $\beta - \alpha$ for the upper side of the wedge model and $\beta + \alpha$ for the lower. However, the local reflecting pattern of a shock wave from a cone at a finite angle of attack in a meridian plane may be influenced not only by the local angle between the shock-propagating direction and the cone ray, γ, but also by the local curvature of the cone surface, a cross-flow from windward to leeward, etc. In Fig. 5, the triple-point trajectories on the cone with $\beta = 30°$ at $\alpha = 0°$ (Case I), on the upper side of the cone with $\beta = 40°$ at $\alpha = 10°$ (Case II), and on the lower side of the cone with $\beta = 20°$ at $\alpha = 10°$ (Case III) are plotted. The symbols show the experimental results and the straight lines are given by the least-squares fitting of them. In all three cases, γ is equal to 30°. In this figure, the difference of the triple-point trajectory angle between case I and case III is negligible, but the differences between case II and the others are not. This finding implies that the local reflecting pattern of a shock wave from a cone at a finite angle of attack cannot be determined only by γ. Hence, we should take account of this fact when investigating the three-dimensional shock-wave reflection phenomena.

4. Conclusion

Results from preliminary experiments for a three-dimensional shock-wave reflection from a cone have been presented. The Mach number of the incident planar shock wave was 1.34. The main conclusions from the present study are as follows: (1) the angle of attack that causes the transition from Mach to regular reflection on the lower side is about 10° for a cone with a semi-apex angle of 30°, it is smaller than that obtained for a similar wedge case; (2) the trajectory angle of the triple point for the cone case is smaller than that for the wedge case; (3) the local reflecting pattern of a shock wave from a cone at a finite angle of attack cannot be determined only by the local angle between the shock-propagating direction and the cone ray.

Acknowledgement. The authors wish to express their appreciation to Prof. O. Igra of Ben-Gurion University of the Negev for his kind advice at an early stage of this study. They would also like to acknowledge the assistance of Mr. Y. Kakuda of Toyama Prefectural University for carrying out the experiments. Thanks are also due to Mr. Y. Usui of Toyama Prefectural University for his technical support to prepare the present paper using $\LaTeX\,2_\varepsilon$.

References

Ben-Dor G (1991) Shock Wave Reflection Phenomena, Springer, New York.

Han Z-Y, Yin X-Z (1992) Shock Dynamics, Kluwer, Dordrecht.

Hookham PA, Rosenblatt M, Takayama K, Watanabe M (1995) Interaction of a planar shock with a cone at an oblique angle: numerical simulation and experiment. In: Shock Waves @ Marseille, Brun R, Dumitrescu LZ (eds), Springer, Berlin, Vol IV, pp 169–174.

Milton BE, Archer RD (1995) Generalized concepts for the internal and external conical Mach reflection of moving shock waves. In: Shock Waves @ Marseille, Brun R, Dumitrescu LZ (eds), Springer, Berlin, pp 157–162.

Suzuki T, Sakamura Y (1995) Shock wave reflections from a cone. Trans Japan Soc Aero Space Sci 38:161–168.

Takayama K, Sekiguchi H (1977) An experiment on shock diffraction by cones. Rep Inst High Speed Mech, Tohoku Univ 36:53–74.

Calculation of focusing of weak shocks using geometrical shock dynamics

J. Cates, B. Sturtevant
Graduate Aeronautical Laboratories
California Institute of Technology
Pasadena, CA 91125 USA

Abstract: The equations of geometrical shock dynamics provide a valuable approximation for many shock-propagation problems. An improved finite difference formulation for geometrical shock dynamics has been developed, and is used to calculate the focusing of plane shock waves reflected from a parabola. The shock pressure jump and the path of the triple point along the shock front were compared to the experimental results of Sturtevant and Kulkarny (1976) for a range of Mach numbers from 1.1 to 1.3. The numerical calculations overestimate pressures near the focus, however, the location of the maximum pressure is accurately predicted for Mach numbers above 1.1. As expected with geometrical shock dynamics, the accuracy deteriorates for weaker shocks.

Key words: Geometrical shock dynamics, Shock focusing

1. Introduction

The equations of geometrical shock dynamics (Whitham 1957) are

$$\nabla \cdot \left(\frac{M}{A(M)} \nabla \alpha \right) = 0, \qquad M = \frac{1}{|\nabla \alpha|}. \tag{1}$$

$\alpha(\mathbf{x}) = a_0 t$ gives the shock position at time t, a_0 being the undisturbed sound speed. $A(M)$ is the Area-Mach-number relation, which gives the relation between the area A of the ray tube and the shock strength M. For a perfect gas, the Area-Mach-number relation is given by

$$A(M) = \exp\left[-\int_{M_0}^{M} \frac{M\lambda(M)}{M^2 - 1} dM \right], \tag{2}$$

where

$$\lambda(M) = \left(1 + \frac{2}{\gamma + 1} \frac{1 - \mu^2}{\mu} \right) \left(1 + 2\mu + \frac{1}{M^2} \right), \tag{3}$$

$$\mu^2 = \frac{(\gamma - 1)M^2 + 2}{2\gamma M^2 - (\gamma - 1)}. \tag{4}$$

These equations form a hyperbolic, second-order partial differential equation describing wave motion of disturbances propagating on the shock front. Discontinuities, called *shock-shocks*, can form along the shock carrying a change in shock angle and Mach number which represent triple points. *Shock-expansions* can also form, such as when the shock diffracts around a sharp corner. At a solid boundary, the boundary condition requires the shock front to be perpendicular to the wall, therefore Mach reflection always occurs and regular reflection is not allowed.

In Cartesian coordinates x, y, Eq. (1) is

$$\frac{\partial(\sigma u)}{\partial x} + \frac{\partial(\sigma v)}{y} = 0, \tag{5}$$

where

$$M = \left[u^2 + v^2\right]^{-1/2}, \qquad \sigma \equiv \frac{M}{A(M)}, \tag{6}$$

and $u = \alpha_x$, $v = \alpha_y$ are the components of the gradient of α, As first noted by Whitham (1959), the equations are analogous to the full potential equation in steady supersonic flow. The extensive numerical methods developed for the supersonic potential equation can be directly adapted to geometrical shock dynamics (Schwendeman 1993).

2. Numerical method

For an arbitrary, non-orthogonal coordinate system defined by $\zeta = \zeta(x, y)$, $\eta = \eta(x, y)$, Eq. (5) can be written in conservative form as:

$$\frac{\partial}{\partial \zeta}\left(\frac{\sigma U}{J}\right) + \frac{\partial}{\partial \eta}\left(\frac{\sigma V}{J}\right) = 0, \tag{7}$$

where U, V are the contravariant velocity components given by

$$U = a_{11}\alpha_\zeta + a_{12}\alpha_\eta, \qquad V = a_{12}\alpha_\zeta + a_{22}\alpha_\eta, \tag{8}$$

and the grid parameters are

$$a_{11} = \zeta_x^2 + \zeta_y{}^2, \qquad a_{12} = \zeta_x\eta_x + \zeta_y\eta_y, \qquad a_{22} = \eta_x^2 + \eta_y^2, \tag{9}$$

$$J = \zeta_x\eta_y - \zeta_y\eta_x. \tag{10}$$

The Mach number is now given by

$$M = \left[U\alpha_\zeta + V\alpha_\eta\right]^{-1/2}. \tag{11}$$

In the solution procedure, we consider ζ to be the marching direction and assume all information is known at the ith and all previous levels. The problem is to advance the solution to the level $i + 1$ and obtain α.

At all levels, the quantities α_ζ, α_η are computed from

$$(\alpha_\zeta)_{i,j} = \frac{\alpha_{i,j} - \alpha_{i-1,j}}{\Delta\zeta}, \qquad (\alpha_\eta)_{i,j} = \frac{\alpha_{i,j+1} - \alpha_{i,j-1}}{2\Delta\eta}, \tag{12}$$

where a backward difference is used for the marching direction ζ. For half-points, the velocities are averaged between neighboring points which gives:

$$(\alpha_\zeta)_{i,j+1/2} = \frac{(\alpha_{i,j+1} - \alpha_{i-1,j+1}) + (\alpha_{i,j} - \alpha_{i-1,j})}{2\Delta\zeta}, \tag{13}$$

$$(\alpha_\eta)_{i,j+1/2} = \frac{\alpha_{i,j+1} - \alpha_{i,j}}{\Delta\eta}. \tag{14}$$

Once these quantities are known, U, V follow directly from Eq. (8) and M from Eq. (11).

Since the ζ direction is the marching direction, the ζ derivative term in Eq. (7) will be backward differenced as

$$\frac{\partial}{\partial\zeta}\left(\frac{\sigma U}{J}\right)_{i+1,j} \approx \frac{(a_1 - \theta b_1)\left\{\left(\frac{\sigma U}{J}\right)_{i+1,j} - \left(\frac{\sigma U}{J}\right)_{i,j}\right\} - \theta b_1\left\{\left(\frac{\sigma U}{J}\right)_{i,j} - \left(\frac{\sigma U}{J}\right)_{i-1,j}\right\}}{a_1\Delta\zeta_1 - \theta b_1(\Delta\zeta_1 + \Delta\zeta_2)}, \tag{15}$$

where

$$a_1 = (\Delta\zeta_1 + \Delta\zeta_2)^2, \qquad b_1 = (\Delta\zeta_1)^2. \tag{16}$$

The parameter θ controls the order: $\theta = 0$ gives first-order accuracy and $\theta = 1$ gives second-order accuracy (Shankar 1982).

The η derivative term in Eq. (7) will be written at the level $i+1$ to make the resulting scheme fully implicit:

$$\frac{\partial}{\partial\eta}\left(\frac{\sigma V}{J}\right)_{i+1,j} \approx \frac{1}{\Delta\eta}\left\{\left(\frac{\sigma V}{J}\right)_{i+1,j+1/2} - \left(\frac{\sigma V}{J}\right)_{i+1,j-1/2}\right\}. \tag{17}$$

To simulate viscosity, the analogy of density biasing (or often called artificial compressibility) will be used. The σ in Eq. (17) will be replaced by $\tilde{\sigma}$ defined to be

$$\tilde{\sigma}_{i+1,j+1/2} = (1-\nu)\sigma_{i+1,j+1/2} + \frac{1}{2}\nu\left(\sigma_{i+1,j+2m} + \sigma_{i+1,j+2m-1}\right), \tag{18}$$

with $m = 0$ when $V_{i+1,j+1/2} > 0$ and $m = 1$ when $V_{i+1,j+1/2} < 0$. The artificial velocity employs upwind differencing based on the characteristic signal propagation, which turns off the upwind differencing when the cross flow is elliptic-type (Shankar 1982). In this case, the artificial viscosity coefficient is given by

$$\nu_{i+1,j+1/2} = \mu\left[1 - \frac{a_{22}a^2}{V^2}\right]_{i+1,j+1/2} \tag{19}$$

with

$$\mu = 0 \quad \text{for} \quad \left(a_{22} - \frac{V^2}{a^2}\right)_{i+1,j+1/2} > 0 \qquad \text{(elliptic cross flow)}, \tag{20}$$

$$= 1 \quad \text{for} \quad \left(a_{22} - \frac{V^2}{a^2}\right)_{i+1,j+1/2} < 0 \qquad \text{(hyperbolic cross flow)}. \tag{21}$$

Given the α values at all previous levels, this formulation results in a set of nonlinear algebraic equations for α_{i+1} which are solved using Newton's method. In general, the artificial viscosity leads to a pentadiagonal system. For most cases, convergence on each step can be achieved while maintaining a tridiagonal system by neglecting derivative terms due to artificial viscosity, i.e., the derivative matrix is calculated as if $\nu = 0$. This often requires more iterations for convergence of each ζ step, but each iteration is significantly faster.

3. Shock-focusing results

The resulting numerical method was used to calculate the focusing of plane shock waves reflected from a parabola. For the initial condition of the calculation, the Mach number of the shock was approximated as the initial Mach number. The computed reflected shock fronts and Mach-number contours for a shock wave with initial Mach numbers $M_0 = 1.3$ and $M_0 = 1.1$ are shown in Fig. 1. The outline of the parabola is shown at the left of each figure, and the focus of the parabola lies on the centerline at $(x, y) = (0, 1)$.

Figure 1a shows an example of the behavior observed for a stronger shock, $M_0 = 1.3$. The initial disturbance from the corner of the parabola travels along the shock front until it reaches the centerline, where a shock-shock discontinuity forms on the shock to enforce the boundary condition. The effect of the discontinuity is clearly shown by the change in the angle of the shock front and the increase in Mach number. The path of the shock-shock propagating outward along the shock front represents the path of the triple point seen experimentally. As the discontinuity propagates outward, it weakens and the Mach number behind the shock-shock decreases.

For weaker shock waves, the shock-shock does not propagate outward, but is swept back toward the centerline as shown in Fig. 1b for $M_0 = 1.1$. Here the shock-shock briefly moves away from the centerline before being pushed back into the centerline. The strength of the shock-shock increases as it is pushed inward, producing a further increase in the centerline Mach number. The strength of the shock-shock continues to increase until the velocity of the shock-shock along the shock front is sufficient for the shock-shock to move outward again. For very weak shocks, the shock-shock is actually pushed to within several grid

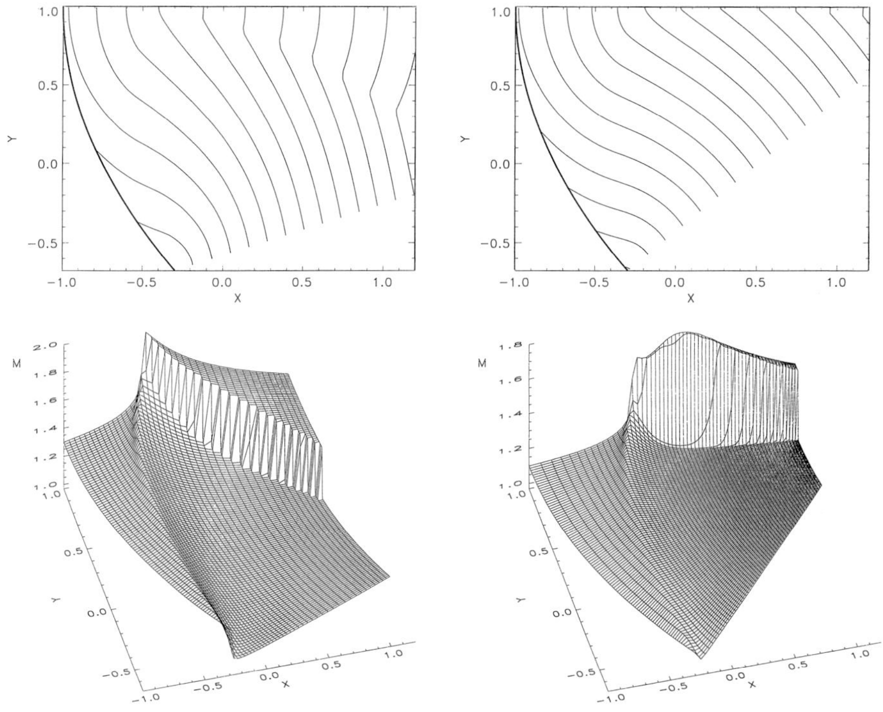

(a) (b)

Figure 1. Shock fronts and Mach-number surface for reflection of planar shock waves from a parabola with initial Mach number (a) $M_0 = 1.3$ and (b) $M_0 = 1.1$.

points of the centerline, and the shock-shock path becomes dependent on the grid spacing. From comparison with experiment, one expects these cases to mark where experiment shows a transition to regular reflection and the front becomes crossed and folded.

In Fig. 2, the computed pressure ratio and shock-shock path are compared with experimental results. The experimental pressure ratio is the maximum pressure ratio, which outside the focal region is not the pressure directly at the shock. Within the focal region, the numerical results correctly predict the location of the maximum pressure for the stronger shocks. For the weaker $M_0 = 1.1$ case, the location of the initial peak agrees with the experimental results, but the additional rise in the Mach number due to the requirement in shock dynamics for Mach reflection at the centerline is artificial. The path of the

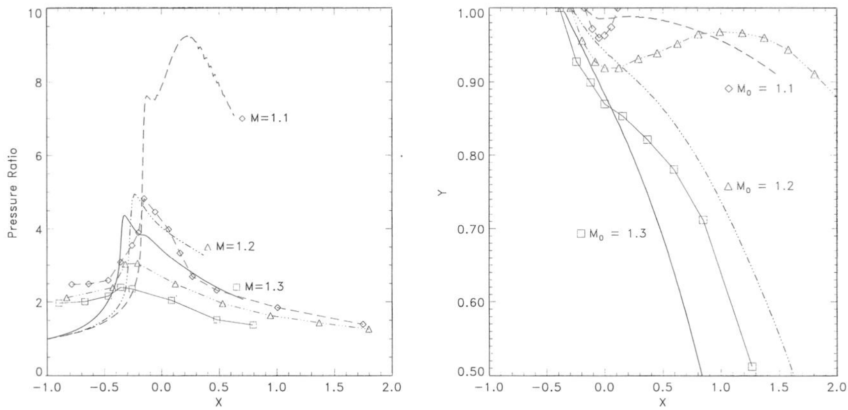

(a) (b)

Figure 2. (a) Pressure ratio and (b) shock-shock path computed for reflection of planar shock waves from a parabola compared with experimental results of Sturtevant and Kulkarny (1976). Experimental pressure ratio is the maximum pressure ratio encountered, which outside the focal region is not the pressure directly at the shock.

shock-shock is well predicted for $M_0 = 1.3$, but for weaker shocks the agreement deteriorates. However, the shock dynamics approximation is able to capture the motion of the shock-shock being towards the centerline for weaker shocks.

The finite-difference method developed for geometrical shock dynamics is useful for efficiently computing problems such as shock focusing where strong shock-shock discontinuities occur. The numerical results qualitatively capture the major effects seen experimentally, however quantitative agreement is poor, especially for weaker shocks.

References

Schwendeman DW (1993) A new numerical method for shock wave propagation based on geometrical shock dynamics. Proc R Soc Lond A 441:331–341.

Shankar V (1982) Conservative full potential, implicit marching scheme for supersonic flows. AIAA J 20:1508–1514.

Sturtevant B, Kulkarny VA (1976) The focusing of weak shock waves. J Fluid Mech 73:651–671.

Whitham GB (1957) A new approach to problems of shock dynamics, part I: two-dimensional problems. J Fluid Mech 2:145–171.

Whitham GB (1959) A new approach to problems of shock dynamics, part II: three-dimensional problems. J Fluid Mech 5:369–386.

The shock reflection process over a non-straight surface

S. Kobayashi, T. Adachi
Department of Mechanical Engineering
Saitama Institute of Technology
1690 Fusaiji, Okabe-machi, Ohsato-gun, Saitama, 369-02 Japan

T. Suzuki
Department of Mechanical Systems Engineering
Toyama Prefectural University
5180 Kurokawa, Kosugi-machi, Imizu-gun, Toyama, 939-03 Japan

Abstract: Shock reflection over a non-straight surface has been investigated. The models used in the present experiment are ordinary circular cylindrical concave and convex wedges as well as step-wise simulations of the circular cylindrical wedges. The step height of the step-like wedges is constant and the equivalent radius of curvature is 40 mm. The reflected-wave structure has been photographed with a schlieren apparatus. The transition wedge angle is compared with the analytical formula based on acoustic theory.

Key words: Shock reflection, Wave structure, Unsteady reflection, Transition

1. Introduction

When a planar shock wave is reflected over a non-straight surface, the incidence angle varies as the shock proceeds and the flow field is truly nonstationary. As a result, such problems are less tractable than steady or pseudo-steady problems. In particular, self-similarity does not hold when a characteristic length is present such as the radius of curvature in the case of circular cylindrical wedges. The unsteady reflection over concave and convex circular cylinders has been investigated experimentally and theoretically (Heilig 1969, Ben-Dor et al. 1987) from various viewpoints. The authors, who have studied during the last decade the shock reflection phenomena over multi-guttered wedges and particulate layers (Adachi et al. 1992, Kobayashi et al. 1992), found that the minute reflected waves from such surfaces are useful in understanding the development of the ordinary reflected wave over a smooth surface as well as that of the precursory reflected wave over non-smooth surfaces. In the present paper, step-like wedges are used to simulate circular cylindrical wedges and to investigate the reflection over them.

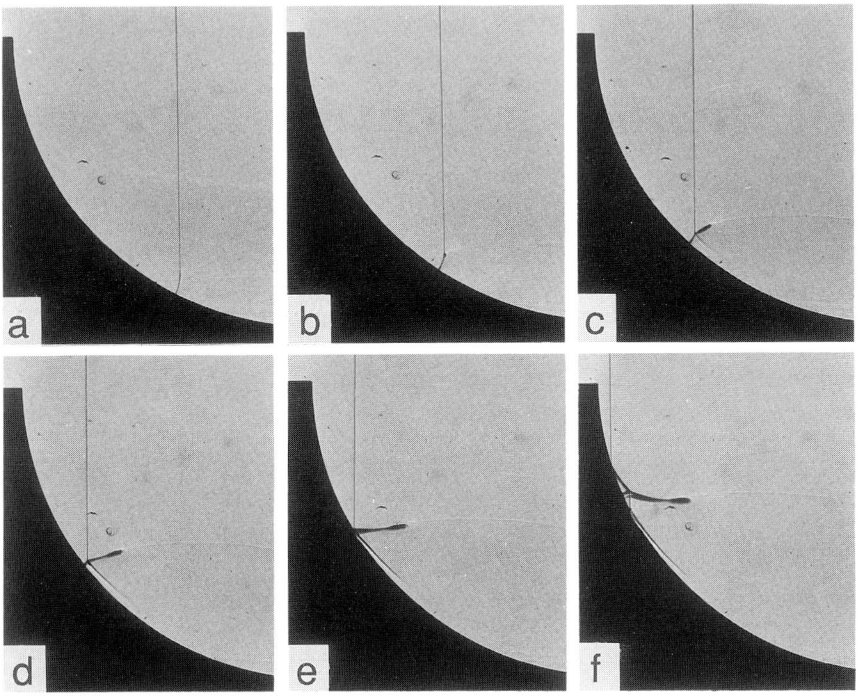

Figure 1. Reflection over a smooth concave wedge (r = 40 mm, M_i = 1.40).

2. Experimental apparatus and method

The experimental setup is the same as that used in our previous paper (Adachi et al. 1992). We used three models with different radii of curvature: r = 20, 30, and 40 mm for the smooth cylindrical wedges. The surface of the step-like wedges is composed of multiple steps with constant height s (1 mm or 2 mm) and an equivalent radius of curvature of 40 mm. The working gas is air and the condition ahead of the incident shock is room temperature and atmospheric pressure. The incident shock Mach number M_i is fixed at about 1.40. The wave structure has been photographed with a schlieren apparatus.

3. Results and discussion

Figure 1 shows shadowgraphs of shock reflection over a cylindrical concave surface. At an early stage, the reflected wave is too weak to be observed clearly. As the incident wave proceeds, a triple point and then a Mach stem appears. The

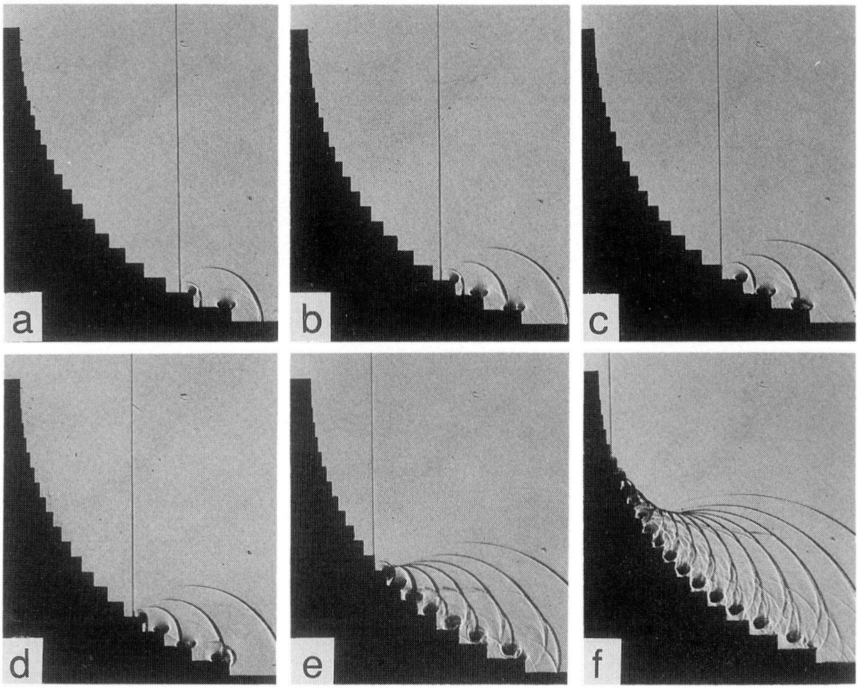

Figure 2. Reflection over a concave step-like model ($r = 40$ mm, $s = 2$ mm, $M_i = 1.41$).

reflected wave is accumulated in the neighborhood of the triple point (Figs. 1c–e). The triple point collides with the surface (Fig. 1e) and the TRR is formed in Fig. 1f. The corresponding reflection over a step-like wedge is shown in Fig. 2. A very weak wave is issued at each step and the "triple point" is barely observable. At an early stage, these weak waves accumulate in the neighborhood of the single triple point, and the triple point can get over the step. However, at a later stage (Figs. 2c–2f), the triple point cannot get over but collides with the step. The resemblance between Figs. 2f and 1f is remarkable. This is more evident in Fig. 3 for $s = 1$ mm. In Fig. 3a, the weak waves accumulate near the foot of the incident shock, which is similar to the accumulated reflected wave in Figs. 1c–e. The difference between the two experiments is that the Mach stem does not grow in the case of the step-like wedge. The reason that the reflected wave over a smooth concave wedge is only restricted near the reflection point is evident in Figs. 2f and 3b. That is to say, the reflected wave over a smooth concave wedge is dispersed away from the reflection point.

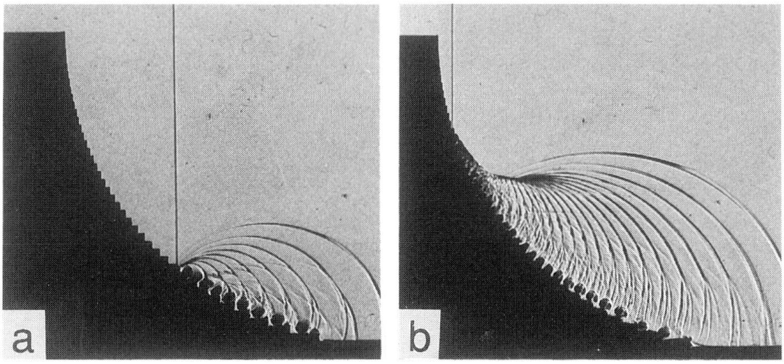

Figure 3. Reflection over a concave step-like model ($r = 40$ mm, $s = 1$ mm, $M_i = 1.41$).

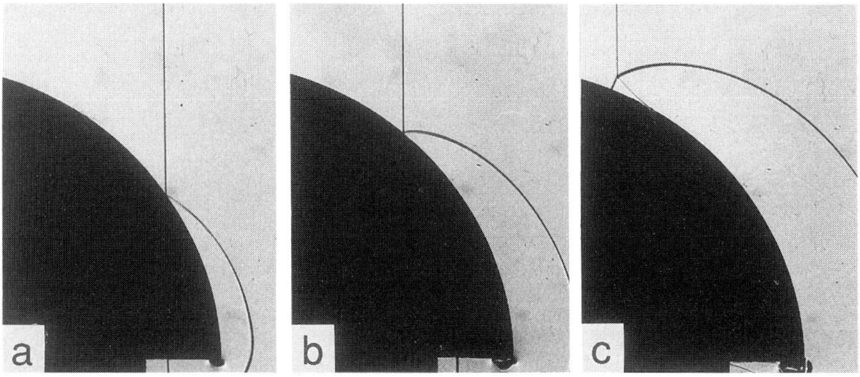

Figure 4. Reflection over a smooth convex wedge (r = 40 mm, M_i = 1.40).

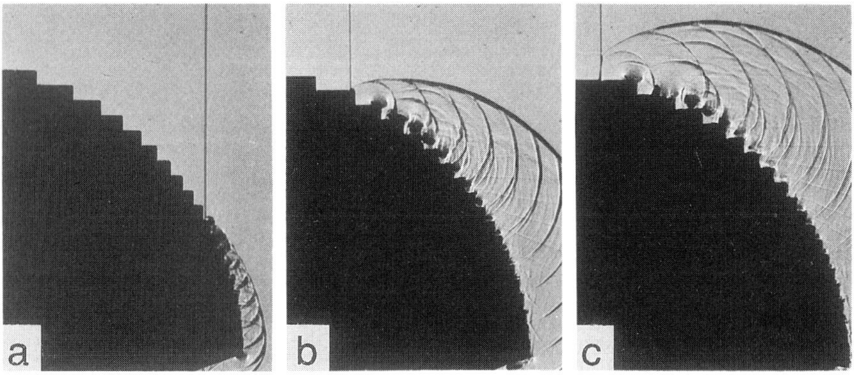

Figure 5. Reflection over a convex step-like model ($r = 40$ mm, $s = 2$ mm, $M_i = 1.41$).

Figure 4 shows the reflection configuration over a cylindrical convex surface. The reflection is RR in Figs. 4a and 4b, and MR in Fig. 4c. As the incident shock propagates (Fig. 4a to Fig. 4b), the angle of reflection increases. The transition from RR to MR has already taken place by Fig. 4c. Figure 5 is the corresponding reflection over a step-like wedge. Contrary to the reflection over the concave step-like wedge, at an early stage, the triple point cannot get over the step until the slope made by the neighboring corners decreases below some value (Fig. 5b). Hereafter, signals issued from each corner catch up with the incident shock near its foot. Consequently, these signals accumulate to form a Mach stem as shown in Fig. 5c. It is noted that the Mach stem is slightly slanted forward at the foot, which is similar to what appears in Fig. 4c.

This experiment suggests that RR cannot persist when the triple point can get over the step. The location of the triple point is the farthest point that the signal generated at the corner can reach. The signal generated at the corner propagates at the sound speed relative to the flow behind the incident shock. Therefore, we have an analytical formula for the transition wedge angle,

$$\theta_{tr} = \tan^{-1}\left\{M_i^{-2}\sqrt{\left[(\kappa-1)\,M_i^2+2\right](M_i^2-1)\,(\kappa+1)^{-1}}\right\}, \tag{1}$$

where κ is the specific heat ratio. When the slope made by the neighboring corners is less than θ_{tr}, the transition from RR to MR occurs. This formula has been derived for the smooth concave wedge as a triple point trajectory angle at glancing incidence (Ben-Dor et al. 1987). However, except for very weak shocks, as shown in Fig. 6 (Takayama et al. 1981), Eq. (1) agrees with the transition wedge angle for convex wedges with surface roughness. For a convex step-like wedge, the transition wedge angle from RR to MR is not determined by a unique value by nature but characterized by upper and lower values defined by two neighboring corners. Thus it is characterized by the angle defined by the corner the triple point can first get over and that by its previous corner. Since the step-like wedge is considered to have surface roughness, it is reasonable the present experimental results over step-like wedges agree with those over rough surfaces.

4. Conclusions

The shock reflection over a non-straight surface has been investigated. Two conclusions may be drawn from this study: (1) We can better understand the development of reflected waves over smooth convex and concave surfaces by considering the reflected wave structure over step-like models. (2) Except for very weak shocks, the analytical formula for the transition wedge angle based on acoustic theory agrees with experiments for convex wedges with surface roughness.

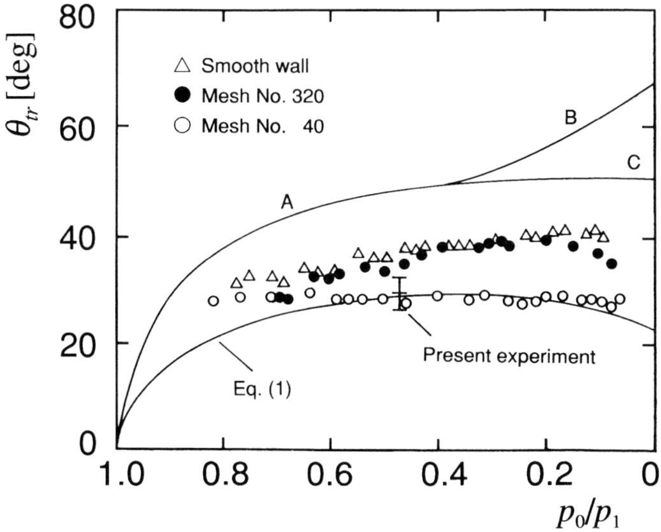

Figure 6. Transition wedge angle vs. inverse pressure ratio across the incident shock. The present experiment: $26° < \theta_{tr} < 32°$ for $s = 2$ mm, $29° < \theta_{tr} < 32°$ for $s = 1$ mm. AB: transition line for steady flow; AC: pseudo-steady flow.

References

Adachi T, Kobayashi S, Suzuki T (1992) An experimental analysis of oblique shock reflection over a two-dimensional multi-guttered wedge. Fluid Dyn Res 9:119–132.

Ben-Dor G, Takayama K, Dewey JM (1987) Further analytical considerations of weak planar shock wave reflections over a concave wedge. Fluid Dyn Res 2:77–85.

Heilig WH (1969) Diffraction of a shock wave by a cylinder. Phys Fluids Suppl 1:154–157.

Kobayashi S, Adachi T, Suzuki T (1992) Unsteady behavior of Mach reflection over a particulate layer. In: Proc 18th Int Symp Shock Waves, pp 239–244.

Takayama K, Ben-Dor G, Gotoh J (1981) Regular to Mach reflection transition in truly nonstationary flows — influence of surface roughness. AIAA J 19:1238–1240.

Oblique shock interactions and the von Neumann paradox

S. Čanić*
Department of Mathematics
Iowa State University, Ames, IA 50011 USA

B. L. Keyfitz†
Department of Mathematics
University of Houston, Houston, TX 77204-3476 USA

Abstract: We present a new theoretical approach to studying weak shock reflection in regimes where shock polar analysis is insufficient. The approach is based on examining the subsonic flow using monotone operator techniques. Our results reveal complicated flow structures in the zone of interaction of waves. Results predicting the region of high velocity gradients in the von Neumann reflection are reported. The precise type and the position of the singularity are described.

Key words: Riemann problem, Shock interactions, von Neumann reflection, Transonic shocks

1. Introduction

When a weak shock wave hits a wedge, different reflection patterns occur depending on the shock strength and the wedge angle. The type of shock interaction that occurs and the transition criteria between different types of reflection have been the subjects of many numerical and experimental studies. The most common theoretical approach is based on shock polar analysis. For certain small Mach numbers and small wedge angles this analysis fails to predict the type of reflection that takes place (the "von Neumann paradox"). The main reason for the inadequacy of this approach is that it assumes that the flow downstream of the reflected wave and near the point of interaction of waves is uniform. However, numerical and experimental studies suggest that in place of a triple-shock configuration dividing the region of uniform flow, as in the case of Mach reflection, for a certain range of parameters, the reflected wave contains a region with high velocity gradients corresponding to a compression wave near the Mach stem (von Neumann reflection).

*Supported in part by the Department of Energy under grant DE-FG02ER25220, and by the NSF under grant DMS-9403598.

†Supported in part by the Texas Advanced Research Program under grant 00365-2102-ARP, and by the Department of Energy under grant DE-FG03-94ER25222.

For this reason we have taken a different approach in order to understand the singularities in the equations that govern the flow, and to obtain a bifurcation diagram corresponding to different types of reflection. Besides shock polar analysis we have used monotone operator techniques to study the subsonic flow in the neighborhood of wave interactions. Equations that approximate the flow in the interaction zone have been obtained (Brio & Hunter 1992, Tabak & Rosales 1994) from the compressible gas dynamics equations. The asymptotic equation is the unsteady transonic small disturbance equation, also known as the 2D Burgers equation:

$$\begin{aligned} u_t + uu_x + v_y &= 0, \\ -v_x + u_y &= 0. \end{aligned} \tag{1}$$

This equation was also found (Morawetz 1994) to be an approximation to the full potential equation in the neighborhood of wave interactions. Here u and v denote scaled, normalized velocity components.

Instead of modeling the initial-value problem directly, we study a *two-parameter family of Riemann problems* associated with shock reflection. The initial data consist of two oblique shock waves separating constant states. The two shock waves correspond to the incident shock, $S_1 : x = ay$, and the reflected wave, $S_2 : x = -by$, at time $t = 0$ when the incident shock hits the wedge. The two parameters a and b are related to the angle of the wedge and to the strength of the incident shock. In principle, the solution to the shock-reflection problem could be recovered from this Riemann problem by determining the parameter b from the flow angle a and the conditions at infinity. The uniform states separated by S_1 and S_2 satisfy Rankine-Hugoniot conditions across the shocks, so these are *oblique-shock Riemann data.* In scaling the equations, one finds that the initial states can be given by $U_0 = (0, 0)$, $U_1 = (1, -a)$, $U_2 = (1 + a/b, 0)$ (see Brio & Hunter 1992). Propagation of the singularity at the origin at $t = 0$ (the point where the two shocks meet) corresponds to a corner signal which gives rise to a region of subsonic flow.

We study self-similar solutions of this initial-value problem. The system of equations (1) written in self-similar coordinates changes type from hyperbolic to elliptic, corresponding to the regions where the flow is supersonic or subsonic. For $t > 0$, the second shock wave, S_2, either overtakes the incident shock, S_1, or the incident shock travels faster than the reflected wave, depending on the values of a and b. Now, the corner signal, which gives rise to a region where the flow is subsonic and therefore not necessarily uniform, may interfere with the reflected and/or the incident shock wave. In order to describe the flow for all choices of a and b we

(1) study hyperbolic wave interactions,

(2) study interaction of shock waves with the subsonic region, and

(3) obtain a bifurcation diagram in (a,b)-space describing different types of solutions, based on the type of the hyperbolic interaction and on the free-boundary solution.

In Section 2 we summarize results (Canic & Keyfitz 1995) on hyperbolic wave interactions relevant to the study of von Neumann reflection. In Section 3 we present results about the subsonic flow that suggest new, interesting structures in the region of wave interaction. Based on the theory given in Sections 2 and 3, we conjecture the structure of the flow in von Neumann reflection. The bifurcation diagram describing other types of reflection can be found in Canic et al. (1994).

2. Hyperbolic wave interactions

In self-similar coordinates ($\xi = x/t, \eta = y/t$) equation (1) becomes

$$\begin{aligned} (u-\xi)u_\xi - \eta u_\eta + v_\eta &= 0, \\ -v_\xi + u_\eta &= 0. \end{aligned} \tag{2}$$

This system is hyperbolic outside and elliptic inside the parabola P_u: $\xi + \eta^2/4 = u$. This parabola replaces the sonic circle of the Euler equations of compressible gas flow. Finding a self-similar solution of the initial-value problem described above reduces to solving a boundary-value problem in physical space (ξ, η), with data given on a curve sufficiently far from the origin (corresponding to data at infinity at $t = 0$). Because the coefficients of system (2) depend explicitly on ξ and η, this system is not standard for the hyperbolic theory of conservation laws. For example, the characteristics in the hyperbolic region, linearized about a constant state, are straight lines tangent to the sonic parabola.

Hyperbolic wave interactions occur for values of a and b for which either the two shock waves, S_1 and S_2, intersect in the hyperbolic region, or the incident shock meets the wall before interacting with either S_2 or the subsonic region. In these situations, new waves are generated as solutions of a quasi-one-dimensional Riemann problem based at the point of interaction, Ξ_I. In Canic & Keyfitz (1995) we prove the following:

Theorem 1 *Given Riemann data (U_0, U_2) at Ξ_I with hyperbolic states U_0 and U_2, let $D(U_0)$ be the downstream wave locus of U_0. Then there are three regions, R_1, R_2, and R_3, in the complement of $D(U_0)$ such that if U_2 belongs to any of the three regions, there exists a unique Riemann solution between U_0 and U_2, qualitatively different for each region. If U_2 belongs to the complement of $R_1 \cup R_2 \cup R_3$, there may be none, one, or two admissible solutions.*

More details about the structure of quasi-one-dimensional Riemann solutions can be found in Canic & Keyfitz (1995) and Canic et al. (1994).

3. Shock interaction with the subsonic region

In order to understand the interaction of a shock with the subsonic region we need to solve a free-boundary problem for a degenerate elliptic equation (2), where the position of a transonic shock is determined by the Rankine-Hugoniot conditions

$$\begin{aligned} \xi'(\eta) &= -\eta/2 \pm \sqrt{\xi + \eta^2/4 - (u+u_0)/2}, \\ \xi'(\eta) &= (v_0 - v)/(u_0 - u), \end{aligned} \tag{3}$$

with $\xi(\eta_*) = \xi_*$ corresponding the point where the shock meets the sonic curve; $\xi = \xi(\eta)$ is the position of the transonic shock, and u and v are the limits of the velocity components along the transonic shock (see Fig. 1a).

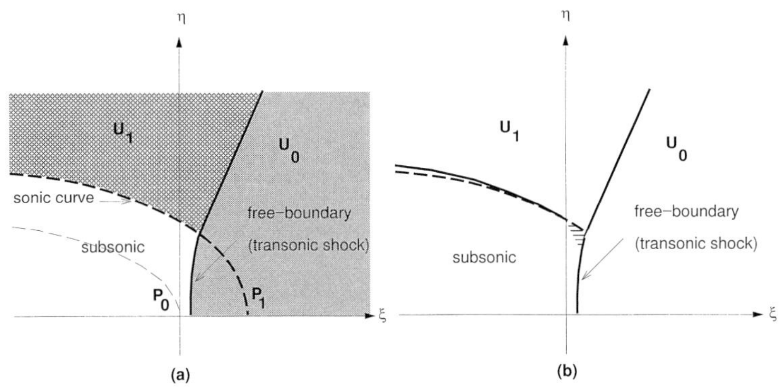

Figure 1. Subsonic boundary-value problems.

In an attempt to capture possible singularities, we first studied a fixed boundary-value problem based on an approximate position of the transonic shock (Canic & Keyfitz 1996a, Canic & Keyfitz 1996b). For this, it is sufficient to solve a second-order, quasi-linear, degenerate elliptic PDE obtained from (2):

$$\left((u-\rho)u_\rho + \frac{u}{2}\right)_\rho + u_{\eta\eta} = 0, \quad \text{where} \quad \rho = \xi + \frac{\eta^2}{4}, \tag{4}$$

and then recover v using the fact that $v = 0$ at the wall. Our results reveal interesting solution structures, as follows.

- The *asymptotic behavior* of a solution is described by the following theorem, proved in Canic & Keyfitz (1996a).

Theorem 2 *If u is constant along the sonic boundary, then there are exactly two possibilities for the asymptotic behavior of a solution at the sonic boundary. Either $u \in C^1$ up to the boundary, and $u_\rho = 1/2$, or $u \in C^\alpha$, where $\alpha \leq 1/2$, and $u = \sqrt{\rho_M - \rho}\; w$, where $w \neq 0$ at the sonic line, $\rho = \rho_M$.*

• *Global existence* of a subsonic solution for an approximate position of the Mach stem is proved in Canic & Keyfitz (1996a) and Canic & Keyfitz (1996b). Using monotone operator techniques adapted for this degenerate elliptic equation we proved that the problem allows two qualitatively different types of solutions, depending on the type of boundary data given along the transonic shock. One solution is *smooth* up to the boundary, another has a *square-root singularity* along the sonic line.

• Another interesting feature of our solution is that the regular subsonic solution (u, v) matches smoothly with the supersonic region, while the singular solution does not (Canic & Keyfitz 1996b). The singular case would require a weak shock wave emerging from the square-root singularity to match the subsonic with the supersonic flow (see Fig. 1b). The existence of a solution of this free-boundary problem has not yet been proved.

• Finally, the *curvature* of the Mach stem is determined by the type of boundary data along the stem. Namely, the first Rankine-Hugoniot condition implies

$$u' = 2\rho'(1 - 2\rho''), \quad \text{where} \quad \rho(\eta) = \xi(\eta) + \eta^2/4. \tag{5}$$

Here u' denotes the derivative of u along the transonic shock $\xi = \xi(\eta)$. Therefore, if the shock steepens monotonically to vertical at the axis, (i.e., $\xi' > 0$), then the shock is convex, (i.e., $\xi'' > 0$ or $\rho'' > 1/2$), if and only if $u' < 0$. Moreover, the higher the gradient of u, the bigger the curvature of the free boundary. Therefore, in order to have a monotonically increasing, convex Mach stem, as observed in the experiments, the data for the approximate free boundary has to be decreasing. By Theorem 2, the solution u must then have a square-root singularity at the point where the Mach stem meets the sonic curve. This argument proves the following:

Theorem 3 *Assume that the incident shock S_1 does not interact with another shock before meeting the sonic curve, and that the Mach stem $\xi = \xi(\eta)$ is an increasing, convex function. Then the subsonic solution must have a square-root singularity at the point where the Mach stem crosses the sonic curve.*

3.1. A conjecture for von Neumann reflection

We conjecture that the following configuration of S_1 and S_2 leads to von Neumann reflection. The incident shock S_1 between the uniform states U_0 and U_1 interacts with the region where U_1 becomes subsonic, giving rise to a free-boundary problem with a convex Mach stem. The flow at the sonic curve has a square-root singularity, which, in turn, gives rise to a weak shock emerging from the singular region. The shock S_2, separating the states U_1 and U_2, interacts with the region where U_2 becomes subsonic, P_2, leading to a free-boundary problem for another transonic shock. The shock generated by the square-root singularity catches up with the transonic shock continuing from S_2 giving rise to the flow pattern shown in Fig. 2.

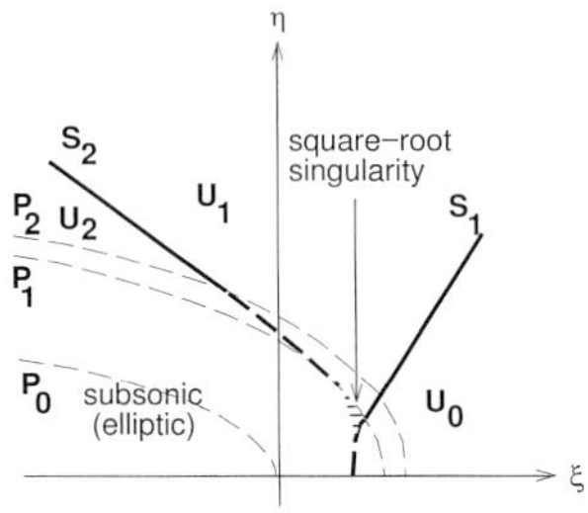

Figure 2. Conjectured structure for von Neumann reflection.

4. Final Remarks

Our results about the square-root behavior of the horizontal velocity component, u, seem to agree with the numerical results described in Tabak & Rosales (1995). Their numerical simulations indicate that the vertical velocity component, v, exhibit's a stronger singularity. In the course of our study to prove the existence of a solution of the free-boundary problem we have come across some indications that this might indeed be the case. Nevertheless, this problem is still open.

References

Brio M, Hunter JK (1992) Mach reflection for the two-dimensional Burgers equation. Physica D 60:194–207.

Canic S, Keyfitz BL (1995) Quasi-one-dimensional Riemann problems and their role in self-similar two-dimensional problems. (In preparation)

Canic S, Keyfitz BL (1996a) An elliptic problem arising from the unsteady transonic small disturbance equation. J Diff Equations (to appear).

Canic S, Keyfitz BL (1996b) A smooth solution for a keldysh type equation. Comm PDE 21:319–340.

Canic S, Keyfitz BL, Wagner DH (1994) A bifurcation diagram for oblique shock interactions in the unsteady transonic small disturbance equation. In: Proc 5th International Conference on Hyperbolic Problems, Theory, Numerics and Applications (to appear).

Morawetz CS (1994) Potential theory for regular and Mach reflection of a shock at a wedge. Comm Pure Appl Math XLVII:593–624.

Tabak EG, Rosales RR (1994) Weak shock focusing and the von Neumann paradox. Physics of Fluids A 6(5):1874–1892.

Tabak EG, Rosales RR (1995) Private communication.

Shock-wave interactions at impingement of impulsive supersonic jet upon an obstacle

T. V. Bazhenova, V. V. Golub, A. M. Shulmeister,
T. A. Bormotova, S. B. Bazarov
High Energy Density Research Center
IVTAN, Russian Academy of Sciences
Izhorskaya, 13/19, Moscow, 127412 Russia

Abstract: The unsteady interaction between an impulsive jet and a flat plate was investigated experimentally and numerically. At small distances between the nozzle and plate, the reflected wave plays an important role, bringing about a rise in pressure. At great distances, the head part of the impulsive jet plays the decisive role; on the plate near the axis of the jet, pressure and gas density rise and at the region of the vortex ring the density drops.

Key words: Impulsive jet, Vortex, Reflection

1. Introduction

The primary stage of an impulsive supersonic jet is characterized by the formation of shock waves and vortex rings. A shock-wave system similar to the barrel structure in underexpanded jets develops in the gas jet with a starting shock wave ahead of the outflowing gas (Golub 1994). The present paper is devoted to the experimental and numerical investigation of the unsteady interaction between an impulsive jet and a flat plate.

2. Experimental set-up

A sonic nozzle with diameter of critical cross-section $d_* = 4$ mm was installed at the end of the shock tube, the plate being mounted in front of the nozzle (at the dimensionless distances $\bar{L} = L/d_* = 6$ to 50. The shock tube and vacuum chamber were filled with nitrogen. The initial temperature of the gas was room temperature. The Mach number M_0 of the incident shocks was varied between 2.0 and 4.4. The pressure P_∞ of the ambient gas in the vacuum chamber was between 4 kPa and 50 kPa. The pressure ratio $n = P_a/P_\infty$ was varied from $n = 8$ to 74, where P_a is the pressure at the nozzle exit and P_∞ is the pressure of the ambient gas. The flow was visualized by means of the shadow device and interferometer with a ruby laser. Five pressure transducers and five thin-film temperature gauges were mounted on the plate.

3. Results

The interferogram and sketch of the interaction between the impulsive supersonic jet and the obstacle are shown in Fig. 1. The starting shock wave SW1

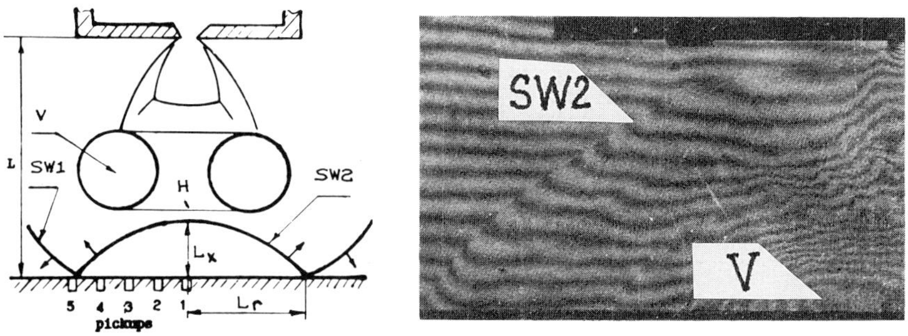

Figure 1. Interaction between the impulsive supersonic jet and the obstacle: sketch (left) and interferogram (right) with $M_0 = 2.75$, $\bar{L} = 10$, and $t = 114$ μs.

(Fig. 1, left) propagating in front of the hot gas and outflowing from the nozzle reaches the plate and is reflected by it, forming shock wave SW2 moving towards the nozzle. The head part of the impulsive jet H and the vortex V are shown in the sketch. The interactions between the high-enthalpy jet flow and vortex V with the reflected shock S2 and with the plate were observed in the later stages of development. In Fig. 1 (right) we present one-half of the axisymmetric interferogram. The vortex ring V touches the plate at this stage, with the reflected shock wave SW2 moving back to the nozzle.

In order to obtain the density distribution, the axisymmetric interferometric fringe distributions were evaluated numerically by solving Abel's integral equation. Since we used finite-width fringe interferometry, there was no need to know the initial values of the density on the shock front. The numerical processing of interferograms is based upon the solution of Abel's integral equation:

$$S(x,y) = 2K/\lambda \int_y^R \frac{(\rho(x,y) - \rho_0)r}{\sqrt{r^2 - y^2}} dr \tag{1}$$

which can be rewritten as an operator formula: $S(x,y) = [A\rho](x,y)$ assuming an axisymmetric jet density field. Here $S(x,y)$ is the fringe shift in the interferogram, K is the Gladstone-Dale constant, λ is the ruby-laser wavelength, x is the coordinate along the jet axis, y is the transverse coordinate, and r is the jet radius at the present cross-section. Abel's equation may be rewritten as:

$$S(y_j) = \frac{RK}{N\lambda} \sum_{i=j}^{N} (\rho_j - \rho_0) a_{ij} \tag{2}$$

where a_{ij} are Pirs's coefficients.

$$\rho(x,y) = [AA^{-1}S](x,y) \tag{3}$$

This formula is precisely the inversion formula of Radon. The density distribution along the symmetry axis of the impulsive jet after impingement was constructed. Figure 2 presents the spatial density distribution for $t = 114$ μs, $L = 40$ mm, and $M_0 = 2.75$. The jet density is normalized by the ambient

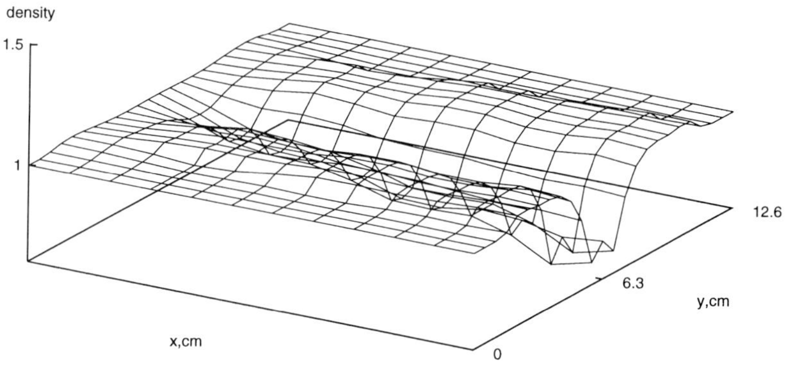

Figure 2. The spatial density distribution for $t = 114$ μs, $L = 4$ cm, $M_0 = 2.75$; $x = 0$ denotes the plate position, with the symmetry axis $y = 6.3$.

gas density ρ_∞ and the coordinates x and y are given in centimeters. At some distance from the axis the density decreases to $0.8\rho_\infty$. Possibly the decrease of the density is related to the interaction of the vortex V with the plate.

The heat flux $q(t)$ into the wall during the impingement process was calculated using the wall temperature history $T(t)$ measured by the thin-film gauges:

$$q(t) = \frac{\lambda}{\sqrt{a\pi}} \int_0^t \frac{dT(t)}{d\tau} \frac{1}{\sqrt{t-\tau}} d\tau \tag{4}$$

where a and λ are coefficients of the thermal conductivity and diffusivity. The pressure and heat-flux histories are shown for two distances between the nozzle exit and the plate, $\bar{L} = 12$ and 38 (Figs. 3a and 3b). The arrival of SW1 at the plate is clearly traced by the initial rise of the pressure recorded by the center pick-ups (Fig. 3a).

After the SW1 reflection process the head part of the impulsive jet H begins to interact with SW2 and after interaction impinges on the plate. The unsteady breaking of the gas jet on the plate leads to the second rise of pressure and heat flux (Fig. 3b). The heat flux rises at the moment of high-enthalpy jet interaction with the plate. The value of the heat flow at $\bar{L} = 38$, $t = 700$ μs rises to $q = 3 \times 10^5$ W/m^2.

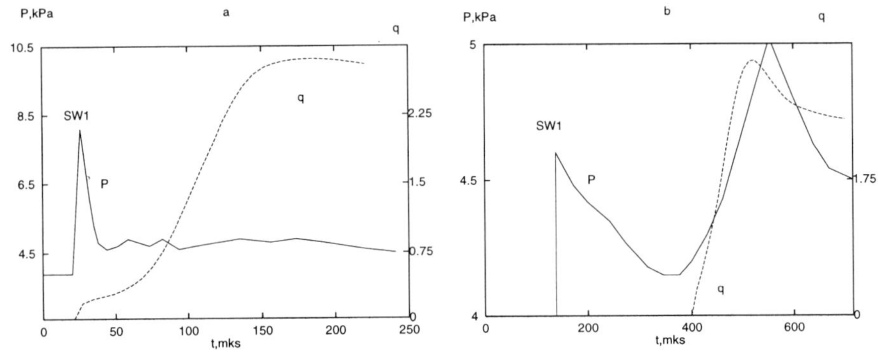

Figure 3. The pressure and heat-flux histories at the obstacle, $M_0 = 4.4$: (a) $\bar{L} = 12$; and (b) $\bar{L} = 38$; q is normalized to 10^{-5} W/m^2.

4. Comparision with the numerical simulation

Numerical simulation was made by solving Euler equations by Godunov's method. The experimental and computed normalized distributions of the pressure and density near the plate are compared. Figure 4a shows the density distribution along the symmetry axis and Fig. 4b shows the distribution along the plate ($M_0 = 2.7$, $\bar{L} = 10$, $t = 114$ μs). As seen in Fig. 4a the calculated density near the plate is in agreement with the experimental data. The density distribution along the plate shows that the main features of the process near the plate are presented both in the calculated data and the experimental data. In the center of the plate the density is slightly increased; at some distance from the center there exists a circular rarefaction zone surrounded by the circular zone of elevated density. This distribution may be interpreted as a result of interaction of the shock wave SW2 and vortex ring V with the plate. The calculated pressure records are in qualitive agreement with the experimental ones.

5. Discussion

The calculated time history of the normalized pressure, density, and gas temperature near the center of the plate is shown in Fig. 5. The first peak on the pressure curve corresponds to SW1 interaction with the plate. It is remarkable that the gas temperature rises slightly at the reflected shock front. Later, the high-enthalpy flow of the jet stagnates near the plate and the gas temperature rises to the value of the stagnation temperature behind the reflected shock wave in the shock tube and then rises higher. Due to unsteady deceleration, the gas temperature near the plate is the higher than the stagnation temperature; this may account for the redistribution of the kinetic energy in nonstationary flows (Stanukovich 1971).

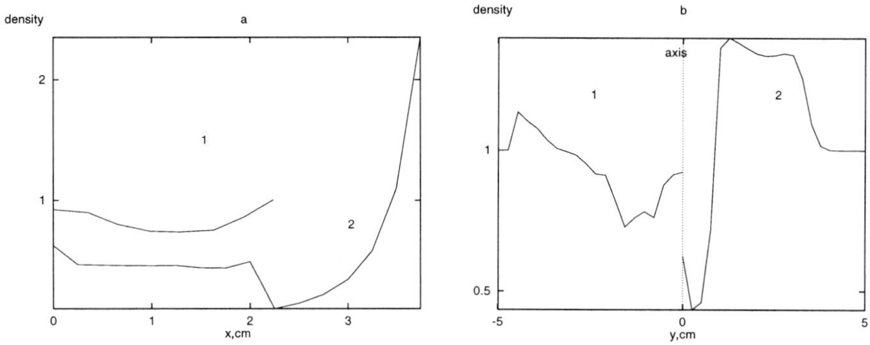

Figure 4. Density distribution along (a) the symmetry axis and (b) the plate; $M_0 = 2.75$, $t = 114$ μs, $L = 4$ cm. (1) corresponds to experiment and (2) to calculation.

Zeldovich and Stanukovich (1947) considered the problem of nonstationary flow interaction with the wall. The flow is generated by the sudden outflowing of gas with stagnation temperature T_{00} through the orifice in the resevoir, mounted opposite the reflecting wall. Using the self-similar solution of the gasdynamic problem for the perfect gas the flow parameters after the interaction with the wall were obtained. The starting shock wave S1 was not taken into account. The one-dimensional case was considered. The nonstationary flow velocity u_0 is equal to $u_0 = 2c_{00}/(\gamma+1)^2$ where c_{00} is the sound velocity at the stagnation temperature, and $\gamma = c_p/c_v$ is the adiabatic index. The temperature T near the reflecting wall after nonstationary flow impingement was found from:

$$c_v T = \frac{u_0^2}{2}\left[\frac{(\gamma^3+\gamma+2)}{(\gamma^2+1)(\gamma+1)}\right]^2 \tag{5}$$

where c_v is the specific heat capacity. The gas-temperature value after impingement related to the stagnation temperature T_{00} is equal to

$$T/T_{00} = [2\gamma/(\gamma-1)]\left[\frac{\gamma^3+\gamma+2}{(\gamma^2+1)(\gamma+1)}\right]^2 \tag{6}$$

In our case $\gamma = 1.4$ which gives a maximum value T/T_{00} equal to 6. Numerical calculation gives $T/T_{00} = 1.25$. The difference is natural because Zeldovich estimation did not take into account the space distribution of the flow energy in the axisymmetric case.

6. Conclusions

(1) At small distances between the nozzle and plate, the reflected wave SW2 plays an important role, bringing about the rise of temperature and pressure

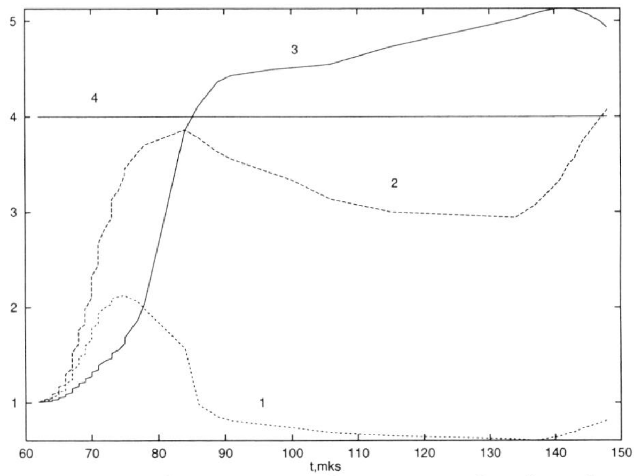

Figure 5. The pressure and temperature histories at the obstacle on the jet axis, $\bar{L} = 10$, $M_0 = 2.75$, (1) corresponds to density, (2) to pressure, (3)to gas temperature, and (4)to stagnation temperature.

(thus, at $\bar{L} = 12$ the pressure in the plate increases 1.5 times). (2) At great distances, the head part of the impulsive jet plays the decisive role. On the plate near the axis of the jet, pressure and temperatures rise and at the region of vortex ring the pressure drops (thus, at $\bar{L} = 38$ the pressure on the plate drops twofold). (3) The gas temperature at the wall at the second stage exceeds stagnation temperature. The increase of the gas temperature at the second stage of interaction leads to an increase in heat flow to the wall.

Acknowledgement. The present research was supported by Grant Nos. 94-01-00218-a and 94-01-01461-a of the Russian Fundamental Research Foundation.

References

Golub VV (1994) Development of shock wave and vortex structures in unsteady jets. Shock Waves 3:279–285.

Stanukovich KP (1971) Unsteady Flows of Continous Media, Nauka Press, Moscow (in Russian).

Zeldovich YB, Stanukovich KP (1947) The reflection of the detonation wave. Doklady Academy Nauk, USSR, 55(7).

Application of the principle of minimum entropy production to the regular-to-Mach-reflection transition in steady flows

H. Li, G. Ben-Dor
Pearlstone Center for Aeronautical Engineering Studies
Department of Mechanical Engineering
Ben-Gurion University of the Negev, Beer Sheva, Israel

Abstract: A new approach based on the principle of minimum entropy production for better understanding the regular-to-Mach-reflection transition in steady flows is presented. To the best of our knowledge the new proposed transition criteria agree with all the available experimental results.

Key words: Minimum entropy production, Shock reflection, Steady flows

1. Introduction

The present understanding regarding the regular $\leftrightarrow$ Mach-reflection transition in steady flows was established by Hornung and Robinson (1982). Since very few experimental works regarding this process were conducted, there were no reasons to discard their conclusions. However, in a recent experimental study which was conducted by Chpoun et al. (1995), results which completely contradict those of Henderson and Lozzi (1975), Hornung and Robinson (1982), and Hornung et al. (1979) were obtained. Hence, there is a need to modify the existing transition criterion to agree with all the available (including the new) experimental results.

Evidently the conservation equations are not sufficient to determine the physical solution which actually takes place. In order to determine the physical solution out of a variety of possible mathematical solutions the irreversibility of the thermodynamic process has to be accounted for. As postulated by Glansdorff and Prigogine (1971) the process must fulfill the principle of minimum entropy production. This principle, which is an a posteriori axiom, has been found to accurately predict the stability of a wide range of phenomena. Since a steady flow field can be regarded as a non-equilibrium steady state of thermodynamics, the principle of minimum entropy production can be adopted to better understand, from a physical point of view, phenomena in which, from a mathematical point of view, multiple solutions are possible.

2. Present study

The principle of minimum entropy production was applied to oblique shock-wave reflections in steady flow. Details of the analysis can be found in Li and Ben-Dor (1996). In the following only the final results are presented. Based on their analysis it was concluded that for given values of γ and M_0 the physical solution must satisfy the condition that $\beta \leq \beta^*$, where, β^* is the angle of incidence between the flow and the deflecting shock wave.

2.1. Application to supersonic-flow deflection

Consider a problem in which a uniform supersonic flow having a Mach number M encounters a wedge having an angle θ. The flow negotiates the wedge by means of a straight attached shock wave. The angle of incidence between the flow and the deflecting shock wave is β. It is well known that the deflecting shock wave is attached provided $\beta \leq \beta_{ext}$. The curves β^* (solid line) and β_{ext} (dashed line) vs. M, for $\gamma = 1.4$, are plotted in the (M,β)-plane shown in Fig. 1. Note that both β^* and β_{ext} approach asymptotic values as $M \to \infty$. $\beta^*(M \to \infty) = 45°$ and β_{ext} $(M \to \infty) = 67.8°$. It is evident from Fig. 1 that the two curves intersect at a point marked by Q. It corresponds to $M_Q = 1.80$ and $\beta_Q = 65.7°$. The shadowed area in Fig. 1, for which $\beta \leq \beta^*$, indicates the domain in the (M,β)-plane in which possible mathematical solutions are physical. An inspection of Fig. 1 indicates the following: (1) for $1 < M < M_Q$ there is a narrow domain in which $\beta_{ext} < \beta < \beta^*$, thus, a "strong" shock solution is physically possible there; (2) for $M > M_Q$ there is a domain in which $\beta^* < \beta < \beta_{ext}$, thus in spite of the fact that this domain lies in the "weak" solution domain, a solution in this domain is aphysical.

It is very important to note here that the facts that the "strong" solution has never been observed in the domain mentioned in (1) and that "weak" solutions have been observed in the domain mentioned in (2) in wind-tunnel experiments should not be used to discard the conclusions mentioned in (1) and (2). This is due to the fact that the analytical results shown in Fig. 1 were obtained under the assumption of "free boundaries" while actual experiments in wind tunnels do not satisfy this constraint.

2.2. Application to regular reflection (RR)

The wave configuration of a steady RR is shown in Fig. 2a. The supersonic flow in state (0) passes obliquely through the incident shock wave i. While passing through i it is deflected by an angle θ_1 and its flow Mach number changes to M_1. The flow in state (1) is then re-deflected by the reflected shock wave r, to become again parallel to the reflecting surface, hence $\theta_2 = \theta_1$. The flow Mach number behind the r is M_2. The angles of incidence are ϕ_1 and ϕ_2.

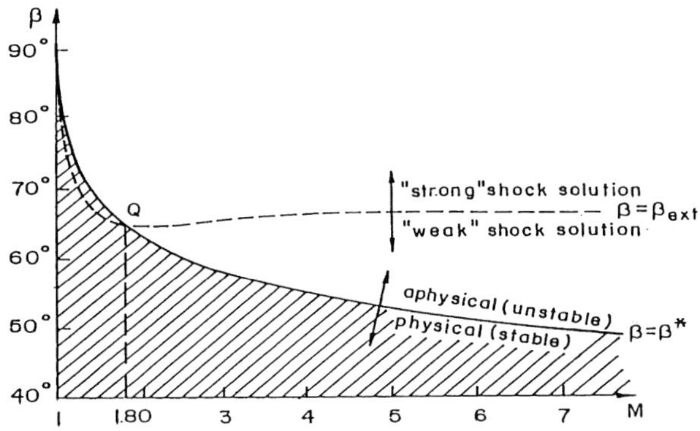

Figure 1. Domains of "strong" and "weak" solutions of the supersonic-flow deflection process and regions of physical and aphysical solution in the (M, β)-plane.

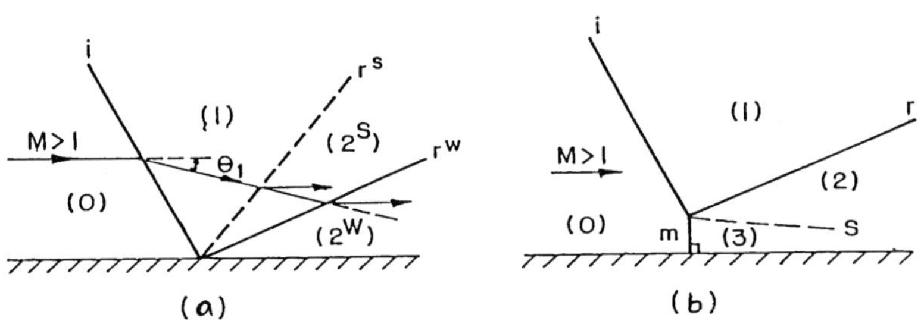

Figure 2. (a) The wave configurations of the "weak" and the "strong" regular reflections (RR). (b) The wave configuration of the Mach reflection (MR).

In order to eliminate the aphysical RR solution out of the two possible mathematical solutions (shown in Figs. 2a and 2b), the principle of minimum entropy production was applied. Owing to the analogy between the RR and the flow-deflection processes, the results shown in Fig. 1 can be used here too. Figure 3 is a reproduction of Fig. 1 for the case of RR. Note the changes in the vertical axis from β to ϕ_2 and in the horizontal axis from M to M_1. Similar to the notation used in Fig. 1, the shadowed area in Fig. 3, which corresponds to $\phi_2 \leq \phi_2^*$, indicates the domain in the (M_1,ϕ_2)-plane in which possible mathematical RR solutions are physical. An inspection of Fig. 3 indicates the following: (1) for $1 < M_1 < M_1^Q$ there is a narrow domain in which $\phi_1^* < \phi_2 < \phi_2^{ext}$, thus a "strong" RR is physically possible there; (2) for $M_1 > M_1^Q$ there is a domain in which $\phi_2^* < \phi_2 < \phi_2^{ext}$, thus in spite of the fact that this domain lies in the "weak" RR domain, an RR solution in this domain is aphysical.

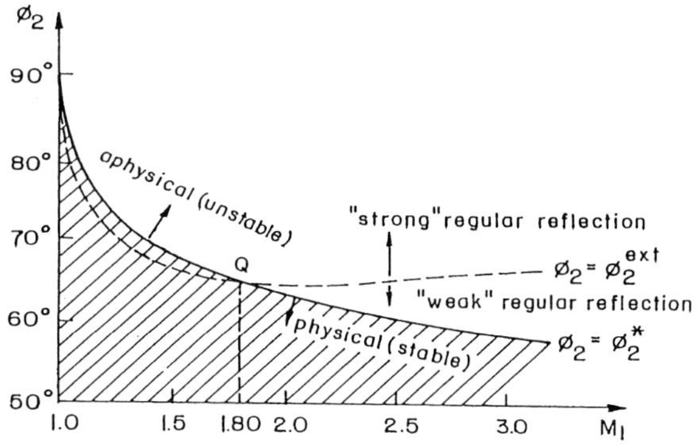

Figure 3. Domains of "strong" and "weak" solutions of a regular reflection and regions of physical and aphysical solutions in the (M_1, ϕ_2)-plane. (Note that M_1 and ϕ_2 are analogous to M and β in Fig. 1.)

The conclusion mentioned in (2) will be discussed following the introduction of the possible RR $\leftrightarrow$ MR transition criteria from in the next section.

2.3. The RR $\leftrightarrow$ MR transition criterion

Figure 4 is a transformed plot of Fig. 3. The curve $\phi_2 = \phi_2^{ext}$ is relabeled as $\phi_1 = \phi_1^{ext}$ in order to stress that it is actually the transition line corresponding to the above-mentioned detachment criterion. In addition to the $\phi_1 = \phi_1^{ext}$ and the $\phi_1 = \phi_1^*$ curves, there is a third curve marked by $\phi_1 = \phi_1^N$ which corresponds to the von Neumann transition criterion. The $\phi_1 = \phi_1^N$ curve originates at point K. For $M_0 < M_0^K$ only the detachment criterion exists and for $M_0 > M_0^K$ both the detachment and the von Neumann criteria are possible. Point K which is a point on $\phi_1 = \phi_1^{det}$, is the origin of $\phi_1 = \phi_1^N$. In addition, $M_0^K = 2.20$ and $M_0^Q = 2.66$. Two more remarks regarding the transformation from the (M_2,ϕ_2)-plane to the (M_0,ϕ_1)-plane should be noted. The first is that the small region in Fig. 3 for which $\phi_2^{ext} < \phi_2 < \phi_2^*$ has vanished owing to the fact that RR is mathematically impossible for $\phi_1 > \phi_1^{det}$. Thus in the range $M_0 < M_0^Q$ the curves $\phi_1 = \phi_1^{det}$ and $\phi_1 = \phi_1^*$ converge smoothly to a single curve. The second remark concerns the range $M_0 > M_0^Q$. Note how the domain $\phi_2^* < \phi_2 < \phi_2^{ext}$ shown in Fig. 3 shrinks down to a very narrow domain for which $\phi_1^* < \phi_1 < \phi_1^{det}$ in Fig. 4. Thus it is practically impossible to distinguish between the $\phi_1 = \phi_1^{det}$

and $\phi_1 = \phi_1^*$. Consequently, the earlier-mentioned conclusion that part of the "weak" RR domain is aphysical is practically meaningless since, as shown in Fig. 4, this domain almost vanishes in the (M_0, ϕ_1)-plane. It should be noted here that based on Fig. 4 the two mathematical solutions are stable since they fulfill the condition that $\phi_1 \leq \phi_1^*$.

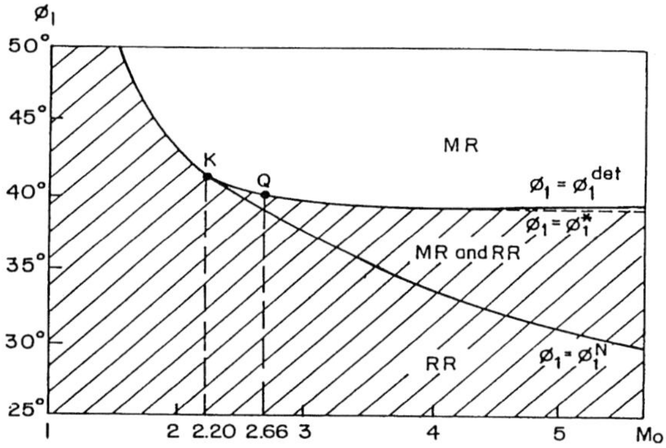

Figure 4. Domains of RR and MR and regions of physical and aphysical solutions.

The suggestion that RR is stable in the $\phi_1^N < \phi_1 < \phi_1^*$ domain disagrees with the state of the art concerning the RR $\leftrightarrow$ MR transition in steady flows which was established by Hornung and Robinson (1982). Their experimental conclusion was that the von Neumann criterion is the correct transition criterion. Furthermore, they noted that Hornung et al.'s (1979) hypothesis of a hysteresis in the RR $\leftrightarrow$ MR transition "could not be confirmed".

However, a recent experimental investigation of the RR $\leftrightarrow$ MR transition in steady flows, which was conducted by Chpoun et al. (1995), clearly validated the existence of the hysteresis phenomenon originally hypothesized by Hornung et al. (1979). Stable RR was recorded in the $\phi_1^N < \phi_1 < \phi_1^* \approx \phi_1^{det}$ domain. Furthermore, their attempts to introduce large disturbances in order to force transition from RR to MR totally failed. The RR in the $\phi_1^N < \phi_1 < \phi_1^*$ domain was perfectly stable. Chpoun et al.'s (1995) results also indicated that the actual angle at which the RR $\rightarrow$ MR transition takes place depends on both the geometrical size of the model and the way the experiment is set up. Using different models and experimental setups the values of the RR $\rightarrow$ MR transition angles were found to be in the range $35° < \phi_1^{tr}(RR \rightarrow MR) \leq 39°$ for $M_0 = 4.96$.

These results, together with the foregoing discussions, led us to propose the following general transition criterion for situations which completely satisfy the "free boundary" constraint under which the principle of minimum entropy production was developed:

For weak shock waves, i.e., $M_0 < M_0^K$:

I. The RR $\leftrightarrow$ MR transition takes place at the detachment criterion.

For strong shock waves, i.e, $M_0 > M_0^K$:

II. The RR $\rightarrow$ MR transition takes place at the detachment criterion.

III. The MR $\rightarrow$ RR transition takes place at the von Neumann criterion.

If the "free boundary" constraint is not satisfied, i.e., if downstream influences can affect the reflection or unsteady effects are present then for strong shock waves, i.e., $M_0 > M_0^K$:

IV. The RR $\leftrightarrow$ MR transition takes place anywhere in the domain $\phi_1^N \leq \phi_1^{tr} \leq \phi_1^* \approx \phi_1^{det}$ and ϕ_1^{tr} depends on the downstream and the unsteady influences.

3. Conclusion

The RR $\leftrightarrow$ MR transition phenomenon was reconsidered with the aid of the principle of minimum entropy production. The reconsideration led to a new set of transition criteria.

References

Chpoun A, Passerel D, Li H, Ben-Dor G (1995) Reconsideration of oblique shock wave reflections in steady flows, part I: experimental investigation. J Fluid Mech 301:19–35.

Glansdorff P, Prigogine A (1971) Thermodynamic Theory of Structure Stability and Fluctuations, John Wiley and Sons, Ltd., London.

Henderson LF, Lozzi A (1975) Experiments on transition of Mach reflection. J Fluid Mech 68:139–155.

Hornung HG, Oertel H Jr, Sandeman RJ (1979) Transition to Mach reflection of shock waves in steady and pseudo-steady flow with and without relaxation. J Fluid Mech 90:541–560.

Hornung HG, Robinson ML (1982) Transition from regular to Mach reflection of shock waves, part 2: the steady-flow criterion. J Fluid Mech 123:155–164.

Li H, Ben-Dor G (1996) Principle of minimum entropy production in the regular Mach reflection transition in steady flows. (Submitted for publication)

Evolution of Courant-Friedrichs structure: self-collimating outflows behind reflected shocks

A. Chernin
Sternberg Astronomical Institute, Moscow University
Moscow, 119899 Russia

P. Voinovich
Advanced Technology Center
P. O. Box 160, St. Petersburg 198103 Russia

Abstract: The evolution of the Courant-Friedrichs structure resulting from the collision of two equal spherical blast waves is simulated numerically on the basis of a locally adaptive unstructured Euler solver. Formation of strong self-collimating axial jets has been demonstrated as a prominent feature of the flow development. The results might supply a possible explanation for some observed astrophysical phenomena.

Key words: Blast waves, Jets, Numerical simulation, Astrophysics

1. Introduction

Almost a half-century ago, Courant and Friedrichs (1948) studied a dynamic structure produced by two identical spherical shock fronts coming into collision. The resulting symmetric structure includes the ring of the Mach front, two surfaces of tangential discontinuity, and two reflected shocks propagating toward the original centers of the blasts. Now we present results of a numerical simulation of the further nonlinear dynamical development of the Courant-Friedrichs structure with the aim of providing a basis to explain the mechanisms that may underlie some observed phenomena.

2. Numerical model

The Sedov-Taylor self-similar point blast solution has been used to set up the initial conditions at a moment in time preceding the collision of shock fronts in a perfect gas with a specific heat ratio of $\gamma = 5/3$. A high-order Godunov-type finite-volume scheme and transiently adaptive unstructured grid approach constitute the principal components of the numerical technique.

Some results are given in the figures in the form of density contour lines. Figures 1, 3, and 5 depict initial conditions for the following images providing

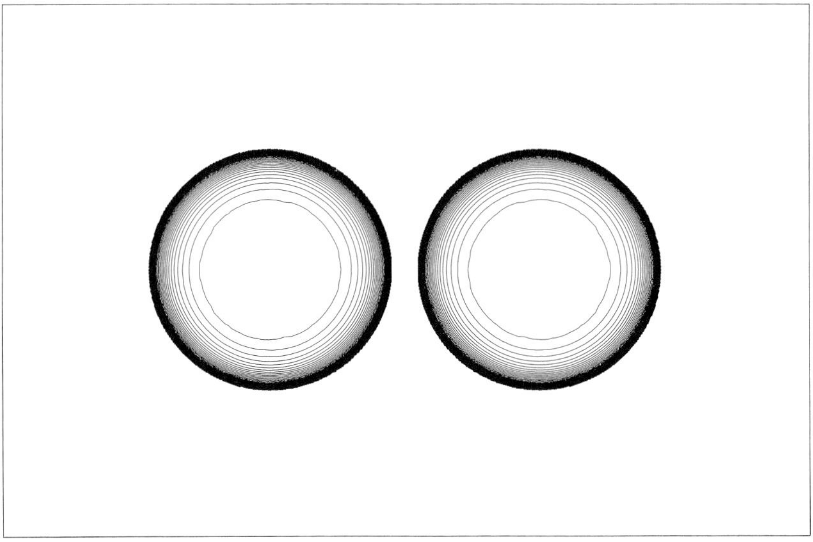

Figure 1. $t = 0.768$.

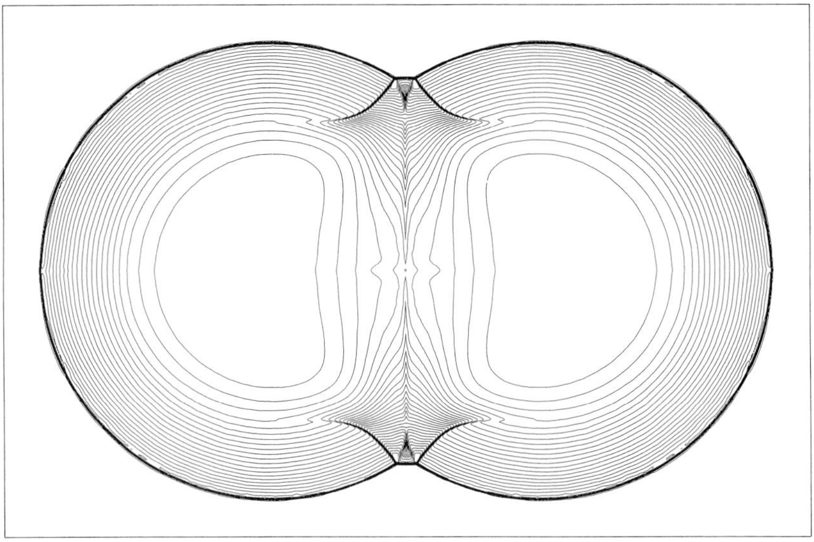

Figure 2. $t = 3.50$.

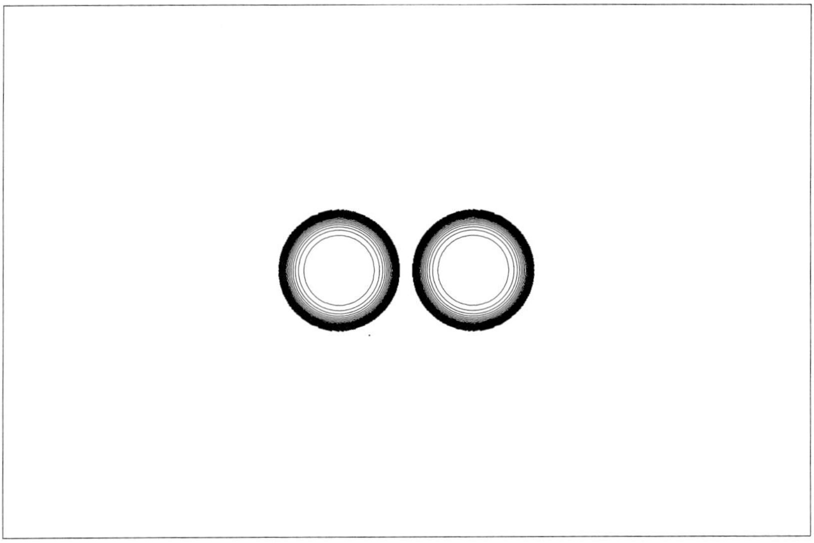

Figure 3. $t = 0.768$.

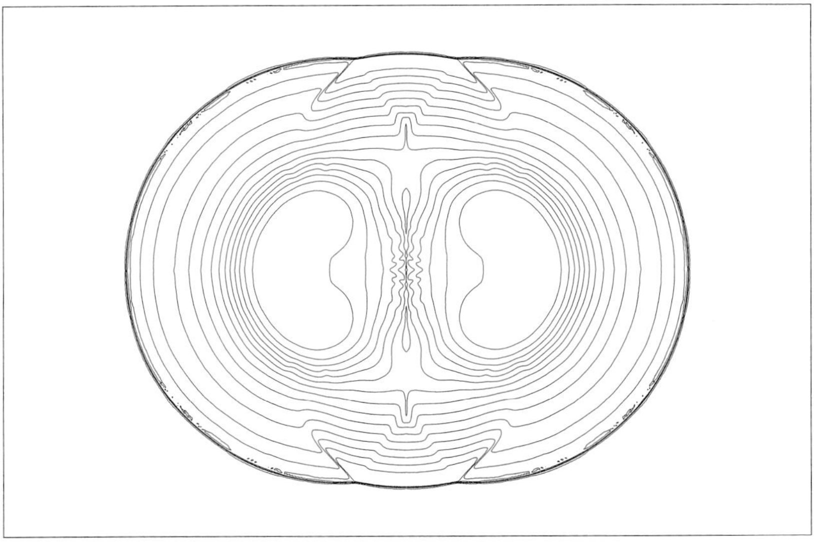

Figure 4. $t = 11.90$.

Figure 5. $t = 0.768$.

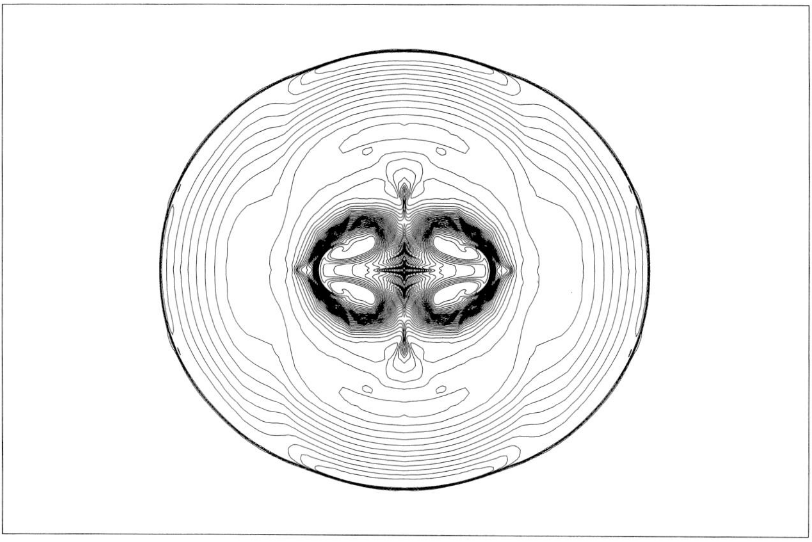

Figure 6. $t = 34.41$.

Figure 7. $t = 865.0$.

a reference scale. The unit time corresponds to the moment of collision of the blast fronts. The contour line spacing is $\Delta\rho = 0.1\rho_0$ for Figs. 1–5, and $\Delta\rho = 0.025\rho_0$ for Figs. 6 and 7, where ρ_0 is the density of ambient gas at rest.

All results were computed on an IBM PC/486 which demonstrates a high performance of the code (for details of numerical technique see Galyukov et al. 1996).

3. Discussion

As can be seen in the figures, the reflected shocks of the Courant-Friedrichs structure are critically affected by the distribution of matter in the area where they propagate. This matter distribution, in turn, is mainly due to the initial shocks that compress the major portion of the gas into two spherical shells behind their fronts. The reflected shocks propagate into the shells and the cavities within shells accelerating in the regions of low gas density.

Another characteristic feature of the flow development is a sort of self-collimation of the outflows behind the fronts of the reflected shocks caused by the rarefaction in the vicinity of the blast centers. As a result, the gas flow behind the reflected shocks tends to transform into directed streams or jets along the symmetry axis. The maximum gas velocity in the jets ranges up to half of the sound speed in the initial gas.

The final stage of the flow development given in Fig. 7 exhibits intensive vorticity generated by shock interactions and passing through a strongly non-uniform density field.

From the point of view of possible astrophysical applications of the process, it can be mentioned that observations provide examples of structures similar to the Courant-Friedrichs configuration; one of them is a system of two giant shells in the Large Magellanic Cloud near 30 Doradus discovered by Meaburn (1980). Collimated binary outflows are also observed in the interstellar medium (Lada 1985).

Acknowledgement. The work was partially supported by the Russian Foundation of Fundamental Investigations (Grant No. 94–01–01456).

The authors express their sincere gratitude to Prof. K. Takayama, Shock Wave Research Center, Institute of Fluid Science, Tohoku University, Sendai, Japan, for valuable support and fruitful discussions.

References

Courant R, Friedrichs KO (1948) Supersonic Flow and Shock Waves, Wiley Interscience, New York.

Galyukov A, Fursenko A, Voinovich P, Saito T, Timofeev E (1996) Efficient 2D and 3D locally adaptive unstructured Euler solvers for unsteady shocked gas flows. In: Shock Waves, Proc of the 20th ISSW, Sturtevant B, Shepherd JE, Hornung HG (eds), World Scientific, Vol 1, pp 691–696.

Lada CJ (1985) Cold outflows, energetic winds, and enigmatic jets around young stellar objects. Ann Rev Astron Astrophys 23:367–318.

Meaburn J (1980) The giant and supergiant shells of the Magellanic Clouds. Mon Not R Astron Soc 192:365–375.

Numerical investigations of the prediction of the Mach-stem height in steady flows

J. Vuillon, D. Zeitoun
I.U.S.T.I. (U.M.R. CNRS 139), Dept. MHEQ, Centre St. Jérome
Université de Provence, 13997 Marseille Cedex 20, France

G. Ben-Dor
Pearlstone Center for Aeronautical Engineering Studies
Department of Mechanical Engineering
Ben-Gurion University of the Negev
Beer Sheva, Israel

Abstract: A two-dimensional supersonic steady flow between a wedge and a symmetry plane was computed by solving the unsteady Euler equations. The numerical method (LCPFCT method) is based on a finite-difference scheme associated with a flux-corrected-transport algorithm. The configuration of the shock-wave reflections is investigated. The numerical results of the Mach-stem height are in good agreement with the experimental data. The results show stable configurations of regular and Mach reflections in the dual-solution domain, and a hysteresis effect in the transition between these two reflections. They also show the influence of the distance h between the wedge trailing edge and the symmetry plane on the RR $\leftrightarrow$ MR transition.

Key words: Steady supersonic flow, Euler equations, Numerical simulation, RR to MR transition

1. Introduction

Generally in a steady flow, as described by Ben-Dor (1991), the wedge-generated shock wave which reflects on a symmetry plane can produce two shock-wave reflection configurations, namely, regular reflection (RR) and Mach reflection (MR). The RR configuration consists of an incident shock wave i and a reflected one r, with a reflection point R which is located on the reflecting surface. The MR configuration contains three shock waves: the incident and reflected ones, as before, the Mach stem m, and in addition, one slipstream s. These four discontinuities interact at a single point called the triple point T, which is located above the reflecting surface. A schematic drawing of these RR and MR wave configurations is shown in Fig. 1. At a fixed freestream Mach number, the transition from one reflection type to another one depends on the angle ω_i between the incident shock wave and the symmetry plane. Two possible tran-

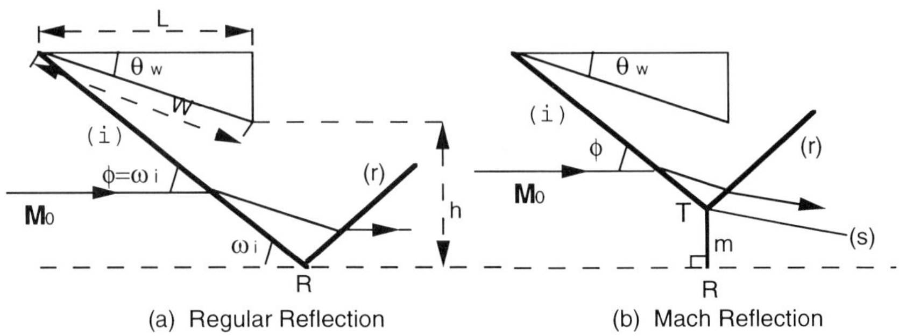

Figure 1. Schematic illustration of the shock-wave configurations between the wedge and the symmetry plane.

sition criteria have been proposed by von Neumann (1963). They are known as the detachment ($\omega_i = {\omega_i}^D$) and von Neumann ($\omega_i = {\omega_i}^N$) criteria which correspond to the values of incident shock-wave angles. The regular reflection does not take place for $\omega_i > {\omega_i}^D$ and the Mach reflection is impossible for $\omega_i < {\omega_i}^N$. In the range of ${\omega_i}^N < \omega_i < {\omega_i}^D$ and for a strong freestream Mach number (i.e., $M_0 > 2.20$), called the dual-solution domain, both reflections are theoretically possible. From their experimental study, Hornung and Robinson (1982) have concluded that: "the RR $\leftrightarrow$ MR transition occurs at the von-Neumann condition and not at the detachment condition." The predicted hysteresis effect in the RR $\leftrightarrow$ MR transition (Hornung et al. 1979) was not confirmed by their results. For this reason, they claimed that the RR configuration in the dual domain was an unstable regular reflection.

Recently, Chpoun et al. (1995) have pointed out this hysteresis effect and have experimentally found a stable RR in the dual domain. Very little numerical work has been devoted to this phenomenon and the aim of the present paper is to numerically study these shock-wave reflections. In order to do that, the unsteady 2D Euler equations have been solved with the LCPFCT method (Oran and Boris 1987). This method is based on a finite-difference scheme associated with a flux-corrected-transport algorithm. The main objectives of this work are firstly, to validate the code with Hornung and Robinson's experimental results; secondly, to put in evidence the hysteresis effect; and finally, to demonstrate that the reflected-wave configurations established inside the dual-solution domain depend on the distance h between the wedge trailing edge and the symmetry plane for given wedge and M_0.

2. Numerical investigations

The numerical description of the problem consists in solving the 2D unsteady Euler equations for describing the supersonic steady flow ($\gamma = 1.4$) between a wedge and a symmetry plane. The LCPFCT code is applied on a cartesian grid, where the ratio between the space steps is equal to the tangent of the wedge angle. The y step is constant and equal to $0.5 \cdot 10^{-2}$ cm in all runs. The computational plane is initialized with the constant upstream conditions and the steady state is supposedly reached when the L2 norm decreases by five orders of magnitude (Vuillon et al. 1995).

2.1. Code validation

For this code validation, the numerical investigations were performed with the values of the nominal freestream Mach number and geometrical data of the wedge used by Hornung and Robinson. In these conditions, the experimental results show MR configurations of the steady flow. Figure 2 represents the evolution of the Mach-stem height versus the incident Mach-number angle for the value of $M_0 = 4.96$. This figure contains Hornung and Robinson's experimental data, Azevedo and Liu's (1993) analytical results, and our computed results. One can notice the good agreement of our results with experimental ones and

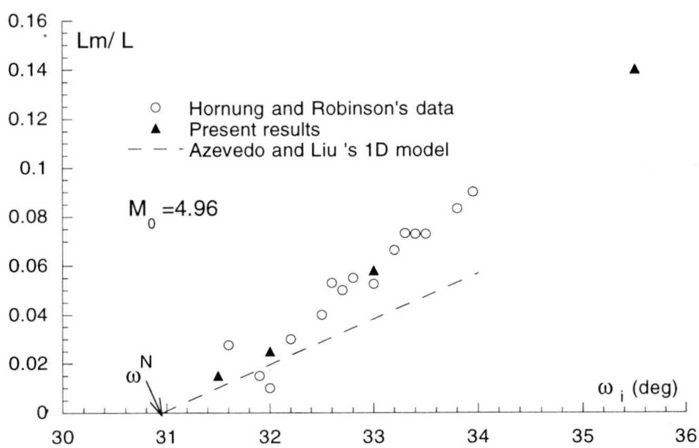

Figure 2. Nondimensional Mach-stem height versus incident shock-wave angle.

the under-prediction given by the 1D analytical model proposed by Azevedo and Liu, certainly due to the strong assumptions listed by those authors. The same agreement has been obtained with lower upstream Mach numbers but are not reported here (see Vuillon et al. 1995).

2.2. Hysteresis effect

In the dual-solution domain (${\omega_i}^N < \omega_i < {\omega_i}^D$), the unsteady computation from an initial constant plane ($M_0 = 4.96$ and $h/w = 0.34$) always leads to a regular reflection. In order to go to an existing stable MR configuration, an over-pressure disturbance is suddenly introduced behind the reflected shock wave (Vuillon et al. 1995). If this is not done, the steady state is always a RR configuration. From this remark, and from an incident Mach-number angle lower than the von Neumann criteria $\omega_i^N = 30.9°$, the wedge angle is increased by 0.5° steps, by a rotation around the wedge trailing edge. Without adding downstream disturbances, stable RR configurations are found until ω_i equals 39°. For $\omega_i = 39.5°$ the RR turns into a stable MR. It is noticed that the detachment criteria is ${\omega_i}^D = 39.3°$. For the value $\omega_i = 40°$, the MR is also obtained, but the reflected shock wave interacts with the wedge trailing edge and the triple point moves upstream until a detached shock wave forms ahead of the wedge (unstarted nozzle flow). Now, from a computed MR configuration as the initial computational plane, it is possible to do the drawback procedure (decreasing the angle from $\omega_i = 39.5°$ by 0.5° steps). Stable MR configurations are obtained until $\omega_i = 31.5°$, with decreasing values of Mach-stem height. The nondimensional Mach-stem height L_m/w is 0.015 for $\omega_i = 31.5°$. For a lower value of ω_i, the Mach stem can not be captured, due to the mesh distribution used, and the steady state is a RR configuration. This behavior is named the hysteresis effect on the RR $\leftrightarrow$ MR transition and is illustrated in Fig. 3, where the Mach-stem height is plotted against the incident shock-wave angle. This effect was also found experimentally by Chpoun et al. (1995).

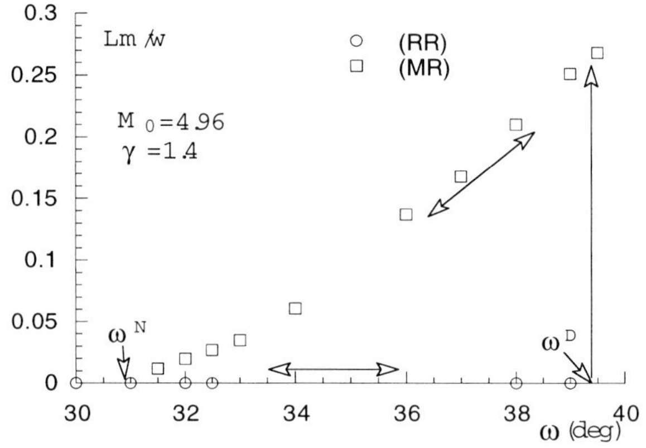

Figure 3. Hysteresis effect in the RR $\leftrightarrow$ MR transition.

2.3. RR $\leftrightarrow$ MR transition function of the h distance

For given ω_i and M_0 in the dual-solution domain, the reflected shock-wave configuration strongly depends on the height h between the wedge trailing edge and the symmetry plane (twice this height defines the exit area of the convergent nozzle). Based on theoretical considerations, this distance is bounded by lower and upper limit values (Vuillon et al. 1995). The lower value, h_{min}, corresponds to the case in which the reflected shock wave of the MR configuration grazes the wedge trailing edge. The upper value, h_{max}, is obtained in a stable RR, when the leading characteristic of the trailing-edge expansion interacts with the incident shock wave on the symmetry plane. The transition between RR and MR occurs for a value of h_{tr} such that $h_{min} < h_{tr} < h_{max}$, and it can be found by numerical simulations. In the dual-solution domain, five test cases are chosen and the numerical results are plotted in the (M_0, h_{tr})-plane for different wedge-angle values θ_w in Fig. 4. One can notice, for a fixed value of θ_w, a decrease of h_{tr} with an increase in M_0; and for the same value of M_0 a decrease of h_{tr} with wedge angle.

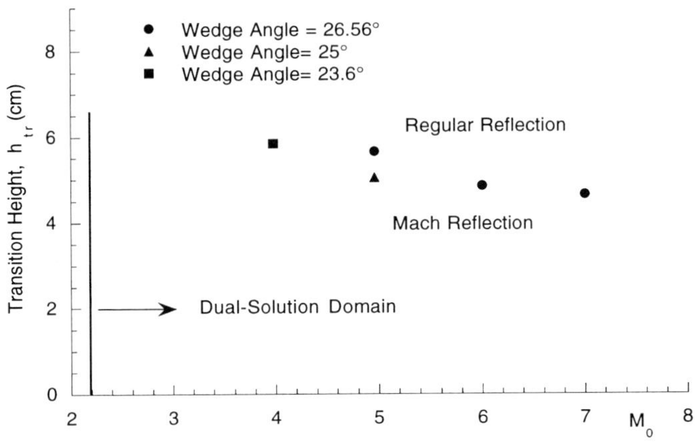

Figure 4. The dependence of h_{tr} versus the incident flow Mach number in the dual-solution domain.

3. Conclusions

The numerical investigations, based on solving the unsteady Euler equations in a cartesian grid, have allowed us to describe the steady shock-wave reflections generated between a wedge and a symmetry plane. The code is validated by the Mach-stem-height data given by Hornung and Robinson. The numerical results have shown firstly, that both regular and Mach reflections are stable in the dual-solution domain; secondly, that the hysteresis effect is produced in the RR $\leftrightarrow$ MR transition; and finally, that the height h between the wedge trailing edge and the symmetry plane influences the RR $\leftrightarrow$ MR transition for given ω_i and M_0. If one takes into account the limiting possibilities of the existing facilities and the strong assumptions of the analytical modeling, the numerical approach is an excellent tool to investigate these shock-wave reflections.

References

Azevedo DJ, Liu SL (1993) Engineering approach to the prediction of shock patterns in bounded high-speed flows. AIAA J 31(1):83–90.

Ben-Dor G (1991) Shock Wave Reflection Phenomena, Springer-Verlag, New York, NY, USA.

Chpoun A, Passerel D, Ben-Dor G (1995) Reconsideration of the state-of-the-art of oblique shock wave reflections in steady flows, part I: experimental investigation. J. Fluid Mech (to appear).

Hornung HG, Robinson ML (1982) Transition from regular to Mach reflection of shock waves, part 2, the steady-flow criterion. J Fluid Mech 123:155–164.

Oran ES, Boris JP (1987) Numerical Simulation of Reactive Flow, Elsevier, New York, USA.

von Neumann J (1963) Collected Works (AH Taub, ed), Vol. 6, Pergamon Press, New York, NY, USA.

Vuillon J, Zeitoun D, Ben-Dor G (1995) Reconsideration of the state-of-the-art of oblique shock wave reflections in steady flows, part II: numerical investigation. J Fluid Mech (to appear).

Interaction of shock waves emitted from a shock tube with a supersonic projectile

Z. Jiang, K. Takayama
Shock Wave Research Center, Institute of Fluid Science, Tohoku University
2-1-1 Katahira, Aoba-ku, Sendai 980-77 Japan

Abstract: This paper reports on shock-wave flows driven by a supersonic projectile released from a shock tube into ambient air. The dispersion control scheme was used for numerical simulations. Two cases were calculated at first and then numerical results were compared with experiments. Three further cases were examined for various Mach numbers of the projectile and different stand-off distances between an initial shock wave and the projectile. It was found that complicated transient phenomena exist in the earlier stages of shock waves, such as interactions between a moving shock wave and a moving body, and induced contact surfaces and moving vortices.

Key words: Supersonic projectile, Moving shock waves, Numerical simulation

1. Introduction

A projectile moving at a supersonic speed in a shock tube drives an initial shock wave in front of it. Once the shock wave is discharged from the open end of the shock tube, shock-wave diffraction and a jet flow develop. Later, when the projectile is released from the shock tube, it interacts with this diffracting shock wave, the jet flow, and also with the initial shock wave. This results in complex moving-shock-wave/moving-body interaction, moving-shock-wave/contact-surface interaction and moving-shock-wave/vortex interaction. Such a transient flow phenomenon shows a significant difference from shock-wave interactions with a body at rest. Therefore, this paper aims at understanding shock-wave/projectile interaction.

A dispersion control scheme based on the dispersion conditions (Jiang 1993, Jiang et al. 1993) was used for solving the Euler equations assuming inviscid and axisymmetric flows. Two cases were calculated first. Then numerical interferograms were constructed from calculated results and compared with experiments to validate the numerical code. Three other cases were carried out with various Mach numbers and different release times of a projectile after an initial shock wave was discharged. Numerical solutions show that interaction among a jet flow, shock-wave diffraction, an initial shock wave, and the bow-shock-wave system around a projectile does produce complicated transient flow phenomena.

2. Governing equations and numerical methods

For axisymmetric problems, a hyperbolic system of conservation laws for perfect gas in cylindrical coordinates can be written as

$$\frac{\partial \mathbf{U}}{\partial t} + \frac{\partial \mathbf{F}}{\partial x} + \frac{\partial \mathbf{G}}{\partial r} + \frac{1}{r}\mathbf{S} = 0, \tag{1}$$

where for the inviscid gasdynamics equations we have

$$\mathbf{U} = \begin{pmatrix} \rho \\ \rho u \\ \rho v \\ e \end{pmatrix}, \quad \mathbf{F} = \begin{pmatrix} \rho u \\ \rho u^2 + p \\ \rho uv \\ (e+p)u \end{pmatrix}, \quad \mathbf{G} = \begin{pmatrix} \rho v \\ \rho uv \\ \rho v^2 + p \\ (e+p)v \end{pmatrix}, \quad \mathbf{S} = \begin{pmatrix} \rho v \\ \rho uv \\ \rho v^2 \\ (e+p)v \end{pmatrix}, \tag{2}$$

where primitive variables in the unknown $\mathbf{U}$ are density ρ, velocity components u and v, and total energy per unit volume e. Fluid pressure is p and the equation of state for perfect gas is given by

$$e = \frac{p}{\gamma - 1} + \frac{1}{2}\rho(u^2 + v^2), \tag{3}$$

where γ, the specific heat ratio, is taken as 1.4. The finite-difference equations of (1) discretized by using the dispersion control scheme are given in the form of half discretion as

$$\left(\frac{\partial \mathbf{U}}{\partial t}\right)_{i,j}^{n} = -\frac{1}{\Delta x}(\mathbf{H}^n_{i+\frac{1}{2},j} - \mathbf{H}^n_{i-\frac{1}{2},j}) - \frac{1}{\Delta r}(\mathbf{P}^n_{i,j+\frac{1}{2}} - \mathbf{P}^n_{i,j-\frac{1}{2}}) - \mathbf{S}^n_{i,j}, \tag{4}$$

where

$$\mathbf{H}^n_{i+\frac{1}{2},j} = \mathbf{F}^+_{i+\frac{1}{2}L,j} + \mathbf{F}^-_{i+\frac{1}{2}R,j}, \qquad \mathbf{P}^n_{i,j+\frac{1}{2}} = \mathbf{G}^+_{i,j+\frac{1}{2}L} + \mathbf{G}^-_{i,j+\frac{1}{2}R}, \tag{5}$$

$$\begin{cases} \mathbf{F}^+_{i+\frac{1}{2}L,j} = \mathbf{F}^+_{i,j} + \frac{1}{2}\text{minmod}(\Delta\mathbf{F}^+_{i-\frac{1}{2},j}, \Delta\mathbf{F}^+_{i+\frac{1}{2},j}) \\ \mathbf{F}^-_{i+\frac{1}{2}R,j} = \mathbf{F}^-_{i+1,j} - \frac{1}{2}\text{minmod}(\Delta\mathbf{F}^-_{i+\frac{1}{2},j}, \Delta\mathbf{F}^-_{i+\frac{3}{2},j}) \end{cases}, \tag{6}$$

$$\begin{cases} \mathbf{G}^+_{i,j+\frac{1}{2}L} = \mathbf{G}^+_{i,j} + \frac{1}{2}\text{minmod}(\Delta\mathbf{G}^+_{i,j-\frac{1}{2}}, \Delta\mathbf{G}^+_{i,j+\frac{1}{2}}) \\ \mathbf{G}^-_{i,j+\frac{1}{2}R} = \mathbf{G}^-_{i,j+1} - \frac{1}{2}\text{minmod}(\Delta\mathbf{G}^-_{i,j+\frac{1}{2}}, \Delta\mathbf{G}^-_{i,j+\frac{3}{2}}) \end{cases}, \tag{7}$$

$$\begin{cases} \Delta\mathbf{F}^\pm_{i+\frac{1}{2},j} = \mathbf{F}^\pm_{i+1,j} - \mathbf{F}^\pm_{i,j} \\ \Delta\mathbf{G}^\pm_{i,j+\frac{1}{2}} = \mathbf{G}^\pm_{i,j+1} - \mathbf{G}^\pm_{i,j} \end{cases}, \qquad \begin{cases} \mathbf{F}^\pm = \mathbf{A}^\pm\mathbf{U} \\ \mathbf{G}^\pm = \mathbf{B}^\pm\mathbf{U} \end{cases}. \tag{8}$$

Steger and Warming flux vector splitting was used in the equations above. Solutions were evolved in time by the second-order Runge-Kutta method. Reflecting boundary conditions were specified both on solid walls and the centerline of a computational domain. Non-reflecting boundary conditions were applied at the inflow and outflow boundaries. Calculations were carried out on an equally spaced grid with 50 grid points along the radius of a shock tube.

3. Numerical results and discussion

3.1. Verification with experiment

The first case is a shock wave discharged in air from a shock tube with $M_s = 1.6$, and the second is a shock wave caught up by a projectile with a speed of $M_p = 2.2$. Numerical interferograms were constructed using techniques described by Babinsky et al. (1995) to compare with the experimental ones shown in Figs. 1 and 2 (Abe 1990) (experimental results on left, numerical on right). It can be seen that computational results agree well with experiments. Hence the accuracy of the present numerical method is well validated.

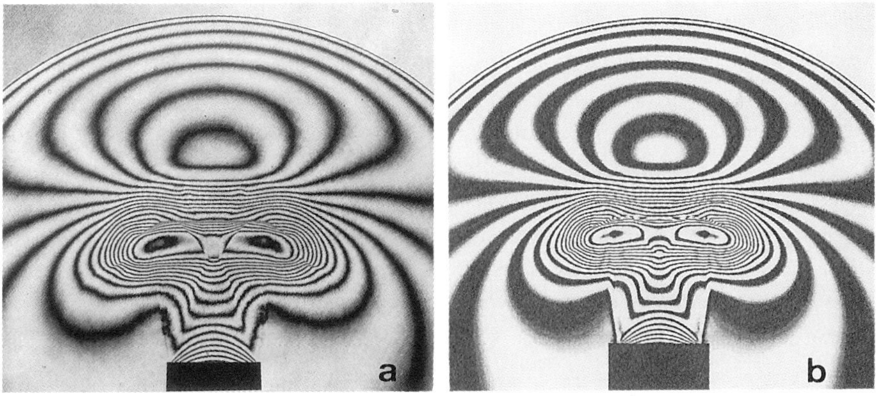

Figure 1. Comparison of numerical and experimental interferograms, $M_s = 1.6$.

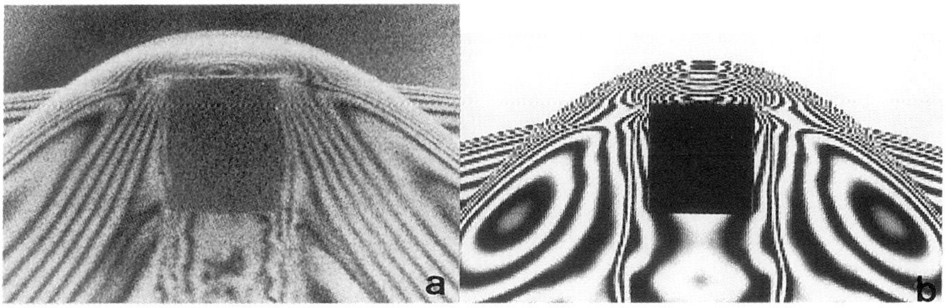

Figure 2. Comparison of numerical and experimental interferograms, $M_p = 2.2$.

3.2. The projectile speed of $M_p = 4.0$

A time sequence of case 3 is shown in Fig. 3 where isobar is plotted in the symmetric lower half and isopycnics in the symmetric upper half (the same display is used in other figures). Figure 3a shows a frame when a projectile is just about to be released and a jet flow develops, in which there are a spherical shock wave with its weak rim sucked into a low-pressure core of a primary vortex ring, and a contact surface between the shock wave and an initial shock wave. Later, as shown in Fig. 3b, the projectile moves through the spherical shock wave and drives out its own bow shock wave. Meanwhile, shock-wave diffraction followed by a contact surface is observed over the rear surface of the projectile due to the body moving through the shock wave that is almost at rest. This diffracting shock wave meets the spherical shock wave in Fig. 3c, distorts the spherical shock wave and changes it into a discontinuous shape consisting of three curved shock waves. The velocity and density discontinuities develop here, and vortices are produced near each discontinuous circle. From the same figure it is found that when the bow shock wave in front of the projectile catches up with the initial shock wave, expansion waves and a contact surface in front of the projectile are produced. Also noticeable is that the contact surface originating from sharp corners of the projectile is going to meet with that of the jet flow. In Fig. 3d, the kink and rolling up of two contact surfaces are observed due to inviscid vortex instability. The similar instability in shear flows appears also near the jet-flow boundary. The contact surface located in front of the projectile is getting thinner and thinner since the fluid flow is leaving the centerline for the side walls of the projectile, and finally vanishes. At this moment, the spherical shock of the jet flow is changed back into a spherical shape again.

3.3. The projectile speed of $M_p = 4.0$ with shorter stand-off distance

The results of the fourth case are shown in Fig. 4 with the same initial conditions as case 3, but a shorter release time of the projectile. From Figs. 3a and 4a, it can be seen that when an initial shock wave arrives at the same position, the projectile in case 4 is going to catch up with its initial shock wave. The contact surface appears in front of the projectile in Fig. 4a where a similar shock-wave diffraction exists over the rear surface of the projectile. Since the contact surface from the projectile meets that of the jet flow earlier, the kink and rolling up of two contact surfaces take on more complicted structures, as shown in Fig. 4b. The flow patterns appear similar to each other in the two cases except for the differences mentioned above.

3.4. The projectile speed of $M_p = 1.25$

Figure 5 shows the shock-wave flow of case 5 with a speed of $M_p = 1.25$. Shock-wave diffraction over the back of a projectile (see Fig. 5a) is found to be similar to the former cases, but the kink of contact surfaces is quite different. Because of the lower Mach number, shock-wave interaction in front of the projectile is invisible in Fig. 5b and the jet flow appears to be the underexpanded one with a repeated expansion-contraction structure.

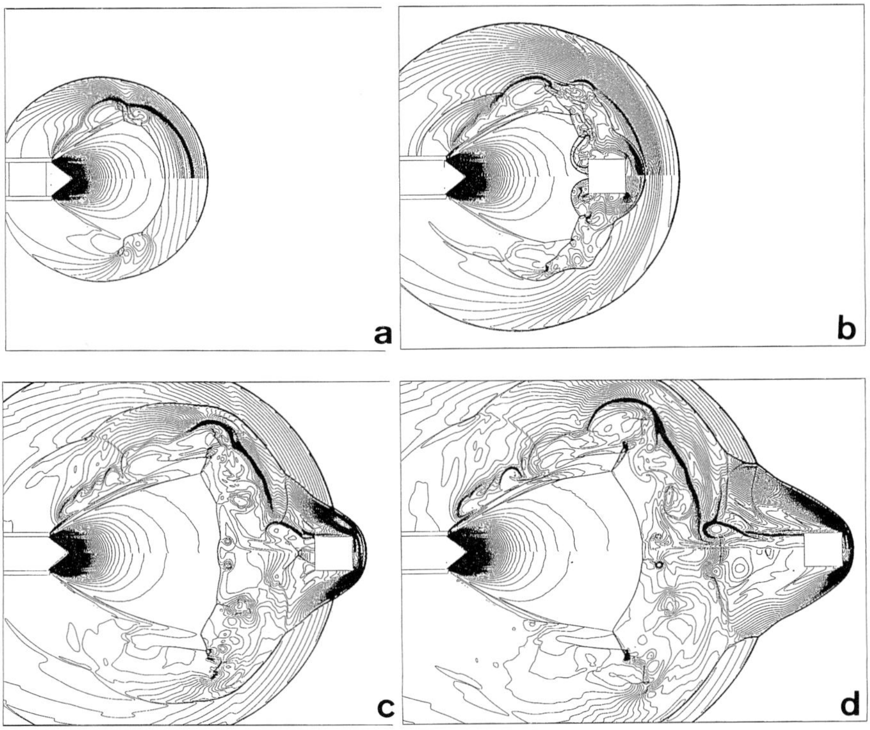

Figure 3. Isobar (lower half) and isopycnics (upper half), $M_p = 4.0$.

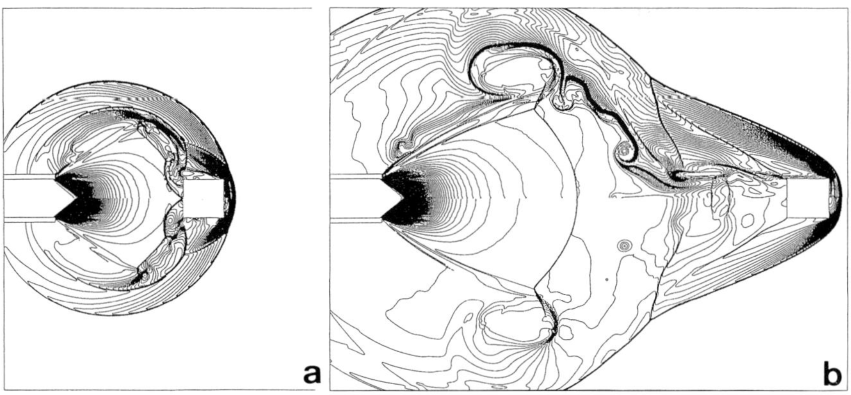

Figure 4. Isobar (lower half) and isopycnics (upper half), $M_p = 4.0$.

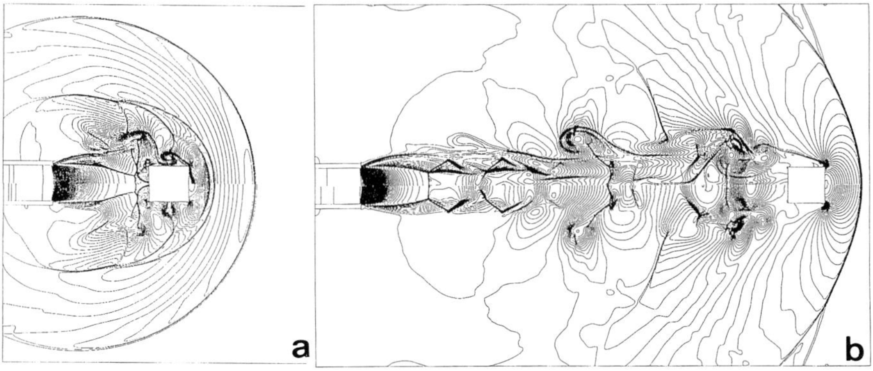

Figure 5. Isobar (lower half) and isopycnics (upper half), $M_p = 1.25$.

4. Conclusions and remarks

Flow patterns change depending mainly on the speed of a projectile because of jet-flow structures varying with the Mach number of the projectile. The stand-off distance between an initial shock wave and a projectile appears to have less effect on flow patterns except for a region behind the projectile. Once they meet, the kink of a contact surface of the jet flow and one produced by a projectile can be observed due to inviscid vortex instability. When the shock-wave system of a projectile catches up with the initial shock wave, the interaction of those two shock waves produces a contact surface and expansion waves behind a new spherical shock wave. In the case of the projectile at a lower speed, this interaction is weaker and invisible.

References

Abe A (1990) Diffraction of Shock Waves Emitted From the Open End of a Shock Tube, Doctoral Thesis, Tohoku University, Sendai, Japan.

Babinsky H, Meguro T, Jiang Z, Takayama K (1995) Numerical visualization of shock wave flow in an expanding tube and comparison with experiment. In: Proc of 4th Symp on Exp and Numerical Flow Visualization.

Jiang ZL (1993) Study on the Finite Difference Theory and Numerical Methods of Week Solution Problems, Doctoral Thesis, Peking University, Beijing, China.

Jiang ZL, Chen YS, Ishii K, Kuwahara K (1993) Numerical dispersion, numerical oscillations, and dispersion control schemes for weak solution probelms. In: Proc of 5th Int Symp on CFD, Sendai, Japan, 1:397.

Numerical investigation of shock-wave reflection problems in steady flows

M. S. Ivanov, S. F. Gimelshein, G. N. Markelov
Institute of Theoretical and Applied Mechanics, Novosibirsk 630090 Russia
A. E. Beylich
Stoßwellenlabor, RWTH Aachen, Aachen 52056 Germany

Abstract: The transition from regular to Mach reflection and back of planar shock waves generated by straight wedges in steady flows was studied numerically with the DSMC method at small Knudsen numbers. The hysteresis effect was observed at increasing and decreasing shock-wave angle. Transformation from one configuration to the other in response to perturbations was shown.

Key words: Mach reflection, Transition, Hysteresis, Stability

1. Introduction

In steady flow, the transition between regular reflection (RR) and Mach reflection (MR) of a wedge-generated shock wave from a symmetry plane can theoretically occur anywhere in the range $\alpha_N < \alpha < \alpha_D$, where α is the angle between the shock and the symmetry plane, α_N is the von Neumann angle, and α_D is the detachment angle. RR is not possible for $\alpha > \alpha_D$ and MR is impossible for $\alpha < \alpha_N$. In the region $\alpha_N < \alpha < \alpha_D$ (dual-solution region), both reflection types are theoretically possible. The dependence of transition angle on the direction of α variation was actively studied from the mid-1970s to the mid-1980s, mostly by experimental methods. The results of such experimental investigations were summarized by Hornung and Robinson (1982) as follows: transition from RR to MR and back in steady flows takes place according to the von Neumann criterion; in the dual-solution domain RR is unstable and a hysteresis effect was not observed. Recent experimental results reported by Ben-Dor et al. (1994), and a numerical study by Ivanov et al. (1995) did, however, exhibit hysteresis. Here we continue the study of Ivanov et al. by considering the influences of flow rarefaction, free-stream, and local perturbations.

2. Method of solution

The Direct Simulation Monte Carlo (DSMC) method for solving the Boltzmann equation is used in the version based on the majorant collision frequency principle (Ivanov and Rogasinsky 1991) in the near-continuum Knudsen-number range, $0.001 \leq \mathrm{Kn} \leq 0.01$. The rectangular computational domain covered one side of the symmetry plane, which was modeled as a specular reflector. Uniform free-stream conditions were imposed on the left and upper boundaries of the computational domain, and specular reflection on the wedge surface.

3. Flow conditions

All results presented here (except for those shown in Fig. 4b) were obtained for the conditions of the argon flow of Hornung et al. (1979), i.e., $\gamma = 5/3$, $M = 16$, $g/w = 0.41$, where g is the distance between the trailing edge of the wedge and the symmetry plane and w is the wedge length. The data in Fig. 4b were obtained for one of the conditions of Hornung and Robinson (1982), i.e., $M = 4.96$, $T_0 = 365$ K, $g/w = 0.34$, $\gamma = 1.4$, nitrogen.

4. Computational results and discussion

4.1. Fixed wedge

Here, the wedge angle was initially fixed such that it generates a shock with α within the range $\alpha_N = 26.16$ and $\alpha_D = 35.37$. A uniform undisturbed flow was assumed in the entire computational domain as the initial condition. Calculations showed that the incident shock flow agreed well with theory. For all α, RR was observed up to $\alpha = \alpha_D$. Figure 1 shows an example of the shock-wave configuration (pressure flow field and streamlines) for the incidence angle $\alpha = 35°$ slightly below α_D. Parameters behind the incident and reflected shock waves are in good agreement with the two-wave solution. When $\alpha > \alpha_D$, MR is observed. An example is shown in Fig. 2 for $\alpha = 36°$ exceeding α_D. The pressure downstream of the Mach stem coincides with the theoretical value. Behind the reflected wave and near the triple point the pressure is up to 10% higher than the pressure downstream of the Mach stem. Thus, RR takes place for all α inside the dual-solution region and MR occurs for $\alpha > \alpha_D$. Similar results were also obtained for nitrogen ($\gamma = 1.4$).

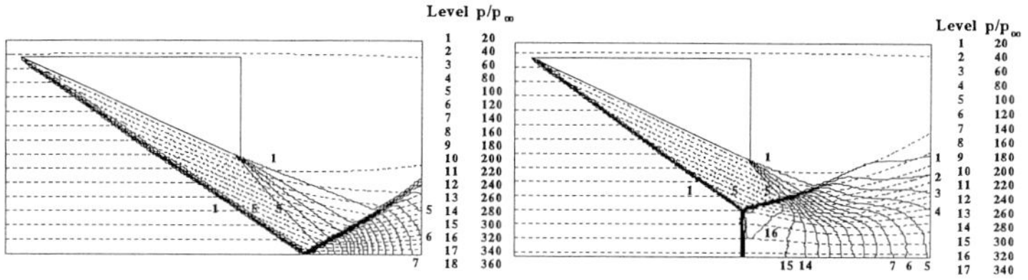

Figures 1 (left) and 2 (right). Pressure field (continuous lines) and streamlines (dashed lines) for uniform flow as initial condition, argon, $M = 16$, Kn $= 0.0025$. Left, $\alpha = 35°$; right, $\alpha = 36°$.

4.2. Moving wedge and the hysteresis effect

Here, initial conditions were chosen to correspond to a calculated MR with $\alpha = \alpha_1 > \alpha_D$. This solution was then used as the initial condition for a calculation with $\alpha = \alpha_2 < \alpha_1$, etc. The angle α was changed by rotating the wedge around its trailing edge. This showed that MR is maintained to values of α well below α_D (see Fig. 3 with $\alpha = 35°$). MR is maintained down to $\alpha = \alpha^* > \alpha_N$. The fact that $\alpha^* > \alpha_N$ may be explained as follows: when $\alpha \to \alpha_N$, the Mach-stem height decreases and becomes comparable with the shock-wave thickness, so MR and RR may no longer be distinguished. Accordingly, α^* decreases with decreasing Kn (see Fig. 4a). DSMC calculations for different flow conditions ($M = 4.96$, $\gamma = 1.4$) also confirmed that the final configuration in the dual-solution domain depends on initial conditions (see Fig. 4b). Thus, numerical solutions exhibit the presence of a hysteresis effect, i.e., of reflection-type dependence on the direction of the incidence angle variation, at α changing through the range $\alpha^* < \alpha < \alpha_D$, both from below and from above (see Fig. 4). The transition from RR to MR takes place at $\alpha = \alpha_D$ when α increases. MR is retained down to $\alpha^* > \alpha_N$ with decreasing α and initial $\alpha > \alpha_D$.

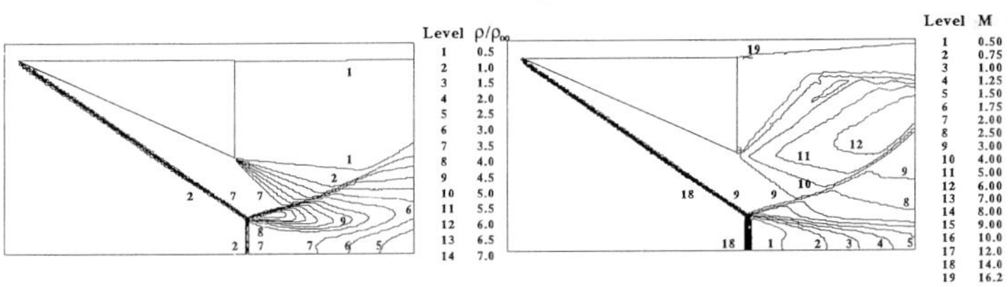

Figure 3. (a) Density flow field and (b) Mach number flow field for the case of Fig. 1.

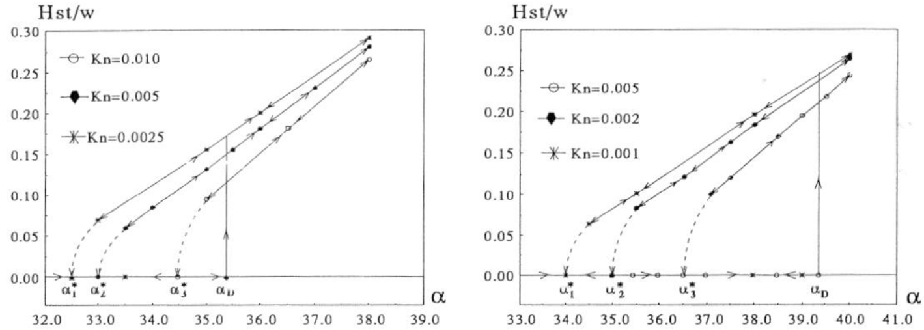

Figure 4. Normalized Mach-stem height vs α: (a) Argon, $M = 16$; (b) Nitrogen, $M = 4.96$.

4.3. MR and RR perturbations

Next, the possibility of MR $\leftrightarrow$ RR transitions was studied in the dual-solution domain and under the effect of external disturbances. MR or RR was prescribed as the initial condition for fixed α and g/w. Then a disturbance was applied during some finite period t_p and the subsequent flow stabilization was observed.

Flow perturbations near the reflection point. RR was prescribed as the initial condition, and a disturbance was introduced in a rectangular region near the reflection point at time $t = 0$, instantaneously stopping all particles present in this region. At $t > 0$, the constraint was removed, and the flow relaxed to a steady state (MR or RR). Figure 5 shows sketches of several perturbation regions for computations with $\alpha = 33.5°$ and Kn $= 0.005$. "RR" and "MR" indicate the final steady configuration. Figures 5a and b show RR is retained when the disturbance region is of the order of the Mach-stem height, H_{st}, for the MR configuration corresponding to a given α ($H_{st}(33.5°) \approx 0.06w$, see Fig. 4a). When the vertical size of this region is increased up to $2H_{st}$, RR holds. However, a larger horizontal size (up to $2H_{st}$) causes MR formation (Fig. 5d). Figure 6 shows some typical perturbed regions for $\alpha = 35°$; RR is retained when the disturbance region is of the order $0.25H_{st}$ ($H_{st}(35°) \approx 0.132w$, see Fig. 4a). However, when the horizontal size is increased up to $0.5H_{st}$, MR appears (Fig. 6b). If this region is located upstream of the incident shock wave, the perturbations exert a weaker effect and RR holds (Fig. 6c). When the vertical size of this region is increased by a factor of 2 (as in Fig. 5a), MR is formed. Thus, increasing α brings about lower RR stability. RR stability is also appreciably reduced with decreasing Knudsen number. For Kn $= 0.0025$, RR transforms into MR already for the perturbed region size of about $0.025H_{st}$ (Fig. 6d). The study of MR stability to velocity perturbations showed that this configuration is much more stable than RR. For example, doubling the particle velocities within the regions of order $2H_{st}$ did not destroy MR.

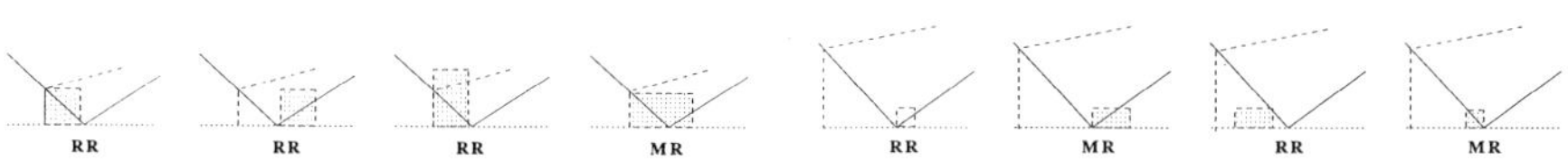

Figures 5 (left) and 6 (right). Shapes and positions of disturbed regions (argon). Solid line – RR configuration, dashed line – MR. Hatched area is disturbed region. On left: $\alpha = 33.5°$, $M = 16$, Kn $= 0.005$. On right: $\alpha = 35°$.

Free-stream perturbations. The effects of free-stream velocity and density perturbations were studied with MR as initial configuration and $\alpha = 33.5°$. For velocity perturbations, the average velocity of the particles entering the computational domain was changed to $U = xU_\infty$ for a period $t = nt_M$, where $t_M = H_{st}/U_\infty$. As seen in Fig. 7a, sufficiently large n and small x cause MR failure and RR formation, while velocity increases cause MR to be maintained. Likewise, density perturbations of the entering particles ($\rho = x\rho_\infty$) were investigated. Figure 7b shows an appreciable density reduction ($x = 0.5$), as in the case of $u = 0.5u_\infty$, which destroys the MR configuration. Computations with RR as initial configuration at $\alpha = 33.5°$ showed this to be more stable. Velocity perturbations with $x \geq 2$ for $n \geq 1$ were necessary to transform it to MR. Figure 8 illustrates the temporal transformation of RR into MR affected by velocity perturbations. Figures 8a–8c show the unsteady flow pattern for fixed time (the fluctuations in the flow field are related to a strong statistical scattering).

The final steady solution is presented in Fig. 8d. The disturbed region consists of a compression subregion (particles with the velocity $u = 2u_\infty$ overtake the particles with $u = u_\infty$) followed by an expansion subregion (Fig. 8a). Interference of these two subregions with the incident shock wave forms the Mach stem (Fig. 8c). An example of MR→RR transition is shown in Fig. 9. Since $x < 1$, the downstream perturbed region contains the expansion subregion first, with the compression subregion to follow. The former subregion smears the MR, and the latter one forms an RR. Computations for large α showed that at $\alpha \to \alpha_D$ MR stability increases whereas that of RR is reduced.

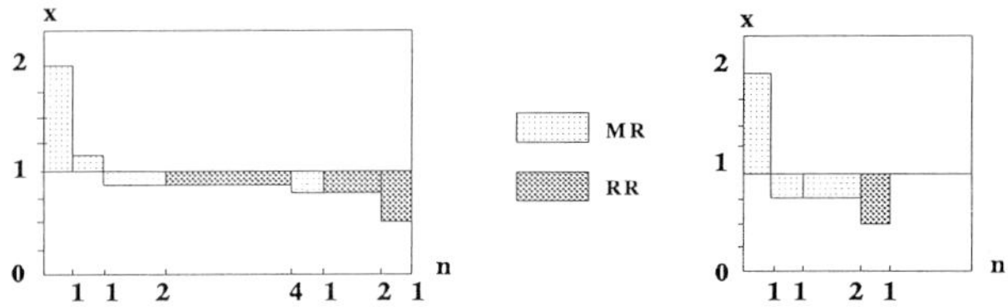

Figure 7. Effect of (a) magnitude and duration of free-stream velocity, and (b) density perturbations on configuration (MR or RR).

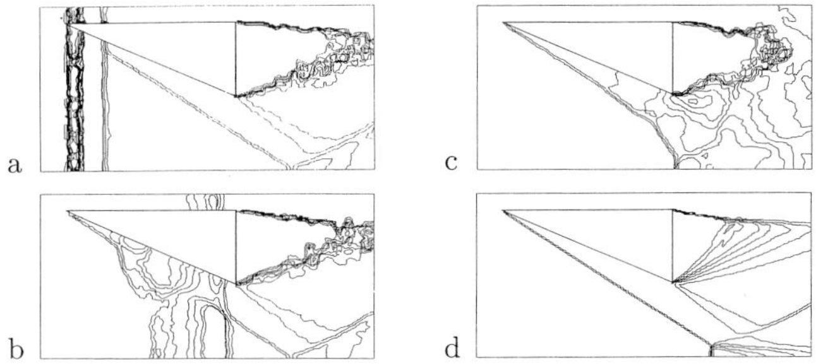

Figure 8. (a) – (d) Free-stream velocity perturbations of RR at t/t_M = 2, 8, 18, and final steady solution.

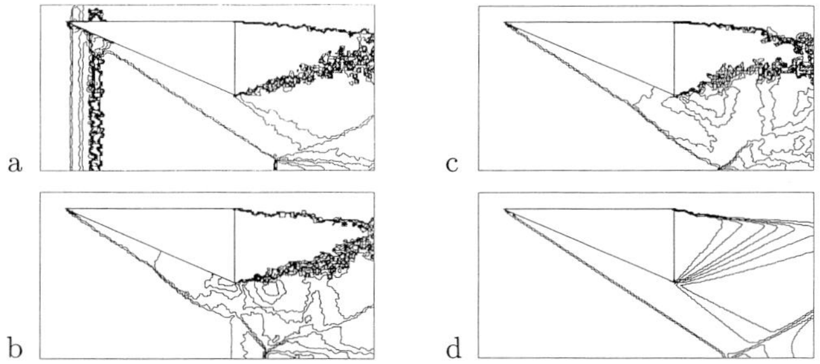

Figure 9. (a) – (d) Free-stream velocity perturbations of MR at t/t_M = 4, 20, 28, and final steady solution.

5. Conclusions

The calculations show that: (1) the reflection type depends on the direction of the incidence angle variation, coming from below or from above $\alpha^\star < \alpha < \alpha_D$, i.e., the hysteresis effect exists; (2) the difference $\alpha^\star - \alpha_N$ is reduced when the Knudsen number decreases, and the dependence of H_{st} on α is close to linear within $\alpha^\star < \alpha < \alpha_D$ for various Knudsen numbers; (3) RR is less stable to the flow disturbances near the reflection point than MR. RR stability to this kind of disturbance is appeciably reduced with the flow rarefaction decreasing. The last fact indicates that experiments on the hysteresis effect should be analyzed very carefully, since even small disturbances can transform RR into MR; and (4) free-stream disturbances can transform RR into MR, and vice versa. MR stability increases with α growing ($\alpha^\star < \alpha < \alpha_D$), whereas that of RR is reduced.

Acknowledgement. The authors wish to acknowledge support from the Deutsche Forschungsgemeinschaft and the International Science Foundation (Grant RBW000).

References

Ben-Dor G, Li H, Chpoun A, Passerel D, Vuillon J, Zeitoun D (1994) Analytical, numerical and experimental investigation of the regular ↔ Mach reflection transition in steady flows. In: 11th Int Mach Reflection Symposium, Univ of Victoria, BC, Canada.

Hornung HG, Oertel H, Sandeman RJ (1979) Transition to Mach reflection of shock waves in steady and pseudosteady flow with and without relaxation. J Fluid Mech 90:541–560.

Hornung HG, Robinson ML (1982) Transition from regular to Mach reflection of shock waves. Part 2. The steady-flow criterion. J Fluid Mech 123:155–164.

Ivanov MS, Gimelshein SF, Beylich AE (1995) Hysteresis effect in stationary reflection of shock waves. Physics of Fluids 7(4):685–687.

Ivanov MS, Rogasinsky SV (1991) Theoretical analysis of traditional and modern schemes of the DSMC method. In: 17th Int Symp on Rarefied Gas Dynamics, Aachen, Germany, pp 629–642.

The interaction of a shock wave discharged from the open end of a shock tube with a flat plate

T. V. Bazhenova, V. V. Golub, A. M. Shulmeister,
T. A. Bormotova, S. B. Bazarov, S. V. Pichugin
High Energy Density Research Center
IVTAN, Russian Academy of Sciences
Izhorskaya, 13/19, Moscow, 127412 Russia

Abstract: This paper is devoted to the investigation of the impingement of a diffracted shock wave on a flat plate and comparison with the numerical simulations.
Key words: Shock-wave diffraction, Vortex ring, Shock-wave impingement on a plate

1. Introduction

During the past four decades, shock waves discharged from the open ends of shock tubes have been studied. The complicated flow field behind a shock wave emitted from the open end of a shock tube was clarified in some details experimentally and numerically (Baird 1987, Abe & Takayama 1989). The purpose of the present paper is to study the impingement of a shock wave discharged from the open end of a shock tube upon a flate plate and the unsteady flow field behind the shock wave. The features of shock-wave reflection after the discharge of a shock of $M_0 = 1.46$ which impinged on the back wall of a square box were reported by Phan & Stollery (1985).

2. Experimental set-up and numerical simulation

A flate plate was mounted opposite to the open end of a shock tube with diameter $d = 20$ mm. The distance L between the open end of the shock tube and the plate was varied from 50 mm to 100 mm. The Mach number of the incident shock wave was varied from 2.0 to 4.7. The low-pressure chamber and vacuum tank were filled with nitrogen to pressures between $P_1 = 4$ and 50 kPa. The flow fields were observed by taking schlieren photographs and interferograms. Computer simulation was based on Euler's equations, which were solved by Godunov's method.

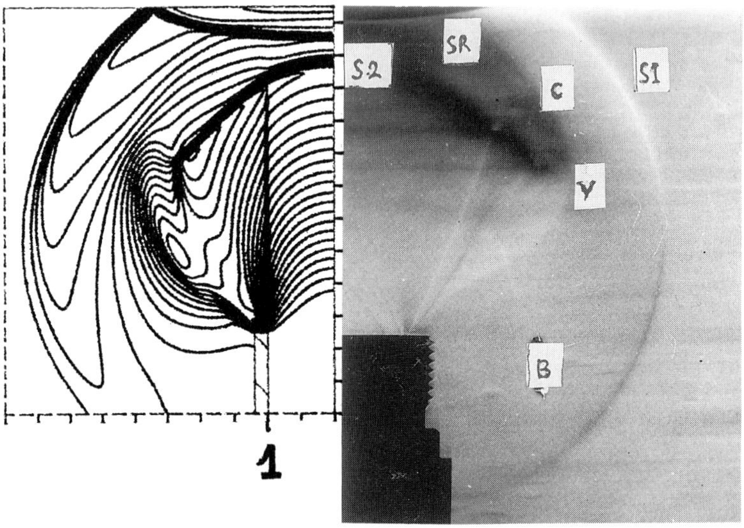

Figure 1. The calculated density contours (left) and schlieren photograph (right) of diffracted shock wave impingement on a flat plate; $M_0 = 3.2$, $P_1 = 4$ kPa, $t = 72$ μs.

3. Results and discussion

The obtained series of shadowgraphs and interferograms have clearly captured the main stages of the process:

- For diffraction into free space of the shock wave discharged from the open end of a shock tube: in the unsteady flow field behind the emitted shock wave there are two kinds of complicated flows depending on the shock-wave Mach number. At low shock-wave Mach numbers, the post-shock flow is subsonic and is dominated by the contact surface and vortex ring with an embedded rearward-facing shock. When the post-shock flow is supersonic, the flow pattern is similar to the barrel structure in the under-expanded stationary jet (Bazarov et al. 1993).

- Figure 1 shows the schlieren photograph and calculated density contours of the impingement of the diffracted shock wave on a flat plate. After the reflection of the spherical shock wave $S1$ from a plate (Fig. 1, right), the reflected shock wave SR arises. The Mach lines of the disturbed flow, the rarefaction fan, and the separation line can be recognized near the open end of the shock tube. The vortex ring V is formed near the head part of the contact surface C of the outflowing gas. The rearward-facing shock $S2$ appears in the outflowing gas. Near the tube exit there occurs a sort of shock B across the boundary of the discharging gas. This shock is seen both in flow discharging in free space and in flow interacting with an obstacle. This shock wave was observed in a previous investigation (e.g. Glass 1974), but the authors don't know where it was discussed. The distance of this shock from the tube exit slightly increases with time.

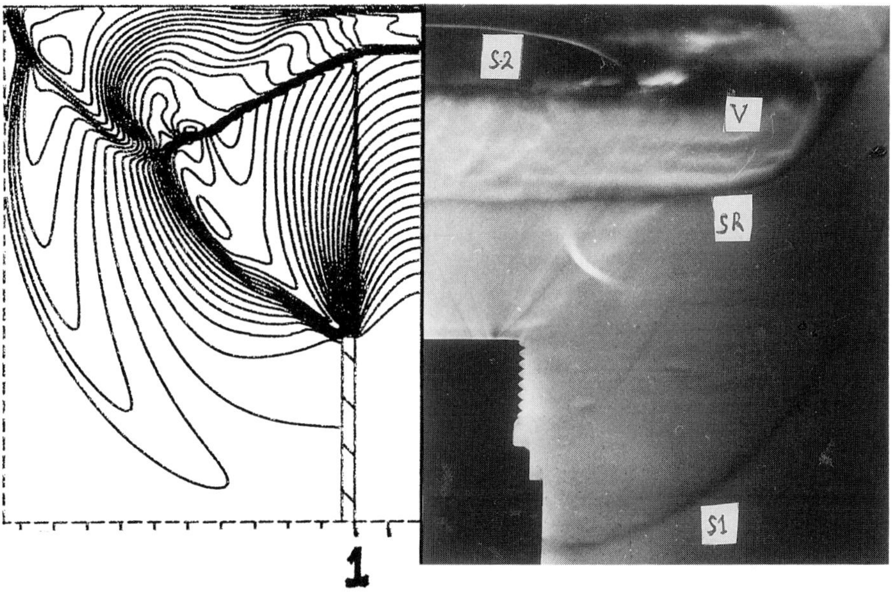

Figure 2. The calculated density contours (left) and schlieren photograph (right) of diffracted shock wave impingement on a flat plate; $M_0 = 4.7$, $P_1 = 4$ kPa, $t = 82$ μs.

As the reflected shock wave SR moves in the upstream direction, it consequently interacts with the contact surface C, secondary shock wave $S2$, and vortex ring V (Figs. 1 and 2). The comparison of calculated density contours and schlieren photographs shows that there is satisfactory agreement between them. Nevertheless, in numerical calculations the bow shock wave B is absent. The schlieren photograph in Fig. 2 shows the reflected shock wave SR in the central part of the outflowing gas, but in the calculated density contours (Fig. 2, left) the SR is seen only in the side part of the outflowing gas. Due to the pressure jump after the shock wave $S1$ reflection near the plate, the velocity of the contact surface moving toward the flate plate is decreased.

The numerical calculation of pressure $\bar{P} = P/P_1$ (curve 1) and density distributions $\bar{\rho} = \rho/\rho_1$ (curve 2) across the flate plate are shown for $t = 72$ μs (Fig. 3a). The pressure and density gain their maxima in the center and are decreased slowly up to the position of the shock wave $S1$ on the plate.

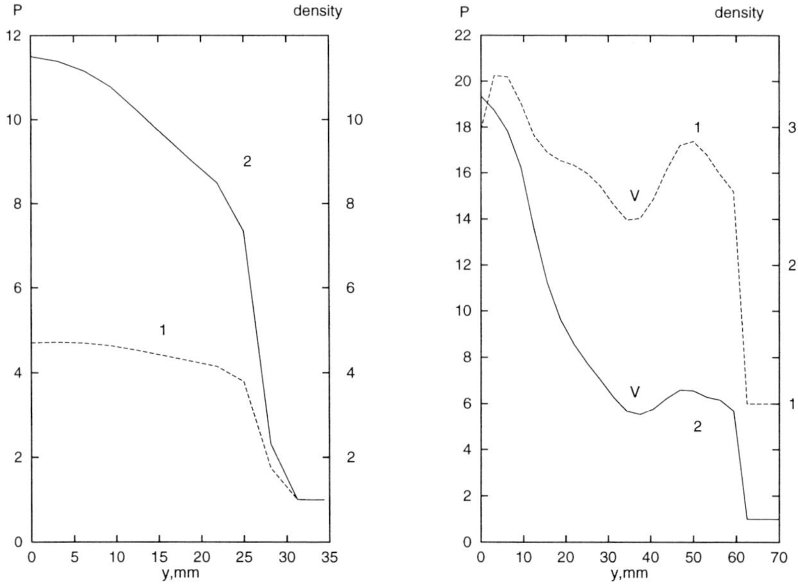

(a) (b)

Figure 3. The calculated pressure and density distribution along the plate after impingement of a diffracted shock wave, $P_1 = 4$ kPa, $L = 50$ mm: (a) $M_0 = 3.2$, $t = 72$ μs; (b $M_0 = 4.7$, $t = 82$ μs. Pressure (1) and density (2) are normalized on their initial values.

The shock wave $S2$ after the interaction with the reflected shock wave SR comes close to the plate and occupies a quasi-stationary position. Thus a flow pattern is formed analogous to that when a supersonic under-expanded flow impinges on a plate.

The schlieren photograph in Fig. 2 (right) shows that after the interaction process between SR and the vortex ring V, the shock wave SR passes through it and SR is seen on the back part of the vortex ring V. The vortex ring V moves toward the plate and its size is increased as it comes closer to the plate. The pressure (curve 1) and density (curve 2) distributions show that in the places where the vortex ring is in contact with the plate (Fig. 3b) there are the local minima in pressure and density.

The main stages of the shock-wave/flat-plate interaction are clearly seen on the pressure $\bar{P}$ (curve 1) and temperature $\bar{T} = T/T_1$ (curve 2) histories in the central point of the plate (Figs. 4a and 4b). The initial impingement of the shock wave $S1$ leads to peaks in pressure and temperature. After that there is subsequent decay of pressure and temperature, but when the outflowing gas comes to the plate there occurs a secondary rise of pressure and temperature. Then the quasi-stationary parameters are formed in the central part of the flat plate. When the primary-shock Mach number is varied from 3.2 to 4.7, the qualitative pattern of the interaction process is the same in both cases, but the quantitative influence of the initial Mach number leads to greater pressure and temperature jumps in the initial stage and greater quasi-stationary parameters on the plate.

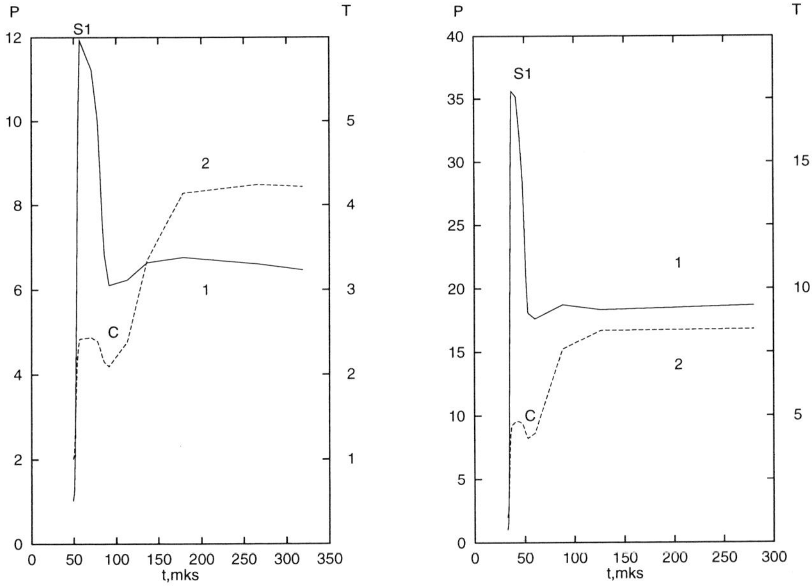

(a) (b)

Figure 4. The calculated pressure and temperature time histories after the impingement of a diffracted shock wave on the center point of the plate for $L = 50$ mm: (a) $M_0 = 3.2$; (b) $M_0 = 4.7$. Pressure (1) and temperature (2) are normalized on their initial values.

The quasi-stationary temperature at $M_0 = 4.7$ is near to the stagnation temperature of the flow behind the incident shock wave; at $M_0 = 3.2$ it is less than that.

4. Conclusions

The complex two-dimensional flow with shock waves and vortex ring which arises after the impingement of a shock wave on a plate was studied by experimental and numerical methods. During this interaction, conditions for different kinds of instability are realized. These instabilities and the vortex produced by shock-wave diffraction play a significant role on the resulting surface-pressure and temperature distributions. The following results were obtained:

- When a diffracted shock wave impinges on a plate, regular reflection accounts for the pressure and temperature distributions in the central part of the plate, but on the side part of the plate Mach reflection accounts for such distributions.
- Pressure and density distributions on the plate after shock impingement on the contact surface are determined by the vortex ring.

Acknowledgement. The present research was supported by Grant No. 94-01-00218-a and by Grant No. 94-01-01461-a of the Russian Fundamental Research Foundation.

References

Abe A, Takayama K (1989) A peculiar shape in the transition process of a shock wave discharged from an open end of a shock tube. In: Proc 1989 National Symp on Shock Wave Phenomena, Japan, pp 41–54.

Baird JP (1987) Supersonic vortex rings. Proc R Soc Lond A409:59–65.

Bazarov SB, Bazhenova TV, Golub VV, Shulmeister AM (1993) Three-dimensional shock ejection from a channel. In: Proc 19th Int Symp on Shock Waves, Vol. 4, Sendai, Japan, pp 135–138.

Glass II (1974) Shock Waves and Man, University of Toronto Institute for Aerospace Studies, Toronto, p 169.

Phan KC, Stollery YL (1985) On the effects of shock wave reflection in a confined space. In: Proc 15th Int Symp on Shock Tubes and Waves, pp 135–145.

Analyses of reflection and diffraction of weak shock waves

Y. Takano, N. Iwamoto, T. Goto
Tottori University
Tottori, 680 Japan

Abstract: In the present paper, analyses are made of two-dimensional flow fields generated by reflection and diffraction of weak shock waves on a wedge at an angle of attack. Similarity solutions are obtained for velocity fields in addition to those for density (pressure) fields which had already been obtained by Keller and Blank (1951). Combining these similarity solutions, analytical solutions are constructed that describe reflection processes of weak shock waves on a triple reflector.

Key words: Analysis, Weak shock wave, Reflection, Diffraction

1. Introduction

Gasdynamic phenomena induced by weak shock waves may be modeled using the linearized gasdynamic equations or acoustic equations. Several mathematical theories have been developed for acoustic shock waves based upon the wave equation, which are reviewed by Friedlander (1958). Keller and Blank (1951) presented analytical similarity solutions for pressure fields induced by reflection and diffraction of plane acoustic shock waves from a wedge or a corner. Takano and Karakawa (1993) compared numerical solutions of the two-dimensional Euler equations for pressure fields induced by incident shock waves with Mach number 1.05 with the analytical solutions of Keller and Blank (1951) and obtained fairly good agreement.

In the present paper, we perform two kinds of analyses that extend the analyses of the wave equation of Keller and Blank (1951). First we analyze the acoustic equations and obtain similarity solutions for velocity fields in addition to those for density (pressure) fields induced by the shock waves incident on a wedge at an angle of attack. Secondly we construct analytical solutions for the density fields that describe reflection processes of weak shock waves on a triple reflector by superposing the similarity solutions of the wave equation.

2. Analyses of the flow field around a wedge

The basic equations for the analysis of flow fields generated by reflection of weak incident shock waves around a wedge are acoustic (linearized gasdynamic)

equations. Introducing the new variable $R = r/t$ to the acoustic equations in polar coordinates (r, θ) with the origin at the apex of the wedge, where t is the time elapsed after arrival of the incident shock waves at the apex, we have

$$R\frac{\partial \rho}{\partial R} = \frac{\partial v_r}{\partial R} + \frac{1}{R}\frac{\partial v_\theta}{\partial \theta} + \frac{v_r}{R}, \tag{1}$$

$$R\frac{\partial v_r}{\partial R} = \frac{\partial \rho}{\partial R}, \qquad R\frac{\partial v_\theta}{\partial R} = \frac{1}{R}\frac{\partial \rho}{\partial \theta}. \tag{2}$$

Here ρ is the density normalized by that in the region of undisturbed stationary gas, and v_r and v_θ are the radial and angular components of the velocity, respectively, which are normalized by the acoustic speed of the stationary gas.

Eliminating the velocity components from Eqs. (1) and (2), we obtain

$$\frac{\partial}{\partial R}\left\{(1 - R^2)\frac{\partial \rho}{\partial R}\right\} + \frac{1}{R}\frac{\partial \rho}{\partial R} + \frac{1}{R^2}\frac{\partial^2 \rho}{\partial \theta^2} = 0, \tag{3}$$

which had been derived from the wave equation in two-dimensional space by Keller and Blank (1951). They had presented analytical solutions for the pressure (density) field which are defined within a circular sector of diffraction around the apex. For example, an analytical solution of the variation of density $\rho(R, \theta)$ due to reflection and diffraction around a wedge at an angle of attack, as is illustrated in Fig. 1a, can be written

$$\rho(R, \theta) = f(q, \omega; \alpha_1) + f(q, \omega; \alpha_2), \tag{4}$$

where

$$f(q, \omega; \alpha) = \frac{1}{\pi}\tan^{-1}\frac{(1 - q^2)\sin\alpha}{(1 + q^2)\cos\alpha - 2q\cos\omega}, \tag{5}$$

$$q = \left[\frac{R}{1 + (1 - R^2)^{1/2}}\right]^\lambda, \qquad \omega = \lambda\theta. \tag{6}$$

Here $\alpha_1 = \lambda\psi_1, \alpha_2 = \lambda(\phi_1 - \psi_2)$, $\lambda = \pi/\phi_1$, $\phi_1 = 2\pi - \phi_0$ and $\phi_0 = \psi_1 - \psi_2$, and the angles ψ_1 and ψ_2 $(0 < \psi_2 \leq \psi_1 < \pi/2)$ are indicated in Fig. 1a.

Figure 1b shows a density-contour diagram of $\rho(R, \theta)$ in the case with $\psi_1 = \pi/4$ and $\psi_2 = \pi/12$. It is observed that the incident shock wave coming from right to left diffracts on the upper surface, reflects on the lower surface, and generates a circular sector of diffraction around the apex.

Integrating Eqs. (2) from the boundary of the circular sector ($R = 1$) to R $(0 < R < 1)$, the velocity within the circular sector can be expressed as

$$v_r(R, \theta) = v_r(1, \theta) + \int_1^R \frac{1}{R}\frac{\partial \rho(R, \theta)}{\partial R}dR, \tag{7}$$

(a) (b)

Figure 1. Wave pattern and polar coordinate (a), and density field (b).

$$v_\theta(R,\theta) = v_\theta(1,\theta) + \int_1^R \frac{1}{R^2}\frac{\partial\rho(R,\theta)}{\partial\theta}dR. \tag{8}$$

For this case the boundary conditions for the formulas for velocity components (7) and (8) are expressed as

$$v_r(1,\theta) = \begin{cases} 2\cos\psi_1\cos\theta, & 0 \le \theta < \psi_1 \\ \cos(\psi_1+\theta), & \psi_1 < \theta < \phi_1 - \psi_2 \\ 0 & \phi_1 - \psi_2 < \theta \le \phi_1 \end{cases} \tag{9}$$

$$v_\theta(1,\theta) = \begin{cases} -2\cos\psi_1\sin\theta, & 0 \le \theta < \psi_1 \\ -\sin(\psi_1+\theta), & \psi_1 < \theta < \phi_1 - \psi_2\ . \\ 0 & \phi_1 - \psi_2 < \theta \le \phi_1 \end{cases} \tag{10}$$

Figure 2 shows the velocity field in the case with $\psi_1 = \pi/4$ and $\psi_2 = \pi/12$. Figure 2a shows vector plots of the flow field combined with the density field and Fig. 2b shows a close-up view near the apex of the wedge. The present analyses predict that the velocity tends to infinity at the apex of the wedge at non-zero angle of attack, although the density converges to a value of $2\pi/\phi_1$ there. It is observed in the close-up view of the velocity field that a flow occurs from the lower surface to the upper surface and a stagnation point exists on the lower surface in the neighborhood of the apex.

3. Analyses of reflection processes from a triple reflector

Figure 3a shows schematic diagrams of reflection processes from a triple reflector consisting of a flat surface and wedges on both sides: (1) an incident shock wave

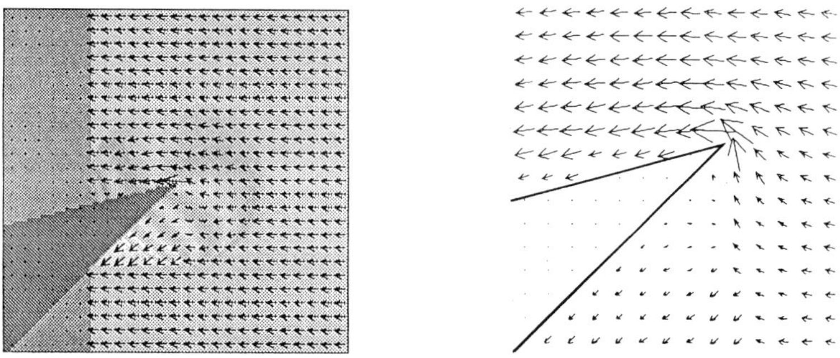

(a) (b)

Figure 2. Velocity field (a) and its close-up view near the apex of the wedge (b).

moving from right to left reflects on the inclined surfaces of the wedges on both sides and circular sectors of diffraction develop from the apexes A and A'; (2) it reflects from the front surface and other circular sectors also develop around corners B and B'. Hence analytical solutions can be constructed for flow fields generated by reflection on a triple reflector by superposing similarity solutions around wedges and corners. As a result of the same kind of analyses as those in the previous section, the pressure (density) field around a corner B between the front surface and the inclined surface at $\phi_0 < \pi/4$ is represented by two combinations of functions $p_1 + p_2$ which are defined as follows

$$p_1(r_B, \theta_B, t_B) = \begin{cases} 0, & r_B/t_B > 1/\cos(\theta_B - \phi_{BS}) \\ 0, & 1 < r_B/t_B < 1/\cos(\theta_B - \phi_{BS}) \text{ and } \phi_{BS} < \theta_B \leq \phi_B \\ 1, & 1 < r_B/t_B < 1/\cos(\theta_B - \phi_{BS}) \text{ and } 0 \leq \theta_B < \phi_{BS} \\ 1 - f(q_B, \omega_B; \alpha_{BS}), & r_B/t_B < 1 \end{cases} \tag{11}$$

$$p_2(r_B, \theta_B, t_B) = \begin{cases} 0, & r_B/t_B > 1/\cos(\theta_B - \phi_{BR}) \\ 1, & 1 < r_B/t_B < 1/\cos(\theta_B - \phi_{BR}) \text{ and } \phi_{BR} < \theta_B \leq \phi_B \\ 0, & 1 < r_B/t_B < 1/\cos(\theta_B - \phi_{BR}) \text{ and } 0 \leq \theta_B < \phi_{BR} \\ f(q_B, \omega_B; \alpha_{BR}), & r_B/t_B < 1 \end{cases} \tag{12}$$

where

$$q_B = \left[\frac{r_B/t_B}{1 + \{1 - (r_B/t_B)^2\}^{1/2}}\right]^{\lambda_B}, \qquad \omega_B = \lambda_B \theta_B. \tag{13}$$

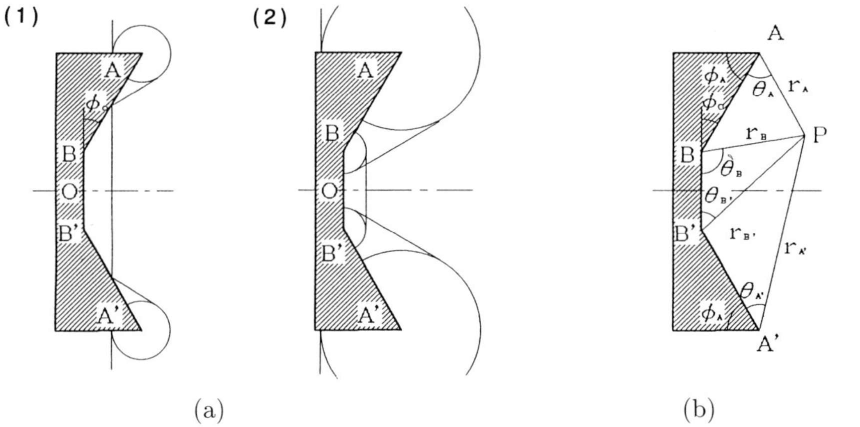

Figure 3. Schematic diagrams of reflection processes from a triple reflector (a) and polar coordinate systems fixed with the reflector (b).

Here a polar coordinate (r_B, θ_B) is defined in Fig. 3b and t_B is the time elapsed after the reflection of the incident shock wave on the front surface, $\phi_{BS} = \pi/2 - 2\phi_0$ and $\phi_B = \pi - \phi_0$ are defined using ϕ_0 indicated in Fig. 3b, $\phi_{BR} = \pi/2$, $\lambda_B = \pi/\phi_B$, $\alpha_{BS} = \lambda_B\phi_{BS}$, and $\alpha_{BR} = \lambda_B\phi_{BR}$. Effects of waves developing around the other corner B' are represented by a function $p_3(r_{B'}, \theta_{B'}, t_B)$ that is a mirror image to $p_1(r_B, \theta_B, t_B)$, and by a function $p_4(r_{B'}, \theta_{B'}, t_B)$ that is the contribution of circular sector around B'. Also, an effect of diffraction from the apex A is expressed as $p_5(r_A, \theta_A, t_A)$. Thus, the pressure field generated by reflection of the incident shock wave on the triple reflector is expressed as

$$p = 1 + p_1(r_B, \theta_B, t_B) + p_2(r_B, \theta_B, t_B) + p_3(r_{B'}, \theta_{B'}, t_B) + p_4(r_{B'}, \theta_{B'}, t_B) + p_5(r_A, \theta_A, t_A). \tag{14}$$

Here a polar coordinate (r_A, θ_A) is defined in Fig. 3b and t_A is the time elapsed after arrival of the incident shock wave at the apex A of the wedge. Figure 4 shows density contour diagrams for the reflection processes on a triple reflector: (a) an incident shock wave moving from right to left reflects on inclined surfaces of the wedges on both sides, and circular sectors of diffraction develop from the apexes; (b) the incident shock just arrives at the front surface; (c) it reflects from the front surface and other circular sectors also develop from both corners; (d) these circular sectors and the normal and oblique reflected shock waves overlap so that complicated wave patterns are observed.

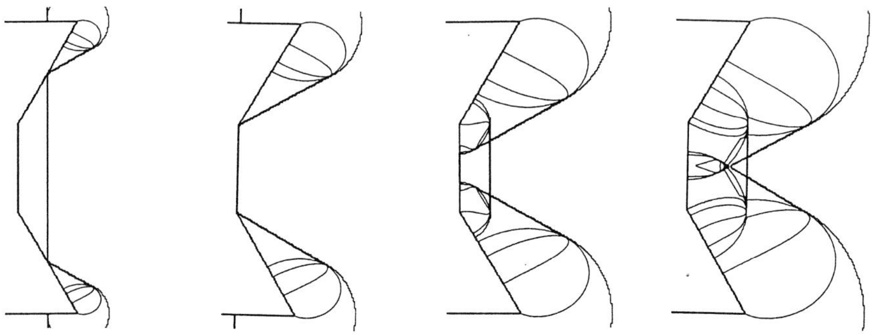

t = - 0.2 (a)

t = 0 (b)

t = 0.2 (c)

t = 0.4 (d)

Figure 4. Density contour diagrams of reflection processes: (a) $t = -0.2$, (b) $t = 0$, (c) $t = 0.2$, (d) $t = 0.4$.

4. Concluding remarks

In the present paper, as a result of analyses that use the two-dimensional acoustic (linearized gasdynamic) equations, similarity solutions are obtained for velocity fields in addition to those for the density (pressure) field generated by reflection and diffraction of weak shock waves on a wedge at an angle of attack. Also, it is shown that analytical solutions can be constructed for reflection processes of weak shock waves on a triple reflector by superposing these similarity solutions.

Acknowledgement. The present investigation is partly supported by the Secom Foundation's Research Grant.

References

Friedlander FG (1958) Sound Pulses, Cambridge University Press.

Keller JB, Blank A (1951) Diffraction and reflection of pulses by wedges and corners. Comm Pure Appl Math 4:75–94.

Takano Y, Karakawa K (1993) Comparisons between simulations and analyses of weak shock waves reflected by a wedge at attack angle. In: 5th Int Symp on Computational Fluid Dynamics, Sendai, Vol III, pp 187–192.

Computation and validation of shock-wave interaction with fluid inhomogeneities

J. Falcovitz
Institute of Mathematics, The Hebrew University
Jerusalem 91904, Israel

K. Yamada, H. Nagoya, K. Takayama
Shock Wave Research Center, Institute of Fluid Science
Tohoku University, Katahira 2-1-1, Sendai, Japan

Abstract: We study the interaction between a shock wave propagating in a homogeneous fluid and an enclosed bubble of a different fluid. Two types of such phenomena are considered: the shock-wave breakup of a helium bubble in air and the interaction of a micro-explosion water shock with an air bubble. Numerical simulations are obtained from the 2D GRP scheme extended to treat multi-fluid flows. The scheme employs "volume-of-fluid" information in cells to approximately track interfaces between dissimilar (and immiscible) fluids. Computational results are compared to experimental visualizations and good agreement is demonstrated.

Key words: Shock/bubble interaction, Multi-fluid scheme

1. Introduction

The encounter between a shock wave propagating in a homogeneous fluid and an enclosed bubble of a different fluid gives rise to diverse interaction patterns. Here we specifically consider two types of such phenomena: (i) shock-wave breakup of a helium bubble in air; (ii) interaction of a shock obtained from a micro-explosion in water with an air bubble. In the following we outline the principal features of the numerical scheme. We then present three instances of multi-fluid shock-wave phenomena, comparing experimental observations with numerical simulations.

2. Material Interface Tracking scheme

The numerical computations were performed using the Generalized Riemann Problem (GRP) scheme for compressible flows in an operator-split 2D mode. An extension of GRP into a scheme for the simulation of multi-fluid shock-wave phenomena was developed and named Material Interface Tracking (MIT). In MIT, pressure and velocity continuity at interfaces are obtained by treating the fluids in 'mixed cells' as a homogenized mixture of fluids, having uniform pressure and velocity. The advection of the correct fluid species is obtained by reconstructing the interface piecewise per cell, using solely volume-of-fluid (VOF) information. In situations where an interface cannot be uniquely reconstructed in a particular cell, the scheme reverts to the less accurate antidiffusive measure of preferentially advecting the fluid type matching that of the receiving cell, so as to maintain sharp interfaces.

3. Experimental validation

As preliminary validation of the GRP/MIT scheme, we considered the $M_s = 1.25$ shock-wave breakup of a spherical helium bubble in air (Fig. 1), which compares well with the experimental shadowgraphs obtained by Haas and Sturtevant (1987). The comparison in this case relied mostly on the air/helium interface, which was visible in the shadowgraphs throughout the interaction.

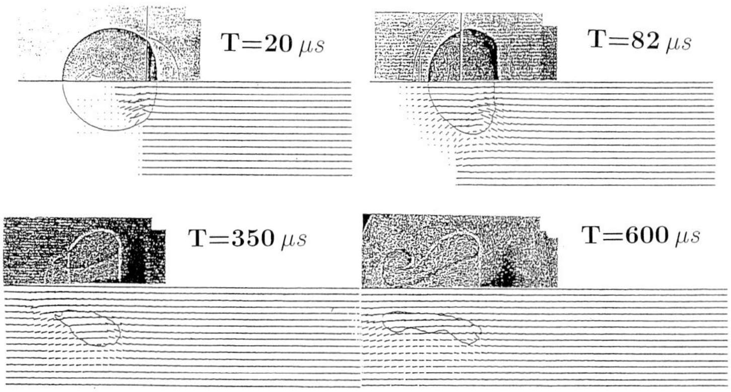

Figure 1. Interaction of $M = 1.25$ air shock with spherical helium bubble (Haas and Sturtevant 1987).

A more detailed observation of the interaction of an $M_s = 1.2$ air shock with a cylindrical He bubble was obtained by recording a series of holographic interferometric pictures at times where the incident shock was passing through the bubble and shortly thereafter (Fig. 2). These interferograms clearly show the following features: shock fronts both in air and within the bubble; the He/air interface; and isopycnics in regions of continuously varying density such as the rarefaction waves arising from the passing of the incident air shock into and around the helium bubble. These features agree well with the GRP/MIT simulation, constituting a physical validation of the method.

The water-shock experiment was conducted as follows. A 10-mg silver azide pellet was immersed in water and was detonated at the moment that an air

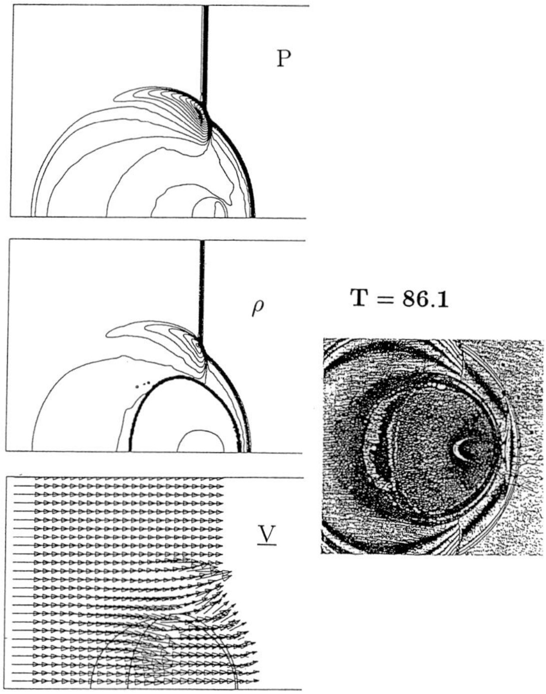

Figure 2. Interaction of $M = 1.20$ air shock with cylindrical helium bubble.

bubble (diameter about 3 mm, initial pressure 0.1 MPa) was located 10 mm away (center-to-center distance). The interaction between the incident shock and the bubble was recorded by a series of shadowgraph pictures taken at 1-μs time intervals (Fig. 3). The shock overpressure at the range of 10 mm was independently measured as about 65 MPa. The following features were observed on the shadowgraphs: the contracting-expanding air bubble; the (nearly spherical) incident wave; a (nearly spherical) rarefaction wave; and a (nearly spherical) rebound shock wave.

The numerical simulation in this case was considerably more involved than in the air/helium case. The (presumably spherical) explosion was computed at a separate 1D computation, starting out from a quiescent compressed gas bubble having the mass and energy densities of burned silver azide, and surrounded by water. At the moment where the water shock overpressure was about 100 MPa, the 1D flow field was recorded and subsequently transformed as initial conditions for the 2D computation. It was found that a simplified scheme where the spherical blast wave was obtained by replacing the silver azide gas by some

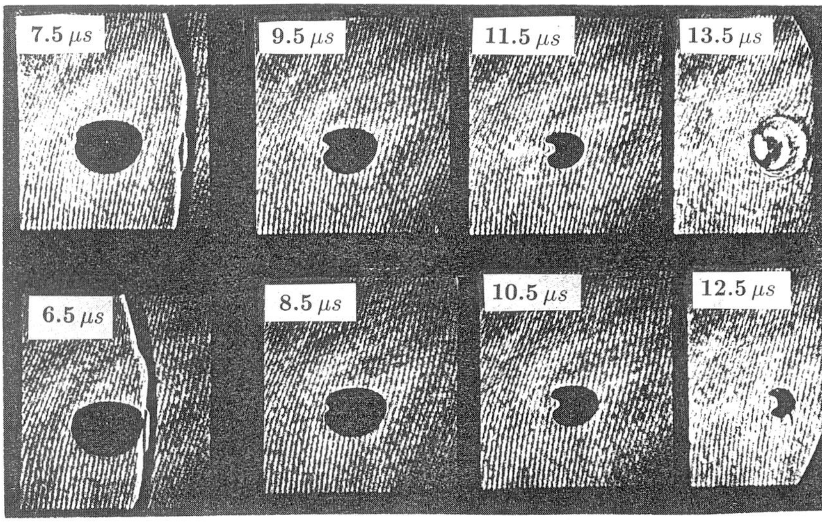

Figure 3. Interaction of shock from 10-mg silver azide explosion in water with air bubble.

equivalent compressed water sphere could be matched to yield the correct shock overpressure. However, the resulting blast wave also contained an incorrect steeper pressure gradient at the shock front. Thus, the silver azide gas bubble had to be maintained in the simulation throughout.

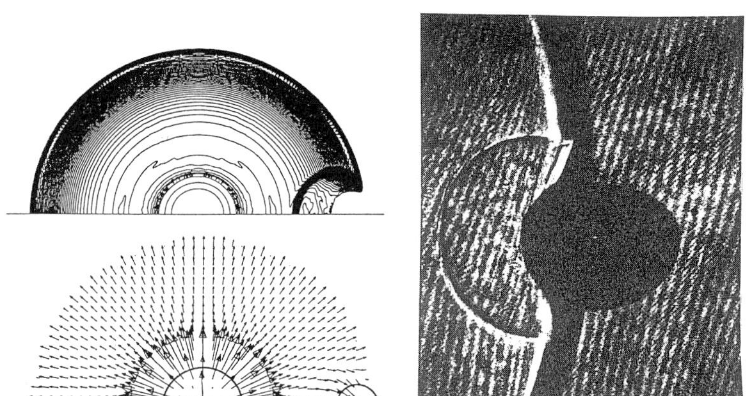

Figure 4. Interaction of shock from 10-mg silver azide explosion in water with air bubble. Incident shock, reflected rarefaction, and bubble at $T = 5.5$ μs.

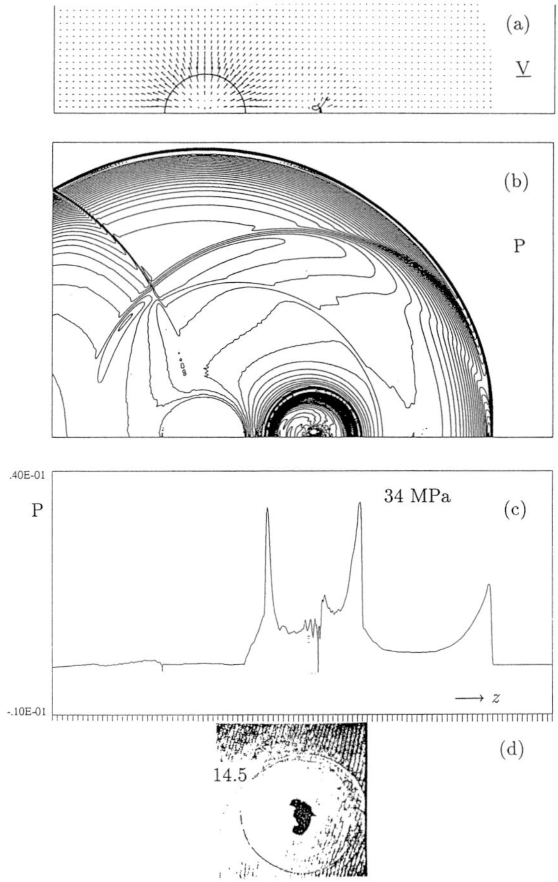

Figure 5. Interaction of shock from 10-mg silver azide explosion in water with air bubble. Rebound shock and bubble at $T = 17.5$ μs: (a) velocity and interface plot; (b) isobars; (c) pressure distribution along axis; (d) shadowgraph (at $T = 14.5$ μs).

Turning to the computation results, all three features previously mentioned were observed in the simulation, with fair — although not complete — agreement with the experimental findings. The rarefaction wave is observed as the incident shock diffracts around the bubble at time 5.5 μs (times are from moment of detonation). A good agreement between the computational and experimental observation is clearly found (Fig. 4). The peak of the computed bubble contraction was observed at about 15 μs. The rebound shock is hence generated from that moment on, and is clearly visible, for example at 17.5 μs (Fig. 5). The shadowgraph shock (Fig. 5) is in qualitative agreement with the computed rebound shock, although it is earlier by about 3 μs. From the recorded time history of average volume and pressure of the bubble, it is found that the peak contraction was at a volume ratio of about 1/300. The corresponding peak bubble pressure was about 320 MPa, whereas by isentropic compression the peak pressure for that contraction ratio would be about 290 MPa. The difference may be attributed to entropy increase during the dynamic compression of the air bubble. A further comparison of computed results with experimental shadowgraphs (the figures are not shown here) indicates the following findings. The first shadowgraph observation of a rebound shock is at about 13.5 μs, which is earlier by about 3 μs than the computation. At about 15.5 μs in the experiment, the rebound shock seems weaker and its front is at a radius of some 6 mm. This front radius is obtained in the computation at about 18.5 μs. Again, the experiment is earlier than the computation by about 3 μs. We do not know the reason for this time discrepancy. But other than that disagreement, the computed and experimental findings agree fairly well.

This study clearly demonstrates the benefit of CFD in enhancing our understanding of the observed phenomena and, furthermore, in estimating effects which cannot be directly measured. In particular, the rebound shock overpressure is computationally estimated as having a peak value of 320 MPa, which is about five times higher than the overpressure of the incident water shock.

Acknowledgement. This study was conducted during the first author's stay as a visiting professor at the Shock Wave Research Center in the Institute of Fluid Science of Tohoku University. The author is grateful to the Center and to its director Prof. K. Takayama for providing a stimulating and pleasant research environment.

References

Ben-Artzi M, Falcovitz J (1984) A second-order Godunov-type scheme for compressible fluid dynamics. J Comp Phys 55:1–32.

Ben-Artzi M, Falcovitz J (1986) An upwind second-order scheme for compressible duct flows. SIAM J Sci Stat Comp. 7:744–768.

Haas JF, Sturtevant B (1987) Interaction of weak shocks with cylindrical and spherical gas inhomogeneities. J Fluid Mech 181:41–76.

The influence of geometric variations on shock focusing in cylindrical cavities

H. Babinsky*, O. Onodera, K. Takayama
Shock Wave Research Center
Tohoku University
Sendai 980 Japan

E. Timofeev, P. Voinovich
Advanced Technology Center
P.O. Box 160, St. Petersburg 198103 Russia

Abstract: A variety of geometric variations to the standard semicircular two-dimensional cavity have been studied experimentally and numerically to investigate the effect on shock-wave focusing at a shock Mach number of 1.4. It was found that introduction of bluntness at the shoulders of the cavity increases the focusing pressures due to significant changes in the shock reflection structure. However, the different flow patterns also change the size and shape of the high-pressure region; thus the suitability of a geometry should be considered with the requirements of a typical application in mind.

Key words: CFD, Unstructured grid, Shock focusing, Interferometry

1. Introduction

During the past few years a number of works have been devoted to the investigation of shock-wave focusing in concave cavities (Takayama 1989). A commonly used geometry is a half-cylindrical cavity with sharp edges pointing towards the incident shock front (see Fig. 1). The aim of the present work is to analyze the influence of some geometric modifications of this shape, seen in Fig. 1, on the focusing process. In all cases the outer edge of the cavity is modified by the introduction of some degree of curvature while the central portion of the cavity remains unchanged. An experimental investigation using holographic interferometry can provide useful information on flow structure and development, however, the most interesting quantity in this problem — the pressure in the focal region — cannot be obtained directly from these results. Nevertheless the experimental data can be used effectively for the validation of a CFD scheme, which in turn offers the whole variety of gasdynamic parameters at any point of interest as well as their development with time.

*Present address: Department of Engineering, University of Cambridge, Cambridge, England.

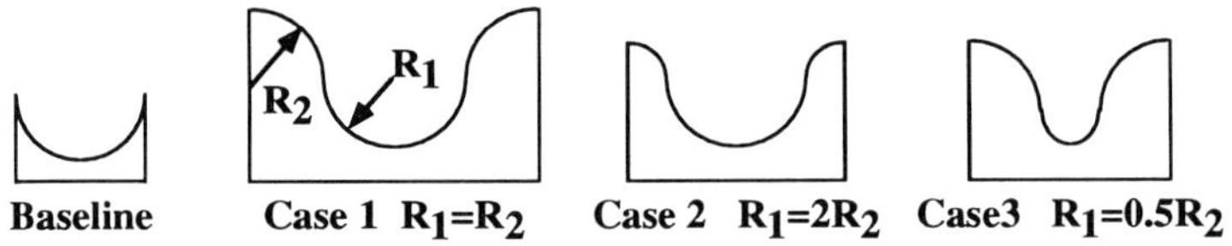

Figure 1. Shock-focusing geometries.

2. Numerical scheme

The problem under investigation involves numerous shocks and slip surfaces of rapidly changing shape combined with a curved model surface. Therefore it is perfectly suited to be simulated by a shock-capturing technique based on locally adaptive unstructured grids. The numerical scheme is described in Galyukov et al. (1996). The grid, consisting of triangles, undergoes a refinement based on the current solution at each time step. The number of total nodes can therefore be kept at a minimum while still providing a solution with very high spatial resolution. The high efficiency of this method and the scheme allow multiple computations on general purpose platforms like workstations and PCs. Some of the results presented here have been obtained in one day of computing on an IBM/486 personal computer. To visualize the results and allow direct comparison with experiments, quasi-experimental images such as interferograms and schlieren photographs have been produced using techniques described in Babinsky et al. (1995). For the purposes of image processing it was required to supply the numerical data on a regular grid. Several approaches have been tried to interpolate the data from the unstructured grid to a rectangular array and it was found that the best results were achieved if the same interpolation routines were used that have also been implemented by the numerical scheme itself, hence the same triangular structure is used for interpolation of data points between nodes. Any other approach leads to unwanted artifacts when visualizing the data. However, the interpolation process is computationally expensive and when many images were created, such as for animations, a Cray T3D supercomputer was used.

3. Results

The experimental results are available as infinite-fringe interferograms, obtained in a shock tube at a shock Mach number of 1.4 in air. The pressure upstream of the shock wave was set to atmospheric conditions and the model width was 60 mm. These parameters have been used in the calculation of numerical interferograms and the two images can be compared directly to evaluate the validity of the numerical scheme. Figure 2 shows such a comparison for case 1, demonstrating the good agreement between CFD and experiment. The maximum disagreement in fringe location was found to be within half a fringe; this is equivalent to a density difference of less than 0.5%. This level is better than the expected experimental accuracy and it can be concluded that the numerical scheme can predict the main flow very satisfactorily. Fine flow structures have been captured well by the present scheme, however, a slight smearing of contact surfaces and slip lines was observed. A similar level of agreement has been found for the other cases investigated.

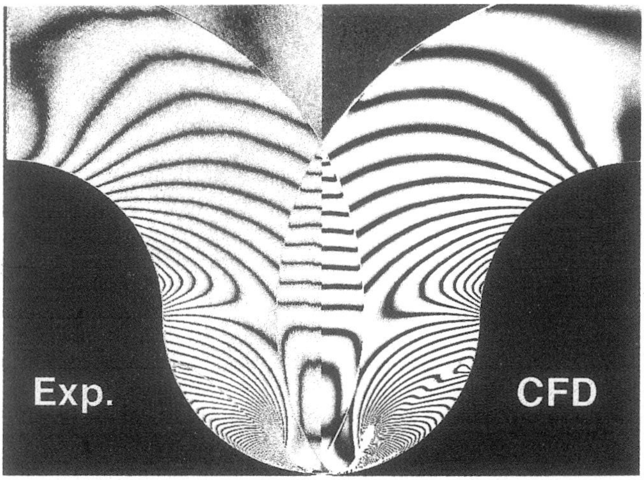

Figure 2. Comparison of experimental (left) and computational (right) interferogram.

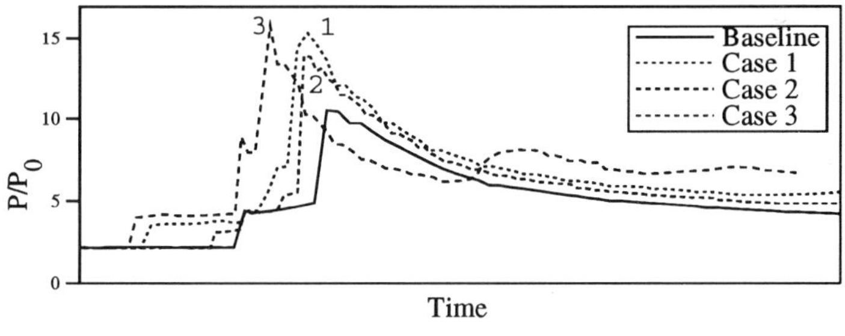

Figure 3. Maximum pressure histories at centerline for all cases.

After establishing confidence in the numerical simulation, the CFD data can be used to compare the various geometries. In Fig. 3 the maximum pressure along the centerline is plotted versus time for all cases. No fixed position was choosen, as the location of the focus on the centerline is initially unknown. It can be seen that the geometrical variations have a significant effect on the achievable pressure levels as well as on the time of focusing. The blunt cases exhibit a higher pressure at the focus and show some additional step-like pressure increases before the actual focusing. A time sequence of shock patterns is given in Figs. 4 and 5, showing shadowgraphs of the flow field together with a schematic of the equivalent location on the pressure curve. In the baseline case, the incoming shock wave undergoes an unsteady shock reflection along the concave side wall of the cavity. Hence, a Mach stem initially appears and then reduces in length. A new shock wave is formed behind the reflection point when the reflection pattern changes from Mach reflection to direct reflection. The focal pressure is achieved when the reflected shock wave and the two new shock waves merge at one point. The blunt cases show a different wave configuration as the incoming shock wave enters the circular part of the cavity. A Mach stem has already formed at this point and the reflected shocks are stronger than in the baseline case. It is this Mach stem which now undergoes inverse Mach reflection and subsequently the wave pattern leading to a focus is different for these cases. Furthermore, the reflected shock waves reach the focal point before focusing takes place, thus also increasing the pressure. The whole history of flow development is significantly different, resulting in a different focusing mechanism and efficiency.

However, the maximum focusing pressure is only one criterion for the selection of a geometry. Other factors include the location of the focus, the sharpness of the pressure peak, and the shape of the high-pressure region. In some applications it may be desirable to have a very sharp focus with low pressures everywhere else in the flow field, whereas other applications may call for a wide area of high pressure. To evaluate the various geometries under these criteria the maximum flow-field pressures, regardless of time, are seen in Fig. 6 for all geometries. It can be seen that the pressure distributions are quite different and the sharpness of the focus strongly depends on the geometry. Of particular importance is the observation that in the blunt cases some very large pressures are observed away from the centerline. These high-pressure regions could easily be missed in an experimental investigation. However, for practical purposes such high-pressure regions in other areas of the flow field are extremely important and may cause the particular geometry to be unsuitable.

4. Conclusions

The combined approach incorporating experimental and CFD methods supplemented by an advanced visualization technique has contributed greatly to the understanding of the presented problem. It was shown that modifications to the geometry of a cavity can have a significant effect on the shock-wave reflection structures observed. In particular these changes have been found to cause large differences in focusing pressure as well as in the shape of the high-

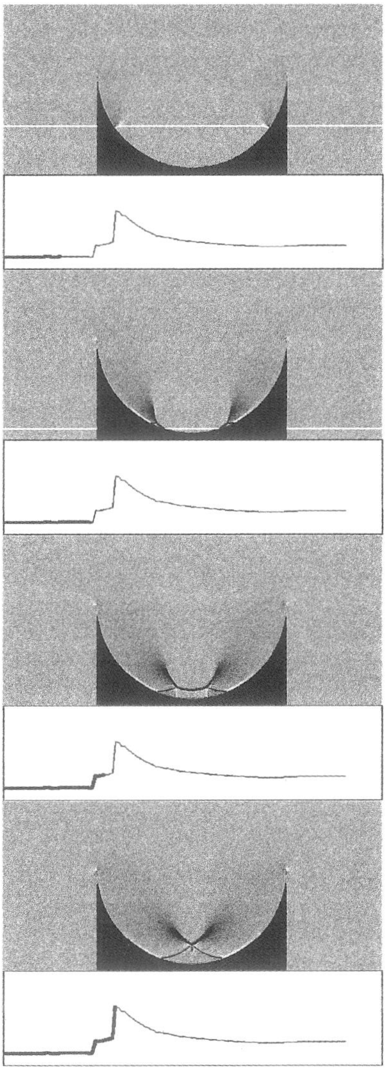

Figure 4. Sequence of numerical schlieren photographs for the CFD solution of the baseline case, together with relevant position on pressure history.

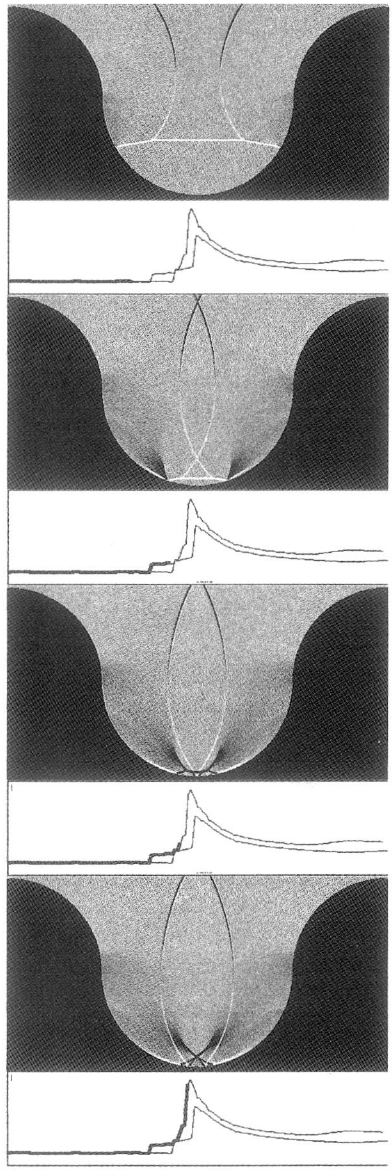

Figure 5. Same as Fig. 4, Case 1.

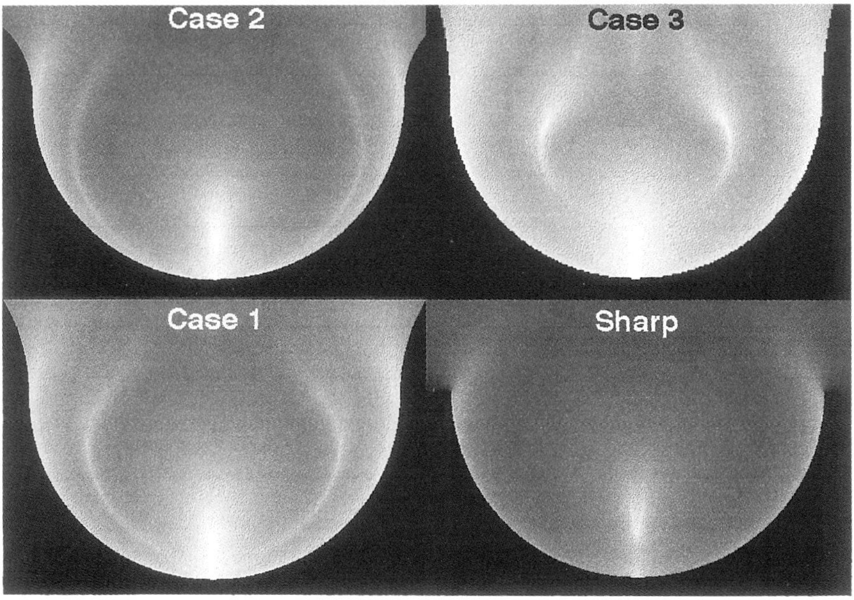

Figure 6. Maximum pressures throughout whole computation time.

pressure region. While blunted cavities can increase the maximum pressure significantly, they also increase the area subject to relatively large pressures and could therefore be unsuitable for some applications. The ideal choice of a focusing geometry cannot be determined generally, but rather depends on the particular requirement of the application.

Acknowledgement. This work was supported in part by the Russian Foundation of Fundamental Investigations (Grant No. 94-01-01456).

References

Babinsky H, Meguro T, Jiang Z, Takayama K (1995) Numerical visualization of shock wave flow in an expanding tube and comparison with experiment. In: 4th ASME/JSME Symp on Exp and Num Flow Vis, Hilton Head, USA, ASME, Vol 218, pp 89–94.

Galyukov A, Fursenko A, Voinovich P, Saito T, Timofeev E (1996) Efficient 2D and 3D locally adaptive unstructured Euler solvers for unsteady shocked gas flows. In: Shock Waves, Proc of the 20th ISSW, Sturtevant B, Shepherd JE, Hornung HG (eds), World Scientific, Vol 1, pp 691–696.

Takayama K (ed) (1989) Proc of Int Workshop on Shock Wave Focusing, Shock Wave Research Center, Inst of Fluid Science, Tohoku University, Japan.

An experimental/numerical approach to shock-wave diffraction

H. Kleine*, Q. Yu, E. Ritzerfeld†, H. Grönig
Shock Wave Laboratory, RWTH Aachen
Templergraben 55
52056 Aachen, Germany

Abstract: Fundamental diffraction processes are investigated by comparing data obtained by numerical simulation with the results of flow visualization. Some small-scale structures and details of the flow pattern that neither experiment nor simulation alone can reveal clearly may be recognized and interpreted by using this combined approach.

Key words: Shock-wave diffraction, Numerical simulation, Visualization techniques

1. Introduction

The process of shock-wave diffraction is usually considered as a fairly well-understood fundamental gasdynamic phenomenon. In fact, for the case of a wave diffracting at a sharp corner most of the essential features were already pointed out in early papers on this issue (Griffith & Brickl 1953, Skews 1967; for further investigations and a review see e.g. Bazhenova et al. 1984 and Hillier 1991). However, even for the apparently most straightforward case of a shock wave diffracting at a sharp 90° corner a number of details had not been discovered in the past. Some of these were discussed in more recent papers (Kleine et al. 1995, Hillier 1995), but a few points still require clarification.

In this contribution, experimental and numerical data are compared to answer some of the open questions concerning the flow patterns that are observed when a shock diffracts at a sharp 90° corner. The same approach is used to describe the shock diffraction around a circular cylinder.

2. Experiments

Most of the experimental data are obtained by applying a number of density-sensitive flow visualization methods, in particular holographic interferometry

*Present address: McGill University, Department of Mechanical Engineering, 817 Sherbrooke Street West, Montreal, PQ, H3A 2K6 Canada.

†Present address: TZN Unterlüß GmbH, Neuensothriether Straße 20, 29345 Unterlüß, Germany.

and color schlieren. For the flows mentioned above and a large number of other gasdynamic processes these techniques provide the almost exclusive means to obtain information about the inner structure of the flow field. The visualization techniques are described in greater detail by Kleine & Grönig (1993) and Kleine (1994). The experiments were conducted in a shock tube with a square cross section of 54.5 mm × 54.5 mm, where shock Mach numbers from $M_S = 1.05$ to $M_S = 3.0$ could be achieved for N_2 or air as test gases.

3. Numerical simulation

Since pressure and temperature behind the incident shock are moderate and since the Reynolds number of the post-shock flow is of the order of 10^5, the problem may be simulated by solving the two-dimensional, time-dependent Euler equations. The case of the 90° corner lends itself directly to using an equally spaced cartesian grid, while for more complex flow fields involving diffraction at curved surfaces, an unstructured, triangular grid with adaptive mesh refinement is chosen. The algorithm for the former case is described by Kleine et al. 1995, while details for the upwind finite-volume scheme used in the latter application are given by Yu & Grönig (1996).

4. Results and discussion

4.1. Shock-wave diffraction at a sharp 90° corner

When a shock diffracts at a sharp 90° corner, the flow field behind it is fully sub- or supersonic, when the shock Mach number is low ($M_S < 1.3$) or sufficiently high, respectively ($M_S \geq M_S^*$; $M_S^* = 2.07$ for $\gamma = 7/5$). The transition regime ($1.3 < M_S < M_S^*$) is characterized by the appearance of secondary shock systems that grow and eventually merge for increasing M_S (Skews 1967). Close to the diffracting corner, the upstream travelling expansion wave accelerates the gas beyond the speed of sound, which is indicated by a number of lambda-shaped shocks on top of the slipstream. The results of numerical simulation, which have been found to be in excellent agreement with experimental data, even suggest that there are supersonic regions embedded in a subsonic flow field where the transition to subsonic flow appears to be shock-free (Kleine et al. 1995). All our subsequent investigations, both experimental and numerical, have led to the same conclusion, although an ultimate proof for the existence of these regions has not been obtained yet.

The existence and the development of the lambda-shaped shock systems on top of the slip stream may be explained by an analogy to the flow patterns observed in stationary transonic flows over solid boundaries (Ackeret et al. 1946). There, for slightly supersonic flow ($1 < M_{flow} \leq 1.2$) the convex curvature of the boundary requires a deceleration of the flow which is established by an

(a) (b) (c)

Figure 1. Holographic interferograms of the flow behind a diffracting shock wave in N_2: (a) $M_S = 1.45$, $p_1 = 1.8$ bar; (b) $M_S = 1.58$, $p_1 = 1.25$ bar; (c) $M_S = 2.14$, $p_1 = 0.6$ bar.

almost normal shock. This shock can interact with the boundary layer and induce a local flow separation. For a laminar boundary layer, the separation region can develop in front of the shock thus causing another shock, an oblique shock, the front leg of the lambda-configuration. This front leg shock usually is rather weak and decays with increasing flow Mach number. When the boundary layer becomes turbulent, the separation region induced by the deceleration shock cannot make its presence felt upstream, consequently, the oblique front leg then vanishes. For the quasi-steady flow following the diffracted shock, the visualized flow fields exhibit almost exactly the same phenomena (Fig. 1). Further experiments, in which the Reynolds number of the post-shock flow was varied by changing the initial pressure p_1 while keeping M_S constant, validated qualitatively the analogy to the stationary case. In the numerical results, the main second shock is predicted correctly in all investigated cases. The preceding front leg, however, does not appear in these calculations since its formation mechanism cannot be simulated by the applied Euler code.

In the supersonic regime ($M_S \geq M_S^*$) it was already observed in early investigations (Skews 1967) that a contact surface develops behind the second shock indicating a non-uniformity of the flow field in this region. This can readily be explained by taking into account that this second shock actually consists of two parts: it is formed by the shock on top of the slip stream (see previous paragraph) which merges with a shock that develops out of a Kelvin-Helmholtz-type instability at the right-hand side of the vortex. The contact surface originates at the intersection of these shocks (Fig. 1c).

Numerical simulations (Uchiyama & Inoue 1995 as well as our own calculations) suggest that the slipstream eventually decays into a number of small vortices. Such a pattern cannot be seen on experimental records, since by the time such a decay might be observed the flow field has usually been influenced already by waves that have been reflected from the solid boundaries of the test section. Some flow pictures, however, indicate an onset of vortices along the slip stream. In addition, the observed local density minimum within the slip stream (Kleine et al. 1995) would let a breakup into several vortices appear likely.

4.2. Shock-wave diffraction at a circular cylinder

In the past, the main points of interest concerning the diffraction of a shock wave at a circular cylinder have been to locate the point of transition from regular to Mach reflection and to determine the variation of shock strength along the subsequently diffracted Mach stem (e.g. Heilig 1970). In addition to this, a combined experimental/numerical approach allows one to investigate the rather complicated flow field that is observed after the two diffracted shocks on both sides of the cylinder have crossed. Having established an excellent agreement with respect to the density contours, the numerical data may again be taken to determine other properties of the flow field that are hard or impossible to obtain experimentally, such as local hot spots or locations of maximum pressure loading. Figure 2 indicates what accuracy can be achieved for experimental and numerical density contours.

In this figure, an advanced stage of the diffraction process is shown. The diffracted portions of the shock on the upper and on the lower side of the cylinder have intersected, eventually causing the formation of a Mach stem in the center of the test section. Numerical and experimental density contours only differ by at most 20% of the density increment ($\Delta\rho = 0.054$ kg/m^3). So, with the possible exception of regions where transport processes like viscous forces or heat transfer may be significant (and our calculations therefore not valid), the density at any point of the flow field may be determined with an accuracy of $\pm$ 0.011 kg/m^3. It seems therefore justified to take numerical data for a further analysis of the flow field.

As already indicated by the density contours and further confirmed by the numerical results, the variation of shock strength along the leading shock front is comparatively moderate (61% for $M_S = 2.33$ and 58% for $M_S = 1.35$). In particular, there are no indications that the Mach stem and its associated triple points cause any distinct hot spots in the flow, although along the shock front they create the highest increase of pressure, density and temperature. The actual values of the thermodynamic variables in this region, however, are fairly below the maxima which are found to be in the stagnation region left of the cylinder.

(a)

(b)

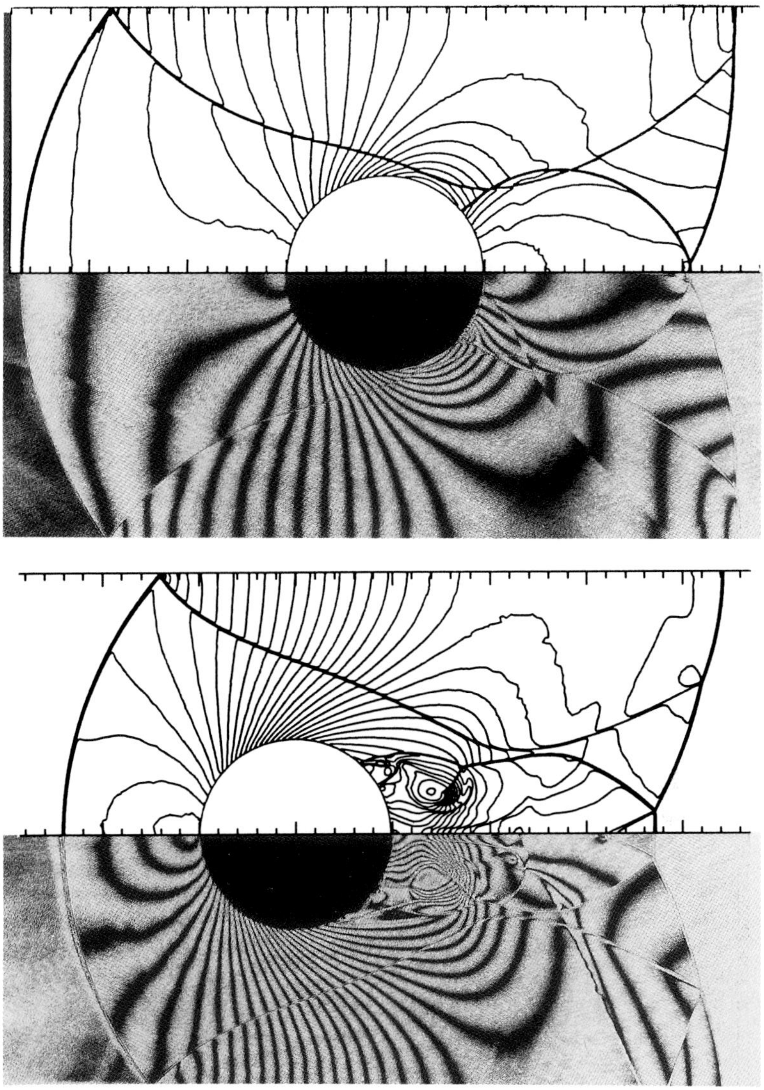

Figure 2. Shock diffraction at a circular cylinder; holographic interferogram and numerically determined density contours: test gas N_2; (a) $M_S = 1.35$, $p_1 = 1$ bar, (b) $M_S = 2.23$, $p_1 = 0.25$ bar.

In the case of $M_S = 2.23$ (Fig. 2b), numerical results indicate that the flow along the cylinder is accelerated to a maximum flow Mach number of $M_f \approx 3.2$ before it is slowed down by the upstream traveling part of the diffracted shock at the cylinder. In the final part (starting at the equator, where $M_f \approx 1.6$), an increase in particle velocity and a decrease in local temperature appear to contribute about equally to this rise of M_f.

References

Ackeret J, Feldmann F, Rott N (1946) Untersuchungen an Verdichtungsstößen und Grenzschichten in schnell bewegten Gasen. Mitt aus dem Inst f Aerodynamik, ETH Zürich.

Bazhenova TV, Gvozdeva LG, Nettleton MA (1984) Unsteady interactions of shock waves. Prog Aerospace Sci 21:249–331.

Griffith WC, Brickl DE (1953) The diffraction of strong shock waves. Phys Rev 89:451–453.

Heilig W (1970) Theoretische und experimentelle Untersuchungen zur Beugung von Stoßwellen an Kugeln und Zylindern. EMI Bericht 2/70, Freiburg.

Hillier R (1991) Computation of shock wave diffraction at a ninety degree convex edge. Shock Waves 1:89–98.

Hillier R (1995) Numerical modelling of shock wave diffraction. In: Brun R, Dumitrescu LZ (eds), Proc 19th Int Symp on Shock Waves, Vol IV, Springer, Heidelberg, Berlin, New York, pp 17–26.

Kleine H, Grönig H (1993) Visualization of transient flow phenomena by means of colour schlieren and shearing interferometry. In: Dewey J, Racca R (eds), Proc 20th Int Congr High Speed Photography & Photonics, Vol 1801, SPIE, Bellingham, pp 400–409.

Kleine H (1994) Verbesserung optischer Methoden für die Gasdynamik, Doctoral thesis, Stoßwellenlabor, RWTH Aachen.

Kleine H, Ritzerfeld E, Grönig H (1995) Shock wave diffraction – new aspects of an old problem. In: Brun R, Dumitrescu LZ (eds), Proc 19th Int Symp on Shock Waves, Vol IV, Springer, Heidelberg, Berlin, New York, pp 117–122.

Skews BW (1967) The perturbed region behind a diffracting shock wave. J Fluid Mech 29:705–719.

Uchiyama N, Inoue O (1995) Adaptive mesh refinement computation of compressible flow. In: Brun R, Dumitrescu LZ (eds), Proc 19th Int Symp on Shock Waves, Vol I, Springer, Heidelberg, Berlin, New York, pp 407–412.

Yu Q, Grönig H (1996) Numerical simulation of the reflection of detonation waves. In: Shock Waves, Proc of the 20th ISSW, Sturtevant B, Shepherd JE, Hornung HG (eds), World Scientific, Vol 2, pp 1143–1148.

Compressibility effects on the propagation and interaction of shock-generated vortices

C. Hébert, M. Brouillette
Département de Génie Mécanique
Université de Sherbrooke
Sherbrooke, Québec, J1K 2R1 Canada

Abstract: This paper examines the vortical flow regimes produced by the impulse of a shock wave emerging from the open end of a shock tube. Spark shadowgraph and schlieren photography were used to characterize the flow field. For shock-wave Mach numbers ranging from $M_s = 1.0$ to 2.0, three distinct flow regimes were identified. From $M_s = 1.0$ to 1.43, conventional-looking vortex rings are produced and are characterized by very thin cores and thus high circumferential instability wavenumbers. From $M_s = 1.43$ to 1.60, a shock wave becomes embedded in the supersonic recirculating region of the vortex ring. From $M_s = 1.60$ to 2.0, a secondary vortex ring forms ahead of the main vortex ring.

Key words: Vorticity, Compressible turbulence

1. Introduction

Compressible vorticity is important in a wide range of applications (Brouillette et al. 1995, Baird 1987, Minota 1993). Unfortunately, little experimental evidence is available to study the propagation and interaction of vortices when compressibility effects are important (Mandella et al. 1986, Baird 1987, Minota 1993, Brouillette et al. 1995). In the incompressible flow regime, vortex rings can undergo transitions from laminar to turbulent propagation, owing to the development of flow instabilities. Krutzsch (1939) observed the core of a ring on which a circumferential instability appeared spontaneously. Those observations where confirmed experimentally by Maxworthy (1972), and analytically by Widnall and Sullivan (1973).

The present paper examines the various vortical flow regimes produced when an impulse is produced by the refraction of a shock wave from the open end of a shock tube. The effects of compressibility are also identified with respect to the strength of the generating shock wave. Spark shadowgraph and schlieren photography are used to visualize the flow field, and fast-response piezoelectric pressure measurements are used to quantify the characteristics of the flow. More specifically, the effects of Reynolds and Mach numbers on the development of the flow field were examined.

2. Apparatus

A shock-tube facility was built specifically for the study of this flow. The driven section has an inside diameter of 38 mm and a length of 1.85 m, of which the last 165 mm, which constitutes the nozzle of the tube, can be replaced with different sections having inside diameters of 6.4 mm and 12.7 mm respectively. The driver section (inside diameter: 38 mm) has an adjustable length, which allows the production of a variable-width pressure pulse at the nozzle end of the shock tube. Compressed nitrogen is used in the high-pressure section, and the low-pressure section is opened to the atmosphere. The diaphragms are punctured on command with a pneumatically actuated needle plunger. Pressure histories recorded at two locations near the end of the driven section provide us with the features of the generated shock; the Mach number of the shock wave is obtained from time-of-flight measurements between those two locations.

To visualize the emerging shock wave and the ensuing vortical flow field, we used a single-spark shadowgraph/schlieren optical system which is built in a folding Z configuration with two 203-mm spherical mirrors, both with a focal length of 1.52 m; the schlieren system used a horizontal knife edge. The images were recorded with a 35-mm camera on Ilford XP2 black-and-white film (400 ASA), which was then developed using the standard C-41 film development process. The light source, a condenser spark-gap, is triggered after an adjustable delay by a pressure transducer mounted near the nozzle of the driven section.

3. Results

3.1. Experimental conditions

With the adjustable driver length of the shock tube, it was possible to change the shape, amplitude, and width of the pressure pulse produced at the nozzle end. The variation of this parameter gave birth to three flow regimes. The first was produced with a driver length greater than a critical value, for which the pressure pulse comprises a positive step function due to the shock arrival, a constant pressure plateau, and a slowly decaying region due to the arrival of the expansion waves from the driver end. The resulting flow shows evidence of a jet-like structure at the nozzle for an extended period of time and the vortex ring produced by this impulse tends to interact with the jet and break up near the nozzle. The second regime is observed using a driver length for which the aforementionned expansion waves interact with the incident shock wave exactly at the exit of the shock tube. For this critical length, the pressure impulse obtained at the nozzle is maximized for a given driver pressure, but the pressure plateau characteristic of the first regime is reduced. This tended to eliminate the jet flow and thus produce cleaner vortex rings that could propagate over longer distances; the results presented here were all obtained using this so-called critical driver length. The third regime is obtained with a driver length shorter than the critical value; in this case the expansion waves start to interact with the incident shock wave before the latter can travel to the end of the driven section, thus producing a flow field qualitatively similar to the second regime but at a lower pressure than with the critical length.

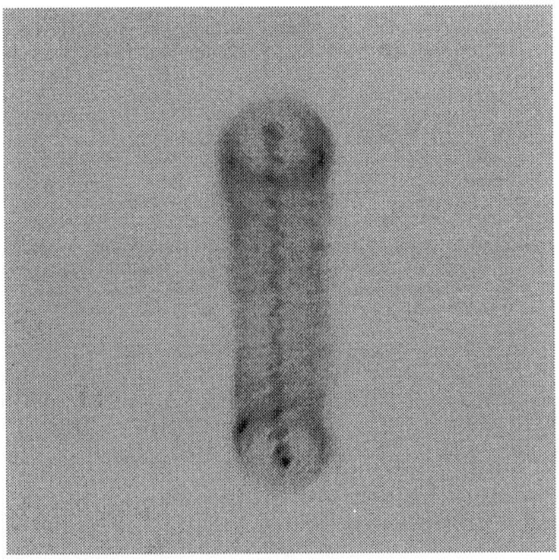

(a)

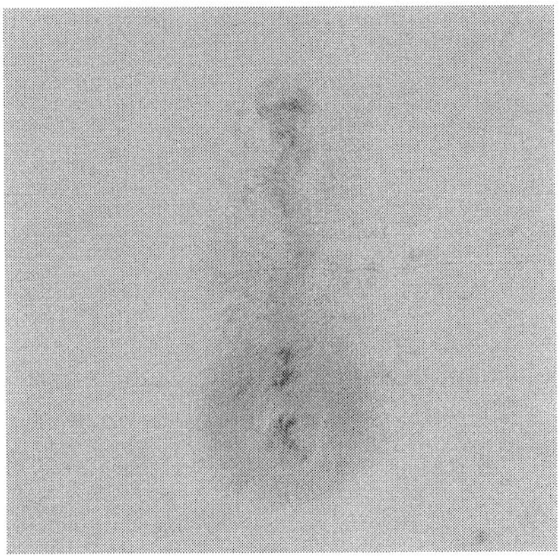

(b)

Figure 1. Spark shadowgraphs of flow field for $M_s = 1.32$: (a) $t = 1.190$ ms; (b) $t = 2.756$ ms.

3.2. Flow regimes

Experiments performed with a generating shock strength between $M_s = 1$ and 2 were able to generate three distinct vortical flow regimes. The first one is illustrated in Fig. 1, showing the flow field produced by a $M_s = 1.32$ shock with a critical driver length. The propagation of the resulting vortex ring is shock-free and, as it procedes downstream, develops a high number of circumferential instabilities around its core region (Fig. 1a). At later times (Fig. 1b), the high wavenumber instabilities seem to have saturated their growth, seemingly increasing the apparent core size of the ring in the process.

Figure 2 shows the second flow regime in which a vortex ring propagates with a rearward-facing shock in the supersonic recirculating region of the ring. The core is no longer observed as a compact structure and is seemingly already highly turbulent. Far downstream of the nozzle (Fig. 2b), the embedded shock seems to disappear as the ring slows down due to the dissipative effects. As the generating shock strength is increased, a secondary vortex ring, with reverse

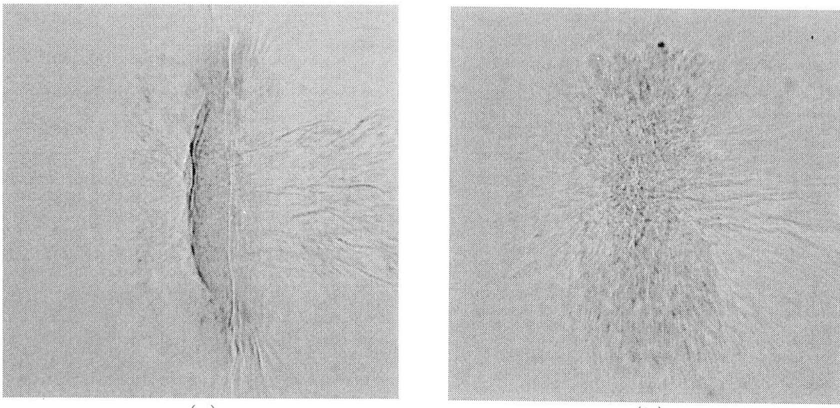

(a) (b)

Figure 2. Spark shadowgraphs of flow field for $M_s = 1.51$: (a) $t = 0.693$ ms; (b) $t = 1.768$ ms.

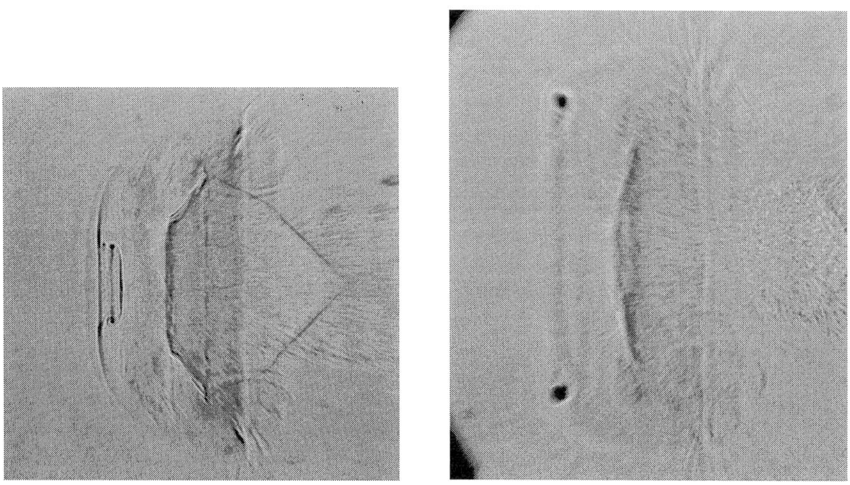

(a) (b)

Figure 3. Spark shadowgraphs of flow field for $M_s = 1.65$: (a) $t = 0.469$ ms; (b) $t = 0.738$ ms.

circulation, is formed in front of the main vortex ring (Fig. 3). The mutual induction between the vortices causes the secondary one to increase in size rapidly as it is entrained around the main vortex. The two then seem to merge in a rather poorly defined vortical structure.

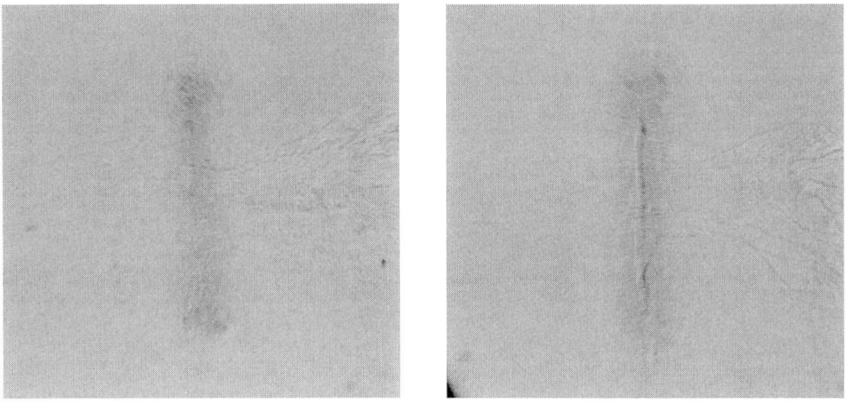

(a) (b)

Figure 4. Spark shadowgraphs of vortex rings at $t = 0.900$ ms: (a) $M_s = 1.42$; (b) $M_s = 1.44$.

3.3. Flow transitions

By varying the strength of the generating shock and using the appropriate critical driver length, it was possible to observe the transition between the three aforementionned flow regimes. Figure 4 shows the first flow transition, between the shock-free compressible regime and the supersonic recirculation regime. This transition is found to be located at $M_s = 1.43$. Previous preliminary studies (Brouillette et al. 1995) had identified this transition at $M_s = 1.34$, but this result was obtained with a driver length larger than the critical length and could thus have been affected by the strong exit-jet characteristic of long drivers.

The onset of the third flow regime takes place when the second counter-rotating vortex forms in the recirculating region of the main vortex ring (Fig. 5b). This secondary counter-rotating vortex results from the displacement of the stagnation point in front of the main vortex because of the total pressure drop across the embedded shock wave. As the shock strength is increased, the formation process is more violent, and the core of the secondary vortex is larger.

4. Conclusions

We have explored the vortical flow field produced by the impulse of a shock wave emerging from the open end of a shock tube. In the $M_s = 1.0$ to $M_s = 2.0$ range, we found that the behavior of the flow can be divided in three regimes: (1) from low supersonic up to $M_s = 1.43$, the vortex ring generated remains shock free, and circumferential instabilities tend to develop around the core region; (2) between $M_s = 1.43$ and $M_s = 1.60$, an embedded shock forms within the

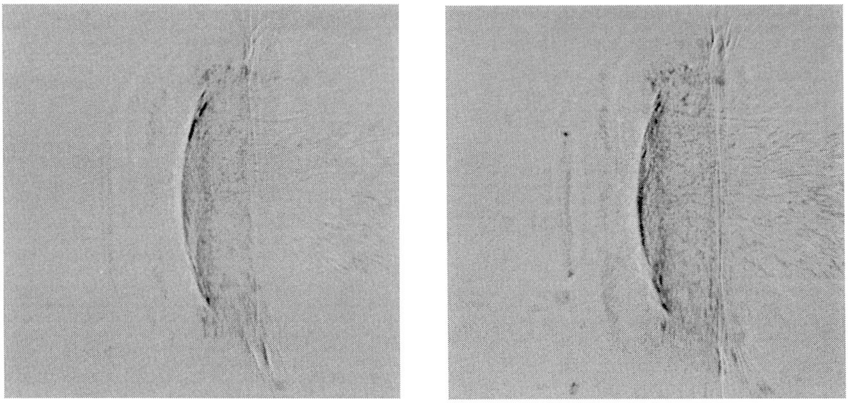

(a) (b)

Figure 5. Spark shadowgraphs of vortex rings at $t = 0.588$ ms: (a) $M_s = 1.58$; (b) $M_s = 1.62$.

recirculating region of the ring, this shock is due to the onset of supersonic flow velocities in the recirculating region; (3) past $M_s = 1.60$, the total pressure drop through the embedded shock causes flow separation which gives birth to a secondary vortex in front of the main vortex.

Acknowledgement. This research is supported by the Natural Sciences and Engineering Research Council of Canada and the Fonds FCAR/Québec.

References

Baird JP (1987) Supersonic vortex rings. Proc R Soc (London) A409:59.

Brouillette M, Tardif J, Gauthier E (1995) Experimental study of shock-generated vortex rings. In: Shock Waves @ Marseille IV, R Brun, LZ Dumitrescu (eds), pp 361–366.

Krutzsch CH (1939) Über eine experimentelle beobachtete Erscheinung an Wirbelringen bei ihrer translatorischen Bewegung in wirklicher Flüssigkeiten. Ann Phys 5:497–523.

Mandella M, Moon YJ, Bershader D (1986) Quantitative study of shock-generated vortex flows. In: Shock Waves and Shock Tubes, Proc 15th Int Symp on Shock Waves, Bershader D, Hanson RK (eds), Berkeley, p 471.

Maxworthy T (1972) The structure and stability of vortex rings. J Fluid Mech 51:15–32.

Minota T (1993) Interaction of a shock wave with a high-speed vortex ring. Fluid Dynamics Research 12:335–342.

Widnall SE, Sullivan JP (1973) On the stability of vortex rings. Proc R Soc Lond A332:335–353.

Shock-wave diffraction by an elastic sphere in water

A. Merlen
Laboratoire de Mécanique de Lille – URA CNRS 1441
Cité Scientifique 59655 Villeneuve d'Ascq, France

P. Pernod, A. Ahyi, A. Kemmou
Institut d'Electronique et Microelectronique du Nord
Cité Scientifique 59655 Villeneuve d'Ascq, France

Abstract: Shock-wave propagation and diffraction on a sphere in water are treated by high-speed cinematography (up to 1 million shadowgraph frames/s), high-power acoustic source and broad-banded hydrophone (from 10 kHz to 10 MHz). The high-power and short-pulse source characteristics are determined, showing the relevance of this kind of source to shock-wave diffraction experiments. Diffraction on an elastic sphere is then investigated emphasizing the easy localization of different phenomena like circumferential (Franz-type) and Rayleigh-type waves. Visualizations show that the flow is far more complicated than the simple diffraction by an infinitely rigid body as it is often supposed in computational fluid dynamics.

Key words: Diffraction, Shocks in liquids, Elastic bodies

1. Introduction

The best way to investigate acoustic diffraction problems is to measure the spatial transient response of the system. The total response of the propagation system may then be simply inferred by convolution of spatial impulse response with the specific signal constituting the source signature. Nevertheless conventional acoustic sources, like piezoelectric transducers, can not be considered as point sources and have generally poor transient responses due to their naturally oscillating behavior. Experimental results can not be easily interpreted because of the great number of diffracting waves and the lack of temporal and spatial resolution. Moreover, the acoustic power radiated by these conventional sources is generally not sufficient to observe the very weak diffusing waves remaining behind the incident wave. We have therefore developed an alternative solution (minisparker), based on underwater electrical discharges, which generates a quasi-spherical impulsive wave front, with a short and simple signature. Moreover, this source, quasi-punctual both in space and time, releases high energy in such a way that a shock wave can be generated.

2. Experimental devices

2.1. Mini-sparker

A coaxial oil-filled capacitor of 18 nF, made for rapid discharges (low inductance of 10 to 40 nH), is charged through resistance and diode protections (Rp, Dp) by a high-voltage power supply adjustable from 1 kV to 10 kV. The discharge is triggered by the use of a 5C22 thyratron tube: a half-period sinusoid of current of 0.4-ms duration and 1000-A maximum is then delivered to the circuit, and a spark is created between the coaxial electrodes tips immersed in water (discharge gap 0.5 mm). As a consequence of this electric arc, a strong acoustic pulse generally followed by an oscillating cavitation bubble is generated (Pernod et al. 1994a, 1994b).

2.2. High-speed shadowgraph system and pressure measurement

Based on the Cranz-Schardin visualization principle, the system consists of 24 light flashes triggered successively after the command with an adjustable time interval (from 100 ns to 1 s). Each flash of 300-ns duration is transformed into a parallel beam by a spherical mirror of 50 cm diameter and a 4-m radius of curvature. The beam then crosses the field of interest where the mini-sparker is placed, and is focused by a second spherical mirror identical to the first one on the 24 lenses of the receiving photographic plate. For its characterization, the pulse emitted by the mini-sparker has been observed by high-speed cinematography and measured by a broad-band hand-made hydrophone (100 kHz to 10 MHz at −7 dB, Nakamura & Otani 1992).

2.3. Characteristics of the mini-sparker emission

The spatial behavior of the emitted wave is shown in the photograph extracted from the 24 photographs (Fig. 1). The pulse presents a well-marked wave front of approximately spherical shape, showing that the source can be considered punctual. The temporal behavior presented in Fig. 2a can be roughly divided into two parts: in the first one, the pressure grows very sharply from 0 to 790 mbar with a rise time of about 100 ns; in the second part, the pressure decreases more slowly in an exponential form, with the total signal duration being about 2 μs. The Fast Fourier Transform of this pulse signature is presented in Fig. 2b. We can see that the frequency spectrum is very large, extending from less than 10 kHz to over than 10 MHz, as can be expected from the steepness of the rising front, and verified by high-pass filtering of the 10-MHz cut-off frequency. Typical maximum frequencies are 2 MHz (−10 dB), 300 kHz (−5 dB), or 200 kHz (−2 dB), the spectrum amplitude being very flat for this last case.

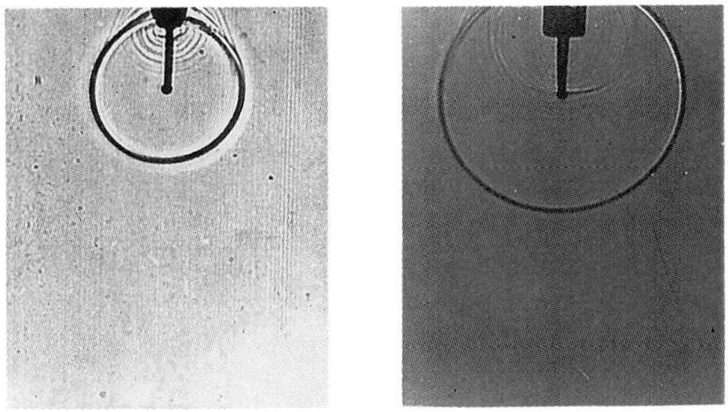

Figure 1. Typical emitted wave.

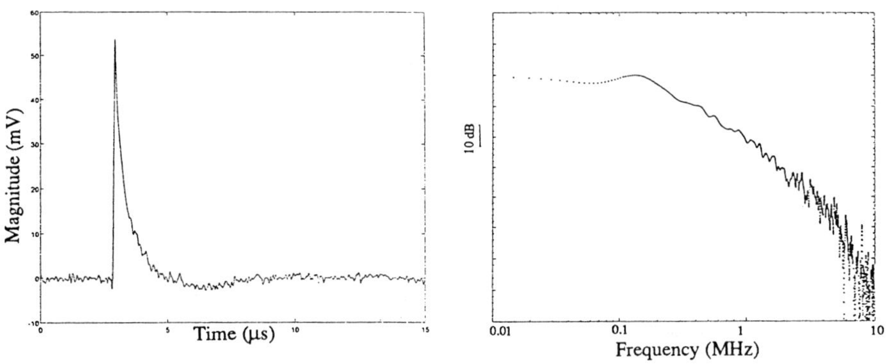

(a) (b)

Figure 2. Pressure measurement and spectrum.

3. Application to diffraction on an elastic sphere

Many papers have been published on investigations of acoustic diffraction of a plane wave by an elastic cylinder or cylindrical shell, which represents a two-dimensional problem. Considering our source properties, we have applied our investigation system to spherical wave diffraction by an elastic sphere, which represents a simple three-dimensional problem owing to symmetry properties.

3.1. Experimental set-up

A 3.5-cm diameter glass sphere is chosen as a diffracting object. It is held by a 0.1-mm thick thread which generates negligible diffraction. The sphere is placed 5 cm above the shock-wave source, as far as possible from the water surface and water-tank sides to avoid echo interferences. Several series of 24 photographs each were used to study the whole diffraction phenomenon with 2-μs intervals.

3.2. Results

Figure 3 shows a series of shadowgraphs of the wave fronts resulting from a pulse coming from below. In our interpretation of these photographs presented in Fig. 4a, we will use the notation (n,m) suggested by Brill and Uberall (1971)

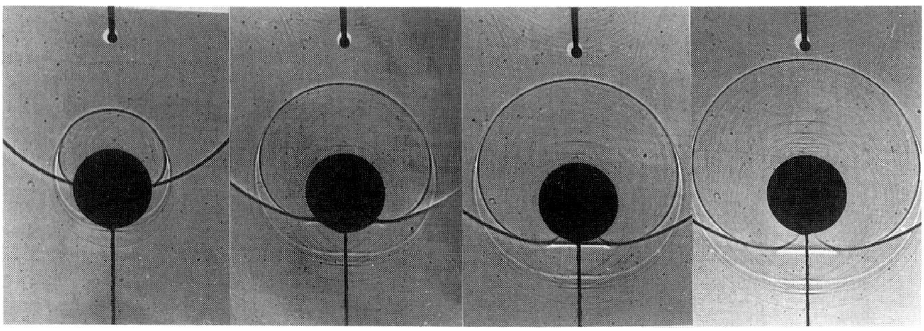

Figure 3. Shock diffraction at different stages. Incident-shock velocity ≈ 1480 m/s.

for identification of the different waves emerging from the target: m is the total number of reflections and n is the number of reflections as a shear wave. Five different types of waves are clearly observed: (1) The strong incident wave appears as a spherical wave front centered on the mini-sparker electrode tips and propagates downward. (2) The specular reflection (0,0) corresponds to the backscattered spherical wave front and joins the incident wave at the contact points with the glass sphere. (3) We can observe with a weaker amplitude the direct longitudinally transmitted wave through the sphere (1,0), which has anticipated the incident wave front owing to the greater speed of propagation in the solid medium. A similar transverse wave through the sphere (1,1) should theoretically be present, but is not observed on the photographs probably due to its very weak amplitude and the limited sensitivity of the visualization system. (4) Far more interesting is the very clear observation, in the shadow region of the sphere, of the Franz wave adjoining and prolonging the reflected wave (0,0). (5) We have grouped in the last category all the waves rotating around the sphere which are due to the multiple internal reflections and refractions, including the mode-converted waves. Their number is theoretically infinite; limitations in observations are due to attenuation and the limited sensitivity of the visualization system. All these waves are identified on the transducer signal (Fig. 4b). Franz waves are clearly visualized in Fig. 5a. Apart from all these identified waves, the photographs also show backscattered waves propagating behind the specular reflection which we have not been able to identify. They form a group with different speeds just like the internal reflection radiation waves (Fig. 5b).

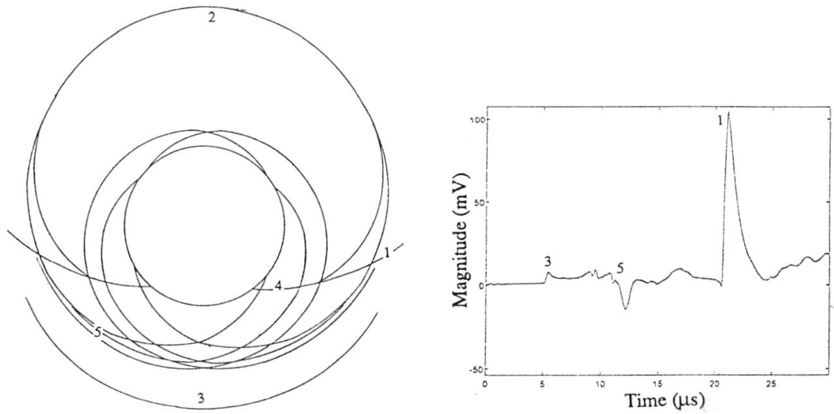

(a) (b)

Figure 4. Wave-front patterns: (a) diagram, (b) signal.

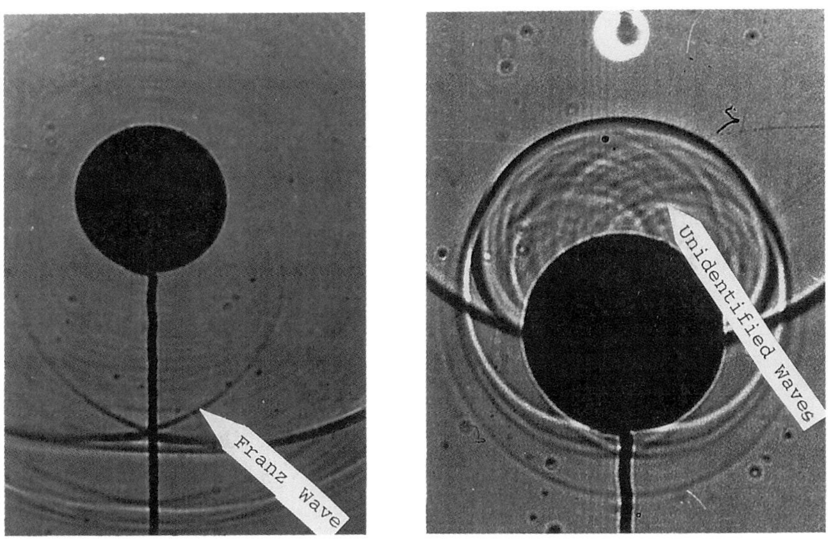

(a) (b)

Figure 5. Visualizations: (a) Franz waves, (b) unidentified waves.

4. Conclusion

A mini-sparker source which generates an optimized spherical wave front with quasi-Dirac-type signature has been developed for impulse-response measurements. Owing to the high acoustic power generated, its association with an ultra-high-speed shadowgraph visualization system is a useful tool for diffraction experiments. Examples with elastic spheres have shown that typical diffused waves reported in the literature (Neubauer & Dragonette 1970, Welton et al. 1980) can be observed with a very high quality of visualization. The same devices can be used for many other wave-structure observations, for example, interactions between shocks and particles, underwater explosions, etc.

References

Brill D, Uberall H (1971) Acoustic waves transmitted through solid elastic cylinders. JASA 50(3):921–939.

Nakamura Y, Otani T (1992) Frequency response of a piezoelectric polymer film hydrophone and an elastic wave induced on the backing surface. Japanese J of Appl Phys 31-1:266–268.

Neubauer WG, Dragonette LR (1970) Observation of waves radiated from circular cylinders caused by an incident pulse. JASA 48(5)2:1135–1149.

Pernod P, Merlen A, Kemmou A, Ahyi A (1994a) Etude par visualisations ultra rapides et mesures de pression d'une source impulsionnelle large bande non directive. Journal de Physique IV, Colloque C5, Supplément au Journal de Physique III, 4:729–732.

Pernod P, Merlen A, Kemmou A, Ahyi A, Delannoy B (1994b) Punctual transient acoustic source for experimental impulse response measurements. In: IMACS, IEEE-SMC, SPRANN 94, IMACS International Symposium on Signal Processing, Robotics and Neural Networks, pp 689–692.

Welton PJ, De Billy M, Hayman A, Quentin G (1980) Backscattering of short ultrasonic pulses by solid elastic cylinders at large k.a. JASA 67(2):470–476.

Physics of a shock-generated freely moving compressible vortex

K. von Ellenrieder, C. T. Kao, D. Bershader*
Stanford University
Stanford, CA 94305 USA

Abstract: This work represents a preliminary study of the physical mechanisms which govern the motion of compressible free vortices. An analysis of results from previous experimental and computational studies suggests that the transport of angular momentum occurring within such vortices can be categorized into three different flow regimes: convective, convective/diffusive, and diffusive. Additionally, the dependence of radial velocity and dissipation in the vortex on unsteady conditions outside of its core are discussed.

Key words: Vortex, Compressibility, Dissipation, Angular momentum

1. Introduction

The two-dimensional motion of an isolated free vortex in an incompressible flow is well understood. Conservation of mass requires that there be no radial velocity in the vortex, and one finds that viscous diffusion is the dominant process governing the fluid's motion. When the compressibility of fluid within the vortex and far from the vortex center are important, an analysis of the flow becomes much more complicated. In studying two-dimensional compressible vortices one must consider the effects of heat conduction, viscous dissipation, compressibility, and convection by radial mass flux on the vortex's motion and structure (Bershader 1995).

This study is based on the vortices examined by Mandella (Mandella 1987 and Mandella & Bershader 1987), which were created by passing a shock of $M \approx 1.8$ over a NACA 0018 airfoil at a 30° angle of attack. The resulting vortex was embedded in a free stream with a Mach number of $M_\infty \approx 0.52$ and a Reynolds number, based on airfoil chord length of 3.7 cm, of $Re_\infty \approx 4.4 \times 10^5$. Using density and pressure measurements of the free vortex, and assuming the radial pressure gradient was balanced by centrifugal forces, Mandella calculated the radial distributions of tangential velocity and angular momentum (Fig. 1). The density distribution was empirically fit to a time-dependent Lorentzian function. During an experimental observation period of several hundred microseconds, the free vortex traveled a distance of more than seven times its

*In memoriam.

own diameter and was convected by the free stream at a remarkably constant velocity of around 1.8×10^2 m/s. This feature, and the rather modest change of vortex structure during the observations, imply that the time scale for associated viscous dissipation effects is substantially longer than experimental times. To see this, consider the classic work by Lamb (1945) concerning incompressible vortices which predicts that vortex decay due to viscosity is exponential:

$$v_\phi = \frac{\Gamma_0}{2\pi r}(1 - e^{-\frac{r^2}{4\nu t}}). \tag{1}$$

Here v_ϕ is tangential velocity, r radial distance from the vortex center, Γ_0 the initial circulation, ν the kinematic viscosity, and t is time. When the exponential term above is the same order as 1.0, or ≥ 0.3, viscous diffusion has significantly affected the vortex's structure. The vortex core size is given by the radius of maximum tangential velocity $r_c \sim 1.0$ mm. When based on r_c and ν the viscous diffusion time is $\sim 1.4 \times 10^{-2}$ s; this is almost two orders of magnitude longer than the experimental observation time ~ 500 μs. Additionally, note that the vortex Reynolds number, based on a circulation strength of $\Gamma \approx 2.5$ m^2/s, is $Re_\Gamma \equiv \Gamma/\nu \approx 1.7 \times 10^5$.

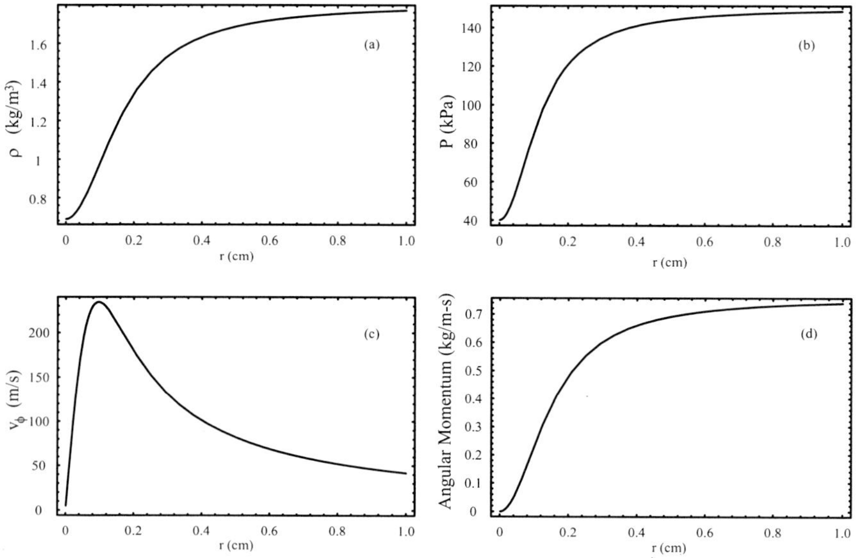

Figure 1. Radial profiles of (a) density, (b) pressure, (c) tangential velocity, and (d) angular momentum within the vortex.

Measurements of pressure and density, denoted as p and ρ respectively, illustrate that the isentropic assumption $p \propto \rho^\gamma$ is not valid within the vortex, and indicate that some sort of dissipation has occurred (Mandella 1987). The key to the observed dissipation may well reside in the radial mass flow associated with unsteady conditions outside of the vortex core.

Colonius et al. (1991) have performed an analytical and computational study in which the pressure, density, and tangential velocity measurements reported by Mandella were used as the initial conditions for a simulation of the free compressible vortex. Flow conditions far from the vortex core were assumed to be constant, and the free vortex was allowed to evolve in time under the action of viscosity. The computational time period used was comparable to the experimental times of Mandella's study, and the resulting radial distributions of pressure, density, and tangential velocity were found to agree with the experimentally measured values. However, the computations predicted much smaller radial velocities than calculated by Mandella.

An analysis of the angular-momentum transport occurring within such compressible vortices suggests that there are distinct flow regimes into which their time evolution can be characterized. Each flow regime is determined by the magnitude of the radial velocity and the core size of the vortex. Additionally, it is shown that small variations in the pressure outside of a vortex's core may influence the radial velocity and dissipation occurring within the vortex.

2. Angular-momentum conservation

For a two-dimensional and cylindrically symmetric fluid flow, the equation governing the conservation of angular momentum, expressed in cylindrical coordinates, is:

$$\rho r^2 \frac{\partial}{\partial t}\left(\frac{v_\phi}{r}\right) + \rho v_r \frac{\partial}{\partial r}(r v_\phi) - \frac{1}{r}\frac{\partial}{\partial r}\left[r^3 \mu \frac{\partial}{\partial r}\left(\frac{v_\phi}{r}\right)\right] = 0, \tag{2}$$

where μ and v_r correspond to viscosity and radial velocity, respectively. The first term in this equation is the local time rate of change of angular momentum, the second its convection by radial mass flux, and the final term is its rate of change due to viscous torque. Using the constants r_c, τ, ρ_∞, μ_0, v_{r0}, and $v_{\phi 0}$, the angular-momentum conservation equation can be cast in the following nondimensional form:

$$\hat{\rho}\hat{r}^2 \frac{\partial}{\partial \hat{t}}\left(\frac{\hat{v}_\phi}{\hat{r}}\right) + \left\{\frac{\tau v_{r0}}{r_c}\right\}\hat{\rho}\hat{v}_r \frac{\partial}{\partial \hat{r}}(\hat{r}\hat{v}_\phi) - \left\{\frac{\mu_0 \tau}{\rho_\infty {r_c}^2}\right\}\frac{1}{\hat{r}}\frac{\partial}{\partial \hat{r}}\left[\hat{r}^3 \hat{\mu} \frac{\partial}{\partial \hat{r}}\left(\frac{\hat{v}_\phi}{\hat{r}}\right)\right] = 0. \tag{3}$$

As shown in Table 1, quantities characteristic of Mandella's experimental results are chosen for r_c, ρ_∞, μ_0, and $v_{\phi 0}$. Note that $v_{\phi 0}$ cancels from the expression, and that analysis yields the values for the time constant of angular-momentum transport τ, and the radial velocity (at r_c) v_{r0}. An examination of the orders of magnitude of the two resulting dimensionless groups (bracketed terms) suggests that there are three flow regimes into which the transport of angular momentum within a vortex can be categorized: convective, convective/diffusive, and diffusive.

Table 1. Physical constants characteristic of the free vortex.

Constant	Value	Description
r_c	1.0 mm	Vortex core radius
ρ_∞	1.8 kg/m^3	Free-stream density
μ_0	2.0×10^{-5} Ns/m^2	Coefficient of viscosity (air)
$v_{\phi 0}$	2.3×10^2 m/s	Tangential velocity at r_c

2.1. Convective: $\tau v_{r0}/r_c \sim 1.0 \gg \mu_0\tau/(\rho_\infty {r_c}^2)$

In this flow regime convection is the dominant process by which angular momentum is transferred within the vortex. By eliminating τ in the inequality, approximate values may be found for the radial velocity: $v_{r0} \gg 1.1 \times 10^{-2}$ m/s or more roughly $v_{r0} > 3.0 \times 10^{-2}$ m/s. Estimates for the time scale of this motion are given by $\tau \sim r_c/v_{r0}$, which yields a time scale of $\tau \ll 9.1 \times 10^{-2}$ s or $\tau < 3.0 \times 10^{-2}$ s. Note that the radial velocities predicted by Mandella $v_{r0} \sim 0.3$ m/s lie in the convective range, and that for the experimental vortex under investigation, when v_{r0} is negative (inward flow) the angular momentum in the vortex is increasing (Fig. 1d).

2.2. Convective/diffusive: $\tau v_{r0}/r_c \sim 1.0 \sim \mu_0\tau/(\rho_\infty {r_c}^2)$

When the magnitude of each dimensionless group is on of the order of 1.0, both the effects of viscosity and convection must be considered in determining angular-momentum transport within the vortex. An approximate value of v_{r0} in this region is given by: 3.0×10^{-3} m/s $\leq v_{r0} \leq 3.0 \times 10^{-2}$ m/s. The radial velocities reported by Colonius et al., $v_{r0} \sim 6.8 \times 10^{-3}$ m/s, lie in this range.

2.3. Diffusive: $\tau v_{r0}/r_c \ll 1.0 \sim \mu_0\tau/(\rho_\infty {r_c}^2)$

In the diffusive regime viscosity is responsible for the spreading of angular momentum. The time scale for these effects is $\tau >\approx 0.09$ s, which is longer than the time scale predicted above for the incompressible vortex, and is the same as the "slow" time scale of Colonius. Radial velocities in the range $v_{r0} \ll 1.1 \times 10^{-2}$ m/s or $v_{r0} < 3.0 \times 10^{-3}$ m/s are expected for this regime.

2.4. Discussion

In the computational study of Colonius et al. the radial velocity found was two orders of magnitude smaller than the radial velocity experimentally inferred by Mandella. This difference is most likely due to unsteady variations in the experimental ambient-flow conditions. In the computational study, the external pressure, density, and temperature were fixed — the only mechanism for changes to occur in the vortex was viscosity. In the experiment, the vortex forms in a complex, unsteady flow field. The vortex is created in the low-pressure region behind an airfoil, and after being traversed by a shock, is convected away from the airfoil. The radial pressure gradient across the vortex is approximately balanced by its tangential velocity. It is likely that small changes in the external pressure cause the tangential velocity profile to change, and that the radial convection of angular momentum is the mechanism responsible.

Dimensional analysis suggests that unsteady ambient pressure may account for the large radial velocity found by Mandella. When the Reynolds Number is large, and viscous effects are negligible, the radial momentum and the angular-momentum conservation equations, respectively, can be written as:

$$\rho\frac{{v_\phi}^2}{r} = \frac{\partial p}{\partial r}, \qquad \text{and} \qquad \rho\frac{\partial}{\partial t}(rv_\phi) = -\rho v_r \frac{\partial}{\partial r}(rv_\phi). \tag{4}$$

Rearranging these expressions, and using the characteristic values listed in Table 1, an estimate for the radial velocity is found to be:

$$v_{r0} \sim \frac{1}{\tau v_{\phi 0}}\left[\frac{{r_c}^3}{\rho_\infty}\frac{\partial p}{\partial r}\bigg|_c\right]^{1/2}. \tag{5}$$

A 5% variation in the vortex core pressure gradient, $\frac{\partial p}{\partial r}\big|_c \sim 3.4 \times 10^7$ N/m^3 (as estimated from Fig. 1b), occurring over a period of $\tau \sim 500$ μs, may be adequate to produce a radial velocity of $\sim 2.7 \times 10^{-1}$ m/s.

3. Dissipation

With the assumption of cylindrical symmetry, a formal statement of the second law governing the creation of entropy in a fluid system is:

$$\frac{\partial s}{\partial t} + v_r\frac{\partial s}{\partial r} - \frac{1}{\rho T r}\frac{\partial}{\partial r}\left(r\kappa\frac{\partial T}{\partial r}\right) + \frac{\kappa}{\rho T^2}\left(\frac{\partial T}{\partial r}\right)^2 \geq 0 \tag{6}$$

where κ is the thermal conductivity $\sim 2.6 \times 10^{-2}$ J/(msK). The left-hand side of the equation is the total production of entropy per unit mass s, which must be positive. The last two terms containing the temperature T represent the outflow of entropy per unit mass, and the first two terms are the net increase in entropy per unit mass. The radial temperature distribution and the temperature gradient may be determined from Figs. 1a and 1b by use of the ideal gas law. Dimensional analysis reveals that the entropy outflow terms are several orders of magnitude smaller than the unsteady and convective entropy terms. Therefore the role of heat conduction within the vortex may be considered a minor one. The unsteady entropy term may be expressed as a function of density and pressure:

$$\frac{\partial s}{\partial t} = \frac{R}{(\gamma - 1)} \left[\frac{1}{p} \frac{\partial p}{\partial t} - \frac{\gamma}{\rho} \frac{\partial \rho}{\partial t} \right], \tag{7}$$

where R and γ are the gas constant for air and the ratio of specific heats, respectively (Colonius et al. 1991). The rate of change of pressure is found by use of the radial-momentum equation:

$$\frac{\partial p}{\partial t} = \frac{\partial p_\infty}{\partial t} - \int_r^\infty \left[\frac{{v_\phi}^2}{r} \frac{\partial \rho}{\partial t} + \frac{2\rho v_\phi}{r} \frac{\partial v_\phi}{\partial t} \right] dr. \tag{8}$$

Note that for constant ambient conditions, the p_∞ term vanishes. However, small variations of $p_\infty \sim 1.5 \times 10^5$ Pa over time scales on the order of Mandella's experiments may have a significant effect on the dissipation produced. For example, a 10% positive change in p_∞ over a time of $\tau \sim 500$ μs, would produce a value of $\rho T \frac{\partial s}{\partial t} \sim 7.5 \times 10^7$ N/m^{2}s. By comparison, in Mandella's experiments, $\rho T v_r \frac{\partial s}{\partial r} \sim 3 \times 10^8$ N/m^{2}s.

4. Conclusions

The analysis above suggests that the dissipation and radial velocities inside compressible vortices are influenced by unsteady processes occurring outside of their cores. It has also been shown that the nature of the angular-momentum transport mechanisms in such vortices varies according to the vortex radial velocity and core size.

Acknowledgement. The authors are grateful for partial support of this work by NASA Consortium Agreement No. NCC2-5002, and the NASA Graduate Student Researchers Program. Additionally, the first author would like to thank Professor Cantwell for his help and advice.

References

Bershader D (1995) Compressible vortices. In: Fluid Vortices, Sheldon Green (ed), Klewer Academic Publishers.

Colonius T, Lele S, Moin P (1991) The free compressible viscous vortex. J Fluid Mech 230:45–73.

Lamb H (1945) Hydrodynamics, 6th Edition, Dover Publications.

Mandella M (1987) Experimental and Analytical Studies of Compressible Vortices, Ph.D. Thesis, Dept of Applied Physics, Stanford University.

Mandella M, Bershader D (1987) Quantitative Study of the Compressible Vortex: Generation, Structure, and Interaction with Airfoils, AIAA Paper 87-0328.

Study of the propagation of shock waves over bodies by holographic interferometry

J. Le, X. Ye, X. Wu, H. Yang
China Aerodynamics Research and Development Center
P.O. Box 211, Mianyang, Sichuan, China

Abstract: The propagation of shock waves of $M_s = 1.8$ over various bodies such as single cylinders, double cylinders, wedges with back step, or plates with sharp-edged slits in a shock tube or in a channel bend was studied experimentally with a holographic interferometer. Quantitative density fields obtained from interferograms are in good agreement with computations for the case of the wedge with back step.

Key words: Shock wave, Holographic interferometry, Shock tube

1. Introduction

The problem of the propagation and diffraction of shock waves past solid bodies has been investigated extensively in the past both experimentally and computationally. Some excellent results of the quantitative measurement of density fields have been obtained with holographic interferograms, such as: fifteen basic cases of oblique shock-wave reflection (Glaz et al. 1986); and shock-wave diffraction through an aligned baffle system (Reichenbach 1994). However, for some problems with complicated flow structure, including shock waves reflected from a solid wall and the interaction of shock waves with vortices, the quantitative measurements of density fields from time sequences of interferogram frames are still very necessary and important for validating computer codes and for studing complicated flow phenomena. For example, only computational results of the propagation of shock waves through double cylinders in a channel bend have been completed (Fursenko and Sharov 1992), thus investigation of this kind of flow using holographic interferometry is of great value in providing insight into the physical phenomena. The purpose of this study is to quantitatively measure the unsteady flow fields of the propagation of shock waves over some typical bodies using a shock-tube technique in which the incident M_s can be precisely controlled and to compare the results of experiment and computation for one of these typical bodies.

2. Experiment

The experiments were carried out in CARDC's small shock tube with a test section of 32 mm × 32 mm, with two pairs of observation windows of 100 mm × 32 mm. The distance between windows is 400 mm. The driver of the shock tube has a circular section with a diameter of 70 mm. Figure 1 shows the shock tube and the optical system of the holographic interferometer. Usually, a 2D model is located in view of the first window and can be easily exchanged for a channel bend where the double cylinders are placed. Their detailed sizes are shown in Fig. 2. The incident-shock Mach number M_s can be precisely controlled by a specially made plastic diaphragm and by high-accuracy measurements of gas pressure both in the driver and in the driven section. The distances from the transducers to the model are arranged to measure the shock speed and to synchronize the ruby laser. For quantitatively measuring the density fields,

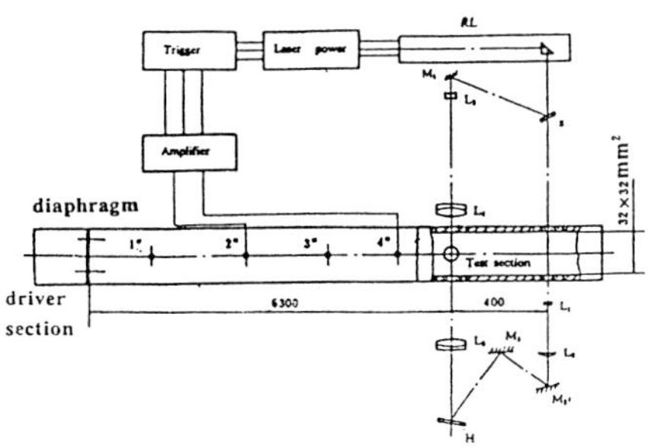

Figure 1. Shock tube and holographic interferometer system.

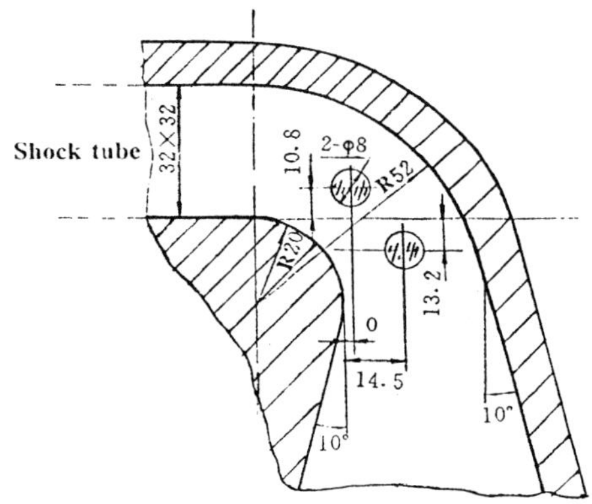

Figure 2. Size of channel bend with cylinders.

double-exposure holographic interferometry was used. The first exposure is performed before the incident shock wave arrives and the second exposure is triggered at the moment at which the shock waves interact with the model. By incrementing triggered delay times (with precison of 1 μs), the frame-by-frome holographic interferograms, separately obtained from a series of tests, can be considered to capture the unsteady process of shock-wave interaction with a body.

3. Results and discussion

Propagation through a slit with sharp edges. The apex angle of the sharp edge is 20° and the distance of two apices is 4 mm. Some of these experimental results are similar to those obtained by shadowgraphs (Amann and Reichenbach 1973). In Fig. 3 are holographic interferograms taken at different instants. After the incident shock wave passes over the model, two intense vortices are formed at the sharp edges but only one diffracting shock wave moves forward. The experimentally determined isopycnics are presented in Fig. 3d. The velocity of the diffracting shock wave along the slit plate can be measured by two instantaneous interferograms (Figs. 3a and b). Then its density ratio (ρ_N/ρ in Fig. 3d) can be calculated by the normal shock-wave relation.

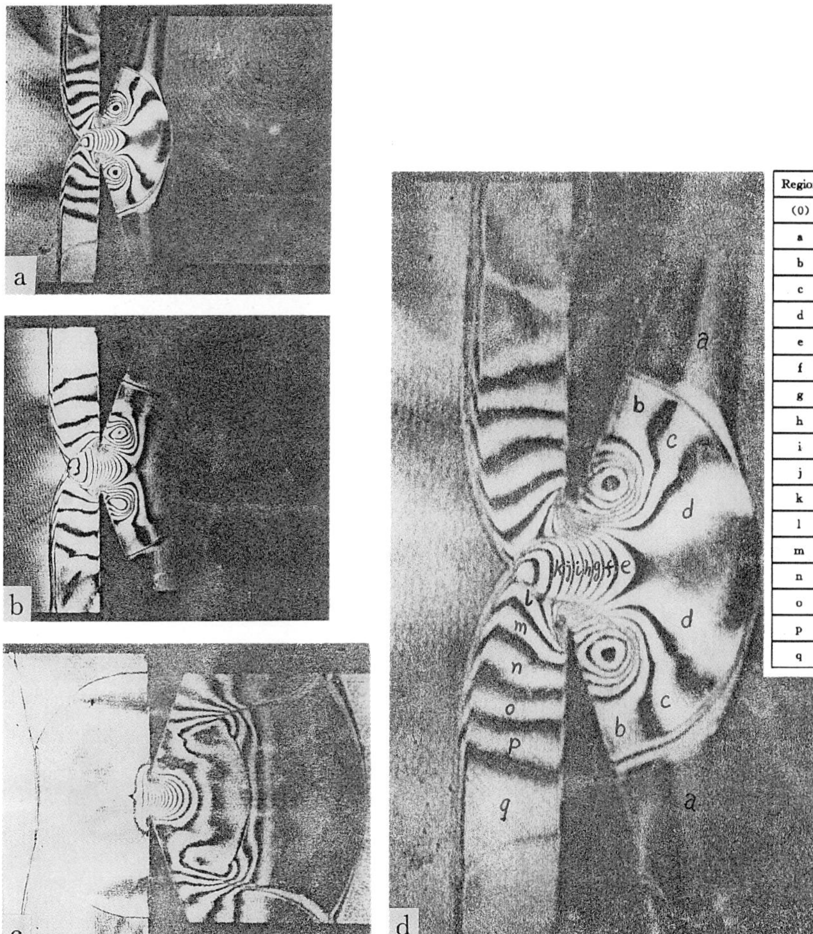

Region	ρ_N/ρ_0
(0)	1.00
a	1.26
b	3.29
c	3.55
d	3.81
e	4.07
f	4.33
g	4.59
h	4.85
i	5.11
j	5.37
k	5.63
l	5.89
m	6.15
n	6.41
o	6.67
p	6.93
q	7.19

Figure 3. Holographic interferograms, $M_s = 1.81$, $P_0 = 32$ kPa, $T_0 = 300.5$ K: (a) $t = 0$ μs; (b) $t = 2$ μs; (c) $t = 30$ μs; (d) experimentally determined density contours of case a.

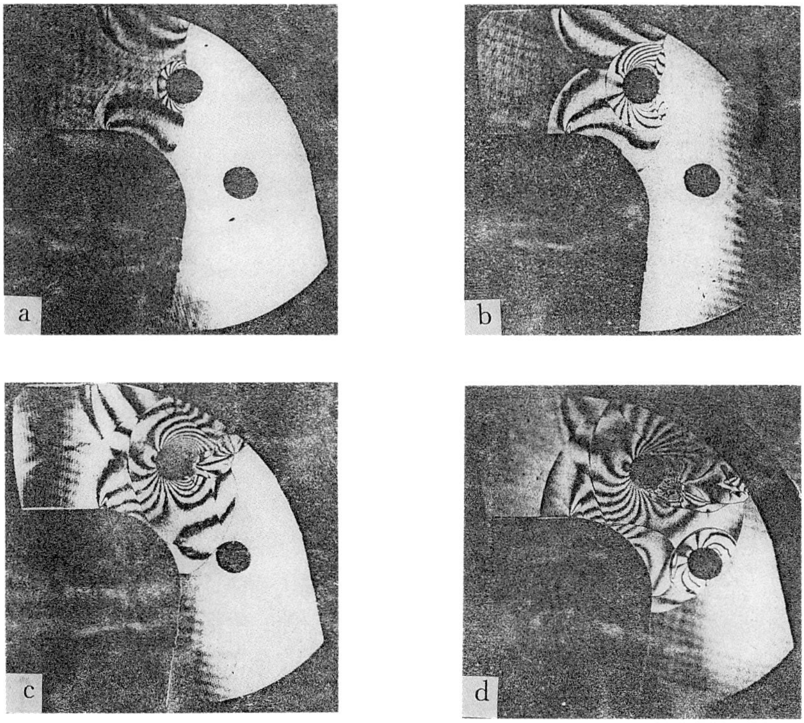

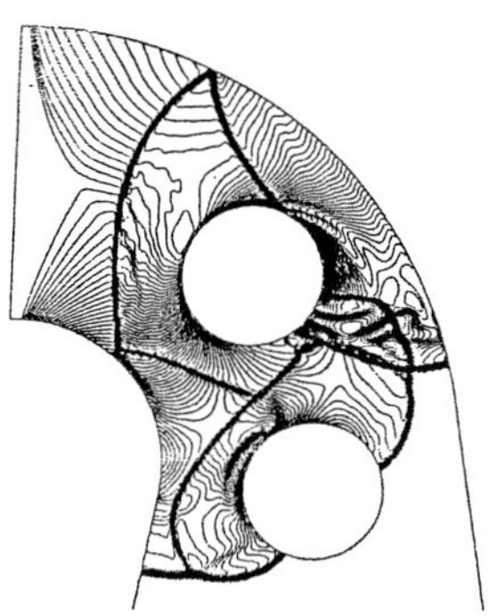

Figure 4. Computed density contours from Fursenko and Sharov (1992); $M_s = 2.1$.

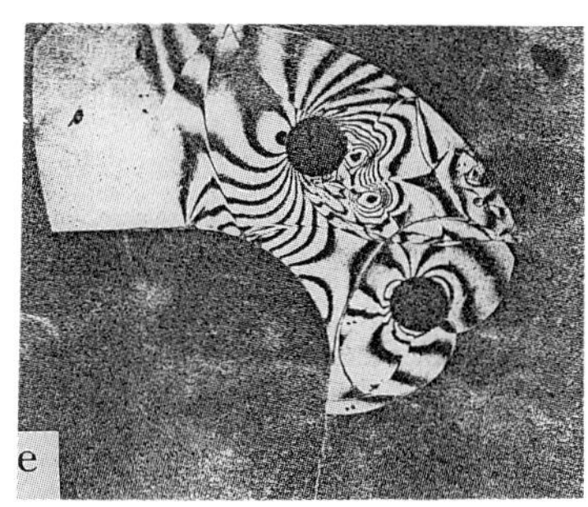

Figure 5. Holographic interferograms from CARDC. $M_s = 1.82$, $P_0 = 32$ kPa, $T_0 = 288$ K: (a) $t = 10\ \mu$s; (c) $t = 24\ \mu$s; (d) $t = 38\ \mu$s; (e) $t = 52\ \mu$s.

Propagation through a channel bend with double cylinders. There is only one paper that computationally treats this unsteady problem based on

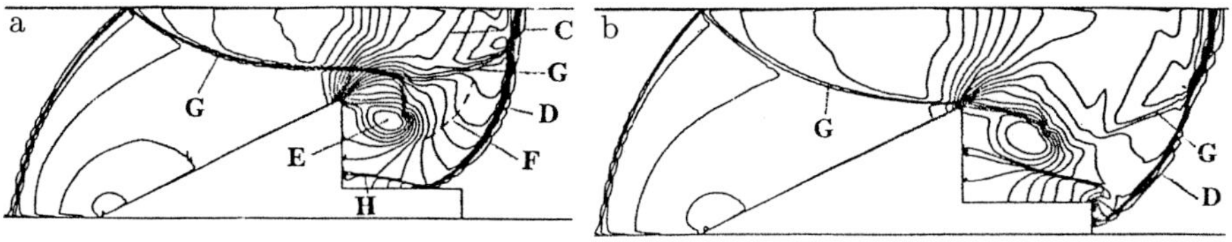

Figure 6. Computed density by ENO scheme. $M_s = 1.80$, $P_0 = 32$ kPa, $T_0 = 285$ K, $Re = 5.3 \times 10^5$, 351×121 grid points: (a) $t = 36.9$ μs; (b) $t = 44.88$ μs.

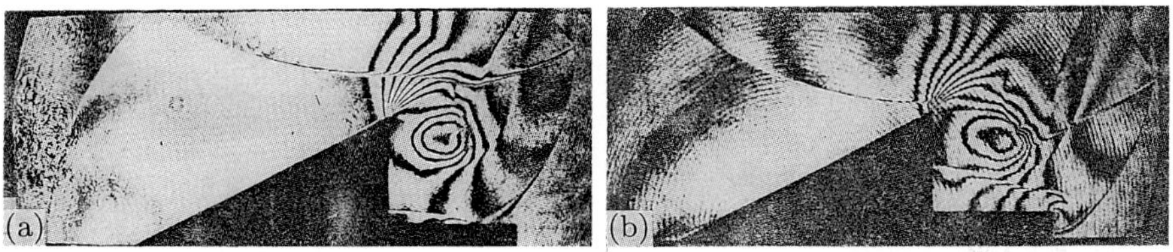

Figure 7. Holographic interferograms. $M_s = 1.80$, $P_0 = 32$ kPa, $T_0 = 285$ K, $\Delta t = 0$ is set at the moment when the Mach stem arrives at the apex of the wedge: (a) $t = 36.9$ μs; (b) $t = 44.88$ μs.

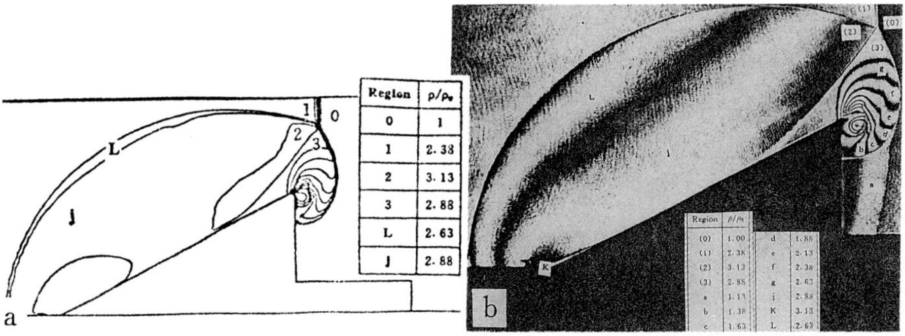

Figure 8. Comparison of density contours from computation and experiement. $M_s = 1.80$, $P_0 = 32$ kPa, $T_0 = 285$ K, $\Delta t = 12$ μs: (a) from computation; (b) from experiment.

the Euler equation with an unstructured finite-volume method (Fursenko and Sharov 1992). Figure 4 shows the result. In contrast with the computation, the experimental results shown in Fig. 5 represent a more complex wave system; in particular, the two vortices formed behind the first cylinder are more extensive. Viscous effects behind so called lambda shock wave are also stronger, thus the flow structure may depend on the Reynolds number.

Propagation past a wedge with back step. Flow fields were computed using the thin-layer approximated N-S equations with second-order ENO schemes (351×121 grid points). The model size and a detailed description of both experiment and computation are given by Le and Li (1994). As shown in Figs. 6 and 7, the vortex (E) has a significant effect on the reflected shock wave (G). The shock wave (G) is distorted a little as it approaches the vortex (shown in Fig. 7a), and then is split into three parts by the vortex (shown in Fig. 7b), its right half is accelerated and its center is completely merged by the vortex. Compared with the experiment, the interaction of the shock wave (G) with the vortex can be more clearly captured by the present computations (shown in Figs. 6a and b). In addition, as shown in Figs. 8a and b, the density contours from experiment and computation are in good agreement.

4. Conclusion

By using the present shock-tube technique, the incident M_s can be precisely controlled and time sequences of holographic interferograms can be taken, making it very ecomomical for us to study quantitatively the problems of unsteady compressible flow. However the accuracy of time measurements is very important and will be further improved.

References

Abe A, Takayama K (1989) Shock wave diffraction from the open end of a shock tube. In: 17th Int Symp on Shock Waves and Shock Tubes, pp 270–275.

Amann HO, Reichenbach H (1973) Unsteady flow phenomena in a shock tube nozzles. In: Proc 9th Int Shock Tube Symp, pp 96–112.

Fursenko AA, Sharov DM (1992) Numerical simulation of shock wave interactions, channel bends and gas nonuniformities. Computers Fluids 21(3):377–396.

Glaz HM, Collella P, Glass II (1986) A Detailed Numerical, Graphical, and Experimental Study of Oblique Shock Wave Reflections, UTIAS Report No. 285.

Le J, Li C (1994) Propagation of shock wave over a wedge. In: Proc 2nd Int Conf of Experimental Fluid Mechanics, pp 786–793.

Reichenbach H (1994) Color interferogram of shock propagation through an aligned baffle system. Shock Waves 4(2).

Accuracy of shock-wave propagation using geometrical shock dynamics

D. W. Schwendeman
Department of Mathematical Sciences
Rensselaer Polytechnic Institute
Troy, NY 12180 USA

Abstract: Shock-wave motion determined by the approximate theory of geometrical shock dynamics is compared with the motion obtained from solutions of the unsteady Euler equations. The problem of two-dimensional shock propagation in a channel with a 90° bend is used to carry out the comparison. Both strong shocks ($M_0 = 2.1$) and weak shocks ($M_0 = 1.1$) are discussed. The best agreement is found in regions where the shock is strengthening. An alternate approach in which the area-Mach-number rule would be replaced by a numerical calculation of the one-dimensional equations for flow in a ray tube is discussed.

Key words: Geometrical shock dynamics, Euler equations, Shock capturing, Shock-wave diffraction

1. Introduction

Geometrical shock dynamics is an approximate theory that determines the motion of the leading shock without an explicit calculation of the flow behind (Whitham 1957, 1959). Numerous investigations have used geometrical shock dynamics to calculate shock-wave motion and in many cases the results have been compared with experiments (e.g. Bryson and Gross 1961 and Henshaw et al. 1986). In general, good accuracy is reported but there is always some error. Geometrical shock dynamics gives an approximation of shock-wave motion as determined by the Euler equations. A detailed study of the accuracy of the approximate theory can be made by comparing numerical calculations of the equations of geometrical shock dynamics with those of the Euler equations. This paper gives some results of this study for the problem of shock-wave diffraction in a channel with a 90° bend.

In geometrical shock dynamics, the shock is moved according to an area-Mach-number rule. The area-Mach-number rule could be obtained from experimental data for a particular problem or set of problems, but in most cases it is taken from an analysis of shock propagation in a channel with gradual area change (Whitham 1958). A study of this problem and the approximations that lead to the area-Mach-number rule give some insight into the accuracy of the geometrical shock-dynamics approximation.

2. Governing equations

The equations of geometrical shock dynamics are

$$\nabla \cdot \left(\frac{M}{A} \nabla \alpha \right) = 0, \qquad M = \frac{1}{|\nabla \alpha|}, \qquad A = A(M), \tag{1}$$

where $\alpha(\mathbf{x}) = a_0 t$ gives the leading shock position $\mathbf{x}$ at a time t and a_0 is the ambient sound speed (Whitham 1959). In (1), $A = A(M)$ is taken to be

$$A = \exp\left(- \int \frac{M\lambda(M)}{M^2 - 1} \, dM \right), \tag{2}$$

where $\lambda(M)$ is a function that varies slowly between 4 when $M = 1$ and 5.0743 when $M \to \infty$ (for $\gamma = 1.4$). The form in (2) was proposed by Whitham following the work of Chester (1954) and Chisnell (1957). The equations in (1) form a hyperbolic system that describes disturbances in M and shock normal $\nabla \alpha / |\nabla \alpha|$ propagating on the leading shock.

The area-Mach-number rule in (2) gives the Mach number $M = U/a_0$ of a shock propagating down a channel with a given cross-sectional area $A(s)$, where s is distance down the channel. It is an approximation of the shock motion as determined by the one-dimensional (hydraulic) equations

$$\begin{aligned} \rho_t + u\rho_s + \rho u_s + \rho u \frac{A'(s)}{A(s)} &= 0, \\ u_t + u u_s + \frac{1}{\rho} p_s &= 0, \\ p_t + u p_s - a^2(\rho_t + u\rho_s) &= 0, \qquad a^2 = \frac{\gamma p}{\rho}, \end{aligned} \tag{3}$$

where ρ, u, p, and a are the density, velocity, pressure, and sound speed in the channel, respectively. The basic assumption that leads to (2) is that the change in $A(s)$ is small, and it can be shown that (2) would be exact if

$$\delta \equiv \left(\frac{1}{U} - \frac{1}{u + a} \right) \left(\frac{p_t + \rho a u_t}{p_t} \right) \tag{4}$$

vanished at the shock. In general, δ is not zero at the shock, and this quantity can be treated as a residual and can be used to study the error in the approximation of (3).

We will consider the motion of the leading shock in two space dimensions $\mathbf{x} = (x, y)$ as determined by a solution of (1). This motion is some approximation of the motion of the shock as determined by a corresponding solution of the unsteady Euler equations

$$\begin{bmatrix} \rho \\ \rho u \\ \rho v \\ E \end{bmatrix}_t + \begin{bmatrix} \rho u \\ \rho u^2 + p \\ \rho v u \\ (E+p)u \end{bmatrix}_x + \begin{bmatrix} \rho v \\ \rho u v \\ \rho v^2 + p \\ (E+p)v \end{bmatrix}_y = 0, \tag{5}$$

where

$$E = \frac{p}{\gamma - 1} + \frac{1}{2}\rho(u^2 + v^2)$$

is the energy. A solution of (5) gives the whole flow field, including the leading shock, and can be used to measure the accuracy of the geometrical shock-dynamics approximation.

3. Numerical methods

To make the desired comparisons, numerical methods are needed to solve (1) and (5). The numerical method used to solve (1) is the same one discussed in Schwendeman (1993). It is a finite-difference formulation of the conservation form in (1). The numerical method used to solve (5) is a second-order Godunov-type finite-difference scheme, similar to the one discussed in Colella (1990).

A numerical scheme is needed also to calculate solutions of (3) for a given $A(s)$, and to calculate the residual (4) at the shock. The desired solution of (3) will always have a shock in the C_+ characteristic, $ds/dt = u + a$, propagating down the channel. For this problem it is convenient to use a numerical scheme based on the characteristic form of (3) for the flow behind the shock and fit the shock in the C_+ characteristic using the normal shock conditions.

4. Results

The problem of two-dimensional shock propagation in a channel with a 90° bend is used to compare the leading shock as determined by a solution of (1) with the motion determined by a solution of (5). This is a good test case because the shock diffraction within the channel produces both regions of shock strengthening and weakening. The gap distance between the inner and outer walls of the channel is normalized to one and the radius of the inner wall is taken to be 0.9368. The channel geometry is the same one discussed in Henshaw et al. (1986) following the experiments in Edwards et al. (1983). The initial shock Mach number is M_0 and the ambient sound speed ahead of the shock a_0 is normalized to one.

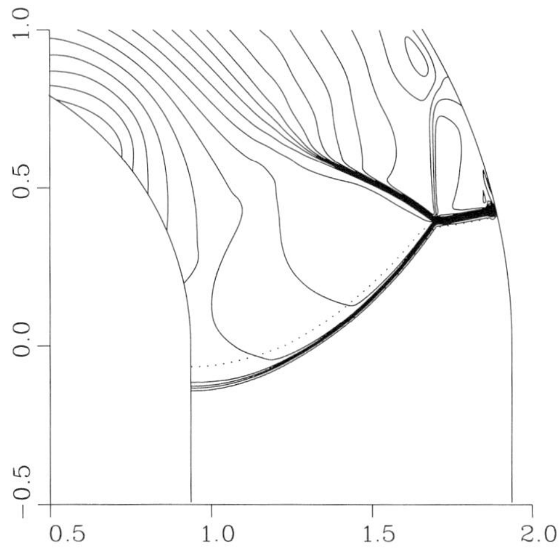

Figure 1. Contours of density (solid curves) and the approximate leading shock position (dotted curve) all at $t = 1$.

Figure 1 shows contours of density in a portion of the channel containing the leading shock at a time $t = 1$. In this example, $M_0 = 2.1$ and $t = 0$ when the shock first meets the 90° bend. The shock weakens near the inner wall and strengthens and forms a Mach stem near the outer wall. The dotted curve in this figure is the leading shock according to geometrical shock dynamics. It is slightly ahead of the Mach stem near the outer wall and shows a significant lag near the inner wall.

The behavior observed in Fig. 1 is consistent with the general belief that geometrical shock dynamics is most accurate in cases when the shock strengthens. This belief is based on a physical intuition of the approximations involved, but is not supported by any mathematical theory. For example, there is no "next correction" in the theory available to assess the error.

One way to make a closer examination of the behavior in Fig. 1 is to calculate $A(s)$ for various rays from the solution of (1) and then determine the motion of the shock along the rays based on solutions of (3). This approach gives an indication of the location of the shock based on a full calculation of (3). The residual in (4) can also be calculated and this quantity gives some measure of the approximation in the area-Mach rule.

Figure 2a shows contours of α in the channel from the numerical solution of (1). These contours give the successive shock positions according to geometrical shock dynamics. The dotted contour curve is $\alpha = t = 1$ ($a_0 = 1$) and this curve

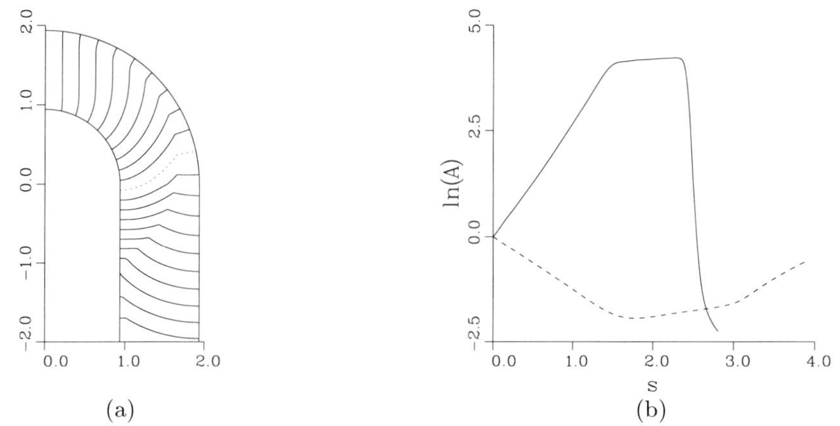

Figure 2. (a) Contours of α and (b) $\ln(A)$ versus s along the channel walls.

is the same one plotted in Fig. 1. From this solution, $A(s)$ can be calculated for various rays and Fig. 2b shows $\ln(A)$ versus s for the ray along the inner wall (solid curve) and the ray along the outer wall (dashed curve).

Numerical calculations of (3) for the two area curves in Fig. 2b give the motion of the shock along the walls without the area-Mach-number rule approximation. The dashed curves in Fig. 3a show this motion and these curves can be compared to the corresponding ones obtained from the solution of (1) which uses the area-Mach-number rule approximation (dotted curves) and the solution of the full Euler equations (solid curves). The set of three curves with the larger values of time t correspond to the inner wall while the other set belongs to the outer wall. Along the outer wall, where the shock strengthens, the agreement is excellent. There is some disagreement in the set of curves belonging to the inner wall and it is interesting to note that the dashed curve given by a calculation of (3) is not closer to the solid curve given by the Euler solution.

Values of δ along the walls are shown in Fig. 3b. The data set with smaller $|\delta|$ belongs to the outer wall and these smaller values are consistent with the better agreement between the dashed and dotted curves in Fig. 3a for the outer wall. The increase in δ near $s = 1.5$ seems to correspond to the change in slope of the area curves in Fig. 2b. It is interesting to note that the values of δ are not particularly small for either wall despite the relatively good agreement between the dashed and dotted for both cases.

A similar study has been carried out for a weak shock case where $M_0 = 1.1$. The general observations noted above for the relatively strong shock still apply

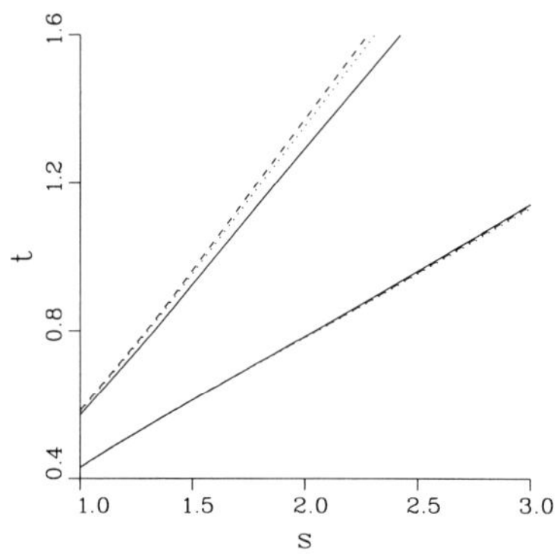

(a)

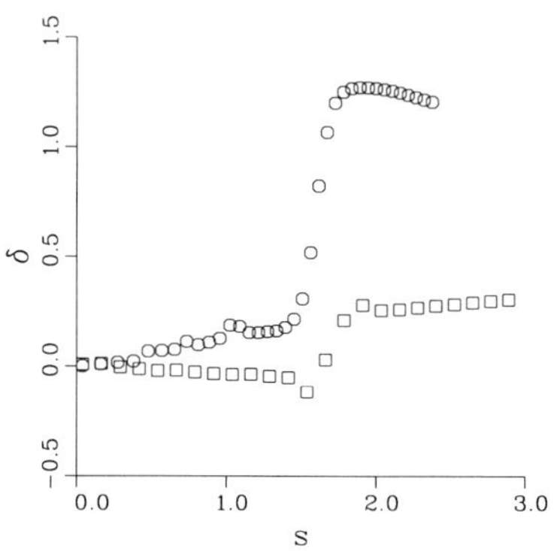

(b)

Figure 3. (a) Shock paths and (b) δ versus s along the channel walls.

and the level of accuracy is about the same in contrast to the general belief that geometrical shock dynamics would be a poor approximation for weak shocks.

Acknowledgement. This work was supported by the National Science Foundation.

References

Bryson AE, Gross RWF (1961) Diffraction of strong shock by cones, cylinders, and spheres. J Fluid Mech 10:1.

Chester W (1954) The quasi-cylindrical shock tube. Phil Mag 45:1293.

Chisnell RF (1957) The motion of a shock wave in a channel, with applications to cylindrical and spherical shock waves. J Fluid Mech 2:286.

Colella P (1990) Multidimensional upwind methods for hyperbolic conservation laws. J Comp Phys 87:171.

Edwards DH, Fearnley P, Nettleton MA (1983) Shock diffraction in channels with a 90° bend. J Fluid Mech 132:257.

Henshaw WD, Smyth NF, Schwendeman DW (1986) Numerical shock propagation using geometrical shock dynamics. J Fluid Mech 171:519.

Schwendeman DW (1993) A new numerical method for shock wave propagation based on geometrical shock dynamics. Proc R Soc Lond A 441:331.

Whitham GB (1957) A new approach to problems of shock dynamics, part I: two-dimensional problems. J Fluid Mech 2:145.

Whitham GB (1958) On the propagation of shock waves through regions of non-uniform area or flow. J Fluid Mech 4:337.

Whitham GB (1959) A new approach to problems of shock dynamics, part II: three-dimensional problems. J Fluid Mech 5:369.

A locally adaptive structured/unstructured Navier-Stokes solver for essentially unsteady shock-obstacle interactions at high Reynolds numbers

D. Ofengeim, E. Timofeev
Ioffe Physical-Technical Institute, Russian Academy of Sciences
Polytekhnicheskaya Str. 26, St. Petersburg 194021 Russia

P. Voinovich
Advanced Technology Center
P.O. Box 160, St. Petersburg 198103 Russia

Abstract: The paper presents a new numerical technique to compute essentially unsteady viscous-inviscid interactions at high Reynolds numbers. Special attention is given to the verification of the method.

Key words: Combined grids, Navier-Stokes equations, Unsteady flow

1. Introduction

The investigation of shock-wave interactions with obstacles is of fundamental and practical interest. The Euler gas model is widely used, providing reasonable results for high Reynolds number flows. Dynamically adaptive unstructured grids and high-resolution schemes contribute greatly to the computational efficiency of the shock-capturing method, as its accuracy is critically affected by resolution at the flow discontinuities. The highly accurate and efficient numerical technique has been developed previously by Fursenko et al. (1993, 1995), however, it cannot accurately predict a flow separation. An extension of the above technique to the Navier-Stokes solver is described in Ofengeim et al. (1995) and Beryozkina et al. (1996). The present paper contains a summary of this approach and focuses on numerical results in comparison with experiments.

2. Numerical model

The numerical model is based on the full Navier-Stokes equations. A thin layer of non-uniform structured orthogonal grid is constructed along the body surface. The rest of the computational domain is covered by an unstructured triangular grid (Fig. 1). In the course of a computation the unstructured grid is adapted to the solution using a hierarchical refinement/derefinement procedure. The structured grid and the adjacent cells of the unstructured grid are fixed.

Nonoverlapping control volumes are established for either grids. An unsplit explicit technique is employed on the unstructured grid. The convective terms are approximated by the second-order Godunov-type scheme (Fursenko et al. 1993), while central differences are applied to the viscous terms. The time-step operator for the structured grid is split according to physical processes. To

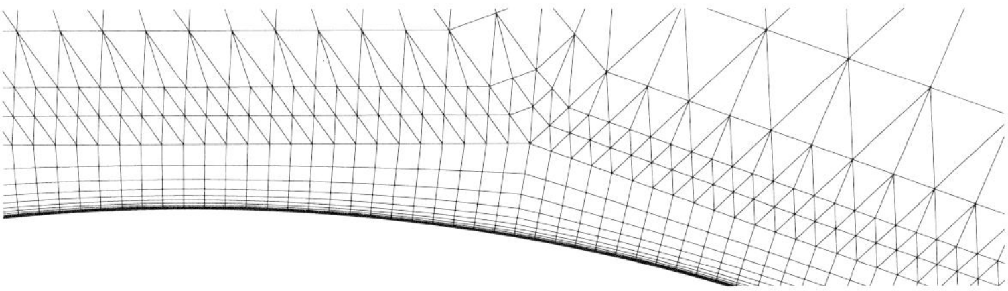

Figure 1. Detail of hybrid structured/unstructured grid.

compensate partially for the severe limitation of the time step, the inviscid operator is also split in space. The respective 1D operators are approximated by a 1D version of the above Godunov-type scheme. A predictor-corrector procedure enhances the temporal accuracy for all operators. The viscous operator on the structured grid is unsplit and contains central differences written implicitly in time for terms with second derivatives along the normal direction to the body surface, and explicitly (a predictor-corrector method) for the remaining viscous terms. The resulting set of linear difference equations with a block tridiagonal matrix of coefficients is solved by the Thomas algorithm.

3. Numerical results and comparison

Verification of the solver presents a problem due to a lack of analytical solutions, reliable computations, and experimental data for unsteady viscous flows with shocks. Consequently, several steady problems have been used as preliminary tests: computations of a boundary layer on a flat plate can be found in Ofengeim et al. (1995); the results for a shock-wave/boundary-layer interaction and for a supersonic flow over a NACA 0012 airfoil are presented below. Shock-wave diffraction over a circular cylinder is also considered as a test problem because of high-quality experimental data by Syshchikova et al. (1990) and Beryozkina et al. (1996) at hand.

All the subsequent computations have been carried out for perfect gases with constant specific heat ratio γ. The Prandtl number is equal to 0.72. The viscosity coefficient is calculated following Sutherland's law.

3.1. Steady tests

The steady-state supersonic $M_\infty = 1.5$, $Re = 10,000$ flow ($\gamma = 1.4$) over an adiabatic NACA 0012 airfoil at zero inclination has been considered as an example without separation. The oblique shock-wave impingement on the laminar flat-plate boundary layer represents another standard benchmark for Navier-Stokes solvers. The flow parameters ensuring separation are: $M_\infty = 2$, $Re = 296,000$, $T_{wall} = 1.7T_\infty$, $\gamma = 1.4$; the primary shock incidence angle is 32.6°.

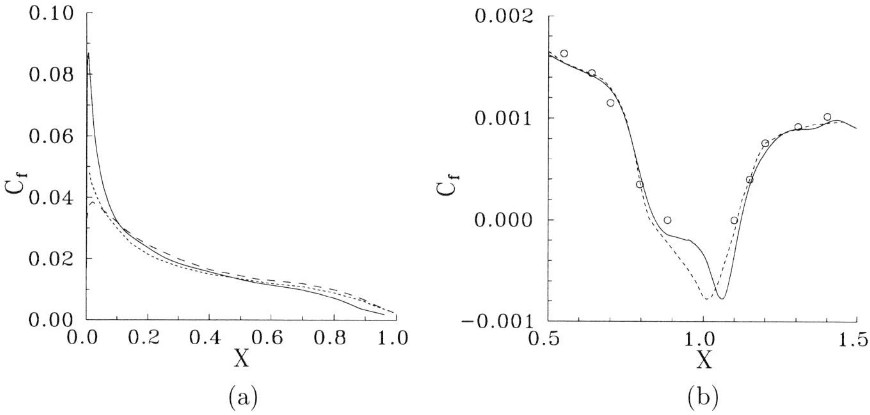

Figure 2. Dependence of the skin-friction coefficient C_f on the longitudinal coordinate X measured from the leading edge: (a) NACA 0012 airfoil; the dotted and dashed lines display computational data by Ivanov et al. (1989) and Schröder & Hänel (1987) respectively; (b) the shock-wave/boundary-layer interaction; the dotted line gives numerical data from Savel'ev (1988), circles are experimental data of Hakkinen et al. (1959). Solid lines represent our calculations.

The solutions are presented in Fig. 2 as skin friction over the surface. The skin friction is known to be very sensitive to both convective and viscous transport, and the good agreement indicates an accurate approximation.

3.2. Shock-wave diffraction over a circular cylinder

The above approach has been applied to the investigation of shock-wave ($M_s = 3.0$) diffraction over a circular cylinder in CO_2 ($\gamma = 1.289$). The Reynolds number based on the cylinder radius r is 80,000; $T_{wall} = 293$ K.

Some results are given in Figs. 3 and 4 for comparison of the shock-tube experiments with Euler computations. The time t_r is measured from the first shock-cylinder collision. The incident shock wave passes a distance of r in one time unit. The number of unstructured-grid nodes varied from 272 for the background grid to 12552 for the final grid at $t_r = 5.8$ (5 refinement levels with

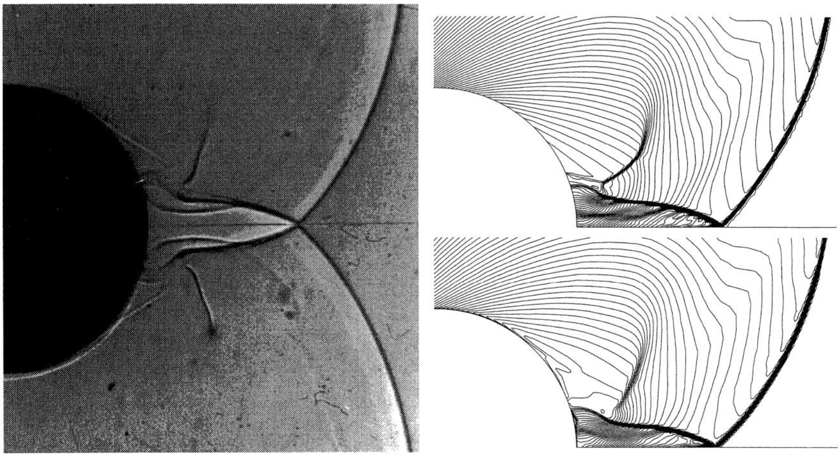

Figure 3. Shock-wave diffraction over a cylinder: the afterbody region for time $t_r = 3.7$. On left, experimental shadowgraph (Syshchikova et al. 1990, Beryozkina et al. 1996); on right, computed density contours: inviscid on top, viscous on bottom.

an initial cell size of $0.4r$). The $0.05r$-thick structured part of the computational grid is equally spaced along the cylinder surface and exponentially stretched in the normal direction with minimal and maximal spacing of $6.8 \times 10^{-5}r$ and $1.6 \times 10^{-2}r$ respectively.

The comparison between experimental data and numerical results shows the good quality of the viscous computations, which describe correctly the wave pattern and the separation location. Pressure histories at fixed points of the cylinder surface also agree with the experiment (see Beryozkina et al. 1996). For $t_r = 3.7$ the Navier-Stokes solver provides correct positions of the rearward-facing stagnation shock wave and the slip surface, and predicts accurately the reflected shock and contact discontinuity near the wall. Additional gasdynamic discontinuities arising near the afterbody, the triple-point configurations at the reflected shock, and the intensity of the latter are reproduced well for $t_r = 5.8$. A fundamental improvement in the solution is obvious compared to the inviscid scheme.

The high efficiency of the proposed adaptive technique made it possible to obtain the presented results using IBM PC/486 personal computers and SUN SPARC workstations while supercomputer power is normally required to achieve the same accuracy with traditional approaches.

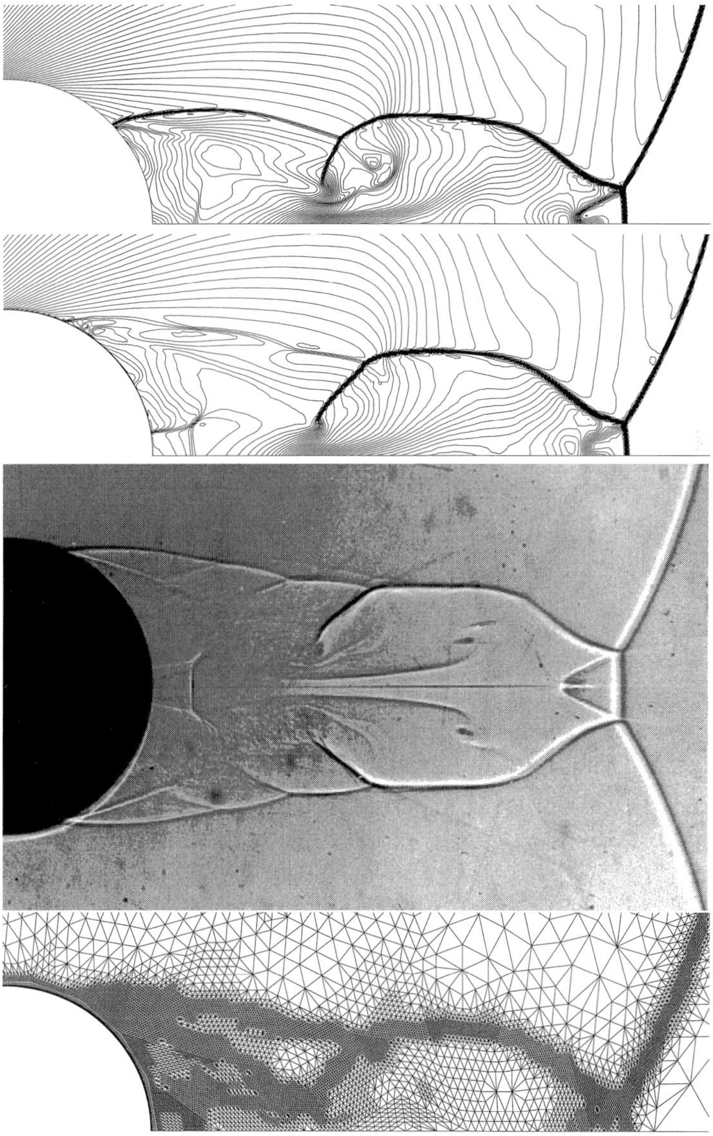

Figure 4. Same as in Fig. 3, $t_r = 5.8$. From top to bottom: density contours for inviscid computation; density contours for viscous computation; experimental shadowgraph (Syshchikova et al. 1990, Beryozkina et al. 1996); grid pattern for the Navier-Stokes solver.

Acknowledgement. Numerous discussions with Drs. M. Syshchikova and M. Beryozkina and their high-quality experimental data are acknowledged sincerely. We thank Prof. Yu. Golovachov and Prof. A. Fursenko for stimulating discussions. E. Timofeev acknowledges the support by Prof. N. Satofuka from Cooperative Research Center, Kyoto Institute of Technology, where the paper was finally prepared for publication. This work was partially supported by the Russian Foundation of Fundamental Investigations (Grants 93-02-14797 and 94-01-01456), Friedrich Alexander University Erlangen-Nürnberg (Contract FAU4001–01IR303) and European Community (COPERNICUS Project COP-94-01490).

References

Beryozkina MK, Ofengeim DM, Syshchikova MP, Timofeev EV, Fursenko AA, Voinovich PA (1996) A combined adaptive structured/unstructured technique for essentially unsteady shock-obstacle interactions at high Reynolds numbers. In: Hyperbolic Problems: Theory, Numerics, Computation, Glimm J, Grove J, Plohr B (eds), World Scientific (to be published).

Fursenko AA, Mende NP, Oshima K, Sharov DM, Timofeev EV, Voinovich PA (1993) Numerical simulation of propagation of shock waves through channel bends. Comp Fluid Dyn J 2:1.

Fursenko AA, Sharov DM, Timofeev EV, Voinovich PA (1995) An efficient unstructured Euler solver for transient shocked flows. In: Shock Waves @ Marseille I, Brun R, Dumitrescu LZ (eds), Springer, pp 371–376.

Hakkinen RJ, Greber I, Trilling L, Abarbanel SS (1959) The interaction of an oblique shock wave with a laminar boundary layer, NASA Memo 2-18-59W.

Ivanov MY, Krupa VG, Nigmatullin RZ (1989) An implicit high-order Godunov scheme for the Navier-Stokes computations. J Numer Math & Math Phys 29:888 (in Russian).

Ofengeim D, Timofeev E, Fursenko A, Voinovich P (1995) A hybrid adaptive structured/unstructured technique for essentially unsteady viscous/inviscid interactions. In: Proc 1st Asian CFD Conf, Hui WH, Kwok Y, Chasnov JR (eds), HKUST, Hong Kong, Vol 1, pp 383–389.

Savel'ev AD (1988) Implementation of Third Order Approximations for Numerical Simulation of Viscous Compressible Gas Flows Near Solid Surfaces, PhD Thesis, Moscow University.

Schröder W, Hänel D (1987) An unfactored implicit scheme with multigrid acceleration for the solution of the Navier-Stokes equations. Comp & Fluids 15:313.

Syshchikova MP, Beryozkina MK, Smirnov IV (1990) The effects upon bodies in the zone of interference of two explosions. In: Unsteady Gas Flows with Shock Waves, Syshchikova MP, Beryozkina MK (eds), Ioffe Physical-Technical Institute, USSR Academy of Sciences, pp 7–29 (in Russian).

Wave Refraction

Vortex-ring impingement on a wall with shock formation

T. Minota
Ariake National College of Technology Omuta, Fukuoka 836 Japan
M. Nishida, M. G. Lee
Kyushu University Fukuoka 812-81 Japan

Abstract: The flow field induced by a high-speed vortex ring approaching a solid plane wall is studied by means of experimental visualization and numerical simulation. As a vortex ring draws close to a wall, an induced boundary layer separates from the wall and forms a wall vortex. A shocklet is generated in the narrow region between the vortex ring core and the wall vortex. Moreover, a new shocklet is formed between the vortex core and the separation point of the boundary layer on the wall. The formation of the former shocklet is by the same mechanism as that in a convergent-divergent nozzle flow. The latter shocklet is caused by the abrupt increase in pressure due to the separation of the flow.

Key words: Vortex ring, Vortex-wall interaction, Formation of shocklet, Flow visualization

1. Introduction

Many studies have been performed to understand flow structure in near-wall turbulence, and have shown that hairpin vortices and bursting phenomena appear in turbulent boundary layers. To understand the physical mechanisms accounting for these phenomena, studies of such fundamental flows as a vortex ring approaching a plane wall have been reported. In an incompressible flow, when a vortex ring approaches a solid wall, secondary and tertiary vortices are generated on the wall owing to viscous effects. The primary vortex is stretched and rebounded by mutual induction with produced vortices (Walker et al. 1987, Orlandi & Verzicco 1993). On the other hand, it is known that in a compressible flow, a turbulent field may be influenced by the shocklets produced and the interactions between them.

The purposes of the present study are to perform optical visualization of a high-speed vortex ring approaching a plane wall and to compare the results with numerical simulations using the axisymmetric Navier-Stokes equations. A shock wave emitted from the open end of a shock tube travels towards wall, hits normal to the wall, and then reflects from it. Immediately after the shock is emitted, the fluid within the shock tube starts to be discharged. A shear layer generated between the emerging and external fluid rolls up, then develops into a strong vortex ring near the open end. Thus the generated vortex ring moves towards the wall at a self-induced velocity, and then interacts with the reflected shock wave traveling towards the open end. After this interaction, the vortex ring approaches the wall, and the boundary layer is induced on the wall, which is followed by a secondary vortex. This wall vortex interacts with the primary vortex ring. The effect of compressibility on this process is studied here.

2. Experimental setup and numerical procedure

A shock wave generated in a shock tube is emitted through a pipe with a 6.0-mm diameter into an atmosphere. The emitted shock wave is followed by a vortex ring as mentioned earlier. A glass plane is placed perpendicularly to the axis of the pipe, at a distance of 16.1 mm from the end of the pipe. Flow visualization was performed by a shadowgraph method with a pulsed dye laser as a light source. The optical system is the same as that described in Minota (1993). Experiments were performed at an incident-shock Mach number of 1.35.

The governing equations for the numerical simulations are the axisymmetric compressible Navier-Stokes equations. They are numerically solved by using the second-order upwind TVD scheme of the Harten-Yee type for convection terms and ordinary central discretization for viscous terms. The computational domain is 21 mm $\times$ 16.5 mm with a grid system of 210 $\times$ 165. No-slip and adiabatic conditions are imposed on the wall.

3. Results and discussion

Figure 1 shows the experimental and computed shadowgraphs at the interaction of the vortex ring and reflected shock, and those after the interaction. In Figs. 1a and b, a slightly bow shock wave connects with the vortex cores and is retarded by the opposing flow due to the vortex. In the computed shadowgraph (Fig. 1b), another shock is embedded in the vortex ring, which is the same rearward-facing shock shown by Baird (1987); but it is not seen clearly in the present experimental shadowgraph (Fig. 1a). A part of the shock wave diffracted in the vortex core travels in the opposite direction and hits the wall. Reflected and intensified shock fronts piled up on the ring surface are seen in Figs. 1c and d. The computation shows that the pressure near the wall at this stage is 1.7 times atmospheric pressure. It may be mentioned that the computed shadowgraphs and experimental ones are in fairly good agreement.

The successive shadowgraphs in Fig. 2 show the development of the wall vortices and the formation of a new shock when the vortex comes close to the wall. The small dark spot shown by "a" on the wall in Figs. 2b–d is a secondary vortex (wall vortex) formed by the concentration of vorticity on the

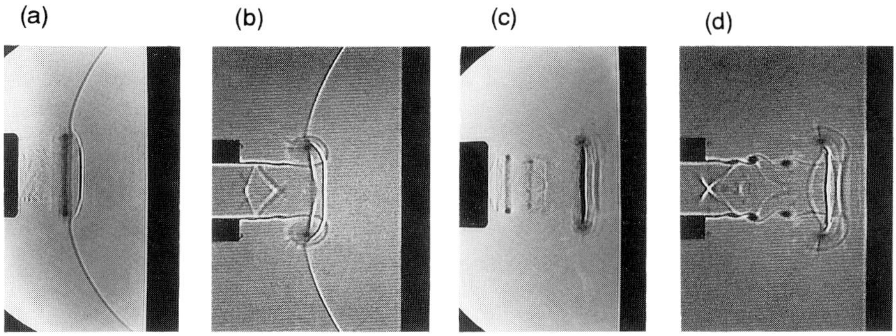

Figure 1. Experimental and computed shadowgraphs: (a) and (c) are experimental shadowgraphs (Mach number of the shock inside the nozzle is 1.35, Reynolds number is 7.1×10^4); (b) and (d) are computed shadowgraphs.

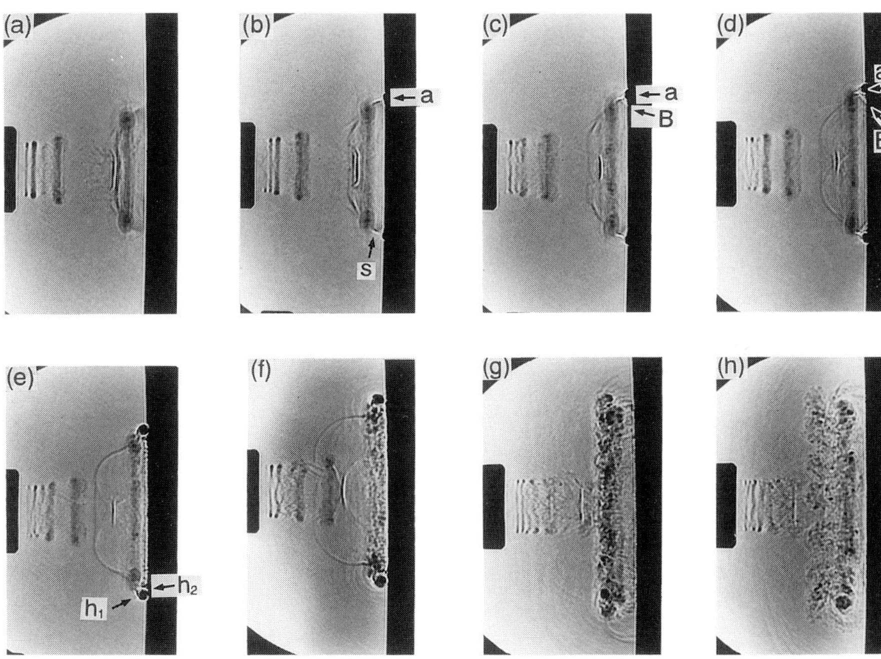

Figure 2. Experimental shadowgraphs showing the development of wall vortices and the formation of shocklets. Times after the stage of Fig. 1a: (a) 66 μs, (b) 68 μs, (c) 70 μs, (d) 74 μs, (e) 82 μs, (f) 89 μs, (g) 114 μs, and (h) 134 μs.

wall. An inward-facing shock shown by "s" is generated between the primary and secondary vortices. In Fig. 2d showing the growth of the secondary vortex, shocklets are formed around the half-circular dark spots (indicated by "a"). In Fig. 2e, these shocklets are observed as two lines "h_1" and "h_2" connecting the primary and the secondary vortex cores. In Figs. 2c and d are seen another new shocklet between the shock "s" and the position B on the wall. This shocklet is also an inward-facing shock. Figures 2f–h illustrate the interaction between the primary and secondary vortex rings. The secondary vortex rises up and winds around the primary vortex. The generated acoustic waves are observed in these figures.

Computational simulations corresponding to Fig. 2 are shown as shadowgraphs, pressure contours, and vorticity contours in Figs. 3 to 5, respectively. The shocklets which are observed in Figs. 2b–d are recognized at almost the same positions in the computed shadowgraphs shown in Figs. 3b–d. As shown in Figs. 4b and c, the shocklet "s" is created from the center of the vortex core towards the wall vortex, and intersects with the shocklet growing from the wall. The vorticity contour line "S" in Figs. 5b and c corresponds to the shocklets "s" in the pressure contours in Figs. 4b and c, respectively.

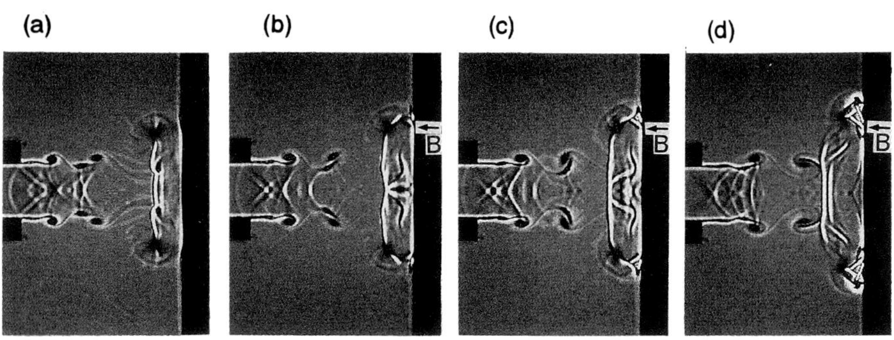

Figure 3. Computed shadowgraph showing the development of the wall vortices and the formation of the shocklets.

The position of the new shocklet on the wall connecting with the shocklet "s" is indicated by an arrow "B" in each figure, which is at $y/R = 0.43$, where y is the radial distance from the central axis and R is the maximum value of y in the computational domain. The point B is the separation point of the boundary layer or the front position of the separation bubble. This can be recognized from the velocity vectors. Figure 6a illustrates the velocity field corresponding to Fig. 3c; the normalized pressure distribution on the wall at the same time is shown in Fig. 6b, where p_1 and γ are the ambient pressure

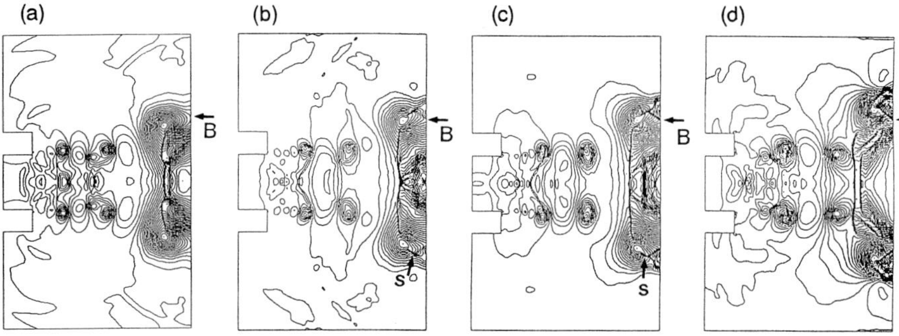

Figure 4. Pressure contours for each step of Fig. 3.

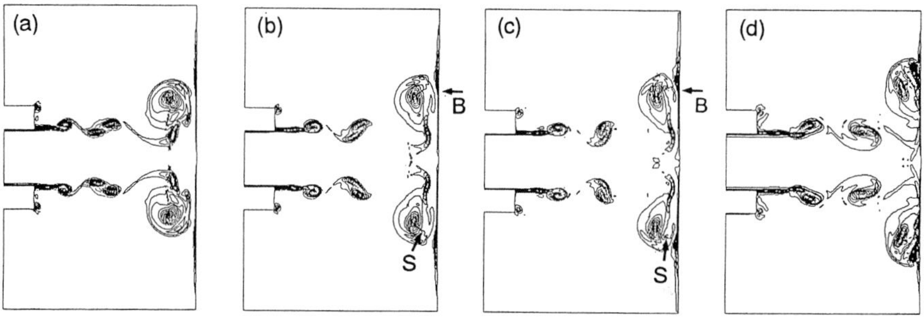

Figure 5. Vorticity contours for each step of Fig. 3.

and the ratio of specific heats, respectively. The fluid impinging on the wall is accelerated while flowing outward along the wall, and the pressure on the wall decreases rapidly. This decrease induces the separation of the boundary layer, then the pressure on the wall rises. Moreover, this abrupt reduction and enlargement cause the formation of the shocklet. The separation bubble generated by such a mechanism becomes a secondary vortex. The arrow "A" in Fig. 6b shows the position of the center of the secondary vortex. On the other hand, the accelerated flow passes through the narrow gap between the primary vortex and the separation bubble or secondary vortex. This is very similar to a convergent-divergent nozzle flow with a normal shock. It has been presumed that the formation of the shocklet "s" between the primary and secondary vortices is owing to the same mechanism as in the convergent-divergent nozzle flow.

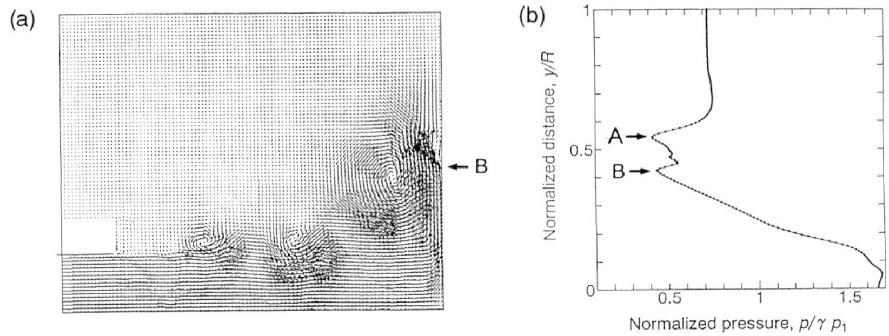

Figure 6. Computed results: (a) velocity field corresponding to Fig. 3c; (b) normalized pressure distribution on the wall.

4. Conclusions

The motion of a high-speed vortex ring which moves towards a plane solid wall has been studied experimentally and numerically. The computed shadowgraphs are in good agreement with the experimental ones. As the vortex ring approaches the wall within a distance comparable to the core diameter, the separation of the boundary layer occurs and a secondary vortex is formed; at the same time a shocklet is generated between both vortices. Moreover new shocklets are generated between the vortex core and the separation point on the wall. The formation of the former shocklet is owing to the same mechanism as that in a convergent-divergent nozzle flow. The latter shocklet is caused by an abrupt increase in pressure due to the separation of the flow. This was confirmed by the simulated results, including pressure contours, vorticity contours, pressure distributions on the wall, and so on.

References

Baird JP (1987) Supersonic vortex rings. Proc R Soc Lond A409:59–65.

Minota T (1993) Interaction of a shock wave with a high-speed vortex ring. Fluid Dynamics Research 12:335–342.

Orlandi P, Verzicco R (1993) Vortex rings impinging on walls: axisymmetric and three-dimensional simulations. J Fluid Mech 256:615–646.

Walker JDA, Smith CR, Cerra AW, Doligalski TL (1987) The impact of a vortex ring on a wall. J Fluid Mech 181:99–140.

Oblique reflection of shock waves in a slow sound-speed fluorocarbon liquid

B. J. P. Mortimer
School of Electrical Engineering, Cape Technikon
PO Box 652, Cape Town, 8000 South Africa

B. W. Skews, L. T. Felthun
School of Mechanical Engineering, University of the Witwatersrand
Private Bag 3 WITS, Johannesburg, 2050 South Africa

Abstract: Very low sound velocities are observed in fluorocarbon liquids. In addition, the nonlinearity parameter B/A for these fluids is exceptionally high. This factor means that the equation of state will be highly nonlinear. This paper presents a fluorocarbon liquid with a sound speed of 655 m/s for use in liquid shock experiments. Analytical, numerical, and experimental results of oblique shock-wave reflection in this medium off of a rigid surface are presented.

Key words: Fluorocarbons, Shock-wave reflection, Nonlinearity parameter

1. Introduction

Research into the propagation and focusing of shock waves in liquids has primarily used water as the medium e.g., Müller (1987), Takayama et al. (1983). Weak shock waves in this medium will travel at close to the acoustic sound speed of 1498 m/s. This value is over four times that of air. The steep rise time of water shocks places several constraints on experimentation. Fluorocarbon liquids have very low sound speeds in comparison to water, being between 470 and 700 m/s at 25°C. The purpose of this paper is to present experimental and numerical results of oblique shock-wave reflection with one of the fluorocarbon liquids, FC-43, as the medium. This liquid has a specific gravity of 1.88, a viscosity of 2.8 centistokes, and a sound speed at 25° of 655 m/s.

The isentropic equation of state for a fluid can be written as:

$$p - {c_o}^2\rho = \frac{c_o^2}{\rho_o}\frac{B}{2A}\rho^2 \tag{1}$$

where p and ρ are the excess pressure and density respectively and A and B are temperature-dependent quantities (Beyer 1974). The term B/A is called the parameter of nonlinearity and is given by:

$$\frac{B}{A} = 2\rho_o c_o \left(\frac{\partial c}{\partial P}\right)_T + \frac{2c_o\alpha T}{\rho_o C_P}\left(\frac{\partial c}{\partial T}\right)_P \tag{2}$$

where T is the temperature in degrees Kelvin, C_P is the specific heat at constant pressure, and α is the coefficient of thermal expansion. For FC-43, B/A has been measured and reported to be 13.2 (Madigosky et al. 1981). This is the highest value of B/A reported thus far. By comparison, water has a B/A of 5.2.

The equation of state can also be derived in a form known as the Tait equation. This form is widely used for water shock computations.

$$p = B\left(\frac{\rho}{\rho_o}\right)^n - B \tag{3}$$

Here B and n are empirically determined constants. The parameter n has been shown to be related to the nonlinearity parameter B/A (Beyer 1974) by:

$$n = B/A + 1 \tag{4}$$

Thus for FC-43, $n = 14.2$ and the constant $B = \rho_o c_o^2/n = 568$ bar.

2. Oblique reflection of shock waves

A mathematical formulation of the reflection of a shock wave from a rigid wall has been considered by von Neuman (1943) and Courant and Friedrichs (1948) for several different configurations. Two types of reflection were found to be possible; firstly, regular reflection and secondly, irregular or Mach reflection.

Regular reflection was found to occur only under restrictive conditions. The analysis can be extended to include water shock reflection (Müller 1987) if a transformation for fluid pressure is introduced (Holl 1981). These equations yield the angle of reflection β as a function of the incident angle α and the incident ratio of pressure P_{21}:

$$\frac{(1 - P_{12}) \cdot t_\alpha}{1 + \mu \cdot P_{12} + (P_{12} + \mu) \cdot t_\alpha^2} = A \tag{5}$$

$$A^2(1-\mu)^2(t_\alpha - t_\beta) + A[(1-\mu)^2 - (t_\alpha - t_\beta)^2 - (\mu + t_\alpha t_\beta)^2] - (t_\alpha - t_\beta) = 0 \tag{6}$$

where:

$$p_{ik} = \frac{P_i + B}{P_k + B}, \qquad t_i = \tan(i), \qquad \mu = \frac{n-1}{n+1} \tag{7}$$

Using this analysis together with the introduction of a suitable transformation for the equation of state of FC-43 yields a plot of the angle of reflection β as a function of the incident angle α and the incident ratio of pressure P_{21} shown in Fig. 1. From this figure it can be seen that β is always greater than α and that this effect increases with increasing shock strength. The reflection is highly

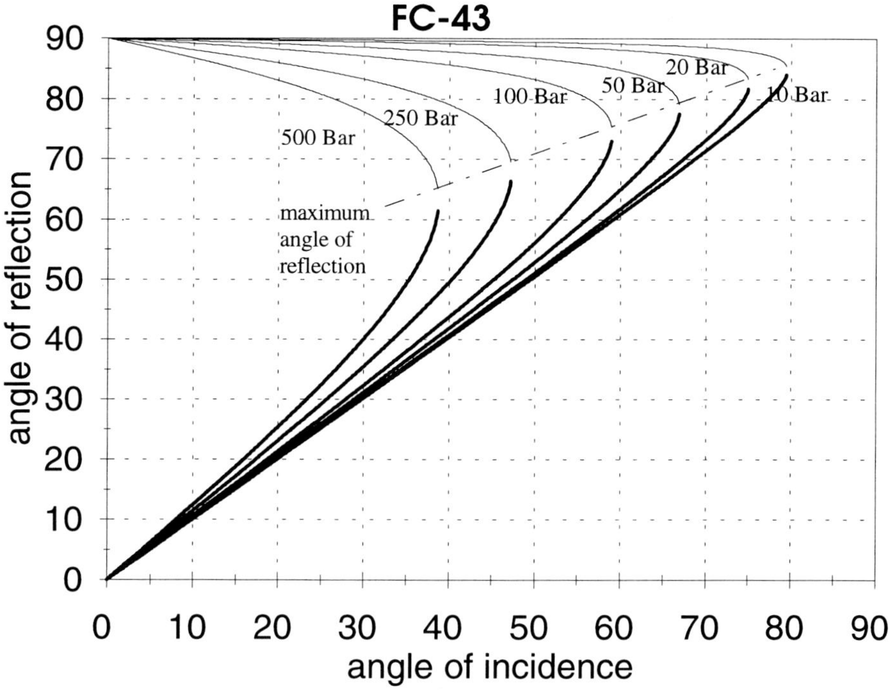

Figure 1. Angle of reflection β as a function of angle of incidence α for different incident shocks.

nonlinear. For a given pressure and incident angle, the deviation in angle of reflection β is greater for FC-43 than for water. For a given pressure, there exists an incident angle at which the reflection will no longer be regular. The reflection then transforms to irregular reflection.

3. Experiments and discussion

The interaction of a shock wave undergoing regular reflection from a rigid surface with FC-43 as the medium has been investigated numerically and experimentally. Wedges were of base angle: 25°, 45°, 75° – 89°, 77° – 87°, and 79° – 85°.

3.1. Computational results

Numerical simulations were conducted using a finite-element code based on the Taylor-Galerkin algorithm (Lohner et al. 1985). This uses a flux-corrected-transport algorithm to maintain high resolution of discontinuities and an adaptive-refinement algorithm to refine the mesh appropriately. In these experiments, the Tait parameters corresponding to FC-43 were used to describe the medium. Shock reflection of various wedges and various incident pressures were simulated. The angle of reflection measured from these experiments is shown in Table 1. Such a simulation is shown in Fig. 2.

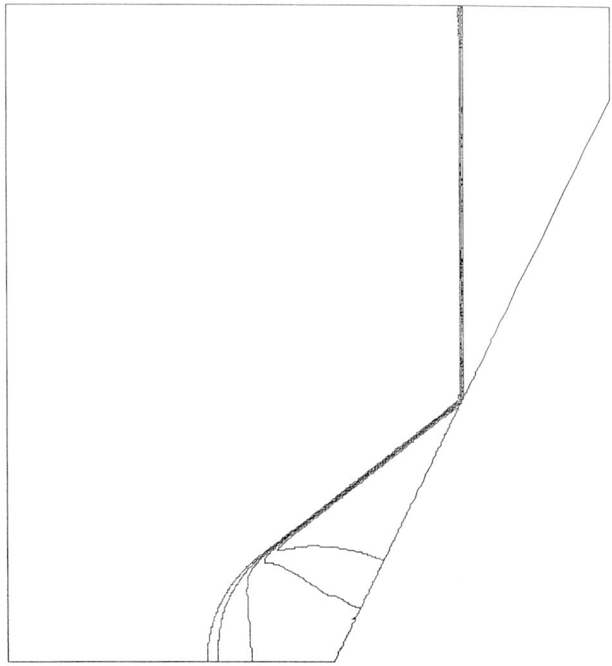

Figure 2. Results from FEM simulation for the reflection of an incident shock of strength 22 bar off a wedge of base angle 25°.

3.2. Shock-tube experiments

Experiments were also conducted using a special vertical shock tube filled with FC-43. The tube is similar to the tube described by Müller (1987), however, the driver section has been modified to facilitate double-diaphragm triggering. A conventional high-pressure air-driver section is pressurized up to 30 MPa. Two pre-scribed metal diaphragms are ruptured and release a shock wave into the air-filled second stage. This shock impinges on a lower test section filled with FC-43. Schlieren photographs were taken of the shock-wave interaction with a wedge mounted at the end of the test section. An example of such an experiment is shown in Fig. 3 — the wave is travelling from top to bottom in the photographs. The light source used in these experiments was a conventional xenon flash of duration of approximately 1 μs.

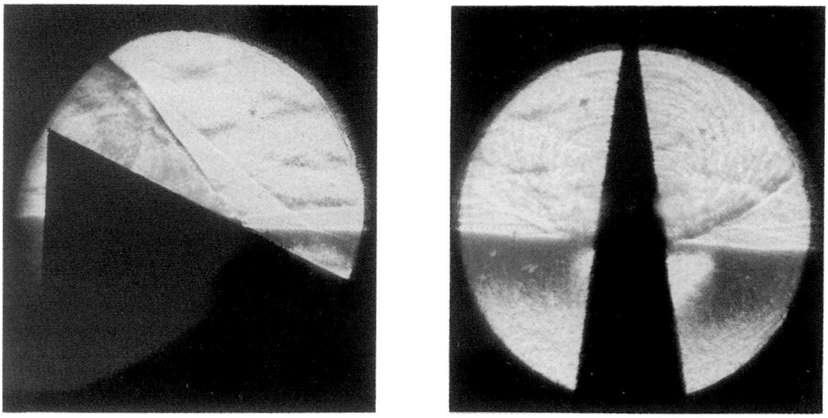

(a) (b)

Figure 3. Schlieren photographs of oblique shock wave reflection off wedges: (a) wedge of 25°, $P_2 = 22$ bar, regular reflection; and (b) wedge of 79° – 85°, $P_2 = 22$ bar, irregular reflection.

For an incident liquid shock of pressure of 22 bar, the Mach number is 1.01. The equivalent gasdynamic inverse pressure ratio is 0.964. These shocks are therefore weak in a gasdynamic sense. The angle of reflection measured from the schlieren photographs for various incident pressures is shown in Table 1.

Table 1. Comparison of the experimental and analytical results.

Wedge Angle	Shock Pressure [bar]	Measured β (degrees) $\beta_{experimental}$	$\beta_{numerical}$	β_{theory}
Wedge: 25°	18	25	25.4	25.33
	22	25.5	25.5	25.41
	28	26	25.6	25.52
Wedge: 45°	18	45.5	45.6	45.8
	22	46	46	45.99
	28	46.5	46.2	46.27
Wedge: 75°	18	80	80	79.95
	22	82	82	irregular
	28	82	82	irregular
Wedge: 79°	10	82	82.5	82.7

The measured angle of incidence for regular reflection with low angles of incidence correspond closely to analytic results. The error in the measurements is less than $0.5°$ for low angles of incidence. As the corner wave interacts with the reflected shock the waves become curved and it becomes more difficult to measure the angles. Several tests were performed where the reflection is theoretically irregular. However no conventional Mach reflection could be resolved for these weak shock experiments.

4. Conclusions

The use of a slow sound-speed fluorocarbon liquid provides several advantages over water for liquid-shock-wave research. Its slow sound speed places fewer demands on experimentation and its high B/A means that nonlinearities in shocks can be obtained for relatively low peak pressures in comparison to water. An equation of state has been derived and analytical and numerical results of oblique shock-wave reflection given. These compare favorably to low-pressure experimental results validating in part the choice of equation of state parameters.

References

Beyer RT (1974) Nonlinear Acoustics, Naval Ship Systems Command, Washington DC, pp 98–101.

Courant R, Friedrichs RO (1948) Supersonic Flow and Shock Waves, Springer-Verlag, New York, pp 329 (reissued in 1976).

Holl R (1981) Stosswellenfokussierung in Fluiden, Thesis, Doktor-Ingenieur, Technischen Hochshule Aachen.

Lohner R, Morgan K, Zienkiewicz O (1985) An adaptive finite element procedure for compressible high speed flows. Comput Meths Appl Mech Eng 51:441–465.

Madigosky WM, Rosenbaum I, Lucas R (1981) Sound velocities and B/A in fluorocarbon fluids and in several low density solids. J Acoust Soc Am 69(6):1639–1643.

Müller M (1987) Stosswellenfokussierung in Wasser, Thesis, Doktor-Ingenieur, Technischen Hochshule Aachen.

Takayama K, Esashi H, Sanada N (1983) Propagation and focusing of spherical shock waves produced by underwater microexplosions. In: Proceedings of the 14th Int Symp on Shock Tubes and Waves, Archer RD, Milton B (eds), Sydney, pp 553–562.

von Neuman J (1943) Collected Works, Volume VI, (Taub AH, General editor), Pergamon Press, Oxford, London (reissued in 1963).

Refraction of a blast wave propagating parallel to a contact discontinuity: a Free-Lagrange numerical simulation

G. J. Ball, R. A. East
Deptartment of Aeronautics & Astronautics
University of Southampton, Southampton, SO17 1BJ UK

Abstract: A recently developed Free-Lagrange code is used to simulate the refraction of planar blast waves propagating parallel to air/propane and air/freon-114 interfaces. The time evolution of the flow fields is described. The rate of decay of the blast wave in the gas with the slower speed of sound is found to be reduced at the expense of an increased rate of decay in the air.

Key words: Blast wave, Shock refraction, Free-Lagrange method

1. Introduction

Interactions between shock waves and contact discontinuities have been the subject of much recent research, focusing both on underlying physical mechanisms and on their role in practical explosion/blast phenomena. These interactions fall into two broad classes: Richtmyer-Meshkov instabilities and shock refraction events. Here we are concerned only with the latter. In a series of papers (see for example Henderson et al. 1991) Henderson and co-workers have described a variety of wave configurations which may occur as a shock passes obliquely through a contact surface across which the speed of sound changes. The wave structure was found to depend strongly on the shock strength and the shock-contact surface incidence angle, α. Ben-Dor and Rayevsky (1991) have examined the behavior of a shock propagating horizontally ($\alpha = 90°$) in air above a layer of gas of higher density, simulating dust-laden air, for a range of density ratios and shock Mach numbers. They observed similar wave structures to those of Henderson et al., and showed that the leading edge of the higher density layer tended to roll-up into a large vortex, providing a strong mechanism for dust entrainment.

In this paper we consider a related problem, in which a weak planar blast wave is propagated parallel ($\alpha = 90°$) to a 'fast-slow' contact surface between air and propane (sound speed ratio 1 : 0.73). The problem may be regarded as an idealization of a flow which might arise following an accidental LPG spillage and explosion. In addition, we examine the more extreme case of refraction at an interface between air and freon 114 (sound speed ratio 1 : 0.36). The use of a blast wave, which decays with time, introduces additional physical effects which are absent in the equivalent shock-wave problem, but which may be significant in accidental explosions.

2. Numerical method

The simulation is performed using a recently developed Free-Lagrange code *Vucalm* (Ball 1996), which solves the two-dimensional unsteady Euler equations on an unstructured Lagrangian grid using a Godunov-type method. The working fluid is divided into discrete packets, between which mass exchange is forbidden. Flow variables are stored at a 'particle' at the center of each packet. The approximate locations of the packet boundaries are determined by constructing a Voronoi mesh, in which each particle is enclosed within a N-sided polygonal Voronoi cell, which forms a control volume for the time integration of the Euler equations. The mesh is reconstructed frequently during the simulation to allow the grid connectivity to change naturally under the influence of shear.

The code employs a three-wave approximate Riemann solver, adapted from Toro et al. (1994), which includes an explicit representation of the contact surface in the local Riemann problem. When implemented in the Lagrangian frame, where the cell boundary and the contact surface are coincident by definition, the solver gives zero numerical diffusion at contact discontinuities, which can therefore be sharply resolved without recourse to an interface-tracking routine. A piecewise-linear representation of conserved variables within grid cells is used to obtain nominal second-order spatial accuracy. A slope limiter of the MUSCL type is used to prevent oscillations in the solution near shock waves.

3. Results and discussion

3.1. Problem specification

The starting geometry for our problem is shown in Fig. 1. The flow occurs in a semi-infinite rectangular shock tube, of which only the first 5 meters are represented in the computational domain. The pressure and temperature in

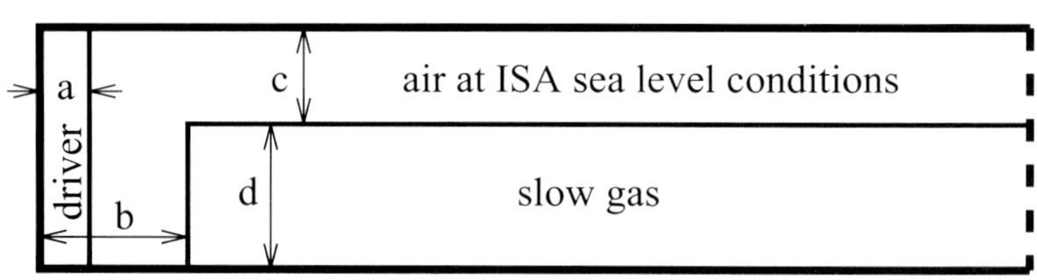

Figure 1. Initial geometry: $a = 0.25\,\text{m}$, $b = 0.75\,\text{m}$, $c = 0.5\,\text{m}$, $d = 0.75\,\text{m}$.

both layers, and in the short driver section, are set to 1.01325 bar and 288.15 K. The particles initially form a square grid, with a pitch of 1/40 m, giving a total of 10^4 cells. To initiate a blast wave, the temperature of the air in the driver section is impulsively increased by a factor of 5. The resulting pressure discontinuity gives rise to a shock propagating to the right and an expansion wave which reflects from the closed end of the tube and then rapidly catches up with the shock to establish a blast wave. In order to characterize the subsequent decay of the blast wave, a calculation was performed using air only. For propagation times between 2 ms and 30 ms, the shock Mach number was found to decay as $M_s = 1 + 0.878t^{-0.475}$ (for t in ms).

3.2. Results for air/propane

Figure 2 shows the flow structure 4 ms after initiation. The contours for this and subsequent plots are of constant $\log_{10}\rho$, with contour interval 0.015. The

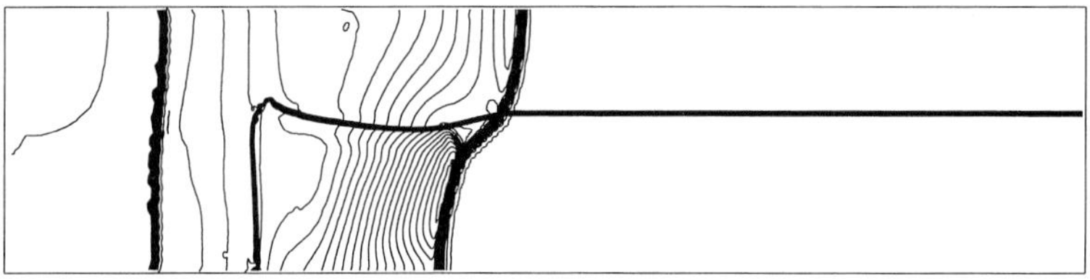

Figure 2. Air/propane solution after 4 ms. Contours are of log ρ.

strong vertical feature to the left is the contact surface marking the edge of the hot driver gas. The shock in the air layer has advanced ahead of that in the propane to form a free precursor shock, and the two waves are linked by an oblique shock, forming a transient Mach reflection in the propane layer. Above the triple point the primary shock in the propane is inclined sharply to the left. Behind the shock structure the pursuing expansion is clearly visible in both layers.

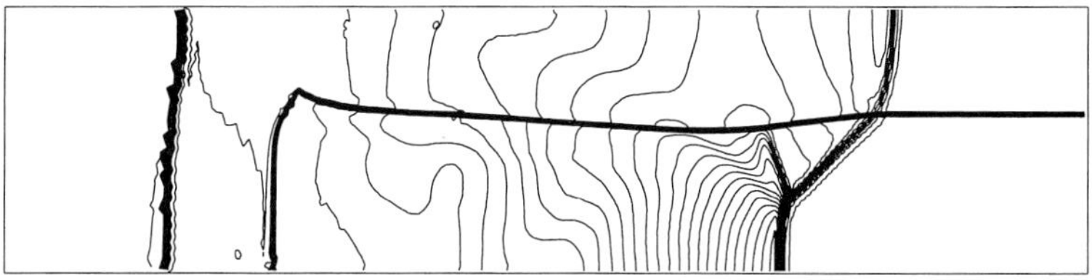

Figure 3. Air/propane solution after 8 ms. Contours are of log ρ.

The subsequent evolution of the flow is shown in Figs. 3 and 4. The triple point moves towards the floor of the tube, while the precursor shock and the

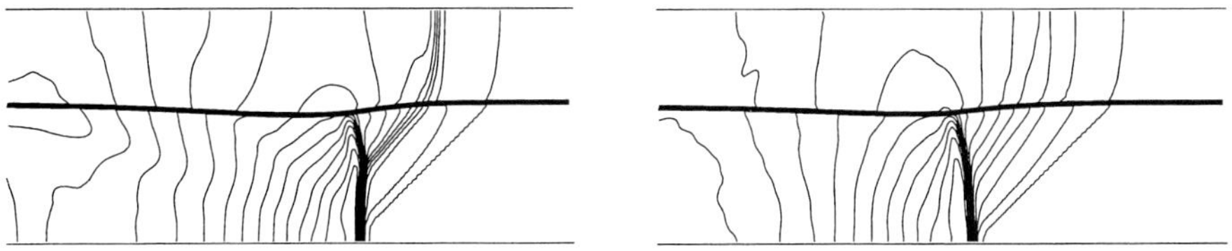

Figure 4. Air/propane solution after 18 ms (left) and 30 ms (right) — log ρ.

linking oblique shock rapidly weaken. An additional weak oblique compression can be seen climbing the rear of the propane shock in Fig. 3. This wave ultimately reinforces the precursor wave (see Fig. 4). Its origin will be discussed in Section 3.3. After 30 ms the precursor wave is little more than a Mach wave, while the propane shock still has a Mach number of about 1.32. At the same stage, the solution for air alone gives $M_s = 1.17$, while for an all-propane case $M_s = 1.18$. It appears that, as a result of the interaction at the contact surface, the blast wave in the gas with the lower sound speed has been sustained at the expense of that in the air, where the rate of decay has been enhanced.

In order to retain the blast wave within the computational domain, the solutions for times beyond 8 ms were obtained in a frame of reference translating at approximately the propagation speed of the wave structure. The domain was truncated to the left at a point where the velocity in the blast had fallen sufficiently to provide a supersonic outflow in the moving frame, thus simplifying the outflow boundary conditions.

3.3. Results for air/freon 114

The initial development of the air/freon flow is qualitatively similar to that of the air/propane case. At 4 ms a transient Mach reflection has been established in the freon (see Fig. 5). The interaction of the blast with the upstream corner of the freon layer has two consequences. Firstly, the difference in post-shock velocities in the two layers causes the freon to begin to roll up into a vortex.

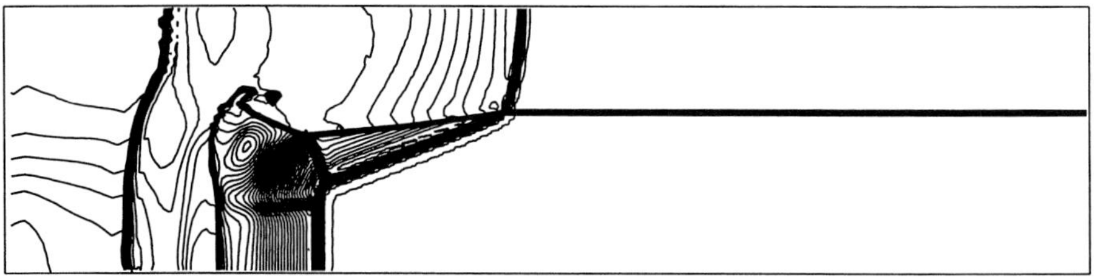

Figure 5. Air/freon 114 solution after 4 ms. Contours are of log ρ.

Secondly, a compression wave appears, aligned almost horizontally and propagating downwards. This latter feature, which is also present, although much weaker, in the air/propane case, seems not to have been observed by other authors.

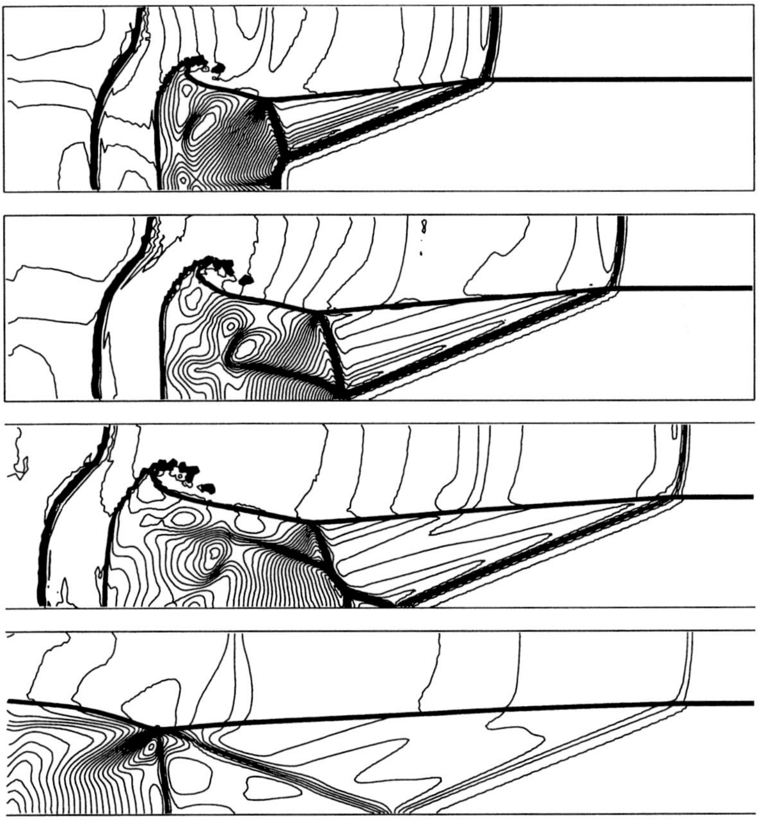

Figure 6. Air/freon-114 solution at (a – d, top – bottom) 6, 8, 10, 16 ms. Reference frame translated at 370 m/s starting $t = 9$ ms. Contours of log ρ.

Subsequent events are shown in Figs. 6a – d. The vorticity at the freon 'corner' is rapidly reduced as the the vortex is traversed by the expansion component of the blast wave, so that the roll-up process is effectively frozen. The freon filament continues to stretch slowly due to the axial velocity gradient. Where the filament thickness is too fine to be resolved on our grid, it breaks into discrete beads. The compression wave reflects from the floor of the tube (Figs. 6a and b), then propagates upwards along the rear of the freon shock (Fig. 6c) and is partially transmitted into the air layer, being refracted in the process so that its propagation direction becomes approximately axial (Fig. 6d). Meanwhile, the precursor shock and the linking oblique shock weaken, the triple point reaches the tube floor, and the transient Mach reflection is replaced by a regular reflection (Fig. 6c). After 16 ms (Fig. 6d) the reflected oblique shock has extended to the contact surface, while the freon shock persists as a nearly normal shock, closely followed by a strong expansion. As in the air/propane case, peak overpressure occurs behind the primary shock in the slower gas, while the pressure excursions in the faster gas rapidly decrease as the wave structure evolves.

4. Conclusions

The flow structures we have observed are in broad accord with those reported by other authors. The use of a decaying blast wave in the present work has two consequences: the vorticity generated at the leading corner of the slow gas layer is rapidly relieved, limiting the extent of entrainment of the slow gas, and the wave structure evolves continuously, exhibiting no self-similar phase, although the rate of evolution falls with time. The rate of decay of the primary blast wave in the slow gas has been shown to be reduced by the interaction, at the expense of an enhanced rate of decay of the precursor shock in the faster gas. An additional compression wave has been observed, originating from the interaction of the blast with the leading corner of the slow gas, which has not apparently been reported elsewhere. It is not clear whether this is particular to blast interactions. The Free-Lagrange code has demonstrated its ability to capture contact surfaces without diffusion, even in the presence of strong pressure waves.

Acknowledgement. The authors are grateful to Mr. D. Webster for his work in developing the graphical user interface for the Vucalm code.

References

Ball GJ (1996) A Free-Lagrange method for unsteady compressible flow: simulation of a confined cylindrical blast wave. Shock Waves 5:311–325.

Ben-Dor G, Rayevsky D (1991) Shock wave interaction with a dense layer. In: Proc International Workshop on Strong Shock Waves, Honma H, Maeno K (eds) Chiba, Japan, pp 145–164.

Henderson LF, Colella P, Puckett EG (1991) On the refraction of shock waves at a slow-fast gas interface. J Fluid Mech 224:1–27.

Toro EF, Spruce M, Speares W (1994) Restoration of the contact surface in the HLL Riemann solver. Shock Waves 4:25–34.

Shock waves generated by the interaction of two vortex rings

T. Minota
Ariake National College of Technology
Omuta, Fukuoka 836 Japan

M. Nishida, M. G. Lee
Department of Aeronautics and Astronautics, Kyushu University
Fukuoka 812-81 Japan

Abstract: This paper describes experimental and numerical studies of the flow field for a head-on collision of two vortex rings and the shock waves generated by it. As the two vortex rings approach, the region between them is compressed, then the pressure becomes high. Moreover, as the rings come close to each other, an inward-facing shock is formed between the two vortex cores. This shock travels together with the stretching of the vortex core line. Computational results of Mach contours and the pressure distribution on the collision plane of the rings show that the formation of this new shock is by the same mechanism as that in a convergent-divergent nozzle flow.

Key words: Vortex ring, Formation of shocklet, Interaction of vortices

1. Introduction

A vortex ring has been used to understand various aspects of vortex motion in a jet flow, vortex interaction, generation of vortex sound, and so on. The motion of interaction between vortex rings exhibits various behaviors depending on initial configuration. Numerous studies of a head-on collision of two vortex rings have been conducted (Shariff 1992). When two vortex rings approach each other, their sizes increase due to mutual induction. In the case of ring collision with shock formation, compressibility plays an important role in the interaction of vortices. However there are not satisfactory studies on the interaction of compressible vortex rings. In a previous work (Minota 1995), the formation and development of a shock connecting with two vortex cores have been observed by means of optical shadowgraphs.

In the present work, the optical visualization and the numerical calculations are carried out under nearly the same conditions. The physical mechanism of the formation of a new shock between the cores is studied by comparison of experimental shadowgraphs with computational results.

2. Experimental setup and numerical calculation

A shock wave generated in a shock tube is separated into two shock waves, and they are emitted at the same time from two nozzles which are placed at a distance of 33.4 mm from each other with a common axis. Two vortex rings formed near the open ends of the nozzles move towards each other at a self-induced velocity. Experiments were performed for an incident-shock Mach number of 1.34. The Reynolds number of the vortex ring in the initial stage is approximately 7.0×10^4. Flow visualization was performed by means of the same shadowgraph method as used in Minota (1995).

The governing equations for the numerical simulations are the axisymmetric compressible Navier-Stokes equations. They are numerically solved by using the second-order upwind TVD scheme of the Harten-Yee type for convection terms and ordinary central discretization for viscous terms. The computational domain is the upper half region, which is 42.9 mm $\times$ 18.5 mm with a grid system of 276 $\times$ 119.

3. Results and discussion

Figure 1 shows the optical shadowgraphs of the flow field of the interaction of shock waves and vortex rings. Computed shadowgraphs corresponding to each moment in Fig. 1 are shown in Fig. 2. A spherical shock wave collides with a vortex ring and is distorted. The bright line (Sd) connecting with the vortex core seen in Fig. 1a shows an impinging shock wave retarded by the opposing flow due to the vortex. In the computed shadowgraphs (Fig. 2a), the shock indicated by "Sd" is the retarded shock mentioned above, and a new rearward-facing shock (Se) is generated. This new shock is not seen clearly in the optical shadowgraph (Fig. 1a). The outer part (Wo) of the impinging shock wave diffracts in the vortex core, and bends as it passes around the core as shown by "Wd" in Fig. 1b. The shock front traveling in the opposite direction interacts with other vortex rings.

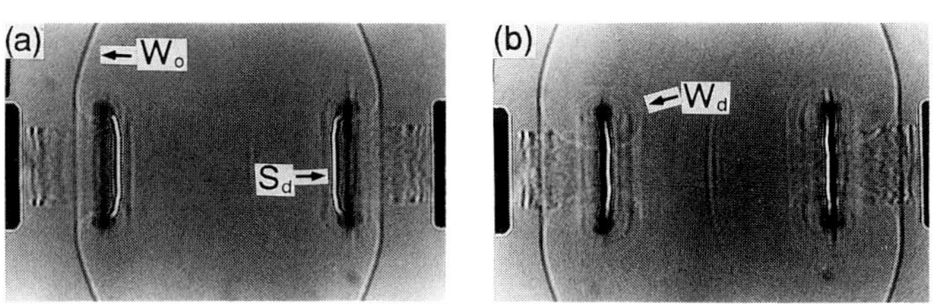

Figure 1. Optical shadowgraphs showing the flow field of the interaction of shock waves and vortex rings. Time: (a) 40 μs; and (b) 50 μs.

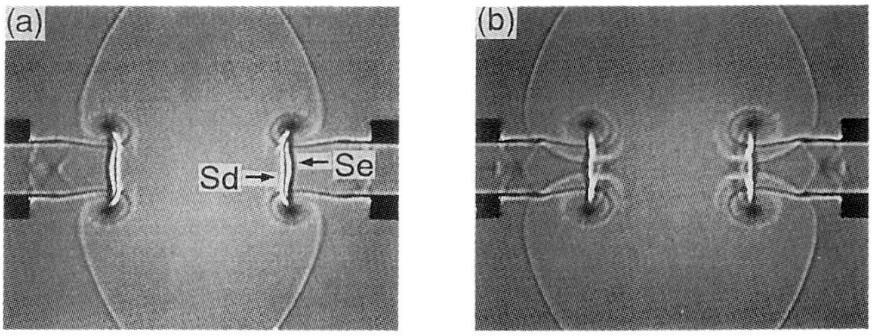

Figure 2. Computed shadowgraphs. Time: (a) 46.5 μs; and (b) 54.2 μs.

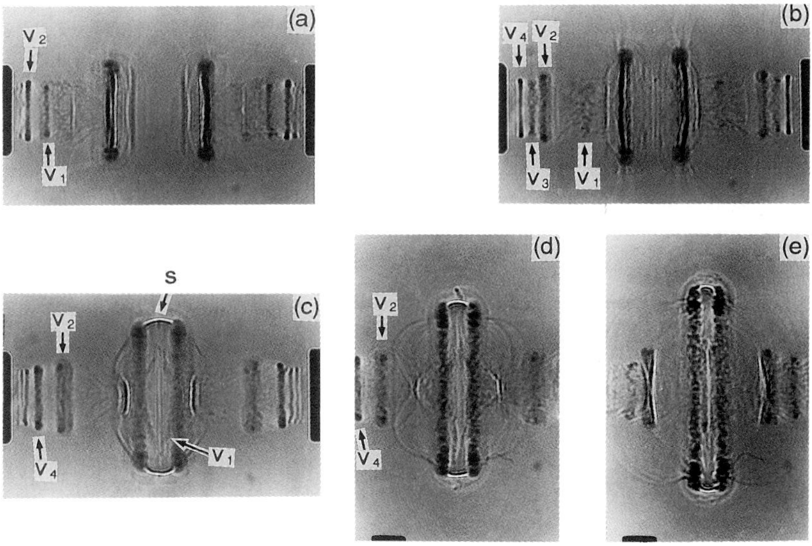

Figure 3. Optical shadowgraphs showing development of the interaction of two vortex rings. Time: (a) 80 μs; (b) 100 μs; (c) 110 μs; (d) 120 μs; and (e) 135 μs.

The successive shadowgraphs in Fig. 3 show the time development of the interaction of two vortex rings. As shown in Fig. 3a, a few shock waves are piled up on the ring surface; these shocks are the waves intensified by the interactions with the vortex. In Fig. 3c, an inward-facing new shock (shown by "s") is seen between the cores of the two vortex rings. The forward motion is blocked and the radial motion is accelerated. Then the vortex core line is stretched and its diameter becomes large. The rate of the diameter development is more than 270 m/s. The inward-facing new shock moves together with the stretching of the vortex core.

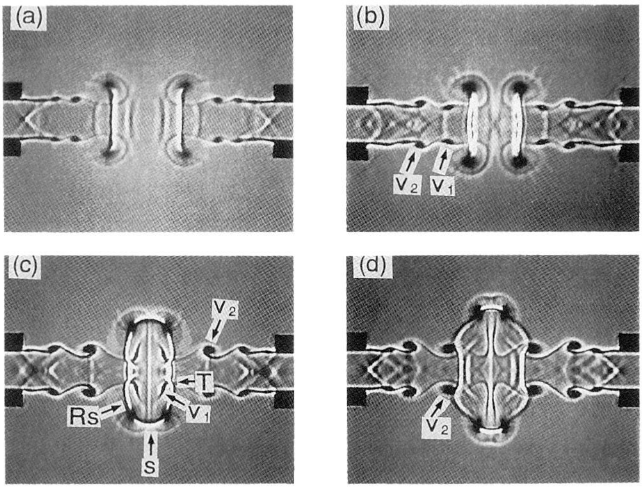

Figure 4. Computed shadowgraphs showing the time development of the interaction of two vortex rings. Time: (a) 80.0 μs; (b) 92.3 μs; (c) 107.6 μs; and (d) 122.0 μs.

A thin shear layer between the jet flow following the vortex ring and the exterior fluid rolls up owing to the Kelvin-Helmholtz instability, and small vortex rings are formed successively. These small vortex rings interact with each other and merge. The small vortex rings "v_1" and "v_2" in Fig. 3a arrive at the positions "v_1" and "v_2" in Fig. 3b, respectively. In Fig. 3b, new rings "v_3" and "v_4" are formed. The ring "v_1" penetrates into the primary vortex, and deforms as indicated by "v_1" in Fig. 3c. The ring "v_3" approaches the ring "v_2", and is merged. Acoustic waves are recognized in Fig. 3e. Computed shadowgraphs corresponding to the optical ones in Fig. 3a–d are shown in Fig. 4. The small vortex "v_1" in Fig. 4b is caught up by the primary vortex in Fig. 4c, and is distorted as shown by "v_1". This is similar to the form of "v_1" in Fig. 3c. Shocks inside the primary ring, indicated by "Rs" in Fig. 4c, have a structure with a triple point shown by "T" through diffraction around the distorted small vortices "v_1". The inward-facing shocks "s" are seen in Fig. 4c–d. These move with the stretching of the vortex ring in a way such as in Fig. 3. The computed shadowgraphs are in fairly good agreement with the optical shadowgraphs concerning the motion of the small vortex rings as well as the process of interaction of the two primary vortex rings and the formation of the new inward-facing shock.

Figure 5 shows the results of computational simulation corresponding to Figs. 4b–d. The top (-A), second (-B), third (-C), and bottom (-D) rows show

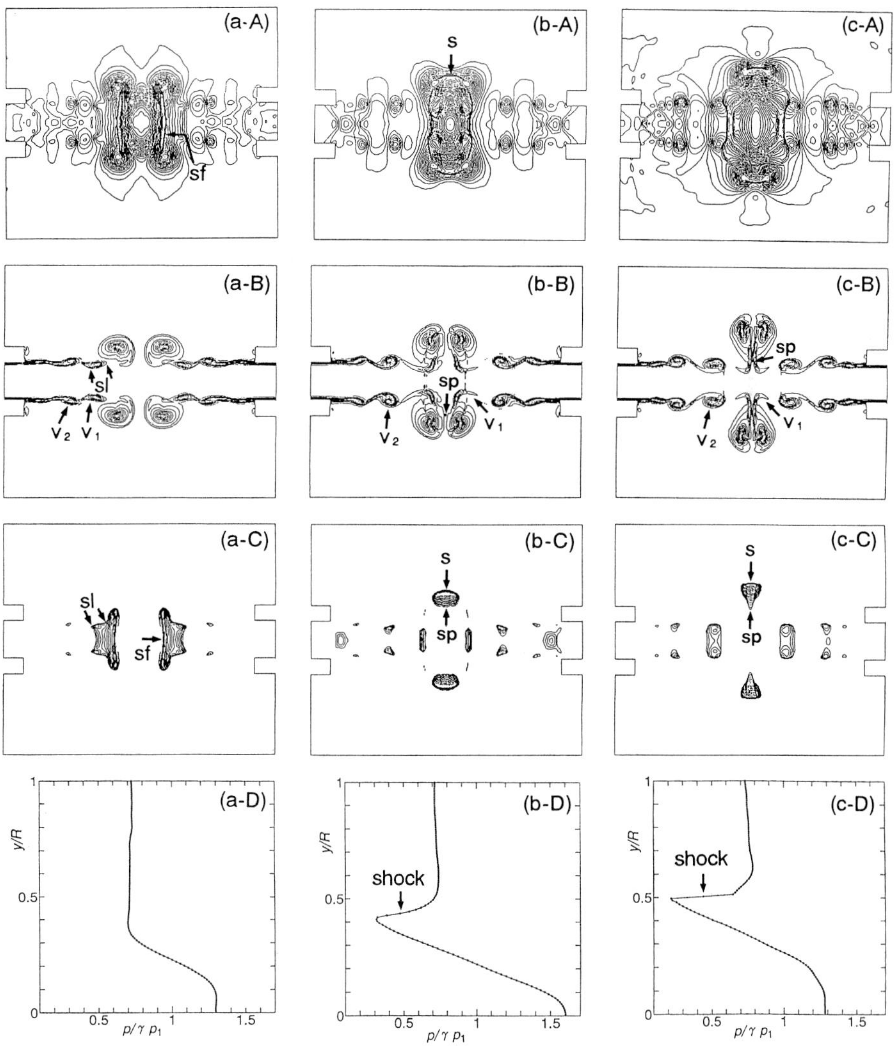

Figure 5. The results of computational simulation corresponding to Figs. 4b–d. Top row shows isobars. Second, third, and bottom rows show vorticity contours, supersonic Mach contours, and the normalized pressure distribution, respectively, where y is the radial distance from the central axis, R the maximum value of y in the computational domain, p_1 the ambient pressure, and γ the ratio of specific heats.

isobars, vorticity contours, supersonic Mach contours, and pressure distribution along the collision plane of the vortices, respectively. The front indicated by "sf" in the supersonic Mach contours of Fig. 5a-C corresponds to outer shock front "sf" in the pressure contours of Fig. 5a-A. The velocity upstream of the shock front "sf" becomes supersonic. The sonic line indicated by "sl" corresponds to the inside segment "sl" of the small vortex "v_1". The penetrating small vortex rings distort as shown in Fig. 5b-B and c-B. These were observed as the vortex "v_1" in the shadowgraphs Fig. 3c and 4c. The inward-facing shock (s) seen in the pressure contours Fig. 5b-A is also recognized as "s" in the Mach-number contours of Fig. 5b-C. The position indicated by "sp" in vorticity contours Fig. 5b-B shows a sonic line "sp" in the Mach contours. The pressure distribution Fig. 5b-D shows that the pressure on the center of the collision plane of the rings is more than 2.2 times the ambient pressure and 1.3 times the initial pressure behind the shock. The pressure of fluid flowing outwards decreases rapidly, and rises suddenly by the formation of a shock. The formation of the inward-facing shock can be explained as follows: the high-pressure fluid on the center of the collision plane is accelerated, and the pressure decreases rapidly. The accelerated fluid passes through the narrow gap surrounded by the two vortices. The flow velocity becomes supersonic downstream of it and a shock is formed in the accelerated flow in order that the low pressure become equal to the ambient pressure, which is higher than the pressure in the accelerated flow. This situation is very similar to a convergent-divergent nozzle flow with a normal shock.

4. Conclusions

The motion of two high-speed vortex rings which approach each other has been studied experimentally and numerically. When the two vortex rings approach each other within a distance comparable to the core diameter, an inward-facing shock is formed between the two vortex rings. This new shock travels together with the stretching of the vortex core line. The formation of the inward-facing shock is by the same mechanism as that in a convergent-divergent nozzle flow. This is confirmed by the computational simulations, i.e., Mach contours and pressure distribution.

References

Minota T (1995) Experiments on shock and vortex interactions. In: Shock Waves @ Marseille, Springer-Verlag, Vol 4, pp 343–348.

Shariff K (1992) Vortex rings. Annual Rev Fluid Mech 24:235–279.

Mechanisms of shock-wave generation resulting from shock-wave/bubble interaction in liquids

K. Yamada, T. Kodama, K. Takayama
Shock Wave Research Center, Institute of Fluid Science
Tohoku University
2-1-1, Katahira, Aoba-ku, Sendai 980 Japan

Abstract: Interactions of a shock wave with a gas bubble were visualized using double-exposure holographic interferometry and high-speed framing photography. The gas bubble was exposed to a spherical shock wave which was generated by detonation of a micro-explosive. The gas bubble contracted by impingement of the shock wave and hence a liquid jet formed; eventually, compression waves or shock waves occurred.

Key words: Flow visualization, Gas bubble, Liquid jet, Shock-bubble interaction

1. Introduction

A flow visualization study was conducted for quantitative observation of interaction between a spherical shock wave and a gas bubble by means of double-exposure holographic interferometry. Processes of shock-bubble interaction from shock impingement to generation of a rebound shock wave were visualized. Four combinations of liquid/gas were examined: silicone-oil/air, water/air, water/helium, and silicone-oil/air-helium (Yamada and Takayama 1995).

2. Experiments

Test liquids were water and dimethylsilicone-oil, and test gases were air and helium. Speed of sound, density, and kinematic viscosity of silicone-oil (Shin-Etsu Chemical Industry Co. 1989; Weissler 1949) were $a_{0s} = 936.5\,\mathrm{m/s}$, $\rho_{0s} = 821\,\mathrm{kg/m^3}$ (293 K), and $\nu_{0s} = 1.0 \pm 0.1 \times 10^{-6}\,\mathrm{m^2/s}$ (298 K), respectively.

An underwater shock wave impinged on a gas bubble released from the bottom of a test tank and rising freely in liquid. A micro-explosive, a 10.0 ± 0.5-mg silver azide (AgN_3) pellet (Chugoku Kayaku Co.), was detonated at a predetermined position in the liquid. The volume of the gas bubble and position of the gas bubble at the moment of initiation of the micro-explosive were specified. The micro-explosive was ignited and subsequently a spherical shock wave interacted with the gas bubble.

3. Results and discussion

Experimental results of silicone-oil/air, water/helium, and silicone-oil/air helium mixed-gas combinations are presented. The shape of the gas bubble before interaction with the shock wave was an oblate spheroidal shape. Typical dimensions of the gas bubbles and overpressures of the incident shock waves are tabulated in Table 1.

Table 1. Typical dimensions of gas bubbles immediately before impingement of an incident shock wave in liquid and overpressure of the incident shock wave. The variables a, b, f, d, and ΔP are major diameter, minor diameter, flattening factor, equivalent diameter of sphere, and overpressure of the incident shock wave, respectively.

Combination of liquid/gas	a (mm)	b (mm)	$f=(a-b)/a$	$d=\sqrt[3]{a^2b}$ (mm)	ΔP (MPa)
Water/helium	3.9	1.5	0.61	2.8	42.2
Silicone-oil/air	2.0	1.0	0.50	1.6	11.5
Silicone-oil/mixed gas	1.2	0.82	0.29	1.0	26.4

Figures 1a–1d show interferograms for the silicone-oil/air combination. In Fig. 1a an expansion wave reflected from an air bubble and the bubble started to contract. In Fig. 1c, because of enhancement of pressure at the side of the bubble where the shock wave impinged, spherical compression waves were generated. Later, the compression waves coalesced into spherical shock waves (Fig. 1d). When the bubble started to expand, its shape appeared to be a jagged one.

The shock-bubble interaction is not a two-dimensional flow and hence fringes on the interferogram do not correspond to isopycnics. However, by assuming an axisymmetric flow, the density distribution can be obtained by analysis of the interferogram. Figure 2a shows pressure distribution as obtained from Fig. 1a. An axis of axisymmetry is a line through the explosion center and the center of the bubble. In Fig. 2c a maximum pressure was obtained at the side of the bubble where the shock wave first hit.

Figures 3a and 3b show sequential shadowgrams of the shock-bubble interaction for the water/helium combination, in which the incident shock wave propagated from left to right.

In Fig. 3a a dark line in frames 1 and 2 shows the incident spherical shock wave. The incident shock wave interacted with a helium bubble, and the wave

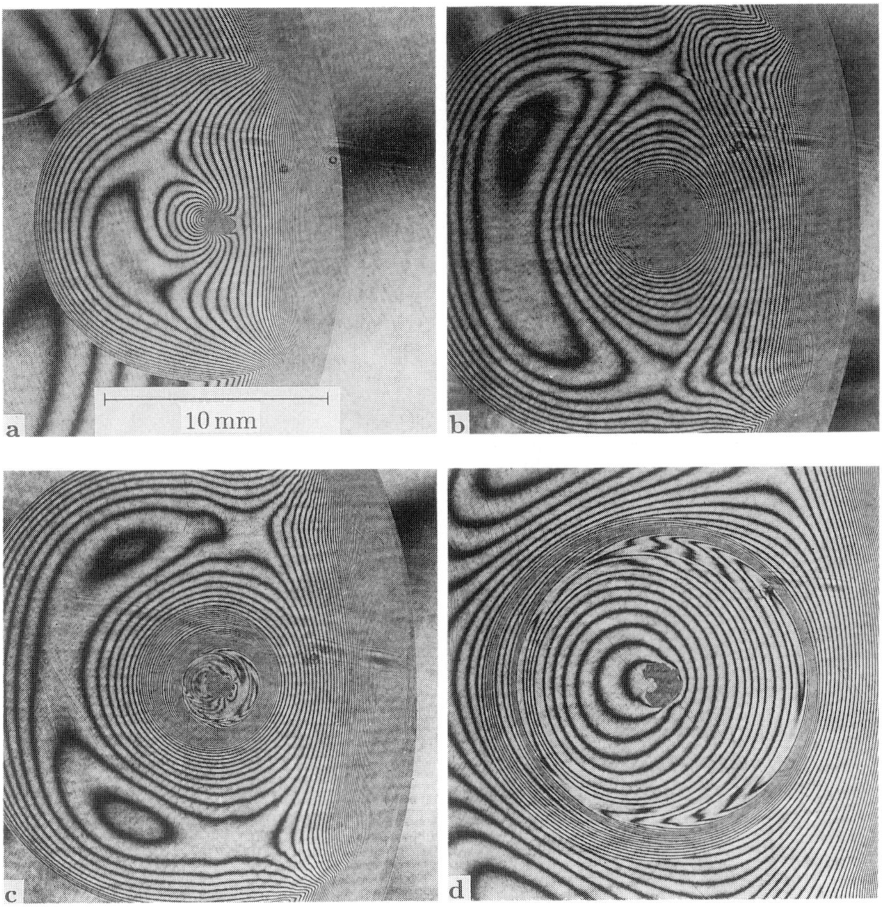

Figure 1. Interferograms of a shock-bubble interaction for the silicone-oil/air combination.

that reflected from the bubble was an expansion wave (frame 2). A liquid jet formed, because liquid particles near the point on the bubble where the shock wave first hit were accelerated due to effect of curvature of the interface and liquid was entrained into the bubble (frames 3–6). The liquid jet penetrated toward the rear surface (frame 6). From these sequential shadowgrams the average speed of the liquid jet was estimated to be $275 \pm 10\,\mathrm{m/s}$ so that stagnation pressure was thus more than 200 MPa. Spherical shock waves were generated, and at this stage the bubble started to expand (frames 6 and 7).

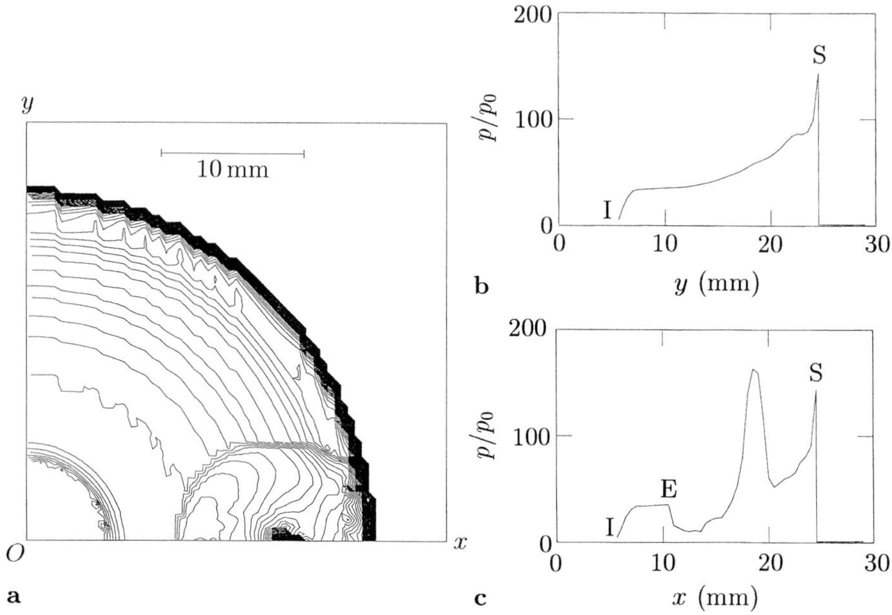

Figure 2. (a) Pressure distribution on the cut-through plane of the axis. The pressure distribution was analyzed from Fig. 1a. Intervals of contours are $5p_0$ where p_0 is atmospheric pressure. (b) Pressure distribution on the line Oy. S – shock front, I – interface of water and explosion product gas. (c) Pressure distribution on the axis Ox. E – wave front of an expansion wave.

In frames 1–3 in Fig. 3b two nearby waves marked A and B were created. In frames 1 and 2, wave A appeares to be stronger at the rear side and weaker at the frontal side. The strength of wave B appeares to be more uniform compared to wave A. The center of wave A is located almost on the rear surface while the center of wave B is in the bubble (frames 1 and 2). Wave A was generated due to water hammer pressure when the liquid jet impinged the rear surface. Wave B is a rebound wave that was generated by expansion of the bubble.

Figure 4 shows shadowgrams of shock-bubble interaction for the silicone-oil/gas consisting air-helium mixture. Concentric spherical waves were generated (frames 3–5). In frame 3 a sonoluminescence can be seen, because temperature was enhanced in the bubble. A counter jet was generated (frames 4–8).

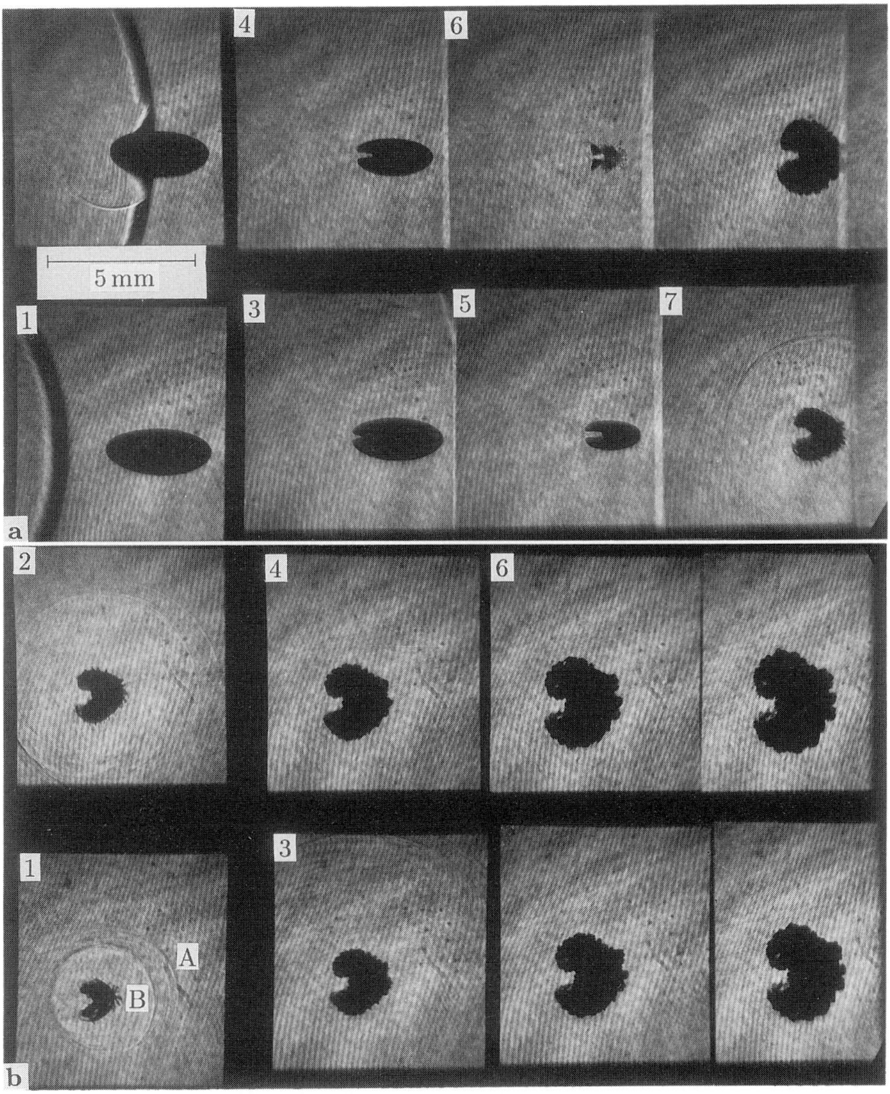

Figure 3. Sequential shadowgrams of a shock-bubble interaction for the water/helium combination. A spherical shock wave propagated from left to right. (a) The interframe time is $2.0\,\mu s$. (b) Later stage than (a). The spherical shock wave already passed. The interframe time is $1.0\,\mu s$. The exposure time is 1/5 of interframe time for all frames.

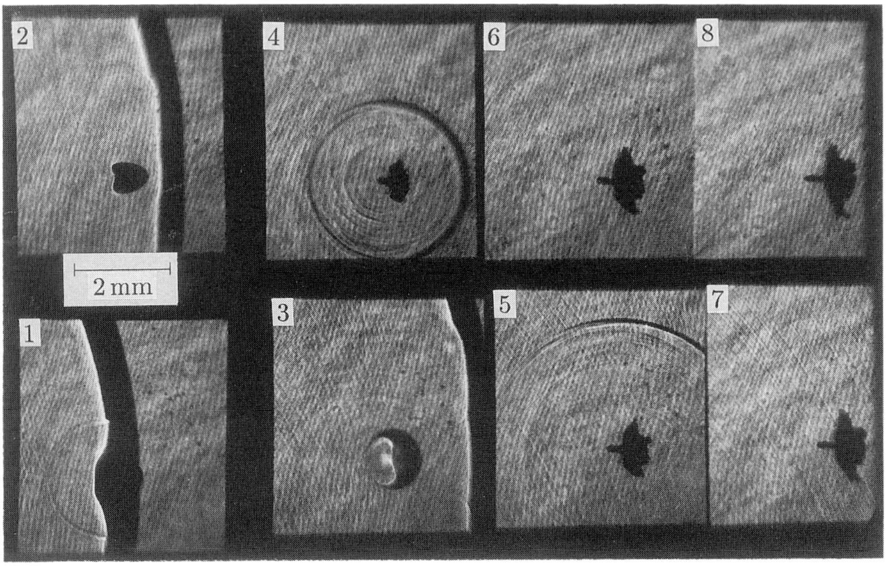

Figure 4. Sequential shadowgrams of a shock-bubble interaction for the silicone-oil/air-helium mixed gas combination. The interframe time is 1.0 μs.

4. Conclusion

Interaction of a shock wave with a gas bubble was investigated experimentally using double-exposure holographic interferometry and high-speed framing photography. Collapse of the gas bubble was well visualized. The pressure profile around the collapsing bubble was analyzed quantitatively.

Acknowledgement. The authors wish to thank to Mr. O. Onodera and Mr. H. Ojima of the Shock Wave Research Center, Institute of Fluid Science, Tohoku University for their assistance in conducting the present experiments.

References

Shin-Etsu Chemical Industry Co. (1989) Explanatory note for KF 96. Shin-Etsu Chemical Industry Co., Japan (in Japanese).

Weissler A (1949) Ultrasonic investigation of molecular properties of liquids. III. Linear polymethylsiloxanes. J Am Chem Soc 71:93.

Yamada K, Takayama K (1995) Interaction of an underwater shock wave with a gas bubble. In: IUTAM Symposium on Waves in Liquid/Gas and Liquid/Vapour Two-Phase Systems, Morioka S, Van Wijngaarden L (eds), Kluwer Academic Publishers, Dordrecht, The Netherlands, pp 161–170.

Passage of two shocks through an interface

S. G. Zaytsev, E. I. Chebotareva, A. N. Aleshin,
E. V. Lazareva, S. N. Titov
The Krzhizhanovsky Power Engineering Institute (ENIN)
Leninsky Pr. 19, Moscow, 117071 Russia

Abstract: We have made an experimental investigation of the evolution of the Richtmyer-Meshkov instability on the passage of incident and reflected shock waves through continuous and discontinuous interfaces. It has been shown that the type of refraction of the reflected shock wave at an interface determines its subsequent evolution.

Key words: Shock wave, Hydrodynamic instability, Turbulence

1. Introduction

We studied the evolution of the mixing zone between gases of different density on the passage of a plane shock wave through the sinusoidal interface modeled by a thin film (Aleshin et al. 1993). It has been shown that the evolution of the interface separating the shock-compressed gases of different density depends upon the type of refraction of the shock. The type of refraction is determined by the following parameters: a_0k, where a_0 and k are the amplitude and the wave number of initial interface, respectively, the Mach number M of the incident shock wave, and the Atwood number $A = (\rho_1 - \rho_2)(\rho_1 + \rho_2)^{-1}$, where ρ_1 and ρ_2 are the densities of the gases separated by the interface. Three typical regimes have been described: soft regular, hard regular, and irregular. For the soft regular regime the growth of the amplitude $a_{\mathrm{K}12}$ of the initial interface K_{12} at initial stage of the evolution for $\rho_1 < \rho_2$ is satisfactorily described by the Richtmyer relation:

$$\frac{da_{\mathrm{K}12}}{dt} = a^*_{\mathrm{K}12}ukA \tag{1}$$

where $a^*_{\mathrm{K}12}$ is the amplitude at the end of refraction, u the shock-induced velocity of K_{12}, and A is the Atwood number. For $M = 3$, the relation (1) for the initial stage of K_{12} evolution is applicable at $a_0k < 2$.

The study of K_{12} evolution for the shock passage through a continuous interface (Zaytsev et al. 1994) has shown that for $a_0k < 2$ the growth of the amplitude a_{k12} at the initial stage of K_{12} evolution is described by the following relation within 30% error:

$$\frac{da_{K12}}{dt} = \frac{a^*_{K12} ukA}{\Psi} \tag{2}$$

The relation (2) was suggested by Brouillette and Sturtevant (1994). The value Ψ is dependent on the thickness Δ of interface, wave number k, and Atwood number; Ψ was calculated according to Duff et al. (1962). To measure the amplitude a_{k23} of the interface K_{23} formed after the interaction of the reflected shock wave S_{e2b} with K_{12}, the following relation was suggested (Brouillette & Sturtevant 1994):

$$\frac{da_{K23}}{dt} = \frac{da_{K12}}{dt} + \frac{a^*_{K12} ukA}{\Psi(k, \Delta, A)} \tag{3}$$

where a^*_{K23} is the a_{K23} amplitude at the end of refraction of S_{e2b} on K_{12}. Based on experiments, the authors concluded that the relation (3) adequately takes into account the influence of K_{12} evolution on K_{23} evolution (Brouillette & Sturtevant 1994).

2. Experimental data

We performed an experimental investigation of the passage of the reflected shock wave S_{e2b} through the interface K_{23}. The interface K_{12} is formed on the passage of the incident shock wave S_{e1a} through the initial interface K_0. The S_{e1a} propagates in a gas of density ρ_1. The reflected shock wave S_{e2b} propagates in a gas of density ρ_2 ($\rho_2 < \rho_1$). The experiments were performed in shock tubes.

In the first series of experiments performed in a shock tube (Aleshin et al. 1993) the initial discontinuous interface was modeled by a thin film. The mechanical effect of the shock and the heat effect of the shocked flows following the refracted shock wave S_{i1b} resulted in formation of the mixing zone K_{12}. K_{12} of sinusoidal shape and finite thickness Δ contained the initial gases and the products of destruction of the film. The refracted shock wave S_{i1b}, reflecting from the end wall of the shock tube, generated the reflected shock wave S_{e2b} which moved toward K_{12} and refracted on it, thus forming the mixing zone K_{23}.

In the second series of experiments we used a continuous interface obtained with a fast removable plate separating the gases before the experiment. In this case the distribution of concentrations was dependent on molecular diffusion.

The x, t-diagram for run no. 583B is given in Fig. 1; $K^{(1)}_{12}$ and $K^{(2)}_{12}$ are the trajectories of experimental points of the front separating the mixing zone K_{12} from the heavy gas. The distance between $K^{(1)}_{12}$ and $K^{(2)}_{12}$ for fixed time t determines the amplitude $a_{K12}(t)$. S_{e1b} and S_{i2a} are the trajectories of the refracted shock waves generated by the interactions of S_{e1a} with K_0 and of S_{e2b} with K_{12}, respectively.

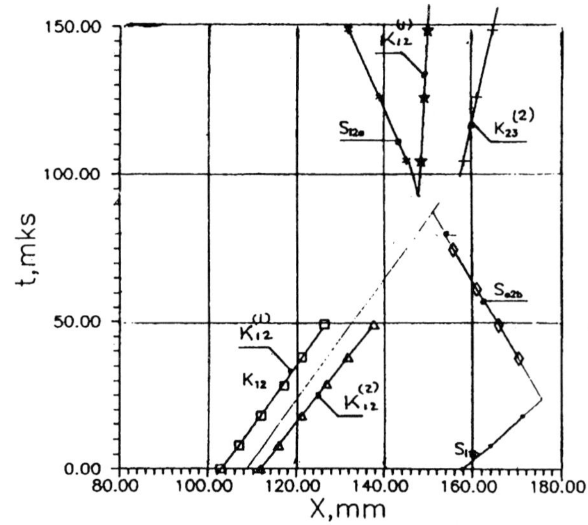

Figure 1. An x, t-diagram for run no. 583B.

The x, t-diagram given in Fig. 2 shows interaction of the incident shock wave S_{e1a} with the continuous interface obtained with a removable plate. The initial conditions and the measured data on K_{12} and K_{23} evolutions are given in Table 1. Numbers of experiments of the first (K_0 – discontinuous interface modeled by a thin film) and the second (K_0 – continuous interface) series are denoted with "B" and "A," respectively. Table 1 gives the experimentally measured values da_{K12}/dt and da_{K23}/dt.

To analyze the evolution of K_{23}, the experimentally measured values da_{K12}/dt are compared to the values calculated from (2) and (3).

$$\chi_1 = \frac{\dfrac{a^*_{K23} ukA}{\Psi(k,\Delta,A)}}{\dfrac{da_{K23}}{dt}} \qquad \chi_2 = \frac{\dfrac{da_{K12}}{dt} + \dfrac{a^*_{K23} ukA}{\Psi(k,\Delta,A)}}{\dfrac{da_{K23}}{dt}}$$

The a^*_{K23} was determined from the trajectories of $K_{12}^{(1)}$, $K_{12}^{(2)}$, $K_{23}^{(1)}$, and $K_{23}^{(2)}$. The $\Psi(k,\Delta,A)$ was calculated according to Duff et al. (1962). The experimentally measured thickness Δ is given in Table 1.

Table 1. Initial conditions and results of experiment.

	583B	**491A**	**560B**	**488A**
A→B→A	Xe→Kr→Xe	Ar→0.5Ar+0.5He→Ar	Ar→He→ Ar	Ar→He→Ar
$\mathbf{P}_0$	0.5	0.5	0.5	0.5
M	3.76	3.22	3.74	3.21
λ, mm	36	11.24	72	4.6
a_0, mm	10		5	
$(da/dt)_1$, mm/μs	0.042	0.017	0.092	0.064
$(da/dt)_2$, mm/μs	0.077	0.067	0.284	0.207
(da/dt), mm/μs	0.114	0.106	1.017	0.632
$a_{K_{12}}(t_R)k$	2.02	1.17	1.49	2.03
A	0.18	0.23	0.76	0.76
regime of refraction	soft	soft	hard	hard
Δ, mm	5.24	3.56	3.08	6.46
χ_1	0.82	0.60	2.90	1.90
χ_2	1.36	0.85	3.30	2.25
Ψ	1.8	2.62	1.2	1.57

3. Discussion

Our previous results (Aleshin et al. 1993) enable us to determine the regime of refraction of S_{e2b} on K_{12} from the experimentally measured interface amplitude $a_{K12}(t_R)$ before interaction of K_{12} with S_{e2b}. Thus, the regime of the refraction of S_{e2b} on K_{12} should be considered as a soft regular for runs 583B and 491A, and as a hard regular regime for runs 560B and 488A. For the runs with a hard regular regime of refraction the amplitude growth rate is expected to follow the relation (3) and hence χ_2 will be 1. This is true, provided that relation (3) adequately describes the influence of K_{12} evolution on K_{23} evolution.

The value χ_1 reflects the ratio between the measured growth rate a_{K23} and the value calculated proceeding on the assumption that K_{23} evolution is independent of K_{12} evolution. Unfortunately, the difference between χ_1 and χ_2, and the differences of each of the values from Table 1 have not allowed us to make a final conclusion whether the relation (3) is adequate or not. However,

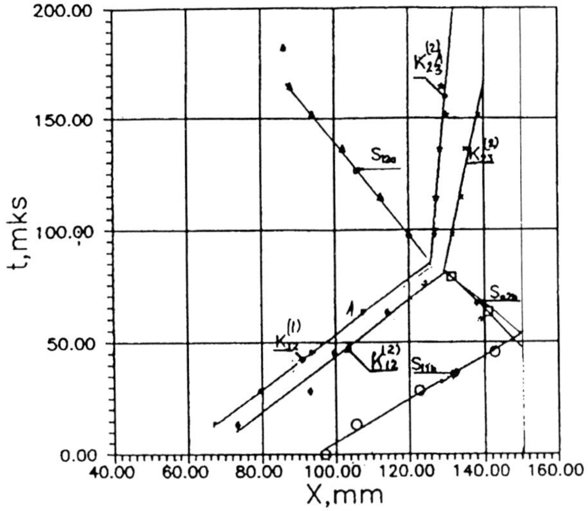

Figure 2. An x, t-diagram for run no. 491A.

the clear excess of χ_1 and χ_2 over 1 confirms the data (Aleshin et al. 1993), indicating that the character of the evolution of the mixing zone is dependent on the type of refraction. Indeed, the amplitude growth rate for hard regular regime is 2–3-fold lower than what is expected from relation (2).

Acknowledgement. The study was partly supported by the Russian Foundation for Fundamental Research (Grant No. 94-01-01144a), by a contract from the French Atomic Energy Commission (CEA/DAM No. 3672), and by the ISTC (Grant No. 94-029).

References

Aleshin AN, Zaytsev SG, Lazareva EV (1993) Experimental and numerical studies on Richtmyer-Meshkov instability. RJCM 1:33–49.

Brouillette M, Sturtevant B (1994) Experiments on the Richtmyer-Meshkov instability: single-scale perturbation on a continuous interface. J Fluid Mech 263:271–292.

Duff RE, Harlow FN, Hirt SW (1962) Effects of diffusion on interface instability between gases. Phys Fluids 5:417–425.

Zaytsev SG, Titov SN, Chebotareva EI (1994) The shock-induced evolution of continuous interface separating gases of different densities. Izv Ross Akad Nauk Mekh Zhidk i Gaza (2):18–26 (in Russian).

Shock-hydrodynamics experiments on the Nova laser

P. L. Miller, T. A. Peyser, P. E. Stry, K. S. Budil
Lawrence Livermore National Laboratory
P.O. Box 808
Livermore, CA 94551 USA

Abstract: We have conducted shock-induced hydrodynamics experiments using the Nova laser at Lawrence Livermore National Laboratory. The laser provides a high-enthalpy source by depositing its energy (about 22 kJ) in a small gold cavity called a *Hohlraum*. The *Hohlraum* serves as a driver section, launching very strong (M $\approx$ 20) shocks into millimeter-scale cylindrical "shock tubes." The flow is imaged radiographically by an electronic framing camera, using a laser-generated X-ray source. Several topics have been addressed with this configuration, including shock-induced mixing at density interfaces (seeded with a variety of perturbations); the development of high-speed, shaped-charge-like jets; the effects of geometry on the planarity of the generated shocks; and shock-shock interactions which develop in the flows. In this paper, we describe the general configuration of our experiments, present an overview of the high-speed jet work, discuss some of our findings, and compare our results with computer simulations we have conducted.

Key words: Jets, Mixing, Richtmyer-Meshkov instability, Nova laser, Compressible flow

1. Introduction

Understanding shock-induced flows, including the Richtmyer-Meshkov instability (Richtmyer 1960, Meshkov 1969) which occurs at shocked density interfaces, is of interest in inertial confinement fusion, astrophysics, and other applications. In this paper, we describe experiments involving high-speed jets generated at a density interface with a hemispherical perturbation. Additional material described in the ISSW oral presentation is not included here, but has been documented in Peyser et al. (1995).

2. Experiment description

The experiments were conducted using the Nova laser facility. Nova has ten separate beam lines, each capable of delivering approximately 2.8 kJ of energy in pulses 1-ns long, at a wavelength of 353 nm. The beams are focused into a 5-m-diameter target chamber, which contains a target positioner holding the target, alignment systems, and various diagnostics.

In our experiments, we use the Nova laser as a high-enthalpy source to drive our miniature shock tubes, and to produce diagnostic X-rays. A small gold cylindrical can, called a *Hohlraum*, serves as the shock-tube driver section (see Fig. 1). The *Hohlraum* dimensions are 3200 μm long by 1600 μm wide, with 1200-μm -diameter holes in both ends and a 45-μm wall thickness. Our miniature shock tube is mounted on a 700-μm hole on the side of the *Hohlraum*. The shock-tube wall is made of beryllium, about 700 μm in internal diameter with 100-μm -thick walls. The working fluid in the shock tube is brominated plastic (2% Br by atomic fraction) in the end closest to the *Hohlraum*, and low-density carbon foam (0.1 g/cc) in the remainder of the tube. The shock-tube length, 2 mm, and external diameter, 0.9 mm, are about the same size as the letter "i" in the word "DIME" on the backside of a U.S. dime.

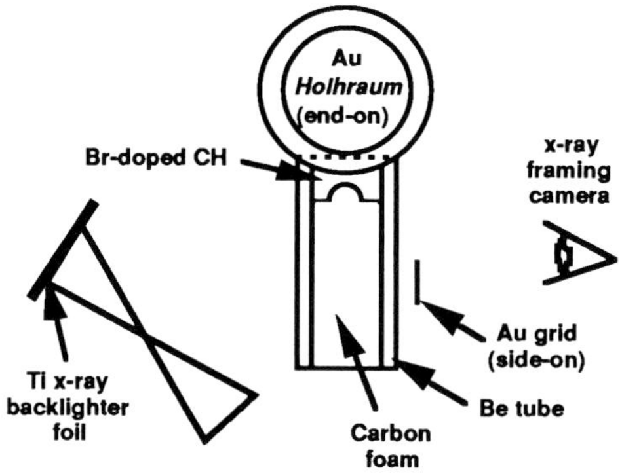

Figure 1. Schematic of the experiment.

The sequence of events in the experiments is as follows. Eight of Nova's beams are focused into the interior of the gold *Hohlraum* can, rapidly heating the interior of the *Hohlraum* to a temperature of about 230 eV (over 2 million K). The plastic-filled end of the shock tube, exposed to the high temperature through the hole in the side of the *Hohlraum* it is mounted over, begins to rapidly ablate, driving a strong shock (tens of Mbars) down the axis of the tube. As the shock propagates down the tube, it crosses the interface between the brominated plastic and the low-density carbon foam. The interface is flat, except for a 150-μm -radius, hemispherical hole in the plastic side (filled with foam), centered on the tube axis. The shock tube and associated parts are made with advanced micro-machining techniques developed at LLNL (Louis et al. 1995).

The primary data collected in these experiments are X-ray images. To provide the X-ray illumination to capture these images, the remaining two of Nova's ten beams, which are not used to heat the *Hohlraum*, are delayed in time and strike a titanium foil located several millimeters behind the shock tube, as viewed from the diagnostic camera. The interaction of the intense laser light with the metal produces X-rays (of about 4.9 keV), providing a backlighting source. The X-ray transmission through the flow is attenuated by a factor which is a combination of path length, type of material, and density. The beryllium tube and unshocked carbon foam are relatively transparent to X-rays of that energy, while the brominated plastic and higher-density materials are more opaque. Therefore, the jet of brominated plastic appears as a dark feature on a lighter background. Images of the resulting flow are captured using an electronic framing camera (Budil et al. 1995), which is basically an X-ray pinhole camera, capturing two-dimensional images with about 100 ps exposures. The data is recorded on film, and later digitized on a densitometer, including wedge corrections.

3. Experimental and numerical results

Two-dimensional, axisymmetric numerical simulations of the jet flow have been conducted using the code CALE. CALE is an arbitrary Lagrangian-Eulerian code that was developed at LLNL. A calculated temporal sequence of the jet behavior, depicting material boundaries, is shown in Fig. 2. Since this is a depiction of material interfaces, densities (and the shock) are not represented. The shock is traveling from top to bottom in the figures. At 0 ns, the initial geometry is evident, and 5 ns after the laser begins to heat the *Hohlraum* the shock is crossing the hemispherical protrusion, causing it to invert. By 10 ns, the hemisphere has fully inverted and is forming a jet, which continues to develop through later times. The sides of the jet, subject to shear, form roll-ups by the Kelvin-Helmholtz instability. Such roll-ups, when viewed in transmission, appear in the data as dark horizontal bands.

Using the known opacities of the materials, the simulation was post-processed to create simulated radiographs, which can then be compared directly with the experimental data. A comparison of data with the results from the simulation is shown in Fig. 3. Cosmetic differences between the two sets of images include the presence of noise, the inclusion of reference grids, and the non-uniformity of the X-ray illumination, all in the experimental data. Features of interest in the experimental data include the jet's bow shock (particularly evident at 10.6 ns), the dark bands of the vortical structure around the head of the jet, swirls of material at 14.8 ns and trailing tendrils from the sides at 19.7 and 21.9 ns, and the growth and thinning of the neck of the jet, particularly at 21.9 ns. Differ-

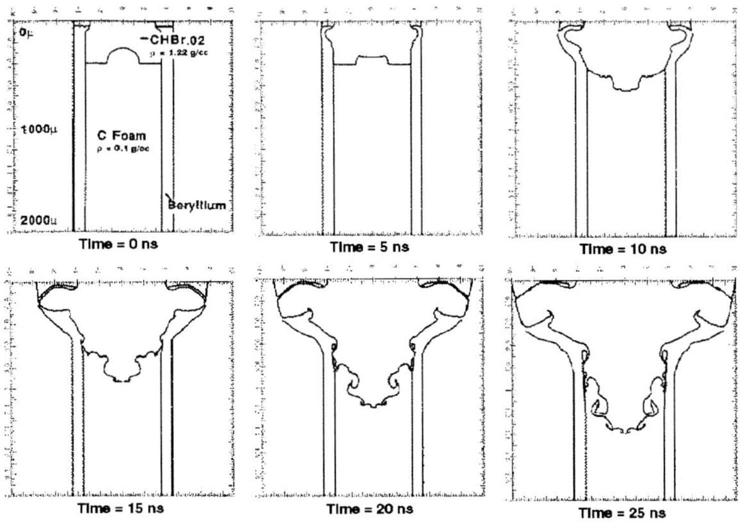

Figure 2. Material boundaries of the high-speed jet experiment, calculated with CALE.

ences between the simulation and the data include faster growth (longer neck) and later development of the swirl features in the simulation.

Using data from a series of separate experiments, four jet features were identified in each picture: the bow-shock location, the tip of the jet material, the center of the dark band from the vortical roll-up, and the "shoulder" or material to the side at the base of the jet. The axial positions of these four features were measured from each data image, and the results are compared in Fig. 4 to the measurements from the simulated X-ray images of the calculation. One data point for the shoulder at 19.5 ns was out of the frame. The time is measured from the beginning of the 1.0-ns laser drive pulse, and the shock crosses the hemispherical protrusion around 5 ns. Axial position is measured from the end of the brominated plastic closest to the *Hohlraum*, with the tip of the hemisphere initially at 250 μm and the flat around it at 400 μm .

The agreement between the simulation and the data is, depending on the context, between excellent and fair. The feature locations from the calculation agree within 5 to 10% of their total travel distance with those from the data, with the exception of the earliest-time data, and the spacing between features at a particular time is also quite similar. The data points at 11.5 ns and 19.5 ns are

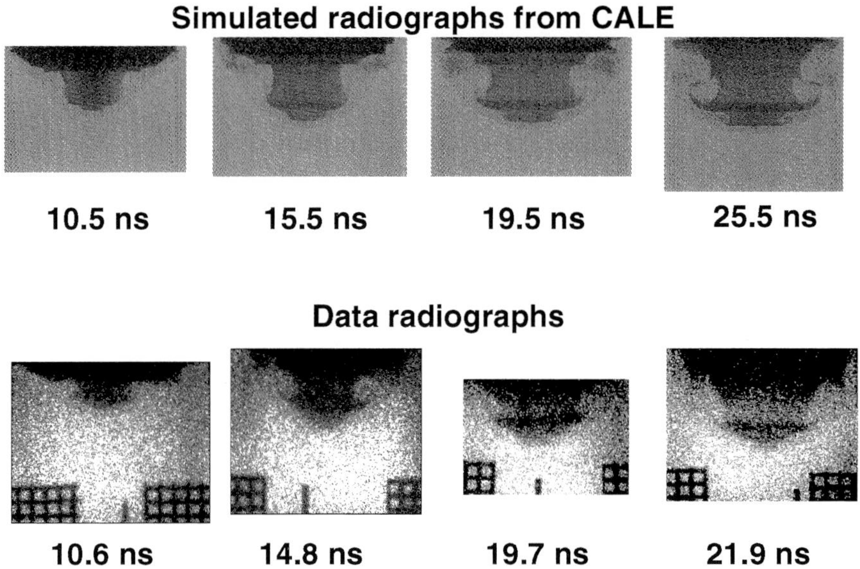

Figure 3. X-ray images from CALE calculation and experimental data.

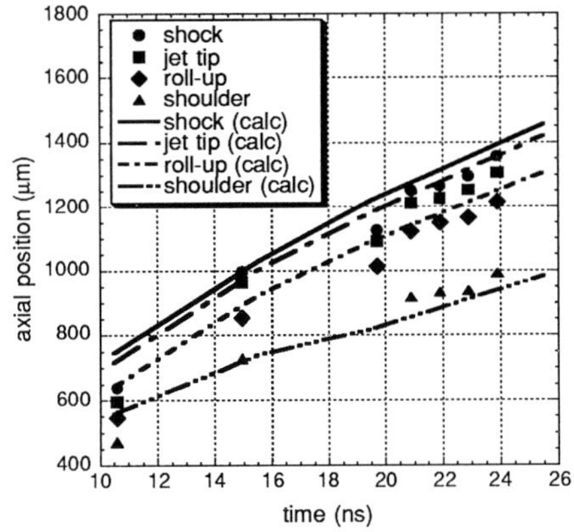

Figure 4. Comparison of jet-feature locations from calculation and experiment.

significantly less advanced than the calculation, however. It should be noted that the drive energy delivered by the laser in each experiment is subject to variations of $\pm 10\%$, or so, and the energies measured for the two lower sets of data were, in fact, lower than the others. One possible approach to correct for this uncertainty in the initial drive would be to assume the shock location is an accurate indicator for the drive condition, and to shift the data points in time to where they agree with the simulated shock location. We do not present such an adjustment here.

4. Conclusions

We have presented results from our shock-induced fluid-dynamics experiments conducted on the Nova laser facility. For this paper, we have focused on our high-speed-jet data and compared it to two-dimensional CALE calculations. The distinct features present in the jets provide good fiducials for comparison with the calculations. The CALE simulations are found to agree in major respects, but differ in some details. We also note that the roll-up observed in the data is possibly the first such direct observation of the Kelvin-Helmholtz instability in this (plasma) regime.

Acknowledgement. This work was performed under the auspices of the U.S. Department of Energy by the Lawrence Livermore National Laboratory under Contract Number W-7405-ENG-48.

References

Budil KS, Perry TS, Bell PM, Hares JD, Miller PL, Peyser TA, Wallace R, Louis H, Smith DE (1995) The flexible x-ray imager. Rev Sci Inst 67(2):1.

Louis H, Demiris A, Budil KS, Miller PL, Peyser TA, Stry PE, Wojtowicz DA, Dimotakis PE (1995) Miniature targets for high-energy density experiments on Nova. Fusion Technology 28(5):1769.

Meshkov EE (1969) Instability of the interface of two gases accelerated by a shock wave. Sov Fluid Dyn 4:101.

Peyser TA, Miller PL, Stry PE, Budil KS, Burke EW, Wojtowicz DA, Griswold DL, Hammel BA, Phillion DW (1995) Measurement of radiation-driven shock-induced mixing from nonlinear initial perturbations. Phys Rev Lett 75:2332.

Richtmyer RD (1960) Taylor instability in shock acceleration of compressible fluids. Commun Pure Appl Math 13:297.

Shock-accelerated heavy-gas layers

J. M. Budzinski, R. F. Benjamin
Los Alamos National Laboratory
Los Alamos, NM 87545 USA
J. W. Jacobs
University of Arizona
Tucson, AZ 85721 USA

Abstract: Shock-accelerated heavy-gas layers are studied experimentally and with front-tracking simulations. Richtmyer-Meshkov instability develops on each interface. Three different patterns form depending on the relative size of the initial perturbations on the upstream and downstream interfaces. The patterns are interpreted in terms of the average vorticity generated on the interfaces.

Key words: Richtmyer-Meshkov instability, Shock acceleration, Shock-induced mixing, Front-tracking

1. Introduction

This study is directed toward a better understanding of the flow of shock-accelerated heavy layers embedded between lower-density fluids. This problem is of fundamental interest because it involves the production, coupling, and transport of vorticity in a system sensitive to initial conditions. Our experimental observations are available in Budzinski et al. (1994), Jacobs et al. (1993), and Jacobs et al. (1995). We report here further analysis of the experimental data and the results of simulations using a front-tracking code.

A fluid interface subjected to a shock acceleration becomes unstable and distorts after the acceleration. This impulsively driven instability is commonly referred to as shock-induced Rayleigh-Talyor instability or Richtmyer-Meshkov (RM) instability. We are studying the impulsive acceleration of two nearby interfaces, and the subsequent growth and coupling of these interfaces. Nearly sinusoidal perturbations are initially imposed on the interfaces. The perturbations grow, distort, and interact, producing distinctive flow patterns. We find that similar initial conditions produce three different flow patterns, identified descriptively as: upstream mushrooms, sinuous, and downstream mushrooms. We report these observations and analysis with simulations using a front-tracking code developed at SUNY Stony Brook (Chern et al. 1986, Grove 1994).

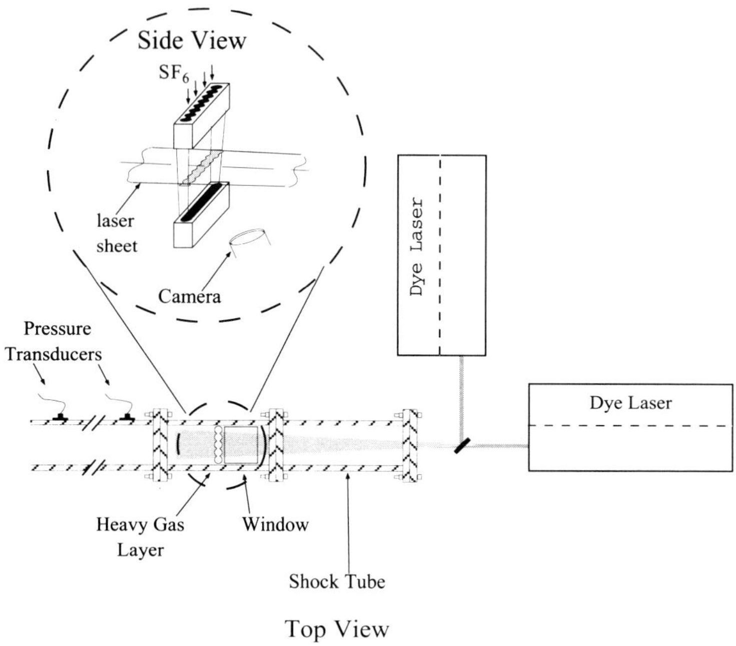

Figure 1. Schematic of shock tube and associated diagnostics. Two images per experiment are acquired: one of the initial conditions of the gas curtain, and the other of the flow pattern at a predetermined time after shock acceleration.

2. Experimental apparatus

Figure 1 shows the experimental apparatus. The layer is generated inside a shock tube by vertically flowing SF_6 through a corrugated nozzle into a test section. In this way, a laminar gas jet (i.e., gas curtain) with a varicose cross section and diffuse boundaries is created with a wavelength of 6 mm and amplitudes up to 1 mm on each side. This flow system has been shown to produce a two-dimensional flow after a shock acceleration (Jacobs et al. 1993, 1995). Small, unsteady, uncontrollable fluctuations in the SF_6 jet cause asymmetries which vary from experiment to experiment. The fluctuations are measured but not controlled. The present experiments use planar laser Rayleigh scattering to capture two frames per experiment, measuring the pre-shock initial conditions and one dynamic condition at varying times after the shock interaction. Two pulsed dye lasers illuminate the test region with coplanar sheets of laser light, and a cooled CCD camera records each pair of images. The images are converted to maps of SF_6 concentration by calibrating the system with scattering from pure air and pure SF_6.

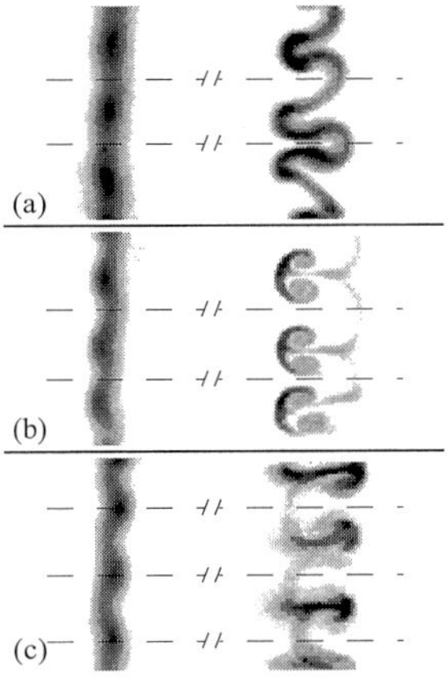

Figure 2. Composite of images from three typical experiments showing the initial and dynamic conditions of each experiment. Each pair of initial and dynamic conditions was recorded on the same event. (a) Upstream mushroom, dynamic image recorded 0.45 ms after the shock interaction. (b) Downstream mushroom, dynamic image at $t = 0.40$ ms. (c) Sinuous, dynamic image at $t = 0.45$ ms.

3. Experimental results

Sample images from the experiments are shown in Fig. 2. We find that asymmetries in the initial conditions correlate with the three post-shock flow patterns in over 90% of the 100+ experiments performed. This correlation is shown in Fig. 3 by plotting the initial upstream amplitude versus the initial downstream amplitude for each experiment, and identifying on the graph which pattern was generated. The post-shock flow patterns are classified visually, and the initial upstream and downstream amplitudes are found from an analysis of the Fourier coefficients of the pre-shock interfaces. On this graph, the experiments which generated similar flow patterns cluster together, with a small amount of overlap between the different clusters. An upstream mushroom pattern develops when the initial perturbations on the upstream side (i.e., on the side first interacting with the shock wave) are larger than the perturbations on the downstream side. A sinuous pattern forms when the initial perturbations on the downstream side are about the same or slightly larger than those on the upstream side. A downstream mushroom develops when the perturbations on the downstream side are much larger than the perturbation on the upstream side.

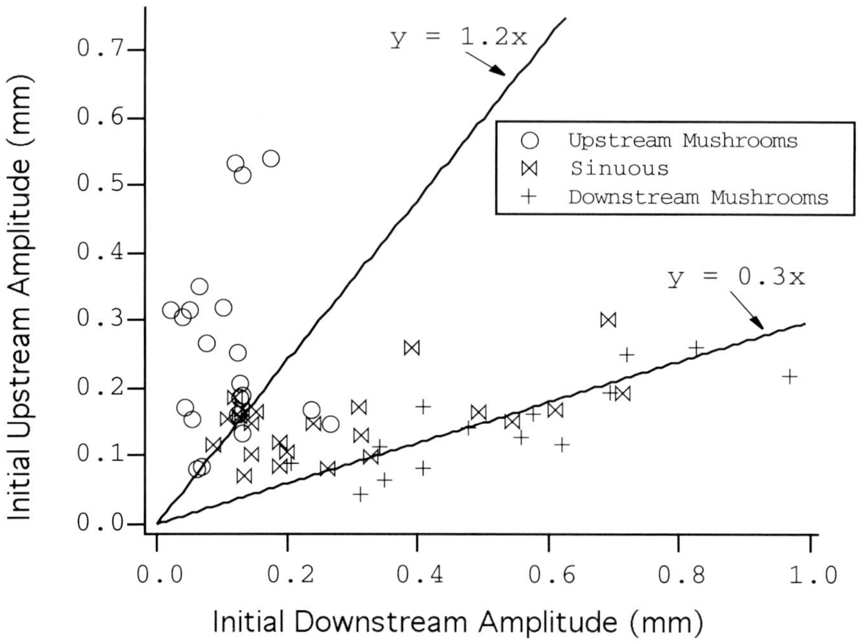

Figure 3. Upstream versus downstream perturbation amplitudes for the experiments show clusters of similar flows.

One way of interpreting the creation of the three patterns is to view the flow evolution within the context of vortex dynamics. The vorticity is generated baroclinically by the shock interaction through the misalignment of pressure and density gradients. The magnitude of vorticity will be greatest where the pressure and density gradients are most misaligned. Assuming the pressure gradients are predominantly from planar waves, most of the vorticity is expected to be in the regions between the peaks and valleys of the initial perturbations. When the perturbation amplitude is greater on one side of the layer, more vorticity is generated on that side. Furthermore, the vorticity is preferentially generated in the lighter fluid. This effect accentuates the offset of the vorticity from the center of the layer. The mushrooms form on the side with the largest initial perturbation amplitude because the heavy fluid is asymmetrically entrained into the side with the most vorticity. The vorticity distribution of a symmetrically perturbed initial layer would be centered on the heavy layer and thus produce a sinuous pattern.

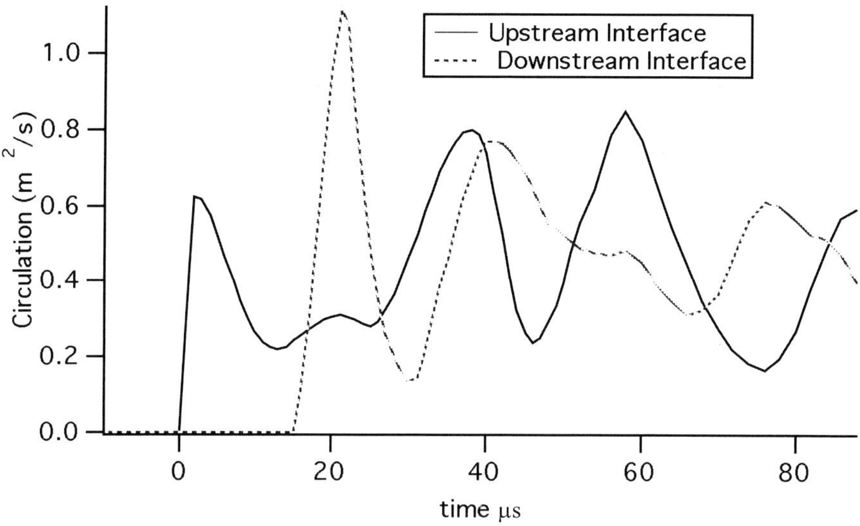

Figure 4. Circulation versus time for a layer with initial symmetric perturbations of 0.33 mm amplitude.

4. Simulations

The simulations, which reproduce all three of the experimental flow patterns, illustrate the complex nature of the wave dynamics and vorticity generation. Every time a pressure wave interacts with a material interface, a transmitted wave and a reflected wave are created. A series of reverberating waves of successively decreasing strengths are generated inside the heavy layer. The simulations show the contribution that each of these waves has on the total circulation generated on the upstream and downstream material interfaces. These contributions are essential to an analysis of the total vorticity generated. The simulations also show the deformation of the shock waves that are reflected and transmitted through the layer. These waves deform to the extent that they break and then create transverse or lateral waves that run along the layer. Interactions of the material interfaces with these lateral waves cause oscillations in the circulation on the upstream and downstream material interfaces. Figure 4 shows the circulation as a function of time for a layer with symmetric perturbations. These oscillations in circulation last for significant periods of time after the first shock interaction. In order to interpret the pattern formation in terms of vorticity on the interfaces, the average of the vorticity over time, or the vorticity after the large oscillations decay, becomes the important parameter.

5. Conclusions

Simulations and experiments are performed on shock interactions with heavy-gas layers. Three flow patterns develop depending on the relative amplitudes of the initial perturbations on the upstream and downstream interfaces. Simulations show large oscillations of the circulation on the interfaces.

Acknowledgement. We are indebted to James Glimm and John Grove for allowing us to use their front-tracking code, and for their expert advice and support. We also would like to recognize the help of Jean-Marc Le Peuvedic in running the front-tracking code. We would like to thank Robert Critchfield for his assistance on the experiments. This work is supported by US DOE, W-7405-ENG-36.

References

Budzinski JM, Benjamin RF, Jacobs JW (1994) Influence of initial conditions on the flow patterns of a shock-accelerated thin fluid layer. Physics of Fluids 6:3510–3512.

Chern IL, Glimm J, McBryan O, Plohr B, Yaniv S (1986) Front tracking for gas dynamics. J Comput Phys 62:83–110.

Grove J (1994) Applications of front-tracking to the simulation of shock refractions and unstable mixing. J Appl Num Math 14:213–237.

Jacobs JW, Klein DL, Jenkins DG, Benjamin RF (1993) Instability growth patterns of a shock-accelerated thin fluid layer. Phys Rev Lett 70:583–586.

Jacobs JW, Klein DL, Jenkins DG, Benjamin RF (1995) Nonlinear growth of shock-accelerated instability in a thin fluid layer. J Fluid Mech 295:23–42.

Experiments on the shock-excited interfacial instability at an air/SF_6 interface

M. Vetter*, B. Sturtevant
Graduate Aeronautical Laboratories
California Institute of Technology
Pasadena, CA 91125 USA

Abstract: Results of flow-visualization experiments of an impulsively accelerated plane interface between air and SF_6 are reported. The shock tube used for the experiments has a larger test section than in previous experiments. The larger extent of uniform test flow relative to nonuniform boundary-layer flow permits unambiguous interpretation of flow-visualization photographs, and the influence of shock-wave/boundary-layer interactions is no longer dominant. The strong wall vortex observed in previous studies is not observed in these experiments. It is found that the thin membrane, which forms the initially plane interface, has a significant influence on the initial growth rate of the interface thickness. However, the measured growth rates after the first reflected shock are independent of membrane configuration and are in good agreement with analytical predictions.

Key words: Richtmyer-Meshkov instability, Flow visualization, Experiment, Shock tube, Turbulent mixing

1. Introduction

The Richtmyer-Meshkov instability occurs when a shock wave impinges upon an interface between two fluids of different densities (Richtmyer 1960, Meshkov 1969). The refraction of the shock wave from spatial perturbations on the interface causes distortion of both the interface and the shock wave. Most of the experiments so far reported in this field have concentrated on the instability of an initially sinusoidal interface with perturbations of wavelength λ and amplitude η_0. Richtmyer (1960) showed that in this case the perturbation amplitude η grows linearly with time. Small-scale perturbations, the subject of this paper, grow rapidly into the nonlinear regime. Perturbations of, for example, $\eta_o/\lambda = 0.1$ and $\lambda = 1$ cm, on an interface which is accelerated impulsively from rest to a velocity of 150 m/s (all values typical for the experiments described here), grow out of the linear regime within 0.7 ms. This finally leads to a tur-

*Presently at: Bayern-Chemie, D-85521 Ottobrunn, Germany.

bulent mixing zone (TMZ) between the two fluids. Mikaelian (1989) adapted results of constant-acceleration experiments of Read (1984) to impulsive acceleration and obtained an analytical expression for the nonlinearly saturated shock-induced thickening of an initially flat interface,

$$\delta = 0.28[u]A't \,, \tag{1}$$

where δ is the interface thickness, $[u]$ is the change of interface velocity induced by the shock wave, A' is the post-shock Atwood ratio, and t is time. It has recently been suggested (Brouillette & Sturtevant 1989, 1993) that, due to the modest-size shock tubes used for the relatively few experiments on this problem to date (Andronov et al. 1976, Andronov et al. 1982, Brouillette & Sturtevant 1989, 1993), shock-wave/boundary-layer effects have had a large influence on the measured growth rates and on the interpretation of flow-visualization pictures. The goal of this work is to reduce this influence by using a larger test facility.

2. Experimental

The experiments were performed in the GALCIT 17-inch diameter horizontal shock tube (Vetter & Sturtevant 1995). A 27-cm square test section is flanged to the end of this tube, and the transition from circular to square cross section is made with a cookie cutter extending 1.5 m upstream into the shock tube (Fig. 1). A thin (0.5 μm) nitro-cellulose membrane, which forms the initial interface, is placed between the shock tube and the test section. Mach numbers above 1.1 are generated by reducing the initial pressure p_1 below atmospheric.

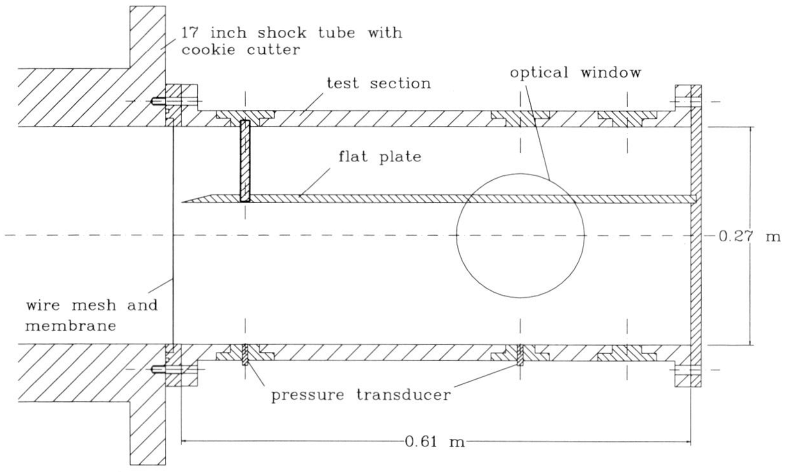

Figure 1. Geometry of the short test section with the mounted flat plate.

For $M_s = 2$, p_1 is as low as 8 kPa. In most of the experiments a wire mesh was installed to support the membrane on the upstream side during the filling procedure. Prior to each run, the membrane was usually pushed slightly into the wire mesh, with an estimated amplitude of about 1 mm. The optical windows for the schlieren visualization system are centered 42 cm downstream of the membrane and are 15 cm in diameter, so the shock-tube wall is not directly visible in the pictures. Therefore, some experiments were performed with a sharp-edged flat plate mounted across the window, extending from the end wall up to a distance of 10 mm downstream of the membrane (see Vetter & Sturtevant 1995). The thickness δ of the interface, defined to be the mean horizontal overall extent of the interface, was measured on both spark-schlieren and high-speed movie photographs. No corrections were made for possible wall effects. The rate of increase of the interface thickness ('growth rate') is obtained by a straight-line least-squares fit to the thickness data.

3. Growth rates

The evolution of the average thickness measured from high-speed motion pictures for four different incident-shock Mach numbers between 1.24 and 1.98 is shown in Fig. 2. The length L of the test section was adjusted in some experiments to insure that the interaction of the reflected shock wave with the interface would be in the field of view. Very small and sometimes no growth

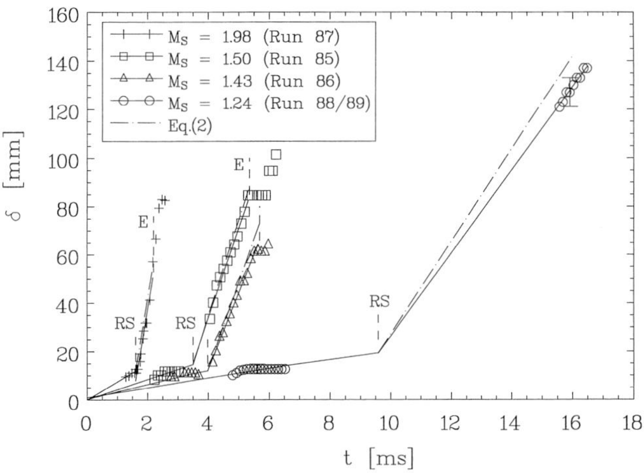

Figure 2. Time evolution of the thickness of the TMZ for the plane discontinuous interface between air and SF_6. Results from high-speed motion pictures (RS: reflected shock; E: expansion).

is observed before the reflected shock interacts with the interface; growth rates are an order of magnitude smaller than predicted by Eq. 1. In experiments in which the wire mesh was placed on the *downstream* side of the membrane to fragment it and enhance mixing, it is observed that the mixing region is up to five times thicker than with the wire mesh on the upstream side. However, this still represents a growth rate only about 2/3 of that predicted by Eq. 1. In this configuration residual wakes of SF_6 from the wire mesh are visible behind the TMZ.

The growth of the interface between the reflected shock and the arrival of the re-reflected expansion wave appears to be linear, and the measured growth rates are in good agreement with Eq. 1 (indicated by the dash-dotted lines in Fig. 2, using the Atwood ratios after the reflected shock), even for the low Mach number with only one datum available. The initial thickness for the post-reflected shock fit is taken from the incident shock fitted result, but nearly the same rate would be obtained if the actual measured pre-reflected shock thickness were used.

Calculated growth rates assuming that the initial thickness is zero are compared with the measurements in Table 1. The index "0" corresponds to the

Table 1. Test conditions and comparison between measured and predicted growth rates of the interface thickness.

	I[a]	II[b]	III[b]	IV[a]	V[a]	VI[b]	VII[b]	
M_s	1.18	1.24	1.43	1.45	1.50	1.50	1.98	–
p_1	55	40	31	29	23	23	8	kPa
L	123	110	62	62	62	62	49	cm
M_T	1.27	1.37	1.67	1.70	1.78	1.78	2.56	–
u_0	56	72	126	126	150	150	287	m/s
$(d\delta/dt)_{0,exp.}$	0.94	2.1	3.0	3.1	4.0	4.2	7.5	m/s
$(d\delta/dt)_{0,Eq.\ 1}$	10.7	14.2	25.5	25.6	30.7	30.7	61.3	m/s
$(d\delta/dt)_{1,exp.}$	19.2	17.0	31.5	35.5	32.6	37.2	74.4	m/s
$(d\delta/dt)_{1,Eq.\ 1}$	20.1	22.2	35.9	37.3	39.8	39.8	75.5	m/s

[a] Results from single spark-schlieren photographs.
[b] Results from high-speed motion pictures.

conditions after the incident shock and the index "1" to those after the reflected shock. The effective initial thickness, due to the starting process, may in fact be finite. Thus the values given for the growth rates after the initial shock are upper bounds. The present results for the growth rate after the reflected shock are summarized in Fig. 3, a plot of growth rate $(d\delta/dt)_1$ normalized by the velocity jump $[u]_1$ induced by the reflected shock vs. the Atwood ratio A'_1 after the reflected shock. The measured growth rates fall in a band 1.5 to 23%

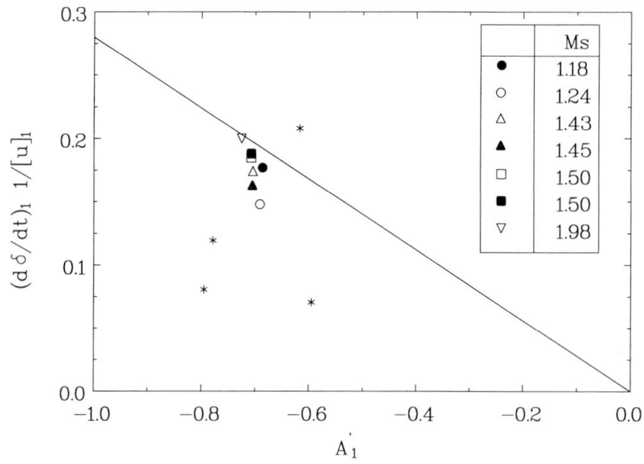

Figure 3. Correlation of reflected-shock growth rates for a discontinuous interface between air and SF_6. Filled symbols: results from single schlieren pictures. Open symbols: results from high-speed motion pictures. —: Mikaelian formula (Eq. 1). ∗: experimental results from Brouillette & Sturtevant (1993).

below the analytical formula of Mikaelian (1989), Eq. 1, with substantially better agreement than in previous investigations. The agreement improves as M_s increases.

4. Discussion and conclusions

Results of growth-rate measurements of impulsively accelerated plane interfaces between air and SF_6 have been presented. Due to the much larger test section than in previous studies, the effects of shock-wave/boundary-layer interaction have been reduced significantly. The resulting increased resolution of schlieren pictures and comparison of pictures taken at the same instant both with horizontal and vertical knife edges permit a better interpretation of the visible flow features. Contrary to previous investigations in smaller facilities, no strong wall vortex was found in these experiments (Vetter & Sturtevant 1995). This suggests that the interactions observed in small tubes may be enhanced by mutual interaction of the wall effects.

The measured growth rate after the incident shock is an order of magnitude less than predicted by Mikaelian (1989), Eq. 1, presumably due to strong damping of mixing by the nitro-cellulose membrane. On the other hand, the growth rate after the reflected shock agrees to within 1.5–23% with Eq. 1, depending on the shock strength, suggesting that the damping effect of the membrane is negated by the reflected shock. The fact that the normalization suggested by Eq. 1 does not remove all of the shock-strength dependence is taken to mean that the growth rate constant 0.28 in Eq. 1 actually depends weakly on initial conditions. Observed growth rates before reshock are not only shock-strength dependent but also apparatus (mesh and membrane) and facility dependent. On the other hand, independently of the pre-reflected-shock conditions, the post-reflected-shock growth rate agrees well with Eq. 1, suggesting that non-linearly saturated Richtmyer-Meshkov mixing is physically similar to saturated Rayleigh-Taylor mixing, from which Eq. 1 is inferred.

Acknowledgement. This research was supported by the U.S. Department of Energy, Lawrence Livermore National Laboratory, under contract agreement DOE W-7405-ENG-48. The first author was also supported in part by Deutsche Forschungsgemeinschaft, Germany.

References

Andronov VA, Bakhrakh SM, Meshkov EE, Mokhov VN, Nikiforov VN, Pevnitskii AV, Tolshmyakhov AI (1976) Turbulent mixing at contact surface accelerated by shock waves. Sov Phys JETP 44:424.

Andronov VA, Bakhrakh SM, Meshkov EE, Nikiforov VV, Pevnitskii AV, Tolshmyakhov AI (1982) An experimental investigation and numerical modeling of turbulent mixing in one-dimensional flows. Sov Phys Dokl 27:393.

Brouillette M, Sturtevant B (1989) Growth induced by multiple shock waves normally incident on plane gaseous interfaces. Physica D 37:248.

Brouillette M, Sturtevant B (1993) Experiments on the Richtmyer-Meshkov instability: small-scale perturbations on a plane interface. Phys Fluids A 5:916.

Meshkov EE (1969) Instability of the interface of two gases accelerated by a shock wave. Sov Fluid Dyn 4:101.

Mikaelian KO (1989) Turbulent mixing generated by Rayleigh-Taylor and Richtmyer-Meshkov instabilities. Physica D 36:343.

Read KI (1984) Experimental investigation of turbulent mixing in Rayleigh-Taylor instability. Physica D 12:45.

Richtmyer RD (1960) Taylor instability in shock acceleration of compressible fluids. Commun Pure Appl Math 13:297.

Vetter M, Sturtevant B (1995) Experiments on the Richtmyer-Meshkov instability of an air/SF_6 interface. Shock Waves 4:247.

Mathematical modeling of gravitational turbulent mixing in non-self-similar problems

V. Y. Neuvazhayev
Russian Federal Nuclear Center – VNIITF
454070, Snezhinsk, Chelyabinsk Region, P.O. 245, Russia

Abstract: Semi-empirical models of different complexity can be applied to describe the turbulent mixing arising as a result of the gas-dynamic instability induced by the action of shock waves and acceleration. In this paper the K and KE models, being the most widely used for scientific applications, are set forth. Considered in detail are four types of problems which can be described, in spite of their complexity, within the framework of the simple models. The significance of the performed investigations lies in fact that all results are obtained in an analytical form which makes it possible to observe general regularities.

Key words: Turbulent mixing, Shock waves, Analytical form

1. The sign-variable acceleration

By compressing the targets in Inertial Confinement Fusion (ICF) problems, the situation in which the acceleration of the boundary changes its sign is typical (Fig. 1). This should be taken into account in turbulent mixing models. The method of separation accounting is based on the addition of the corresponding transport terms (Neuvazhayev 1993a) and is used within the framework of the known K and KE models. In the case of incompressible fluids, the equation for the density profile for the K model is used in the following form:

$$\frac{\partial \rho}{\partial t} = \frac{\partial}{\partial x}\left[0.3L\overline{v}\frac{\partial \rho}{\partial x} - \sqrt{a_s}\sqrt{\frac{2|g|L}{(\rho_1-\rho_2)(\rho_1+\rho_2)}}(\rho_1-\rho)(\rho-\rho_2)\right] \tag{1}$$

$$\text{(a)}\quad \frac{dL}{dt} = 3.06\overline{v} \qquad \text{(b)}\quad \frac{d\overline{v}^2}{dL} + (0.5+0.23A^2)\frac{\overline{v}^2}{L} = 0.11\pi g A \tag{2}$$

Equations (2a) and (2b) for turbulent velocity are deduced for the mean velocity $\overline{v}$ over the mixing zone L, with the width of the mixing zone chosen effectively considering the law of mixed mass conservation. $A = \frac{\rho_1-\rho_2}{\rho_1+\rho_2}$, where ρ_1 and ρ_2 are the initial densities of heavy and light matters, respectively.

The specific coefficient value 0.07 is also presented here: under constant acceleration $L/2 = 0.07 g_o A t_o$; and with attenuation, $L \sim t^{2/7}$ when the acceleration goes to zero.

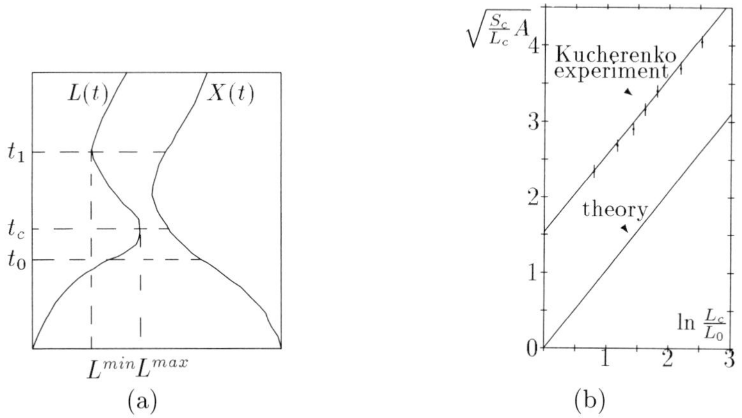

Figure 1. (a) Typical picture of the behavior of the boundary $X(t)$ between light (ρ_2) and heavy (ρ_1) matter under change of acceleration sign: at t_0 and t_1, the acceleration becomes zero; L is the mixing-zone width. (b) Dependence of the value $(S_c/L_c)^{1/2}$ on $\ln(L_c/L_0)$, where L_c is the width of a continuous density profile, L_0 is the initial roughness, and S_c is the displacement of the ampoule if $L = L_c$.

It is essential to note under what conditions the transport term is entered into (1) — the second component on the right side under the differential sign. The transport term is added only in the stable flow stage when the increment $gA < 0$. It is added not at once, but after the period during which the turbulent velocity $\overline{v}$ reduces from $\overline{v}_o$ to some critical value $\overline{v}_c$ according to Eqs. (2a) and (2b): with $t = t_c \overline{v}_c^2 = N \overline{v}_o^2$. By $\overline{v}_o$ we represent velocity at the moment when the acceleration sign changes. Value N is left indeterminate for now. It can be determined by the experiment or calculation.

Equations (2a) and (2b) are used only until critical time t_c, that is, when $t \le t_c$. When $t \ge t_c$, it would be well to present the new equation for $\overline{v}$ but in so far as it is not clear how to do it, we assume that when $t \ge t_c$, velocity $\overline{v}$ is constant: $\overline{v} = \overline{v}_c$. So, when $t \ge t_c$ in (1), $\overline{v} = \overline{v}_c$ and the mixing-zone width is determined effectively by the density profile. After changing sign, if the acceleration is kept constant as $t \to \infty$, the stationary density profile is elaborated and calculated from the equation

$$\frac{\partial \rho}{\partial x} = \frac{1}{3\overline{v}_o \sqrt{NL}} \sqrt{\frac{2|g|}{(\rho_1 - \rho_2)(\rho_1 + \rho_2)}} (\rho_1 - \rho)(\rho - \rho_2)$$

The limiting value of the mixing-zone width L_∞ is connected with the problem parameters in the following way: $L_\infty = 3.3N|\frac{g_o}{g_1}|L_o$. Notice that setting $N = 0$ means the complete separation $L_\infty = 0$ is taking place (Neuvazhayev 1993b). The experiments of Kucherenko et al. (1994) made it possible to determine one more constant, $\alpha_s = 0.01$.

2. The delay in the turbulent mixing process

Let there be two incompressible fluids with densities ρ_1 and ρ_2. The constant acceleration g is directed perpendicular to the boundary so that the Rayleigh-Taylor instability is taking place. It is known that by the initial continuous boundary there appears a delay in the turbulent mixing process. This problem was studied theoretically by LeLevier et al. (1955), Zababakhin and Ptizin (1988), and others. This problem was studied experimentally by Ardashova et al. (1988), where the relationship between the delay and the value of initial perturbations was set for Atwood number 0.6. The perturbations were preset in the middle of the continuous layer by the particles of matter with intermediate density. The theoretical studies of the delay phenomenon on the basis of the K model of turbulent mixing, conducted by Neuvazhayev (1994) lead to the determination of the analytical relationship between the initial roughness L_o and the mixing delay L_c determined by transference of system S_c at the instant when the turbulent mixing-zone width L is equal to L_c, i.e., to the width of the initial continuous boundary. This formula has the form

$$\sqrt{\frac{L_c}{L_o}} = \exp\left(0.48\sqrt{\frac{S_c A}{L_c}}\right) \tag{3}$$

It results from integrating the equations

$$\frac{\partial \rho}{\partial t} = \frac{\partial}{\partial x} D \frac{\partial \rho}{\partial x} \qquad \frac{d\overline{v}^2}{dL} + (0.5 + 0.23A^2)\frac{\overline{v}^2}{L} = \frac{\pi}{16} L\gamma^2 \tag{4}$$

$$\gamma^2 = \begin{cases} \frac{1.78gA}{L_c} & \text{if} \quad L \le L_c \\ \frac{1.78gA}{L} & \text{if} \quad L \ge L_c \end{cases} \tag{5}$$

Integrating the obtained equations with (2a) leads to the solution for mixing-zone width:

$$\sqrt{L} = \begin{cases} \sqrt{L_o}\exp\left(0.638\sqrt{\frac{gA}{(3.5+0.11A^2)L_c}}t\right) & \text{if} \quad L < L_c \\ \approx \sqrt{L_c} + 0.319\sqrt{\frac{gA}{(3.5+0.11A^2)L_c}}(t - t_c) & \text{if} \quad L > L_c \end{cases} \tag{6}$$

It must be emphasized here that the special feature of the developed solution — the exponential law of mixing in the beginning — is determined by the increment γ which takes the limited value with $L < L_c$. Formula (3) results from (6) if we determine the delay moment t_c by the equality $L = L_c$ and go from time t_c to the transference of the system S_c:

$$S_c = \int_o^t \left(\int g dt \right) dt$$

Then (3) will be true also in case of smoothly changing acceleration g.

Let us discuss the application of formula (3). First we note that when $L_o \to 0$, $S_c \to \infty$, i.e., infinitely small initial roughness corresponds to infinitely large delay in mixing. This result agrees well with experimental data (Ardashova et al. 1988) in which mixing was not observed unless initial perturbations were given. Formula (3) allows one to process experimental data in the plane of variables (x, y): $x = \ln \frac{L_c}{L_o}$; $\quad y = \sqrt{\frac{S_c}{L_c}}$.

In Fig. 2, experimental results are plotted and the slope of the straight line is determined:

$$y = (1.02/\sqrt{A})x + 1.54 \tag{7}$$

The indicated distinction disappeares when setting $L_o = 0.3L_o^{exp}$.

3. The evaporation front

Taking into account turbulent mixing in the evaporation front is a complex problem because it has a number of peculiarities. One of them is the front transference over the matter. This is known to cause stabilization of short wave perturbations. Takabe and Yamamoto (1991) suggested the following formula for increment γ:

$$\gamma = \sqrt{kg} - \beta k v_o \tag{8}$$

where $\beta = 3.5$, Atwood number $A = 1$, v_o is the transference velocity of the eveporation front, and k is the wave number related to the mixing-zone width by the relationship $k = 3/L$. It is obvious that the increment in the form (8) leads to the limitation of the rise of short-wave perturbations. This limitation can be stated in the following way

$$\gamma = \begin{cases} \frac{0.81}{4\beta v_o} g & \text{if} \quad L < L_{cr} \\ 0.9\sqrt{\frac{3}{L} g} - \frac{3}{L}\beta v_O & \text{if} \quad L > L_{cr} \end{cases} \tag{9}$$

In (9) L_{cr} is the critical value of the mixing-zone width when the increment is at its maximum, $L_{cr} = (14.8\beta^2 v_o^2)/g$. Furthermore, much as it was in the previous section, Eqs. (4), (5), and (2a) are integrated and as a result the solution for the mixing- zone width L is obtained. To make the formulas simpler, Eq. (2) in its simplest form was used here without the first term. Then we have the simple expression for the velocity $\overline{v}$: $\overline{v} = \frac{\sqrt{\pi}}{4} \frac{L\gamma}{\sqrt{5+0.23A^2}}$; the equation for mixing-zone width is integrated and the solution has the following form

$$\sqrt{L} = \begin{cases} \sqrt{L_o} \exp\left(\sqrt{\frac{0.07g}{4L_{cr}}} t\right) & \text{if} \quad t \leq t_{cr} \\ \sqrt{0.07g}(t - t_{cr}) + \sqrt{L_{cr}} \left(1 - \ln\sqrt{2\sqrt{\frac{L}{L_{cr}}} - 1}\ \right) & \text{if} \quad t \geq t_{cr} \end{cases} \tag{10}$$

$$t_{cr} = \sqrt{\frac{L_{cr}}{0.07g}} \ln \frac{L_{cr}}{L_o}$$

Classically, in the case without evaporation, $V_o = 0$, and the solution has the following form

$$\sqrt{L} = \sqrt{L_o} + \sqrt{0.07g}\ \ t \tag{11}$$

Comparing the formulas (10) and (11), let us draw attention to the various dependencies on the initial roughness L_o. In the classic case, (11) with zero roughness, the nontrivial solution takes place; but in case (10), when the evaporation front moves over the matter, only the trivial solution corresponds to zero roughness (Neuvazhayev 1994).

4. Thin initially turbulized layer

In many experiments on the boundary, the membrane is present, and the function, especially for gases, is difficult to construct. There are many reasons for the initial turbulization of a thin layer, for example, the removal of a barrier between the gases. The complex targets in the ICF problem have usually thin interlayers.

The problems concerning the atmosphere and the ocean which deal with the determination of the laws of initially turbulized admixture propagation lead to setting up the thin-layer problem.

In accordance with the admixture-density environment-density relationship, several special situations can take place. If the admixture density has an intermediate value, then in accordance with the position of light and heavy media, the intensive mixing by the square law (heavy matter above) or slow mixing with the degree $\frac{2}{7} - \frac{2}{6}$ (light matter above) takes place over the course of time. The admixture spreads out, up, and down according to nearly the same law. If the admixture is heavier than the environment, then in the case when the heavy medium is above, mixing with constant rate directed only to the heavy matter takes place. If the admixture is lighter than the environment, the same situation takes place, and the mixing is directed to the light matter which is below and occurs with the constant rate, i.e., according to the linear law in time.

All these cases are studied on the basis of semi-empirical models. The special case of a thin layer immersed into a heavy one within the framework of the K model was considered by Belenikii and Fradkin (1965). The generalized consideration of this problem can be found in (Neuvazhayev 1983, 1988, 1993a), where the thin layers at the bottom and lid of an accelerating vessel, and the self-simulating problem on spreading of a turbulent mixture layer, were studied.

5. Conclusion

The complexity of the four problems discussed shows that rather simple K and KE models can deal with a fairly wide range of applications quite well.

References

Ardashova RI, Balabin SI, Voloshin NP, Kucherenko YA, Ptizina NV, Chitaikin VI (1988) The experimental investigation of perturbation propagation in gravitationally unstable system with continuous density spreading. Vopr At Nauki Tech Ser Teor Prikl Fiz 1:20–27.

Belenkii SZ, Fradkin ES (1965) Theory of turbulent mixing. In: Proceedings of FIAN, 29:207–237.

Kucherenko YA, Neuvazhayev VE, Pylaev AP (1994) Behavior of gravitational turbulent mixing region under conditions leading to separation. DRAN, 334, 4:445–448.

LeLevier R, Lasher G, Bjorklund (1955) Effect of Density Gradient on Taylor Instability, Report UCRL-4459, Univ. of California Radiation Laboratory.

Neuvazhayev VE (1983) The model properties of the turbulent mixing boundary of accelerating fluids of various densities. Appl Mech & Tech Physics 5:81–88.

Neuvazhayev VE (1988) Blurring of the turbulent mixing layer. Appl Mech & Tech Physics 2:46–51.

Neuvazhyaev VE (1993a) Asymtotic behaviour of a thin interlayer after passing a shock wave. In: Proc 4th Int Workshop on the Physics of Compressible Turbulent Mixing, Linden PE, Youngs PL, Dalziel SB (eds), Cambridge University Press, England, pp 461–477.

Neuvazhyaev VE (1993b) The properties of the turbulent mixing model based on the binary two-velocity model. The separation addition to diffusion models. VNIITF, Snezhinsk Preprint 48.

Neuvazhyaev VE (1994) Dependence of turbulent mixing on the initial roughness in the evaporation front with continuous density profile at the interface. Phys Rev E 50(2):1394–1398.

Takabe H, Yamamoto A (1991) Reduction of turbulent mixing at the ablation front of fusion targets. Phys Rev A44:5142–5152.

Zababakhin EI, Ptizin AR (1988) Convective instability limitation. Vopr At Nauki Tech Ser Teor Prikl Fiz 1:4–6.

Numerical simulation of Richtmyer-Meshkov instability

I. G. Lebo, V. B. Rozanov
Lebedev Physics Institute (FIAN)
Leninsky Pr. 53, Moscow 117924 Russia
V. F. Tishkin, V. V. Nikishin, I. V. Popov
Institute of Mathematical Modelling (IMM), Miusskaya Sq. 4, Moscow 125047
A. P. Favorsky
Keldish Appl. Mathematics Institute (IAM), Miusskaya Sq. 4, Moscow 125047

Abstract: The paper presents results of 2D and 3D Eulerian code "NUT" simulations of Richtmyer-Meshkov instability in shock-tube experiments and laser-driven targets.

Key words: Richtmyer-Meshkov instability, Numerical simulation, Shock wave

1. Introduction and short description of the NUT code

Richtmyer-Meshkov instability (RMI) (Richtmyer 1960, Meshkov 1969) develops near the contact boundary when a shock wave comes through the disturbed interface between two gases. Numerical modeling of RMI is presented for 2D and 3D contact-boundary perturbations. To describe the physical processes in the shock tube, a model of nonviscous, non-heat-conducting ideal gas is used which is presented by a system of Euler equations.

For the approximation of the system of gas-dynamic equations we propose a nonlinear conservative quasi-monotonic difference explicit scheme of a high order of accuracy (analogous schemes for 2D codes have been considered in Vjaznikov et al. 1989). The boundary conditions in this system are given according to the law: $x = 0$, $x = L$, then $u = -u$ (reflection condition). The shock wave moves along the 0Z axis, $z = Z_{boundary}$; the parameters of the shock wave and shocked gas are calculated over the given Mach number and unshocked gas density and pressure. We also solve the equation of thermal conductivity. This effect is very important for the modeling of laser plasma phenomena (see section 3). We have solved it numerically using an implicit scheme.

2. Modeling and comparison with shock-tube experiment

Using the NUT code we performed a series of test simulations modeling the shock-tube experiments made at G. M. Krzhizhanovsky Power Institute, Moscow (Aleshin et al. 1990). First we compared our results with the results of the

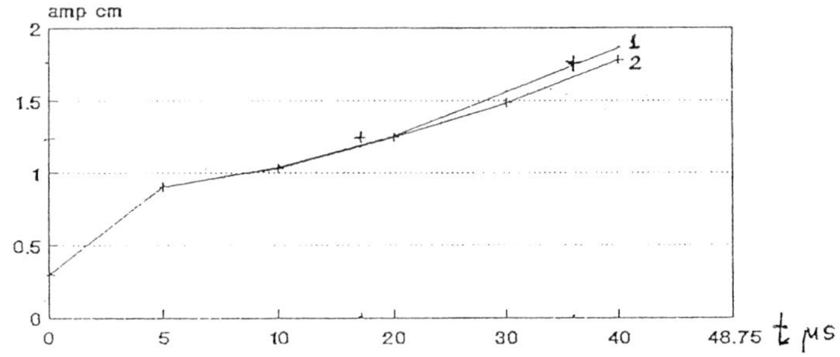

Figure 1. Comparison of perturbation growth velocity: (1) NUT code simulation; (2) results obtained at Moscow Physics and Technology Institute.

group of scientists from the Moscow Institute of Physics and Technology (Belotserkovskii et al. 1994). Figure 1 illustrates the growth of complete amplitudes of perturbations of the contact boundary (a so-called "mushroom structure") between helium and xenon gases (a "complete amplitude" means the distance from the "bottom" to the "top of the spike" of the contact boundary). The initial pressure was 0.5 bar, the Mach number of the initial shock wave was 2.5, the initial disturbance wavelength $\lambda = 8$ mm (these parameters suit the case of a late nonlinear stage, $ka_0 \geq 3$, where k is the wavenumber; $k = 2\pi/\lambda$; a_0 is the initial amplitude of a sinusoidal perturbation; and a_0 is equal to half of the complete initial amplitude). However, the perturbation growth in our case is somewhat higher. We assume that the reason for such a discrepancy lies in the fact that our code is typical of lower approximative viscosity. In the second test, the shock wave goes through the contact boundary between argon and xenon gases. The initial pressure is 0.5 bar, the Mach number, 3.5. Figure 2 shows the calculated growth of the complete amplitude for the perturbation wavelengths $\lambda = 36$ mm, $a_0 = 0.5$ mm. The + symbols denote the perturbation amplitudes obtained in the shock-tube experiments at G. M. Krzhizhanovsky Power Institute (Moscow). The calculations showed good agreement between the complete amplitude growth velocity and the experimental data.

A series of 3D simulations has been carried out. The initial perturbation of the contact boundary (argon-xenon) had the form: $Z_c^{pert} = Z_c^0 + a_0 * \cos(k_x x)\cos(k_y y)$ where $a_0 = 0.5$ mm, $k_x = 2\pi/\lambda_x$, $k_y = 2\pi/\lambda_y$, $\lambda_x = \lambda_y = 36$ mm. An incident shock wave was propagating along the axis OZ, and its velocity corresponded to the Mach number 3.5. As follows from Richtmyer (1960), at a linear stage the perturbations in 2D and 3D geometry have the same velocity provided the condition is fulfilled: $k_{2D} = 2\pi/\lambda_{2D} = k_{3D} = \left((2\pi/\lambda_x)^2 + (2\pi/\lambda_y)^2\right)^{1/2}$ where $\lambda_{3D} = \sqrt{2}\lambda_{2D}$ if $\lambda_x = \lambda_y = \lambda_{2D}$. From this it follows that in 3D geometry the perturbations with $\lambda_{3D} = 36$ mm should be compared in 2D geometry to the perturbations with $\lambda_{2D} \approx 24$ mm. Figure 3

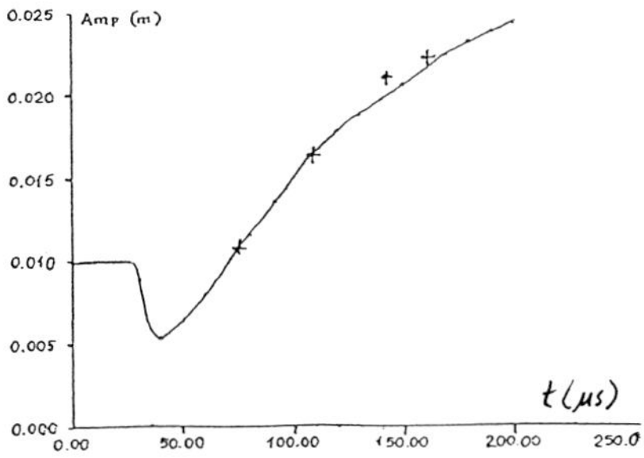

Figure 2. Contact boundary perturbation amplitude growth versus time; + denote perturbation amplitudes obtained in shock-tube experiments.

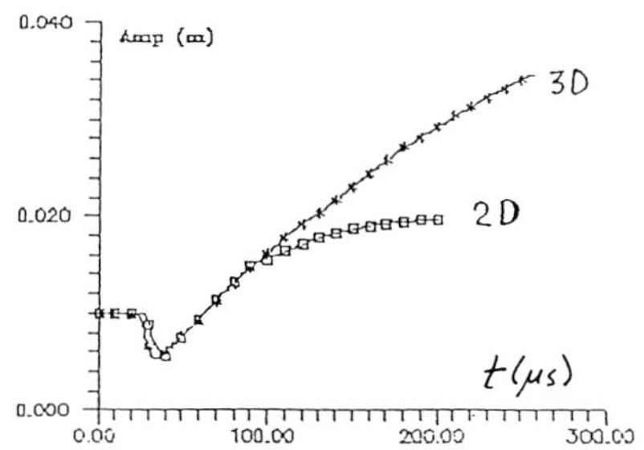

Figure 3. Contact boundary perturbation amplitude growth versus time for 3D ($\lambda = 36$ mm) and 2D ($\lambda = 24$ mm) perturbations, $a_0 = 0.5$ mm.

shows the growth of complete perturbation amplitudes for 3D and 2D geometry. It is seen that at the nonlinear stage ($k^*a > 5.0$, where k is the wavenumber and a the current complete amplitude) the perturbations in 2D geometry grow more slowly. This may be explained by the fact that due to the growth of Kelvin-Helmholtz instability near the top of the spike the processes of vortex-structure formation and "blurring" of such a spike are faster in 2D geometry (see Rozanov et al. 1992).

In order to investigate the influence of the boundary conditions on the development and formation of the vortex structures we have performed a calculation of the Ar-Xe contact boundary perturbation development where: (a) the initial perturbation is a sinusoid with the wavelength $\lambda = 24$ mm, $a_0 = 0.5$ mm; (b) the calculated region of OX axis is twice as high, and the initial perturbation has the following form:

$$Z_c(x,y) = \begin{cases} Z_0 + a_0\cos(k_x x), & \text{if } 0 \le x \le \lambda_x/2; \\ 0, & \text{if } \lambda_x/2 \le x \le \lambda_x \end{cases}$$

that is, "an absolutely elastic wall" (i.e., $u = -u$) is placed at $x = \lambda = 24$ mm; and (c) there is "free flow through a wall" ($du/dx = 0$ at $x = \lambda/2$). The dependence of the perturbation complete amplitude growth is close to that which is observed when the "wall" is located at point $x = \lambda/2$, but there are some differences.

The shape of the perturbation (case b) at $t = 170$ μs is shown in Fig. 4a; Fig. 4b illustrates the amplitude development of the contact boundary corresponding to cases (a), (b), and (c). One can see that a "solitary perturbation" has a somewhat bigger amplitude at $t = 200$ μs. The next series modeled a shock-wave passage through the contact boundary of krypton and xenon gases in 2D geometry. As observed previously, the perturbation of the contact bound-

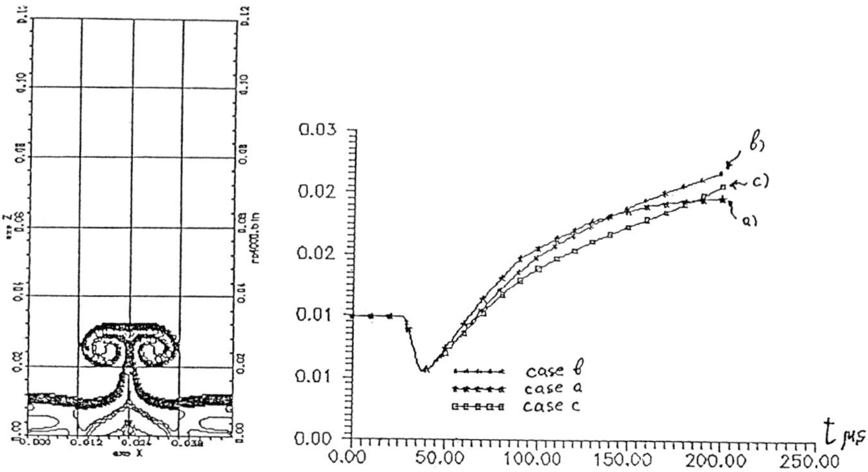

Figure 4. (a) Density isolines near Ar-Xe contact boundary for "solitary perturbation" (case b). (b) Contact boundary perturbation amplitude growth for cases (a), (b), (c).

ary had a sinusoidal shape with wavelength $\lambda = 3.6$ cm and amplitude $a_0 = 0.5$ cm (we compared our results with the results of American scientists from the University of New York at Stony Brook). The initial density in Kr is $\rho = 1.87$ mg/cm^3, and in Xe $\rho = 2.95$ mg/cm^3. In the calculations we set the initial velocity of the unperturbed gas flow $w = -55.5E+4$ m/s, which allowed us to restrict the calculated region. Below we give the calculation results.

Figure 5 illustrates the amplitude perturbation growth of the contact boundary. A satisfactory agreement is observed between our results and the American results (+ symbols). At the late nonlinear stage of the perturbation growth the amplitude obtained in our calculations is somewhat larger than the American results. We assume that the reason for such a discrepancy is similar to that for test 1. However, this question demands further investigation.

3. Richtmayer-Meshkov instability in a laser target

In contrast to the shock-wave experiments, the development of instabilities in laser targets is essentially influenced by electron (and radiative) heat conductivity. A one-temperature approximation that is true in the dense compressed target (where the electron-ion exchange time is small due to high plasma density) is considered. In order to calculate the development of perturbations in the target, we used as the initial data the 2D Lagrange code "Atlant" data (Lebo et al. 1994). Note that most of the Lagrange codes used in the target compression calculations would not allow us to obtain a "mushroom-like" structure at the contact boundary, due to a strong deformation of the calculation cells.

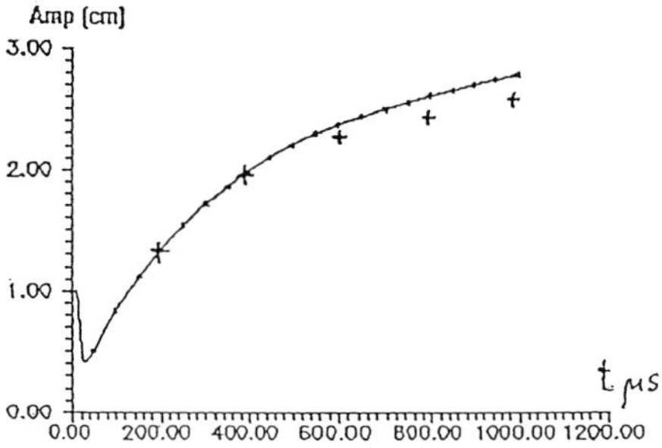

Figure 5. Contact boundary perturbation amplitude growth for Kr-Xe case; + show results of American group from Stony Brook.

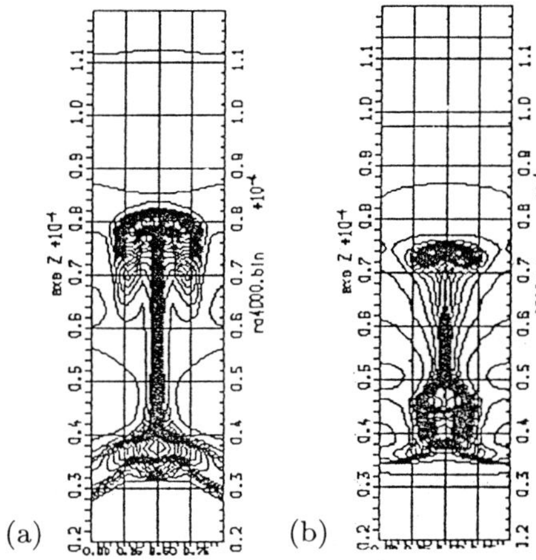

Figure 6. Plasma density isolines near "shell-fuel" and "shell-corona" contact boundaries: (a) without heat conductivity; (b) with heat conductivity.

In order to calculate this problem using the NUT code we have set three subregions: (1) Subregion I with the DT fuel with density $\rho_F = 0.05$ g/cm^3. (2) Subregion II with the nonevaporated part of the shell. The averaged density (ρ_S) and the region thickness (L) have been changed in the calculations so as to obtain $\rho_S * L = 10$ mg/cm^2. (3) Subregion III is the corona. In the calculations the following values of the average corona density were set: $\rho_C = 3$ mg/cm^3 and $\rho_C = 150$ mg/cm^3. The initial temperature in the corona was equal to 1 keV and 0.02 keV. The temperature in subregions I and II was calculated in such a way as to make the pressure at $t = 0$ constant within the whole region. The shock wave moved from the DT fuel to the shell. The gas velocity behind the shock front was $w = 250$ km/s (in one of the variants, 500 km/s; in this case when $\lambda = 5$ μm the "mushroom-like" structure was blurred completely). We have performed 2D calculations of the "shell-fuel" contact boundary perturbation for different wavelengths, $\lambda = 5$, 10, and 20 μm, and initial amplitude $a_0 = 2$ μm. We have performed calculations also with $w = 250$ km/s, but with a higher density of the shell (variant 2: $\rho_S = 0.2$ g/cm^3, variant 3: $\rho_S = 0.8$ g/cm^3, variant 4: $\rho_S = 0.8$ g/cm^3, $\rho_C = 150$ mg/cm^3). In variant 4, the corona density was increased, and the temperature, correspondingly, was decreased. As a result, the heat flow from the corona toward the shell was suppressed. The instability development is strongly affected by the heat conductivity. First, the heat flow from the corona to the shell leads to the formation of another shock wave moving from the corona to DT fuel. As a result, we obtain density and temperature profiles which differ from calculations without heat conductivity (no jump in the density at the contact boundary). Second, the heat conductivity leads to the "blurring" of the vortex structure near the top of a spike. Figure 6a illustrates the isolines of the shell density at $t = 0.237$ ns without heat conductivity; Fig. 6b shows the isolines with allowance for this effect at $t = 0.237$ ns. The vortex structures in the vicinity of the spike top are blurred, and the amplitude of the perturbation is smaller than that in case (a).

4. Conclusions

(1) The NUT code allows one to numerically model the development of such a complicated phenomena as the Richtmyer-Meshkov instability up to its late nonlinear stage, when vortex structures, destruction of juts, and the formation of turbulent mixing zones occur. The comparison of numerical calculations with results from shock-tube experiments has demonstrated good agreement. (2) The 2D perturbations at the nonlinear stage are destroyed faster compared to the 3D perturbations. (3) We started to study the multi-mode perturbation development. (4) The development of instability at the "shell-fuel" contact boundary has some peculiar features in the high-temperature plasma of the laser targets. They are due to the influence of nonlinear heat conductivity. The heat flow at the high temperature of the fuel suppresses the small-scale vortex structures in the vicinity of the spike top, and can reduce the velocity of the perturbation development.

Acknowledgement. The authors are grateful to S. G. Zaytsev and his colleagues for fruitful discussion of the calculation results. This work was supported by the ISTC, Grant No. 29.

References

Aleshin AN, Lazareva EV, Zaytsev SG, Rozanov VB, Gamaly EG, Lebo IG (1990) Dokladi Akademii Nauk SSSR, Fizika 310(5) (in Russian).

Belotserkovskii OM, Demchenko VV, Oparin AM (1994) Doklady Russian Academii of Science 39(2):118 (in Russian).

Lebo IG, Popov IV, Rozanov VB, Tishkin VF (1994) Journal of Russian Laser Research 15(15):136 (in Russian).

Meshkov EE (1969) Izv. Akad Nauk SSSR, Mekhanika Zhidkosti i Gaza (5):151 (in Russian).

Richtmyer RD (1989) Comm Pure and Appl Math 13(2):297.

Rozanov VB, Zmitrenko NV (1992) Preprint FIAN, No. 16 (in Russian).

Vjaznikov KV, Tishkin VF, Favorsky AP (1989) Matematicheskoe modelirovanie 1(5):95 (in Russian).

The effect of refraction regime on the evolution of perturbation for Richtmyer-Meshkov instability

S. G. Zaytsev, A. N. Aleshin, E. V. Lazareva
Krzhizhanovsky Power Engineering Institute (ENIN)
Leninsky Pr. 19, Moscow, 117071 Russia

Abstract: We studied the evolution of an interface after the plane incident shock S_0 passed through it, moving from light to heavy gas. It has been found that the increased a_0k (where a_0 and k are the amplitude and the wave number of the initial two-dimensional interface K_0, respectively) resulted in reduced depth of penetration of one gas into the other at the final turbulent stage of the interface evolution, and this was due to the change in the type of refraction of S_0 on K_0.

Key words: Shock wave, Hydrodynamic instability, Turbulence

1. Introduction

The passage of the shock S_0 from light (ρ_1) to heavy (ρ_2) gas through a two-dimensional sinusoidal interface K_0 is accompanied by formation of a mixing zone K which separates the shocked flows formed by the refracted (S) and reflected (R) shock waves. At the end of refraction the shapes of S, R, and K are similar to that of K_0 (Fig. 1). Because the generated shock waves S and R are curvilinear, the subsequent shocked flow acquires the component of velocity U_y which is perpendicular to the direction of the motion of S_0 (axis X). Collision of the two flows behind the converging sections of S and R (S_c and R_c, respectively) results in formation of the high-pressure regions P_S^+ and P_R^+ behind these sections, and at the same time, the low-pressure regions P_S^- and P_R^- are formed behind the diverging sections of S and R (S_d and R_d). The properties of the resultant distribution of pressure determine the evolution of the interface K and the shock waves S and R (Richtmyer-Meshkov instability, RMI). These properties are dependent on: (i) the angle α between the normal to S_0 and the tangent to K_0; (ii) the density ratio for the contacting gases; and (iii) the intensity of the incident shock.

After the end of refraction the amplitude a_K of the interface K is increased while the amplitudes a_S and a_R of the shock waves S and R, respectively, are decreased. Visualization and numerical calculations show that when the curvature of K_0 (for unchanged ρ_1, ρ_2, and Mach number M_0 of the incident shock)

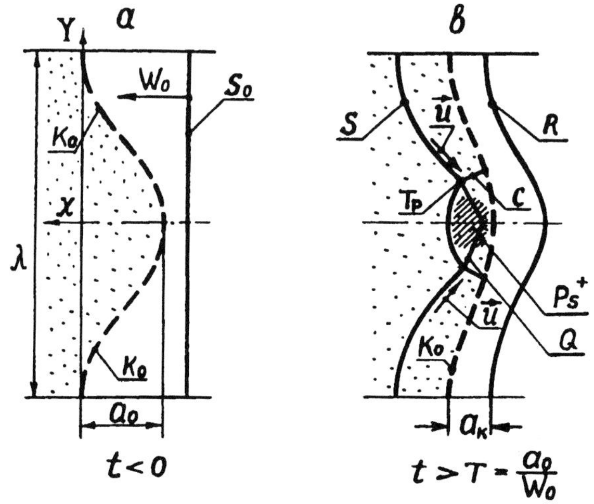

Figure 1. The scheme of excitation of Richtmyer-Meshkov hydrodynamic instability for the hard regime of refraction: (a) dispositions of S_0 and K_0 before the interaction, and (b) dispositions of the refracted (S) and reflected (R) shock waves and the interface K after the end of refraction.

overcomes its critical value, transversal shock waves C appear in the shock-compressed flow behind S, and the interaction of C with S results in a triple point T_p (Fig. 1).

For $\alpha = \pi/2$, the parameters of S, R, and K are determined by one-dimensional equations and denoted by *. For $\alpha_1 < \alpha < \pi/2$ the regime of refraction is regular, i.e., S_0, S, and R meet in one point (Henderson & Macpherson 1968). For $\alpha = Const$, parameters of the flows in the vicinity of the point of intersection of S_0, S, and R remain unchanged during the motion of S_0 through K_0. For $\alpha_1 < \alpha$, the point of intersection of S_0 with R moves aside from K_0, forming a Mach configuration. This is an irregular regime of refraction. These regimes have been studied for $\alpha = Const$ (Henderson & Macpherson 1968).

2. Results of experimental and numerical investigations

Experiments were performed in a shock tube; the internal section of the driven was 72×72 mm^2. Different couples of the inert gases He, Ar, Kr, and Xe were used. The a_0k varied from 0.87 to 7.96, and the initial pressure p_0 was 0.5 atm. The Mach number of the incident shock S_0 varied from 2.5 to 4.0. Numerical calculations were performed using the code of Dr. V. V. Demchenko (Moscow Physico-Technical Institute).

The amplitudes $a_K(S)$ of the interface K versus the distance S passed by K are given in Fig. 2. Experimental points 1–4 and 5 are related to regular and irregular refraction, respectively. The experimental and numerical investigations of RMI on a 2D sinusoidal interface K_0 allowed us to find the correlation between the observed changes in the interface evolution and different types of refraction of S_0 on K_0. Thus we can distinguish three regimes of the evolution: soft regular, hard regular, and irregular.

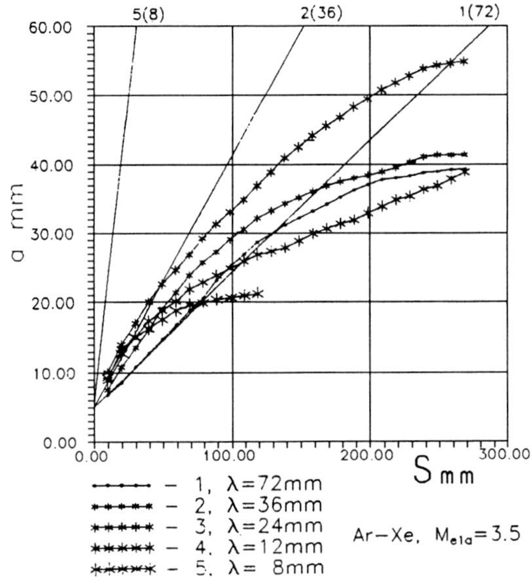

Figure 2. Amplitude $a_K(S)$ of the interface K versus the distance S passed by K. Lines 1(72), 2(36), and 5(8) correspond to the values of a_K calculated from the Richtmyer relation for K_0 with λ of 72, 36, and 8 mm, respectively.

Soft regular regime. During formation of the high-pressure regions and their subsequent "spreading," the flow between the shock front and interface remains continuous. For this regime the process of K evolution after the refraction of S_0 at K_0 may be divided into three typical stages preceding the formation of a turbulent mixing zone: linear, nonlinear, and transitional. For this regime the maximum of pressure behind the converging section S_C does not exceed 1.2P* (See Fig. 1). For the soft regular regime the K evolution starts from the linear stage which is sufficiently described by the Richtmyer relation:

$$\frac{da_K(t)}{dt} = a_K^* ukA$$

where a_K^* is the amplitude at the end of refraction, u is the velocity given to K by the shock, k is the wave number, and $A = (\rho_2 - \rho_1) \cdot (\rho_1 + \rho_2)^{-1}$.

Hard regular regime. The maximum of pressure behind the converging section S_C does not exceed $2P^*$ (See Fig. 1). For this regime, in the flow between S and K the transversal shock waves C are generated. Interaction of C with S results in a triple point T_p on the S front. During the refraction the interface transforms into the spikes of heavy gas penetrating into the light gas and bubbles of light gas penetrating into the heavy gas. This shape of K is typical for a nonlinear stage. Thus, for the hard regular regime the interface evolution after the end of refraction starts with a nonlinear stage.

Irregular regime. For the experiments corresponding to the points 5 in Fig. 2 the numerical calculations indicate the irregular type of refraction. For this regime, the vortex structures on the heavy gas spikes typical for the transitional stage are formed during the refraction.

The calculated data for the shock passage from He to Xe ($M_0 = 2.5$, $a_0 = 10$ mm, $\lambda = 8$ mm) are given in Fig. 3. At the end of refraction ($t = 4$ μs), a vortex structure (Q zone) appeared on the spike of the heavy gas (Xe) penetrating into the light gas (He). By $t = 9.8$ μs, this vortex structure developed into the typical mushroom-like structure. For this regime of refraction, the maximum pressure in P_S^+ is $3.3P^*$. The interface evolution after the end of refraction starts with a nonlinear stage.

3. Boundaries for different regimes of refraction

The position of the boundary to distinguish the regular and irregular types of refraction for a sinusoidal interface is approximately determined as follows. The process of refraction is presumably dependent on the character of refraction on the passage of S_0 through the central part of K_0, where the change in the angle α between the direction of S_0 motion and the tangent to K_0 will be relatively small. Here α is close to $\alpha_{\min}$ (the vicinity of K_0 inflection point).

The numerically calculated conservation equations for the refraction of a plane shock front on a plane interface arranged at the angle α allow one to determine the wave image of refraction. Thus we determine the boundary between regular and irregular regimes in the (α, M_0)-plane for a fixed A. In Fig. 4 the boundary position is denoted by line 2 in the (a_0k,A)-plane for a fixed M_0. The angle α is counted as $\alpha_{\min}$ for a sinusoid with the corresponding a_0k. The points in Fig. 4 correspond to the experiments in which Topler pictures demonstrate a specific network of lines resulting from the interaction of the transversal shock wave C with the reflected shock wave. Experiments performed for the irregular regime recorded the confluence of adjacent mushroom-like structures. Numerical calculations performed for the regimes disposed above line 2 (Fig. 4) demonstrated formation of mushroom-like structures in a heavy gas spike penetrating into light gas. Thus, irregular regimes are disposed above line 2 (Fig. 4).

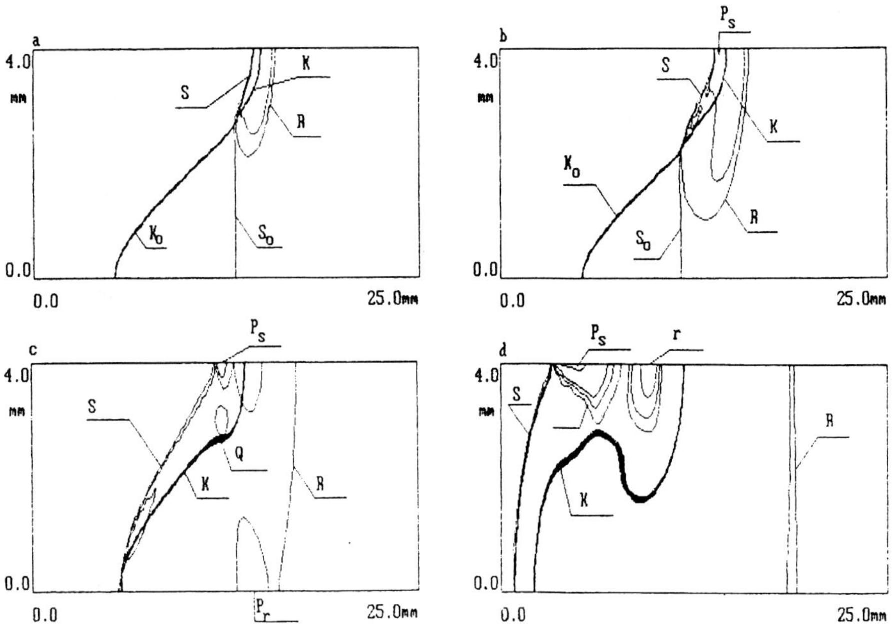

Figure 3. Results of calculation for the passage of the incident shock S_0 from He to Xe through sinusoidal interface K_0 ($\lambda = 8$ mm, $p_0 = 0.5$ atm, $M_0 = 2.5$): (a) $t = 1.0$ μs, (b) $t = 1.6$ μs, (c) $t = 4.0$ μs, and (d) $t = 9.8$ μs.

Line 1 is obtained from numerical calculation using the following empirical condition: $U_y \geq 0.2c$, where c is the local speed of sound (Aleshin et al. 1993). The area between lines 1 and 2 corresponds to the hard regular regime. In the shock-compressed gas behind S, Topler pictures show a specific network of lines of tangential breaks frozen in the flow. The breaks in the S front (triple points) are due to the interaction of transversal shock waves with the S front. The C fronts move along the shock-compressed flow.

For the regimes disposed below line 1, Topler pictures show neither breaks on S front nor a specific network of lines frozen in the flow described above. This area corresponds to soft regimes.

4. Conclusions

The experimental and numerical investigations of the Richtmyer-Meshkov instability for a single-mode two-dimensional initial interface K_0 have shown that when time intervals are large enough for the turbulent mixing zone to be formed, the depth of penetration of one gas into the other decreases with growth of a_0k

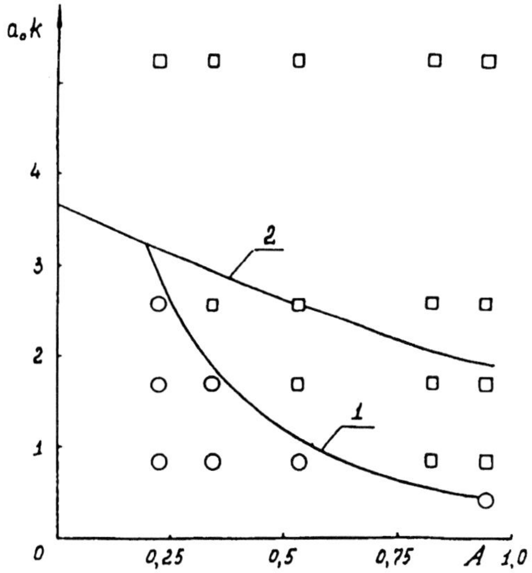

Figure 4. Boundaries of different regimes of refraction in the (a_0k,A)-plane. The irregular regimes of refractions are disposed above line 2. The soft regular regimes are disposed below line 1. The area between lines 1 and 2 corresponds to hard regular regimes.

(a_0 and k are the amplitude and wave number of K_0, respectively). This is caused by the reduced time of transition to the turbulent stage because of the altered evolution of interface which is due to the change in type of refraction. At the turbulent stage the rate of penetration of one gas into the other slightly depends on initial conditions.

Acknowledgement. The study was partly supported by the Russian Foundation for Fundamental Research (Grant No. 94-01-01144a) and by the ISTC (Grant No. 94-029).

References

Aleshin AN, Zaytsev SG, Lazareva EV (1993) Experimental and numerical studies on Richtmyer-Meshkov instability. RJCM 1:33–49.

Henderson LF, Macpherson AK (1968) On the irregular refraction of a plane shock wave at a Mach number interface. J Fluid Mech 32(I):185–202.

A numerical study of shock acceleration of a diffuse helium cylinder

J. A. Greenough, J. B. Bell, P. Colella
Center for Computational Sciences and Engineering
Lawrence Livermore National Laboratory*
Livermore, CA 94550 USA

E. G. Puckett
Department of Mathematics and Institute of Theoretical Dynamics
University of California
Davis, CA 95616 USA

J. W. Jacobs
Department of Aerospace and Mechanical Engineering
University of Arizona
Tucson, AZ 85721 USA

Abstract: The development of a shock-accelerated diffuse helium cylindrical inhomogeneity is investigated using a new numerical method. The new algorithm is a higher-order Godunov implementation of the so-called multi-fluid equations. This system correctly models multiple-component mixtures by accounting for differential compressibility effects. This base integrator is embedded in an implementation of adaptive mesh refinement (AMR) that allows efficient increase in resolution where the computational effort is concentrated when high accuracy, or increased resolution, are required. Qualitative and quantitative comparisons with previous experimental data are excellent. The simulations show that counter-sign vortex blobs are deposited in the jet core by baroclinic generation of the curved shock wave as it traverses the jet. This vorticity deposition occurs over timescales that scale with the shock passage time ($\sim 10\ \mu$s). Three phases of development are identified and characterized. The first is the weak-deformation (WD) phase, where there is weak distortion of the helium jet due to weak vorticity-induced velocity effects. The second phase is the strong-deformation (SD) phase where there is large distortion for the jet and the vortex blobs due to large induced-velocity effects. The last is a relaxation/reorganization (RR) phase where the vorticity field reorganizes into a point-like vortex pair.

*Work performed under the auspices of the US DOE by the Lawrence Livermore National Laboratory under contract W-7405-Eng-48. Additional support under contract W-7405-Eng-48 by the Applied Mathematical Sciences Program and the HPCC Grand Challenge Program of the Office of Scientific Computing at DOE, Defense Nuclear Agency under IACRO 94-831, and the Center for Advanced Fluid Dynamics Applications under contract B295120.

Key words: Shock-accelerated diffuse interfaces, Multi-fluid methods, Adaptive methods

1. Introduction

The evolution of interfaces accelerated by shock waves has been the focus of investigations for many years. This class of problem has applications in such disparate fields as inertial confinement fusion (ICF) and high-speed combustion. The literature is rich in studies on the stability of plane interfaces (Richtmyer 1960, Meshkov 1969, Zaitsev et al. 1985, and Brouillette & Sturtevant 1988). The stability of curved interfaces has received similar attention culminating with the study of Haas & Sturtevant (1987), where cylindrical or spherical shapes, constructed from soap bubbles or nitrocellulose membranes, were filled with either a light or heavy gas and then accelerated by a shock wave. This last study and the previous works relied on such membranes to separate gases. A different class of experiment on acceleration of gaseous cylinders was developed by Jacobs (1992, 1993) that was free from the effect of membranes and used a planar laser-induced fluorescence (PLIF) technique that allowed for excellent visualization of the flow and quantitative information regarding mixing.

The present simulations use a high-resolution adaptive method designed for multi-fluid flows with initial mixed layers. In this paper, the development of the helium mole fraction and vorticity field will be examined. Extensive comparisons with experimental data have also been completed as well as development of a new conceptual model describing this flow, but these cannot be described here due to space limitations.

2. Multi-fluid system and numerical method

The equations used for this study are the so-called multi-fluid equations as developed by Collela et al. (1995) and also described in Puckett & Saltzman (1992). This system is based on a volume-of-fluid approach, where the fraction of a cell volume occupied by a distinguished fluid is followed during the flow evolution. The basic assumptions in this approach are that there is pressure equilibrium within a cell, a single velocity vector for both fluids, and that all changes of state are adiabatic. The equations are given as

$$\frac{\partial f^\alpha}{\partial t} + \nabla \cdot (\boldsymbol{u} f^\alpha) = f^\alpha \frac{\hat{\Gamma}}{\Gamma^\alpha} \nabla \cdot \boldsymbol{u} \tag{1}$$

$$\frac{\partial}{\partial t}(f^\alpha \rho^\alpha) + \nabla \cdot (\boldsymbol{u} f^\alpha \rho^\alpha) = 0 \tag{2}$$

$$\frac{\partial \rho \boldsymbol{u}}{\partial t} + \nabla \cdot (\boldsymbol{u}\boldsymbol{u}\rho) + \nabla p = 0 \tag{3}$$

$$\frac{\partial}{\partial t}\left(f^\alpha \rho^\alpha E^\alpha\right) + \nabla \cdot \left(\boldsymbol{u} f^\alpha \rho^\alpha E^\alpha\right) + p f^\alpha \frac{\hat{\Gamma}}{\Gamma^\alpha} \nabla \cdot \boldsymbol{u} + f^\alpha \frac{\rho^\alpha}{\rho} \boldsymbol{u} \cdot \nabla p = 0 \tag{4}$$

where f^α, ρ^α, and E^α are the volume fraction, density, and total energy density of fluid component α. The volume fraction is defined as $f^\alpha = \Lambda_\alpha/\Lambda$ where Λ is the volume of the cell and Λ_α is the volume of the cell occupied by fluid α. Γ^α is the sound speed gamma for fluid α, and $\hat{\Gamma} = 1/\sum_\alpha \left(f^\alpha/\Gamma^\alpha\right)$ represents fraction weighted Γ for the mixture.

The pressure that appears in the above system is defined to be a thermodynamically consistent pressure and defined as $p = \sum_i \hat{\Gamma}\left(f^i p^i/\Gamma^i\right)$, where p^α is the partial pressure of component α. Note that the formulation is sufficiently general to allow real-gas equation of state (EOS) systems described by pressure given as a function of density and internal energy.

The solution procedure for the system is a higher-order Godunov method following that of Colella (1984) and detailed in Greenough et al. (1995). This base integrator for the multi-fluid system is embedded within an implementation of adaptive mesh refinement (AMR). This methodology is based on the original work of Berger & Oliger (1984) and later Berger & Colella (1989). This is a means for managing a refined grid hierarchy composed of logically rectangular grid patches that allows for efficient increases in resolution by focusing the computational accuracy where errors are deemed large, or in regions where high resolution is required.

3. Problem setup

The initial conditions for this problem are taken from Jacobs (1993). The shock-wave Mach number is 1.094 in air. In Budzinski (1992), a Rayleigh scattering technique is used to measure the actual molar concentrations of helium. It shows the initial profile is Gaussian in shape with a core concentration of approximately 80% helium. The initial half radius is set as $r_{\text{half}} = 0.2$, where it is defined as the distance where the helium mole fraction is half the center value. The density ratio between air and helium is $\rho_{\text{helium}}/\rho_{\text{air}} = 0.138$.

The computational domain is 9 cm wide by 27 cm long with a base grid of 36 cells wide by 108 cells long. There are two levels of refinement with the first level a factor of 4 more refined than the base grid and the second level a factor of 8 more refined than the first level of refinement. This gives an effective grid resolution of $\Delta x_{\text{fine}} = \Delta y_{\text{fine}} = 0.0078125\ cm$. The shock is initialized in air at a distance of $3r_{\text{half}}$ from the jet. All boundaries are inflow/outflow type.

4. Results and discussion

4.1. Flow visualization

Figures 1 and 2 show the evolution of the mole fraction of helium and the development of the vorticity field, respectively. Figure 1a shows the initial conditions and the initial mixed layer. At 88 μs, Fig. 1b, there is a flattening of the jet and a generally weak deformation of its original circular structure. From here out to about 400 μs, spanned in time by Figs. 1c through f, there is strong deformation of the jet as it is inverted and transformed into the vortex pair. After 400 μs, from Fig. 1g on, the flow evolves as a well-defined vortex pair. Comparing these results with those given in Jacobs (1993), the agreement is exceptional.

Based on these observations and for ease and clarity of later exposition, we define three phases of development. The first is the weak-deformation phase (WD) that occurs over short times out to about 90 μs. The second is the strong-deformation phase (SD) that occurs after WD out to about 400 μs. The last is the relaxation/reorganization (RR) phase that occurs after SD.

In the vorticity time sequence, Fig. 2a shows the shock, still essentially planar, leaving positive (in the upper half plane) vorticity behind it. By 66 μs, Fig. 2b, the shock is well out of the jet core and two counter-sign circular vortex blobs are seen located in the jet core. During the WD phase, the vortex blobs remain essentially undistorted. This is due to the fact that vorticity-induced velocity effects are very weak during the WD phase. Also, the blobs are deposited in the jet core over a timescale that scales with the shock passage time ($\sim$ 10 μs). During the SD phase however, there are large induced-velocity effects that distort and stretch the vortex blobs into something like continuous finite-thickness sheets. The underlying vortical field swirls these sheets so that they are reorganized, at later times, into a vortex pair. Note that during the SD phase, there is strong counter-sign vorticity appearing also oriented in sheets, ahead of the distorted blob. Although not shown here, this counter-sign production acts to modulate the overall circulation as it relaxes to a near constant value.

The development of this flow is seen to be determined by the interaction and evolution of the vortex blobs. Once they are present in the flow, they at first interact weakly (during WD), then interact strongly (during SD) as they change from blobs to more point-like vortices (during RR). During the course of the transformation in the vorticity field, the helium jet undergoes a similar transformation as shown. From a modeling point of view, this flow could be accurately depicted by the development of a vortex blob pair in an essentially circular helium jet (slightly flattened on the upstream side in reality). The

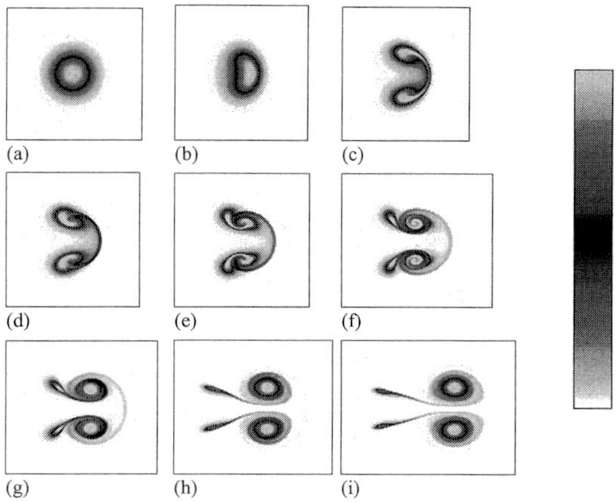

Figure 1. A sequence showing the computed helium mole-fraction evolution. Note that the shock arrival time at the jet is approximately 40 μs and all times reported are relative to the start of the calculation: (a) the initial jet, (b) 88 μs, (c) 202 μs, (d) 259 μs, (e) 318 μs, (f) 436 μs, (g) 561 μs, (h) 943 μs, (i) 1182 μs. The shock moves from left to right. The color bar [sic] to the right varies highest to lowest values from top to bottom.

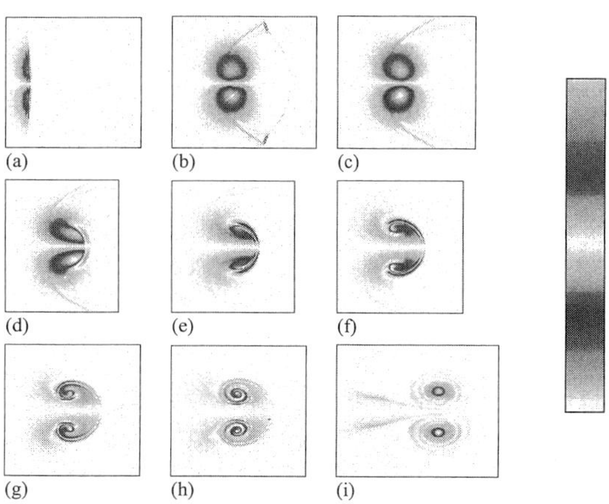

Figure 2. A sequence showing the vorticity evolution. Note that the shock arrival time at the jet is approximately 40 μs and all times reported are relative to the start of the calculation: (a) 44 μs, (b) 66 μs, (c) 88 μs, (d) 141 μs, (e) 202 μs, (f) 259 μs, (g) 318 μs, (h) 436 μs, (i) 1182 μs. The shock moves from the left to right. The color bar [sic] to the right is centered with respect to the data. Values vary highest positive to highest negative from top to bottom.

subsequent development could be predicted by their interaction and reorganization/relaxation to point vortices.

References

Berger MJ, Colella P (1989) Local adaptive mesh refinement for shock hydrodynamics. J Comput Phys 82:64–84.

Berger MJ, Oliger J (1984) Adaptive mesh refinement for hyperbolic partial differential equations. J Comput Phys 53:484–512.

Brouillette M, Sturtevant B (1988) Shock induced Rayleigh-Taylor instability at a continuous interface. Abstract submitted to the Int Workshop on the Physics of Compressible Turbulent Mixing, 24–27 October 1988, Princeton, New Jersey.

Budzinski JM (1992) Planar Rayleigh Scattering Measurements of Shock Enhanced Mixing, Doctoral dissertation, California Institute of Technology.

Colella P (1984) A direct Eulerian muscl scheme for gas dynamics. SIAM J Sci Stat Comput 6:104–117.

Colella P, Ferguson RE, Glaz HM (1995) Multifluid algorithms for Eulerian finite difference methods. In manuscript.

Greenough JA, Bell JB, Colella P (1995) An Adaptive Multifluid Interface-Capturing Method for Compressible Flows in Complex Geometry, AIAA Paper 95-1718.

Haas J-F, Sturtevant B (1987) Interaction of weak shock waves with cylindrical and spherical inhomogeneities. J Fluid Mech 181:42–76.

Jacobs JW (1992) Shock-induced mixing of a light-gas cylinder. J Fluid Mech 234:629–649.

Jacobs JW (1993) The dynamics of shock accelerated light and heavy gas cylinders. Phys Fluids A 5:2239–2247.

Meshkov EE (1969) Instability of the interface of two gases accelerated by a shock wave. Izv Akad Nauk SSSR Mekh Ahidk Gaza 4:1151–157; Russian: Izv Acad Sci USSR Fluid Dyn 4:101–104.

Puckett EG, Saltzman JS (1992) A 3-D adaptive mesh refinement algorithm for multimaterial gas dynamics. Physica D 60:84–104.

Richtmyer RD (1960) Taylor instability in shock acceleration of compressible fluids. Comm Pure Appl Math 23:297–319.

Zaitsev SG, Lazareva EV, Chernukha VV, Belyaev VM (1985) Experimental investigation of the hydrodynamic instability of the interface between media of different density in an acceleration field. Translated from Teplofzika Vysokikh Temperatur 23:535–541.

Hypervelocity Richtmyer-Meshkov instability

R. Samtaney*, D. I. Meiron
Firestone Flight Sciences Laboratory
D. I. Pullin
Graduate Aeronautical Laboratories
California Institute of Technology
Pasadena, CA, 91125 USA

Abstract: The Richtmyer-Meshkov instability is investigated for shocks sufficiently strong enough to produce active dissociation chemistry. To model the interaction of the flow with chemistry typical of hypervelocity flows, Lighthill's ideal dissociating gas model is employed. While dissociation causes significant changes in density and temperature, the change in growth of the perturbations is small when compared to frozen flow.

Key words: Hypervelocity, Ideal dissociating gas, Growth rate

1. Introduction

Hitherto, most investigations of the Richtmyer-Meshkov (RM) (Richtmyer 1960, Meshkov 1969) instability have been confined to shocks of Mach number $M \leq 4$. In this paper, we present results of *hypervelocity* RM instability. The term "hypervelocity" is used in contrast with the term "hypersonic." Hypervelocity implies not only high Mach number ($M \gg 1$) but also large velocities and therefore high total enthalpy. Thus, in hypervelocity RM instability, the shocks are of strength sufficient to activate dissociation/recombination chemistry and related gas-chemistry interactions. This is the topic of the present paper.

2. Physical parameters, equations and chemistry model, and numerical method

2.1. Physical parameters

The physical domain is a two-dimensional rectangular shock tube of dimensions $[x_l, x_r] \times [0, \lambda/2] \equiv [-0.1, 4.9] \times [0, 0.5]$. A vertical interface separating two gases is perturbed sinusoidally with wavelength $\lambda \equiv 1$, and initial amplitude $A_0 = 0.1$ (see Fig. 1). We use a hydrogen-nitrogen interface, initially in thermal ($T_0 = T_b = 298$ K) and mechanical ($p_0 = p_b = 0.1$ atm) equilibrium.

*Also affiliated with the Graduate Aeronautical Laboratories.

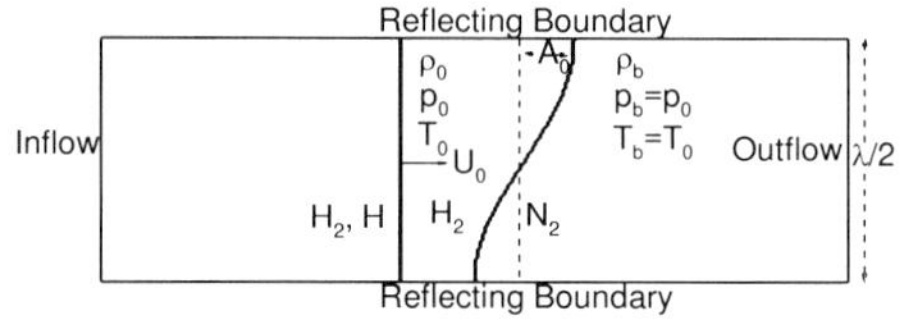

Figure 1. Schematic of the physical domain and parameters.

Although for chemically active flows the Mach number is an ill-defined quantity, for convenience we define a Mach number by $M \equiv U_0/c_0$, where U_0, c_0 are the shock speed and the frozen sound speed in the unshocked hydrogen. In fact, for $T_0 = T_b = 298$ K the frozen and equilibrium sound speeds are identical.

2.2. Equations and chemistry model

Consistent with previous studies of the RM instability, the governing equations of motion are the compressible Euler equations in two dimensions. In this paper we examine two limits: (a) *frozen* (Damkohler number $\Omega = 0$); and (b) *equilibrium* ($\Omega \to \infty$). For $\Omega \to \infty$, we assume that the equilibrium chemical reactions are are adequately modeled by the Lighthill Ideal Dissociating Gas (IDG) model (Lighthill 1957). The law of mass action for the IDG model is

$$\frac{\alpha_k^2}{1-\alpha_k} = \frac{\rho_{d,k}}{\rho_k} \exp\left(\frac{-\theta_{d,k}}{T}\right) \tag{1}$$

In Eq. (1), ρ_k is the density of the kth gas ($k = 1, 2$); α_k is the mass fraction of the monoatomic constituent of the kth gas; $\theta_{d,k}$ is the dissociation temperature of the kth gas, and $\rho_{d,k}$ is the characteristic density. The IDG properties for hydrogen and nitrogen are: $\rho_d = 1800, 130000$ kg/m^3 and $\theta_d = 51900, 113200$ K. Note that in the IDG model the gas is always in a state of vibrational excitation which implies an incorrect limit of specific heat ratio, $\gamma \to 4/3$ at low temperatures.

2.3. Numerical method

The numerical method employed here is a second-order accurate Equilibrium Flux Method (EFM) (Pullin 1980). For equilibrium chemistry computations, we employ operator splitting i.e., the hydrodynamical equations are solved and then the law of mass action (for $\Omega \to \infty$, Eq. 1) for each gas is solved in every computational cell. The physical domain is subdivided into a uniform 2000×200 grid. A message-passing parallel code was implemented on the 512 Intel Paragon at Caltech.

3. Simulation results

In Figs. 2 and 3 density and vorticity contours are shown at "early" and "late" times during the interaction. The incident shock refracts at the interface

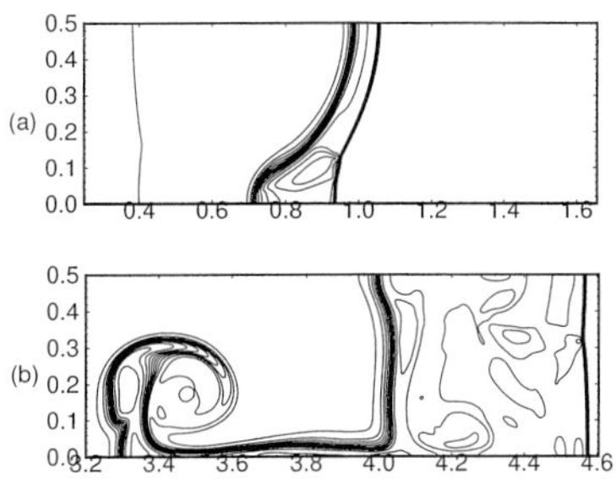

Figure 2. Density contours for an $M = 10$ shock interaction with a H_2-N_2 interface at $t = 1.38 \times 10^{-4}, 6.9 \times 10^{-4}$ for frozen (left column, $\rho = (0.05, 1.0)$) and equilibrium (right column, $\rho = (0.07, 1.8)$) chemistry.

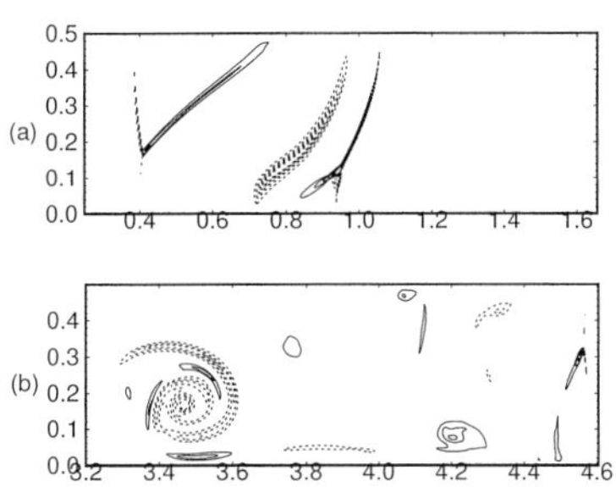

Figure 3. Vorticity contours (dashed negative), $\omega = (-10^5, 10^5)$ for frozen (left column) and equilibrium (right column) chemistry.

and the reflected and transmitted waves are shocks for this low-high acoustic impedance interface. The mean velocities of the reflected and the transmitted shocks, as well as the contact surface are larger for the equilibrium case than the frozen chemistry case. However, the transmitted and the reflected shocks lie closer to the interface for the equilibrium case. During the refraction process, negative vorticity is generated due to *baroclinicity.* As the curved reflected and transmitted shocks straighten, triple points form on the shock fronts due to the nonlinear interactions. Associated with these triple points are shear layers with positive vorticity. As time increases, these triple points traverse the reflected (or transmitted) shock fronts and further reflections from the top and bottom boundaries cause the appearance of shear layers (with alternating signs of vorticity). The contact rolls up and gives rise to the familiar mushroom-shaped morphology of the interface. The bubble rapidly assumes a nearly flat shape. This implies that the growth of the perturbation is mostly due to the relative velocity of the intense spike. The main effect of the endothermic dissociation reaction is to increase the gas density and lower the temperature. The ratio of the peak density (temperature) in the equilibrium chemistry to the frozen chemistry case is approximately 1.7 (0.4). The maximum dissociated fraction of hydrogen (nitrogen) is about 20% (36%). At any instant, the growth rate dA/dt is different for both cases, while the amplitude A of the perturbation shows a very small reduction due to dissociation (see Fig. 4). The growth rate given by the impulse model (Richtmyer 1960) over-predicts the growth rate significantly. In fact, no "linear" period of growth is observed for this case or even for the case with $A/\lambda = 0.01$ (not shown). The first slope change in Γ, Γ_i, (see

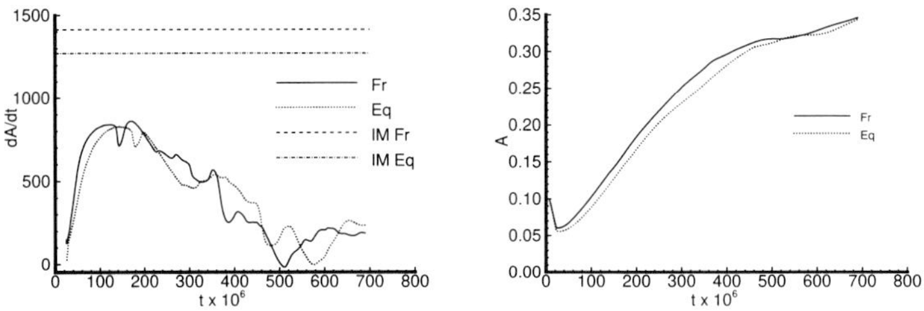

Figure 4. Growth rate (left) and amplitude (right) of the perturbation as a function of time. Eq: Equilibrium, Fr: Frozen chemistry, and IM: Impulse Model.

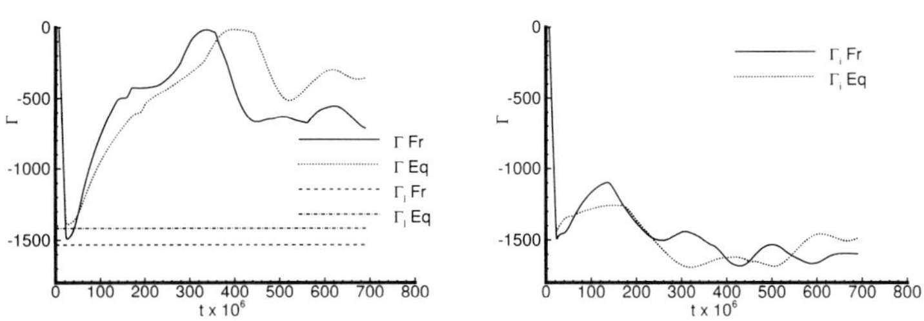

Figure 5. Total circulation (left) and interfacial circulation (right) as a function of time. Γ_l is circulation from a local analysis.

Fig. 5) occurs when the incident shock has completely traversed the interface. The circulation at this point may be determined by means of a local analysis (see Section 4). The circulation Γ shows strong variation with time due to the generation of the shear layers as explained above. The interfacial circulation changes in large part due to the strong compression waves interacting with the interface. At late times, the difference in Γ_i in the two limiting cases is small.

3.1. Mach-number scaling

The hydrodynamic equations are invariant under the following transformation:

$$t \to tM, \; p \to p/M^2, \; u \to u/M, \; E \to E/M^2 \tag{2}$$

In Fig. 6, the amplitude is plotted as a function of tM. While the frozen cases exhibit good scaling behavior, observe the monotonic decrease in A with increase in M for the equilibrium case.

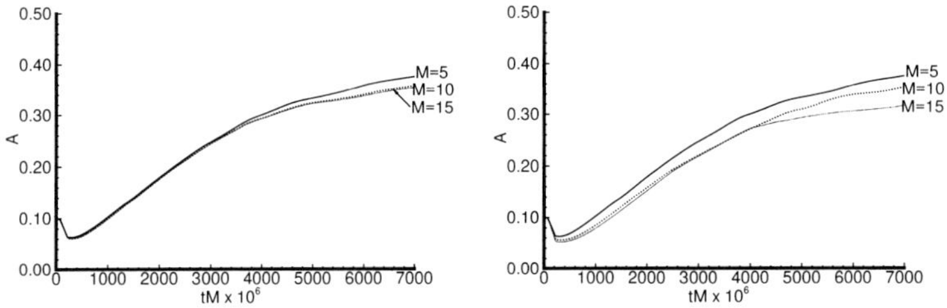

Figure 6. Perturbation amplitude as a function of scaled time, tM, for frozen (left) and equilibrium (right) chemistry.

4. Local analysis

Consider a sawtooth perturbation of a vertical fast-slow interface, where the angle β_0 between the incident shock and the interface is constant. Extending previous work (Samtaney and Zabusky 1994) for $\Omega = \infty$ we obtain the solution for small β_0 in a small neighborhood of the node where all the waves meet. For $\beta_0 = 30°$ and a H_2-N_2 interface, ρ and T behind the reflected and transmitted shocks, and σ, the vortex sheet density, are plotted in Fig. 7. The differences due to dissociation occur at about $M = 6$. Note that the difference in σ between equilibrium and frozen chemistry is small. For $M > 15$, all the diatomic species have dissociated completely and the resulting gas behaves as a monoatomic perfect gas. The initial baroclinic-circulation generation on a sinusoidal interface is obtained by integrating the σ over the length of the interface. This result was plotted as the horizontal dotted line in the circulation plot (Fig. 5).

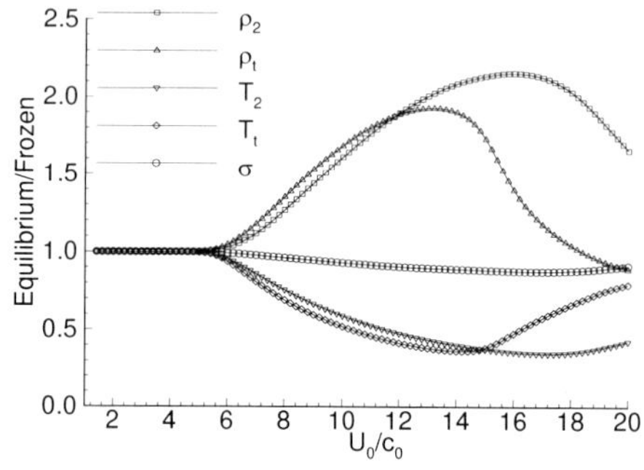

Figure 7. Local solution for equilibrium dissociated flow normalized by the frozen solution for a $\beta_0 = 30°$, H_2-N_2 interface. Subscript '2' ('t') is for values between the reflected (transmitted) shock and contact surface; σ is the vortex sheet strength.

5. Conclusion

In this paper, we examined the effects of dissociation in the interaction of hypervelocity shocks with sinusoidally perturbed density interfaces. As far as the perturbation amplitude is concerned, the chemistry plays a minor role. For frozen cases, Mach-number scaling works well for the entire duration of the simulation. For equilibrium chemistry this scaling is questionable only when dissociation is significant.

Acknowledgement. This work was funded by the Center for Advanced Fluid Dynamics Applications, Lawrence Livermore National Laboratory under grant No. W-7405-ENG-48 Subcontract B295121 and by AFOSR under grant F49620-93-1-0338. RS acknowledges the help of Berna Massingill who provided the mesh archetypes for the Intel Paragon and Thanh Phung who helped optimize the code.

References

Lighthill MJ (1957) Dynamics of a dissociating gas. Part 1. Equilibrium flow. J Fluid Mech 2:1.

Meshkov EE (1969) Instability of the interface of two gases accelerated by a shock wave. Fluid Dynamics 4:101.

Pullin DI (1980) Direct simulation methods for compressible ideal gas flow. J Comput Phys 34:231–244.

Richtmyer RD (1960) Taylor instability in shock acceleration of compressible fluids. Comm Pure and Appl Math XIII:297.

Samtaney R, Zabusky NJ (1994) Circulation deposition on shock-accelerated planar and curved density stratified interfaces: models and scaling laws. J Fluid Mech 269:45–78.

Numerical investigation of Richtmyer-Meshkov instability using front tracking

J. W. Grove, R. L. Holmes*
Department of Applied Mathematics and Statistics
University at Stony Brook, Stony Brook, NY 11794-3600 USA
D. H. Sharp
Complex Systems Group, Theoretical Division
Los Alamos National Laboratory, Los Alamos, NM 87545 USA

Abstract: Front-tracking simulations of Richtmyer-Meshkov instability produce significantly better agreement with experimentally measured growth rates than obtained in previous computations. Careful analysis of the early stages of the shock-acceleration process show that nonlinearity and compressibility play a critical role in the behavior of the shocked interface and are responsible for the deviations from the linear theories.

Key words: Front tracking, Richtmyer-Meshkov instability

1. Introduction

The quantitative understanding of Richtmyer-Meshkov (shock-accelerated) interface instability has challenged scientists for over 30 years. Interest in this instability is motivated by its importance in subjects such as inertial confinement fusion and supernovae dynamics.

The growth of perturbations on a shocked material interface was first studied in Richtmyer's seminal 1960 paper. Richtmyer found that after an initial rapid increase, the growth rate in the solution to a linearization of the Euler equations oscillated about a limiting value. He proposed a formula for this limit, known as the impulsive model, that agreed quite well with the computed limit of the linearized equations. Later shock-tube experiments (Meshkov 1970, Benjamin 1992, Benjamin et al. 1993) produced growth rates that were significantly smaller by as much as a factor of two than the predictions of the impulsive model. In sharp contrast however, many numerical simulations of experiments agreed quite well with the impulsive model (Benjamin et al. 1993, Cloutman & Wehner 1992, Meyer & Blewett 1972). Numerous articles on this instability can be found in the Proceedings of the International Workshops on the Physics of Compressible Turbulent Mixing (Besnard et al. 1991, Dannevik et al. 1992,

*Present address: Courant Institute of Mathematical Sciences, New York University, 251 Mercer St., New York, NY 10012.

Linden et al. 1993), and interesting exceptions are found in Mikaelian (1993) and Yang et al. (1994).

Recent front-tracking simulations of experiments of Benjamin et al. (1993) have substantially improved agreement between experiment, theory, and computation, producing growth rates that are in excellent agreement with those found in Benjamin's experiments. Our computations show that the nonlinear simulation and the linearized theory agree for small initial amplitude perturbations and sufficiently early times, but that there is an abrupt departure of the nonlinear theory from the linear theory at later times caused by a re-acceleration of the interface by shocks produced by nonlinear self-interactions of the reflected and transmitted waves.

2. Experimental configurations

Benjamin's experiments were conducted in a rectangular shock tube where air and sulfur hexafluoride (SF_6) were separated by a thin membrane with an imposed single-wavelength sinusoidal perturbation (Fig. 1 shows a schematic of the geometry). A Mach 1.2 plane shock ruptures the membrane producing a transmitted shock and either a reflected shock or rarefaction wave. The interface perturbation had an amplitude of 0.24 cm and a wavelength of 3.75 cm. The instability consists of the growth of perturbations on the material interface.

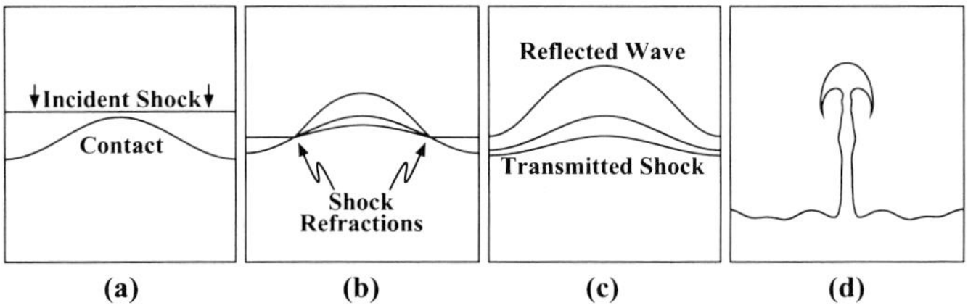

Figure 1. A schematic of the geometry in Richtmyer-Meshkov experiments.

The ambient pressure at the Los Alamos National Laboratory is 0.8 bar and the densities of air and SF_6 are 0.95 g/l and 4.85 g/l, respectively. The gases are assumed to be perfect with $\gamma_{air} = 1.4$ and $\gamma_{SF_6} = 1.09$. The rear wall of the tube was located sufficiently far from the interface that measurements of the growth rate could be made before the arrival of the shock reflected off this rear wall.

We compared the computed amplitude of the perturbations on the interface, $a(t)$, defined to be one-half the total inter-penetration width, and the amplitude growth rate, $\dot{a}(t)$, with Benjamin's measurements. We also measured the flow velocities at the tips of the heavy gas spike and the light gas bubble. Note that $\dot{a}(t) = (v_{spike} - v_{bubble})/2$, where v denotes the velocity at the interface.

3. Comparison of the linear and nonlinear theories

Richtmyer's impulsive model is based on an analogy of the shock acceleration process to an impulsive body force in an incompressible Rayleigh-Taylor instability. If the perturbation is sinusoidal with wave number k, Richtmyer's formula for the perturbation-amplitude growth rate is

$$\dot{a}(t) = kA\Delta u a(0+). \tag{1}$$

Here $A = (\rho_2^* - \rho_1^*)/(\rho_2^* + \rho_1^*)$ is the Atwood ratio using post-shock densities ρ_i^*, Δu is the change in translational velocity of the interface due to the action of the shock, and $a(0+)$ is the shock-compressed amplitude. Equation (1) is assumed hold for $ka(t) \ll 1$. Note that this formula explicitly assumes that compressibility is negligible after the initial shock refraction. Meyer & Blewett (1972) proposed a modification of Eq. 1 for reflected rarefactions in which $a(0+)$ is replaced by the average of the pre- and post-shock amplitudes.

For $ka \ll 1$ the Euler equations can be linearized about the solution of the Riemann problem for a planar shock refracting through a planar material interface. Note that this linearization makes no incompressibility assumption.

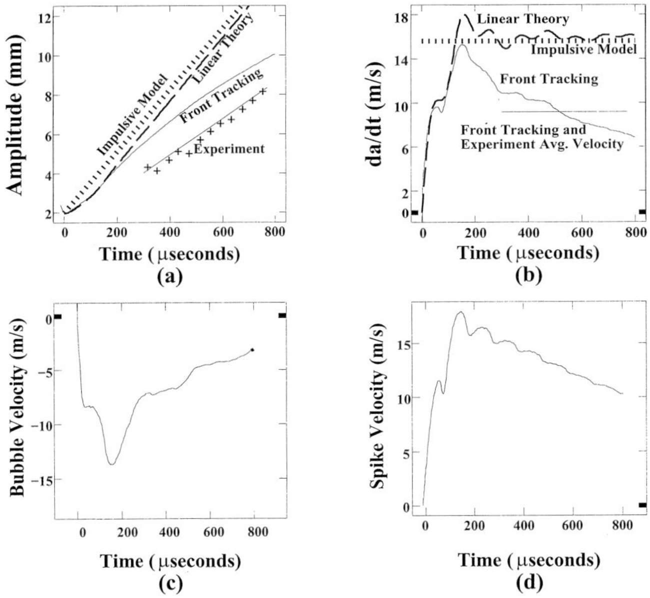

Figure 2. Perturbation amplitude, $a(t)$, and amplitude growth rate, $\dot{a}(t)$, of a shocked air-SF_6 interface.

Figures 2a and b show amplitude and amplitude growth rates from the impulsive model, numerical solution of the linear theory, nonlinear simulations, and experimental results. Also shown is a least-squares fit to the front-tracking amplitude data over the period of experimental observation. We focus on two regions of the growth-rate curves where the linear and nonlinear theories diverge. The first occurs at $t \approx 70$ μs where the growth rate $\dot{a}(t)$ reverses and then begins a second period of rapid increase. Subsequently, at $t \approx 200$ μs, $\dot{a}(t)$ saturates near its maximum value for the linear solution while the nonlinear solution experiences a marked deceleration.

The rapid deceleration of the shocked interface at 200 μs is induced by a series of re-accelerations of the material interface by secondary shocks generated from self-interactions at the reflected and transmitted wave edges. At 33 μs these waves are clearly seen near the edges of the reflected shock (Fig. 3a). As the interaction proceeds, the compression fronts steepen into shocks that collide near the tip of the spike at about 68 μs (Fig. 3b), producing a high-pressure pulse that causes the growth rate to decrease. This deceleration corresponds to the first blip in the growth-rate graph in Fig. 2b. At approximately 207 μs (Fig. 3c) a strong pulse again crosses the interface, leading to the severe (and permanent) deceleration of the bubble. That this decrease is due primarily to action at the bubble side of the interface is clear from Figures 2c and d, which show respectively the separate bubble and spike velocities. By time $t = 344$ μs the strong wave action has moved away from the interface region and the growth-rate curve has smoothed considerably.

Figures 3e and f show the pressure field of the solution to the linearized equations and of the solution to the nonlinear Euler equations at $t = 195$ μs, near the time of the most extreme divergence of the two solutions. The linearized equations of motion do not allow focusing of characteristics and constrain the geometry of the wave fronts to be sinusoidal. These two restrictions prevent the production of the cascading shocks that are present in the nonlinear solution. The linear solution displays a cascading series of acoustic waves, but these do not sharpen into shocks, which greatly reduces their effect on the growth rate of the interface.

4. Comparison to experiments

Comparison of simulations to experiments results in excellent agreement for the growth rates (see Figs. 2a and b, and Grove et al. 1993). However, there is a significant offset in the value of the absolute amplitude of the perturbations. It is believed that this offset is due to the delayed rupture of the membrane used in the experiments to separate the two gases. A detailed discussion of membrane effects is beyond the scope of this paper. We note, however, that a

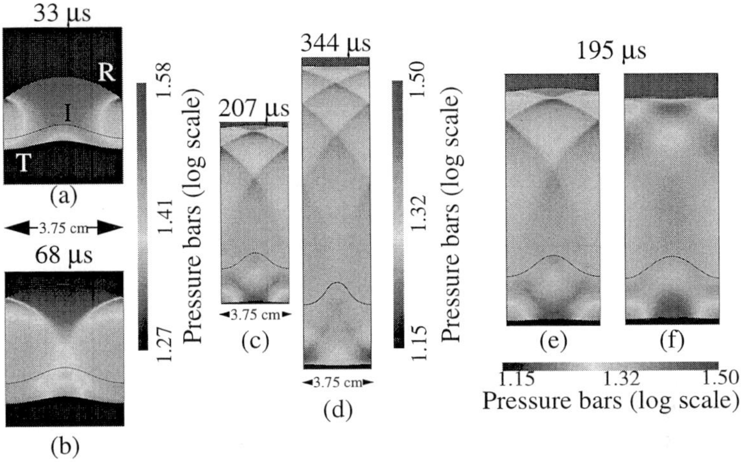

Figure 3. Grey scale plots of the pressure field for the nonlinear simulation and linearized solution of Richtmyer-Meshkov instability. Here T = transmitted shock, I = material interface, R = reflected shock.

computational approach to estimating membrane effects would be complicated. Analyzing the deformation and fragmentation of a thin elasto-plastic membrane would add severe demands to an already demanding computation. There is no established and benchmarked methodology for modeling the fragmentation of such a membrane. Even if one did come up with a model of membrane dynamics, it would be very difficult to know if it was correct, since such a model is likely to require several phenomenological parameters.

5. Conclusion

Simulations of Richtmyer-Meshkov instability show that compressibility and nonlinearity play an important role in the development of Richtmyer-Meshkov instability, up to and including the times measured in experiments. These wave interactions are not captured in the linearized theories commonly used to estimate perturbation growth rates.

Front tracking leads to improved agreement between experiment and simulation perturbation growth rates. It appears worthwhile to study the effect of mass diffusion and membrane effects on the early-time growth of the perturbations to aid in understanding the still unresolved difference in amplitudes.

Acknowledgement. We give our thanks to James Glimm for his encouragement and guidance. This work has been supported by the U.S. Army Research Office and the Mathematical Sciences Institute of Cornell University under subcontract to SUNY Stony Brook, ARO Contract number DAAL03-91-C-0027, the National Science Foundation, Grant number DMS-9057429, the National Science Foundation, Grant number DMS-9201581, the U.S. Army Research Office, Grant number DAAL03-92-G-0185, and the U.S. Department of Energy.

References

Benjamin R (1992) Experimental observations of shock stability and shock induced turbulence. In: Advances in Compressible Turbulent Mixing, WP Dannevik, AC Buckingham, CE Leith (eds), National Technical Information Service, U.S. Dept of Commerce, 5285 Port Royal Rd, Springfield VA 22161, pp 341–348.

Benjamin R, Besnard D, Haas JF (1993) Shock and Reshock of an Unstable Interface, LANL Report LA-UR 92-1185, Los Alamos National Laboratory.

Besnard D, Cavailler C, Dagens L, Figeac P, de Gliniasty M, Haas JF, Holstein PA, Montigny J, Parisot C, Rupert V, Sitt B, Wilke N (1991) Proceedings of Third International Workshop on the Physics of Compressible Turbulent Mixing at Royaumont, France, CEA DAM.

Cloutman LD, Wehner MF (1992) Numerical simulation of Richtmyer-Meshkov instabilities. Phys Fluids A 4:1821–1830.

Dannevik WP, Buckingham AC, Leith CE (1992) Advances in Compressible Turbulent Mixing, National Technical Information Service, U.S. Department of Commerce, 5285 Port Royal Rd, Springfield VA 22161.

Grove JW, Holmes R, Sharp DH, Yang Y, Zhang Q (1993) Quantitative theory of Richtmyer-Meshkov instability. Phys Rev Lett 71(21):3473–3476.

Linden PF, Youngs DL, Dalziel SB (1993) Proceedings of the 4th International Workshop on the Physics of Compressible Turbulent Mixing, Cambridge University Press, Cambridge, England.

Meshkov EE (1970) Instability of a Shock Wave Accelerated Interface Between Two Gases, NASA Tech Trans F-13:074.

Meyer KA, Blewett BJ (1972) Numerical investigation of the stability of a shock-accelerated interface between two fluids. Phys Fluids 15:753–759.

Mikaelian KO (1993) Growth rate of the Richtmyer-Meshkov instability at shocked interfaces. Phys Rev Lett 71(18):2903–2906.

Richtmyer RD (1960) Taylor instability in shock acceleration of compressible fluids. Comm Pure Appl Math 13:297–319.

Yang Y, Zhang Q, Sharp DH (1994) Small amplitude theory of Richtmyer-Meshkov instability. Phys Fluids 6(5):1856–1873.

Richtmyer-Meshkov mixing-zone deformation induced by shock-tube wall boundary layers

G. Jourdan, L. Houas
IUSTI (U.M.R. CNRS 139)
Université de Provence, Marseille, France

Abstract: An experimental investigation of the gaseous mixing zone that originates from Richtmyer-Meshkov instability has been undertaken in a square-cross-section shock tube. Mass concentration fields have been determinated within the mixing zone for the cases when the shock wave passes from a heavy gas to a light one, from one gas to another of close density, and from a light gas to a heavy one before and after being compressed by the reflected shock. The diagnostic method is based on simultaneous three-directional laser absorption of one of the two constituents of the mixing zone. It is shown that the mixing zone is strongly deformed by the wall boundary layer when the latter is turbulent, and becomes more homogeneous when the Atwood number tends to zero.

Key words: Richtmyer-Meshkov instability, Shock/interface interaction, Turbulent mixing zone, Shock-tube experiments, CO_2 laser absorption

1. Introduction

The instability of a shock-accelerated interface separating two fluids of different densities was first studied theoretically by Richtmyer (1960) and experimentally by Meshkov (1970). When the interface is subjected to a shock wave parallel to it, the random small-scale defects due to the making of the interface itself lead to the creation of a mixing zone between the two fluids. Therefore, the density gradients at the interface and the pressure gradient caused by the shock passage are not collinear anywhere and random small-scale vortices are produced.

Experiments in a shock-tube environment are performed in order to understand some of the mechanisms occurring in the implosion of inertial-confinement fusion targets as a consequence of Richtmyer-Meshkov instability. Many experiments for low Mach numbers and atmospheric initial pressure have been undertaken to study this phenomenon. But when the Mach number increases and the initial pressure decreases simultaneously, shock/interface interaction leads to turbulent three-dimensional flow which is difficult to describe. Recently, a multi-directional laser absorption technique has been developed in our laboratory and temperature and species concentration profiles within the mixing zone have been experimentally obtained within nine imaginary identical regions of the square-cross-section shock tube (in the four corners, near the four walls and in the center). Previously, we used several experiments to explore one abscissa with two detectors; now we use three detectors in the same run. The shock-tube wall boundary-layer deformation of the mixing zone has been clearly pointed out with the help of the 3D shape reconstruction of the mixing zone.

2. Experiments

Experiments are performed in a 8.5-cm by 8.5-cm square-cross-section double diaphragm shock tube of about 9 meters total length. The cases when the shock wave passes from the heavy gas to the light one (CO_2/He), from one gas to another of close density (CO_2/Ar), and from the light gas to the heavy one (CO_2/Kr) have been studied for the same initial conditions. The Mach number in the CO_2 is about 3.1 which corresponds to Mach number 2.2, 3, and 3.5 in helium, argon, and krypton respectively, and the initial pressure for the gases is 1500 Pa. CO_2 is used for its spectroscopic properties. The gases are initially separated by a thin membrane (1.5-μm thick mylar) resting over a square grid (of 5×5 wires, diameter of 0.2 mm, and spacing of 17 mm), which allows a regular rupture of the membrane. Figure 1 presents the sketch of the experimental set-up.

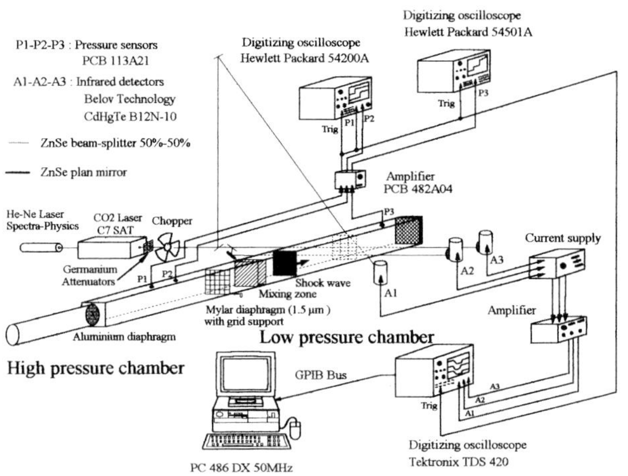

Figure 1. Experimental set-up.

The diagnostic method is a simultaneous three-directional laser absorption technique previously described by Fortes et al. (1994). This method allows the determination of the density and the temperature of one of the mixing constituents (CO_2), and therefore of the concentration. Measurements have been realized at four different abscissa before the return of the first reflected shock, but only the results of the last abscissa are presented in this paper. After compression, the quasi-taylorization of the CO_2/Kr mixing zone does not permit a complete measurement because of the infinite time of the absorption signals, and thus, only CO_2/He and CO_2/Ar results are presented. The accuracy of the concentration and density results is estimated to be about 10%. Concerning the 3D representations, the error bars are evaluated to be about ± 1 to 4 mm as a function of the errors on the velocity of the mixing zone.

3. Results and discussion

Figures 2, 3, and 4 show the concentration fields and the 3D reconstructions of the three tested mixing zones. Each iso-concentration surface corresponds to a plan cut of the mixing zone perpendicular to the direction of its propagation and is delimited by the shock-tube walls. L is the total thickness of the mixing zone. The simultaneous use of three detectors per run has permitted us to obtain a two-dimensional visualization of the mixing-zone shape because we can estimate the absolute delay in time between the three absorption signals.

These visualizations are given before (Figs. 2a, 3a, and 4) and after (Figs. 2b and 3b) the compression by the reflected shock wave. The first comment that we can make is that each 3D representation of the mixing zone is in good agreement with the corresponding concentration fields. Before the interaction with

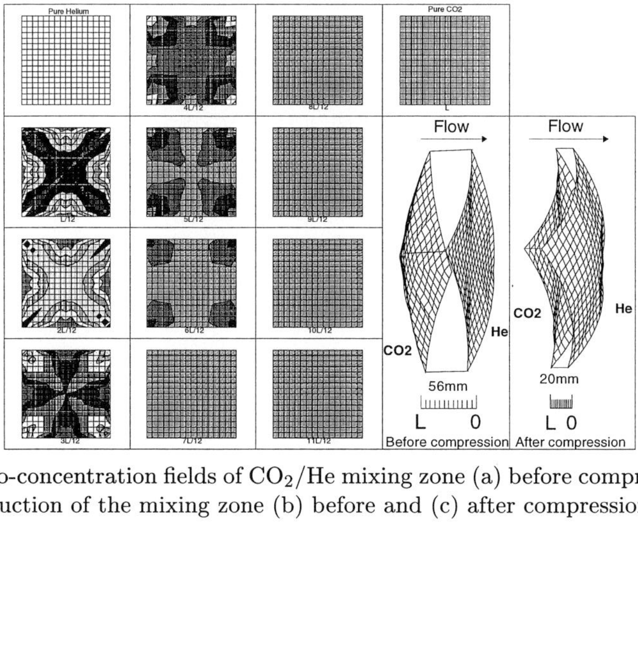

Figure 2. Iso-concentration fields of CO_2/He mixing zone (a) before compression; and 3D reconstruction of the mixing zone (b) before and (c) after compression.

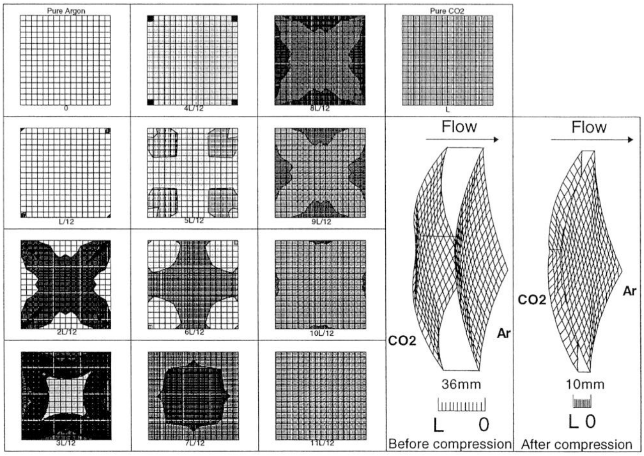

Figure 3. Iso-concentration fields of CO_2/Ar mixing zone (a) before compression; and 3D reconstruction of the mixing zone (b) before and (c) after compression.

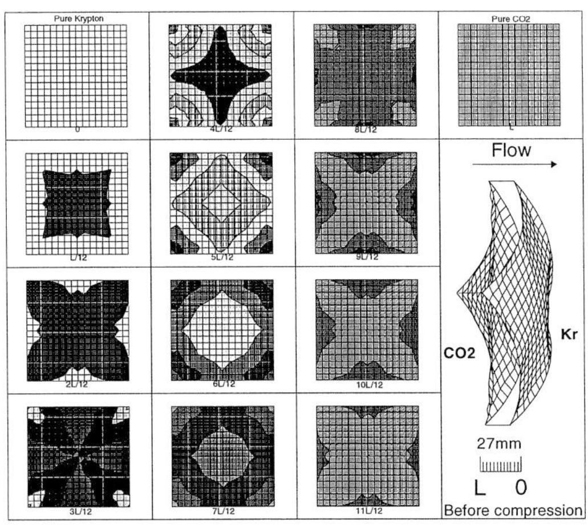

Figure 4. Iso-concentration fields of CO_2/Kr mixing zone (a) before compression; and (b) 3D reconstruction of the mixing zone before compression.

the reflected shock, the CO_2/He mixing zone seems to be the least deformed, since the mixing is present for the first slice $\frac{L}{12}$ in all shock-tube cross sections. Calculations, based on the work of Mirels (1984), have shown that the mixing zone propagates with a boundary layer for CO_2/He which is laminar, transitional for CO_2/Ar, and turbulent for CO_2/Kr. This can explain the different mixing-zone deformations.

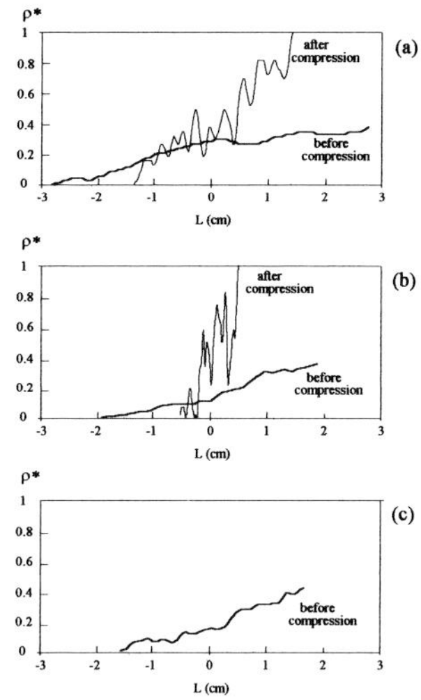

CO_2/He	A	ΔU	$\frac{\rho_{CO2}}{\rho_{He}}$	Time
After incident shock	-0.73	850 m/s	# 10	915 μs
After reflected shock	+0.79	-560 m/s	9.3	60 μs

CO_2/Ar	A	ΔU	$\frac{\rho_{CO2}}{\rho_{He}}$	Time
After incident shock	-0.31	640 m/s	# 1	1215 μs
After reflected shock	+0.5	-585 m/s	3	100 μs

CO_2/Kr	A	ΔU	$\frac{\rho_{CO2}}{\rho_{He}}$	Time
After incident shock	-0.05	535 m/s	# 0.5	1450 μs
After reflected shock	+0.22	-528 m/s	1.5	-

Figure 5. Adimensional density profiles for (a) CO_2/He; (b) CO_2/Ar; and (c) CO_2/Kr mixing zones. A is the Atwood number and ΔU is the mixing-zone velocity jump.

The iso-concentration fields have to be considered as qualitative results. However, some information can be gleaned from their observation. First, the CO_2/He mixing zone (heavy/light with a high Atwood number) is the least homogeneously mixed since the iso-concentration variations reach 70% per slice. The CO_2/Ar and CO_2/Kr mixing zones seem to be more homogeneous and the concentration variation does not exceed 30% per slice. On the other hand, one can say that when the density of the second constituent increases (He, Ar, or Kr), its relative penetration into the CO_2 also increases. As we can observe from Figs. 2a, 3a, and 4, approximately 50% of the rear part of the CO_2/He mixing zone is highly composed of CO_2 (90%). For the CO_2/Ar and CO_2/Kr cases, it is only about 20% of the rear part. Therefore, this is a confirmation that the homogeneity of the mixing increases when the Atwood number tends to zero and also that the mixing zone develops towards the CO_2 in our experimental configuration (for CO_2/He we obtain 60% of the rear part with more than 90% of CO_2 at 164 mm and 335 mm and 50% at 550 mm).

After compression (no result for the CO_2/Kr case) the mixing seems to become more homogeneous (iso-concentration fields are not shown owing to space limitations). The compression seems to be more intense for the CO_2/Ar case and this can be understood since we are nearer taylorization. (Note that each experimental 3D reconstruction of the mixing zone is obtained from a different run; the fact that the CO_2/He and CO_2/Ar mixing zones are bulged in the same direction, before and after the compression by the reflected shock, is coincidental.) This result is quite visible on Fig. 5, where we have plotted the adimensional CO_2 density before and after the compression by the reflected shock for the CO_2/He and CO_2/Ar cases (Figs. 5a and 5b) and only before the compression for CO_2/Kr (Fig. 5c). We also observe that the reflected shock passage has considerably increased the small-scale perturbations which were present within the mixing zone. The density ratio across the shock passage seems to be the same (2.5 for the two cases), but the thickness is decreased by a factor of 2 for CO_2/He and 4 for CO_2/Ar. The difference in the thickness compression can be attributed to the difference in homogeneity of the two mixing zones, or the proximity to the taylorization for the CO_2/Ar case, or both.

Acknowledgement. This study was supported by the C.E.A. Centre d'Etudes de Vaujours-Moronvilliers, on contract No. 304 843/00 D1.

References

Fortes J, Ramdani A, Houas L (1994) CO_2 laser absorption measurements of temperature and density in shock induced Richtmyer-Meshkov mixing zone. Physical Review E 50(4):3041–3049.

Meshkov EE (1970) Instability of a Shock Wave Accelerated Interface Between Two Gases, NASA TT F-13, 074.

Mirels H (1984) Turbulent boundary layer behind constant velocity shock including wall blowing effects. AIAA J 22:1042–1047.

Richtmyer RD (1960) Taylor instability in shock acceleration of compressible fluids. Comm Pure Appl Math 13:297–319.

Vortex methods for early and intermediate time Richtmyer-Meshkov environments

J. Ray, N. J. Zabusky
Department of Mechanical and Aerospace Engineering
P. O. Box 909, Rutgers University
Piscataway, NJ 08855-0909 USA

R. Samtaney
Applied Mathematics
California Institute of Technology
Pasadena, CA 91125 USA

Abstract: The *vortex paradigm* (Hawley and Zabusky 1989) has been used to develop reduced vortex models that predict the time rate of change of the amplitude of harmonics on planar interfaces. Our work represents a substantial improvement over the linearized studies of this environment, e.g. as first conducted by Richtmyer in a pioneering analysis.

The physical parameter domain of our models embraces $1.05 \leq M \leq 4.0$, $\eta = 1.54, 3.0, 5.04$, and $A/\lambda = 0.025, 0.05, 0.1$, where M is the Mach number of the incoming planar vertical shock, η is the density ratio across the interface, and A/λ is the amplitude-to-wavelength ratio of the perturbation to a planar vertical interface. Note, that if the shock moves from a low acoustic impedance to a higher one, $\eta > 1$ and vice versa.

Our models were validated against simulations of the compressible Euler equations.

Key words: Visiometrics, Reduced models, Growth rate

1. Introduction

A linear stability analysis associated with shocks striking a weak, long wavelength harmonic perturbation, ($x = A \cos ky, Ak \ll 1$), on an interface between two gases, was first done by Richtmyer (1960). His "impulse" model states

$$\dot{A} = kA|_{0+} \Delta U \frac{\eta - 1}{\eta + 1} \tag{1}$$

where k the wavenumber, η is the density ratio, $A|_{0+}$ is the amplitude of the harmonic, and ΔU is the velocity of the interface from a 1D interaction, after shock passage. Yang et al. (1994) elucidated Richtmyer's linear analysis.

The vortex paradigm is based on the evolution equation in inviscid flows:

$$\frac{D\omega}{Dt} = \frac{\nabla\rho \times \nabla p}{\rho^2} + \omega \cdot \nabla\vec{u} - \omega\nabla \cdot \vec{u} \tag{2}$$

As the shock traverses the interface, the misalignment of ∇p and $\nabla\rho$ deposits a vortex sheet on the interface which dominates the largely incompressible evolution of the interface, especially for weak shock and low-η cases.

2. Equations, numerics, and parameters

2.1. Numerical method

Consistent with previous numerical studies, we employ 2D compressible Euler equations. We track the interface using a level-set method. The numerical method is a second-order accurate Godunov scheme and is similar to the Eulerian MUSCL scheme.

The physical domain is a rectangular shock tube. The governing parameters are M, η, and Ak. The undisturbed incident gas is initialized with unit density and pressure; the shock is initialized as per the Rankine-Hugoniot conditions.

The test gases are air–CO_2 ($\eta = 1.54, \gamma_0 = 1.4, \gamma_b = 1.297$), air–R22 ($\eta = 3.0, \gamma_0 = 1.4, \gamma_b = 1.172$), and air-$SF_6$ ($\eta = 5.04, \gamma_0 = 1.4, \gamma_b = 1.0935$).

2.2. Convergence of interfacial circulation

We conduct a simulation with $M = 1.5, \eta = 3.0$, and $A/\lambda = 0.05$ on a 1600×80 grid with $\Delta x = \Delta y = 1$. The results are shown in Figs. 1–3. Figure 1a shows excellent convergence of interfacial circulation for the last two grids.

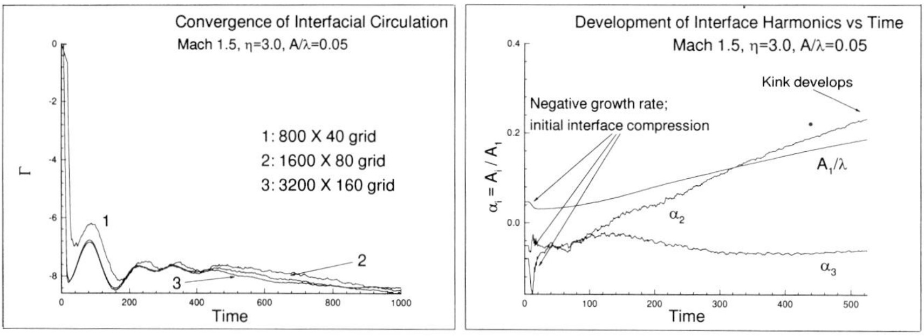

Figure 1. (a) Resolution study; (b) development of interface harmonics.

3. Vortex models

We propose models — analytical and numerical — for the different regimes. The numerical models require minimal inputs from numerical simulations. The growth rate is defined as half the difference between the velocities of the x-extrema of the interface.

3.1. Qualitative aspects

After shock passage, variations in Γ (Fig. 1a) are caused by secondary waves sweeping over the interface. Over time, higher harmonics of the perturbation develop (Fig. 1b) and the vortex sheet rolls-up into a patch (Fig. 2a). Nonlinearity and the rate of roll-up depend on $\Gamma(M, \eta, A/\lambda)$.

The evolution can conveniently be split into four regimes (Fig. 3a): (1)Very Early Time (VET), when the incident shock and dominant compression waves sweep over the interface; (2) Early Time (ET), when the evolution can be modeled by an evolving incompressible vortex sheet; (3) Intermediate Time (IT), when the contact discontinuity becomes a multivalued function of y; and (4) Late Time (LT), when the evolution is nearly incompressible and driven by a turbulent-like collection of vortices. The transition times depend specifically on the details of the run.

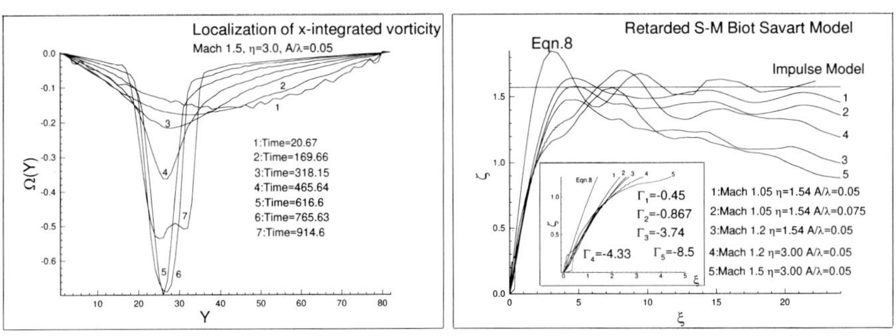

Figure 2. (a) Localization of vorticity; (b) very early time behavior.

The model for VET is corrected by a factor $2\sqrt{\eta}/(\eta + 1)$, which appears in the growth rate of a perturbation on a uniform vortex sheet located on a stratified interface (Chandrasekhar 1961).

3.2. Numerical Biot-Savart model (NBS)

We use this to validate the assumption of incompressibility. We integrate the instantaneous vorticity distribution over the instantaneous interface shape — both extracted from the numerical simulation — to obtain the interface growth rate.

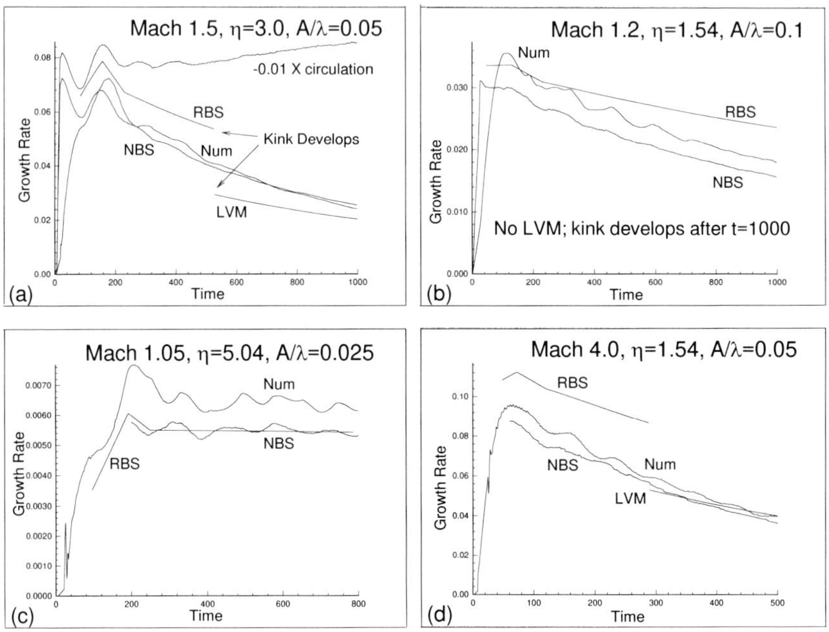

Figure 3. (a)–(d): comparision of models vis-a-vis numerical simulations.

3.3. Reduced Biot-Savart model (RBS)

This Biot-Savart-based model, applicable in ET, assumes incompressibility, a single-mode representation of the perturbation, and a frozen spectral content of the vorticity distribution. We get (Samtaney and Zabusky 1994a)

$$u(x,y,t) = -\frac{1}{4\pi}\int_S \frac{(y-y')\Omega(y')dy'}{|\vec{x}-\vec{x'}|^2}; \quad \dot{A} = \frac{\Gamma(t)k}{4\pi}\int_{-\infty}^{\infty}\frac{\xi\sin(\xi)d\xi}{k^2A(t)^2(1-\cos\xi)^2+\xi^2} \tag{3}$$

where $\Omega(y)$ is the x-integrated vorticity on the interface and $\xi = ky$.

To capture the initial growth rate, $\Gamma(t)$ is obtained by linearly interpolating between $\Gamma(t_1)$ and $\Gamma(t_2)$ for $t_1 \le t \le t_2$, $\Gamma(t_2)$ and $\Gamma(t_3)$ for $t_2 \le t \le t_3$ and $\Gamma(t) = \Gamma(t_3)$ for $t \ge t_3$, where $\Gamma(t_i)$ are the numerical (simulation) values of interfacial circulation at times of the first three local extrema after the shock has crossed the interface.

3.4. Retarded, Stratification-Modified Biot-Savart model (RSMBS)

We derive a model for VET from SMBS under the assumptions of $M \approx 1$ and $Ak \ll 1$; consequently $\Gamma(t) \approx const$. We will approximate the circulation on the interface by the first term of the sine expansion (Samtaney and Zabusky 1994b). The effect of compressibility is introduced with a retarded time, as in aeroacoustics (Kambe 1986, Crow 1970). For $M \approx 1$, P/ρ is almost unchanged after shock passage and we take $c_s = \sqrt{\gamma}$. The growth rate is given by

$$\dot{A} = \frac{2\sqrt{\eta}}{\eta+1}\frac{\Gamma k}{4\pi}\int_{-\infty}^{\infty}\frac{(\xi'\sin(\xi'))H(t-r'/c_s)d\xi'}{k^2A(t)^2(1-\cos\xi')^2+(\xi')^2} \tag{4}$$

where $(kr')^2 = k^2A(t)^2(1-\cos\xi')^2+(\xi')^2$. If $Ak \ll 1$, then $r' = c_st \approx y'$ and the expression simplifies to

$$\dot{A} = \frac{2\sqrt{\eta}}{\eta+1}\frac{(\Gamma_1' kA|_{0-})\mathrm{Si}(\xi)}{\pi}, \quad \xi = kc_st = \sqrt{\gamma}kt \tag{5}$$

where $\mathrm{Si}(\xi)$ is the Sine Integral function.

Instead of Γ_1' we use its analytical scaling result (Samtaney and Zabusky 1994b) in Eq. (5) to get

$$\dot{A} = \frac{4\sqrt{\gamma}}{\gamma+1}\left(1+1/M+2/M^2\right)(M-1)\frac{\sqrt{\eta}-1}{\eta+1}\frac{kA|_{0-}}{\pi}\mathrm{Si}(\xi) \tag{6}$$

We plot $\zeta = \dot{A}/(\Gamma_{sc}A|_0k)$ from various numerical simulations and $\mathrm{Si}(\xi)$ versus ξ in Fig. 2b. The departure from the model with increasing interfacial circulation Γ is also evident from Fig. 2(b).

Physically, the oscillations in $\mathrm{Si}(\xi)$ are due to the arrival of signals from vortex images required to enforce the no-flow boundary condition, $\vec{u}\cdot\vec{n} = 0$, at $y = 0$ and $y = \lambda/2$. As $t \to \infty$, $\dot{A} \to \dot{A}_{Richtmyer}$.

3.5. Constant-Symmetric Line Vortex model (LVM)

This incompressible model applies at intermediate times and is analytic. We assume an array of line vortices of constant strength $(-1)^n\Gamma_L$ and location $y = (2n+1)\lambda/4$, $n = \ldots -2, -1, 0, 1, 2, \ldots$, where Γ_L is the first maximum in circulation and can be obtained analytically from Samtaney and Zabusky (1994b). This captures the solenoidal component of the velocity, the compressible component being small for weak shocks and at later times.

Summing the contribution from all the vortices,

$$\dot{A} = -\frac{\beta}{\cosh kA(t)} \quad \text{or} \quad \sinh kA(t) = -\beta kt + \sinh kA(t_s) \tag{7}$$

where $\beta = \Gamma_L k/2\pi$.

For late time and large amplitude ($Ak \gg 1$, $\tanh(Ak) \approx 1$) we get

$$\dot{A} = \frac{1}{kt + C} \qquad C = \frac{\sinh kA|_0}{-\Gamma_L k/2\pi}$$

which is similar in form to Hecht's model for v_{bubble} (Hecht et al. 1994).

4. Conclusion

Our vortex models provide a robust way of estimating the growth of a single-mode perturbation in Richtmyer-Meshkov environments. For $M \approx 1$ and $kA \ll 1$, there exists an analytical model for the growth rate.

Acknowledgement. This work was funded in part by a DOE grant, DE-FG02-93ER25179.A000, monitored by Dr. Daniel Hitchcock. The computations were performed on the CM5 at NCSA, University of Illinois, Urbana-Champaign.

References

Chandrasekhar S (1961) Hydrodynamic and Hydrodynamic Stability, Oxford University Press.

Crow, S (1970) Aerodynamic sound emmision as a singular perturbation problem. Stud Appld Math 49:21–44.

Hawley JF, Zabusky NJ (1989) Vortex paradigms for shock-accelerated density-stratified interfaces. Phys Rev Lett 63:241-1244.

Hecht J, Alon U, Mukamel D, Shvarts D (1994) Simple potential flow models of Rayleigh-Taylor and Richtmyer-Meshkov bubble-fronts. Phys Fluids 6:4019.

Kambe T (1986) Acoustic emissions by vortex motions. J Fluid Mech 173:643–666.

Richtmyer RD (1960) Taylor instability in shock acceleration of compressible fluids. Comm Pure and Appl Math XIII:297.

Samtaney R, Zabusky NJ (1994a) Vorticity Generation and Growth Rate in Richtmyer-Meshkov Instabilities, Caip Tech. Rept. No. 187, CAIP, Rutgers University, Piscataway, NJ 08855-1390 (submitted Phys Flds).

Samtaney R, Zabusky NJ (1994b) Circulation deposition on shock-accelerated planar and curved density stratified interfaces: models and scaling laws. J Fluid Mech 269:45.

Yang Y, Zhang Q, Sharp D (1994) Small amplitude theory of Richtmyer-Meshkov instability. Phys Fluids 6(5):1856.

Visualization of shocked mixing zones using differential interferometry and X-rays

G. Rodriguez, I. Galametz, M.-H. Thorembey, C. Rayer, J.-F. Haas
Commissariat à l'Energie Atomique / Vaujours-Moronvilliers
B.P. 7, F 77181 Courtry, France

Abstract: We compare the behavior of continuous and discontinuous heavy-light interfaces between two gases, sulphur hexafluoride (SF_6)-air or xenon (Xe)-air, subjected to the acceleration by an incident upward propagating shock wave (Mach 1.45 in SF_6 and 1.4 in Xe) and the deceleration by up to three weaker reflected shock waves in a vertical shock tube. Continuous interfaces, obtained after the retraction of a separating diaphragm, have a smooth density profile with an initial thickness of 2.4 and 2.6 cm for SF_6-air and Xe-air respectively. Discontinuous interfaces are initially created by a plastic membrane 0.5 μm thick torn by the passage of the incident shock wave. The images are obtained by differential interferometry and, in the Xe-air case, by X-rays. The visualization illustrates the different development of shock-induced instabilities for the two types of interfaces, and different modes of reflected shock propagation in a polyatomic gas (SF_6) and monatomic gas (Xe). From the X-ray method, profiles of Xe partial density are obtained in the planar region of the mixing zones for the two types of interfaces at all times. In contrast, the differential interferometry provides partial-density profiles for continuous interfaces only and as long as the turbulent scales on the walls do not perturb the signature from the flow in the center. The continuous interfaces do not thicken and never turbulize. The discontinuous interfaces lead to turbulent mixing zones which thicken between reflected shocks. The shapes of average partial-density profiles are similar for the two types of interfaces, for the two gas pairs and, in the continuous case, for the two diagnostics.

Key words: Richtmyer-Meshkov instability, Turbulent mixing, Density measurement, X-rays, Differential interferometry

1. Experimental set-up and diagnostics

The vertical shock tube has been designed for the measurement of the Richtmyer-Meshkov (RM) instability of continuous heavy-light and discontinuous gas interfaces subjected to weak shock waves (Cavailler et al. 1992). It has a square cross section (8 by 8 cm^2) and the distance between the initial interface position and the end plate has been set to 30 cm in the experiments described here. For the shadowgraph or schlieren visualization of the mixing zones arising from the RM instability we normally use a Cordin high-speed framing camera. Images are recorded at times ranging from 1.1 ms, shortly before the arrival of the first reflected shock waves, to 2.7 ms, after passage of the third reflected shock (Fig. 1). For higher quality visualization and measurement of the density averaged along the optical axis, we have adapted to the gas mixing problem (with strong density gradients) the differential interferometry method developed for compressible aerodynamics (Desse & Pegnaux 1993). Figure 2 schematizes the optical set-up for the white-light differential interferometer. As it provides pictures in which the colors correspond to optical path differences proportional to the gradient of optical index, this set-up can be considered a quantitative schlieren system. As we are interested in mostly horizontal shock waves and mixing zones, the two Wollaston biprisms have the same angle of 1 or 0.5° and are rotated so that the vertical density gradient is measured. The single color image recorded per shot is digitalized using a triple CCD camera (for the colors red, green, and blue). The colors are then automatically matched to the optical index and density gradient. By integration from a position of known density, we deduce the partial density and mass concentration in the mixture. The X-ray technique (Rodriguez et al. 1994), for which only Xe significantly absorbs the low-energy ($\leq$ 60 keV) X-ray photons, has been used previously to obtain the Xe partial density. As both techniques integrate the three-dimensional flow field on a two-dimensional image, wall effects are always present and perturb the density measurement at late times.

2. Visualization

The mixing zones arising from the shock interaction with discontinuous (using thin membranes) or continuous (diffusive zone) interfaces for the gas pairs (Xe-air or SF_6-air) have been visualized using the differential interferometer and the X-rays. As X-ray pictures were presented previously (Rodriguez et al. 1995), we show only a few interferometric images for the case SF_6-air. Unfortunately they appear here in black and white.

Figure 3a shows the SF_6-air continuous interface recorded at time denoted 1 in the x-t diagram (Fig. 1). The discontinuous interface at time 1 is shown on Fig. 3b. One sees from top to bottom the downward-propagating reflected shock wave in air and the upward-moving mixing zone with SF_6 on the bottom. The horizontal layers in Fig. 3a indicate that the diffusion zone is still laminar and the grainy pattern below it is the signature of fine-scale mixing in the boundary layer. The contact surface seen on Fig. 3b displays a dark line which is the signature of the ruptured membrane. The turbulent mixing seen a little above and much further below probably lies along the walls because the membrane

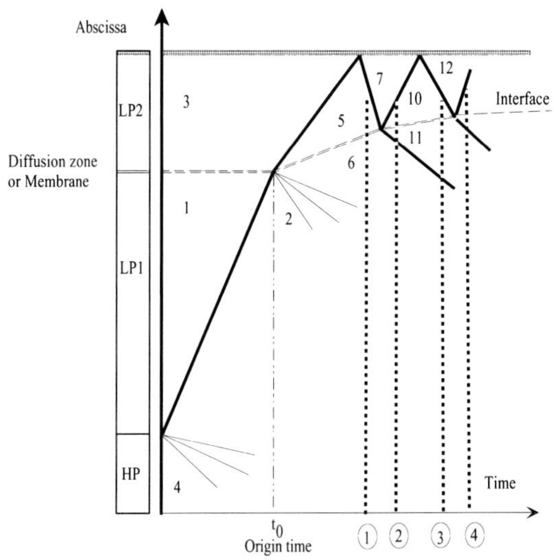

Figure 1. An x-t diagram of the experiment.

has ruptured there at shock passage. Far from the walls the membrane may be still intact and thus prevents gas mixing at this early stage. Figures 4a and 4b recorded at time 2 show the bifurcation of the reflected shock as it refracts into the polyatomic SF_6. A *bubble* can be seen on the walls between the shock and mixing zone which is locally deformed by the interaction. The diffusion zone has been compressed by the reflected shock and the discontinuous interface has been turbulized. Later, the wall bubble seems to turbulize even in the case of the continuous interface. The interferometer photo shown on Fig. 4c was recorded after passage of the second reflected shock on the continuous interface (time 4). The shock appears locally deformed by the wall structures of the mixing zone. It is in fact more disturbed in the discontinuous case (not shown) because the mixing zone is turbulent everywhere. Below this second reflected shock, the turbulent wall layer due to the first reflected shock can be seen. The continuous mixing zone is still laminar but the colored layers tend to be erased by the signature of the small scales on the wall. The measurement of the density profile becomes difficult at this stage.

3. Density profiles

Because the random optical signature of the small scales of turbulent mixing on the walls complicates the data-reduction process, the differential-interferometry method can be used only to obtain concentration and partial-density profiles

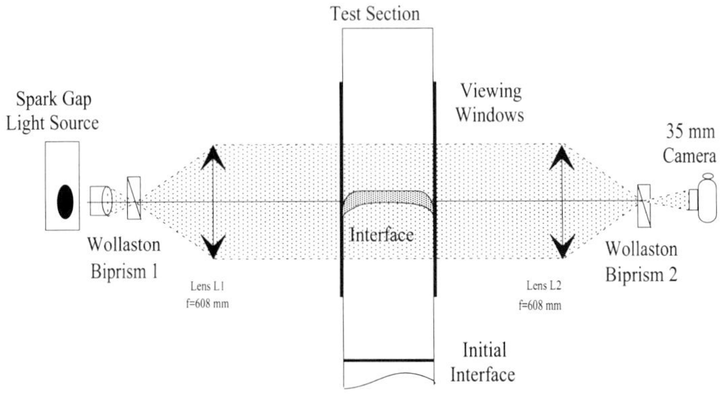

Figure 2. Optical set-up for the differential interferometer.

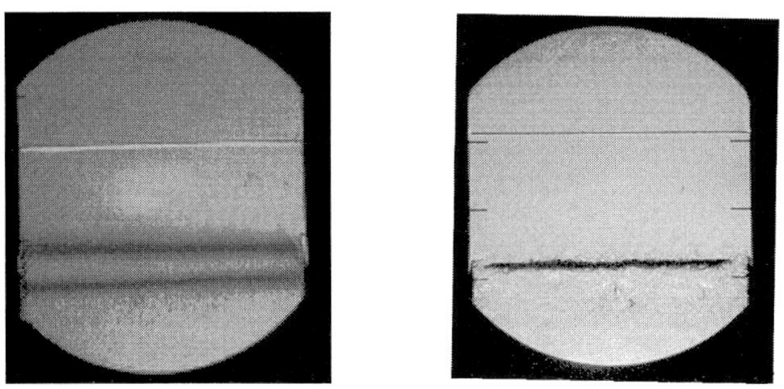

(a) (b)

Figure 3. SF_6-air interface time 1: (a) continuous; (b) discontinuous.

for the continuous mixing zone until time 3 (just before the interaction with the second reflected shock). An additional difficulty in the case of discontinuous interfaces is the pertubating signature of the membrane at early time or its fragments at late time.

Figure 5 shows the partial-density profiles of SF_6 in the SF_6-air mixing zone at times 1, 2, and 3. The profiles are typical of the molecular diffusion process for which the concentrations follow the error function law. Between times 1 and 2, the profiles are steepened due to the hydrodynamic compression by the first reflected shock. There is no widening, however, between times 2 and 3 indicating that no rapid diffusion has happened in this time interval. This is consistent with the absence of turbulent scales in the core of the mixing zone. A comparison of the two diagnostics is possible for the continuous Xe-air mixing zone. The partial densities of Xe (not shown) at times 2 and 3

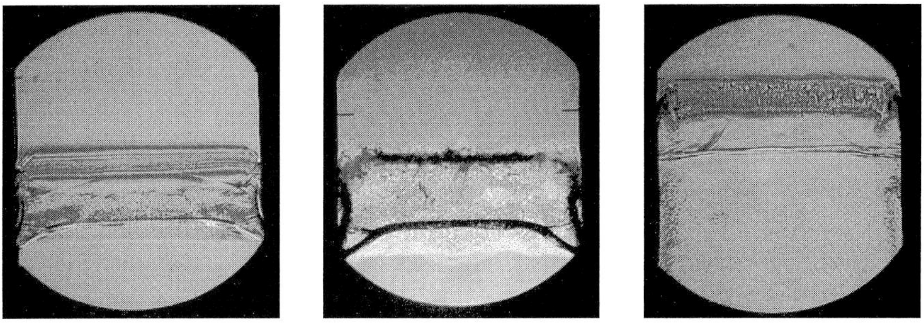

(a) (b) (c)

Figure 4. SF_6-air interface: (a) continuous time 2; (b) discontinuous time 2; (c) continuous time 4.

(between the passage of the first and second reflected shock wave) are similar for the two methods. The X-ray method could be used for the discontinuous Xe-air interfaces, because the membrane (made of low atomic-number elements) is transparent to the X-ray photons and because the absorption process integrates the random scales along the optical axis to yield an optical density on the radiographic film. Figure 6 shows the partial Xe density profiles at times 1, 2, 3, and 4. The profiles widen between times 2 and 3, in accordance with a process of turbulent diffusion.

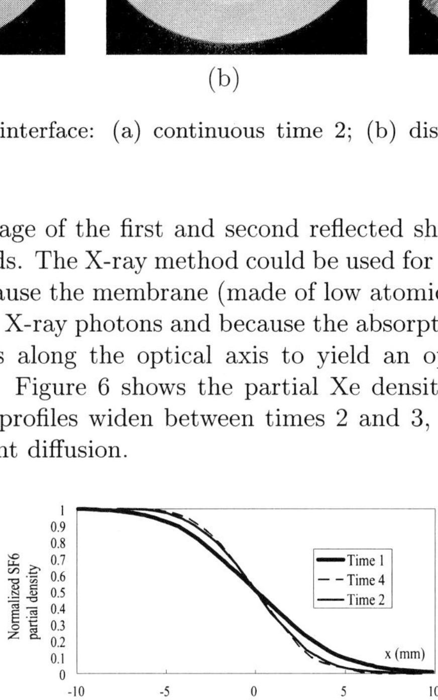

Figure 5. Density profiles for SF6-air continuous interface.

4. Conclusions

With differential interferometry, high-quality visualization of the mixing zones subjected to shock interaction can be obtained, as with conventional schlieren techniques. Our method has provided density measurements for mixing zones which remain essentially controlled by molecular diffusion and as long as turbulent scales on the walls do not perturb the optical signature, i.e. for continuous interfaces until the arrival of the second reflected shock. The results are similar to the radiography and applicable to any gas pairs. The partial-density pro-

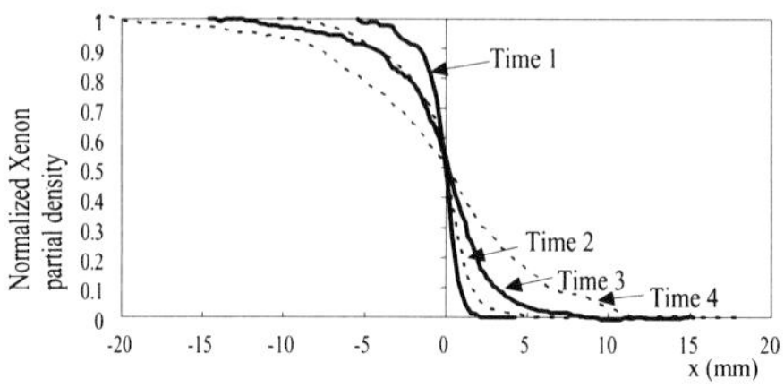

Figure 6. Density profiles for Xe-air discontinuous interface.

files can be used to determine a mixing-zone thickness based on quantitative criteria. In the present experimental conditions with a slow blade retraction for the continuous interface, the mixing zone never turbulizes and its thickness does not increase. The mixing zones created by the shock interaction with a discontinuous interface thicken due to turbulent diffusion and become wider than the continuous interface after the second reshock. There is no significant shape difference between the twice-averaged density profiles of planar diffusive or turbulent mixing zones.

Acknowledgement. We thank G. Delouis for his precious help in the laboratory.

References

Cavailler C, Croso H, Haas JF, Rodriguez G (1992) A new vertical shock shock tube for Rayleigh-Taylor instability studies. In: Proceedings of the 3rd International Workshop on the Physics of Compressible Turbulent Mixing, Royaumont, Wilke N (ed), CEA/DAM, p 27.

Desse JM, Pegneaux JC (1993) Direct measurement of the density field using high speed differential interferometry. Exp in Fluids 15:452.

Rodriguez G, Galametz I, Croso H, Haas JF (1994) Richtmyer-Meshkov instability in a vertical shock tube. In: Proceedings of the 4th International Workshop on the Physics of Compressible Turbulent Mixing, Linden PF, Youngs DL, Dalziel SB (eds), Cambridge University Press, Cambridge, England, p 260.

Rodriguez G, Galametz I, Croso H, Haas JF (1995) Richtmyer-Meshkov instability in a vertical shock tube. In: Shock Waves @ Marseille, Brun R, Dumitrescu L (eds), Proceedings of the 19th International Symposium on Shock Waves, Springer Verlag, Heidelberg, Vol 4, p 275.

Interaction of a turbulent shear flow with a Mach 1.5 normal shock wave

K. Mahesh, S. K. Lele, P. Moin
Department of Mechanical Engineering
Building 500, Stanford University
Stanford, CA 94305 USA

Abstract: Results from direct numerical simulation of the interaction of a Mach 1.5 normal shock with a turbulent shear flow are presented. Interaction with the shock is seen to increase the level of turbulent kinetic energy and thermodynamic fluctuations in the shear flow. The Reynolds shear stress, $\overline{uv}$ is observed to decrease across the shock wave. The observed trends are in agreement with linear analysis in which the influence of upstream anisotropy and entropy fluctuations on the evolution of the turbulence was considered.

Key words: Shock/turbulence interaction, Direct numerical simulation, Linear analysis, Turbulent kinetic energy, Reynolds shear stress, Thermodynamic fluctuations

1. Introduction

Interaction with a shock wave is known (Fernholz and Finley 1981) to alter significantly the mean flow and turbulent fluctuations in a turbulent boundary layer. In recognition of the complex nature of shock wave/boundary layer interaction, recent experiments and computations have studied simpler problems such as the interaction of isotropic turbulence and free shear layers with shock waves. This paper presents results from direct numerical simulation of one such problem: the interaction of a normal shock with an idealized shear flow.

A schematic diagram of the computed flow is shown in Fig. 1. Note that the shock wave is stationary in the mean. The mean velocity at the inflow of the computational domain varies linearly across streamlines while the mean pressure is uniform. The mean density and temperature vary such the mean Mach number is uniform, i.e.

$$\overline{U}_1(y) = U_0 + S(y - y_{\min}), \qquad \overline{P}_1(y) = \frac{1}{\gamma}, \qquad \overline{T}_1(y) = \frac{\overline{U}_1^2(y)}{(\gamma - 1)M_1^2}. \tag{1}$$

The results presented in this paper correspond to a Mach 1.5 shock wave, i.e., $M_1 = 1.5$.

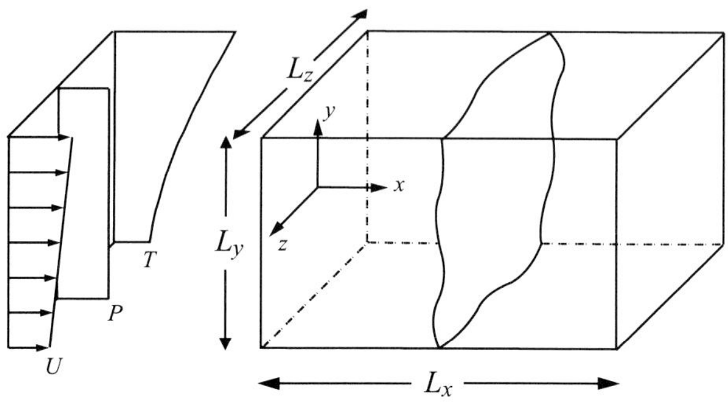

Figure 1. Schematic diagram of the computational setup.

2. Numerical method

The governing equations are the three-dimensional compressible Navier-Stokes equations in conservative form. The computational mesh is uniform in the directions transverse to the shock wave. A non-uniform mesh is used in the streamwise direction to cluster points in the vicinity of the shock. A combination of the sixth-order Padé scheme (Lele 1992) and the sixth-order ENO scheme (Shu and Osher 1989) is used to compute spatial derivatives. In order to reduce the computational cost, the shock-capturing scheme (ENO) is only applied in the shock-normal direction in the vicinity of the shock wave. Time advancement is performed by the third-order Runge Kutta scheme (Wray 1986). Periodic boundary conditions are imposed at the z boundaries while approximately non-reflecting boundary conditions are specified at the y boundaries. Non-reflecting boundary conditions (Poinsot and Lele 1990) are specified at the outflow boundary while turbulent fluctuations are superposed onto the mean flow of Eq. (1) at the inflow boundary.

These turbulent fluctuations are obtained from a separate temporal simulation. The mean flow in the temporal simulation is given by Eq. (1). The initial turbulent field is a superposition of a solenoidal velocity field and entropic thermodynamic field. The initial velocity field is isotropic with $E(k) \sim k^4 \exp(-2k^2/k_0^2)$. The initial pressure fluctuations are set to zero while the initial density fluctuations are set proportional to $-v$; i.e., the initial correlation coefficient between ρ and v equals -1. This ensures that the correlation between u and T will be negative (as in a boundary layer) for a transient period of time. The initial field is advanced in time until the velocity field attains the anisotropy that is typical of a turbulent shear flow. An instantaneous field is then taken and used to generate inflow turbulence in the spatial computation.

3. Results

Table 1 lists the relevant non-dimensional parameters that characterize the upstream turbulence. All quantities are measured at the centerline immediately upstream of the shock. The turbulence upstream of the shock has a low Reynolds number ($R_\lambda = u_{\rm rms}\lambda_1/\nu$), very low fluctuation Mach number ($M_t = \sqrt{\overline{u_i u_i}}/\bar{c}$), and anisotropic Reynolds stresses. The intensity of pressure fluctuations is an order of magnitude lower than those of the density and temperature. Also, the correlation coefficient between the density and temperature fluctuations is -0.99. The upstream turbulence therefore is essentially composed of an anisotropic velocity field along with entropy fluctuations of comparable intensity. The entropy fluctuations are correlated with the velocity field such that the correlation coefficient between u and T is negative.

Table 1. Relevant non-dimensional parameters of the upstream turbulence.

R_λ	M_t	$\overline{u^2}/q^2$	$\overline{v^2}/q^2$	$\overline{uv}/q^2$	$q_{\rm rms}/U_1$	$\rho_{\rm rms}/\overline{\rho_1}$	$T_{\rm rms}/\overline{T_1}$	$p_{\rm rms}/\overline{p_1}$
29.0	0.03	0.53	0.23	-0.22	0.025	0.03	0.03	0.002

Statistics reveal the jump in the mean flow across the shock wave to be very close to that predicted by the Rankine-Hugoniot relations for a laminar shock. For example, the jumps in mean velocity, density, and temperature at the centerline are 0.54, 1.86, and 1.31 respectively, while the corresponding values for a laminar shock ($\gamma = 1.4$) are 0.54, 1.86, and 1.32.

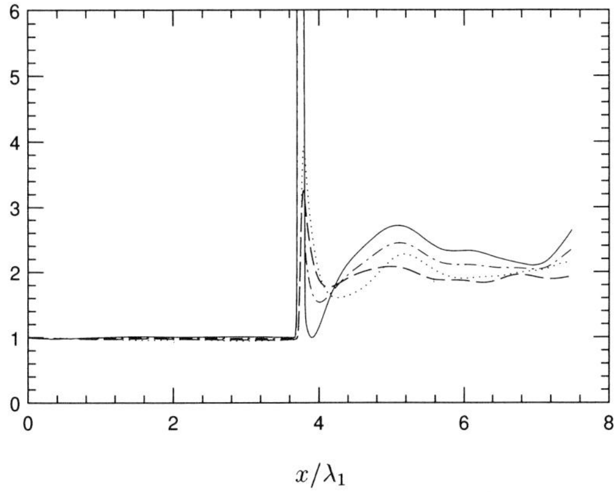

Figure 2. Variation of kinetic energy along the center streamline. All curves are normalized by their values at the inflow; —— $(\overline{u^2})$, ---- $(\overline{v^2})$, ········ $(\overline{w^2})$, — — — (q^2).

3.1. Turbulent kinetic energy

Interaction with the shock wave significantly enhances the kinetic energy of the turbulent fluctuations. The evolution of kinetic energy along the centerline is shown in Fig. 2. Note that λ_1 denotes the streamwise Taylor microscale at the inflow. The high levels in the vicinity of the shock wave are due to intermittency caused by motion of the shock front. Rapid increase in kinetic energy is observed immediately downstream of the shock wave. Similar behavior is observed in the interaction of a shock with isotropic turbulence (Lee et al. 1994). Linear analysis of the isotropic turbulence problem (Mahesh et al. 1995a) reveals rapid conversion of potential energy $(\overline{p^2})$ to turbulent kinetic energy immediately behind the shock, leading to the rapid rise in kinetic energy.

Using downstream values outside the region of rapid variation, $\overline{u^2}, \overline{v^2}$, and $\overline{w^2}$ are amplified by 2.3, 1.9, and 2 respectively, across the shock. These levels of amplification in the shear flow are noticeably higher than that of isotropic turbulence. Linear analysis (Ribner 1954) predicts $\overline{u^2}, \overline{v^2}$, and $\overline{w^2}$ to be amplified by 1.6, 1.38, and 1.38 across a Mach 1.5 shock if the upstream turbulence is isotropic. The higher levels of amplification observed in the shear flow are attributed to two factors: upstream anisotropy and upstream entropy fluctuations with negative $\overline{uT}$.

Incompressible Rapid Distortion Theory was used by Mahesh et al. (1994) to study the influence of upstream anisotropy. The analysis predicted higher levels of kinetic energy amplification across a shock wave in a shear flow than in isotropic turbulence. The relevant parameters responsible for this trend were identified as upstream $E_{11}(k_1)$ and $\overline{u^2}/q^2$ (note that x was assumed to be the shock-normal direction).

Table 1 reveals the presence of significant level of upstream entropy fluctuations with negative correlation between u and T. Linear analysis of the influence of upstream entropy fluctuations (Mahesh et al. 1995b) shows that this would further enhance the amplification of kinetic energy. A preliminary computation of a Mach 1.2 shock wave exhibited a similar strong influence of entropy fluctuations. The sign of $\overline{uT}$ in the Mach 1.2 simulation varied across streamlines upstream of the shock. Kinetic energy on the streamlines with negative upstream $\overline{uT}$ was seen to be amplified by a factor of about 2 while no noticeable change in kinetic energy was observed on the streamlines where $\overline{uT}$ was positive.

3.2. Reynolds shear stress

While kinetic energy is amplified, the Reynolds shear stress, $\overline{uv}$ is seen to *decrease* in magnitude across the shock. The streamwise variation of $\overline{uv}$ and $\overline{uv}/q^2$ on the centerline is shown in Fig. 3. The decrease in magnitude of shear stress

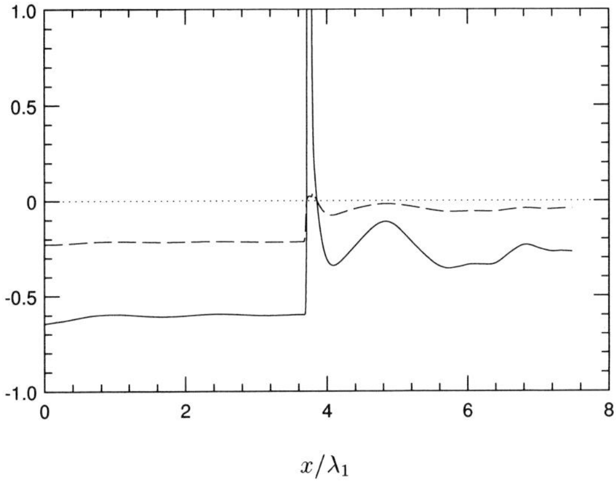

Figure 3. Evolution of shear stress along the center streamline; $\overline{uv}$ is normalized by the value of $u_{\rm rms}v_{\rm rms}$ at the inflow. —— ($\overline{uv}$), ---- ($\overline{uv}/q^2$).

is apparent. This observation is in agreement with the predictions of Mahesh et al. (1994) using Rapid Distortion Theory. Both normal and oblique shocks were considered in this study, which predicted a decrease in shear stress across normal shocks and a slight increase across oblique shocks at sufficiently large angles of obliquity.

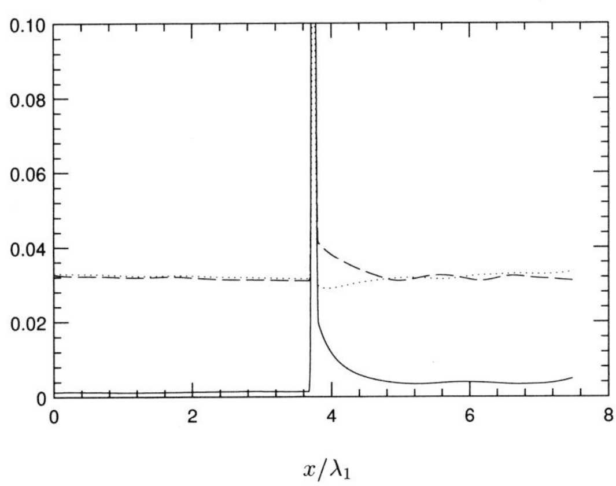

Figure 4. Variation of thermodynamic fluctuations along the centerline; —— ($p_{\rm rms}/\gamma\overline{P}$), ---- ($\rho_{\rm rms}/\overline{\rho}$), ········ ($T_{\rm rms}/\overline{T}$).

3.3. Thermodynamic fluctuations

Thermodynamic fluctuations are seen to be amplified across the shock wave. Figure 4 shows the streamwise evolution of the *intensities* of density, pressure, and temperature. Note that pressure fluctuations are amplified across the shock wave and rapidly decay downstream. This decay is similar to that observed in the interaction of isotropic turbulence with a shock and is inviscid in character. The intensity of pressure fluctuations in the near-field is comparable to those of density and temperature while far-field levels are much smaller. Furthermore, the correlation coefficient between density and temperature in the far-field is nearly -1, i.e., while acoustic fluctuations are significant immediately behind the shock wave, entropy fluctuations dominate in the far-field.

Acknowledgement. This work was supported by the Air Force Office of Scientific Research under Contract F49620-92-J-0128 with Dr. Leonidas Sakell as technical monitor. Computer resources were provided by NASA Ames Research Center.

References

Fernholz FF, Finley PJ (1981) A Further Compilation of Compressible Boundary Layer Data with a Survey of Turbulence Data, AGARDograph 263.

Lee S, Lele SK, Moin P (1994) Interaction of Isotropic Turbulence with a Strong Shock Wave, AIAA Paper 94:0311.

Lele SK (1992) Compact finite difference schemes with spectral-like resolution. J Comput Phys 103:16.

Mahesh K, Lele SK, Moin P (1994) The response of anisotropic turbulence to rapid homogeneous one-dimensional compression. Phys Fluids 6:1052.

Mahesh K, Lee S, Lele SK, Moin P (1995a) The interaction of an isotropic field of acoustic waves with a shock wave. J Fluid Mech 300:383–407.

Mahesh K, Lele SK, Moin P (1995b) The Interaction of a Shock Wave with a Turbulent Shear Flow, report in preparation, Thermosciences Division, Mechanical Engineering Department, Stanford University.

Poinsot TJ, Lele SK (1990) Boundary conditions for direct simulation of compressible viscous reacting flow. J Comput Phys 101:104.

Ribner HS (1954) Shock-turbulence Interaction and the Generation of Noise, NACA TN 3255.

Shu CW, Osher S (1989) Efficient implementation of essentially non-oscillatory shock-capturing schemes, II. J Comput Phys 83:32.

Wray AA (1986) Very Low Storage Time-advancement Schemes, Internal Report, NASA Ames Research Center, Moffett Field, CA.

Waves, General

Converging shock waves: modeling and analysis of structures

H. Wallus, F. Demmig
Universität Hannover
Institut für Plasmaphysik
Callinstraße 38 D-30167 Hannover, Germany

Abstract: We developed a model for the description of converging shock waves, especially for studying the process of fragmentation that occurs during Mach reflection. Nonequilibrium processes like vibrational, dissociative, and ionizational relaxation have been taken into account. In order to give a detailed physical interpretation of specific experimental results, a variety of flow-visualization techniques have been utilizied. A number of numerical methods have been applied to compute converging shock waves.

Key words: Converging shock waves, Time-resolved visualization, Model computation, Pattern recognition, Fragmentation wave system

1. Snapshot model of converging shock waves

Two features due to experimental conditions affect investigations of converging cylindrical shock waves: the shock contour is not perfectly symmetrical and the focus of convergence is not exactly on the symmetry axis of the experimental set up. This can be seen most clearly from time-resolved experiments (Kleine 1986). Our two-dimensional model computations aim to describe shock-wave propagation and reflection, with special attention to the process of fragmentation which occurs as a consequence of the initial shock-front deformation, i.e., its deviation from an ideal circular form. In order to give a detailed interpretation of specific time-resolved experimental results we obtained the initial shock configuration from the first snapshots as initial configuration of the shock entering the test section of the shock tube. The computation then yields snapshots of the converging shock wave, which can be evaluated in the form of isopycnics, shock-front contours, etc., for direct comparison with the corresponding experimental snapshots. As has been explained in more detail in Demmig et al. (1995) the experimental records are digitized, and then are represented by a curve fit. Usually, the first two photographs of the shock just having entered the test section of the tube provide the main information for the formulation of the initial conditions, namely the angular position $r_S(\phi)$ of the shock and the shock Mach number $M_S(\phi)$.

2. Numerical methods

Our preferred method is the UNO scheme (Harten and Osher 1987). In order to reduce computional effort it is desirable to have a fine mesh about the center of convergence and a wide mesh in the rather uniform region behind the initial shock. This meshing can be achieved by the use of polar coordinates quite well, although the coordinate singularity at $r = 0$ has to be treated suitably. For symmetrical shocks, the singularity can be avoided by inserting a thin axis around $r = 0$, which, in this case, is the convergence point. For unsymmetrical shocks the convergence point is offset from the origin, hence the coordinate system has to be modified. Meier (1994) found that the least-complicated solution was to shift the grid by half of a radial step size along a specific diameter, such that the operator functions need not to be evaluated at $r = 0$.

A finite-element method (FEM), which has the advantage of using unstructured grids, is under construction. The matrices of the FEM equation are calculated analytically using bilinear elements with three nodes. A third-order dissipative term is added to suppress oscillations behind the shock front. The λ-scheme of Moretti (1979) might turn out to be a good compromise between difference schemes and methods of bicharacteristics, which are being developed by Meier (1993). A very promising first step toward utilizing the λ-scheme in computations of converging shock waves has been done by vom Endt (1995).

3. Flow visualization

Converging shock waves in molecular gases develop rather complex flow features, especially if nonequilibrium processes are appreciable. This complexity is evident when shock fragmentation occurs resulting in a Mach-reflection configuration and the formation of additional structures after reflection. Without suitable flow visualization aimed at the specific features in question, a physical interpretation is not possible. In addition to the widely used contour maps and 3D plots of relevant physical quantities, we utilized and also developed flow-visualization methods that belong to pattern-recognition techniques. (Demmig and Petersen 1991). The latter allow us to locate and depict the fragmentation wave system that consists of shocks, compression waves, and slip lines. Sometimes it is useful to combine related physical quantities in one plot, as e.g., in Figs. 1c, 2c, and 3c, where the arrows represent the velocity field and their gray tone represents the value of the local energy density.

The process of convergence and reflection of an initially slightly deformed shock wave is illustrated by Fig. 1 through 3 in three large time steps. The isopycnics of Fig. 1a already indicate shock fragmentation. As a consequence of the disturbed three-fold symmetry of the original experimental shock contour, the fragmentation wave system found by the pattern-recognition program looks irregular. In Fig. 1b, full circles denote shock-front points, open circles denote compression waves, and grey points denote slip lines. As can be seen also from Figs. 1a and 2c, the fragmentation is more pronounced in the lower part. Figure 2 shows the shock wave immediately before reflection. Clearly, the convergence 'point' does not coincide with the center of the observation window and the wave system is considerably distorted. The complexity of the flow field behind the reflected shock manifests itself in Fig. 3b. However, in conjunction with Figs. 3a and 3c, a closer interpretation can be given, since behind the reflected shock we observe vortices that originate in the fragmentation wave system during reflection. For comparison, Fig. 4 displays isopycnics from a model computation and an enlarged picture of an experimental snapshot taken at the same instant. Though the agreement of the minute structures cannot be expected to be complete, it is surprisingly good. The model-based computations can obviously facilitate the interpretation of experimental snapshots.

4. The effect of nonequilibrium processes

The propagation of converging shock waves in molecular gases can be affected by nonequilibrium processes like vibrational and dissociative relaxation. The temperature reached in the region near convergence depends on the initial Mach number of the shock wave. Consequently, the fragmentation wave system and, more distinctly, the post-reflection phenomena, will show the influence of nonequilibrium processes. Whether the nonequilibrium effects can be observed depends on the experimental resolution, because the converging shock eventually strengthens and the relaxation time may approach the time characterizing the variation of hydrodynamical quantities. Although vibrational relaxation does not play a decisive role in experiments with Mach number ~ 2 as have been carried out, e.g., by Takayama et al. (1987), an influence of vibrational and dissociation relaxation is already visible for shocks with initial Mach numbers of $2 - 3$ (Klingspohn 1994). The formation of the fragmentation wave system will be markedly affected for shocks with initial Mach numbers in the range of $4 - 5$. If oxygen is taken as the test gas instead of air these phenomena can be observed already for shocks of Mach 2. The analysis of nonequilibrium effects in real gases can be supported by plots of relaxation variables like the vibrational energy (e.g., in Fig. 5, a gray-shade plot of fractional vibrational energy). The initial shock contour had been chosen to have four-fold symmetry. Figure 5, taken well after reflection, clearly shows regions of increased vibrational energy, which underline the shape of the above-mentioned vortices. The fractional shock front that was determined by pattern-recognition methods is represented by a broken black line. Similar plots can be used in the study of ionizational relaxation, which is neglible) for converging shock waves with an initial Mach number smaller than 6.

5. Conclusion and outlook

Our concept of using the instantaneous shape and velocity of a converging shock wave to predict its future propagation including fragmentation and reflection has been elaborated further through a number of computational and flow-visualization tools. A detailed physical interpretation of features like the fragmentation wave system and structures behind the reflected shock can be given. The model could be refined, if more experimental data were avaible, especially regarding the downstream condition of the incoming shock. We recommend experiments with converging shock waves with initial Mach numbers in the range of 3 – 7, where nonequilibrium processes play a role. These experiments are worthwhile from both basic physical and applied standpoints. A fully established fragmentation wave system is a prerequisite for the study of the physical aspects of this interesting shock-wave flow.

References

Demmig F, Grönig H, Kleine H, Wallus H (1995) Experiments and model computation of cylindrical shock waves with time-resolved deformation and fragmentation. In: Shock Waves @ Marseille, R Brun, L Dumitrescu (eds).

Demmig F, Petersen J (1991) Two-dimensional model computation of converging shock wave propagation, fragmentation, and reflection accounting for nonequlibrium processes. In: Int Workshop Strong Shock Waves, Chiba, pp 221–257.

vom Endt A (1995) Dip. Thesis, Ein Differenzenverfahren mit Stoßfronteinpassung und dessen Anwendung zur numerischen Berechnung konvergenter Stoßwellen in Molekülgasen, U Hannover.

Harten H and Osher S (1987) Uniformly high-order accurate nonoscillatory schemes. SIAM J Numer Anal 24:2:279–309.

Kleine H (1986) Dip. Thesis, Kurzzeitphotographie der Fokussierung zylindrischer Stoßwellen, RWTH Aachen.

Klingspohn M (1994) Dip. Thesis, Ausbreitung, Fragmentation und Reflexion zweidimensionaler konvergenter Stoßwellen in Molekülgasen, U Hannover.

Meier J (1994) Dip. Thesis, Konvergente Stoßwellen im Bereich der Fragmentation und Reflektion, U Hannover.

Meier U (1993) Dip. Thesis, Die Lösung instationärer Stoßwellenprobleme mit zweidimensionalen Charakteristikenverfahren, U Hannover, and unpublished papers.

Moretti G (1979) The λ-scheme. Computers and Fluids 7:191–205.

Takayama K, Kleine H, Grönig H (1987) An experimental investigation of the stability of converging cylindrical shock waves in air. Exp Fluids 5:315–322.

Wallus H (1993) Dip. Thesis, Modellrechnungen zur Ausbreitung konvergenter Stoßwellen auf der Grundlage zeitaufgelößter Experimente, U Hannover.

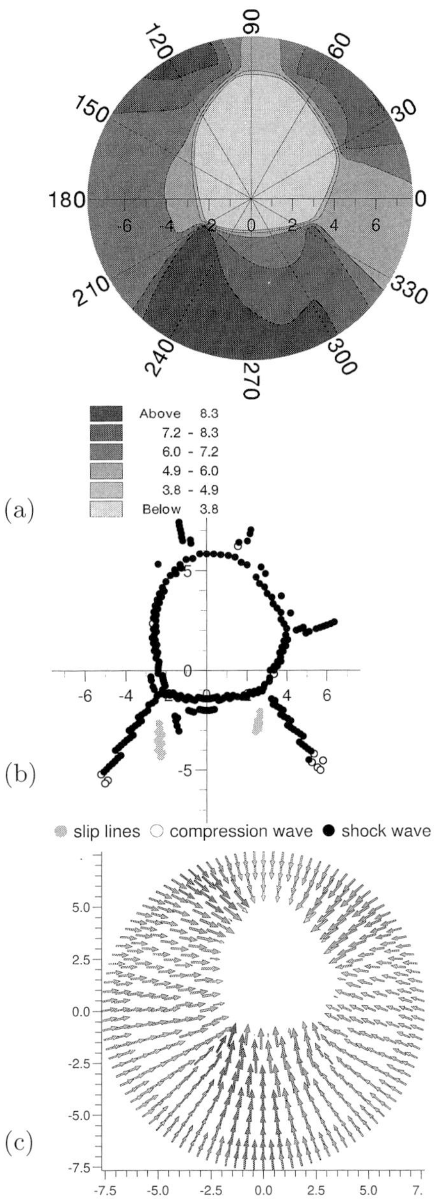

Figure 1. Isopycnics, fragmentation wave system, and velocity field of converging shock before reflection.

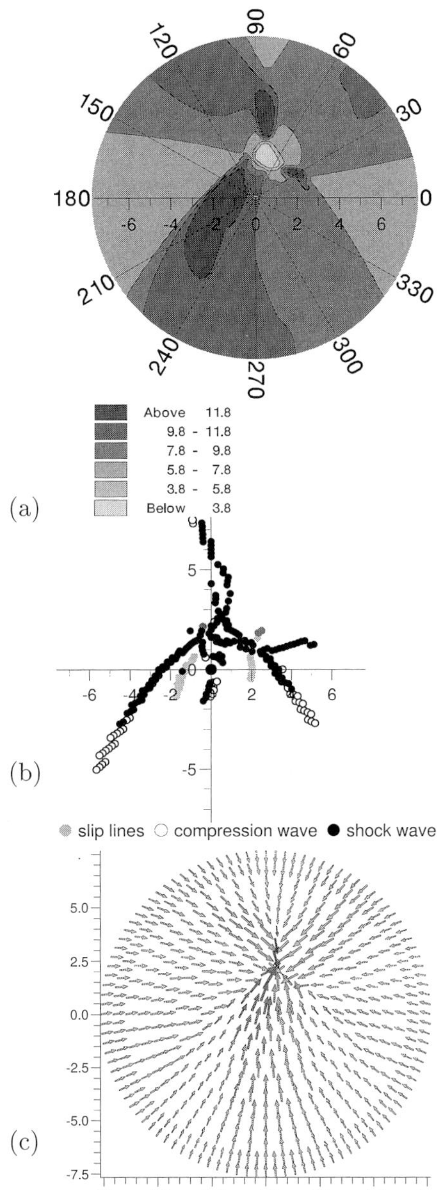

Figure 2. Cf. Fig. 1. Immediately before shock reflection.

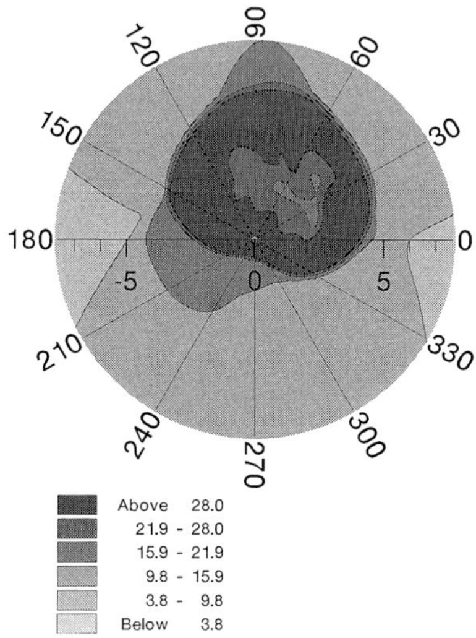

(a)

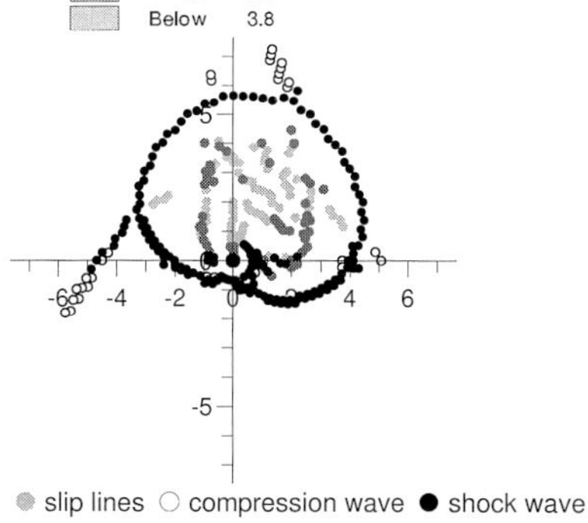

(b)

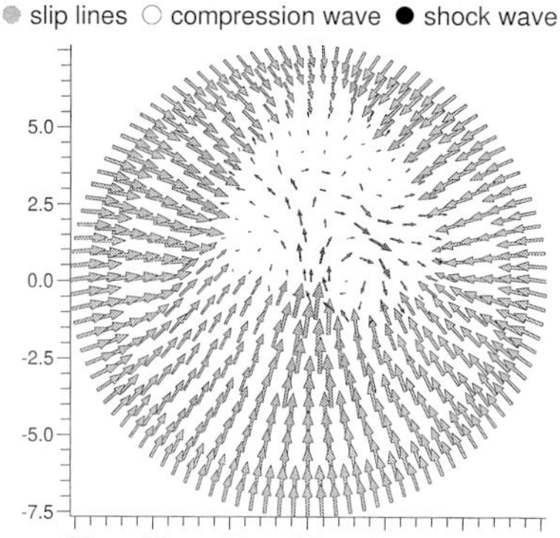

(c)

Figure 3. Cf. Fig. 1. After reflection. Vortices fully established.

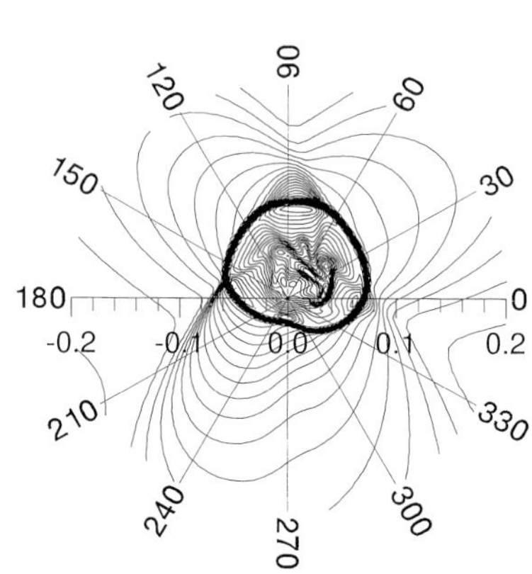

(a)

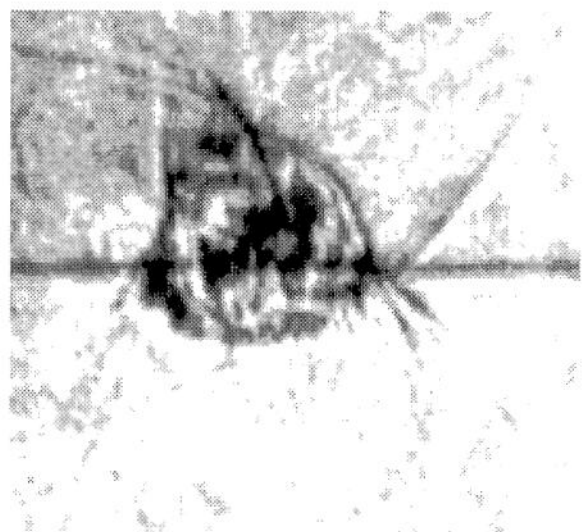

(b)

Figure 4. Isopycnics of model computation and enlarged experimental photo. After reflection.

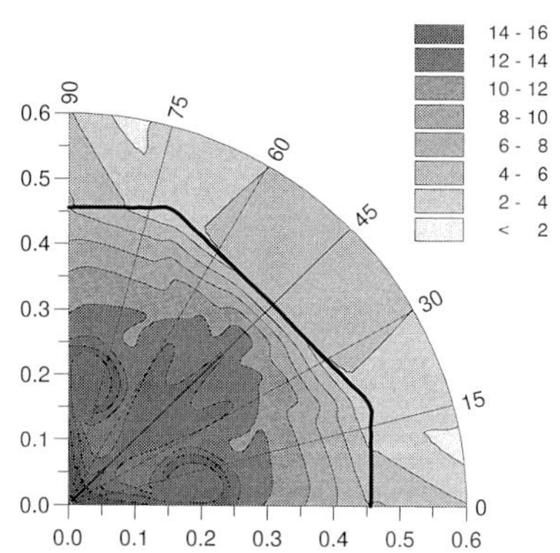

Figure 5. Isoline of vibration energy (in percent). M_S=4. After reflection.

Propagation and stability of strong shocks in gases

M. Mond, I. Rutkevich, E. Toffin
Pearlstone Center for Aeronautical Engineering Studies
Department of Mechanical Engineering
Ben-Gurion University of the Negev, Beer Sheva, Israel

Abstract: The steady state as well as the stability properties of strong shocks propagating in cold monatomic gases are investigated. Excitation of the electronic levels of the atoms as well as ionization processes behind the shock are taken into account in constructing the Hugoniot curve and in reformulating the shock's stability conditions. A parametric study is carried out in order to find domains in parameter space in which shocks become unstable by spontaneously emitting sound- and entropy-vortex waves. Available experimental results agree reasonably well with the theoretical calculations.

Key words: Ionization, Excitation, Shock instability

1. Introduction

Shock waves with high Mach numbers provide a rich source of phenomena and problems of fundamental scientific nature. As the shock's Mach number rises so does the importance of various inelastic processes occurring behind it. In the current work, attention is focused on excitation of the electronic energy levels of the atoms and on ionization. These two processes result in lowering the temperature of the gas behind the gas-dynamical shock and in its subsequent compression beyond the asymptotic classical limit. The steady-state properties of the strong shocks are studied by investigating their Hugoniot curves. It is shown that these latter possess a maximum that represents the maximal degree of compression that is achievable by a shock for given initial conditions.

A deep interest in the problem of shock instability rose during recent years due to some experimental evidence of irregular flows behind strong shocks (Griffiths et al. 1976, Glass and Liu 1977; Glass et al. 1978). The problem of shock instability ranks together with the well-known Kelvin-Helmholtz and Rayleigh-Taylor instabilities as one of the fundamental problems of fluid mechanics. The criteria for shock instability were first obtained by Dyakov (1954) and until recently their straightforward application failed to explain the unstable behavior behind strong shocks. Mond and Rutkevich (1994) have pointed out the importance of the ionization kinetics and, accordingly, modified the criterion for spontaneous emission (SE) of acoustic as well as entropy-vortex waves from shocks. The modified criterion resulted in thresholds for SE that are in reasonable agreement with available experimental data. In the current work, the effects of excitation of the electronic energy levels of the atoms on the steady-state and stability properties of strong shocks are examined. In addition, a parametric study is carried out in order to define the SE domains in the space of parameters like the shock Mach number and the initial pressure of the gas.

2. Steady-state properties of strong shocks in gases

A planar shock wave is considered which propagates with a constant velocity in a cold monatomic gas. It is convenient to describe the shock in a frame of reference that is moving with the shock's normal velocity. As the shock's Mach number rises so does the temperature behind it. Subsequently, excitation of the electronic-energy levels of the atoms as well as their single and multiple ionizations become progressively important as the temperature behind the shock increases to higher and higher values. The classical gas-dynamical shock, with its sharp surface of discontinuity (whose thickness is of the order of a few molecular free-path lengths) gives way to a more complex flow pattern. Thus, the steady-state flow field is divided into four distinct areas: (1) The region ahead of the shock in which the gas is in a cold equilibrium. Due to the low equilibrium temperature in this region it is assumed that all the atoms are in their ground state (region 1). (2) The gas-dynamical shock wave. (3) The region along which the excitation and ionization processes take place, accompanied by a drop in the temperature and a consequent increase of the density. This region is called the relaxation zone (RZ). (4) The region in which excitation and ionization equilibrium is maintained (region 2). The number of free electrons as well as the distribution of excited states reach a constant value in this region. It should be noted that the flow variables in this region are only approximately constants due to the radiation from the hot gas. However, changes caused by radiation occur over a length scale that is much longer than the RZ so the effect of radiation may be neglected.

In the current work, a unified picture is employed in which the gas-dynamical shock together with the RZ constitute an effective surface of discontinuity which is treated as a single unified shock wave. The properties of such shocks and, in particular, their Hugoniot curves are studied by writing the conservation equations that relate both constant-state sides (i.e., regions 1 and 2). The continuity equation and momentum equation have the same form as those for a classical gas-dynamical shock. The energy equation, however, is modified due to the excitation and ionization processes and is

$$e_1 + \frac{p_1}{\rho_1} + \frac{v_1^2}{2} = e_2 + \frac{p_2}{\rho_2} + \frac{v_2^2}{2} + Q_{ion} + Q_{exc} \, , \tag{1}$$

where ρ, v, p, and e are the density, velocity, pressure, and internal energy per unit mass of the gas, respectively, and Q_{ion} and Q_{exc} are the the amounts of energy lost by the gas per unit mass due to ionization and excitation of electronic levels, respectively. The indices 1 and 2 denote the values of the parameters ahead of the shock (region 1) and behind it (region 2), respectively. For a monatomic gas the internal energy per unit mass is

$$e = \frac{p}{(\gamma - 1)\rho}, \tag{2}$$

where γ is the specific heat ratio. The energy-loss terms are

$$Q_{ion} = \frac{NI\alpha}{\mu}, \quad Q_{exc} = \frac{N\bar{E}_a(1-\alpha)}{\mu}, \tag{3}$$

where N is the Avogadro number, μ is the atomic weight of the gas, I is the ionization potential, $\bar{E}_a$ is the average electronic energy level, and $\alpha = n_e/(n_i + n_a)$ is the ionization degree, where n_e, n_i, and n_a are the concentrations of free electrons, ions, and atoms, respectively. The total pressure is determined by the sum of the partial pressures of all the species,

$$p = kT(n_e + n_i + n_a). \tag{4}$$

The average electronic level is calculated according to Boltzmann statistics that yield

$$\bar{E}_a(T) = \frac{1}{D_a} \sum_j g_a^j E_a^j \exp(-E_a^j/kT), \tag{5}$$

where D_a is the atom partition function, and g_a^j and E_a^j are the degeneracy and the value of the jth electronic energy level in the atoms. The partition function is

$$D_a = \sum_j g_a^j \exp(-E_a^j/kT). \tag{6}$$

The ionization degree, α, is calculated through the Saha equation that takes the form

$$\frac{\alpha^2}{1-\alpha} = \frac{G}{n_i + n_a} \left[\frac{m_e k T_2}{2\pi\hbar^2}\right]^{3/2} e^{-I/kT_2}, \tag{7}$$

where $G = 2D_i/D_a$, m_e is the electron's mass, and $\hbar$ is Planck's constant. The quantity D_i is the partition function of the single-charge ions and is calculated in an equivalent manner to Eq. (6). Defining the dimensionless parameters $\theta = T_2/T_1$, $\eta = \rho_2/\rho_1$, and $P = p_2/p_1$, the equations that describe the Hugoniot curve, i.e., $\eta(P)$, may be cast into the implicit parametric form

$$\eta(\theta) = \frac{1 + \Gamma P(\theta)}{\Gamma + P(\theta) - \varsigma\alpha(\eta(\theta), \theta) - \xi(\theta)[1 - \alpha(\eta(\theta), \theta)]}, \tag{8}$$

where $\Gamma = (\gamma+1)/(\gamma-1)$, $\varsigma = 2I/kT_1$, $\xi(\theta) = 2\bar{E}_a(T)/kT_1$, and α is

$$\alpha(\eta,\theta) = -\frac{1}{2}\psi + \sqrt{\frac{1}{4}\psi^2 + \psi}, \tag{9}$$

where

$$\psi = b(p_1,T_1)\theta^{3/2}\eta^{-1}e^{-\varsigma/2\theta} \quad \text{and} \quad b(p_1,T_1) = G\left(\frac{m_e}{2\pi\hbar^2}\right)^{3/2}\frac{(kT)^{5/2}}{p_1}. \tag{10}$$

Before turning to the description of the Hugoniot curve, note that the right-hand sides of Eqs. (5) and (6) represent diverging infinite series for free atoms. However, for atoms in a partially ionized plasma only those energy levels should be taken into account that lie within the Debye sphere (Drellishak et al. 1963, Biberman et al. 1986). Thus, the sums in Eqs. (5) and (6) are truncated after $N_{max} = (Ir_D/e^2)^{1/2}$ terms, where $r_D = (kT/4\pi n_e e^2)^{1/2}$ is the Debye radius. A close inspection of Eq. (8) reveals that the Hugoniot curve possesses a maximum for a certain pressure ratio, $P = P_{max}$. This is so because on the one hand, the Hugoniot curve is close to its classical counterpart for low pressure ratios. On the other hand, in the range $P \gg \varsigma$, where the ionization degree is practically 1, the following asymptotic relationship is obtained:

$$\eta \approx \Gamma\left(1 + \frac{\varsigma}{P}\right), \tag{11}$$

which describes a decreasing dependence of η on P. Due to the energy losses during ionization as well as electronic excitation, the degree of maximal compression exceeds its classical asymptotic value. Thus, for example, numerical calculations show that argon at an initial pressure of 5 torr and temperature of 300 K may be compressed by a shock up to 12 times its initial density for shocks with pressure ratio of $P_{max} = 1400$. A universal law for the maximal density ratio behind ionizing shocks was given by Rutkevich and Mond (1994).

3. Stability of strong shocks in gases

The problem of stability of planar shock waves was first investigated by Dyakov (1954). In particular, he and later Kontorovich (1957) formulated the conditions under which acoustic- and entropy-vortex waves are spontaneously emitted from the shock. This phenomenon was termed spontaneous emission (SE) and the criteria for its occurrence are

$$h_c < h < 1 + 2M_2, \quad h_c = \frac{1-(\eta+1)M_2^2}{1+(\eta-1)M_2^2}, \quad h = -\frac{j^2}{\rho_2^2}\left(\frac{\partial \rho_2}{\partial p_2}\right)_H, \tag{12}$$

where M_2 is the Mach number behind the shock, j is the mass-flow rate across the shock, and the subscript H signifies the derivative along the Hugoniot curve.

A straightforward application of criterion (12) to shocks in ideal gas results in stability of the shock. On the other hand, a series of experiments (Griffiths et al. 1976, Glass and Liu 1977, Glass et al. 1978) revealed irregularities in the flow behind strong shocks in the form of spatial oscillations in the mass density and in the electrons' concentration when the shock's Mach number exceeded a certain threshold.

This apparent discrepancy between theory and experiment has puzzled many researchers over the past twenty years (McIntyre et al. 1991) and was recently resolved by Mond and Rutkevich (1994) who pointed out the crucial role of the electrons' kinetics in the occurrence of SE. It was shown that a frequency range exists for which, on one hand, the perturbations in the electrons' temperature differs from that of the heavy particles while, on the other hand, ionization equilibrium is maintained. This window of thermal non-equilibrium between the electrons and the heavy particles is defined by all wave frequencies ω that satisfy $\nu_\varepsilon \ll \omega \ll 2\nu_{ion}$ where ν_{ion} is the characteristic frequency of ionization by electron impact and ν_ε is the inverse of the relaxation time of the temperature of the heavy particles due to their collisions with the electrons. The latter is proportional to the sum of the elastic-collision frequencies of the electrons with the atoms and the ions. The Coulomb interactions of the electrons with the ions become important as the Mach number rises and the ionization degree approaches unity. Both collision frequencies and ν_{ion} are calculated according to expressions given in Hoffert and Lien (1967). Consequently, the definition of h is replaced by

$$h = -\frac{j^2}{\rho_2^2}\left[\left(\frac{\partial \rho_2}{\partial p_2}\right)_{\rho_1, p_1, \alpha} + \frac{\delta\alpha}{\delta p_2}\left(\frac{\partial \rho_2}{\partial \alpha}\right)_{\rho_1, p_1, p_2}\right]. \tag{13}$$

The ratio $\delta\alpha/\delta p_2$ differs from the derivative $d\alpha/dp_2$ along the Hugoniot curve since the ionization behind the shock is mainly due to electron impact and δT and δT_e are not necessarily equal. In addition, the sound velocity for waves in the non-equilibrium frequency window is also modified in order to take into account the thermal non-equilibrium between the electrons and the heavy particles. A preliminary examination of criterion (12) with the modified h (Eq. 13) and the modified sound velocity for shocks in argon and krypton at initial pressures in the range of 5–10 torr resulted in the occurrence of SE for Mach numbers above 10–15. These results are in reasonable quantitative and qualitative agreement with the experimental results of Glass and Liu (1977) and Glass et al. (1978).

Further numerical investigations show that taking into account the energy losses due to excitation of the electronic energy levels in the atoms affects both the Hugoniot curve and the threshold for SE very little. In addition, an extensive parametric study conducted for argon at room temperature shows that the threshold for SE does not depend on the initial pressure. However, for initial pressures below 11 torr it is impossible to maintain ionization equilibrium without thermal equilibrium in a certain range of the shock Mach numbers. In that regime $\nu_{ion} \ll \nu_\varepsilon$ and hence the classical criterion should be used, resulting in the non-occurrence of SE. These circumstances lead to a significant lowering of the upper limit for SE for low initial pressures. For initial pressures above 11 torr no such restriction exists and SE occurs for Mach numbers above 11.5 (independent of the initial pressure) and below 40–42 (very slight dependence on the initial pressure). It should be noted, however, that for very high Mach numbers such as near the upper limit of the SE, effects like second ionization may play an important role in determining the Hugoniot curve as well as the conditions for SE. For example, the Hugoniot curve is expected to have two maxima and thus to significantly alter the stability condition.

References

Biberman LM, Vorobjev VS, Yakubov IT (1986) Kinetics of Non-Equilibrium Low Temperature Plasmas, Plenum, New York, USA.

Dyakov SP (1954) On the stability of shock waves. Zh Eksp Teor Fiz 27:288–295.

Drellishak KS, Aeschliman DP, Cambel AB (1963) Partition functions and thermodynamic properties of argon plasma. Phys Fluids 6:1280–1284.

Glass II, Liu WS (1977) Effect of hydrogen impurities on shock structure and stability in ionizing monatomic gases: part 1, argon. J Fluid Mech 84:55–77.

Glass II, Liu WS, Tang FC (1978) Effect of hydrogen impurities on shock structure and stability in ionizing monatomic gases: part 2, krypton. Can J Phys 55:1269–1279.

Griffiths RW, Sandeman RJ, Hornung HG (1976) The stability of shock waves in ionizing and dissociating gases. J Appl Phys 9:1681–1691.

Hoffert M, Lien H (1967) Quasi-one-dimensional, nonequilibrium gas dynamics of partially ionized two-temperature argon. Phys Fluids 10(8):1769–1777.

Kontorovich VM (1957) On the stability of shock waves. Zh Exsp Teor Fiz 33:1525–1526.

McIntyre TJ, Houwing AFP, Sandeman RJ, Bachor HA (1991) Relaxation behind shock waves in neon. J Fluid Mech 227:617–640.

Mond M, Rutkevich I (1994) Spontaneous acoustic emission from strong ionizing shocks. J Fluid Mech 275:121–146.

Rutkevich I, Mond M (1994) Similarity relation for maximal gas compression by strong ionizing shocks. J Propulsion and Power 10:906–908.

Blast parameters from unconfined gaseous detonations

S. B. Dorofeev, V. P. Sidorov, A. E. Dvoinishnikov
Applied Chemical Physics Institute
Russian Research Centre "Kurchatov Institute"
Moscow, 123182 Russia

Abstract: Experimental and theoretical data for parameters of the air blast from gaseous detonations are analyzed. Comparison is made on the basis of the main blast parameters themselves and the corresponding TNT-equivalents of the gaseous explosions.

Key words: Gaseous detonations, Blast parameters, TNT-equivalent

1. Introduction

Numerous experimental and theoretical studies of the air blast from gaseous detonations have been conducted. These studies make it possible to proceed with data analysis in order to define the most reliable description of the air blast. Such an analysis has been undertaken earlier (Dvoinishnikov et al. 1993a, 1993b) for the blast waves from HE explosions. One of the important problems in the estimation of explosion effects is the application of the TNT-equivalence concept. This problem has been discussed in different studies (Fishburn 1976; Brossard et al. 1983; Borisov et al. 1985), but many questions are still unresolved. Different values of the TNT-equivalent and different functions of distance have been presented by the different authors. The aim of the second part of this study is to re-evaluate the concept of TNT-equivalent on the basis of selected data for air blasts from high explosives and gaseous detonations.

2. Blast parameters

The comparison of the blast parameters was carried out using Sachs dimensionless variables for unconfined spherical explosions. The data for positive overpressure (P^+) and impulse (I^+) are collected in Fig. 1. A relatively small number of measurements is available for parameters of negative phase (Brossard et al. 1984; Lannoy 1984; Dorofeev et al. 1993). They are not presented here, but these data show good correlation and the comparison was presented by Dorofeev et al. (1993). Figure 1 shows experimental data and the results of calculations. Corresponding experiments have been made with fuel-air and fuel-oxygen mixtures in laboratory and large-scale experiments. Except for the data by Kogarko et al. (1965) and calculations by Fishburn (1976), quite good correlation is observed. Mutual agreement is slightly better for overpressures than for impulses. The comparison shown in Fig. 1 suggests that the data of Brossard et al. (1984) and Dorofeev et al. (1993) are the most reliable. They are based on independent experiments and reveal good mutual agreement. Another important feature of their functions of blast parameters on distance is their refusal to apply $1/R$ behavior for overpressure in the far field. This feature permits reasonable predictions of overpressures at large distances from the explosive cloud to be made. At the same time, the means of representing blast-wave parameters are different in these studies. The only reservation about these data concerns the data for positive impulse that were presented by Brossard et al. (1984) for distances $\overline{R} > 5$. The reason for the behavior shown in the plot needs some explanation or additional validation.

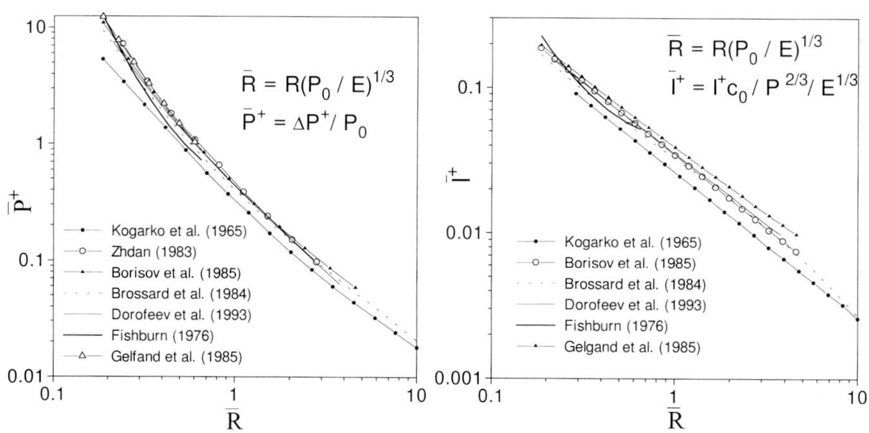

Figure 1. Positive overpressure and positive impulse in air-blast wave versus distance.

3. TNT-equivalents

The concept of TNT-equivalent is supported by the convenience of describing the blast effect of non-ideal explosions by that of some equivalent amount of high explosive. However this convenience does not have a firm basis, since the explosives themselves differ greatly in their intrinsic characteristics. The TNT-equivalent of a gaseous explosion should be defined as the ratio of the energy of the TNT explosion to that of gaseous explosion which has the same value of a given parameter at a fixed distance. In view of this, the TNT-equivalent should be considered (i) as a function of distance and (ii) separately for each parameter of the blast wave. This shows that the usage of some average TNT-equivalent (Brossard et al. 1983; Lannoy 1984) is either incorrect or should be specially validated. In addition, different parameters of air blast are known with different degrees of confidence. Their relative contribution to the blast effect differs itself. Although no need exists to use a TNT-equivalent to describe air-blast parameters from gaseous explosions, an analysis of TNT-equivalents permits a better understanding of the difference between blast waves from HE and gaseous detonation. The following analysis is based mostly on the data selected in the present paper (gaseous detonation) and on the data chosen for high explosives (Dvoinishnikov et al. 1993a, 1993b). Results of calculation of the TNT-equivalents are presented in Figs. 2 and 3. Figure 2 shows that the TNT-equivalent for $\overline{P}^+$ depends strongly on distance. Within the range $\overline{R} = 0.3$ to 2, it increases from 0.3–0.4 to 0.6–0.7 with the $\overline{R}$ increasing for all the data used. Beyond $\overline{R} \approx 2$, which corresponds approximately to overpressures less than 0.1, the analysis is more complicated in view of the lack of data and their limited accuracy. Nevertheless, the behavior of most of the functions indicates the maximum in the range $\overline{R} = 1$ to 3. The choice of data for calculation of TNT-equivalent appears to be very significant. The asymptotic behavior of $\overline{P}^+$ at the far field should be realistic and the same for gaseous and HE explosions. Two examples shown are the data 1 and 3, where the realistic function $1/R^{4/3}$ is used, and the data 6 and 7, where the wrong asymptotic $1/R$ results in an unreliable prediction that the TNT-equivalent approaches 1. While the data for $\overline{P}^+$ show a clear trend, the data for $\overline{P}^-$ do not. The choice of data is very limited and their mutual agreement is poor. The main problem concerns the description for $\overline{R} < 2$. It is not connected with the data for gaseous detonations, which agree well, but with that for HE explosion. In any case, gaseous explosions produce blast waves with relatively stronger negative phase than those of HE with the same positive overpressure. The important question, which remains unclear, is whether it is a factor of 2 or 8.

Figure 3 shows the TNT-equivalent for the impulse. The data used for HE include the mean function obtained by Dvoinishnikov et al. (1993a, 1993b), and the data by Baker (1983). All the values appear to be within the range 0.5–0.6 up to $\overline{R} \approx 4$. The decrease of the equivalent beyond $\overline{R} = 4$ is due to the data for gaseous explosion from Brossard et al. (1984). As was mentioned above, this decrease seems to be unrealistic. Additional analysis is required to clarify this important question. The TNT-equivalent presented by Fishburn (1976), where the comparison was made with ideal point source (Brode 1955), seems also not reliable. The situation with the lack of data for detailed analysis of the TNT-equivalent $\overline{I}^-$ is similar to that for negative overpressure. This was the reason for using also the data for point-source explosions in ideal (Brode 1955) and real gases (Brode 1956). We should notice that all functions grow with distance. The preferable estimate seems to be in between curves 1 and 3 and curves 2 and 3. Again, the negative impulse produced by gaseous detonation

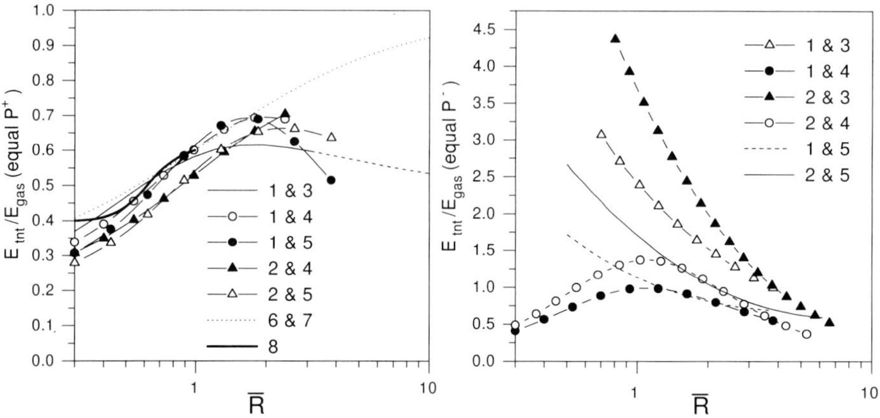

Figure 2. TNT-equivalent for positive and negative overpressure versus dimensionless distance defined by the energy of the gaseous explosion. $\overline{P}^+$: 1 — Dorofeev et al. (1993); 2 — Brossard et al (1984); 3 — Dvoinishnikov et al. (1993b) (spherical charge); 4 — Sadovsky (1952); 5 — Goodman (1960); 6 — Borisov et al. (1984); 7 — Lewis (1981); 8 — Fishburn (1976). $\overline{P}^-$: 1 — Dorofeev et al. (1993); 2 — Brossard et al. (1984); 3 — Dvoinishnikov et al. (1993a) (cylindrical charge); 4 — Lewis (1981); 5 — Dvoinishnikov et al. (1993b) (spherical charge).

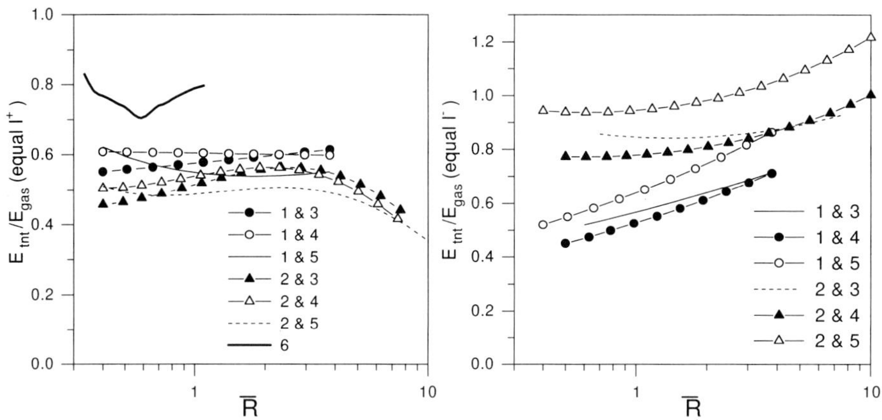

Figure 3. TNT-equivalent for positive and negative impulse. $\overline{I}^+$: 1 — Dorofeev et al. (1993); 2 — Brossard et al. (1984); 3 — Dvoinishnikov et al. (1993a) (mean values); 4 — Dvoinishnikov et al. (1993a) (cylindrical charge); 5 — Baker (1983); 6 — Fishburn (1976). $\overline{I}^-$: 1 — Dorofeev et al. (1993); 2 — Brossard et al. (1984); 3 — Dvoinishnikov et al. (1993a) (cylindrical charges); 4 — Brode (1955); 5 — Brode (1956).

appeared to be greater than that from HE, with the positive zone being the same. This difference, however, is less than that for $\overline{P}^-$.

4. Conclusions

The analysis of the data for the parameters of air blast from unconfined gaseous detonations shows that the description of the air blast from gaseous detonations is at the same level as that for HE explosions. Moreover, the data for the negative phase that is available in the literature seems to be more reliable for the case of gaseous detonation. At the same time, the lack of data for negative phase should be emphasized. In these circumstances the application of the TNT-equivalence concept for estimation of an effect of a gaseous explosion seem to be useless. Nevertheless, the analysis of TNT-equivalents as a function of distance and air-blast parameters can be still useful to clarify the differences in character of air blasts from gaseous detonations. The values of TNT-equivalents still interest many practical engineers. Energetic equivalents are estimated to be within the ranges 0.3–0.7 (P^+), 0.5–0.6 (I^+), 0.5–0.9 (I^-). The mean value of 0.2 ($E_{GAS} = 5E_{TNT}$) presented by Brossard et al. (1983) and Lannoy (1984) is certainly underestimated, probably due to incorrect usage of HE blast parameters. The data for P^- (especially for HE) do not permit the confident evaluation of TNT-equivalent.

References

Baker WE, Cox PA, Westine PS, Kulesz JJ, Strehlow RA (1983) Explosion Hazards and Evaluation, Elsvier, Amsterdam, Oxford, New York.

Borisov AA, Gelfand BE, Tsyganov SA (1985) On modelling of pressure waves forming by detonation and combustion of gas mixtures. J Physics of Combustion and Explosion, Russian Acad of Sci 2:90.

Brode HL (1955) Numerical solutions of spherical blast waves. J Appl Phys 6:766.

Brode HL (1956) Point Source Explosion in Air, Research Memorandum RM-1824-AEC, RAND Corporation, Santa Monica, California.

Brossard J, Leyer JC, Desbordes D, Garnier JL, Hendrickx S, Lannoy A, Perrot J, Saint Clouds JP (1983) Experimental analysis of unconfined explosions of air-hydrocarbon mixtures. characterization of the pressure field. In: 4th Symp Int on Loss Prevention and Safety Promotion in the Process Industries, European Fed of Chem Eng Event No 290 1:D, pp 10–19.

Brossard J, Leyer JC, Desbordes D, Saint Clouds JP, Hendrickx S, Garnier JL, Lannoy A, Perrot J (1984) Air blast unconfined gaseous detonations. Progress in Astron and Aeron 94:556.

Dorofeev SB, Sidorov VP, Dvoinishnikov AE, Alekseev VI, Kuznetsov MS (1993). Experimental study of air blast parameters from fuel rich mixtures detonation. 14th Int Colloquium on Dynamics of Explosions and Reactive Systems, (ICDERS) 2:E1.4.1.

Dvoinishnikov AE, Dorofeev SB, and Gelfand BE (1993a) Analysis of blast wave data from HE explosions. In: Shock Waves @ Marseille, Brun R, Dumitrescu LZ (eds), Vol 4, pp 407–412.

Dvoinishnikov AE, Dorofeev SB, Gelfand BE (1993b) Analysis of the Data on Air Blast Parameters from HE Explosions, Preprint IAE-5605/13, Russian Research Centre "Kurchatov Institute," Moscow.

Fishburn BD (1976) Some aspects of blast from fuel-air explosives. Acta Astronautica 3:1049.

Gelfand BE, Gubin CA, Mikhalkin VN, Shargatov VA (1985) Computation of shock wave parameters by detonation of combustible gaseous mixtures of a variable composition. J Physics of Combustion and Explosion, Russian Acad of Sci 3:92.

Goodman HJ (1960) Compiled Free Air Blast Data on Bare Spherical Pentolite, BRL Report 1092, Aberdeen Proving Ground, Maryland.

Kogarko SM, Adushkin VV, Liamin AG (1965) Investigation of spherical detonation on gas mixtures. Combust Explos Shock Waves USSR 1-2:22.

Lannoy A (1984) Analyse des explosions air-hydrocarbures en milieu libre: etudes deterministe et probabiliste du scenario d'accident. Prevision des effets de surpression. Bulletin Direct Etudes et Recherches EDF, A4.

Lewis DJ (1981) Condensed phase explosions and their blast characteristics. Euromech-139, Aberystwyth.

Sadovsky MA (1952) Mechanical action of blast waves on data of experimental studies. In: Physics of Explosion, USSR Acad of Sci Moscow, Vol 1, pp 70–110.

Zhdan SA (1983) Calculation of gas mixture explosion with regard for shift of chemical equilibrium products. J Physics of Combustion and Explosion, Russian Acad of Sci 1:131.

Numerical simulation of the blast in the World Trade Center

J. D. Baum, H. Luo
Science Applications International Corporation
1710 Goodridge Drive, MS 2-3-1
McLean, VA 22102 USA

R. Löhner
George Mason University
Fairfax, VA 22030 USA

Abstract: This paper describes the application of FEFLO96, a recently developed three-dimensional, adaptive, finite-element, edge-based, ALE shock-capturing methodology on unstructured tetrahedral grids, to the simulation of blast diffraction inside the B-2 level of the World Trade Center garage. The geometry of the B-2 garage level was modeled based on the building blueprints and included all important geometrical features. The shock-diffraction simulation was initiated using the results of a one-dimensional detonation code for a one-thousand pound ANFO charge. The resolution and fidelity of the simulated shock-wave diffraction phenomena, performed via a solution of the transient compressible Euler equations, were enhanced by the application of the classic H-refinement/coarsening grid-adaptation scheme. A large number of shock waves was produced due to shock diffraction around walls, cars, and ramps. Excellent shock adaptation and resolution were obtained at all times. Several phenomena such as three-dimensional shock-diffraction processes, the formation of "hot spots" due to three-dimensional wave focusing at corners, shock reverberation within the parking level, and the formation of three-dimensional Mach stems, were identified and discussed.

Key words: Numerical blast simulation, World Trade Center

1. Introduction

The evaluation of civilian-structure vulnerability to terrorist attacks has become a timely and critical issue after the terrorist attack on the World Trade Center, and the failed attacks on other civilian targets in New York City. Since testing large structures for blast resistance is very expensive, and testing real targets is simply not practical, detailed numerical simulations offer a unique, accurate, and cost-effective approach to assessing vulnerability of these structures to internal explosions.

The numerical simulation of an explosion within a confined environment requires simulating blast initiation and evolution, shock-wave diffraction about many structural components, structural response to the blast, structural failure, and blast transmission through the fragmented structural elements. This list indicates that methodologies from widely disparate fields such as computational fluid dynamics (CFD), computational structural dynamics (CSD), and material modeling must be combined to completely assess structure vulnerability to internal blasts. This coupling has been recently completed. Nevertheless, it was decided that for the first simulation it would be prudent to eliminate the structural coupling and treat all surfaces as non-responding.

Over the past five years we have developed a general CFD methodology capable of efficiently simulating transient shock-shock and shock-structure interactions for realistic, complex, engineering-type, three-dimensional geometries. The unstructured grid approach (Löhner et al. 1987) was chosen as the most suitable for generating surface and volume grids for complex-geometry bodies. Since typical areas of interest require simulation of both steady and transient strong shocks, a high-resolution monotonicity preserving algorithm for unstructured grids, called FEM-FCT was developed. Finally, since the flow fields are typically smooth (except for a few regions where strong gradients appear), efficient adaptive-refinement techniques for transient problems are required (Baum and Löhner 1994).

2. The numerical methodolgy

The primary CFD tools incorporated in the FEFLO methodology include the CAD tool (FECAD), the background-grid generator (PRE-BACK), the advancing-front grid generation (FRGEN3D), the flow solver (FEFLO96), and the post-processing methodology (FEPOST3D). The FEFLO96 flow solver is a new 3D ALE edge-based hydro-solver based on FEM-FCT (Löhner and Baum 1990; Luo et al. 1994). H-refinement is the preferred approach for grid adaptation (Baum et al. 1994, Löhner and Baum 1990). The high-order scheme used is the consistent-mass Taylor-Galerkin algorithm. Combined with a modified second-order Lapidus artificial-viscosity scheme, the resulting scheme is second-order accurate in space, and fourth-order accurate in phase. The critical parameter for the refinement/deletion criteria is density. The recently implemented edge-based data structure reduced memory requirements by about a factor of 2.5 and CPU requirements by about 2.5 (Luo et al. 1994).

3. Numerical results

Solid-body definition. In the absence of CAD data, all wall locations, ramps, exhaust plenum spaces, mechanical rooms, stairways, toll booths, curbs, A/C, and supply and elevator shafts were modeled accurately from the structural and architectural drawings (Fig. 1a). Ten typical car groups were modeled, parked at the correct parking locations.

Blast-wave initiation and diffraction. Blast initiation assumed a vehicle parked near the right curb of the exit lane, leading to the northbound exit to West Street (ramp B). The 1D detonation code modeled a charge of 1000 pounds of ANFO HE placed on the floor, and integrated the solution until the leading shock was only a short distance from the south wall of tower A (the north tower). The solution was then overlayed on a 3D mesh.

The numerical simulation of shock diffraction within the B-2 floor of the World Trade Center is especially suited for an adaptive refinement methodology. The large dimensions of the computational domain (20,249 cm long, 17,743 cm wide, and 580 cm high) are in contrast to the requirement to model the shock as a sharp discontinuity (or at least, minimize shock thickness), and model the smallest physical features of interest (i.e., while modeling the complete domain, also model shock diffraction about a wall that is less than 20 cm thick, a ratio of 1:1000). These contrasting requirements can be fulfilled only by using an adaptive methodology. Still, some compromises on mesh size and total number of nodes had to be made. We initiated the simulation by modeling only the zone fairly close to blast point, while allowing only two levels of refinement. Thus, the initial mesh before refinement had only about 370,000 points and 1.95 million elements. At $t = 25.7$ ms, when the leading shock was close to the boundary of the modeled domain, we remeshed, modeling a larger portion of the garage. We then interpolated the solution from the old mesh to the new mesh, and adapted the solution. By this time we gained access to a Cray YMP/M-92 with 256 MWords static memory, so our mesh size was about 780k points and 4.1M elements. Finally, at $t = 108.9$ ms, we modeled the complete domain. The nonadapted mesh included about 1.54M points and 8.0M elements. After solution adaptation, the final mesh included about 3.52M points and 18.1M elements executing on a new Cray YMP/M-92 with 1 GWords static memory.

The complex shock evolution within the structure is described in great detail in Baum et al. (1994). Due to severe space limitations we will include only two snapshots and demonstrate the complexity of shock evolution in such an environment by analyzing these results.

Examination of the leading shock at $t = 12.51$ ms (Fig. 1b) shows on the right (east) the diffracted wave about the south-east corner room of tower A, and shock diffraction about a column. Closer to the center of the domain we observe a nicely detailed shock diffraction about a column, with almost symmetric rarefaction waves on both sides, and a reflected shock upstream of the column. A similar diffraction process is observed about a column toward the left of the expanding shock. Shock diffraction about the cars situated between these columns is more complex. The reflected pressure amplitude depends on the car profile (i.e., percentage of blocked area from floor to ceiling), car width, and shock incident angle. The impacting shocks were partially transmitted under the cars (wheels were not modeled here).

The dominant shocks on the west (left) side at $t = 95.81$ ms (Fig. 1c) are the shock reflections from the walls of ramps A and B, shock reflection from the second row of cars (reflected in a northwest direction), and shock reflection from the wall of ramp N. The primary shock was partially reflected from the side wall of ramp N, and the row of cars north of ramp T, expanded around the wall surrounding ramp N, and about the vehicles parked west of ramp T. The front propagating south shows, from left to right: a large-amplitude shock and a Mach stem formed near the west wall (between the wall and the adjacent cars); a lower-amplitude shock diffracting about the parked cars; the incident, large-amplitude shock in the traffic lane, followed closely by the reflection from the side wall of ramp N; the broad-side impact of this reflected shock on the row of cars north of ramp T; shock diffraction about the southwest corner of ramp T; and the impact of the shocks reflected from ramps A and B on the central two rows of cars and on the row of cars north to tower N. Naturally, each shock impact produces a reflected shock, plus an assortment of shocks that reverberate between the cars. At the center, shock stagnation at the north wall of ramp T (and the resulting high pressure) explains the significant damage to the cars parked nearby, and the collapse of the wall. Since the cars to the left (west) and at the center of the wall are parked normal to the wall, the shock front is fairly uniform. On the right (east), however, the two vehicles parked diagonally perturbed the flow sufficiently to produce a curved (almost spherical) reflected shock. At the center-right we observe the sharp shock transmitted in the traffic lane between ramp T and the west wall of tower B, the multiple reflections from rooms near the northwest corner of tower B, and from the broad-sided parked cars. At $t = 95.8$ ms, the primary shock diffracts about the south wall of ramp T and the transfer shaft room near the west wall of tower B. The reflected shock from the north wall of ramp T expands at this time near the ramp entrance. In the Secret Service area, we observe the evolution of the primary curved shock and the multiple diffraction processes about the columns. The shock reflected from the two central columns at $t = 89.55$ ms and interacted at $t = 95.81$ ms with the walls of the pipe room attached to the north wall of tower B, to produce a strong reflection, an expansion, and a Mach stem. The weakened incident shock expanded to the north, and hit the PATH wall at $t = 106$ ms. The shock front impacted the presidential limousines at $t = 115.91$ ms. The large-amplitude, spherically reflected shock near the north wall of tower B, with

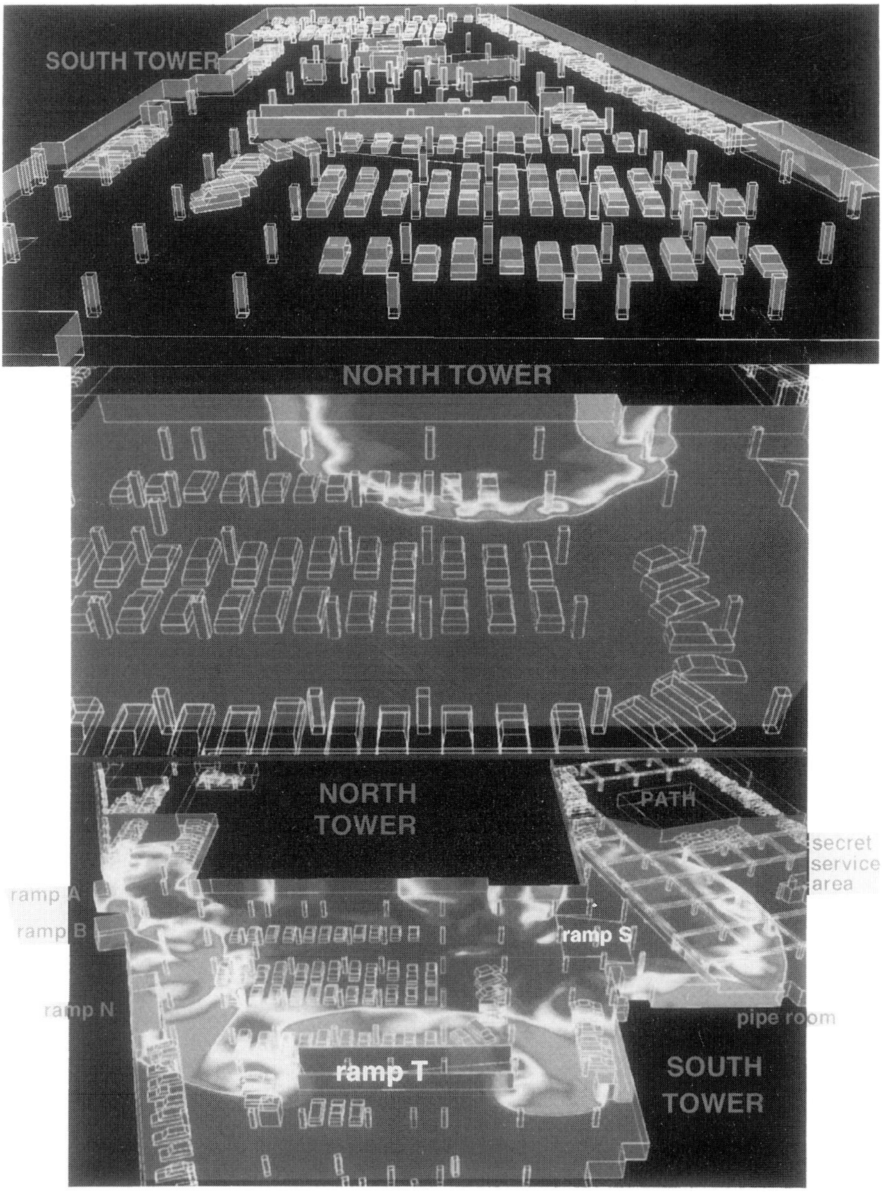

Figure 1. CFD surface mesh and pressure contours: top – surface mesh of central parking area; middle – expanded view of pressure contours at $t = 12.51$ ms; bottom – expanded view of pressure contours at $t = 95.81$ ms.

a circular imprint on the floor, resulted from shock reflection from a corner of the support beams.

The solution was integrated to $t = 287.73$ ms, at which time the shocks have filled the complete computational domain.

4. Summary and conclusions

This paper described the application of a recently developed three-dimensional, adaptive, finite-element, edge-based, ALE shock-capturing methodology on unstructured tetrahedral grids (FEFLO96) to the simulation of blast diffraction inside the B-2 level of the World Trade Center. The evolution of the primary shock and the multitude reflections were studied in great detail. Several shocks were tracked over time. Several interesting 3D shock-diffraction processes were identified and discussed. Among these are the formation of "hot spots" due to three-dimensional wave focusing, shock reverberation between cars, shock diffraction about cars and columns, and the formation of three-dimensional Mach stems. The results demonstrate excellent shock adaptation at all times. All shocks were captured as sharp discontinuities without producing pre- or post-shock oscillations.

Acknowledgement. Funding for this research work was provided by the Defense Nuclear Agency. Drs. Michael E. Giltrud and Douglas A. Sunshine served as the technical program monitors.

References

Baum JD, Löhner R (1994) Numerical simulation of shock-elevated box interaction using an adaptive finite element shock capturing scheme. AIAA J 32(4):682.

Baum JD, Luo H, Löhner R (1994) Numerical Simulation of Blast in The World Trade Center, SAIC Report 1994-23.

Löhner R, Baum JD (1990) Numerical Simulation of Shock Interaction with Complex Geometry Three-Dimensional Structures Using A New Adaptive H-Refinement Scheme on Unstructured Grids, AIAA-90-0700.

Löhner R, Morgan K, Peraire J, Vahdati M (1987) Finite element flux-corrected transport (FEM-FCT) for the Euler and Navier-Stokes equations. Int J Num Meth Fluids 7:1093.

Luo H, Baum JD, Löhner R (1994) Edge-based finite element schemes for the Euler equations. AIAA J 32(6):1183.

Blast-wave interaction with multiple structures

L. H. J. Absil, A. C. van den Berg, J. Weerheijm
TNO Prins Maurits Laboratory
P.O. Box 45, 2280 AA Rijswijk, The Netherlands

Abstract: This paper presents an experimental and numerical study of blast propagation and interaction with multiple urban structures. The main objective of this research is to provide detailed information on blast propagation through urban and industrial areas so as to improve guidelines for the determination of safety distances around potentially hazardous industries.

Key words: Blast waves, Blast interaction with structures, Blast damage assessment, Explosion effects, Safety distances

1. Introduction

Accidental explosions form a threat to communities in the vicinity of hazardous industries. An important aspect of an explosion is the generation of a blast wave, which is responsible for most of the damage at large distances from the explosion source. In the past much effort has been put into developing an understanding of what mechanisms cause explosions. Little attention, however, has been given to assessing the characteristics of the blast wave as it propagates away from the explosion source. Because of this lack of knowledge, current risk analysis is based on the strength of the undisturbed pressure wave: i.e., damage and risk contours are related to the free-field decay of the peak overpressure with distance from the explosion source. Due to the shielding effect of houses on the edge of a city, however, the blast will be attenuated resulting in a reduction of damage to the rest of the city. This modification of the blast wave due to the interaction with surrounding urban structures has not been accounted for in present-day risk analysis. This deficiency has a significant effect on damage-assessment and the determination of safety distances.

On behalf of the Dutch Ministry of Housing, Physical Planning and Environment, a research program was started comprising an experimental and numerical study of blast propagation and interaction with multiple urban structures. The main objective is to generate data that will form the basis for new guidelines for blast damage assessment to urban areas. The first phase of this study was to investigate what parameters play a major role in the shielding effect. Variables that were expected to mainly influence the degree of shielding are the

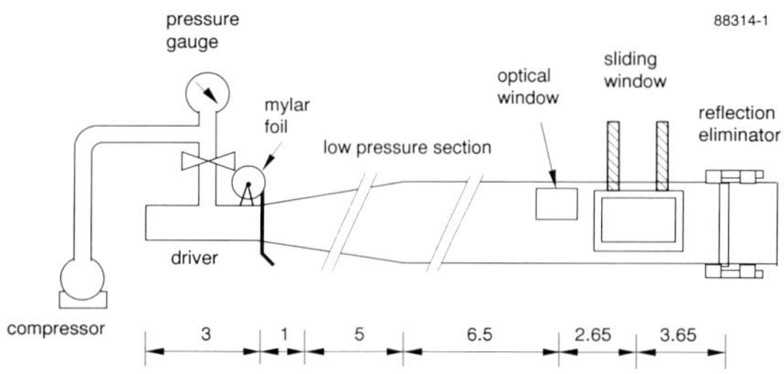

Figure 1. A schematic drawing of the blast simulator (distances given in meters).

peak overpressure and duration of the blast wave, the structure geometry and size, the height of the first structure, the number of structures, and the distance between them.

2. Experimental set-up

The experiments were conducted in a shock tube with a square test section of 40 $\times$ 40 cm^2, see Fig. 1. The 3-m long driver section can be filled with compressed air, up to a maximum of 30 bar, generating blast waves with a peak overpressure of up to 110 kPa and a positive phase duration of 30 ms. The strength of the shock wave can be regulated by varying the initial pressure in the driver section. As scale models, two-dimensional rectangular (5 $\times$ 5 cm^2) cylinders were used and loaded with shock waves of 1 and 10 kPa overpressure: 1 kPa is a typical damage criterion for glass failure, while partial demolition of houses can occur at 10 kPa. These relatively low overpressure levels were selected because the emphasis in the present study is on far-field conditions, where shielding effects can reduce the amount of damage inflicted considerably. The size of urban housing and the mutual distance between houses in a city is typically of the order of 10 meters. Hence, in the experiments the geometry was scaled by a factor of 200. According to the Hopkinson scaling law (Baker et al. 1973) the duration of the blast wave should be scaled proportional to the length scale. The shock-tube driver section was shortened to its smallest size, i.e. 7 cm, to limit the positive phase duration of the blast wave to 4.2 ms, which corresponds to a full-scale duration of about 800 ms. This duration was rather long with regard to the duration of a vapor-cloud explosion blast at far-field conditions, which is typically of the order of 100 ms. Because of the long duration of the blast wave generated in the shock tube, a few experiments were performed with electric detonators to generate the blast wave. With these detonators, blast-wave durations of the order of 0.5 ms could be achieved which correspond to the desired duration of 100 ms at full scale. Highly sensitive piezo-resistive pressure transducers were used to record the pressure-time history at the centers of the front, top, and rear faces of the structures. A Wollaston-prism shearing interferometer (Absil et al. 1992) and a holographic interferometer were used to visualize the shock-wave flow.

3. The BLAST code

For the simulation of a three-dimensional blast-wave/structure interaction, the Computational Fluid Dynamics code BLAST has been developed at TNO-PML (Van den Berg 1990). The code solves the Euler equations which describe inviscid compressible flow. The Flux-Corrected Transport scheme is used for optimum description of shocks and contact discontinuities. The code has the capability to calculate the pressure, density, and temperature distribution around structures for a given infalling blast wave. The BLAST code was used to simulate the flows experimentally studied.

4. Results

A typical example is shown in Fig. 2, which shows the interaction of a blast wave with two rectangular structures. The excellent agreement between the measured interferograms (left) and the isobar pattern as obtained by the numerical simulation (right) indicates that the BLAST code gives a fairly good description of the main features of the shock-wave flow. The first row shows the vertical shock wave striking the structure from the left. After the shock strikes the first structure, part of the shock is reflected backwards, as shown in the second row. The overpressure in the reflected shock is substantially greater, by about a factor of 2, than that of the incident shock. This pressure difference causes a flow from the region of high to low pressure, resulting in the formation of circular expansion waves at the corners of the structure. The reflected frontal overpressure is relieved by the rarefaction wave that travels down the front surface. At the edge of the structure, viscous separation occurs generating a rolled-up vortex featuring a steep density gradient. In the third row, the shock wave has passed the rear face of the first structure, and the compression wave that bridges the pressure difference between the induced air flow above the buildings and the undisturbed flow in between is shown. The compression wave travels downward across the rear face of the first building and subsequently reflects at the earth's surface. The fourth row shows the complicated wave pattern that develops in between the two buildings. The front of the second building will be hit by the weaker diffracted wave, which indicates that it is shielded against the blast wave by the first building.

Next, the infalling wave on the second building is reflected back and the rear face of the first building is hit again, as shown in the last row. This explains why the rear face of a building, which is closely surrounded by other buildings, also shows considerable damage after an accidental explosion.

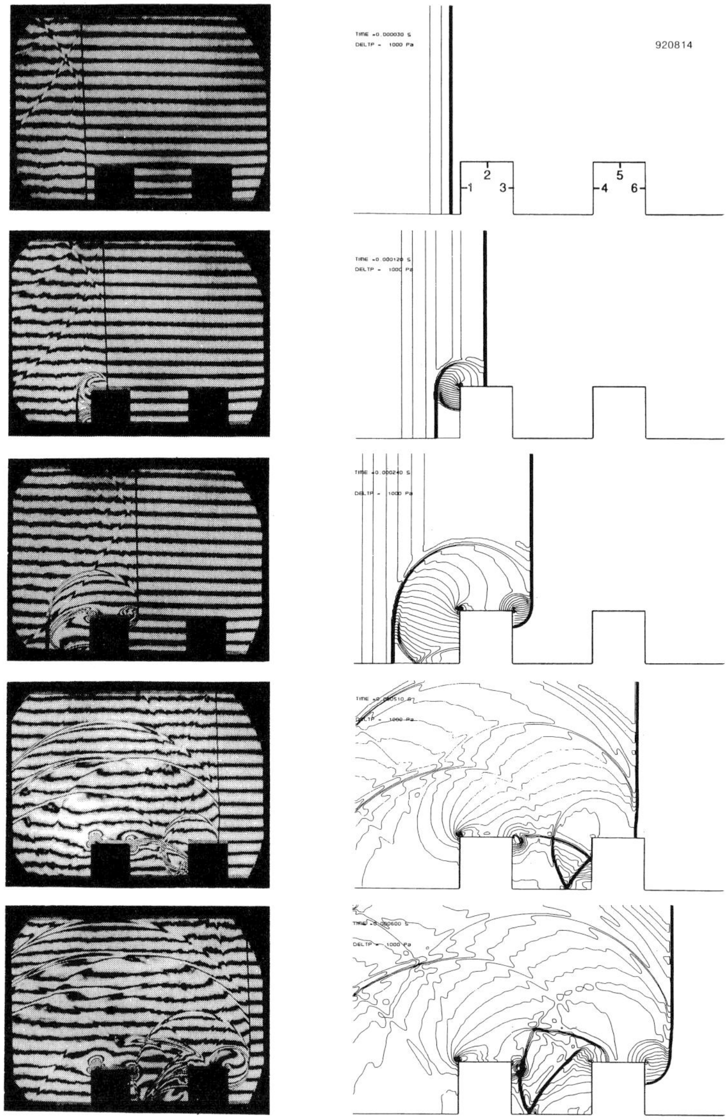

Figure 2. Comparison of interferograms (left) and calculated isobar patterns (right) for a blast wave of 10 kPa and 60-ms duration.

The overpressure versus time records as calculated with BLAST at the mid-points of the front, top, and rear faces of the structures are presented in Fig. 3.

This calculation was made for a blast wave of 10 kPa and 60-ms duration falling in at 10-m high buildings located 15 m apart. In these overpressure signals the complicated wave pattern as seen in Fig. 2 can be readily recognized. The overpressure signal predicted at location 3, at the back of the first building, for instance, shows a sequence of four shock phenomena which can be traced in the flow patterns shown in Fig. 2. The first peak is due to the passage of the infalling blast wave. The gradual pressure build-up is the result of the diffraction of the shock around the first structure. The second pressure jump corresponds with the reflection of the primary wave by the ground. The third pressure increase corresponds with the shock wave reflected directly from the second building. This shock wave is immediately followed by a fourth wave, which is the result of the reflection by the second building and subsequently by the ground. Because the second structure is hit by the weaker diffracted wave, the blast load on the front of the second structure will be lower than that on the front of the first structure. The computation shows that the blast overpressure experienced by the front of the second building is about two times lower than the reflected overpressure of the undisturbed blast wave endured by the front of the first building. A similar reduction in overpressure was found in the experiments using electric detonators. This shows that a considerable shielding effect can occur under realistic conditions.

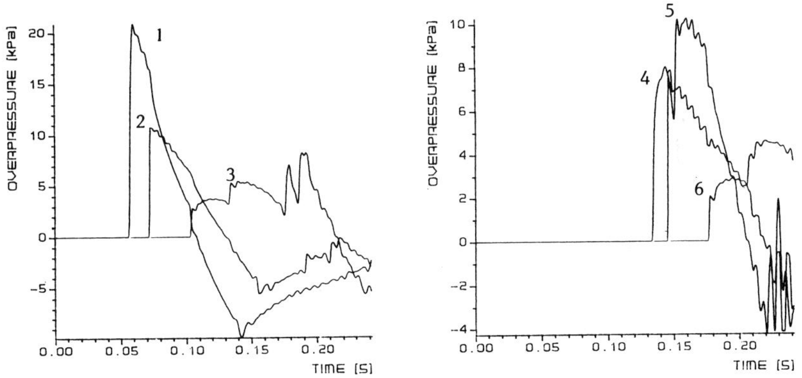

Figure 3. Pressure vs. time signals at stations 1 to 6 (see Fig. 2) as calculated for a blast-wave (10 kPa, 60 ms) interaction with two buildings (10-m high).

The findings of the parametric study can be summarized as follows:

- the shielding effect was found to behave linearly for these rather low overpressures, i.e., nearly the same shielding effect was found for the 1-kPa blast wave as for the 10-kPa blast wave;

- the shielding effect hardly increases after the second structure, i.e., the pressure load on the second, third, and fourth structure is nearly the same;

- the overpressure build-up in between the buildings and the shielding effect were found to be highly dependent on the blast-wave duration and the spacing between the buildings. As the blast wave sustains, multiple wave reflections from the surrounding structures cause a pressure increase on the front and rear faces of the structures;

- the degree of shielding strongly depends on the height of the first structure;

- the geometry of the roof (i.e., flat versus pitched) only had a minor effect on the pressure-load on the structures.

5. Conclusions

The present paper has demonstrated that a significant shielding effect can occur in realistic situations. This shielding effect strongly depends on the duration of the blast wave, the height of the first building, and the distance between the buildings. Good agreement between measurements and numerical simulation was found, indicating that the BLAST code gives a good description of the main features of the diffraction process.

Acknowledgement. The authors are indebted to M. Dirkse for performing the measurements and D. van Leeuwen for performing the calculations. This research was funded by the Dutch Ministry of Housing, Physical Planning and Environment.

References

Absil LHJ, Bakkum EA, van der Meer B, Weerheijm J (1992) Shearing interferometry and diffraction of weak shock waves by scale models. In: Proceedings of the 20th International Congress on High Speed Photography and Photonics, Victoria, Canada.

Baker E, Westine P, Dodge F (1973) Similarity Methods in Engineering Dynamics, Spartan Books, New Jersey.

Van den Berg AC (1990) BLAST 2D — A Code for Numerical Simulation of Two-Dimensional Blast Effects, TNO-PML Internal Report.

Efficient 2D and 3D locally adaptive unstructured Euler solvers for unsteady shocked gas flows

A. Galyukov, A. Fursenko, P. Voinovich
Advanced Technology Center
P. O. Box 160, St. Petersburg 198103 Russia
T. Saito
Cray Research Japan Ltd, Supercomputer Center, Institute of Fluid Science
Tohoku University, 2-1-1 Katahira, Aoba-ku, Sendai 980-77 Japan
E. Timofeev
Ioffe Physical-Technical Institute, Russian Academy of Sciences
Polytekhnicheskaya str. 26, St. Petersburg 194021 Russia

Abstract: This paper presents a set of efficient 2D and 3D Euler solvers for strongly unsteady shocked gas flows. The key points of the numerical methods are: a high-order predictor-corrector Godunov-type finite-volume scheme and local refinement/derefinement procedures supported by the flexible data structures of unstructured triangular (2D) and tetrahedral (3D) grids. The grid generation is performed based on the Watson incremental 2D/3D triangulation providing control over the grid spacing to enhance the resolution in the regions of special interest.

Key words: Unstructured Euler solver, Local grid refinement, Grid generation

1. Introduction

The recent trend toward extensive use of unstructured grids in computational fluid dynamics (CFD) is due to their remarkable adaptability to complex geometries and possible dynamic grid adaptation to the flow pattern. The latter feature appears to be particularly attractive when applied integrally with a shock-capturing scheme, as the accuracy in that case is limited by the resolution at discontinuities, i.e., shock waves, contact surfaces, and slip surfaces. An unstructured grid allows the natural combination of a rather large grid spacing adequate for smooth flow approximation with a very fine grid at discontinuities. When only a small part of the computational domain is covered by the fine grid the local grid refinement leads to high computational efficiency of the method. However, the development of an unstructured code is labor intensive and involves numerous points offering a broad spectrum of alternatives.

2. Grid generation

The principal stage of pre-processing in CFD is the grid generation. Numerous recent papers have been devoted to this subject (see e.g., the review by Barth 1992), but the totally automatic unstructured grid generation for an arbitrary domain remains a challenge.

In the 2D case our grid generation procedure includes the following basic stages: the major boundary nodes are allocated to define the domain geometry; additional nodes are introduced optionally within pre-determined spacing at the boundaries and/or into the domain; the whole set of nodes is triangulated utilizing the technique by Watson (1981) and following the Delaunay criterion; neighboring nodes located prohibitively close are united; the grid can be incrementally refined using the above triangulation technique to meet additional criteria; edge swapping is applied to improve the grid locally; a global smoothing is performed to maintain gradual variation in grid spacing. Some of the steps are invoked iteratively, either individually or in combinations.

The 3D case presents a fundamental problem at the stage of the domain boundary definition; CAD systems are used in many practical engineering applications to create the domain bounding surfaces. For simple cases we utilize a modified version of our 2D grid generator for the surface triangulation. As the visual quality control becomes impractical for 3D unstructured grids, statistical data provided by the grid generation procedures are used extensively.

The basic principles of grid generation in 3D domains are the same as described for two dimensions. A 3D modification of Watson's algorithm performs tetrahedrization of the domain. Nodes positioned too closely to the domain boundary due to grid smoothing are deleted, and the polyhedral cavity surrounding a deletable node is triangulated by the advancing front method (Tanemura et al. 1983).

3. The finite-volume scheme

Constructing the numerical method we proceed from the integral conservation equations applied to node-centered control volumes. In 2D and axisymmetric cases the control volume about a node is confined by the respective parts of medians of the triangular grid elements sharing the vertex. The 3D control volume is bounded by planes passing through the tetrahedron center and the centers of its faces and edges (see e.g., Fursenko 1993).

A Riemann solver is used to compute the mass, momentum, energy, and species fluxes at the control volume interfaces. Linear interpolation within elements (triangles in 2D and tetrahedra in 3D) enhances the spatial accuracy. A limiter is applied when calculating average gradients of primitive gasdynamic variables at each node to prevent spurious oscillations. The second-order accuracy with respect to time is achieved by the two-step predictor-corrector technique. The predictor step of the scheme involves no Riemann solver which contributes to the efficiency of the method. For a detailed description of the scheme see Fursenko (1993).

4. Adaptive grid refinement

To improve the resolution of discontinuities, a grid adaptation procedure is invoked periodically to introduce new grid nodes in the vicinity of shocks, contact surfaces, and slip surfaces and to remove previously inserted nodes where the solution becomes smooth. The classical hierarchical refinement technique (Löhner 1987) is used together with the 2D/axisymmetric solvers. A bisectional refinement procedure based on the method by Rivara & Levin (1992) supplements the solver in the 3D case. A variety of sensors and criteria are used in different applications depending on the specific problem and flow features of interest. The most typical and universal sensor is based on the second derivative of density. The grid nodes are labeled for either refinement or derefinement in two dimensions, while edges are labeled for refinement and nodes for derefinement in three dimensions. Limited gradients at the grid nodes and Riemann solver solutions are optionally used to assign gasdynamic values to the newly inserted nodes. As extra nodes are introduced only at local singularities of the solution, the total amount of grid nodes is normally one to two orders of magnitude lower than for a uniform grid providing the same accuracy. This essentially increases the computational efficiency of the methods.

Some specific features should be emphasized regarding the adaptive grid refinement for transient flow simulations. First, the grid refinement should be reversible; second, the grid rearrangement should be fast compared to the solver; and third, the same data structure should support both the solver and grid refinement/derefinement routines.

5. The data structures

The required memory and efficiency of the grid generators, solvers, and grid adaptation procedures are dramatically affected by the underlying data structures which establish the connectivity of the grid objects. An extremely compact (however suited exclusively for scalar processing) data structure which contains no other objects but nodes has been developed for the 2D unstructured solver. The cross references of nodes are sorted and positioned so that the neighbors of a given node follow each other counter-clockwise. Twelve fixed reference positions are reserved for each node. This is sufficient for all newly inserted nodes and for nodes of the basis grid which originally have 6 or less neighbors. Basis-grid nodes with more neighbors are assigned a doubled reference space.

The data structure for 2D grid generation is also based on nodes, but since the number of neighbors can vary unpredictably references are stored in unidirectional linked lists.

The 3D grid generator, solver, and adaptation procedures utilize the same data structure with tetrahedra as leading elements. A tetrahedron refers to the neighboring tetrahedra, to the edges, and to the vertices (nodes). The edges in turn refer to the nodes they connect. Extra references from nodes to two "parent" nodes are used to support reversibility of the refinement procedure.

The indirect addressing of the grid objects is used in all cases on the basis of dynamic lists of occupied/spare vacancies to avoid expensive procedures for collecting memory units released by the grid derefinement.

6. Vector processing

As mentioned before, the 2D code was developed for efficient operation on scalar single-processor hardware (workstations and PCs). The grid size (in terms of the number of grid objects) in 3D cases normally requires supercomputer resources.

The main loops of the 3D solver which contain recursions are the loops over tetrahedra and over edges. The control volumes, gradients of primitive gasdynamic variables, and increments of conservative variables for the boundary grid nodes are computed in the loops over tetrahedra; the time step limit and increments of conservative variables for internal grid nodes are computed in loops over edges. For efficient vector processing the above loops are split into vectorizable subloops so that within any particular subloop a quantity associated with a specific grid node is updated only once. The tetrahedra (edges) partitioning is performed as follows: zero binary attributes are assigned to all nodes. For each tetrahedron (edge) the attributes of the involved nodes are analyzed and the position of the first common zero bit denotes the subloop number within which the tetrahedron (edge) is to be treated; the respective bit in the attributes is then set to 1.

7. Post-processing

Post-processing for CFD usually implies an appropriate flow visualization technique applied to the computed flow fields. Contour lines are most popular when presenting 2D results; the same method is used in three dimensions, displaying the flow characteristics at the domain bounding surfaces and/or in a set of cross-sections.

For the purpose of quantitative comparison with experimental data, the computed results can be presented in an experiment-like form, e.g., as interferograms, schlieren, and shadowgraphs (see Fursenko et al. 1993a, 1993b).

In some cases extensive data treatment and acquisition is required for multiple time moments, for example, to prepare flow-field animations. For a shock-focusing problem the pressure data reduction was performed with respect to time, i.e., the complete computation for a separate case resulted in a map of maximal pressure values whenever achieved during the strongly transient process of shock interactions (Babinsky et al. 1995). As the respective procedures involved an expensive data conversion into a fixed structured grid independently for every time step, it was accomplished on the massively parallel platform CRAY T3D.

8. Concluding remarks

A high level of automation in pre-processing is achieved and practical experience is gained mostly for 2D and axisymmetric problems. Numerous flows of those classes have been simulated during the past years, including shock-wave reflection, diffraction and focusing, blast-wave interactions, and some engineering applications (see bibliography by the authors given below). The numerical investigation of a shock-wave interaction with the material interface separating two different gases and analysis of the resulting Richtmyer-Meshkov instability are currently in progress.

The described methods and developed software have been shown to be extremely efficient not only in the use of computer resources, but also in terms of the time required to set up a new problem and get reliable data. Preliminary 3D results suggest that respective software can provide comparable labor efficiency on the basis of modern supercomputer architectures in the near future.

9. A viscous extension

The flexible data structures of the unstructured Euler solvers make it possible to combine them with solvers based on alternative grid systems. For the analysis of viscous flows with shocks and transient separations the 2D unstructured solver has been upgraded to approximate the Navier-Stokes equations and coupled with an implicit structured Navier-Stokes solver to compute the flow in the boundary layer at high Reynolds numbers. Results are reported in Ofengeim et al. (1995).

To simulate strong viscous-inviscid interactions over realistic 3D geometries the unstructured solver was combined with a Navier-Stokes solver written for a prismatic grid attached to the body surface.

Acknowledgement. The authors express their sincere gratitude to Profs. F. Durst, Yu. Golovachov, N. Satofuka, M. Schäfer, and K. Takayama, who contributed greatly to the arrangement of this work. We appreciate helpful assistance of Drs. H. Babinsky, Yu. Makarov, and D. Sharov. The sponsorship by the Russian Foundation of Fundamental Investigations (Grants 93–02–14797 and 94–01–01456), Friedrich Alexander University Erlangen–Nürenberg (Contract FAU4001–01IR303), European Community (COPERNICUS Project COP–94–01490), and Ministry of Education, Science and Culture, Japan, is kindly acknowledged.

References

Babinsky H, Onodera O, Takayama K, Timofeev E, Voinovich P (1995) The influence of geometric variations on shock focusing in cylindrical cavities. In: Shock Waves, Proc of the 20th ISSW, Sturtevant B, Shepherd JE, Hornung HG (eds), World Scientific, Vol 1, pp 495–500.

Barth TJ (1992) Aspects of Unstructured Grids and Finite-Volume Solvers for the Euler and Navier-Stokes Equations, AGARD Report 787.

Fursenko AA (1993) Unstructured Euler solvers and underlaying algorithms for strongly unsteady shocked flows. In: Proc 5th Int Symp on CFD, Daiguji H (ed), JSCFD, 1:227–234.

Fursenko AA, Mende NP, Oshima K, Sharov DM, Timofeev EV, Voinovich PA (1993a) Numerical simulation of propagation of shock waves through channel bends. CFD J 2:1.

Fursenko AA, Sharov DM, Timofeev EV, Voinovich PA (1993b) An efficient unstructured Euler solver for transient shocked flows. In: Shock Waves @ Marseille, Brun R, Dumitrescu LZ (eds), I:371–376.

Löhner R (1987) The Efficient Simulation of Strongly Unsteady Flows by the Finite Element Method, AIAA Paper 87–0555.

Ofengeim D, Timofeev E, Voinovich P (1995) A locally adaptive structured/unstructured Navier-Stokes solver for essentially unsteady shock-obstacle interactions at high Reynolds numbers. In: Shock Waves, Proc of the 20th ISSW, Sturtevant B, Shepherd JE, Hornung HG (eds), World Scientific, Vol 1, pp 537–542.

Rivara MC, Levin C (1992) A 3D refinement algorithm suitable for adaptive and multi-grid techniques. Commun in Appl Num Methods 8:281.

Tanemura M, Ogava T, Ogita N (1983) A new algorithm for three-dimensional Voronoi tesselation. J Comput Phys 51:191–207.

Watson DF (1981) Computing the n-dimensional Delaunay tesselation with application to Voronoi polytopes. Computer Journal 24:167.

Viscous Effects

Boundary-layer transition on slender blunt cones at hypersonic speeds

M.A. Zanchetta, R. Hillier
Department of Aeronautics
Imperial College of Science, Technology and Medicine
Prince Consort Road, SW7 2BY, London, UK

Abstract: Experimentally measured heat-transfer data are presented for hypersonic transitional boundary layers. The experiments are performed on a 5° spherically blunted cone in the Imperial College No. 2 gun tunnel, a Mach 9, high Reynolds number facility. The data presented include time-resolved heat-transfer-rate traces, from which information concerning the structure of the transitional hypersonic boundary layer is extracted. The effects of roughness are examined in the reversal regime.

Key words: Transition, Hypersonics, Intermittency

1. Introduction

Kinetic heating poses severe problems for the design of hypersonic vehicles. Regions of high thermal loads include stagnation points, attachment and re-attachment zones, shock and boundary-layer interactions, shock impingement areas, and boundary layer transition regions. Design proposals use nose blunting to reduce the thermal loads experienced in the stagnation regions. The degree of the blunting, via the effect of the resultant entropy layer, strongly influences the location and the extent of the transition region. This important topic is the subject of this paper.

2. Techniques

The test geometry is a 5° cone. Two models are used. The first is a 700-mm chord length model with ten interchangeable spherically blunted nose cones. This model is equipped with 65 thin-film platinum gauges. The second, a 350-mm chord model, has a 25-mm nose radius and is reserved for the purpose of flow visualization. The test facility is a nominal Mach 9 intermittent gun tunnel. Three fully calibrated conditions offer unit Reynolds numbers, Re_∞/m, between 6.5 and 55 $\times 10^6$, with total temperature in excess of of 1050 K. Run times are typically 20 ms with a 4 ms steady flow window.

3. Results

Thirty different test cases are examined. Each test case is characterized by the nose radius, r_n and the unit Reynolds number, Re_∞/m, or more generally by the nose radius Reynolds number, $Re_n = r_n \times Re_\infty/\mathrm{m}$. Results, shown in Fig. 1, indicate that small amounts of nose blunting, $Re_n < 300,000$, cause a delay in

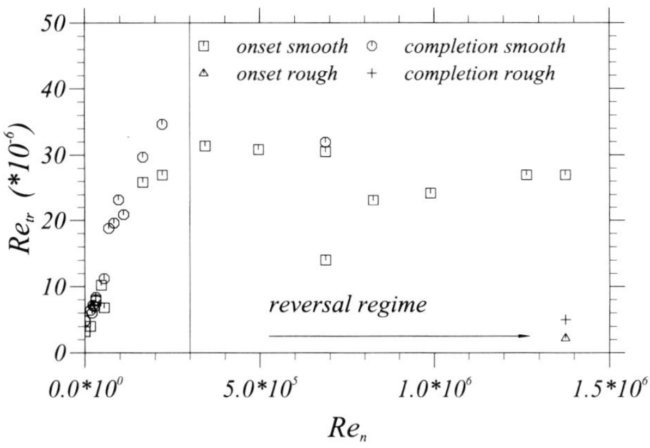

Figure 1. Transition Reynolds number against nose Reynolds number.

transition onset and an increase in the length to transition completion, $Re_{\Delta tr}$. For $Re_n > 300,000$, experiments indicate that the transition behavior is highly sensitive to surface roughness. This regime is denoted the transition reversal regime. A roughness Reynolds number, Re_r, is defined as the product of the roughness height, h, and the unit Reynolds number, Re_∞/m: $Re_r = h \times Re_\infty/\mathrm{m}$. Testing shows that for $Re_r < 500$, the boundary layer is laminar for surface Reynolds numbers in excess of 32×10^6. At $Re_r > 500$, transition completion occurs at 4.2×10^6.

3.1. The structure of the hypersonic transitional boundary layer

In Fig. 2, five heat-transfer-rate time traces are presented. The data are gathered over seven runs with the 3 mm nose at $Re_\infty/\mathrm{m} = 55 \times 10^6$, $Re_n = 165,000$. Due to the nose-blunting entropy-layer effects, there is a significant delay in transition onset, ($Re_{tr} = 25.9 \times 10^6$ cf. $Re_{tr} = 3.2 \times 10^6$ for the sharp nose) and an increase in the length to transition completion ($Re_{\Delta tr} = 3.8 \times 10^6$ cf. $Re_{\Delta tr} = 1.62 \times 10^6$ for the sharp nose). The transition region spans the entire array of gauges. The most upstream gauge, (gauge 02, $s = 0.3505$ m), is wetted by a laminar boundary layer which is characterized by the presence of 'spikes', connected with turbulent or transitional events convecting over the gauge. The

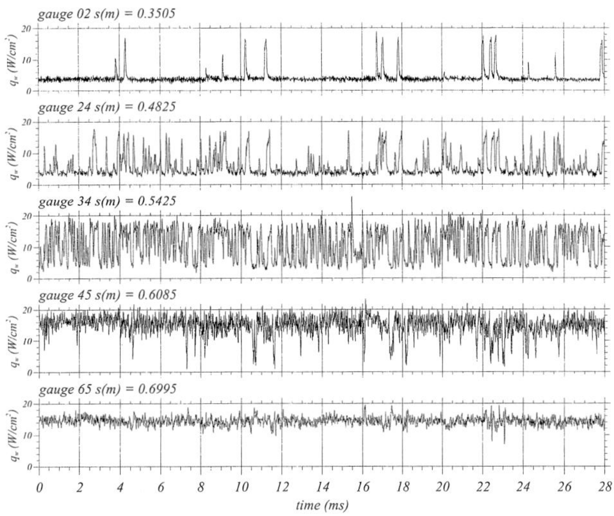

Figure 2. Heat-transfer-rate time histories for five selected gauges; $r_n = 3$ mm, $Re_\infty/\mathrm{m} = 55 \times 10^6$, $Re_n = 165,000$.

duration of the events is 1/10 ms, indicating that they are elongated structures 150δ in length. The structures are tracked as they convect and evolve downstream. Convection speeds calculated between gauge 02 and gauge 24, ($s = 0.4825$ m), indicate that the leading edges and trailing edges of the events travel at 98% and 68% of the local edge velocity, respectively. Further event formation and growth are seen at gauge 34, ($s = 0.5425$ m), where the sensor is witnessing in equal measure laminar and turbulent flow. Due to the speed differential between the leading edge and the trailing edge of the disturbances, successive events merge, leaving only 'stubborn' regions of laminar flow, as seen in the time trace for gauge 45 ($s = 0.6085$ m). Further downstream, the transition process is completed and the boundary layer exhibits purely turbulent behavior, as seen at gauge 65 ($s = 0.6995$ m).

For each gauge the normalized probability density function is determined using 3500 data samples, Fig. 3. The two features visible are the peaks in correspondence to the mean laminar and turbulent heat-transfer-rates, 4 W/cm^2 and 15 W/cm^2 respectively. The distribution at gauge 02 is biased to the 4-W/cm^2, laminar value. Moving downstream, gauges 24, 34, and 45, the effect of boundary-layer transition is to redistribute the PDF from the laminar value

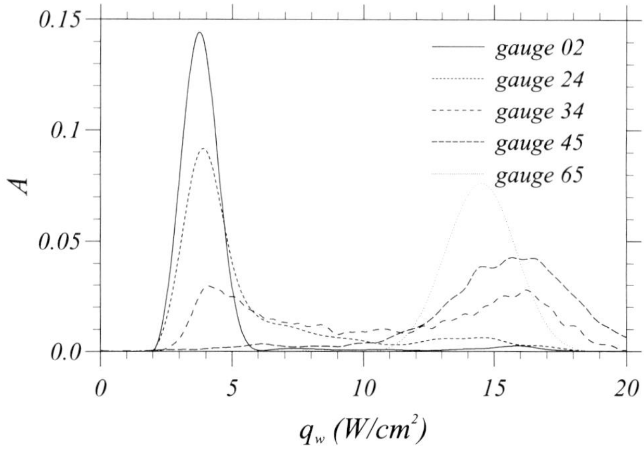

Figure 3. Normalized probability density functions for five selected gauges; $r_n = 3$ mm, $Re_\infty/\mathrm{m} = 55 \times 10^6$, $Re_n = 165,000$.

to the turbulent value. The PDF for gauge 65 is centered purely around the turbulent value of 15 W/cm^2.

It is possible to extract from the time histories the percentage of time each gauge witnesses turbulent flow, represented by the intermittency function, γ. By definition, $\gamma = 0$ for laminar flow and $\gamma = 1$ for turbulent flow. The intermittency distribution against wetted distance for the present case can be found in Fig. 4.

3.2. The reversal regime

The experiments performed in the reversal regime illustrate the boundary-layer sensitivity to roughness in the hemispherical nose region. Liquid-crystal surface-temperature visualization experiments have been performed to study roughness effects. Liquid crystal paints are composed of encapsulated thermochromic particles. When applied to the model surface, they form a covering 10 μm to 20 μm thick. The roughness Reynolds number of the paint, $Re_r < 200$, is insufficient to promote transition and the boundary remains laminar over the entire model length, $Re_{tr} > 22 \times 10^6$, Fig. 5. Controlled application of the paints leads to the formation and deposition of 50 μm particles on the model surface. This enhanced roughness environment, $Re_r = 500$, provides sufficient disturbance to promote transition, Fig. 6. The transition process is characterized by the presence of streak-like structures visible downstream of the sphere–nose junction. The streaks propagate 1.5 nose radii downstream before becoming turbulent. The onset of turbulence is marked by an abrupt change in spreading angle.

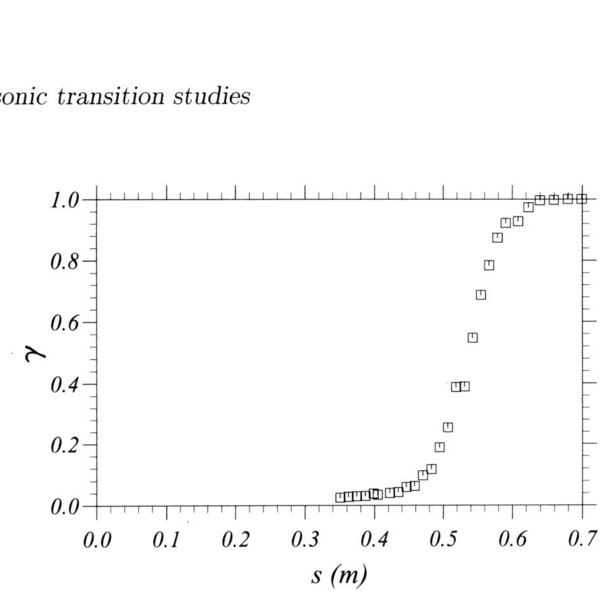

Figure 4. Intermittency function γ, against wetted distance s; r_n = 3 mm, Re_∞/m = 55 $\times 10^6$, $Re_n = 165,000$.

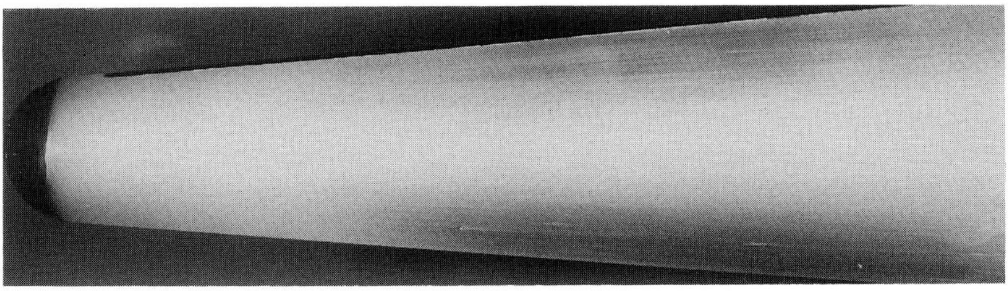

Figure 5. Liquid-crystal surface-temperature visualization, smooth nose configuration; r_n = 25 mm, Re_∞/m = 55 $\times$ 10^6, $Re_n = 1,350,000$.

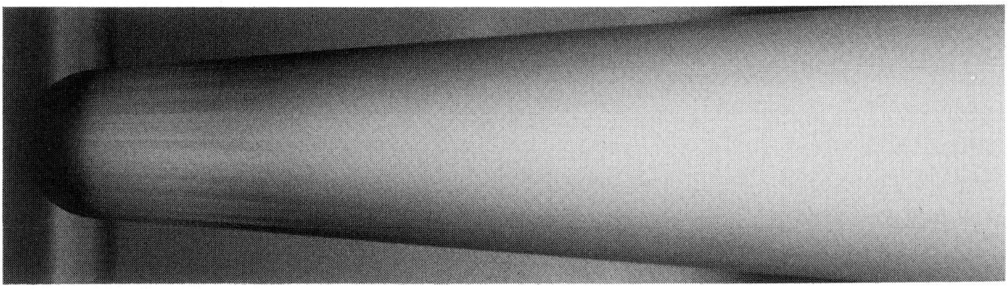

Figure 6. Liquid-crystal surface-temperature visualization, rough nose configuration; r_n = 25 mm, Re_∞/m = 55 $\times$ 10^6, $Re_n = 1,350,000$, $Re_r = 500$.

Due to close circumferential spacing of the streaks, they merge readily once turbulent, giving $Re_{tr} < 5 \times 10^6$.

The sensitivity to isolated roughness elements is explored using 0.13-mm high, 1-mm side length, equilateral triangle trips, located on the sphere–cone junction, ($Re_s = 2.15 \times 10^6$, $Re_r = 1300$.) The effect, Fig. 7, is to cause a highly

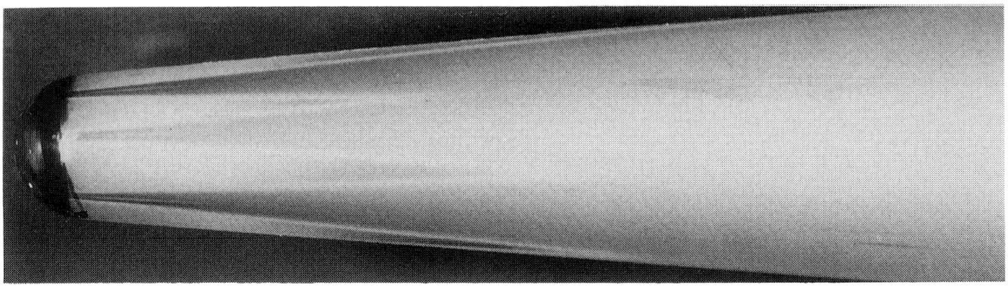

Figure 7. Liquid-crystal surface-temperature visualization, triangular trip elements applied; $r_n = 25$ mm, $Re_\infty/\text{m} = 55 \times 10^6$, $Re_n = 1,350,000$, $Re_r = 1300$.

organized wake, bounded on either side by vortices shed from the rear corners of the trip. The vortices are unstable and undergo transition after travelling two nose radii downstream, giving a transition Reynolds number for the wake of 5×10^6. Transition in other regions of the boundary layer is caused by contamination from the rapidly spreading turbulent wake.

4. Conclusion

The transition location is strongly affected by the presence of the entropy layer associated with nose blunting. The transition location has been experimentally measured for 30 test cases. Transition is delayed by small degrees of nose blunting. Further blunting promotes roughness instability mechanisms. The critical roughness Reynolds number has been determined. The transition process is governed by the formation and growth of turbulent structures. The convection speeds of the disturbances have been calculated and the intermittency distribution has been found.

Acknowledgement. This work is supported by the Defence Research Agency. The authors acknowledge Dr. M. Collyer, Dr. I. Wrisdale, Dr. M. Gilmore, and Dr. T. Caine of DRA Farnborough; and Mr. R. Hutchins and Mr. S. Johnston of Imperial College.

Reflected-shock/side-boundary-layer interaction in a shock tube

M. Nishida, M. G. Lee
Department of Aeronautics and Astronautics, Kyushu University
Fukuoka 812-81 Japan

Abstract: This paper describes the numerical analysis of a reflected-shock/side boundary-layer interaction in a shock tube. The present shock-tube flow is considered as two dimensional and the interaction considered here is such that an initially plane shock reflected from the end wall interacts with the side boundary layer. The computations have been conducted for various Reynolds numbers and incident-shock Mach numbers. Reynolds number of 5×10^5 and incident-shock Mach number of 3.0 were used as typical conditions. Shadowgraphs made from numerical data based on the principle of optical shadowgraph clearly show the bifurcation of a reflected shock wave and the rolling up of a slip line on the shock-tube end wall. Furthermore, the interaction of the shock/boundary layer with suction from the side wall was also investigated to control the interaction.

Key words: Shock/boundary-layer interaction, Shock tube

1. Introduction

It is well known that a reflected shock wave interacts with a side boundary layer in a shock tube which leads to the bifurcation of the reflected shock wave. Concerning the shock/boundary-layer interaction, a number of experimental studies have been conducted. According to these experiments, the flow structure of the shock/boundary-layer interaction is such as shown in Fig. 1 and the simple theory of Mark (1957) can predict that such bifurcation may easily occur for gas with a low ratio of specific heats. On the other hand, numerical simulations of the shock/boundary-layer interaction also have been attempted (Takano 1988, Hataue 1989, Kleine et al. 1992). Takano (1988) did the simulation for nitrogen and argon gases using the thin-layer approximation by means of the flux-corrected-transport method and Beam-Warming scheme as solvers. His result did not show clear shock bifurcation. Although Hataue's simulation using the TVD scheme captured a bifurcated shock, he did not obtain a vortex sheet (Hataue 1989). Kleine et al. (1992) numerically simulated the flow field of this interaction using the MacCormack scheme. They showed the bifurcated reflected shock, but the vortex sheet was not obtained by their numerical simulation.

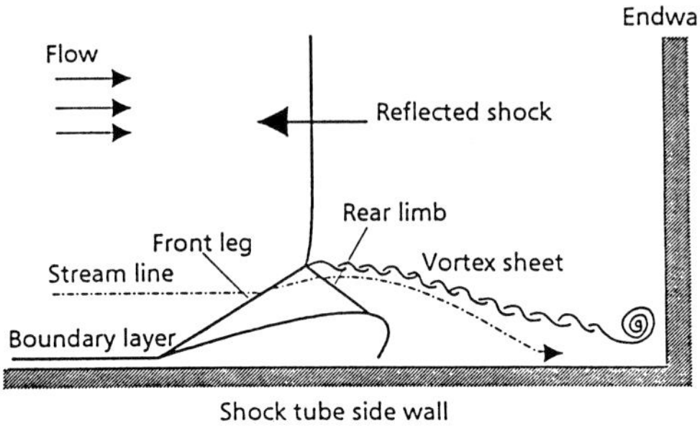

Figure 1. Sketch of a shock/boundary-layer interaction flow field.

In the present work, the numerical calculations of the shock/boundary-layer interaction were carried out for various incident-shock Mach numbers and Reynolds numbers. Furthermore, the interaction of the shock/boundary layer with suction from the side wall was also investigated to control the interaction. The calculations of the shock/boundary-layer interaction with suction were done for various suction speeds. It is the purpose of the present study to investigate the structure of a bifurcated shock wave, the effects of Reynolds number and incident-shock strength on shock/boundary-layer interaction, and the effect of suction on the development of shock bifurcation.

2. Numerical calculation

The two-dimensional flow field treated here is such that an initially plane shock is reflected from the shock-tube end wall and the reflected shock interacts with the side-wall boundary layer. The gas treated is air.

Governing equations for the present problem are two-dimensional Navier-Stokes equations. The computational domain is the lower half region of the duct composed of the left free boundary, lower side wall, upper duct axis, and right duct end wall. The boundary conditions imposed are: no slip and adiabatic wall conditions on the wall, symmetry conditions on the axis, and inflow/outflow condition on the free boundary. Numerical calculations were carried out using the second-order upwind TVD scheme of the Harten-Yee type (Yee 1989).

3. Numerical results

3.1. Structure of bifurcated shock

Computational results for $M_{s1} = 3.0$ and $Re = 5 \times 10^5$ are shown as isopycnics in Fig. 2. The time t' in the figure has been normalized by R/a_1, and $Re = \rho_1 a_1 R/\mu_1$, where R is a half of the duct height, a is the speed of sound,

ρ is the density, μ is the viscosity and the subscript 1 denotes the quantity ahead of an incident shock wave. In this computation, a 400×150 grid system was employed. Like past experimental results, the bifurcation of a reflected shock can be observed in the present numerical result, namely a front-leg shock (indicated by F.L.) and a rear-limb shock (indicated by R.L.) are captured. In addition, it can be clearly seen in the isopycnics that a slip line runs from a triple point towards the end wall and rolls up on the end wall. Strangely, a new discontinuity appears between the front-leg shock and rear-limb shock which has not been shown in any past experiments or computations of reflected shock/side boundary-layer interaction. We have no satisfactory explanation for whether this discontinuity is physical or numerical.

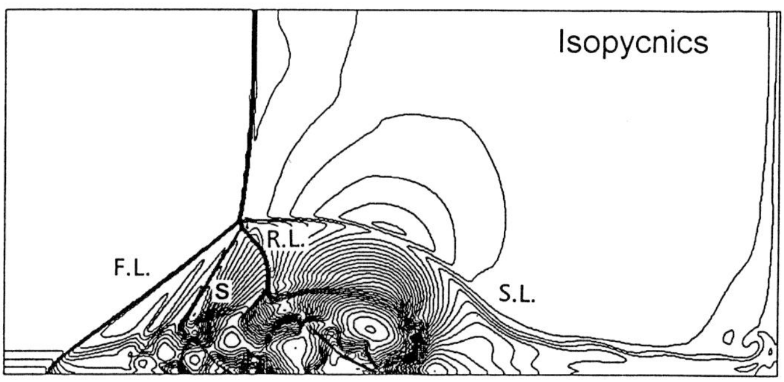

Figure 2. Isopycnics for $M_{s1} = 3.0$ and $Re = 5 \times 10^5$.

3.2. Development of bifurcation in reflected shock

The computed shadowgraph of Fig. 3 illustrates the development of the bifurcation of the reflected shock for $M_{s1} = 3.0$ and $Re = 5 \times 10^5$. This illustration has been developed using the principle of optical shadowgraph. The front-leg shock and rear-limb shock can be clearly identified at $t' = 1.28$. The aforementioned new discontinuity between the front-leg shock and rear-limb shock appears only at $t' = 1.91$ and it is not observed before this time. At $t' = 1.91$, a slip line runs towards the end wall; especially interesting is that this slip line forms a swirl on the end wall. This is very similar to the colored schlieren picture of Kleine et al. (1992). There have been no computational results which succeeded in numerical capture of a vortex sheet.

3.3. Effects of Reynolds number and incident-shock strength

The effect of Reynolds number on the development of a bifurcated shock wave was experimentally investigated by Fokeev and Gvozdeva (1990). Here this

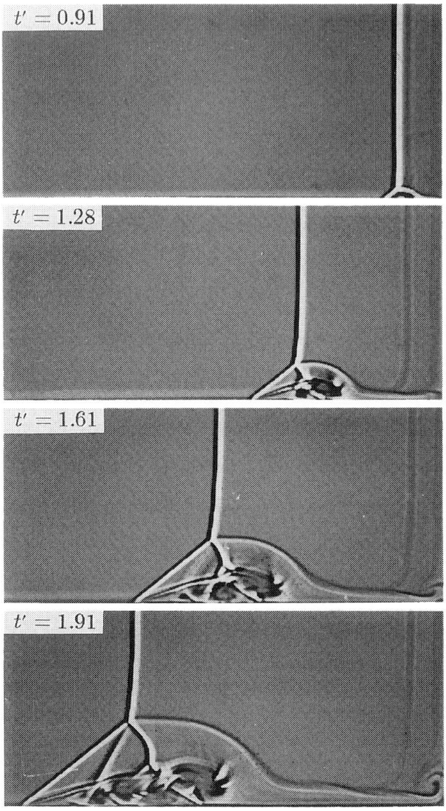

Figure 3. Computed shadowgraphs for M_{s1} = 3.0 and Re = 5 × 10^5.

problem has been numerically studied. Computations were carried out for M_{s1} = 3.0 and Re = 5.0 × 10^4, 1.5 × 10^5, 5.0 × 10^5. The height of the bifurcation is almost linearly proportional to the distance from the end wall and slightly increases with a decrease in Reynolds number (see Fig. 4).

The effect of incident-shock strength on bifurcation of a reflected shock wave was also investigated. The calculations were carried out for Re = 5 × 10^4, and M_{s1} = 2.0, 2.5, 3.0, and 3.5. The results are shown in Fig. 4, where the height of a triple point versus distance from the end wall is also illustrated for the four incident-shock Mach numbers. It is seen from the figure that the height of the triple point is increased with incident-shock strength.

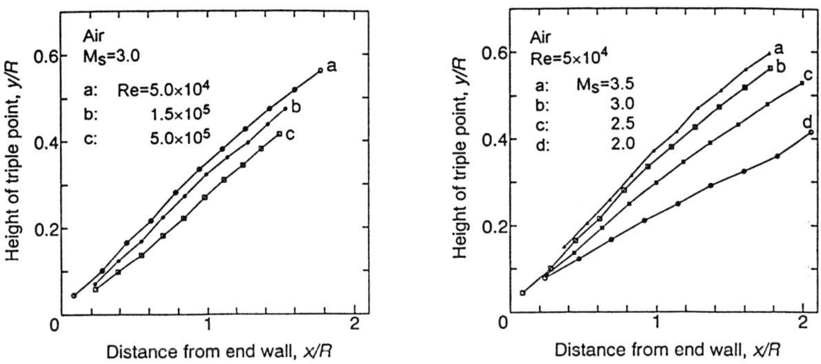

Figure 4. Effects of Reynolds number and incident-shock strength.

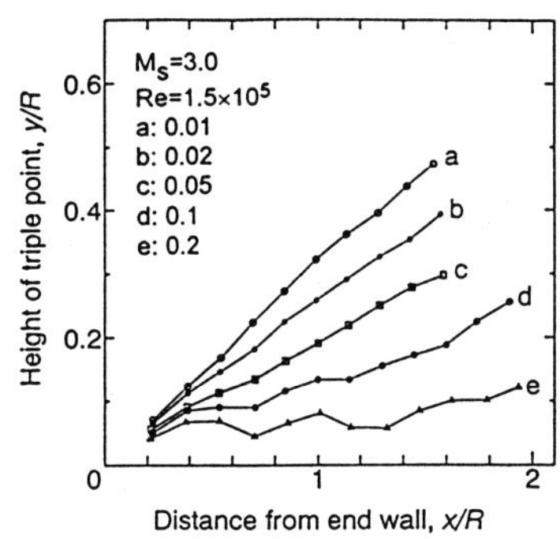

Figure 5. Control of bifurcated shock wave with boundary-layer suction.

3.4. Effect of boundary-layer suction

Active control of the bifurcation of a reflected shock wave is very important for usage of high-pressure and high-temperature gas behind a reflected shock as reservoir gas, because premature contamination is caused by the wall jet in the vicinity of the side wall and as a result, operation time of a shock tunnel is reduced. Xu et al. (1988) experimentally confirmed that the suction of the boundary layer induced by an incident shock wave can control the height of a triple point and reduce the wall jet.

In the present work, the flow field of the bifurcated shock wave with boundary layer suction was numerically analyzed for normalized suction speeds $U_{\rm suc}/a_1$ of 0.2, 0.1, 0.05, 0.02, and 0.01, where a_1 is the speed of sound ahead of an incident shock wave. The height of the triple point versus a distance from the end wall is shown in Fig. 5, where the values for lines a–e are the ratio U_{suc}/a_1. As expected, the height of the triple point is decreased with an increase in suction speed.

4. Concluding remarks

In the present study, the flow field of reflected shock/side-boundary-layer interaction was numerically analyzed using full Navier-Stokes equations. The present computation captured clear bifurcated shock structure and a slip line rolling up on the end wall; these are very similar to experimental results. Next, in order to reveal effects of Reynolds number and incident-shock strength on the bifurcated shock structure, the height of the triple point was estimated for various Reynolds numbers and various incident-shock Mach numbers. The results show that the height of the triple point is almost linearly increased with distance from the end wall. It is remarkably dependent on incident-shock Mach number, but the effect of Reynolds number is relatively small. Finally, the effect of boundary-layer suction on the bifurcated shock structure was investigated and it was found that the height of the triple point decreased with increasing suction speed.

References

Fokeev VP, Gvozdeva LG (1990) Study of bifurcation of reflected shock waves in channels of various cross-sections. In: Current Topics in Shock Waves, American Institute of Physics, pp 862–866.

Hataue I (1989) Computational study of the shock-wave/boundary-layer interaction in a duct. Fluid Dynamics Research 5:217–234.

Kleine H, Lyakhov VN, Gvozdeva LG, Grönig H (1992) Bifurcation of a reflected shock wave in a shock tube. In: Proc 18th Intl Symp on Shock Waves, pp 261–266.

Mark H (1957) The interaction of a reflected shock wave with the boundary layer in a shock tube. J Aero Sci 24:304–306.

Takano Y (1988) Simulations for effects of side-wall boundary-layer on reflected shock flowfields in shock tubes. In: Proc 16th Intl Symp on Shock Tubes and Waves, pp 645–651.

Xu L, Gourlay CM, Stalker RJ (1988) Reduction of driver gas contamination in a reflected shock tube by boundary layer suction. In: Proc 16th Intl Symp on Shock Tubes and Waves, pp 637–643.

Yee HC (1989) A Class of High Resolution Explicit and Implicit Shock-Capturing Methods, NASA TM–101088.

Effects of shear-layer transition and leading-edge bluntness on control-flap effectiveness

D. Kumar, J. L. Stollery
College of Aeronautics, Cranfield University
Cranfield, Bedfordshire, MK43 0AL UK

Abstract: The effects of shear-layer transition and leading-edge bluntness on flap-control effectiveness have been studied experimentally. A quasi-two-dimensional flat plate equipped with a full-span trailing-edge flap was employed for tests. Leading-edge bluntness effects were simulated by introduction of a hemicylindrical leading edge. Flow structure was studied using high-speed schlieren photography and surface-pressure and heat-transfer measurements. Analytical theories have been developed to estimate the flap pressure and laminar and turbulent heat-transfer levels for the sharp and blunt configurations. These are compared against experimental measurements.

Key words: Control flap, Laminar separation, Transition, Bluntness

1. Introduction

Large flap deflections produce extensive regions of separated flow. At low hypersonic Mach numbers, the instability of the separated shear layer promotes transition of the initially laminar boundary layer and leads to either transitional or in severe cases, fully turbulent reattachment. Shear-layer transition produces a substantial increase in heat transfer and hence imposes severe aerothermal loads on the flap. The need to relieve stagnation-point heat transfer at hypersonic speeds dictates the intentional blunting of leading edges. The introduction of bluntness produces a considerable change in the aerodynamic flow structure from that of an equivalent sharp leading edge body. The primary effect is the formation of a bow shock standing off from the leading edge. The shock creates a region of high temperature, low-density flow confined to a region known as the "entropy layer." The increase in leading-edge shock strength together with its highly curved structure give rise to gradients in flow properties (confined to the entropy layer), such as Mach number and unit Reynolds number along and normal to the body surface. The changes in the body flow field due to leading-edge bluntness considerably change the performance of a submerged control flap.

2. Analytical study

Analytical theories employed for estimating the flap heat transfer on both sharp and blunt configurations are deacribed by Kumar & Stollery (1994). In the present work, the leading edge was taken as the origin for the laminar solution whilst the hingeline was used as a virtual origin for the turbulent solutions.

3. Wind-tunnel facility

The Cranfield hypersonic gun tunnel was used; tests were conducted with air as the working gas. An axisymmetric contoured nozzle provided a useful test jet diameter of 15.0 cm at a freestream Mach number of 8.2. The tunnel was equipped with a single-pass schlieren system which used a high-intensity microsecond-duration argon spark source to illuminate density gradients in the flow. Pressure measurements were made using Kulite XCS-190 series pressure transducers. Surface heat-transfer measurements were made using thin-film platinum gauges mounted on a Macor substrate. The temperature history of the gauges was integrated using a low-noise, high-bandwidth analog integrator circuit. The models used are shown in Fig. 1.

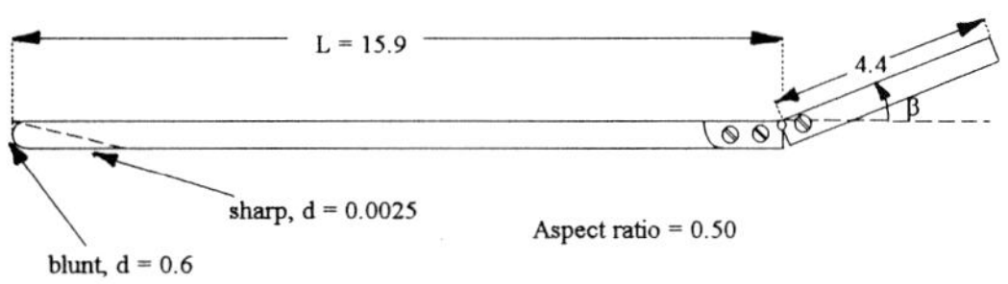

Figure 1. Diagram of control flap model (all dimensions in cm).

4. Results and discussion

In Fig. 2, the single hingeline shock for the $\beta = 5°$ configuration confirms the flow as attached. The compression associated with this shock increases the pressure on the flap. The increase starts at the hingeline and is typical of the pressure distribution for an attached flow configuration. Whilst the flow is attached for the $\beta = 5°$ configuration, the plateau in the pressure distribution for $\beta = 10°$ indicates a well-separated flow structure. This is confirmed by the dual-shock interaction structure in the corresponding schlieren photograph with separation located at $x/L = 0.88$. Further increasing the flap deflection angle to $\beta = 25°$ results in separation occurring at $x/L = 0.6$. There is a small upstream movement in reattachment. The sharp flat-plate heat-transfer measurements of Kumar & Stollery (1994) have shown the flat-plate boundary layer to be laminar. For all the sharp leading-edge configurations shown in Fig. 2, the decrease in the measured heat-transfer distributions below the flat-plate level at the start of their respective interactions is characteristic of laminar interactions. For $\beta = 5°$, the agreement of the measured flap heat-transfer rates with the predictions of laminar reference temperature theory confirms the flap boundary layer as laminar. Thus, transition is not promoted by this flap deflection. For $\beta = 10°$, the increase in the measured flap heat-transfer rates

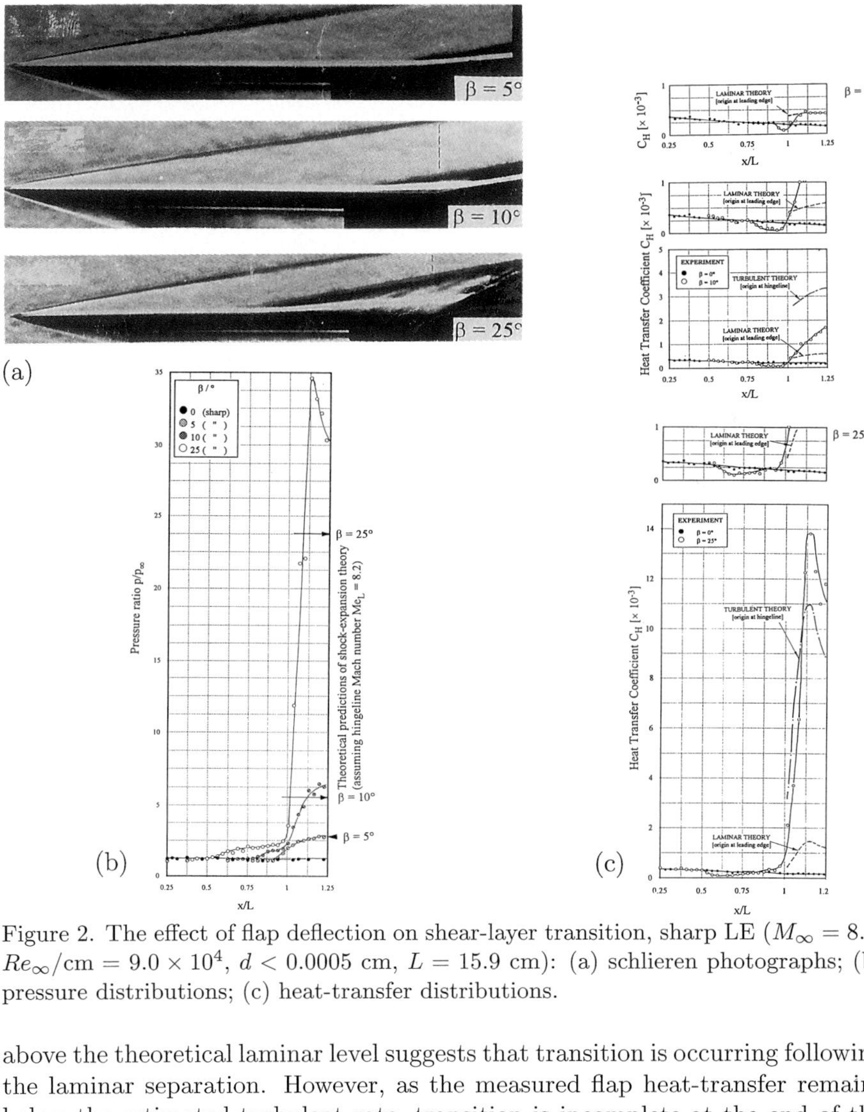

Figure 2. The effect of flap deflection on shear-layer transition, sharp LE ($M_\infty = 8.2$, $Re_\infty/\text{cm} = 9.0 \times 10^4$, $d < 0.0005$ cm, $L = 15.9$ cm): (a) schlieren photographs; (b) pressure distributions; (c) heat-transfer distributions.

above the theoretical laminar level suggests that transition is occurring following the laminar separation. However, as the measured flap heat-transfer remains below the estimated turbulent rate, transition is incomplete at the end of the flap. For $\beta = 25°$, the decrease in heat transfer at separation confirms the local flow as laminar. However, the subsequent increase in heat transfer from within the interaction region suggests rapid transition of the shear layer. This is also evident from the corresponding schlieren photograph which shows shock waves emanating from the edge of the disturbed shear layer. The agreement between the measured flap heat-transfer rates and the predictions of turbulent theory suggest that transition is complete prior to reattachment.

In Fig. 3, the measured flap heat-transfer rates at $x/L = 1.19$ are compared with the predictions of laminar and turbulent reference temperature theory. This clearly shows that flap deflection promotes transition, and in severe cases

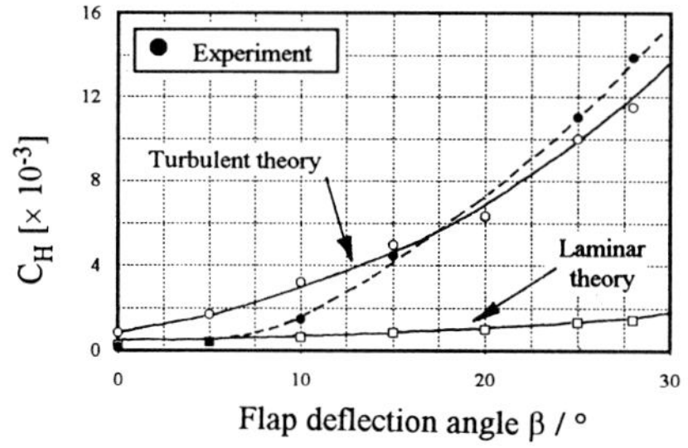

Figure 3. The effect of shear-layer transition on flap heat-transfer rates, sharp LE ($M_\infty = 8.2$, $Re_\infty/\text{cm} = 9.0 \times 10^4$, $x/L = 1.2$ cm).

leads to fully turbulent conditions being realized on the flap. The promotion of separation together with the increased flap adverse pressure gradient are the primary agencies responsible for the promotion of transition with flap deflection. The pressure distribution for the blunt configurations is shown in Fig. 4. The blast effect of the blunt leading edge increases the flat-plate pressure level above that for a sharp leading-edge flat plate. For both $\beta = 5°$ and $\beta = 10°$, the increase in pressure from the hingeline together with the single hingeline shock in the corresponding schlieren photograph confirms the flow on these configurations as attached. The attached flow on the latter configuration contrasts with the well-separated flow structure observed on the corresponding sharp configuration. Thus, for this configuration, bluntness completely surpresses separation. For well-separated flows, ($\beta = 25°$) the delaying of separation from $x/L = 0.6$ on the sharp configuration to $x/L = 0.77$ on the blunt configuration is evident by comparing the pressure distributions and schlieren photographs, shown in Figs. 2 and 5 respectively. The pressure distributions also show that bluntness considerably reduces the pressure recovered on the flap. The calculated flap pressure levels on the blunt configuration, made by assuming the flap to be submerged in an effective supersonic freestream, are shown in Fig. 5. The agreement of these estimates with the measured flap pressure level immediately downstream of reattachment confirms that the loss of stagnation pressure across the leading edge bow shock is responsible for the reduction in flap pressure by bluntness. In a classical theoretical study, Cheng et al. (1961) found that the relative effects of leading-edge bluntness and the viscous displacement effect are measured by the bluntness-viscous interaction parameter $\chi_\epsilon/\kappa_\epsilon^{2/3}$. The sep-

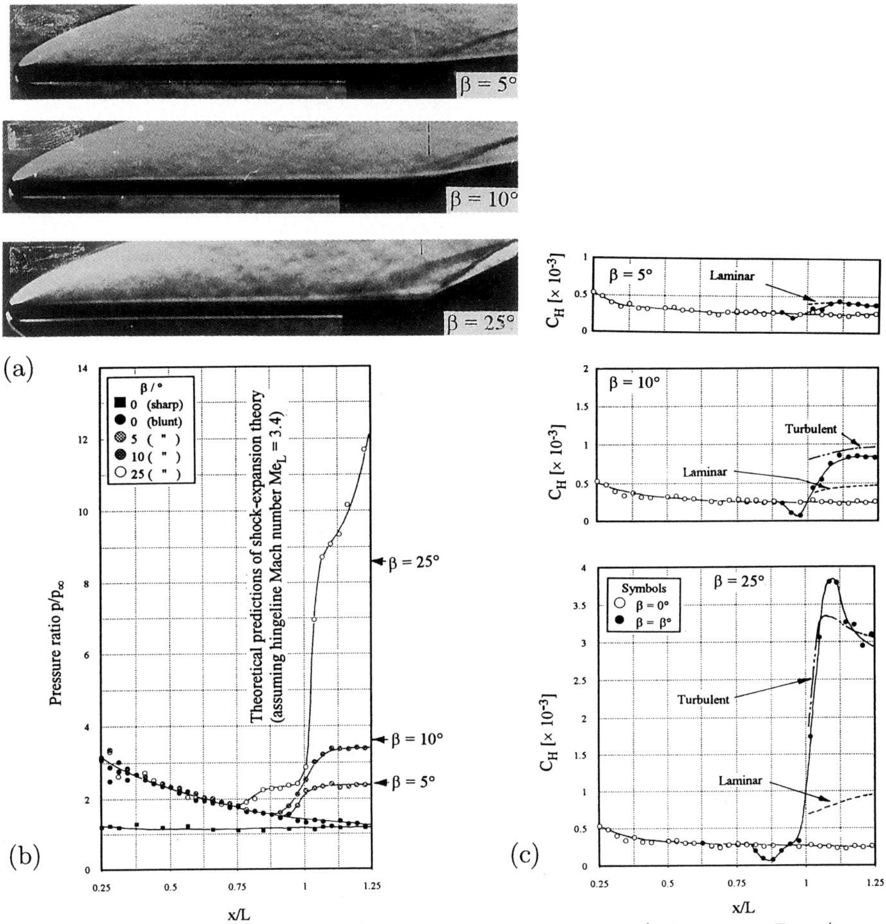

Figure 4. The effect of leading-edge bluntness on separation ($M_\infty = 8.2$, $Re_\infty/\text{cm} = 9.0 \times 10^4$, $d = 0.6$ cm, $L = 15.9$ cm: (a) schlieren photographs; (b) pressure distributions; (c) heat-transfer distributions.

aration lengths and reattachment pressure ratio from the present study and from that of Holden & Moselle (1992) are correlated with Cheng's parameter in Fig. 5. The figure clearly shows two branches — one associated with "small" bluntness and the other associated with "large" bluntness. For the former, changes in the separated flow length occur due to the changes in boundary-layer edge conditions arising within the entropy layer generated by leading-edge bluntness. For "large" bluntness, the correlation shows a substantial decrease in the reattachment pressure ratio with bluntness. This decreases the length of separated flow. The differences between the present study and that of Holden and Moselle's may be due to the transitional nature of the former.

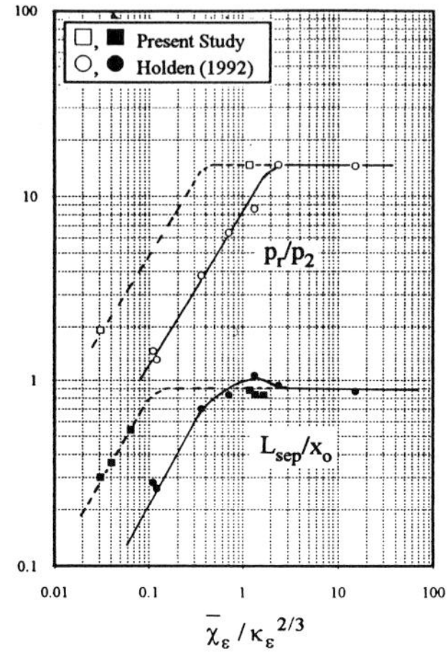

Figure 5. The effect of bluntness on separated flows.

5. Conclusion

Flap deflection promotes separation of laminar boundary layers. The instability of the separated shear layer, together with the adverse pressure gradient promotes transition of an initially laminar boundary layer. Large flap deflections produce extensive regions of separated flows. This results in the onset and sometimes the completion of transition within the separated flow interaction region. Large bluntness substantially delays separation. This is due to the reduction of the pressure recovered on the flap. The reduction in flap pressure by bluntness also alleviates flap heat transfer for both laminar and turbulent flap boundary layers.

References

Ames Research Staff (1953) Equations, Tables and Charts for compressible Flows, NACA-R-1135.

Holden MS, Moselle JR (1992) A Database of Aerothermal Measurements in Hypersonic Flow for CFD Validation, AIAA Paper 92-4023.

Kumar D, Stollery JL (1994) Hypersonic control flap effectiveness. In: Proc of the 19th ICAS Congress, Anaheim, California.

Lukasiewicz J (1961) Hypersonic Flow Blast Analogy, AEDC TN-61-158.

Needham DA (1965) Laminar Separation in Hypersonic Flows, PhD Thesis, University of London (see also Needham DA and Stollery JL, AIAA Paper 66-455, 1966).

Statistical simulation of nonequilibrium hypersonic flows over concave bodies

M. S. Ivanov, S. F. Gimelshein, G. N. Markelov
Institute of Theoretical and Applied Mechanics, Novosibirsk 630090 Russia
Y. E. Gorbachev
International Institute for Interphase Interactions
St. Petersburg 194291 Russia

Abstract: Hypersonic rarefied flows over 2D bodies (a plate, a plate with a flap, an ellipse, an ellipse with deflected rear part) are calculated by the Direct Simulation Monte Carlo (DSMC) method. Various models of vibrational-translational energy exchange and chemical reactions are considered and the influence of the model on the simulated flow is examined.

Key words: Chemically reacting hypersonic rarefied flows, DSMC method

1. Introduction

The study of space-vehicle aerodynamics at the altitudes 80–120 km becomes more and more important. This is primarily caused by the development of a new generation of space transportation systems, for which the mentioned altitudes are determining factors. This requires more comprehensive and detailed understanding of the problems of hypersonic aerodynamics at high flight altitudes. The principal tool for solving the problems of rarefied gas aerodynamics is the Direct Simulation Monte Carlo (DSMC) method. At present, this method is frequently used for studying the flow over space vehicles at altitudes of about 80 km where the real-gas effects are essential, i.e., nonequilibrium excitations of internal (rotational and vibrational) degrees of freedom of molecules and chemical transformations. Many models have been suggested to describe the excitation of internal degrees of freedom and chemical reactions by the DSMC method, varying in their labor intensity and physical validity. The variety of models and algorithms leads to one of the most interesting problems: to what extent does the choice of this or that new model affect the simulated aerodynamics of various bodies. In the present paper this problem is considered.

2. Flow conditions and gas models

The following set of bodies was considered: (a) a flat plate, (b) a plate with a flap, (c) 10% elliptic profile, (d) 10% elliptic profile with deflected rear part. The length of the deflected part (cases b and d) was $0.2L$ where L is the total body length, and the angle of flap deflection was $\delta = 30°$. The flap was deflected towards the free stream. In all cases the angle of attack was 40°. The free-stream velocity and temperature were $U_\infty = 7600$ m/s and $T_\infty = 189$ K, respectively. The wall temperature was assumed constant and equal to $T_w = 1000$ K. The Knudsen number determined from characteristic length L was $Kn = 0.01$.

The calculations were made by the use of the DSMC numerical schemes based on the majorant principle (Ivanov and Rogasinsky 1991). For the intermolecular potential, the Variable Hard Sphere model with an exponent of 0.75 for viscosity-temperature dependence was chosen. The diffuse law with complete accommodation of translational and internal energies was suggested for gas-surface interaction. The Larsen-Borgnakke phenomenological model was employed for simulating the exchange between translational and rotational energy degrees of freedom. Rotation-relaxation collision number Z_r was assumed to be 5. Molecular oxygen was selected as a free-stream gas.

Models used to describe the vibrational-translational (VT) energy exchange:

V1. The Larsen-Borgnakke model with vibration-relaxation collision number $Z_v = 50$. It is the model with the fastest vibrational relaxation.

V2. A model taking into account the quantum nature of vibrations (Ivanov 1993); molecules are considered as anharmonic oscillators (AHO) and only one-step vibrational transitions are involved. It is a model with the slowest vibrational relaxation.

V3. A model for VT exchanges with multi-step transitions (Gimelshein et al. 1994). It employs the energy-dependent form for VT-transition cross sections found using quasiclassical theory for the Morse intermolecular potential.

Models used to describe the gas-phase chemical reactions:

C1. A model with reaction probabilities that depend only on the total energy (i.e. the sum of translational, rotational, and vibrational energies) of a colliding pair. The collision dynamics and algorithm for this case were described in detail in the paper of Gimelshein and Ivanov (1994). This model was used together with model V1 for vibrational excitations.

C2. A model of chemical reactions which takes into account the vibration-dissociation coupling (VDC) (Ivanov 1993). This model was used together with models V2 and V3 for vibrational excitations.

3. Results and discussion

Flat plate. The influence of the gas model on the pressure distribution C_p along the plate is shown in Fig. 1 for various models of nonreacting gas. Here, a TR gas is a gas with translational and rotational degrees of freedom; a TRV gas means the vibrational mode is also considered. Accounting for vibrations leads to a slight C_p reduction along the plate. The choice of vibrational model is not significant for this case (the maximum discrepancy between V1 and V2 models is about 7%).

Figure 2 presents the comparison of C_p distribution along the plate for TRV gas and various models of chemically reacting oxygen. The values of C_p for the V1+C1 model differ appreciably from those for other models on the first half of the plate. This is due to the fact that V2+C2 and V3+C2 models take into account the VDC effect. Since the vibrational mode has no time to be excited in the shock wave near the plate tip, C_p values for these models are almost the same as those for nonreacting gases. Multilevel VT transitions in the V3 model accelerate noticeably the VT relaxation (and hence, chemical reaction rates), compared with the V2 model, and C_p coincides on the lower part of the plate for the V1+C1 and V3+C2 models. Note that the shock wave is attached to the plate and its inclination angle varies depending on the gas model. Inclination-angle reduction causes lower C_p values along the plate.

Plate with a flap. The C_p distribution along the plate with flap for various vibrational models is shown in Fig. 3. The effect of the model in this case is evidently much more significant than for a flat plate. C_p values coincide for various gas models near the body tip but their difference is about 30–50% in the middle of the body and on the flap. The choice of the model for chemical reactions exerts also a considerable effect on C_p, though it is not as significant as the VT model. C_p near the plate tip is slightly lower for V1+C1 than for VDC models, as it was for a flat plate. The maximum difference is observed on the flap.

The influence of vibrational-relaxation and chemical-reaction rates for various models is clearly illustrated by the fields of atomic oxygen mole fraction (Fig. 5). For the V1+C1 model (Fig. 5a) a high concentration of atomic oxygen is observed on the plate tip. The highest increase in concentration for this model occurs in the bow shock wave and is due to rapid growth of translational temperature there. The chemical-reaction rate for VDC models is primarily determined by the vibrational temperature, which increases much slower than translational and rotational temperatures. Therefore, chemical-reaction rates for VDC models are fairly high in the region between the shock and the body. Note that in all above cases, the shock wave was attached to the body.

Elliptic profile. Unlike the case for a plate, the shock wave induced by an ellipse is detached. This is probably the main reason for the smaller influence of the vibrational and chemical-reaction models on the aerodynamic coefficients. As is seen from Figs. 6 and 7, the change of vibrational model hardly influences C_p, and the effect of the chemical-reaction model is 4–5%. The most prominent difference is observed near the body tip.

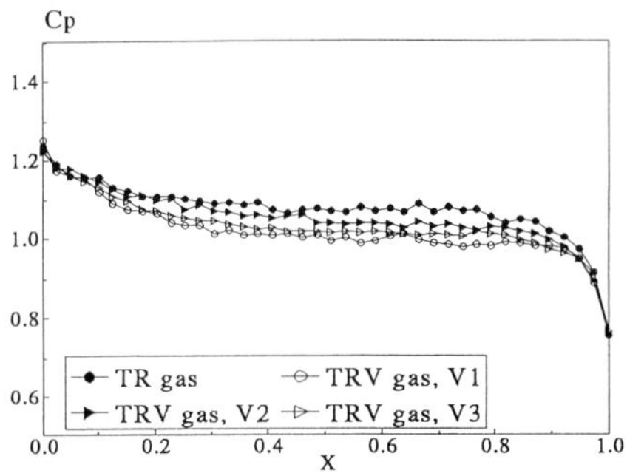

Figure 1. Pressure coefficient over a plate with a flap for various models of molecular vibrations.

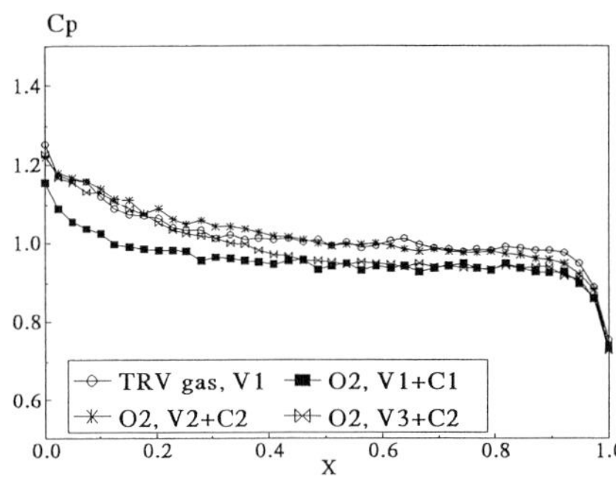

Figure 2. Pressure coefficient over a flat plate for various models of chemical reactions.

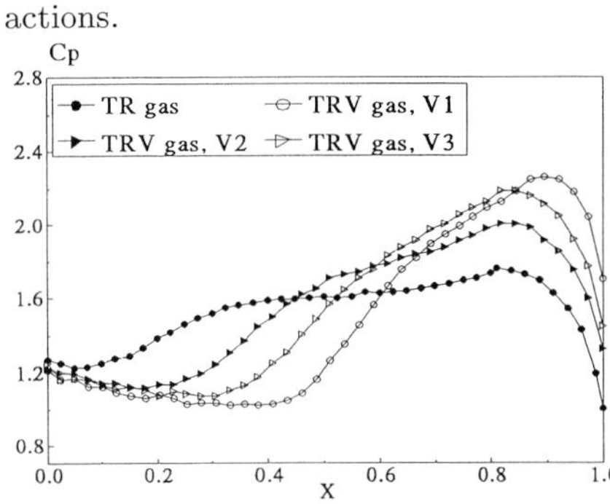

Figure 3. Pressure coefficient over a plate with a flap for various models of molecular vibrations.

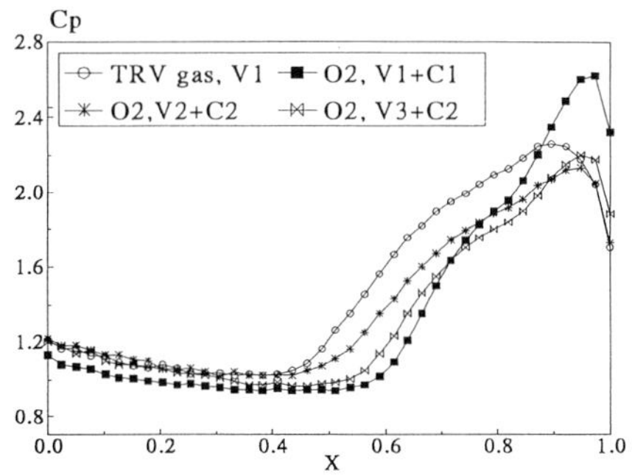

Figure 4. Pressure coefficient over a plate with a flap for various models of chemical reactions.

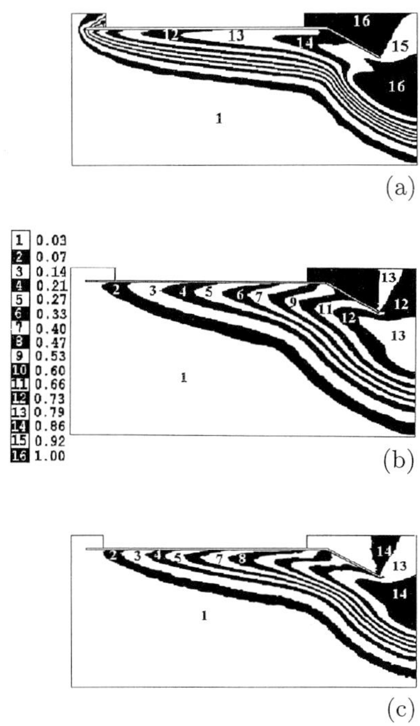

Figure 5. Atomic-oxygen mole fraction about a plate with a flap for model combinations: (a) V1 + C1; (b) V2 + C2; (c) V3 + C2.

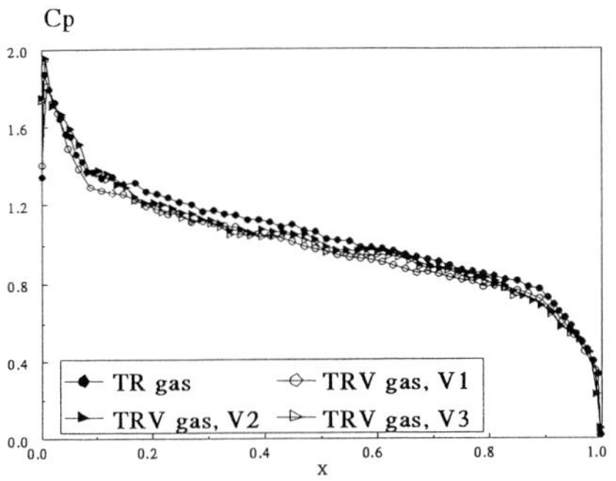

Figure 6. Pressure coefficient over an ellipse for various models of molecular vibrations.

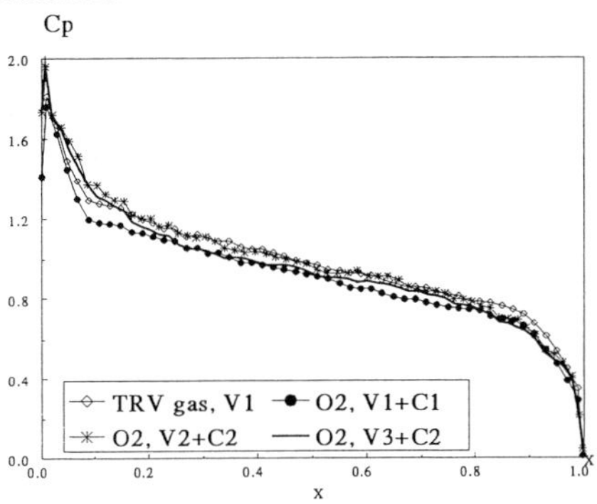

Figure 7. Pressure coefficient over over an ellipse for various models of chemical reactions.

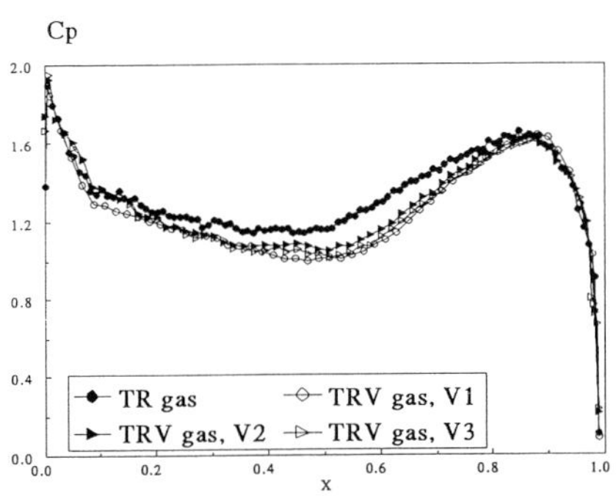

Figure 8. Pressure coefficient over over an ellipse with a flap for various models of molecular vibrations.

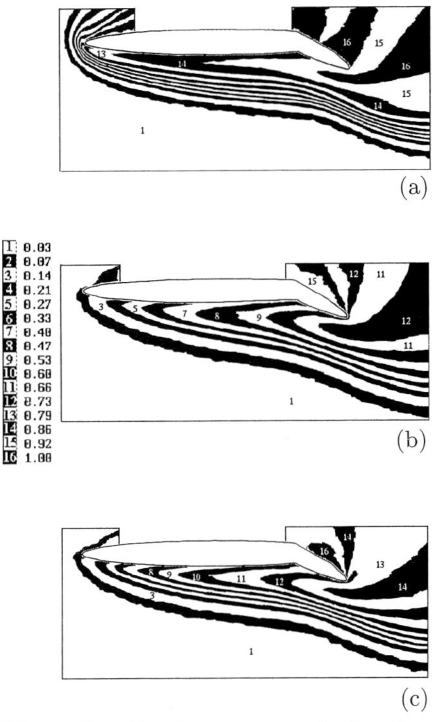

Figure 9. Atomic-oxygen mole fraction about an ellipse with a flap for model combinations: (a) V1 + C1; (b) V2 + C2; (c) V3 + C2.

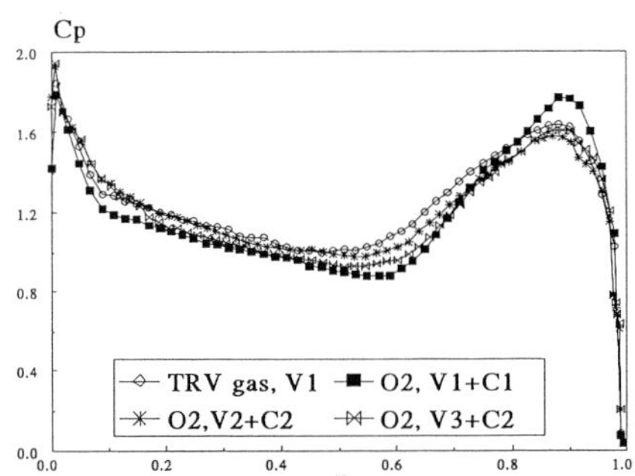

Figure 10. Pressure coefficient over over an ellipse with a flap for various models of chemical reactions.

Elliptic profile with deflected rear part. The deflection of the rear part of the ellipse causes a C_p increase on the second half of the body (Fig. 8). Excitation of the vibrational mode reduces C_p, whereas the choice of vibrational model scarcely affects it, and the maximum discrepancy amounts to 3–5%.

The effect of the chemical model is most conspicuous in the fields of atomic oxygen mole fraction (Fig. 9). However, despite an appreciable difference in the flow structure, the choice of chemical model weakly affects the pressure distribution for the elliptic profile with deflected rear part (Fig. 10).

4. Conclusions

The flow over simple 2D bodies was calculated by the DSMC method for various models for vibrations and chemical reactions. Concave bodies are much more sensible to the gas model variation than convex ones. The choice of the gas model is most significant for a plate with a flap.

For the flow over a plate and a plate with a flap, the choice of the vibrational model makes a larger contribution to aerodynamic coefficients than the chemical model. The reverse situation occurs for elliptic profiles. The determining factor here is whether the bow shock is detached or attached to the body.

Acknowledgement. The authors wish to acknowledge support from the International Science Foundation (Grant No. RBW000).

References

Gimelshein SF, Gorbachev YE, Ivanov MS, Kashkovsky AV (1994) Real gas effects on the aerodynamics of 2D concave bodies in the transitional regime. In: 19th Int Symp on Rarefied Gas Dynamics, Oxford, UK.

Gimelshein SF, Ivanov MS (1994) Simulation of chemically reacting gas flow using majorant frequency scheme of DSMC. In: Rarefied Gas Dynamics: Theory and Simulations (BD Shizgal and DP Weaver, eds), Vol. 159, Progress in Astronautics and Aeronautics, AIAA, Washington, DC, pp 218–233.

Ivanov MS (1993) Transitional regime aerodynamics and real gas effects. In: Proc Int Symp on Aerospace and Fluid Science, Sendai, Japan, pp 77–101.

Ivanov MS, Rogasinsky SV (1991) Theoretical analysis of traditional and modern schemes of the DSMC method. In: 17th Int Symp on Rarefied Gas Dynamics, Aachen, Germany, pp 629–642.

Transition of unsteady boundary layer induced by propagating compression wave

T. Aoki, N. Kondoh, K. Matsuo
Kyushu University
Kasuga, Fukuoka, 816 Japan
S. Mashimo
West Japan Railway Co.
Kita-ku, Osaka, 530 Japan

Abstract: The transition of an unsteady boundary layer induced by a propagating compression wave has been investigated by means of a recently developed laser differential interferometry (LDI) technique using Wollaston prisms. LDI can be made 2 to 3 orders of magnitude more sensitive to optical path change than classical interferometry. The sensitivity was high enough to record the density variations of a compression wave. Measurements were made of time-dependent profiles that gave detailed information on the process of formation of the unsteady boundary layer induced by propagating compression waves in a constant-area duct. It was found that the laminar-to-turbulent transition in the unsteady boundary-layer flow exits describing the density profiles.

Key words: Transition, Unsteady boundary layer, Compression wave, Laser differential interferometry

1. Introduction

The discharged wave from the exit of a tube radiates toward the surrounding area as an impulsive wave and causes an impulsive noise. We can easily understand that this sort of noise due to a discharged wave is similar to the compression wave emerging from the exit portal of a high-speed railway tunnel (Ozawa 1979).

According to the aeroacoustic theory, the magnitude of an impulsive noise from the exit of a tunnel is proportional to the rate of pressure change of the compression wave front arriving at the exit (Ozawa 1979). Thus, we must understand the mechanism of the attenuation and distortion of a compression wave front propagating along a tunnel.

The attenuation and distortion of a compression wave are considered to be related the fact that whether the boundary layer of the flow induced by the compression wave is laminar or turbulent (Matsuo et al. 1991, Aoki et al. 1993). Therefore, in order to estimate the magnitude of an impulsive noise correctly, the transition of the unsteady boundary layer induced by a propagating compression wave in a tunnel is examined and experimental information on the transition is required.

The final objective of our study is to understand the mechanism of the attenuation and distortion of propagating compression waves in a tunnel. In the present paper, experimental investigations are carried out on the transition of the unsteady boundary layer induced by a propagating compression wave in a model tunnel by means of a recently developed laser differential interferometry technique.

2. Experimental apparatus

The open-ended shock tube with a fast-opening gate valve instead of a diaphragm is described in Aoki et al. (1993). Laser differential interferometry (LDI) using Wollaston prisms (Smeets & George 1973) was used for recording the density distribution. A schematic diagram of the optical arrangement is

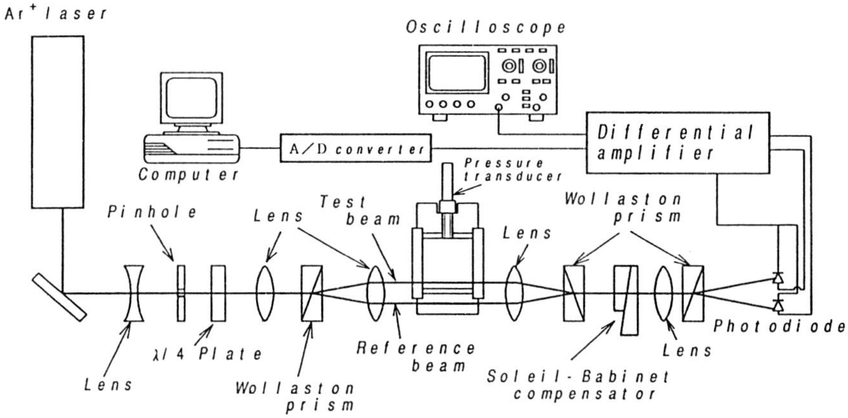

Figure 1. Optical arrangement of laser differential interferometer.

shown in Fig. 1. A continuous argon gas laser is used as a light source and photodiodes are used for detecting the difference of light intensities. The difference of the two detector signals from LDI will therefore be essentially free from vibrations and laser output variations. LDI can be made more sensitive to optical path change by 2 to 3 orders of magnitude than classical interferometry. The sensitivity was high enough to record density variations of a compression wave. This LDI has been used to detect the transition region in the unsteady wall boundary layer induced by a compression wave at $x = 3.15$ m.

Figures 2b and b show schematically two LDI setups, A and B. In the case of setup A, one of the interfering laser beams (test beam) passes through the unsteady boundary layer at a height h. The other beam (reference beam) travels through the main stream of the flow. Therefore the output signal of setup A shows the difference of density between the boundary layer and the main stream. Setup B was used to measure the spatial difference history of density in an unsteady boundary layer at height h.

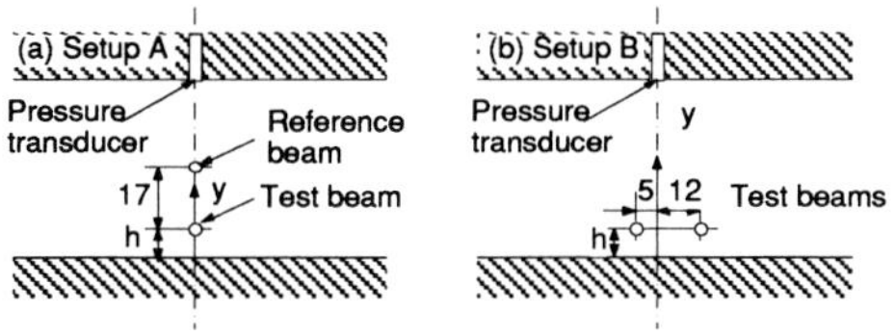

Figure 2. Schematic diagram of optical setup of laser beams.

3. Results and discussion

Typical experimental pressure-time records at $x = 0$ and 3.15 m are shown in Fig. 3a. Let the pressure rise by a compression wave be denoted by $\Delta p_c \equiv p_2 - p_1$, where p_1 and p_2 represent the pressures before and after a compression wave, respectively. The maximum rate of pressure rise $(dp/dt)_{\max}$ at the compression wave front is defined as shown Fig. 3a. Then it is seen that $\Delta p_c = 19$ kPa and $(dp/dt)_{\max} = 6$ kPa/ms. It is clear from the figure that the compression wave form is well controlled by a fast-opening gate valve.

The output signals of the laser differential interferometry at the point $x = 3.15$ m are shown in Fig. 3b (same conditions as Fig. 3a). The figure shows that the both signals for setup A and B detect well at the edge of the boundary layer and the transition. Density distributions of the boundary layer can be calculated by using the signal for the setup A as shown in Fig. 2a. A typical experimental density-time record, the difference of density $\rho - \rho_\infty$ between the boundary layer and the main stream, is shown in Fig. 4, where the compression wave is same as in Fig. 3 and the height of the test beam from the wall is $h = 0.8$ mm. The dotted and dashed lines represent the theoretical results of density profiles in laminar and turbulent boundary layers induced by a shock wave, respectively (Mirels 1956). This experimental data gives detailed information on the process of formation of the unsteady boundary layer induced by a propagating compression wave. The point a denotes the edge of the laminar boundary layer at $h = 0.8$ mm. It can be also concluded from Fig. 4 that the laminar-to-turbulent transition in the unsteady boundary layer flow exits describing the density profiles at point b. It should be noted that the transition time of point b in Fig. 4 coincides with the minus peak on the signal for setup B in Fig. 3b and LDI can be used to locate the unsteady boundary-layer transition induced by a propagating compression wave. The transition time t_r of point b such as

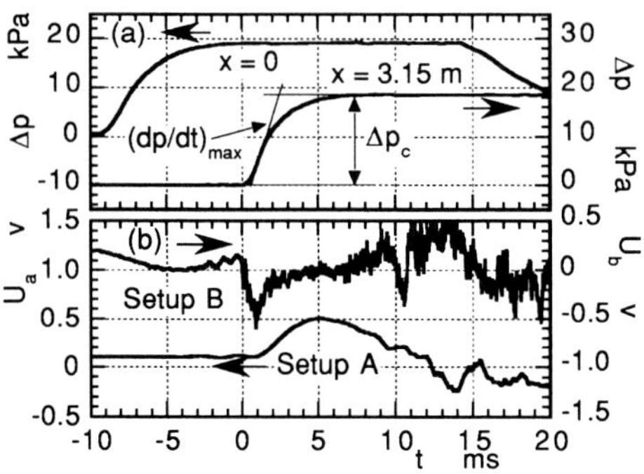

Figure 3. Typical pressure-time record of compression wave and output signals from LDI.

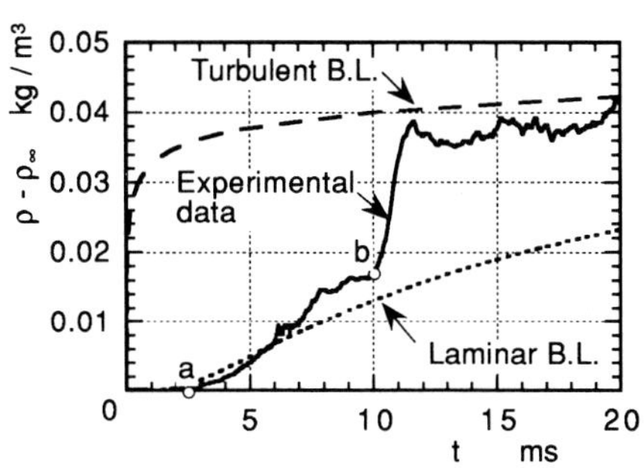

Figure 4. Measured density profiles in unsteady boundary induced by compression wave.

shown in Fig. 4 is determined for every record. The results are plotted in Fig. 5 for the range $\Delta p_c = 15$, 17, 19 kPa, $(dp/dt)_{\max} = 5$, 6 kPa/ms. The value of t_r decreases slightly with Δp_c. To clarify the spatial structure of the unsteady boundary layer, the location of the edge and transition on the boundary layer was measured for $h = 0.4 - 1.6$ mm. Figure 6 shows the experimental results against the length $x_c = at$ (where a is the sonic velocity in front of compression wave) from the wave front for $\Delta p_c = 19$ kPa, $(dp/dt)_{\max} = 5$ kPa/ms. The solid line represents the theoretical results for the edge position in a laminar boundary layer induced by a shock wave (Mirels 1956). Figure X? shows that the solid line agrees well with the experimental point. It looks like the transition propagates from the edge of the boundary layer to the wall. This is a subject for future study.

The transition Reynolds number R_{etr} defined by the wave-fixed coordinate system (Hartunian et al. 1960) is determined for every record. The results are compared with the experimental values for weak shock waves (Asbridge 1959, Hartunian et al. 1960, Thompson & Emrich 1967) as shown in Fig. 7, where the horizontal axis represents the density ratio across the wave front and the shock Mach number. It is seen that the transition Reynolds number of the compression wave agrees well with that of the shock wave. It may be the fact that the characteristic length of transition is relatively longer than that of wave front of the compression wave.

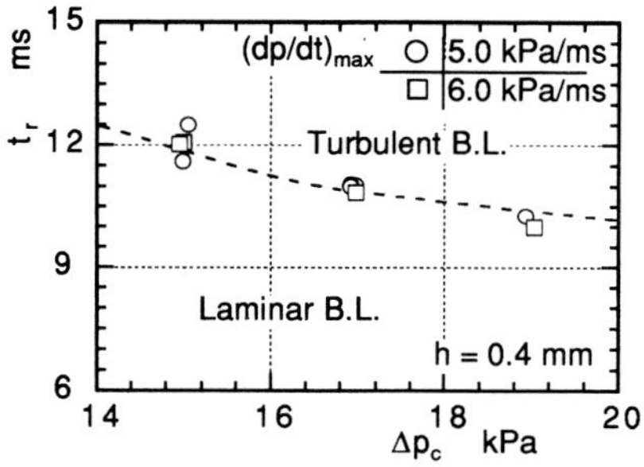

Figure 5. Transition time from laminar to turbulent boundary layer.

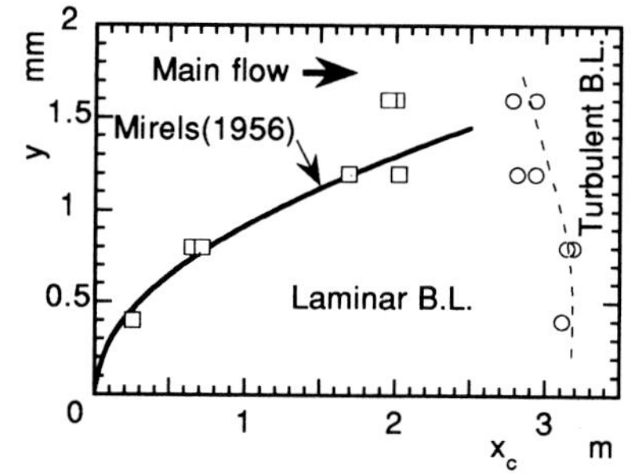

Figure 6. Occurrence of laminar-to-turbulent transition in unsteady boundary layer.

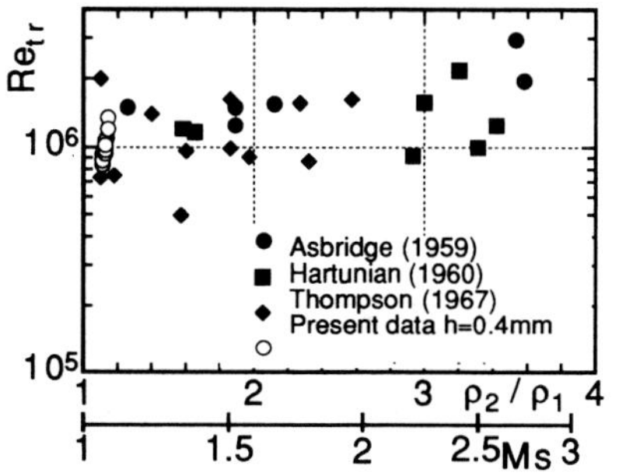

Figure 7. Transition Reynolds number of boundary layer induced by compression wave.

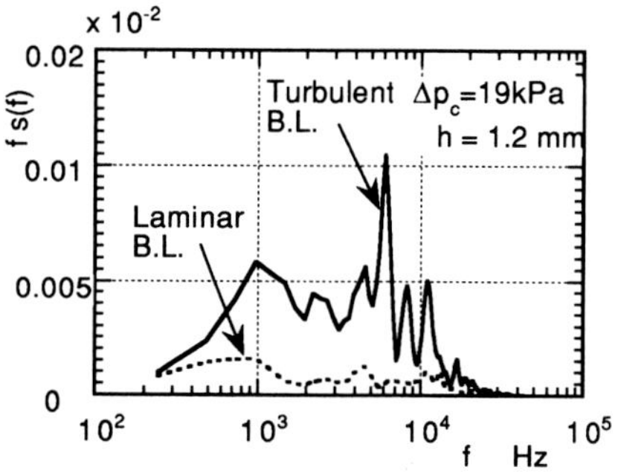

Figure 8. Frequency spectrum of laminar and turbulent boundary layer.

Finally, to obtain information on the boundary-layer structure in the frequency domain, frequency spectrums $fs(f)$ of laminar and turbulent boundary layers measured by the output signal of LDI (setup B) are shown in Fig. 8. Figure 8 indicates that the spectrum curve of the turbulent boundary layer has more peak than that of laminar flow for frequencies between 1 k and 10 kHz. This means that LDI can be used successfully to detect the boundary-layer transition.

4. Concluding remarks

The high-sensitivity interferometry technique, called LDI, was used to record the density variations of compression waves. The measurements were made of time dependent profiles and gave detailed information on the process of formation of the unsteady boundary layer induced by a propagating compression wave. It was found that the laminar-to-turbulent transition in unsteady boundary-layer flow exits describing the density profiles.

References

Aoki T, Matsuo K, Hidaka H, Noguchi Y, Morihara S (1993) Attenuation and distortion of propagating compression wave in high-speed railway model and real tunnels. In: Shock Waves @ Marseille, Vol 3, pp 347–352.

Asbridge JR (1959) An Interferometric Study of Shock Tube Boundary Layer, Lehigh Univ Ins Res Tech 14.

Hartunian RA, Russo AL, Marrone PV (1960) Boundary-layer transition and heat transfer in shock tube. J Aero Space Sci 27(8):587–594.

Matsuo K, Aoki T, Kashimura H, Kawaguchi M, Takeuchi N (1991) Attenuation of compression waves in a high-speed railway tunnel simulator. Proc 7th Intl Symp on Aerodynamic and Ventilation of Vehicle Tunnels, pp 239–252.

Mirels H (1956) Boundary Layer Behind Shock or Thin Expansion Wave Moving into Stationary Fluid. NACA Tech Note 3712, pp 1–53.

Ozawa S (1979) Studies of micro-pressure wave radiated from a tunnel exit. In: Railway Tech Res Rept of Japanese National Railways 1121, pp 1–92 (in Japanese).

Smeets G, George A (1973) Investigation of shock boundary layers with a laser interferometer. In: 9th Int Shock Tube Symposium, pp 431–438.

Thompson WP, Emrich RJ (1967) Turbulent spots and wall roughness effect in shock tube boundary layer transition. Phy Fluids 10(1):17–20.

Structure of multiple shock-wave/ turbulent boundary-layer interactions in a rectangular duct

H. Sugiyama, T. Arai, S. Hongo, T. Kawase, N. Uno
Mechanical Systems Engineering
Muroran Institute of Technology
Muroran, 050 Japan

Abstract: The structure and the turbulence phenomena of multiple shock-wave (called pseudo-shock wave, or PSW)/turbulent boundary-layer interactions in a square duct were investigated using flow-visualization methods and a two-component laser Doppler velocimeter (LDV). First, the PSW was visualized by schlieren photography and laser holographic interferography. Second, the time-mean and fluctuating velocities in the PSW were explored in detail using LDV. Spatial distributions of turbulence intensity, Reynolds shear stress, and turbulence kinetic energy are presented.

Key words: Shock-wave/turbulent boundary-layer interaction, Pseudo-shock wave, Turbulence, LDV, Flow visualization

1. Introduction

When supersonic or hypersonic flow in ducts interacts with downstream blockage devices which could be a flow plug, nozzle, or combustion heat release, a multiple-shock-wave system is produced as the result of shock-wave and duct-wall boundary-layer interactions. A multiple-shock-wave system is called as pseudo-shock wave (PSW) or shock train. The study of PSW has important implications for the design and operation of scramjet/combustor isolators, supersonic gas ejectors, and supersonic wind tunnel diffusers, and so on (Bement et al. 1990).

So far, the characteristics of PSW in ducts have been studied by many researchers — for instance Ikui et al. (1974), Om & Childs (1985), Sugiyama et al. (1988a, 1988b, 1992a, 1992b), and Carroll & Dutton (1990, 1992) — and the characteristics of the PSW were clarified. However, the turbulence phenomena in the PSW have not been fully clarified. This paper describes the turbulence phenomena in the PSW in a square duct that were measred using a two-component laser Doppler velocimeter (LDV).

2. Experimental apparatus and method

A suction-type intermittent supersonic wind tunnel was used in this experiment (Sugiyama et al. 1992a). The test section is a square duct of width $D = 50$ mm and length $L = 1030$ mm. The location of the PSW was fixed in the duct by making small adjustments of the shock location controller. The mean and fluctuating velocities were measured using a fiber-optic two-component LDV system (TSI SYSTEM 9110). The measurement volume diameter was about 0.12 mm and the effective measurement volume length was about 2.4 mm. All of the measurements were made on the mid-plane of the wind tunnel. The two-channel coincidence windows were adjusted to 1 μs. At each measurement location, 2048 velocity samples were taken.

3. Results and discussion

3.1. Overall structure of the PSW

Figure 1 shows the time-mean wall static pressure distributions along the duct measured when the locations of the PSW formation were at $X_f = 140$ mm and 400 mm, respectively. The ordinate is a non-dimensional pressure P/P_a, where P_a is the atmospheric pressure, and X is the distance from the duct entrance. From Fig. 1 it is seen that the pressure rise by the PSW increases as the location of the PSW moves upstream.

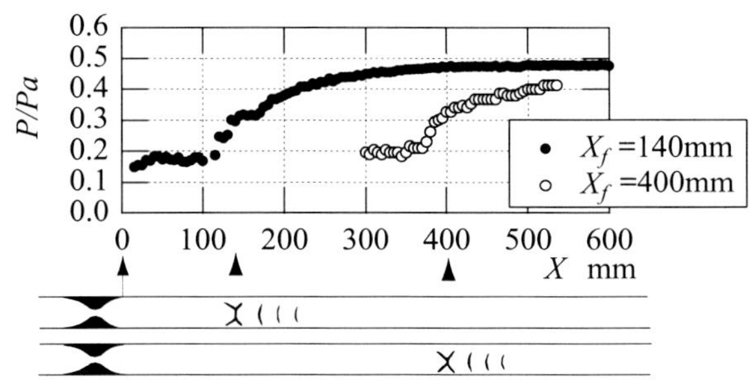

Figure 1. Wall static-pressure distributions along the duct.

Figures 2a and b show a schlieren photograph and laser holographic interferogram of the PSW in a square duct. The location of the PSW was $X_f = 8.0D = 400$ mm. The flow Mach number, boundary-layer thickness and unit Reynolds number just prior to the PSW were $M_\infty = 1.78$, $\delta_\infty = 0.3h$ (where $h = D/2$), and $Re = 13{\times}10^6$ m^{-1}, respectively. The first shock of the PSW is bifurcated while the following second and third shocks are not. The bifurcated shock consists of a leading oblique shock, a nearly normal shock, and a trailing oblique shock, with the three intersecting at the bifurcation point. A thickening of the boundary layer is observed under the leading oblique shock.

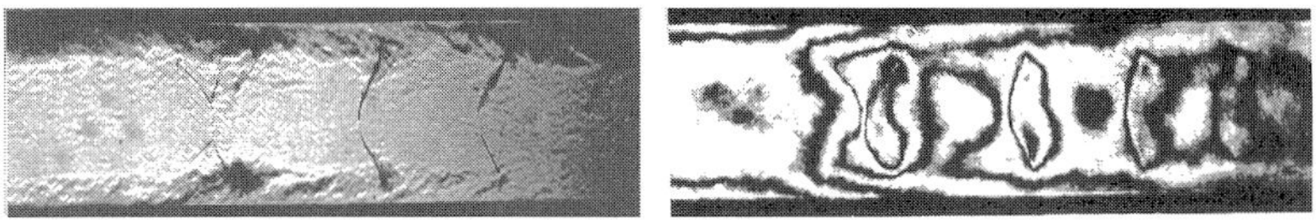

(a) (b)

Figure 2. Visualization of the PSW ($X_f = 8.0D$, $M_\infty = 1.78$, $\delta_\infty/h = 0.3$, $Re = 13\times10^6\ \mathrm{m}^{-1}$): (a) schlieren photograph, (b) holographic interferogram.

3.2. Mean streamwise velocity profiles in the PSW

Detailed LDV measurements were made of the PSW shown in Fig. 2. Vertical traverses were made at nine stations between the incoming undisturbed region and third shock. LDV measurement stations are shown in Fig. 3. At stations 2 and 3, the leading oblique portion of the first bifurcated shock has reduced the near-wall velocity below that of the incoming boundary layer. At stations 4 and 5, a peak in the velocity occurs at the boundary layer edge (Sugiyama et al. 1992a,b). The core flow reaccelerates following the first shock, and a nearly uniform flow exists outside the boundary layer at station 6.

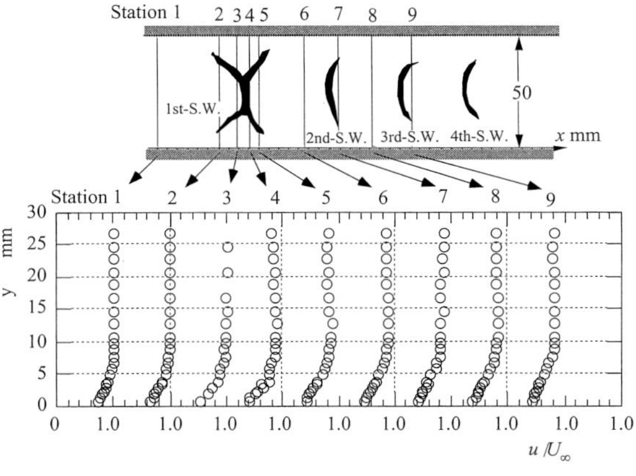

station	X mm	X/D
1	360	7.20
2	387	7.74
3	395	7.90
4	400	8.00
5	405	8.10
6	425	8.50
7	440	8.80
8	455	9.10
9	473	9.46

Figure 3. Mean streamwise velocity profiles in the PSW.

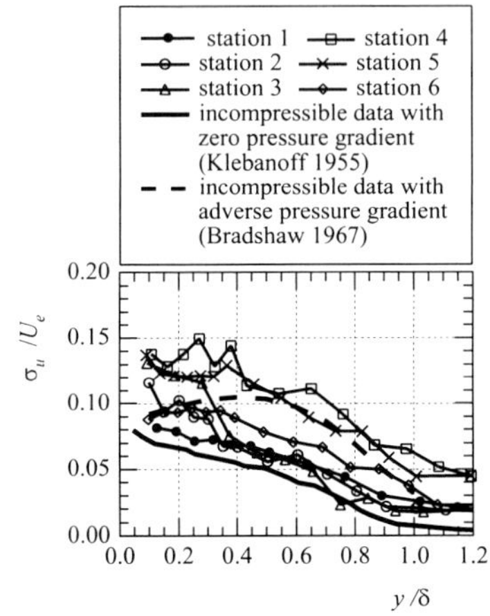

Figure 4. Streamwise turbulence intensity profiles in the PSW.

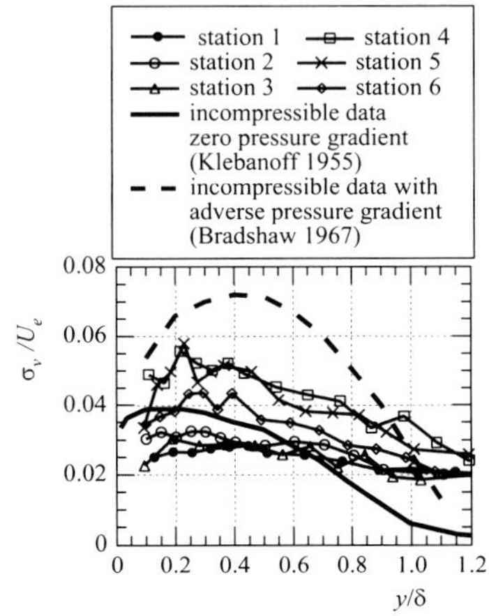

Figure 5. Transverse turbulence intensity profiles in the PSW.

3.3. Turbulence data

Turbulence intensity. Figure 4 shows the standard deviation σ_u of the streamwise velocity u, i.e., turbulence intensity. Here, U_e is the velocity at the outer edge of the boundary layer and δ the boundary layer thickness, which is defined as the distance from the wall for which $u = 0.99U_e$. The incompressible data with zero and adverse pressure gradients over flat plates (Klebanoff 1955, Bradshaw 1967) are also included in Fig. 4. At station 1 in the undisturbed boundary layer, the distribution of turbulence intensity is similar to that of the incompressible flat plate case. At stations 2, 3, and 4, a large increase is caused by the first bifurcated shock to a maximum of $\sigma_u/U_e = 0.15$ for $0.2 < y/\delta < 0.4$. At station 6 in the reacceleration region between the first and second shocks the turbulence intensity decreases. Figure 5 shows the turbulence intensity of transverse velocity, σ_v. At stations 1, 2, and 3, the transverse turbulence intensities are nearly constant across the boundary layer. At stations 4 and 5 under the bifurcated first shock σ_v increases, and at station 6 in the acceleration region σ_v decreases.

Reynolds stress. Figure 6 presents the dimensionless kinetic turbulence shear stress, or Reynolds shear stress, defined as $- < u'v' > /U_e^2$. An increase in the Reynolds stress is caused by the bifurcated first shock. The location of the peak Reynolds stress becomes higher away from the solid wall. At station 6 the Reynolds stress decreases.

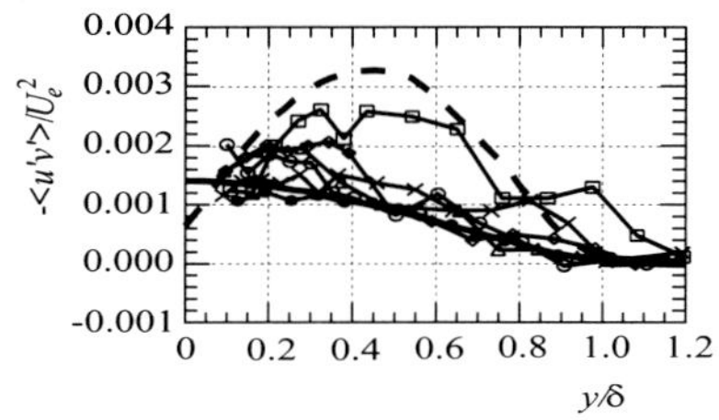

Figure 6. Reynolds shear stress, $-<u'v'>/U_e^2$, in the PSW.

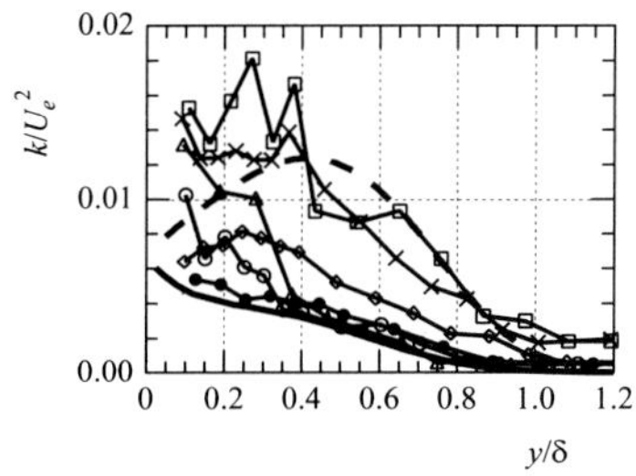

Figure 7. Turbulence kinetic energy, k/U_e^2, in the PSW.

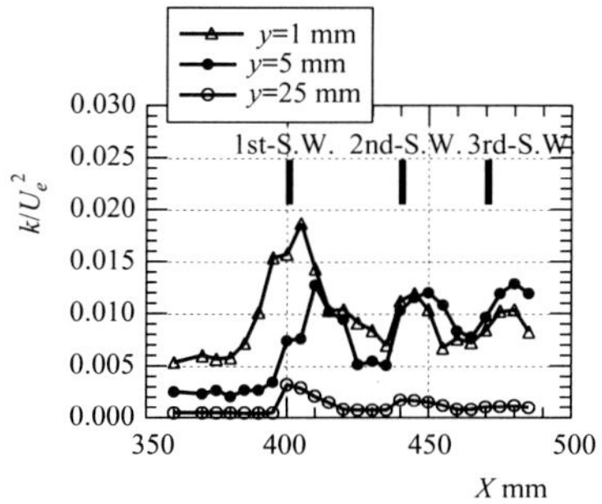

Figure 8. Streamwise distributions of turbulence kinetic energy.

Turbulence kinetic energy. Figure 7 shows the dimensionless turbulence kinetic energy k/U_e^2, where $k = (\sigma_u^2 + \sigma_v^2 + \sigma_w^2)/2$. In this study, the spanwise normal stress σ_w^2 was estimated as $\sigma_w^2 = (\sigma_u^2 + \sigma_v^2)/2$, making $k = 0.75(\sigma_u^2 + \sigma_v^2)$. At station 1 in the incoming, undisturbed boundary layer, the turbulence kinetic energy has almost the same level as in the incompressible case. A large increase in k is caused by the first shock, and the location of the peak moves away from the wall, ranging from $y/\delta = 0.1$ to 0.4. At station 6 in the reacceleration region, the peak level of k decreases. Figure 8 shows the streamwise distribution of k. In the core flow (along the center line of the duct $y = 25$ mm), k is very small and it increases at the first shock because of high frequency oscillation of the multiple shock system. It is clearly seen that in the PSW k increases at each shock, and that k decreases in the reacceleration region between each shock.

4. Conclusions

The internal structure and the turbulence phenomena of multiple shock-wave (PSW)/turbulent boundary-layer interaction in a square duct were investigated experimentally. First, the internal structure of the PSW was visualized by schlieren photography and laser holographic interferography. It was found that the boundary-layer thickness increases at the leading oblique shock of the first bifurcated shock and decreases behind the trailing shock of the first shock, and that there exists a flow reacceleration region between the first and second shocks of the PSW. Next, the time-mean and fluctuating velocities in the PSW were measured in detail using LDV. Spatial distributions of the turbulence intensity, Reynolds shear stress and turbulence kinetic energy were presented. It was found that the turbulence kinetic energy in the boundary layer in the PSW increases under the bifurcated first shock and decreases in the flow reacceleration region between the first and the second shocks.

References

Bement DA, Stevens JR, Thompson MW (1990) Measured operating characteristics of a rectangular combustor/inlet isolator. In: AIAA/SAE/ASME/ASEE 26th Joint Propulsion Conference, AIAA-90-2221, July 16–18, Orlando, FL.

Bradshaw P (1967) The turbulence structure of equilibrium boundary layers. J Fluid Mech 29(4):625–645.

Carroll BF, Dutton JC (1990) Characteristics of multiple shock wave/turbulent boundary-layer interactions in rectangular ducts. J Propulsion and Power 6(2):186–193.

Carroll BF, Dutton JC (1992) Turbulence phenomena in a multiple normal shock wave/turbulent boundary-layer interaction. AIAA J 30(1):43–48.

Ikui T, Matsuo K, Nagai M (1974) The mechanism of pseudo-shock waves. Bulletin of JSME 17(108)731–739.

Klebanoff PS (1955) Characteristics of Turbulence in a Boundary Layer with Zero Pressure Gradient, NACA Report 1247, pp 1135–1153.

Om D, Childs ME (1985) Multiple transonic shock-wave/turbulent boundary-layer interaction in a circular duct. AIAA J 23(10):1506–1511.

Sugiyama H, Takeda H, Zhang J, Abe F (1988a) Multiple shock wave and turbulent boundary layer interaction in a rectangular duct. In: Proc 16th Int Symp on Shock Tubes and Waves, Grönig H (ed), Aachen, pp 185–191.

Sugiyama H, Takeda H, Zhang J, Okuda K, Yamagishi H (1988b) Locations and oscillation phenomena of pseudo-shock waves in a straight rectangular duct. JSME Int J Series II 31(1):9–15.

Sugiyama H, Arai T, Uno N, Takahashi T (1992a) Three-dimensional structure of pseudo-shock waves in a rectangular duct. In: Proc 18th Int Symp on Shock Waves, Takayama K (ed), Sendai, pp 631–636.

Sugiyama H, Arai T, Uno N, Matsumoto T, Homareda M (1992b) Visualization of a pseudo-shock wave in a rectangular duct. In: Proc 6th Int Symp on Flow Visualization, Tanida Y, Miyashiro H (eds), Yokohama, pp 623–626.

Asymptotic defect boundary layer applied to nonequilibrium laminar hypersonic air flows

S. Séror, E. Schall, D. Zeitoun
Lab. I.U.S.T.I. (U.M.R. CNRS 139) Dept. MHEQ, Centre St. Jérôme
Université de Provence, 13397 Marseille Cedex 20, France

J. P. Brazier
ONERA CERT Department of Aerothermodynamics
2, Avenue E. Belin, BP 4025 31055 Toulouse, France

Abstract: The laminar, viscous, nonequilibrium boundary-layer flow around an axisymmetric blunt body has been computed to predict the heat-flux transfer. The extension of the defect formulation to two-dimensional hypersonic flows out of equilibrium including chemical and vibrational nonequilibrium phenomena has been done. This method ensures a smooth matching of the boundary-layer solution with a vortical external flow. Using matched asymptotic expansions, a new set of first-order boundary-layer equations has been derived. The air was considered as a mixture of the five species N_2, O_2, NO, N, O, in which the diatomic species N_2, O_2 are in vibrational nonequilibrium. Comparison between defect calculations and Navier-Stokes calculations (Zeitoun et al. 1994) for the nonequilibrium reacting flow over a sphere-cone shows good agreement. Results of heat-flux predictions are compared to experimental data.

Key words: Boundary layer, Hypersonic flow, Reactive flow, Blunt body, Defect formulation, Asymptotic expansion, Chemical non-equilibrium, Vibrational non-equilibrium

1. Introduction

For hypersonic re-entry flows, the bow shock wave upstream of the spacecraft is the cause of an inviscid vortical flow in the shock layer. Prandtl's equations cannot take into account normal gradients outside of the boundary layer, whereas Van Dyke's second-order matched asymptotic expansions approach (Van Dyke 1962) no longer gives a good matching of the viscous and inviscid solutions when the boundary layer is thick as it is the case for hypersonic re-entry flows. Thus no correct prediction of the influence of the external vorticity on the wall heat flux or the skin friction can be achieved with Van Dyke's approach. To remedy this problem, a defect approach coupled with asymptotic expansions had been proposed (Brazier 1990), but only perfect gas flow had been considered up to now. Moreover, during hypersonic re-entry, an important transfer of the high kinetic energy to the internal energy modes of the gas takes place through this shock wave. This transfer causes a very strong increase in temperature and, therefore, significant modifications of the composition and of the internal energy distribution of the gas. In this work, with regard to the range of Mach numbers investigated, only the chemical and vibrational nonequilibrium processes are taken into account. Energy is thus supplied to the vibration mode and the molecules dissociate. We propose in this paper an extension of the defect approach in the context of hypersonic flows including chemical and vibrational relaxation phenomena. Only steady, two-dimensional planar or axisymmetric laminar compressible reactive flows will be addressed here.

2. Physical modeling

For the considered stagnation conditions, ionization is neglected. The dissociated air can be modeled by a mixture of five species, namely N_2, O_2, NO, N, O. The model assumes seventeen chemical reactions, whose reaction rates are given by Gardiner (1984). The diatomic species N_2 and O_2 are in thermal nonequilibrium and since the vibrational relaxation time for nitrogen monoxide is very small, NO is taken at vibrational equilibrium. The harmonic-oscillator assumption is used and the vibrational exchanges take into account only the translation-vibration (V-T) transfers through the Landau-Teller relaxation theory. A global relaxation time models the exchanges of the molecules with the other species of the mixture. The various relaxation times between species can be found in Druguet et al. (1992).

Only the three major modes, namely translation, rotation, and vibration are accounted for when modeling the partition of internal energy for the various species. Translational and rotational modes are always considered in full equilibrium and thus are characterized by a single temperature T, while vibrational modes may depart from equilibrium. Therefore each of the diatomic species N_2 and O_2 is assumed to have a separate vibrational temperature distinct from the mixture temperature T. Each species of the mixture can be assumed to behave as a perfect gas. The pressure of the mixture is determined from Dalton's law.

The basic equation set is the Navier-Stokes equations together with relaxation equations for the chemical species and the vibration energy. The mass diffusion flux for the I-species are modeled according to the Fick's law using a single diffusion coefficient.

The species viscosity are calculated from the kinetic theory of gases, and Wilke's mixing rule is used to obtain the mixture viscosity. Species translation/rotation thermal conductivities are computed from the Eucken relationship, with mixture values again obtained using Wilke's law. The vibration thermal conductivities of the molecules are expressed in terms of the single diffusion coefficient calculated assuming a constant Lewis number.

3. The defect approach

The proposed strategy is to extend the boundary-layer theory, as developed by Van Dyke (1962) using matched asymptotic expansions, to account for variations of the inviscid flow solution throughout the boundary layer. A way to do it is to use a defect formulation, i.e., not to deal with physical variables in the boundary layer but with the difference between the viscous and the inviscid solutions (Brazier 1990). Therefore applying the method to the Navier-Stokes equations for two-dimensional planar or axisymmetric laminar flows, plus the species conservation equations and the vibrational energy relaxation equations for N_2 and O_2 molecules, new sets of boundary-layer equations have been obtained. The first-order approximation, in the sense of matched asymptotic expansions, includes the usual Prandtl equations plus higher-order terms which ensure a smooth merging of the boundary layer into the inviscid flow.

Thanks to the small-perturbations approach, the calculations of external flow and boundary layer are uncoupled and can be performed separately provided that a specified sequence is respected. First-order external problems must be solved first, then first-order internal. The defect boundary-layer equations are parabolic and can be solved by space marching at a very low cost, like the standard Prandtl equations. The stagnation-point solution is obtained with the help of series expansions with respect to the curvilinear coordinate ξ.

4. Flow past blunt body in the F4 wind tunnel

We have performed a computational study of the high-enthalpy air flow past the blunt body *Electre* defined by the E.S.A for the 1994 workshop. The geometry is essentially a sphere-cone of nose radius $R_0 = 0.035$ m, cone half angle $\alpha = 4.66°$, and overall length $L = 0.4$ m. The wall is assumed to be either fully catalytic or non-catalytic, and at a fixed temperature $T_w = 300$ K. Concerning the boundary conditions relative to the vibrational temperatures, we have considered two extreme cases. First we have assumed that the flow was in thermal equilibrium at the wall, then on the other hand, that there were no vibrational energy exchanges at the wall, i.e., we consider that the wall vibrational heat flux is null. The characteristics of the oncoming flow are given below from the experiment carried out in the F4 hypersonic wind tunnel of ONERA: $\mathcal{M}_\infty = 14.98$, $U_\infty = 5612.4$ m/s, $\rho_\infty = 6.8714\ 10^{-4}$ kg/m^3, $T_\infty = 302$ K, $P_\infty = 71.34$ Pa, $Tv_{N_2\infty} = 5682.5$ K, $Tv_{O_2\infty} = 3618.2$ K, $Y_{N_2\infty} = 0.784812$, $Y_{O_2\infty} = 0.002582$,

$Y_{NO\infty} = 0.002812$, $Y_{N\infty} = 0.003876$, $Y_{O\infty} = 0.205918$. This case corresponds to a Reynolds number based on upstream conditions and nose radius of 900, i.e., typically a low Reynolds-number flow.

5. Discussion of the results

Boundary-layer profiles are displayed in Fig. 1 for a location in the nose region. Longitudinal velocity profiles are plotted in Fig. 1a, and the atomic oxygen mass-fraction profiles are plotted in Fig. 1b. Only the defect approach can match the inviscid solution with a first-order approach. The agreement with the Navier-Stokes solution is fair. Fig. 2a shows the evolutions of the translational and vibrational temperatures along the stagnation streamline in the case of a non-vibrationally catalytic wall. We can notice a strong thermal nonequilibrium in the boundary layer since Tv_{N_2} is about 8000 K at the wall and Tv_{O_2} equals 3300 K. Moreover in the outer flow the vibrational non-equilibrium region of each diatomic species is clearly visible. In the core flow, the relaxation zone of

O_2 is shorter than that of N_2. Figure 2b shows the freezing of the vibrational relaxation process in the expansion zone of the flow. The mass-fraction evolutions along the stagnation streamline are plotted in Fig. 3a. One can notice a strong recombination of atomic species in the boundary layer since the wall is fully catalytic. The evolution of the wall heat flux is given in Fig. 3b for a

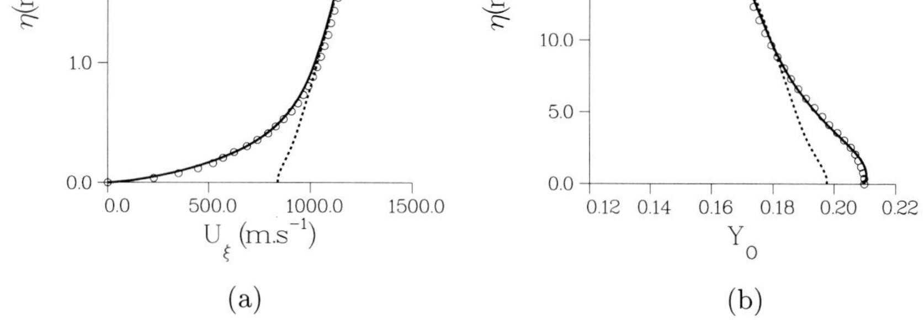

Figure 1. Boundary-layer profiles, $\xi = 2$ cm, non-catalytic wall and wall vibrational heat flux $= 0$: (a) longitudinal velocity profiles, (b) Y_O mass fraction.

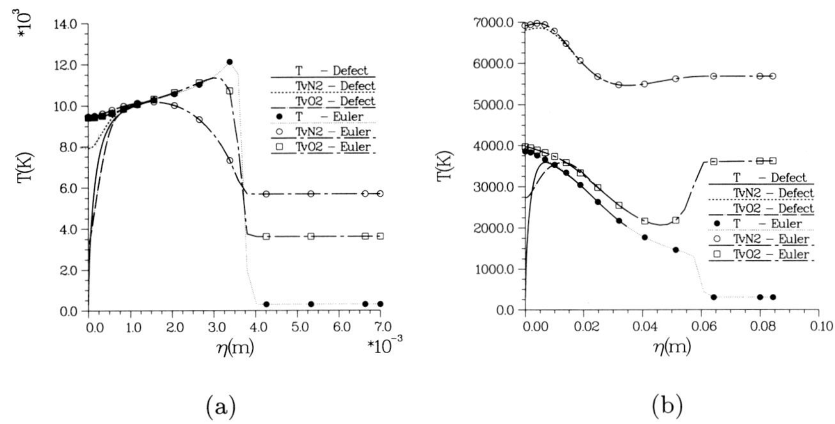

Figure 2. Temperature profiles for the catalytic wall with a wall vibrational heat flux $= 0$: (a) $\xi = 0$, (b) $\xi = 20$ cm.

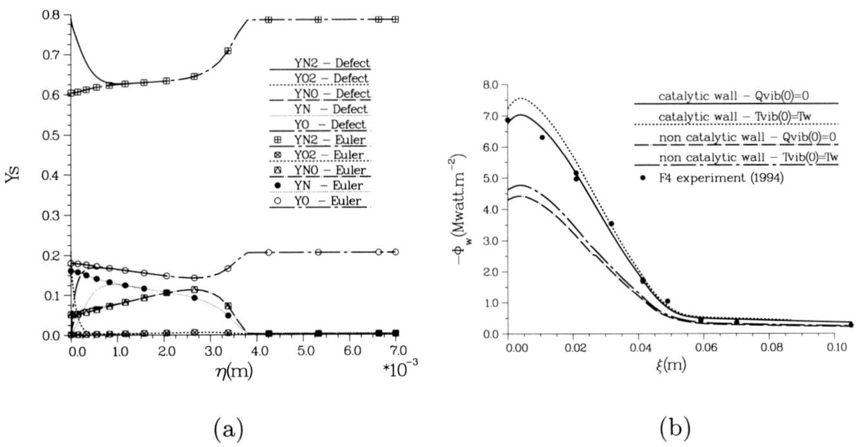

Figure 3. (a) Mass-fraction profiles along stagnation line for the catalytic wall with a wall vibrational heat flux $= 0$, (b) wall heat-flux predictions.

fully catalytic wall without vibrational energy exchanges at the wall. This case seems to be closer to the experimental results. The influence of the boundary condition used for the vibration energy equations is clearly visible. For a catalytic wall the assumption of a total deactivation of the vibrational mode of N_2 and O_2 molecules at the wall induces a nose heat flux 13% higher than for a wall without vibrational energy exchanges. For a non-catalytic wall, the deviation in the nose heat flux between these two extreme conditions is 10%. These results confirm the previous one obtained by Tirsky (1993) concerning wall heat-flux predictions for the Buran shuttle.

6. Conclusions

The defect approach has been extended to derive boundary-layer equations from the Navier-Stokes equations for laminar viscous flows out of equilibrium. Together with matched asymptotic expansions methodology, it ensures a correct matching of the inviscid flow with the first-order boundary-layer solution. In this case the agreement with the Navier-Stokes solution is fair. Wall heat-flux predictions for various boundary conditions have shown that the best agreement with the experimental data is achieved for a fully catalytic wall assuming that there are no vibrational energy exchanges at the wall.

Acknowledgement. The authors would like to thank the "Centre de Calcul de l'I.M.T." for allowing access to the Cray YMP-2E computer.

References

Brazier JP (1990) Etude asymptotique des équations de couche limite en formulation déficitaire, Ph.D. Thesis, Ecole Nationale Supérieure de l' Aéronautique et de l' Espace, Toulouse.

Druguet MC, Zeitoun D, Brun R (1992) Inviscid hypersonic nozzle flows in chemical and vibrational nonequilibrium state. In: Computing Methods in Applied Sciences and Engineering, R Glowinski (ed), Nova Science, New York, pp 729–738.

Gardiner WC Jr (1984) Combustion Chemistry, Springer-Verlag.

Tirsky GA (1993) Up to date gasdynamic models of hypersonic aerodynamics and heat transfer with real gas properties. Annual Review of Fluid Mechanics 25:151–181.

Van Dyke M (1962) Second order compressible boundary-layer theory with application to blunt bodies in hypersonic flow. In: Hypersonic Flow Research, FR Ridell (ed), Academic Press, Vol 7, pp 37–76.

Zeitoun D, Schall E, Burtschell Y, Druguet MC (1994) Vibration-dissociation coupling in nonequilibrium hypersonic viscous flows. AIAA Journal 33(1):79–85.

Analysis of nonequilibrium, hypersonic blunt-body flow with the local-similarity method

A. Klomfass, S. Müller, J. Ballmann
Lehr- und Forschungsgebiet für Mechanik, RWTH Aachen
Templergraben 64, 52056 Aachen, Germany

Abstract: A method for the calculation of the stagnation-streamline quantities and the stagnation-point heat transfer in hypersonic flows about spheres or cylinders is suggested which is based on the local similarity of the flow field. Comparisons with two-dimensional Euler and Navier-Stokes calculations and experiments are presented, which include perfect-gas flows and nonequilibrium air flows.

Key words: Nonequilibrium, Stagnation streamline, Local similarity

1. Introduction

A spacecraft entering the atmosphere generates a strong bow shock that leads to the excitation of molecular vibration and dissociation. A nonequilibrium flow prevails if the relaxation times of these processes are of the order of the fluid-dynamic time scale. Such flows are described mathematically by the conservation laws and additional balance equations for the component densities and the vibrational energies. Physical models provide relations for the chemical production rates, the rates of vibrational energy transfer, and the molecular fluxes. Flows about spheres or cylinders with high aspect ratios are suitable for the investigation and validation of these models. The stagnation-streamline quantities and the stagnation-point heat flux are of major interest in this context. We therefore consider a method for the fast calculation of these quantities.

2. The local-similarity method

The stagnation-streamline quantities are usually obtained from an asymptotic steady-state solution of the two-dimensional, unsteady flow problem. However, as two-dimensional hypersonic flows about spheres or cylinders exhibit an approximate local similarity in the stagnation-streamline area, the number of independent spatial variables can be reduced from two to one (Kao 1964). The dimensionally reduced Navier-Stokes equations (DRNSE) are derived from the Navier-Stokes equations in spherical or cylindrical coordinates. In our empirical approach the flow quantities are represented as products of unknown functions of the radial coordinate r and prescribed, assumed functions of the angular coordinate θ, which is here measured from the stagnation streamline. The DRNSE are then obtained by carrying out all differentiations with respect to θ, equating coefficients of like functions of θ, and letting $\theta \to 0$. The velocity components, the temperatures, the mass fractions, and the pressure are here assumed as

$$v_r = \overline{v}_r(r)\cos\theta, \quad v_\theta = \overline{v}_\theta(r)\sin\theta, \quad T^i = \overline{T}^i(r), \quad x_\alpha = \overline{x}_\alpha(r), \tag{1}$$

$$p = p_\infty + \overline{p}(r)\cos^2\theta\,. \tag{2}$$

These relations are motivated by the boundary conditions and Newtonian flow theory, which provides an approximation for the surface pressure. The resulting DRNSE are formally given as

$$\overline{\mathbf{u}}_{,t} + \overline{\mathbf{F}}_{,r} + \overline{\mathbf{G}} = \mathbf{S}\,, \tag{3}$$

with $\overline{\mathbf{u}} = (\rho, \rho\overline{v}_r, \rho\overline{v}_\theta, \rho e^t, \rho_\alpha, \rho_\alpha e^v_\alpha)^T$ and corresponding dimensionally reduced fluxes $\overline{\mathbf{F}}$, and metric terms $\overline{\mathbf{G}}$, which both contain viscous and inviscid contributions. The nonequilibrium source terms $\mathbf{S} = (\mathbf{0}, \omega_\alpha, \Omega^{vib}_\alpha)$ are not affected by the transformation of the equations. The DRNSE are subject to boundary conditions that specify the undisturbed supersonic parallel free stream, a no-slip or slip wall, an isothermal or adiabatic wall, and a fully catalytic or non-catalytic wall.

3. The numerical method

An implicit time-integration method is applied to obtain steady-state solutions of the DRNSE, which are approximated by the finite-difference equations

$$\frac{\overline{\mathbf{u}}_i^{n+1} - \overline{\mathbf{u}}_i^n}{\Delta t} + \frac{\overline{\mathbf{F}}_{i+1}^{n+1} - \overline{\mathbf{F}}_{i-1}^{n+1}}{\Delta r} + \overline{\mathbf{G}}_i^{n+1} = \mathbf{S}_i^{n+1}\,, \tag{4}$$

with $\Delta t = t^{n+1} - t^n$ and $\Delta r = r_{i+1} - r_{i-1}$. The viscous parts of the fluxes and the metric terms still contain radial derivatives of certain flow quantities. These are approximated by second-order accurate central-, left-, or right-sided finite differences as functions of the respective values at points $i-1, i$, and $i+1$. The unknowns $\overline{\mathbf{u}}^{n+1}$, $\overline{\mathbf{F}}^{n+1}$, $\overline{\mathbf{G}}^{n+1}$, and $\overline{\mathbf{S}}^{n+1}$ at time t^{n+1} are extrapolated from their values at t^n by Taylor expansions. For $\mathbf{f}$ being any of these quantities

$$\mathbf{f}^{n+1} = \mathbf{f}^n + \left.\frac{\partial\,\mathbf{f}}{\partial\,\overline{\mathbf{w}}}\right|^n \Delta\overline{\mathbf{w}}^n\,, \quad \Delta\overline{\mathbf{w}}^n = \overline{\mathbf{w}}^{n+1} - \overline{\mathbf{w}}^n, \quad \overline{\mathbf{w}} = \left(\overline{v}_r, \overline{v}_\theta, \overline{p}, \overline{T}^{tr}, \overline{x}_\alpha, \overline{T}^{vib}_\alpha\right)^T. \tag{5}$$

The introduction of the variables $\overline{\mathbf{w}}$ as calculational quantities simplifies the incorporation of the boundary conditions and the calculation of the Jacobians

$\partial \mathbf{f} / \partial \overline{\mathbf{w}}$ considerably. Introducing (5) into (4) and rearranging terms renders the time-integration scheme

$$\mathbf{a}_i \, \Delta \overline{\mathbf{w}}_{i-1}^n + \mathbf{b}_i \, \Delta \overline{\mathbf{w}}_i^n + \mathbf{c}_i \, \Delta \overline{\mathbf{w}}_{i+1}^n = -\left(\overline{\mathbf{F}}_{i+1}^n - \overline{\mathbf{F}}_{i-1}^n + \Delta r \, \overline{\mathbf{G}}_i^n - \Delta r \, \overline{\mathbf{S}}_i^n \right) , \quad (6)$$

where the matrices $\mathbf{a}_i, \mathbf{b}_i$, and $\mathbf{c}_i$ are given in terms of the Jacobians. The system of equations for the interior points $i = 2, \ldots, N-1$ of the computational domain can be written as $\mathbf{A}^n \Delta \overline{\mathbf{w}}^n = \mathbf{RHS}^n$, with a block-tridiagonal matrix $\mathbf{A}$. The products $\mathbf{a}_2 \Delta \overline{\mathbf{w}}_1^n$ and $\mathbf{c}_{N-1} \Delta \overline{\mathbf{w}}_N^n$ are included on the right-hand side and the boundary conditions are used to eliminate the unkowns $\Delta \overline{\mathbf{w}}_1$ and $\Delta \overline{\mathbf{w}}_N$. Calculations are initialized with suitable linear distributions of $\overline{\mathbf{w}}$. Steady-state solutions are obtained asymptotically for $t \to \infty$. Due to the implicit formulation, time steps with high CFL numbers can be realized. The Fortran routine NBTRIP of R. Chakravarthy is used to solve $\mathbf{A}^n \Delta \overline{\mathbf{w}}^n = \mathbf{RHS}^n$ for $\Delta \overline{\mathbf{w}}^n$. The approximation of $\overline{\mathbf{F}}_{,r}$ by central differences in (4) is appropriate for flows with thick shock waves. A shock-fitting strategy is used for flows with thin shock waves. In this case only the shock-layer part of the stagnation streamline is the computational domain. The free-stream boundary conditions are then replaced by the post-shock conditions. Since the steady-state shock position is a priori unknown, it is calculated simultaneously with the shock-layer flow. Starting from an initial guess, the shock position at t^{n+1} is determined from the shock speed, which is calculated from the Rankine-Hugoniot relations and a compatibility equation (cf. Klomfass 1995). The computational grid is fitted to the new shock position after each time step.

4. The gas physical model

Air is considered as a mixture of 5 components (N_2, O_2, NO, N, O). The energy is calculated on the basis of independent Boltzmann distributions, where a common temperature $T^{(tr)}$ is assigned to all translational, rotational, and atomic-electronic modes. Individual temperatures $T_{N_2}^{(vib)}, T_{O_2}^{(vib)}$, and $T_{NO}^{(vib)}$ are assumed for the vibrational and electronic modes of the molecular components. The model of a rigid rotor and truncated harmonic-oscillator molecule is used. The vibrational energy exchange in non-reactive molecular collisions consists of VT and VV contributions. The volume specific rates are

$$\Omega_{\alpha\beta}^{(VT)} = \rho_\alpha c_\beta \, k_{\alpha\beta}^{VT} \left(1 - \exp\left(-\theta_\alpha^{(vib)} / T^{(tr)}\right)\right) \left\{ e_\alpha^{(vib)}(T^{(tr)}) - e_\alpha^{(vib)} \right\} \quad (7)$$

$$\begin{aligned} \Omega_{\alpha\beta}^{(VV)} &= \rho_\alpha c_\beta \, k_{\alpha\beta}^{VV} \frac{1}{R_\beta \theta_\beta^{(vib)}} \left\{ \left(e_\alpha^{(vib)} + R_\alpha \theta_\alpha^{(vib)} \right) e_\beta^{(vib)} \exp\left(-\frac{\theta_\alpha^{(vib)} - \theta_\beta^{(vib)}}{T^{(tr)}} \right) \right. \\ & \qquad \left. - \left(e_\beta^{(vib)} + R_\beta \theta_\beta^{(vib)} \right) e_\alpha^{(vib)} \right\} . \end{aligned} \quad (8)$$

The fundamental rate constants $k_{\alpha\beta}^{VT}$ are here calculated from curve fits to experimental data, while the $k_{\alpha\beta}^{VV}$ are obtained from the heuristic relation

$$k_{\alpha\beta}^{VV} = \sqrt{k_{\alpha\alpha}^{VT}\, k_{\beta\beta}^{VT}\, \exp\left\{(\theta_\alpha^{(vib)} - \theta_\beta^{(vib)})/T^{(tr)}\right\}}\,, \tag{9}$$

which fits the few available experimental data. The chemical-production rates ω_α are calculated from the usual rate equations, where the 17 major chemical reactions are considered. The preferential dissociation model of Treanor and Marrone is applied for the rate constants $k^{(r')}$ of the endothermic reactions; the rate constants $k^{(r'')}$ of the exothermic reactions are deduced from the principle of detailed balance:

$$k^{(r')} = \Psi\, \tilde{k}^{(r')}(T^{(tr)}) \qquad \text{(endoth. reactions)} \tag{10}$$

$$k^{(r'')} = \tilde{k}^{(r')}(T^{(tr)}) \,/\, K_r(T^{(tr)}) \qquad \text{(recombination)} \tag{11}$$

$$k^{(r'')} = \Psi\, \tilde{k}^{(r')}(T^{(tr)}) \,/\, K_r(T^{(tr)}) \qquad \text{(exoth. exchange reactions).} \tag{12}$$

Here $\tilde{k}^{(r')}$ denotes the equilibrium rate constants; the preferential dissociation parameter $\Psi = \Psi(T^{(tr)}, T_\alpha^{(vib)})$ accounts for the effect of vibrational nonequilibrium of the respective molecular reactant. The correlations of Park (1990) are used for $\tilde{k}^{(r')}$. The equilibrium constants K_r are calculated from partition functions. The vibrational energy loss in reactive molecular collisions is deduced from the preferential dissociation theory. The volume-specific rates of vibrational energy loss and gain are then

$$\Omega_\alpha^{(vib,diss)} = \omega_\alpha^{(r')}\, e_\alpha^{(vib)}(T_\alpha^*)\,, \qquad \frac{1}{T_\alpha^*} = \frac{1}{T_\alpha^{(vib)}} - \frac{1}{T^{(tr)}}\,, \tag{13}$$

$$\Omega_\alpha^{(vib,form)} = \omega_\alpha^{(r'')}\frac{1}{2}\left(e_\alpha^{(0)} - R_\alpha \theta_\alpha^{(vib)}\right)\,, \tag{14}$$

with $e_\alpha^{(0)}$ denoting the (positive) dissociation energy. Equation (14) is derived from (13) and the equilibrium conditions $\Omega_\alpha^{(vib,diss)} = \Omega_\alpha^{(vib,form)}$, $\omega_\alpha^{(r')} = \omega_\alpha^{(r'')}$, and $T_\alpha^{(vib)} = T^{(tr)}$. It states that the formation of a molecule takes place with equal probability for each vibrational level. This process of random recombination is assumed to occur in nonequilibrium, too. Details about the thermochemical model are given in Klomfass (1995). The theories of Newton, Fourier, and Fick are applied to the molecular fluxes. Bulk viscosity, and thermo- and pressure diffusion are as yet neglected. The species viscosities are calculated according to the gas kinetic theory, where the collision integrals are calculated from curve fits (Metz 1995). The mixing rule of Herning and Zipperer is used to approximate the mixture viscosity μ. The conductivities and the diffusion coefficient are approximated by constant Prandtl and Schmidt numbers, where $Pr = Sc = 0.7$ is assumed. Details are given in Klomfass et al. (1993) and Metz (1995).

5. Results and discussion

Several calculations have been carried out to assess the accuracy of the DRNSE. In the following we consider flows about spheres and focus on the shock stand-off distance δ and the stagnation-point heat flux q_{st}. For flows of a perfect gas with $\gamma = 1.4, R = 289\,\text{J/kgK}, \mu = 1.5 \cdot 10^{-5}\,\text{kg/ms}$, and $Pr = 0.7$ the following table shows the comparisons with the results of 2D-Euler calculations and the boundary-layer analysis of Fay and Riddell. These test cases are without

Flow conditions (Wall: 300 K)						DRNSE		2D-Euler/Fay&Rid.	
R_{sph} [m]	V [$\frac{m}{s}$]	p [Pa]	T [K]	Ma	Re	q_{st}	δ/R_{sph}	q_{st} [$\frac{kW}{m^2}$]	δ/R_{sph}
0.152	697	1000	300	2	1.6e5	11.2	0.245	10.7	0.345
0.152	1742	1000	300	5	4.1e5	151	0.139	141	0.156
0.152	3484	1000	300	10	8.1e5	980	0.123	957	0.135
0.152	6967	1000	300	20	16.3e5	6030	0.120	6160	0.129

physical relevance. However, as there is no question concerning the thermo-chemical equilibration of the boundary layer, the Fay and Riddell results are reliable. As the table above shows, the shock stand-off is significantly underestimated for larger values of δ/R_{sph}. The deviation decreases as δ/R_{sph} becomes smaller. The stagnation-point heat fluxes are acceptably accurate for all cases. Not shown here is that the calculated stagnation pressures and adiabatic stagnation temperatures match the Euler results perfectly. For the chemical and vibrational nonequilibrium flow specified below, Figs. 1 and 2 give a detailed comparison with a full 2D Navier-Stokes calculation.

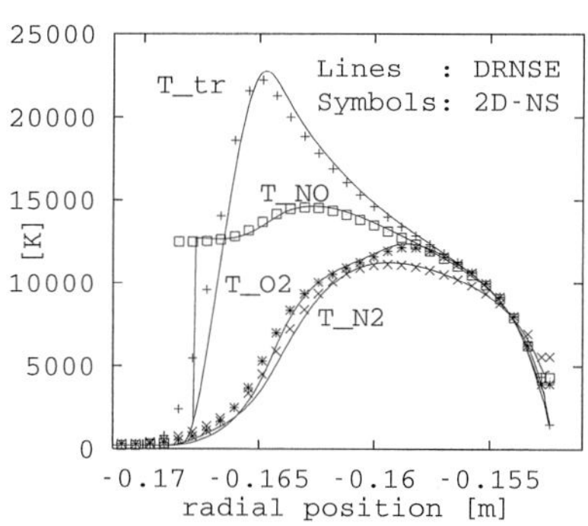

Figure 1. Temperatures.

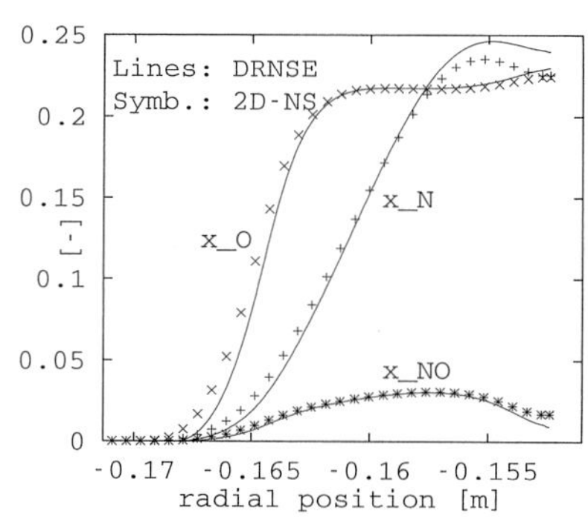

Figure 2. Mass fractions.

Flow conditions (Wall: 1500 K, non-catalytyic)						DRNSE		2D NS	
R_{sph}	V [$\frac{m}{s}$]	p [Pa]	T [K]	Ma	Re	q_{st}	δ/R_{sph}	q_{st} [$\frac{MW}{m^2}$]	δ/R_{sph}
0.152m	7600	4.17	212	26	11400	1.12	≈ 0.081	0.97	≈ 0.085

The deviation of the atomic-nitrogen mass fraction near the wall may be attributed to the different resolution, which is 35×160 points for the 2D and 200 points for the DRNSE calculations. The potential of the DRNSE can be assessed from the computing times, which were about 3 h on a Cray Y/MP for the 2D and about 15 min on a standard workstation for the DRNSE calculation. The following examples are flows with higher Reynolds numbers, where thin boundary layers exist. A good agreement with experimentally determined heat fluxes

Flow conditions (Wall: 300 K, non-catalytic)						DRNSE		Exp.	2D-Euler
R_{sph}[m]	V $[\frac{m}{s}]$	p [Pa]	T [K]	Ma	Re	q_{st}	δ/R_{sph}	q_{st} $[\frac{MW}{m^2}]$	δ/R_{sph}
0.0254	3929	11180	1371	5.5	108000	14.8	0.094	15.26	0.096
0.0254	5301	13090	2359	5.3	61000	34.3	0.084	34.51	0.085

(Wen and Hornung 1995) is found. The calculations revealed near-equilibrium boundary layers. In such flows a very high spatial resolution is required to achieve convergence of the surface concentrations and the heat fluxes. About 200 boundary-layer grid points were necessary to resolve the atomic-oxygen recombination near the stagnation point. This resolution could so far only be realized with the DRNSE approach or boundary-layer methods.

6. Conclusions

A useful approximation for the stagnation-streamline quantities in hypersonic blunt-body flows can be obtained from the numerical integration of the dimensionally reduced Navier-Stokes equations. The presented numerical method provides a helpful tool for the investigation of nonequilibrium flows, as it allows a fast, high-resolution calculation of the stagnation-streamline quantities.

Acknowledgement. This work is funded by the Deutsche Forschungsgemeinschaft.

References

Kao HC (1964) Hypersonic viscous flow near the stagnation streamline of a blunt body. AIAA J 2:1892–1897.

Klomfass A (1995) Hyperschallströmungen im Thermodynamischen Nichtgleichgewicht, Dissertation, RWTH Aachen.

Klomfass A, Müller S, Ballmann J (1993) Modelling of Transport Phenomena for the Hypersonic Stagnation Point Heat Transfer Problem, AIAA Paper 93-5047.

Metz S (1995) Numerische Verfahren zur Berechnung des Stoff- und Impulstransports in Wasserstoff-Luft und Methan-Luft Gemischen, Dissertation, RWTH Aachen.

Wen C, Hornung H (1995) Experiments on hypervelocity dissociating flow over spheres. In: Shock Waves @ Marseille, Vol 2, Springer.

Formation of standing shock waves due to evaporation from a spherical condensed phase

Y. Onishi, J. W. Fuchs
Department of Applied Mathematics and Physics
Tottori University, Tottori 680 Japan

Abstract: Transient motions of a vapor due to evaporation processes at a spherical condensed phase placed in an infinite expanse of the vapor have been considered. The Boltzmann equation of BGK type has been solved numerically by a difference scheme subject to the diffusive boundary condition at the interface together with an initial condition associated with sudden change in temperature of the condensed phase. Unlike problems with simple plane geometry, there exists a certain range of parameters in which the evaporation process causes strong expansion near the condensed phase, accelerating the flow to locally supersonic. A standing shock wave is, then, formed in order to connect the supersonic and subsonic regions within the flow field.

Key words: Kinetic theory, Evaporation, Spherical condensed phase, Shock wave, Standing shock wave

1. Introduction

Transient motions of a vapor due to phase-change processes at the condensed phase are of fundamental importance in gas dynamics in the sense that a clear understanding may be obtained as to how the flow fields approach their final states through propagations and continuous interactions of various waves, e.g., shock waves, contact regions, and sometimes expansion waves, produced by these phase-change processes. Problems of this kind, however, are the ones to which the ordinary continuum-based fluid dynamics is not applicable because of the existence of a nonequilibrium region called the Knudsen layer in the vicinity of the condensed phase. It is this nonequilibrium region that is responsible for the phase-change processes and its existence cannot be neglected in any problems, even in the continuum limit, as long as these processes are involved. Therefore, the analysis for flow problems of this kind must necessarily be based on kinetic theory.

The present study is concerned with one of these flow fields, i.e., the transient motions of a gas due to evaporation processes from a spherical condensed phase placed in an infinite expanse of its vapor and the consequent propagation of shock waves through a three-dimensional region. The formation of a standing shock wave may be expected in the flow field during the transition process to its final state. The Boltzmann equation of BGK type (Bhatnager et al. 1954) has been solved numerically for the present study subject to the boundary condition of diffusive reflection at the interface. It may be noted here that the steady state of this problem in the presence of an inert gas has been treated analytically by Onishi (1986) for weak phase-change cases and by Takeuchi & Onishi (1995) for slightly strong cases. Also the numerical study of the steady evaporation from a spherical phase into a vacuum has been given by Sone & Sugimoto (1993).

2. Equations, initial and boundary conditions

Consider a spherical condensed phase with its radius R_1 in an infinite expanse of its vapor at rest. Initially, the gas is in complete equilibrium with the condensed phase at temperature T_∞. The pressure and number density of the gas at this state are P_∞ and N_∞, respectively. At time $t = 0$, the temperature of the condensed phase is suddenly raised to a temperature T_W ($T_W > T_\infty$), which leads to the onset of evaporation from the condensed phase. The Boltzmann equation of BGK type (Bhatnager et al. 1954) in terms of the spherical coordinates $(r,\ \theta,\ \phi)$ may be written for the present problem with spherical symmetry as

$$\frac{\partial f}{\partial t} + \xi\cos\alpha\,\frac{\partial f}{\partial r} - \frac{\xi\sin\alpha}{r}\frac{\partial f}{\partial \alpha} = N\nu_c\,(F_e - f) \tag{1}$$

$$F_e = \frac{N}{(2\pi RT)^{3/2}}\exp\left\{-\frac{\xi^2 - 2U\xi\cos\alpha + U^2}{2RT}\right\} \tag{2}$$

$$\begin{bmatrix} N \\ NU \\ 3NRT \end{bmatrix} = 2\pi\int_0^\pi\int_0^\infty \begin{bmatrix} 1 \\ \xi\cos\alpha \\ \xi^2 - 2U\,\xi\cos\alpha + U^2 \end{bmatrix} \xi^2\sin\alpha\, f\, d\xi\, d\alpha \tag{3}$$

$$P = N\mathrm{k}T \tag{4}$$

$$E = 2\pi\,\frac{m}{2}\int_0^\pi\int_0^\infty \xi^5\cos\alpha\sin\alpha\, f\, d\xi\, d\alpha \tag{5}$$

$$\xi_r = \xi\cos\alpha, \quad (\xi_\theta^2 + \xi_\phi^2)^{1/2} = \xi\sin\alpha \qquad 0 \le \xi < \infty \quad 0 \le \alpha \le \pi$$

where t is the time; r is the radial coordinate; ξ is the magnitude of the molecular velocity vector $\boldsymbol{\xi}$ ($\xi = |\boldsymbol{\xi}|$) with its components $(\xi_r,\ \xi_\theta,\ \xi_\phi)$; α is an angle between the radial vector $\mathbf{r}$ and the molecular velocity vector $\boldsymbol{\xi}$; f, N, U, T, P, and E are, respectively, the distribution function, the number density, the mean flow velocity in the r direction, the temperature, the pressure, and the energy flux. F_e is the local Maxwellian distribution. $R = \mathrm{k}/m$ is the gas constant per unit mass, k being the Boltzmann constant and m the molecular mass; ν_c is a constant ($N\nu_c$: local collision frequency) related to the transport coefficients of the gas, for example, $\nu_c = P_\infty/(N_\infty \mu)$, μ being the viscosity of gas at the initial equilibrium state.

Now, the initial distribution function is given by

$$f = \frac{N_\infty}{(2\pi R T_\infty)^{3/2}} \exp\left\{-\frac{\xi^2}{2RT_\infty}\right\} \qquad \text{for} \quad 0 \le \alpha \le \pi \tag{6}$$

everywhere in the flow field ($R_1 \le r < \infty$). At time $t = 0$, the temperature of the condensed phase is suddenly changed to a different constant temperature T_W. Since the boundary condition for the distribution function is assumed to be of diffusive type, i.e., molecules leaving the condensed phase after their interactions with it obey a Maxwellian distribution characterized by the velocity (zero here) and the temperature of the condensed phase, the distribution function f in the present case takes for $t \ge 0$

$$f = \frac{N_W}{(2\pi R T_W)^{3/2}} \exp\left\{-\frac{\xi^2}{2RT_W}\right\} \qquad \text{for} \quad 0 \le \alpha \le \frac{\pi}{2} \qquad \text{at} \quad r = R_1 \tag{7}$$

$$f = \frac{N_\infty}{(2\pi R T_\infty)^{3/2}} \exp\left\{-\frac{\xi^2}{2RT_\infty}\right\} \qquad \text{for} \quad \frac{\pi}{2} \le \alpha \le \pi \qquad \text{at} \quad r \to \infty \tag{8}$$

where N_W is the number density of the molecules of gas leaving the interface after their interaction with the condensed phase. N_W is a unique function of T_W and is taken here to be the saturated vapor number density corresponding to T_W. From the Clapeyron-Clausius relation (see e.g., Landau & Lifshitz 1980), it is given by

$$N_W = \frac{P_W}{\mathrm{k}\, T_W} \qquad \frac{P_W}{P_\infty} = \exp\left\{-\Gamma\left(\frac{T_\infty}{T_W} - 1\right)\right\} \qquad \Gamma = \frac{h_L}{RT_\infty} \tag{9}$$

where P_W is the saturated vapor pressure corresponding to T_W and h_L is the latent heat of vaporization per unit mass. It may be noted that the values of Γ are usually larger than unity, for example, $\Gamma = 19$ at $T_\infty = 25°\mathrm{C}$ and $\Gamma = 13$ at $T_\infty = 100°\mathrm{C}$ for H_2O.

Appropriate nondimensionalization of the present problem gives the following parameters which characterize the motions of the gas:

$$Kn = \frac{l_\infty}{R_1} = \left(\frac{8\gamma}{\pi}\right)^{1/2} \frac{Ma}{Re} \qquad \frac{T_W}{T_\infty} \qquad \Gamma \quad \left(\text{or} \quad \frac{P_W}{P_\infty}\right) \tag{10}$$

where Kn, Re, and Ma are the Knudsen number, Reynolds number, and Mach number, respectively, at the initial state; l_∞ is the molecular mean free path defined by $l_\infty = (8RT_\infty/\pi)^{1/2}/(N_\infty \nu_c) = (\mu/P_\infty)(8RT_\infty/\pi)^{1/2}$. Note that $Ma = 1$ in the present case because the sound speed $c_\infty = (\gamma R T_\infty)^{1/2}$ with $\gamma = 5/3$ has been used here as the reference velocity. Furthermore, a time scale $\tau_\infty = (\gamma/2)^{1/2}(R_1/c_\infty)$ has also been introduced for the present numerical calculations.

3. Results and discussion

The fluid dynamic quantities for various sets of the parameters are obtained as functions of r/R_1 and t/τ_∞. Some of the results are shown in Figs. 1–3. The steady solutions to the Euler equations (with the boundary conditions given by the present numerical analysis) are shown by dashed lines in Fig. 1, whereas in Fig. 2 the steady solutions (Takeuchi & Onishi 1995) to the set of macroscopic equations of Navier-Stokes type and the appropriate boundary conditions derived from the kinetic theory analysis (Onishi 1993) are shown also by dashed lines. Figure 3a shows the weakening process of the initial shock wave and Fig. 3b the local Mach-number distributions at the virtually steady state.

Figure 1 shows the propagation of an initial shock wave followed by a contact region and fairly strong expansion (virtually adiabatic) at an early stage promoted by the three-dimensional expanse. This strong incipient expansion causes a supersonic flow near the condensed phase, which leads to the gradual formation of the standing shock wave at a finite distance from the condensed phase as time goes on. The occurrence of this standing shock wave is necessary in order for the flow field to adjust itself to the subsonic state established behind the initial shock wave propagating toward infinity. When the evaporation is not strong (see Fig. 2), the steady behavior is not of a supersonic-shock-subsonic-type, but simply of a smoothly changing weak subsonic, as can be expected.

Acknowledgement. We acknowledge the finacial support of the Association for Promotion of the Tottori Prefectural Engineering Technology.

References

Bhatnager PL, Gross EP, Krook M (1954) A model for collision processes in gases. Phys Rev 94:511–525.

Landau LD, Lifshitz EM (1980) Statistical Physics, Pergamon Press, New York, pp 251–257.

Onishi Y (1986) The spherical-droplet problem of evaporation and condensation in a vapour-gas mixture. J Fl Mech 163:171–194.

Onishi Y (1993) Kinetic theory based nonlinear treatment for motions of a binary gas mixtures involving slightly strong evaporation and condensation processes II — A system of macroscopic equations, the boundary conditions and the Knudsen-layer corrections. ZAMP 44:827–864.

Sone Y, Sugimoto H (1993) Kinetic theory analysis of steady evaporating flows from a spherical condensed phase into a vacuum. Phys Fluids A 5:1491–1511.

Takeuchi S, Onishi Y (1995) Evaporation and condensation from a spherical condensed phase in a vapor-gas mixture. In: Proc of 33rd Symp of the Chuugoku-Shikoku Branch of Japan Soc Mech Eng (in Japanese), pp 53–54.

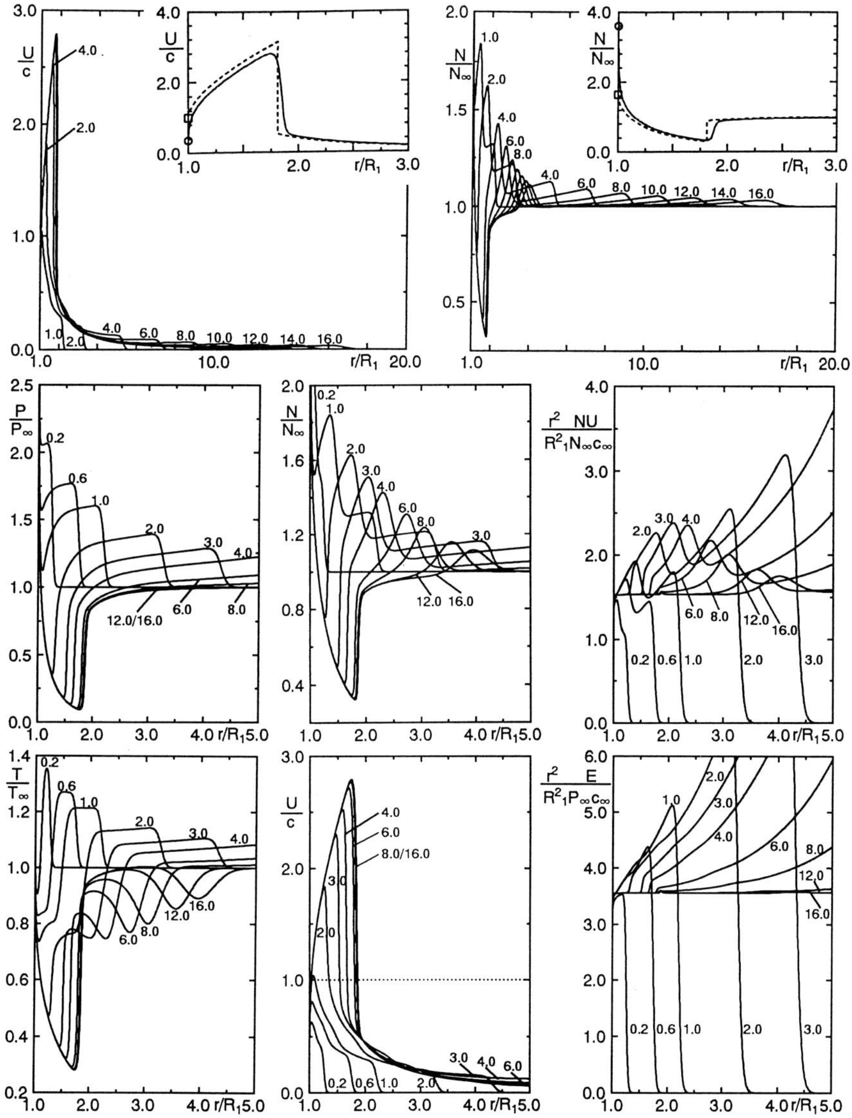

Figure 1. $Kn = 0.01$, $T_W/T_\infty = 1.2$, $\Gamma = 11$ ($P_W/P_\infty = 6.2547$). The numbers in the graphs indicate the values of time t/τ_∞; c is the local sound speed. Dashed lines indicate the steady-state solutions of the Euler equations. The values of $P(1)/P_\infty$ and $T(1)/T_\infty$ at $t/\tau_\infty = 16.0$ are 3.7319 and 1.0375, respectively.

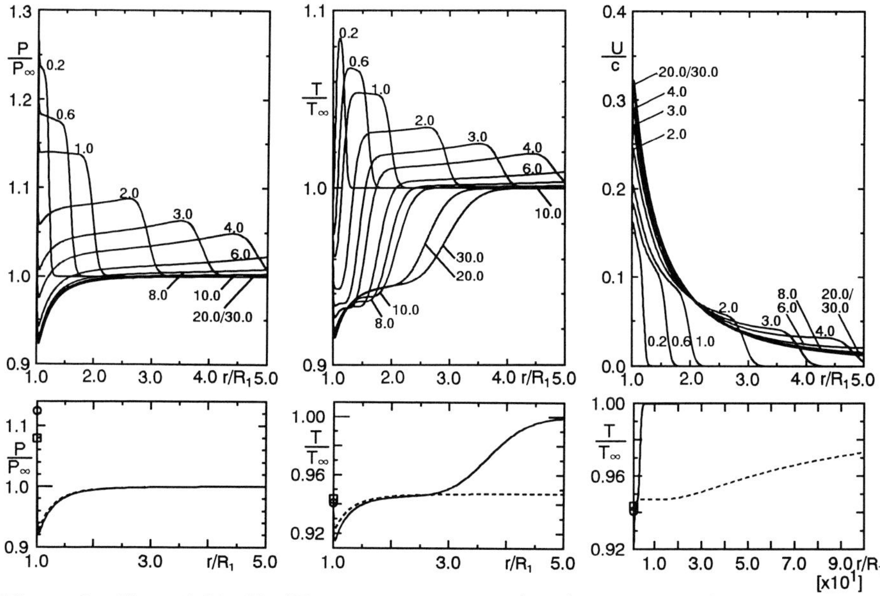

Figure 2. $Kn = 0.01$, $T_W/T_\infty = 1.05$, $\Gamma = 11$, $(P_W/P_\infty = 1.6884)$. The numbers in the graphs indicate the values of time t/τ_∞; c is the local sound speed. Dashed lines indicate the steady solutions (Takeuchi & Onishi 1995) obtained from Navier-Stokes type of equations, to which the present numerical solutions (solid lines) should approach as time goes on.

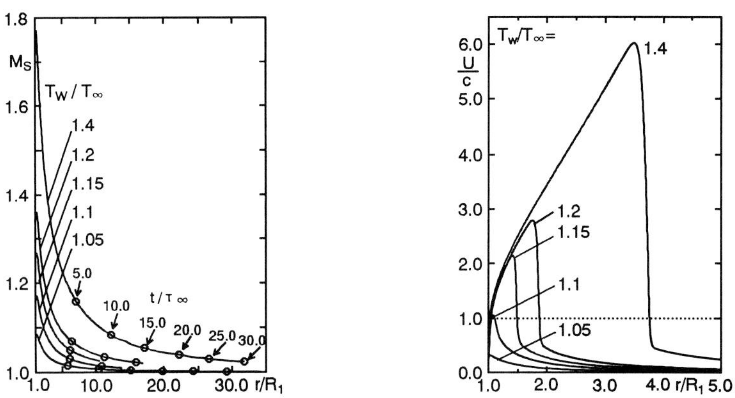

Figure 3. (a) The variations of the initial shock Mach number Ms with the distance from the condensed phase; (b) the local Mach-number distributions at the virtually steady state.

Knudsen-number effects in high-speed microchannel flows

T. X. Nguyen, C. K. Oh, X. N. Saint-Martin-Tillet,
J. V. Piana, E. S. Oran
Naval Research Laboratory, Code 6404
Laboratory for Computational Physics and Fluid Dynamics
Washington, DC 20376-5344 USA

Abstract: The effects of high Knudsen number on the flow in a microchannel are investigated through a series of numerical simulations. Both the Navier-Stokes and Direct Simulation Monte Carlo approaches are used to simulate a high-speed flow of He gas in a two-dimensional microchannel. The results show that the NS leading-edge shock angle is larger than that calculated by the DSMC, causing the two leading-edge shocks to interact sooner. Also, DSMC simulations predict substantial velocity slip and temperature jump at the wall near the leading edges of the channel.

Key words: High Knudsen-number flow, Direct Simulation Monte Carlo, Navier-Stokes, Microchannel flows

1. Introduction

The flow of gas through microdevices, for which the characteristic dimension L is near the order of the molecular mean free path λ, belongs to the transitional regime of fluid dynamics. In this regime, the Knudsen number, $Kn = \lambda/L$, is on the order of 0.1, the value for which the continuum Navier-Stokes (NS) formulation becomes invalid. Molecular approaches must be used to predict accurate slip velocities and temperature jumps at the solid surfaces and their effects on the flow-field structure.

One of the most promising methods applicable to the high Kn regime is the Direct Simulation Monte Carlo (DSMC) technique (Bird 1976), a method that has been used extensively for rarefied gas flows. Recently, DSMC has been used to simulate microchannel flows, in which the relevant channel dimension is on the order of microns and the densities may be relatively high (see, for example, Oh et al. 1995a).

In this work, a series of numerical simulations is used to examine the effects of high Kn on the flow structure in a two-dimensional microchannel. DSMC and NS computations are performed for the same physical system and their results are compared. The DSMC calculations are performed with the DSMC-

MLG algorithm (Cybyk 1994) which was recently optimized on the Thinking Machines CM-5 (Oh et al. 1995b, Nguyen 1995). The second type of calculation is performed with a NS code which has also been optimized on the CM-5 (Weber 1994) and used to study shock/boundary-layer interactions.

2. Numerical methods

2.1. DSMC-MLG approach

The MLG (Boris 1986), a method for managing very large data structures, ensures that particles that are close in physical space are also close in array index space. The MLG achieves this by requiring the spatial coordinates of the particles to satisfy monotonicity constraints.

The combined DSMC-MLG method abandons the spatially fixed grid used in conventional DSMC calculations. Instead, it subdivides the computational domain into templates, which are defined by blocks of nearest-neighbor particles. The locations of the boundary particles define the boundaries of the template. As the fluid particles move during the course of the flow, the particles are resorted into MLG order. The result is a template grid which automatically adapts to the continually changing local number densities of the gas. The combined DSMC-MLG was parallelized on the CM-5, resulting in a very efficient code (80% parallel efficiency achieved) and significant reductions in computational time (Oh et al. 1995b).

2.2. Continuum approach

The NS simulations were performed using a parallelized code based on Flux-Corrected Transport (FCT). FCT (Boris and Book 1976) is a high-order algorithm for integrating the convective transport terms while maintaining conservation, monotonicity, and positivity. The diffusive transport terms are calculated using a central-difference, finite-volume approach, and these terms are combined with the convective transport as source terms in FCT.

3. Results and discussion

The simulations are performed for a 6 μm $\times$ 1.2 μm channel with a Mach 5 inflow of He atoms at 298 K and 1 atm. For this system, $Kn = 0.14$. In the DSMC-MLG simulations, the He atoms are modeled as hard spheres. A total of 110,250 simulated particles are divided into 150 $\times$ 15 time-varying computational templates of 7 $\times$ 7 particles. The solid walls are represented by the fully diffuse boundary conditions and the molecules are reflected according to the Maxwellian distribution based on the wall temperature. In the NS simulations, a rectangular 256 $\times$ 64 grid is used. At the solid walls, no-slip conditions are applied. For both types of calculations, the wall temperature is equal to the free-stream temperature. Figure 1 is a schematic of the flow configuration.

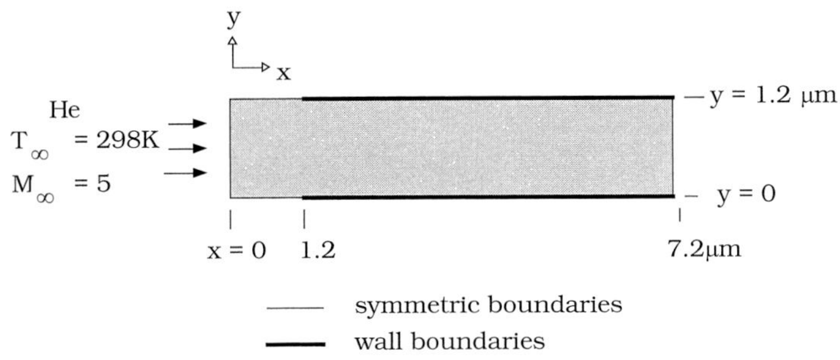

Figure 1. Schematic of the microchannel flow.

Figures 2 and 3 show that NS and DSMC contours of the pressure, p, and viscosity coefficient, μ, differ both qualitatively and quantitatively for the two cases. In the NS calculation, the leading-edge shocks are sharper and more

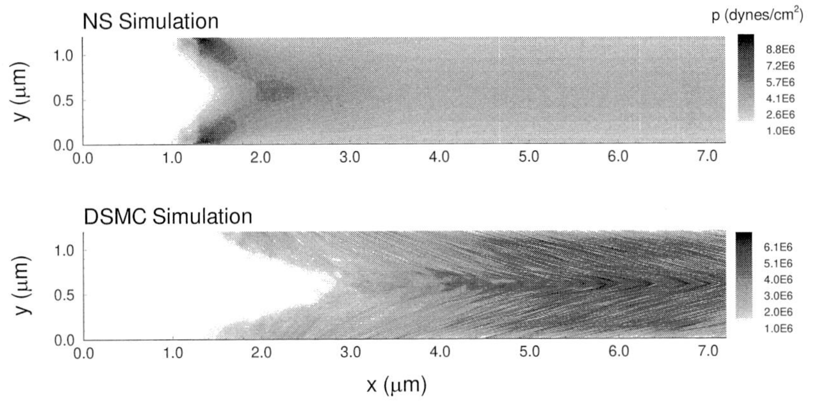

Figure 2. Contours of pressure in the microchannel. The DSMC result represents an instantaneous solution. The streaks in the DSMC result are statistical fluctuations typical of DSMC simulations.

highly curved than the DSMC shocks, indicating the presence of strong viscous interactions between the boundary layer and the outer shock. Furthermore, the

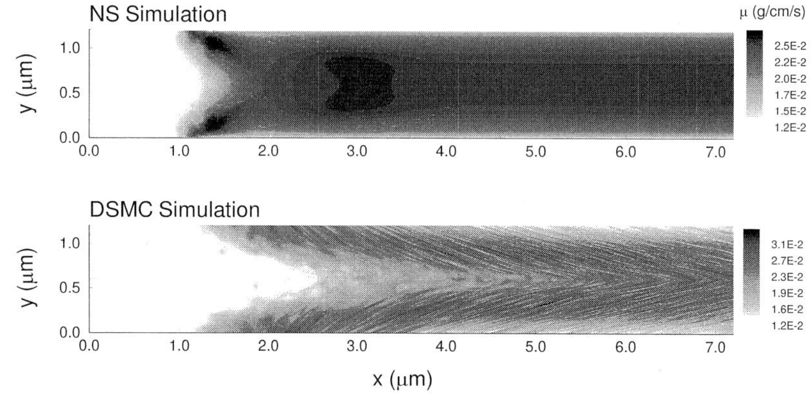

Figure 3. Contours of viscosity coefficient in the microchannel. The DSMC result represents an instantaneous solution.

NS shock angle, on the average, is larger than that calculated by the DSMC method. The shock angle varies between near normal at the wall to $\sim 50°$ near the channel centerline for the continuum case, whereas it is only $\sim 20°$ for the DSMC calculation. The larger shock angle in the NS case means that the two leading-edge shocks interact sooner. The NS shock is also much stronger than the DSMC shock, indicated by the respective pressure ratios of 60 and 7.5 across the shock at the wall.

Figure 4 shows the velocity and temperature profiles at several locations along the length of the microchannel obtained from the DSMC simulation. These velocity profiles show the development of the viscous boundary layer, including the slip velocity generated at the channel walls. The slip velocity is higher near the channel entrance and decreases with distance downstream. The thermal boundary layer, indicated by the temperature profiles, shows a jump in temperature at the walls. The slip velocity and temperature jump along the lower wall of the channel are shown in Fig. 5. The slip velocity is higher near the channel entrance and decreases with distance downstream. Beyond $x = 3.5\ \mu$m, the velocity slip does not change significantly. The temperature jump at the wall initially increases near the channel entrance, reaches a maximum at approximately $x = 2.2\ \mu$m, then decreases further downstream. Thus the DSMC simulations predict the expected velocity slip and temperature jump at the microchannel walls typical of high-Kn flows where the continuum formulation is invalid.

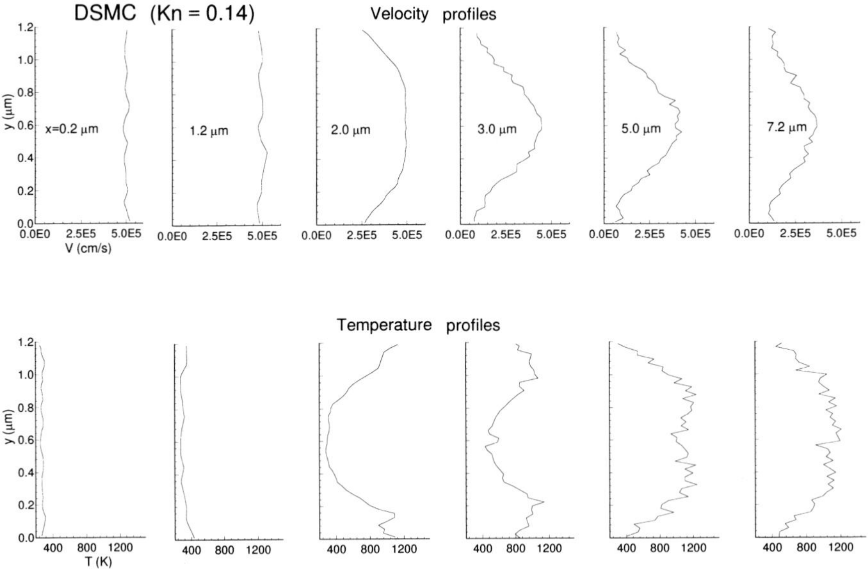

Figure 4. DSMC-computed velocity and temperature profiles at selected locations.

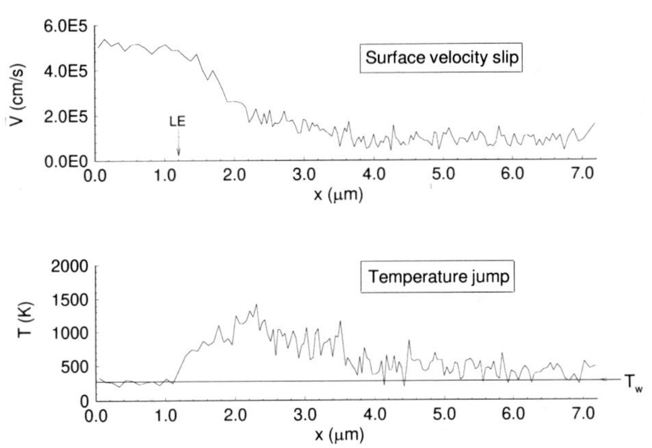

Figure 5. Velocity slip and temperature jump along the lower wall of microchannel.

4. Conclusion

These DSMC results represent one instantaneous solution which requires approximately 7 hours of cpu time on the CM-5 to complete. Even more extensive time-averaging is required to reduce the statistical fluctuations in the solution to an acceptable level, further increasing the computational cost. Each NS simulation requires only $\sim$ 5 minutes of cpu time on the CM-5 for convergence. However, the continuum approach is not valid for $Kn > 0.03$. It therefore becomes important to develop ways of modifying NS calculations to simulate some of the high-Kn effects. These computations will be used as a basis for the derivation of boundary conditions and constitutive relations that extend the validity of the NS simulations.

Acknowledgement. This work was sponsored by ASEE, NRC, ARPA, ONR/NRL, SNECMA, and EM2C-CNRS.

References

Bird GA (1976) Molecular Gas Dynamics, Clarendon Press, Oxford.

Boris JP (1986) A vectorized "near-neighbor" algorithm of order n using a monotonic logical grid. J Comp Phys 66:1–20.

Boris JP, Book DL (1976) Solution of the continuity equation by the method of flux-corrected transport. Methods in Computational Physics 16:85–129.

Cybyk BZ (1994) Combining the Monotonic Lagrangian Grid with Direct Simulation Monte Carlo; a New Approach for Low-Density Flows, PhD Dissertation, University of Maryland.

Nguyen TX (1995) Enhancements of the Parallelized DSMC-MLG Method for Applications to Complex Hypersonic Transitional-Regime Flows, PhD Dissertation, University of Maryland.

Oh CK, Oran ES, Cybyk BZ (1995a) Microchannel Flow Computed with the DSMC-MLG, AIAA-95-2090.

Oh CK, Sinkovits RS, Cybyk BZ, Oran ES, Boris JP (1995b) Massive Parallelization of DSMC Combined with the Monotonic Lagrangian Grid, AIAA-95-0409.

Weber YS (1994) The Numerical Simulation of the Reflected-Shock/Boundary-Layer Interaction in Shock Tubes, PhD Dissertation, University of Maryland.

Studies of separated flow structure and oscillating separation shock waves in shock-wave/turbulent boundary-layer interaction regions induced by a blunt protuberance

S. Maekawa, S. Aso
Department of Aeronautics and Astronautics, Kyushu University
Hakozaki 6-10-1, Higasiku, Fukuoka 812 Japan
K. Karashima, K. Sato
Institute of Space and Astronautical Sciences
Yoshinodai 3-1-1, Sagamihara, Kanagawa 229 Japan

Abstract: The three-dimensional shock-wave/turbulent boundary-layer interaction region induced by a blunt protuberance is examined. In the experiments, oscillating shock waves in the interaction region are also experimentally investigated. Experiments are performed with freestream Mach numbers of 2, 3, and 4 and almost adiabatic wall conditions.

Key words: Oscillating shock wave, Blunt protuberance, Flow visualization

1. Introduction

The shock-wave/turbulent boundary-layer interaction is a very interesting problem and arises in many practical flows. Representative studied were made by Korkegi (1971) and Sedney (1973). One of these phenomena is the oscillating shock-wave motion generated by the shock-induced turbulent boundary-layer separation. This unsteadiness would appear to be an inherent feature of the shock-induced separation, and has been investigated by many researchers (e.g. Sedney and Kitchens 1977 and Dolling and Bogdonoff 1981).

A multi-stage rocket such as H-II is designed to store the maximum amount of fuel, hence pipelines are located outside the rocket body. In such a configuration, the bow shock wave induced by a pipeline fairing interacts with the turbulent boundary layer that is developed on the main body, causing severe aerodynamic heating phenomena and oscillating shock waves in the interaction region.

2. Experimental apparatus and procedures

The experiments are performed in a supersonic wind tunnel of the Institute of Space and Astronautical Sciences with a square test section of 600 × 600 mm and variable Mach number capability over the range of 2–4. A portion of a side wall is an optical window which provides excellent optical access so that flows may be visualized in detail. The boundary-layer thickness is about 50 mm at a freestream Mach number of 4, estimated from schlieren and pitot-pressure measurements (cf. Aso and Maekawa 1993). A schematic of the flow field is shown in Fig. 1.

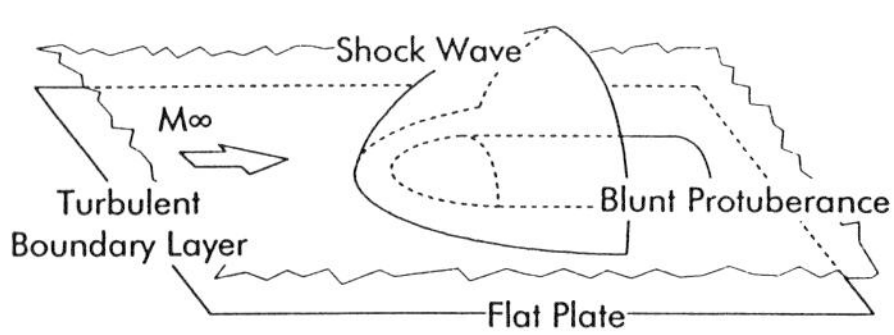

Figure 1. Schematic diagram of the flow field.

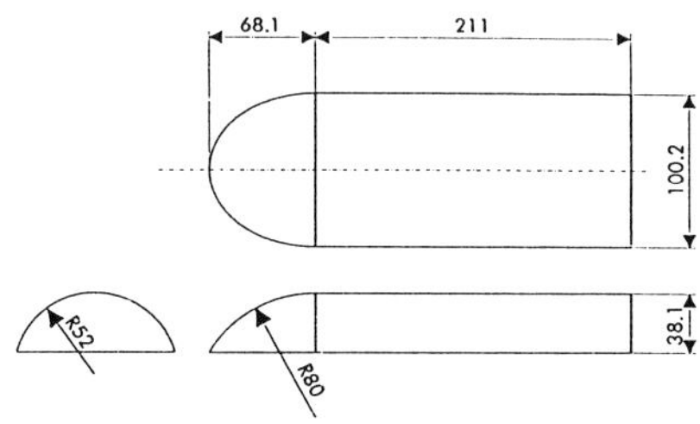

Figure 2. Blunt protuberance model and coordinate system.

The configuration of the blunt protuberance is shown in Fig. 2. The x-axis is the main flow direction, the y-axis is lateral, and the z-axis is normal to the side wall. The origin is the tip of the blunt protuberance. The blunt protuberance is located on the side wall for oil-flow visualizations and pressure-distribution measurements or is attached on the flat plate installed in the wind tunnel for visualization at the foot of the body. Experiments are performed under almost adiabatic wall conditions. Total pressure is 0.29 MPa at a freestream Mach number of 2, 0.49 MPa at Mach 3, and 0.59 MPa at Mach 4.

3. Results and discussion

3.1. Visualizations by oil-flow technique

A representative picture and a sketch of an oil-flow pattern are shown in Fig. 3 for Mach number 4. The locations of the primary separation lines, which represent the extent of the interaction, with various Mach numbers are shown in Fig. 4. In Fig. 3, S_1, S_2, and A are the primary separation line, secondary separation line, and attachment line, respectively. The primary separation lines are located at almost the same position irrespective of freestream Mach number. This indicates that the main governing parameter of the interaction regions is not freestream Mach number, but the size of the shock generator, i.e., the blunt

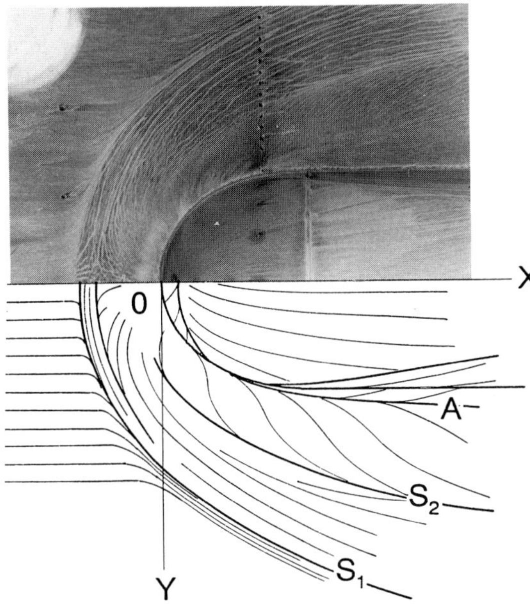

Figure 3. Picture and sketch of an oil-flow pattern for $M_\infty = 4$.

protuberance. Between a secondary separation line and an attachment line, an inverse-S type pattern is observed, which indicates the existence of a separation vortex. Another separation region arises from the shoulder of the blunt protuberance. For the Mach 4 case, just behind the primary separation line, a new separation region exists, suggesting that new vortices in the interaction region are produced.

3.2. Pressure measurements

Pressure measurements on the centerline and $x = 46$ mm are shown in Fig. 5. Pressures are normalized by the value of the undisturbed pressure in front of the interaction region. A pressure plateau is formed close to the secondary separation line due to the separated vortex in the interaction region. The local pressure minimum lies just upstream of the blunt protuberance, due to the small separated vortex located before the protuberance. The maximum pressure is much higher than the pressure plateau. However, the locations at which the pressure starts to rise and the maximum pressures are nearly independent of Mach number. The same tendency is observed for pressure distributions at $x = 46$ mm. Away from the centerline the pressure decreases rapidly, and a

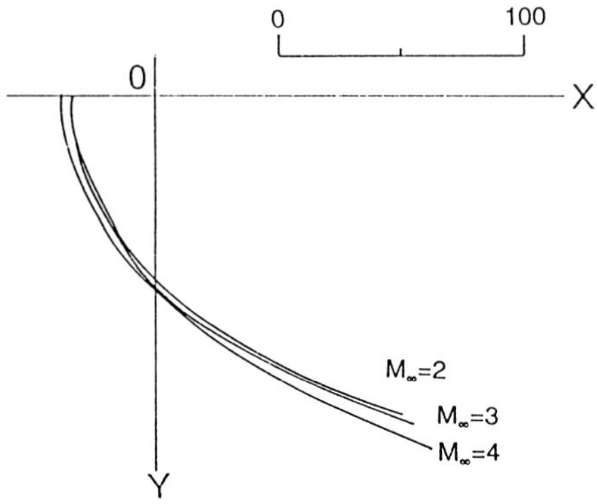

Figure 4. The locations of the primary separation lines.

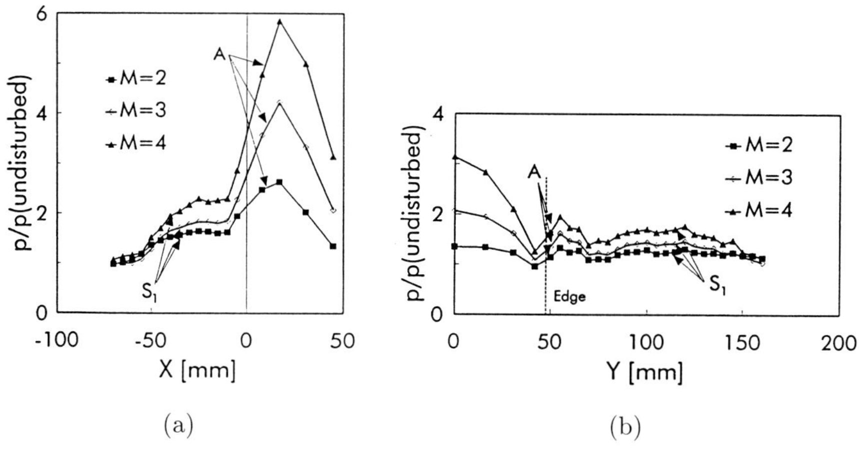

Figure 5. Measurements of pressure distributions: (a) pressure distributions on the centerline, and (b) pressure distributions on $x = 46$ mm.

local pressure minimum is observed upstream of the maximum peak, due to the separated vortex at the corner.

3.3. Visualization by shadowgraph technique

Shadowgraphs are instantaneously taken as randomly shot sequences, shown in Fig. 6. These pictures show the unsteady characteristics of the interaction region. In these pictures, oscillating separated shock waves are observed. Also a slip layer is observed at the intersection point of the separation shock wave and

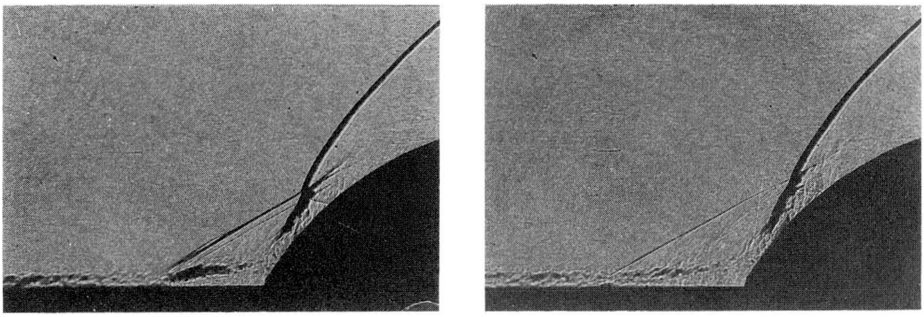

(a) (b)

Figure 6. Instantaneous shadowgraph at $M_\infty = 3$: (a) strong shock separation, and (b) weak shock separation.

the bow shock wave. When a strong-separation shock wave is formed (Fig. 6a), the generating point of the separation shock wave is located slightly further downstream than in the weak-separation shock-wave case (Fig. 6b), and the interaction region decreases. In the strong-separation shock-wave case, a strong slip layer impinges on the blunt protuberance surface.

4. Conclusions

1. The primary separation lines are located at almost the same position irrespective of freestream Mach number. Thus the main governing parameter of these interaction regions is the size of the blunt protuberance.

2. For a freestream Mach number of 4, just behind the primary separation line a new separation region appears, suggesting that new vortices in the interaction region are produced. The position of the primary separation and attachment lines varied very little with Mach number.

3. As the freestream Mach number increases, the maximum pressure becomes higher, but the locations of the start of the pressure rise and the maximum pressures are nearly independent of freestream Mach numbers.

4. The oscillating separated shock waves have been observed by optical flow visualization. The separation phenomenon is highly unsteady. Also there are two significant cases of weak separation and strong separation.

References

Aso S, Maekawa S (1993) Intensive Studies on Unsteady Secondary Separations and Oscillating Shock Waves in Three-Dimensional Shock Waves/Turbulent

Boundary Layer Interaction Region Induced by Sharp/Blunt Fins, AIAA Paper 93-2939.

Dolling DS, Bogdonoff SM (1981) An Experimental Investigation of the Unsteady Behavior of Blunt Fin-Induced Shock Wave Turbulent Boundary Layer Interactions, AIAA Paper 81-1287.

Korkegi RH (1971) Survey of viscous interaction associated with high Mach number flight. AIAA J 9:5:771–784.

Sedney R (1973) A survey of the effects of small protuberances on boundary layer flows. AIAA J 11:6:782–792.

Sedney R, Kitchens CW (1977) Separation ahead of protuberances in supersonic turbulent boundary layers. AIAA J 15:4:546–552.

A numerical study of shock-wave reflections in nonequilibrium chemically reacting flows

S. Aso, T. Mitomo
Department of Aeronautics and Astronautics
Kyushu University, Japan

T. Fujiwara
Department of Aeronautics
Nagoya University, Japan

M. Hayashi
Nishinippon Institute of Technology
Japan

Abstract: In the present study the full two-dimensional Navier-Stokes equations are solved in order to investigate unsteady aerodynamic heating phenomena induced by shock impingement on a ramp surface. The effects of chemical reactions on aerodynamic heating are investigated. The results show that the effects of chemical reactions on shock-induced aerodynamic heating phenomena are quite significant at high temperature.

Key words: Aerodynamic heating, Nonequilibrium flow, Chemical reaction

1. Introduction

Recently, high-speed winged vehicles have been studied. One of the most important problems for designs of such vehicles is the severe aerodynamic heating and pressure rise caused by the impingement of shock waves on vehicle surfaces. In particular, the unsteady aerodynamic heating caused by shock-wave reflections at high shock Mach number has been investigated and the peak heating due to a Mach stem and the second peak heating due to a slip layer have been observed by the present authors (Aso et al. 1989, Ohyama et al. 1993).

In the present study the full two-dimensional Navier-Stokes equations are solved in order to investigate unsteady aerodynamic heating phenomena induced by shock impingement on a ramp surface. When the shock is strong, nonequilibrium chemical reactions occur in the flow. The effects of chemical reactions on aerodynamic heating are investigated. The results are compared with experiments and calculated results without reactions.

2. Numerical methods

In the numerical calculations, nonequilibrium chemically reacting flows are considered. The governing equations are the two-dimensional full Navier-Stokes equations. In conservation form, these are

$$\frac{\partial \boldsymbol{U}}{\partial t} + \frac{\partial \boldsymbol{E}}{\partial x} + \frac{\partial \boldsymbol{F}}{\partial y} = \boldsymbol{J}$$

where each vector is expressed as follows:

$$\boldsymbol{U} = \begin{bmatrix} \rho \\ \rho u \\ \rho v \\ \rho(E + \frac{V^2}{2}) \\ \rho c_i \end{bmatrix}, \qquad \boldsymbol{E} = \begin{bmatrix} \rho u \\ \rho u^2 + p - \tau_{xx} \\ \rho u v - \tau_{xy} \\ \rho(E + \frac{V^2}{2})u + pu - q_x - u\tau_{xx} - v\tau_{xy} \\ \rho c_i u - \rho D_{im}\frac{\partial c_i}{\partial x} \end{bmatrix}$$

$$\boldsymbol{F} = \begin{bmatrix} \rho v \\ \rho u v - \tau_{yx} \\ \rho v^2 + p - \tau_{yy} \\ \rho(E + \frac{V^2}{2})v + pv - q_y - u\tau_{yx} - v\tau_{yy} \\ \rho c_i v - \rho D_{im}\frac{\partial c_i}{\partial y} \end{bmatrix}, \qquad \boldsymbol{J} = \begin{bmatrix} 0 \\ 0 \\ 0 \\ 0 \\ \dot{w}_i \end{bmatrix}.$$

Assuming Stokes' hypothesis of $\lambda = -\frac{2}{3}\mu$, viscous stress tensors are expressed as follows:

$$\tau_{xx} = \frac{2}{3}\mu\left(2\frac{\partial u}{\partial x} - \frac{\partial v}{\partial y}\right), \quad \tau_{yy} = \frac{2}{3}\mu\left(2\frac{\partial v}{\partial y} - \frac{\partial u}{\partial x}\right), \quad \tau_{xy} = \tau_{yx} = \mu\left(\frac{\partial u}{\partial y} + \frac{\partial v}{\partial x}\right).$$

Heat-conduction terms are expressed as follows:

$$q_x = k\frac{\partial T}{\partial x} + \sum_i \rho D_{im}\frac{\partial c_i}{\partial x}h_i, \qquad q_y = k\frac{\partial T}{\partial y} + \sum_i \rho D_{im}\frac{\partial c_i}{\partial y}h_i .$$

For chemical reactions of oxygen gas the following elementary reactions are assumed:

$$O_2 + O_2 \rightleftharpoons 2O + O_2, \qquad O_2 + O \rightleftharpoons 2O + O .$$

For convective terms a Harten-Yee upwind TVD scheme is used and for viscous terms a conventional central difference is used. For boundary conditions, non-slip conditions are applied for the ramp surface and zero derivatives along the freestream are assumed at the incoming and downstream boundaries. Since in the flows with nonequilibrium chemical reactions the density, temperature,

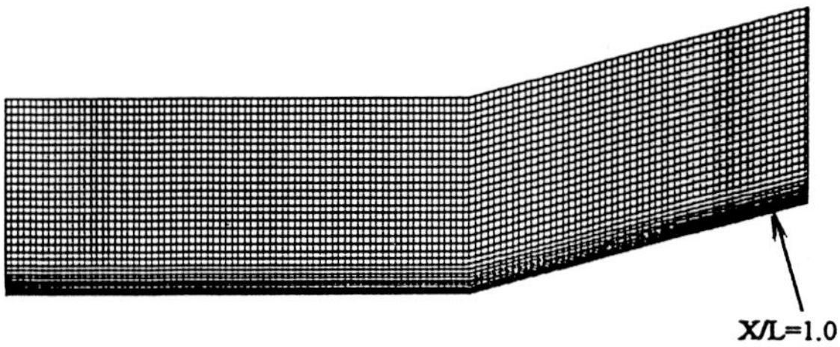

Figure 1. Representative mesh system. Ramp angle is 15 degrees; 515 × 161; every fourth grid shown.

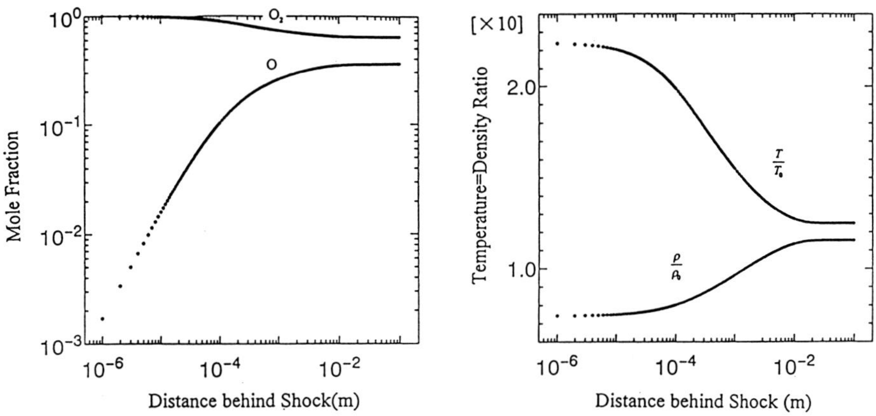

Figure 2. Distributions of chemical species (left) and temperature behind incident shock wave (right); O_2, $M_s = 11.7$, $\theta = 15°$, $p_\infty = 15$ torr, and $T_\infty = 298.0$ K.

and chemical compositions behind the shock change along the distance from the shock front, the one-dimensional, steady Euler equations for nonequilibrium chemically reacting flows are solved first in order to obtain the initial conditions for the incident shock wave. Also, zero physical derivatives normal to incoming flow are imposed for the upper boundary in order to keep the incident shock wave normal. For the energy equation a constant wall temperature condition is assumed. Both the fully catalytic wall case and non-catalytic wall case are considered. Thermal equilibrium is assumed. A representative mesh system is shown in Fig. 1.

3. Numerical results and discussion

Shock-reflection processes in O_2 gas at $M_s = 11.7$, $\theta = 15°$, $p_\infty = 15$ torr, and $T_\infty = 298.0$ K are calculated. The chemical-composition and temperature distributions behind the incident shock wave, which are calculated from the one-dimensional steady Euler equations for nonequilibrium chemically reacting flows, are shown in Fig. 2.

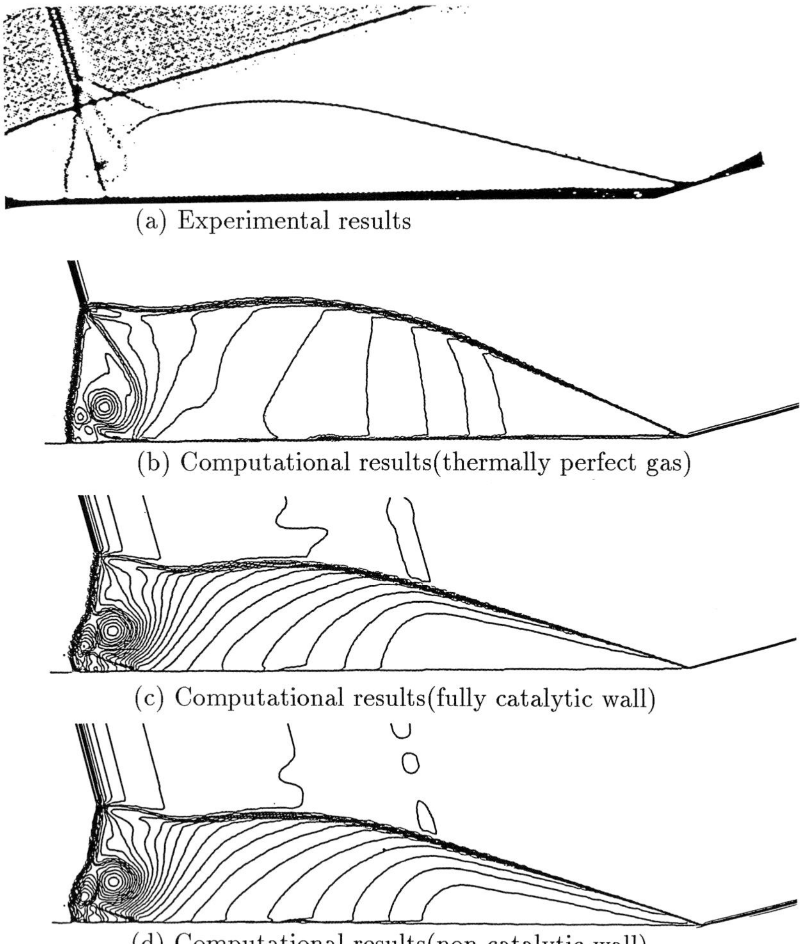

(a) Experimental results

(b) Computational results(thermally perfect gas)

(c) Computational results(fully catalytic wall)

(d) Computational results(non-catalytic wall)

Figure 3. Comparison of calculated results with experiments: (a) experimental results; (b) computational results (thermally perfect gas); (c) computational results (fully catalytic wall); and (d) computational results (non-catalytic wall). O_2, $M_s = 11.7$, $\theta = 15°$, $p_\infty = 15$ torr, and $T_\infty = 298.0$ K.

The calculated results for thermally perfect gases and chemically non-equilibrium gases for fully catalytic and non-catalytic wall conditions are shown in Fig. 3. The calculated shock patterns for chemically non-equilibrium gases with catalytic and non-catalytic wall conditions show quite good agreement with experiments of Weynabts (1968).

The calculated shear-stress and surface heat-flux distributions are shown in Figs. 4 and 5. Though little change in the shear-stress distributions is observed among the three models, quite drastic change in the surface heat-flux distributions is observed. In the fully catalytic wall case, enormous heat flux is seen because all dissociated oxygen recombines at the wall and gives enormous amounts of energy to the wall. These results show that wall catalycity is very important in the estimation of precise aerodynamic heating levels in chemically reacting flows.

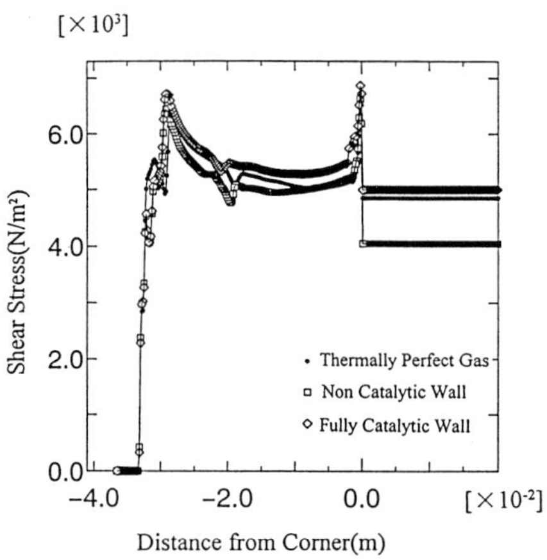

Figure 4. Calculated shear-stress distribution; O_2, $M_s = 11.7$, $\theta = 15°$, $p_\infty = 15$ torr, and $T_\infty = 298.0$ K.

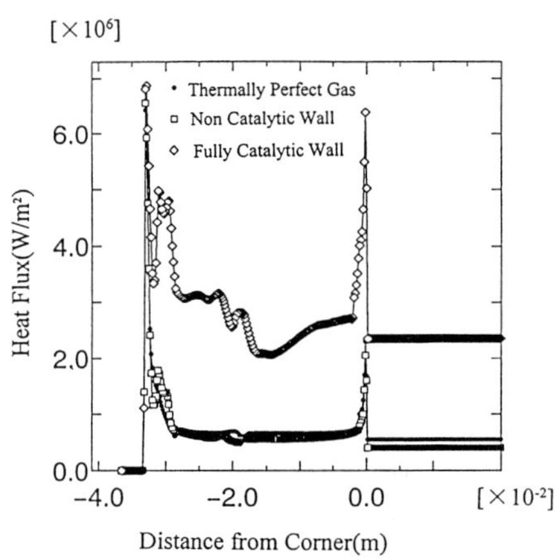

Figure 5. Calculated surface heat-flux distribution; O_2, $M_s = 11.7$, $\theta = 15°$, $p_\infty = 15$ torr, and $T_\infty = 298.0$ K.

4. Conclusions

In the present study the full two-dimensional Navier-Stokes equations are solved in order to investigate unsteady aerodynamic heating phenomena induced by shock impingement on a ramp surface. When the shock is strong, nonequilibrium chemical reactions occur in the flows. The effects of chemical reactions on aerodynamic heating are investigated. The results show the effects of chemical reactions on shock-induced aerodynamic heating phenomena are quite significant at high temperature.

References

Aso S, Hayashi M, Tan A, Tanahashi Y (1989) Surface heat transfer change induced by unsteady shock reflections and numerical simulations on shock reflection. In: Proc 17th ISWST, pp 630–635.

Ohyama K, Aso S, Hayashi M (1993) Numerical simulation of shock induced unsteady aerodynamic heating with highly dense mesh system. In: Shock Waves @ Marseille, R Brun and L Dumitrescu (eds), Springer, pp 395–400.

Weynabts RR (1968) Experimental Investigation of Shock Wave Diffraction over Compression and Expansion Corners, UTIAS TN No.126.

Turbulent spots in boundary layers in a free-piston shock-tunnel flow

D. J. Mee, C. P. Goyne
Department of Mechanical Engineering
The University of Queensland
Brisbane 4072 Australia

Abstract: In this paper, experiments to detect turbulent spots in transitional boundary layers formed on a flat plate in a free-piston shock-tunnel flow are reported. Experiments indicate that thin-film heat-transfer gauges are suitable for identifying turbulent-spot activity and can be used to identify parameters such as the convection rate of spots and the intermittency of turbulence.

Key words: Turbulent spots, Transition, Shock tunnel

1. Introduction

Boundary-layer transition has been observed to take place through formation, growth, and merger of turbulent spots in both incompressible and compressible flows (e.g., Emmons 1951; Clark et al. 1994). In studying transition in high-enthalpy impulse facilities, the frequency of spot production and the growth rate of spots may influence the interpretation of averaged heat-transfer and skin-friction data if an insufficient averaging period is used. Also, recent theoretical analysis by Doorly and Smith (1992) has indicated that there may be differences in spot dynamics at high Mach numbers. Experimental work by Clark et al. (1994) has supported this for Mach numbers up to 1.86.

The aims of the present paper are to demonstrate that turbulent spots can be detected using fast-response heat-transfer instrumentation in a high-enthalpy impulse facility, to obtain an experimental indication of the rate of spot production and spot size, and to investigate the suitability of the present instrumentation for study of spot dynamics in high-speed flows.

2. Detection of turbulent spots

A simplified model of a spot, based on the observations of Schubauer and Klebanoff (1955), as shown in Fig. 1, can be used to determine whether spots might be detectable in hypervelocity flows. The speeds of the leading and trailing edges of the spots are u_t and u_b respectively.

Let the freestream flow speed be u_∞ and the transition length be l_t. Consider a spot initiated at the start of the transition region which passes over a detector

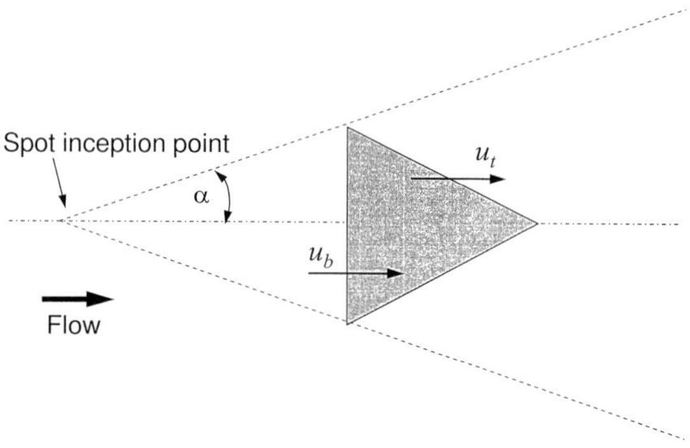

Figure 1. Model planform of a turbulent spot.

located a fraction, f, of the distance l_t from the start of transition. The time between when the apex and when the base of the turbulent spot pass over the detector, t_t, will be given by

$$t_t = \frac{fl_t}{u_\infty} \frac{\frac{u_t}{u_\infty} - \frac{u_b}{u_\infty}}{\frac{u_t}{u_\infty}\frac{u_b}{u_\infty}}. \tag{1}$$

For a typical Mach 5.6 condition in the T4 free-piston shock tunnel at 10-MJ/kg nozzle-supply enthalpy, the transition length on a flat plate with zero pressure gradient is approximately 400 mm and the flow speed is 4000 ms^{-1} (see Section 4). As a first approximation, we take typical low-speed values for $\frac{u_t}{u_\infty}$ and $\frac{u_b}{u_\infty}$ to be 0.88 and 0.50 respectively (Schubauer and Klebanoff 1955). For a gauge at the middle of the transition region, $f = 0.5$ and thus $t_t = 43$ μs. Taking the intermittency to be 0.5 at the middle of the transitional region, the period of laminar flow between the turbulent bursts can be expected to be of a similar duration. Surface-mounted thin-film gauges, such as those used in T4, should be capable of detecting turbulent spots of this size and frequency.

3. Experiments

Experiments were performed in the T4 free-piston shock tunnel at The University of Queensland (Stalker and Morgan 1988). An axisymmetric contoured nozzle was used. The test surface was a flat plate 1500 mm long. This formed one of the inner walls of a duct which captured the flow from an opening 120 mm wide and 60 mm high. Adjustable divergence on the three non-instrumented walls enabled pressure gradients on the test surface to be minimized. Flush-mounted thin-film heat-transfer gauges were used to measure surface heat-transfer rates. These gauges consisted of a platinum film, 1 mm long × 0.3 mm wide × 1 μm thick, painted onto a 2.1-mm-diameter fused quartz cylinder. The frequency response is estimated to be 100 kHz. Locations for gauges were available every 50 mm from a distance of 370 mm to 1470 mm from the sharp leading edge. Not all locations were used. Static pressure measurements were taken simultaneously using PCB type 112A22 piezoelectric pressure transducers located at the same axial location but at a different spanwise location from the corresponding thin-film gauge. The thin-film gauge signals were multiplexed and the signal from each transducer was sampled at 3- or 4-μs intervals. One gauge was sampled at 1-μs intervals for comparison.

4. Results

The present instrumentation was set up to test skin-friction gauges in turbulent boundary layers in the T4 tunnel (Goyne 1995). The experiments reported here were performed at the end of a series of tests for the purpose of seeing if turbulent-spot activity could be detected in T4. The tests were made at a nominal nozzle supply enthalpy of 10 MJ/kg and the flow Reynolds number was varied. The test conditions are given in Table 1.

Table 1. Conditions for experiments.

Shot	Enthalpy MJ/kg	T_∞ K	ρ_∞ kg/m^3	u_∞ m s^{-1}	M_∞	Unit Re m^{-1}
4686	9.9	1260	0.021	3900	5.6	1.6×10^6
4687	9.3	1220	0.031	3850	5.6	2.5×10^6
4688	10.2	1360	0.034	4000	5.6	2.7×10^6

Typical nozzle-supply pressure and test-section Pitot-pressure time histories are shown in Fig. 2a. These results have been processed by a 50-μs moving average filter. The time scales on this and subsequent figures are arbitrary. The flow arrives at the Pitot probe located in the test section 200 μs after the rise in the nozzle-supply pressure. With appropriate time delay, the ratio of Pitot to supply pressure is constant 500 to 600 μs after flow arrival. Previous tests indicate that, subsequently, there is a similar period before driver gas contaminates the test gas (Skinner 1994). During this test time, the pressures remain constant to within about 10%. After the test time, the pressures slowly decrease. The measured static-pressure distributions at the three test conditions, averaged over 500-μs periods during the test time, are plotted in Fig. 2b. The pressures are approximately constant down the test surface for each condition.

Heat-transfer results averaged over a 500-μs period starting at 600 μs after the arrival of flow for the three test conditions are shown in Fig. 3. Shown also are laminar and turbulent levels determined using Eckert's reference temperature method (White 1991) and the theory of Spalding and Chi (1964),

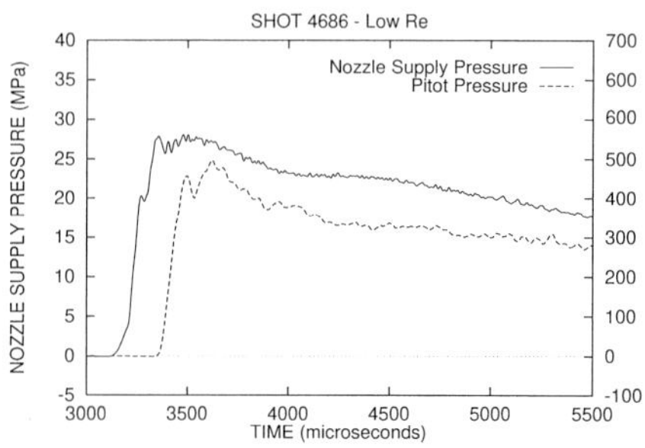

Figure 2(a). Pressure time histories.

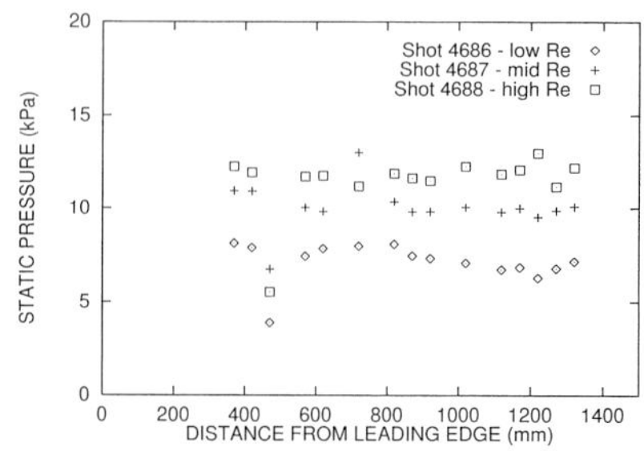

Figure 2(b). Pressure distributions.

respectively. (Sutherland's formula has been used for viscosity and the origin of the turbulent boundary layer has been taken as the leading edge of the plate.) Regions of laminar, transitional, and turbulent flow are evident.

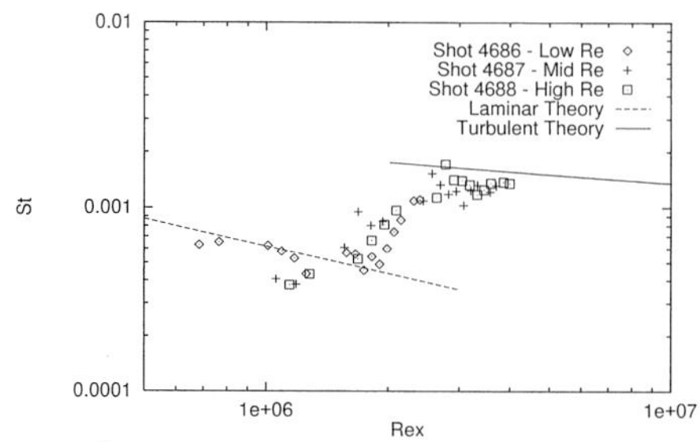

Figure 3. Heat transfer measurements averaged over a 500-μs period.

The time histories of the heat-transfer rates measured using the gauges are plotted for the three test conditions in Fig. 4. The zero level for each trace is plotted in proportion to the distance of the gauge from the leading edge of the test surface. (Thus the plot gives an indication of the development of the flow and enables the propagation of spots down the surface to be identified.) The arrival of flow at the gauge location can be seen from the sudden rise in heat-transfer level, and it can be seen that the flow arrives at later times for gauges progressively further down the plate. Note that results are presented from before the arrival of flow to well beyond the end of the test flow. No filtering or averaging has been applied to the data.

At the low-Re condition, the boundary layer remains laminar to 1120 mm from the leading edge when a turbulent spot appears at time 4300 μs. The spots are characterized by a rapid increase in the heat-transfer rate from the

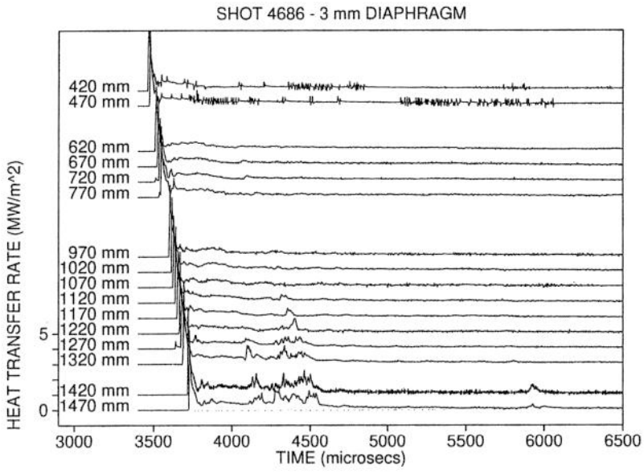

Figure 4(a). Low Re (4686).

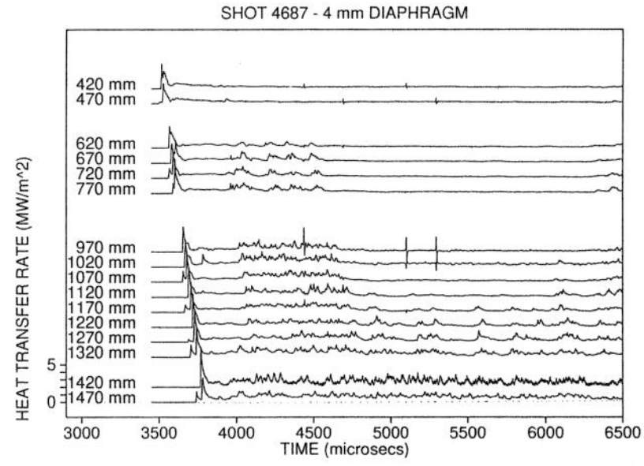

Figure 4(b). Mid Re (4687).

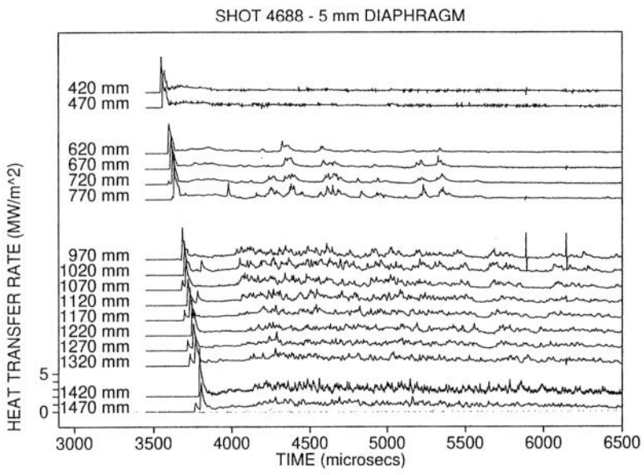

Figure 4(c). High Re (4688).

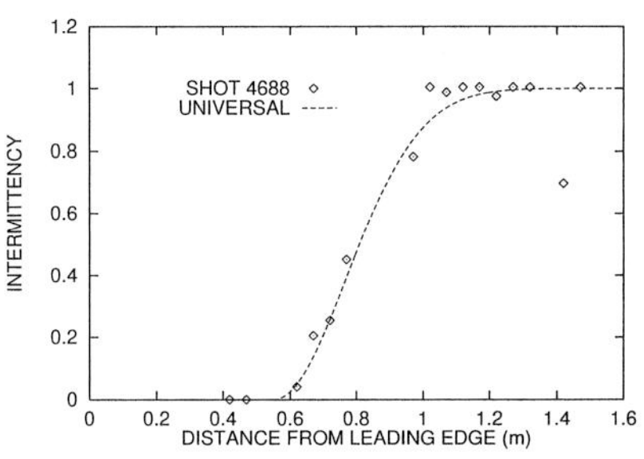

Figure 5. Shot 4688 intermittency.

lower laminar level to the higher turbulent level. This spot grows and others form and the boundary layer is still transitional at the last gauge. The gauge at 1420 mm is the one sampled at the higher rate. The apparent noise on the first two gauges is due to a sampling discretization problem.

At the mid-Re condition, the transition process is clearer. The first spots appear at 620 mm from the leading edge and the growth and merger of spots up to the gauge at 770 mm can be identified. At the gauge at 970 mm, the flow is almost fully turbulent during the test time. After the test time (beyond about 1.5 ms after arrival of flow), the boundary layer towards the front of the plate re-laminarizes before transition occurs once again, now further downstream. From this shot, the speed of the trailing edge of the spots has been estimated from the propagation of the last spot before relaminarization to be approximately 60% of the freestream flow speed. This compares with the low-speed flow value of 50% (Schubauer and Klebanoff 1955) and values between 53% and 56% in the work of Clark et al. (1994) for higher speed flows.

The time histories at the high-Re condition show similar trends to the mid-Re condition but with less distinct evidence of the end of the test flow. The durations of the spots are similar to those expected from the simple analysis in Section 2. For the high-Re condition a numerical intermittency meter has been made along the lines of Hedley and Keffer (1974). The distribution of intermittency down the plate for the high-Re condition, based on a 1-ms sample period, are compared with Narasimha's (1957) "universal" intermittency distribution in Fig. 5. Quite good agreement is obtained.

5. Conclusions

The detection of turbulent spots in transitional boundary layers in hypervelocity flows using thin-film heat-transfer gauges has been demonstrated in tests in the T4 shock tunnel. The formation, growth, and merger of spots can be studied under such conditions. It has been demonstrated that factors such as spot-propagation rates and the intermittency of turbulence can be measured. Further tests are planned to study turbulent-spot activity over a range of hypervelocity flow conditions.

References

Clark JP, Jones TV, LaGraff JE (1994) On the propagation of naturally-occurring turbulent spots. J Eng Math 28:1–19.

Doorly DJ, Smith FT (1992) Initial-value problems for spot disturbances in incompressible or compressible boundary layers. J Eng Math 26:87–106.

Emmons HW (1951) The laminar-turbulent transition in a boundary layer. J Aero Sci 18:490–498.

Goyne CP, Paull A, Stalker RJ (1995) A Skin Friction Gauge for Impulsive Flows, AIAA Paper 95-3152.

Hedley TB, Keffer JF (1974) Turbulent/non-turbulent decisions in an intermittent flow. J Fluid Mech 64:645–678.

Narasimha R (1957) On the distribution of intermittency in the transition region of a boundary layer. J Aero Sci 24:711–712.

Schubauer GB, Klebanoff PS (1955) Contributions on the Mechanics of Boundary-Layer Transition, NACA TN 3489.

Skinner K (1994) Mass Spectrometry in Shock Tunnel Experiments of Hypersonic Combustion, PhD Thesis, The University of Queensland, Australia.

Spalding DB, Chi SW (1964) The drag of a compressible turbulent boundary layer on a smooth flat plate with and without heat transfer. J Fluid Mech 18(1):117–143.

Stalker RJ, Morgan RG (1988) Free Piston Shock Tunnel T4 — Initial Operation and Preliminary Calibration, NASA CR-181721.

White FM (1991) Viscous Fluid Flow, McGraw-Hill, New York, p 511.

Influence of reflected-shock/boundary-layer interaction on driver-gas contamination in high-enthalpy shock tunnels

R. S. M. Chue, K. Itoh
National Aerospace Laboratory, Kakuda Research Center
1 Kimigaya, Kakuda, Miyagi, 981-15 Japan

Abstract: A numerical analysis has been carried out to examine the interaction of the reflected shock with a turbulent boundary layer and its role in driver-gas contamination in a high-enthalpy reflected shock tunnel. The results show that wall jetting is responsible for the premature arrival of driver gas in the tailored and over-tailored tunnel conditions, but it does not play a dominant role in the under-tailored case. The effects of nozzle flow at the end plate on driver-gas contamination are also assessed.

Key words: Reflected-shock/boundary-layer interaction, Driver-gas contamination, High enthalpy, Shock tunnel, Test time

1. Introduction

The contamination of the test gas by driver gas is an important problem in high-enthalpy shock-tunnel facilities, as the premature arrival of driver gas severely limits test time. The interaction of the reflected shock with viscous boundary layers has been recognized as a major candidate to cause driver-gas contamination. Mark (1958) established the first theoretical treatment of the resulting bifurcated shock and the transport of boundary-layer gas by wall jetting towards the shock-tube end plate. Based on this mechanism, analytical models were developed to predict the premature arrival of the driver gas (Davies & Wilson 1969, Stalker & Crane 1978). Recently, numerical simulations have provided more detailed information about the contamination process. Wilson et al. (1993) examined the leakage of driver gas for an over-tailored case with laminar boundary layer. However, the tailored condition, which is more relevant to actual tunnel operations, was not examined. Indeed, the effect of the interaction on driver-gas contamination has yet to be quantitatively assessed in order to compare with other mechanisms. The reflected-shock/boundary-layer interaction is therefore analyzed to examine its role in causing driver-gas contamination and the loss of test time for different tunnel operating conditions.

2. The computational model

The present study concentrates on high shock Mach numbers ($M_s \simeq 10$) and high shock-tube pressures that are typical for high-enthalpy facilities. The shock tube investigated has an internal diameter of 30 mm, a length of 4 m, and shock-tube-to-driver-tube area ratio of 0.09. The driver and test gases are helium and air, respectively, which are assumed to be perfect gases. The helium has specific heat ratio of 5/3 and molecular mass of 4.003 kg/kmol. The test gas is assumed to be high-temperature air having a specific heat ratio of 1.25 and molecular mass of 25.9 kg/kmol. The wall boundary layers are assumed turbulent and the algebraic eddy viscosity model of Baldwin & Lomax (1978) is used. Initial conditions are taken at shock reflection with the initial flow field calculated using the boundary-layer theory of Mirels (1964). To isolate the effect of reflected-shock/boundary-layer interaction, the contact region is initially taken to be a plane discontinuous surface free of the effects of diaphragm rupture.

The computational code was originally developed by Takahashi et al. (1994) using the second-order KRC scheme, which is a pointwise nonoscillatory shock capturing method (Itoh et al. 1992). Time integration was done using the second-order Runge-Kutta method. The computational domain covers 26 cm of the end of the shock tube with a 433×121 grid that clusters in the wall regions. The upstream inflow condition and the tube-wall temperature are held fixed. Two sets of computations have been calculated — one for a closed-end shock tube and another for a shock tunnel with a choked nozzle. The shock Mach number is fixed at 10 with initial shock-tube pressure (p_1) at 10 kPa. The total enthalpy behind the reflected shock is 13 MJ/kg.

3. Results and discussion

Figure 1 displays the sequence of temperature and driver-gas mass fraction contours as the reflected shock interacts with the turbulent boundary layer in the closed-end shock tube. The tailored interface condition is enforced in this calculation (i.e., the ratio of transmitted shock pressure to the reflected shock pressure, p_7/p_5, equals 1). The formation of the bifurcated shock and the resulting mechanism of wall jetting of "cold" boundary-layer fluid towards the end plate, which generates a vortex at the corner region, is illustrated in Fig. 1a. The distortion of the contact surface and the leakage of driver gas into the test gas through the bifurcated shock foot can be seen in Fig. 1b. At 50 μs, a stream of driver gas near the edge of the boundary layer and the triple point of the bifurcated shock is driven towards the tube wall and is then partially transported towards the end plate. However, the vortical motion at the contact surface has a counter-clockwise direction which serves to retard further driver-gas leakage. The driver gas transported by wall jetting arrives at the end plate at about 95 μs after shock reflection.

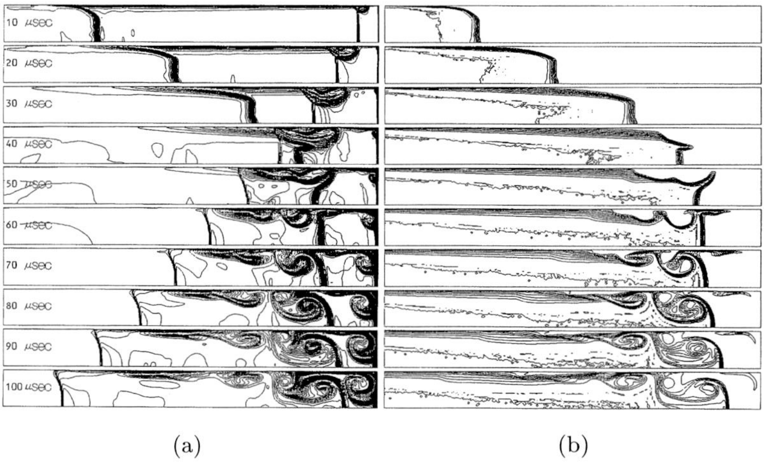

Figure 1. (a) Temperature and (b) driver-gas mass fraction distributions in the shock tube.

The effects of off-tailored tunnel conditions are compared in Fig. 2. The values of p_7/p_5 are, respectively, 0.8 and 1.1 for the under- and over-tailored conditions. The tailored case is redisplayed in Fig. 2b for comparison. Wall jetting is most severe in the over-tailored case and the driver gas arrives at the end plate the earliest, although the overall shape of the contact surface remains largely undistorted. On the other hand, the wall jet in the under-tailored case is too weak to transport the driver gas towards the end plate. However, the contact surface is severely distorted with a large protruded region of driver gas near the center line that flows toward the end plate.

Besides the contamination by "cold" wall fluid and driver gas, Fig. 3 shows that the pressure at the center of the shock-tube end plate becomes quite non-steady with a large pressure peak occurring near 80 μs for both the tailored and off-tailored cases. The quality of the freestream flow would then be affected if the gas is allowed to discharge through a nozzle at the end of the shock tunnel.

An initial effort is made to simulate the effect of nozzle flow at the end of the shock tunnel by assuming the nozzle to be an effective "sink" with choked flow. The nozzle-throat diameter is taken to be 10 mm. The purpose of this simple model is to provide a qualitative picture of how the driver gas may exhaust into the nozzle. Figure 4 shows the temperature distributions and the transport of

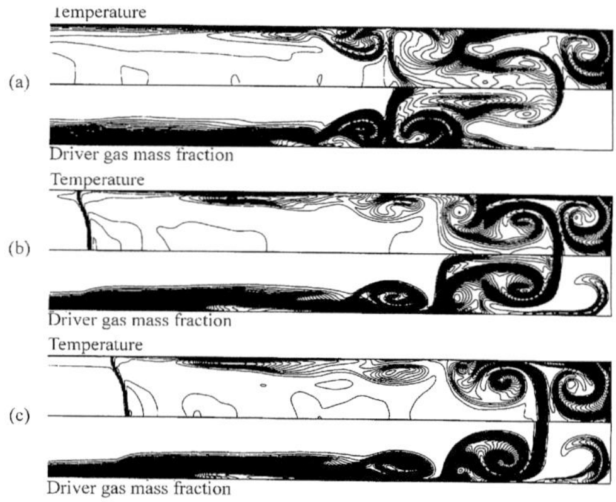

Figure 2. Effect of tunnel operating condition on driver gas contamination at $t = 100\ \mu s$: (a) under-tailored, (b) tailored, and (c) over-tailored condition.

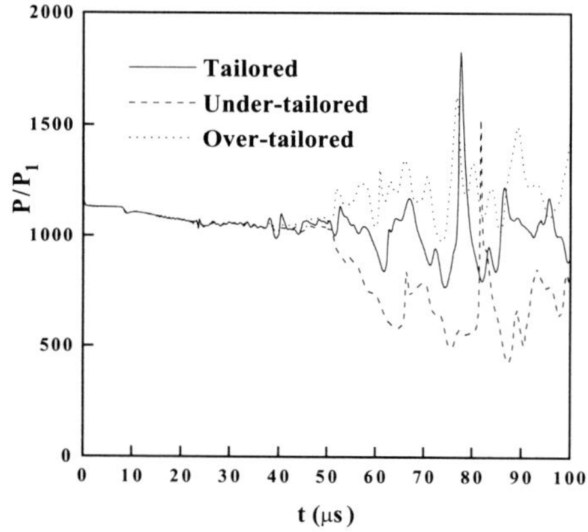

Figure 3. Pressure (p/p_1) history at the center of the shock-tube end plate.

driver gas towards the nozzle for the tailored condition. The early events are quite similar to those without the nozzle, although the corner vortex is stronger with nozzle flow (c.f. Fig. 1). From 80 μs the volume of driver-gas leakage at the end plate appears to be entrained in a vortex close to the end plate with a clockwise motion. As a result, only part of leakage gas is exhausted into the nozzle with the rest being recirculated in the shock tube (100 μs). The time of arrival of driver gas at the nozzle is not significantly different from the closed-end case. A set of off-tailored conditions has also been recalculated with the nozzle flow, and the results are again quite close to those for the closed-end conditions presented in Fig. 2.

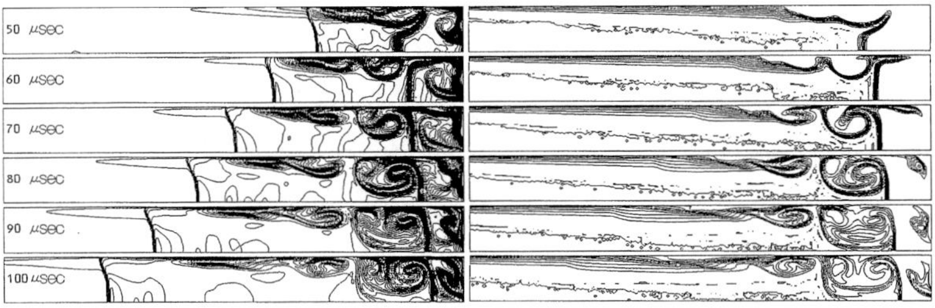

(a) (b)

Figure 4. Effect of nozzle flow on driver gas transport, tailored condition: (a) temperature and (b) driver-gas mass fraction.

Table 1. Time of arrival of driver gas at the end wall.

	Time after shock reflection (μs)
Closed-off end wall	95
With nozzle	90
Analytical approx. by Davies	62
"Ideal" drainage time[a]	226

[a]Where $t_{drainage} = \frac{4M}{A^*} \frac{1}{\rho_5 a_5} \left(\frac{\gamma+1}{2}\right)^{\frac{\gamma+1}{2(\gamma-1)}}$, M = mass of test slug, a = sound speed, and A^* = nozzle throat area. Subscript 5 denotes the condition behind the reflected shock.

The arrival time of the driver gas at the end plate is summarized in Table 1 for the tailored condition. Also included in the table are the values calculated using the analytical approximation of Davies & Wilson (1969), based on the mechanism of wall jetting, and the "ideal" drainage time for reference. The drainage time is the time required to drain the test gas with consideration of mass loss due to boundary-layer effects prior to shock reflection. The approximation of Davies obtained a value on the conservative side and can be a useful tool for predicting test time in actual tunnel operations if the effect of wall jetting is dominant. However, as wall jetting may not always be the cause of driver-gas contamination, as illustrated earlier in the under-tailored condition, the formulation may not always be adequate. In any case, the premature arrival of the driver gas can significantly reduce test time. For the shock tunnel considered in the present study, the arrival time of driver gas is less than 50% of the drainage time.

4. Concluding remarks

The interaction of a reflected shock with a turbulent boundary layer and the contact region has been analyzed. While wall jetting is demonstrated to be an important mechanism to cause contamination of the test gas, it may not necessarily play the dominant role. In particular, the mechanism for contamination is quite dependent on the tunnel operating condition, with under-tailoring appearing to be affected more by the distortion of the contact surface than by wall jetting. This also points out the need to examine other mechanisms, such as non-ideal diaphragm rupture and mixing, before a more precise prediction of driver-gas contamination can be achieved.

Acknowledgement. Special thanks are due to M. Takahashi for his help and discussions, and to G. Eitelberg of DLR-Göttingen for suggesting the calculation with nozzle flow. R. S. M. Chue is supported by the STA Fellowship program of Japan.

References

Baldwin BS, Lomax H (1978) Thin Layer Approximation and Algebraic Model for Separated Turbulent Flows, AIAA–78–257.

Davies L, Wilson JL (1969) Influence of reflected shock and boundary-layer interaction on shock-tube flows. Phys Fluids Supplement I:I-37–43.

Itoh K, Takahashi M, Hiraiwa T (1992) Pointwise non-oscillatory shock capturing scheme. In: Proc of 6th Symp on Computational Fluid Dynamics.

Mark H (1958) The Interaction of a Reflected Shock Wave with the Boundary Layer in a Shock Tube, NACA TM 1418.

Mirels H (1964) Shock tube test time limitation due to turbulent-wall boundary layer. AIAA J 2:38–54.

Stalker RJ, Crane KCA (1978) Driver gas contamination in a high-enthalpy reflected shock tunnel. AIAA J 16:277–279.

Takahashi M, Itoh K, Tani K, Tanno H (1994) Numerical analysis of the driver gas contamination in a reflected shock tunnel. In: Proc of Symp on SHOCK WAVES, JAPAN '94, pp 211–214.

Wilson GJ, Sharma SP, Gillespie WD (1993) Time-Dependent Simulations of Reflected-Shock/Boundary Layer Interaction, AIAA–93–0480.

Suppression of shock-induced separation in dense gases

M. S. Cramer, S. Park
Engineering Science and Mechanics Department
Virginia Polytechnic Institute and State University
Blacksburg, VA 24061 USA
L. T. Watson
Computer Science Department
Virginia Polytechnic Institute and State University
Blacksburg, VA 24061 USA

Abstract: We consider the reflection of oblique compression waves from a two-dimensional, steady, flat-plate laminar boundary layer. The full Navier-Stokes equations are solved numerically through use of a dense-gas version of the Beam-Warming implicit scheme. In the classical perfect-gas theory the compression waves are necessarily shocks. As is well known, the adverse pressure gradient carried by these shocks can lead to boundary-layer separation.

In the present investigation we consider the behavior of Bethe-Zel'dovich-Thompson (BZT) fluids in the context of this classical benchmark problem. BZT fluids are ordinary gases which have specific heats large enough to cause the fundamental derivative of gasdynamics to be negative over a finite range of temperatures and pressures in the single-phase vapor regime. The purpose of the present study is to determine whether the unique dynamics of BZT fluids can suppress boundary-layer separation. We make direct comparisons between the behavior of BZT fluids and more conventional fluids such as steam. The primary result of interest is that this suppression of separation can indeed be achieved. The physical mechanism for the suppression is discussed as are strategies for realizing the suppression in practical configurations.

Key words: Gasdynamics, Shock waves, Shock-induced separation

1. Introduction

An important loss mechanism in both turbomachinery and aerodynamic flows is caused by the shock-induced separation of boundary layers. The physical mechanism behind shock-induced separation is caused by the strong adverse pressure gradients carried by compression shocks which originate either at neighboring blades in the case of turbomachinery, other portions of the fuselage or aerodynamic surfaces in the case of supersonic aircraft, or which form on the wing or fuselage itself in the case of transonic flow. For a given pressure rise, shock waves are more likely to separate the boundary layer because the pressure gradient of the incoming signal is infinite, at least in the context of the inviscid theory. In most flows of practical interest, regions of flow compression are inevitable and in supersonic and transonic flows of perfect gases these compressions inevitably steepen to form compression shocks which in turn carry the strong adverse pressure gradients leading to flow separation.

In recent years it has been shown that such steepening is not inevitable if the class of fluids considered is extended to include high-pressure gases. By high pressure we mean pressures which are so large that the ideal-gas law no longer provides an accurate description of the gas behavior. At such pressure levels more complicated equations of state, such as the van der Waals equation, are required. We will also refer to such high-pressure conditions as dense-gas conditions. It is the principal goal of the present study to demonstrate that shock-induced separation can in fact be suppressed through use of a class of commercially available fluids operating at moderate pressure levels.

The fluids considered here are of the Bethe-Zel'dovich-Thompson (BZT) type. These are ordinary gases which have specific heats which are so large that the fundamental derivative of gasdynamics

$$\Gamma \equiv \frac{a}{\rho} + \left.\frac{\partial a}{\partial \rho}\right|_s = \left.\frac{V^4}{2a}\frac{\partial^2 p}{\partial V^2}\right|_s \tag{1}$$

becomes negative over a finite range of pressures and temperatures in the single-phase vapor regime. Here ρ, p, s, and a are the fluid density, pressure, entropy, and thermodynamic sound speed. The quantity $V \equiv \rho^{-1}$ is referred to as the specific volume. We also refer to the thermodynamic quantity (1) as the intrinsic gasdynamic nonlinearity parameter. For descriptions of the nonclassical dynamics of BZT fluids we refer the reader to the introductions and summaries of Bethe (1942), Zel'dovich (1946), Thompson (1971), Thompson and Lambrakis (1973), Menikoff and Plohr (1989), and Cramer (1991). Here we simply point out that the well-known compression discontinuities of the perfect-gas theory violate the entropy inequality if $\Gamma < 0$ at all points in the neighborhood of the proposed discontinuity. If such an inadmissible discontinuity is introduced into a $\Gamma < 0$ flow, say by a wedge inserted in a supersonic $\Gamma < 0$ flow, it disintegrates into a centered compression reminiscent of the well-known Prandtl-Meyer fans of the perfect-gas theory. This disintegration process was first discussed in the context of steady supersonic flows by Thompson (1971).

The potential advantages of BZT fluids with respect to shock-induced separation should now be obvious. Although the disintegration of the compression discontinuity does not change the polarity of the signal, i.e., it remains a compression, it does in fact significantly reduce the gradient. That is, the incoming compressive signal arriving at the boundary layer is of nonzero length rather than the zero length associated with shock waves. Intuition based on the perfect-gas or even incompressible-flow theory strongly suggests that such a reduction in the adverse pressure gradient experienced by the boundary layer should result in a reduction or even the elimination of shock-induced separation. In the present investigation, we demonstrate that such a suppression of shock-induced separation can indeed occur.

If such a reduction or suppression of separation can be realized in BZT fluids, it is natural to ask whether these nonclassical dynamics can be exploited in technology. One of the most promising applications of BZT fluids is as working fluids in organic Rankine cycles. A general discussion of the nature and the advantages of organic Rankine cycles has been given by Curran (1981) and Angelino and Invernizzi (1993). In order to save space we will restrict our comments to pointing out that BZT fluids tend to be a subclass of retrograde fluids and turn out to be excellent candidates for use as working fluids in such power systems. Not only do BZT fluids have all the advantages of other retrograde fluids, but they also have the potential to reduce or eliminate shock-induced separation in the turbine stage of the cycle.

2. Formulation

To illustrate the suppression of separation we consider the classical benchmark problem of the reflection of an oblique shock from a laminar flat-plate boundary layer. The flow conditions are chosen so that the reflection is regular. As a result the flow is supersonic everywhere in the inviscid part of the flow. The governing equations are taken to be the two-dimensional steady-flow Navier-Stokes equations. These are solved numerically by the well-known Beam-Warming implicit scheme (Beam and Warming 1978), which has been generalized to account for dense-gas effects. A full description of the scheme used and the battery of consistency checks performed have been given by Park (1994).

The equation of state is the state-of-the-art model developed by Martin and Hou (1955). The dense-gas viscosity and thermal-conductivity models chosen were those developed by Chung et al. (1988). Each of these models, the treatment of bulk viscosity, and the physical data for the fluids employed here are described in detail by Park (1994).

Finally, we note that the flows considered here are all single phase. The saturated vapor line correlation of Riedel (1954) was used to ensure that all computed points in the flow field correspond to the single-phase gas regime.

3. Results

In the present study we have compared the behaviors of two fluids. The first is ordinary steam which was chosen because of its widespread use in Rankine power cycles and the fact that is not a BZT fluid, i.e., $\Gamma > 0$ for all pressures and temperatures. The second fluid is the commercially available BZT fluid perfluoro-trihexylamine ($C_{18}F_{39}N$) and will be referred to by its trade name FC-71. As pointed out in Section 2, a full discussion of the physical-property data for each fluid is given by Park (1994) and the references therein.

We have attempted to make direct comparisons between the fluids. For each fluid the freestream pressure, temperature, Mach number, and impingement Reynolds number was fixed at 8.55 atm, 646.15 K, 2.0, and 2.96×10^5, respectively. We note that the freestream pressure of 8.55 atm is only about 4% of the pressure at steam's thermodynamic critical point. As a result the flow of steam can be regarded as that of a perfect gas. For both fluids the incident shock carries a flow deflection angle of 3°. Thus, the incident-shock strength can be regarded as essentially the same for each fluid.

The resultant variation of the plate skin-friction coefficient c_f is depicted in Fig. 1. There x is the coordinate measured parallel to the undisturbed flow direction. The leading edge of the plate was taken to be at $x = 0$ and the shock impingement point is given by $x = L_s$. The circles denote the variation for steam. The separation is seen to be well developed with two local minima in c_f. The + symbols denote the variation of c_f for the BZT fluid FC-71. Here $c_f > 0$ for all x and therefore no separation is observed. A check on the flow details reveals that the values of $\rho\Gamma/a$ immediately upstream and downstream of the incoming compression are -0.04 and -0.16. The theory developed in the references by Bethe, Zel'dovich, Thompson, etc. can be used to show that the initial compression discontinuity in FC-71 is inadmissible and therefore has disintegrated into a centered fan. As a result the adverse pressure gradient is too weak to separate the boundary layer. An examination of a number of shock/boundary-layer interactions and pure boundary-layer flows in dense gases reveals that the reaction of the viscous boundary layer is essentially classical. As discussed by Park (1994), the primary physical mechanism for this suppression is the disintegration of the shock and the resultant weakening of the adverse pressure gradient of the incoming inviscid signal.

Finally, we have also included the computed variation of c_f for FC-71 at the same freestream temperature and Mach number, the same impingement Reynolds number, and the same incoming shock strength as in the previous examples. This variation is given by $\times$ symbols in Fig. 1. The only new feature in this case is that the freestream pressure was taken to be 1 atm instead of 8.55 atm. As a result, the FC-71 behaves as a nearly perfect gas and $\Gamma > 0$

everywhere in this flow. The initial compression discontinuity remains intact as a shock and, as seen in Fig. 1, separates the flow. In fact, the skin-friction variation is very similar to that of steam. The point of this additional calculation is that it supports the idea that the suppression of separation is primarily due to the differences in the incoming inviscid signal and not due to differences in some other detail of the viscous flow. Furthermore, this demonstrates how one can eliminate separation simply by increasing the pressure to bring the state from the ideal-gas conditions to one corresponding to a negative Γ flow.

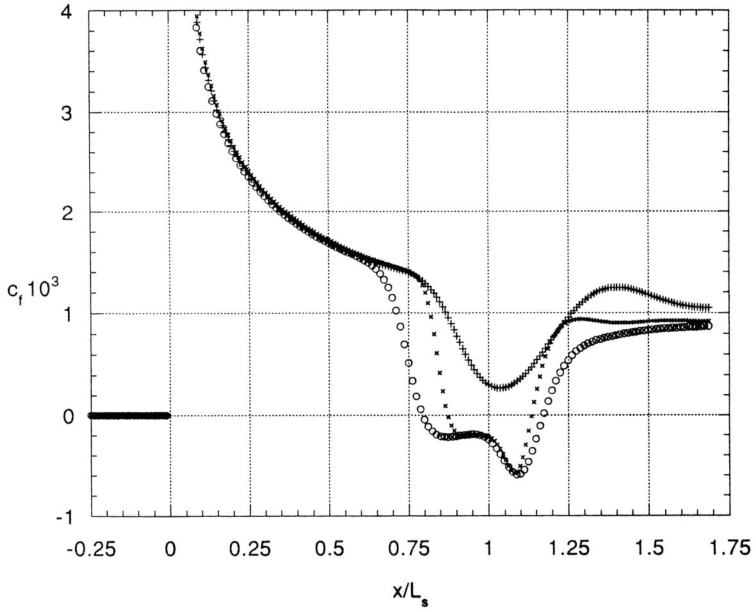

Figure 1. Computed skin-friction coefficients. The open circles denote steam at a freestream pressure of 8.55 atm, the + symbols denote FC-71 at a freestream pressure of 8.55 atm, and the $\times$ symbols denote FC-71 at a freestream pressure of 1 atm.

4. Summary

We have employed a dense-gas version of the Beam-Warming implicit scheme to compare the viscous-inviscid interactions of steam to those for a BZT fluid at the same freestream conditions, the same shock-impingement Reynolds number, and the same incoming-signal strength. It is found that the natural dynamics of BZT fluids in their $\Gamma < 0$ regime leads to a disintegration of the initial compression discontinuity. As a result of this reduction in the adverse pressure gradient, shock-induced separation was suppressed in the BZT fluid. The authors feel that the discovery that the suppression is primarily due to the inviscid dynamics can lead to considerable simplification in attempts to exploit this behavior. One can employ perfect-gas intuition when considering the viscous parts of the flow and restrict any optimizations to the inviscid flow regions.

Acknowledgement. This work was supported by the National Science Foundation under Grant # CTS-8913198. The first two authors would also like to thank Professor S. A. Ragab for many useful discussions and suggestions in the course of this research. We would also like to thank L. T. Watson for making facilities of the Computer Science Department available to S. Park.

References

Angelino G, Invernizzi C (1993) Cyclic methylsiloxanes as working fluids for space power cycles. J Solar Engr 115:130.

Beam RM, Warming RF (1978) An implicit factored scheme for the compressible Navier-Stokes equations. AIAA J 16(4):393.

Bethe HA (1942) The Theory of Shock Waves for an Arbitrary Equation of State, Office Scientific Resesarch & Development Report No. 545.

Chung TH, Ajlan M, Lee LL, Starling KE (1988) Generalized multiparameter correlation for nonpolar and polar fluid transport theories. Ind Eng Chem Res 27:671.

Cramer MS (1991) Nonclassical dynamics of classical gases. In: Nonlinear Waves in Real Fluids, Kluwick A (ed), Springer-Verlag, New York.

Curran HM (1981) Use of organic working fluids in Rankine engines. J Energy 5:218.

Martin JJ, Hou YC (1955) Development of an equation of state for gases. AIChE J 1(2):142.

Menikoff R, Plohr B (1989) Riemann problem for fluid flow of real materials. Rev Mod Phys 61:75.

Park S (1994) Viscous-Inviscid Interactions of Dense Gases, Ph.D. dissertation, Virginia Polytechnic Institute and State University.

Reidel L (1954) Eine neue Universelle Dampfdruckformel-Untersuchungten über eine Erweitung der Theorems der übereinstimmenden Zustände Teil 1. Chem Ing Tech 26:83.

Thompson PA (1971) A fundamental derivative in gas dynamics. Phys Fluids 14:1843.

Thompson PA, Lambrakis K (1973) Negative shock waves. J Fluid Mech 16:187.

Zel'dovich YB (1946) On the possibility of rarefaction shock waves. Zh Eksp Teor 4:363.

Shock waves diffracting over rigid ramps in a viscous, heat-conducting gas (argon)

R. J. Virgona*
Department of Mechanical Systems Engineering
Tokyo Noko University
Naka-cho 2-24-16
Koganei, Tokyo 184 Japan

W. Y. Crutchfield
Center for Computational Sciences and Engineering
Lawrence Livermore National Laboratory, P.O. Box 808
Livermore, California 94550 USA

L. F. Henderson†
I.U.S.T.I.-M.H.E.Q.
Saint-Jérôme Centre
Université de Provence, 13397 Marseille, France

Abstract: Shocks were diffracted over symmetrical steel wedges in argon. The shock Mach number M_i averaged 2.33, so the shocks were strong. The experimental data were simulated by numerically integrating the 2D, compressible, unsteady Navier-Stokes equations for a viscous, heat conducting, polytropic gas. It is found that real-gas effects delay both the onset of Mach reflection (MR), and its evolution to a self-similar state.

Key words: Shock wave, Diffraction, Navier-Stokes, Experiment, CFD

1. Introduction

When a plane incident shock i diffracts over a rigid ramp of apex angle θ either a regular (RR) or an irregular (IR) wave system appears. The criteria which determine when transition from RR to IR (RR $\rightleftharpoons$ IR) takes place have been discussed previously for the perfect gas (von Neumann 1943, Hornung 1986, Henderson 1987, Ben-Dor 1992). Here a study is made of the transition between RR and IR on a ramp for a real gas. A real gas is defined as one which has a finite shear viscosity μ and thermal conductivity k. In a real gas it is found that RR on a ramp generally persists beyond all the perfect-gas transition criteria into regions were it is theoretical impossible (Bleakney & Taub 1949; Kawamura & Saito 1956; Henderson & Lozzi 1975, 1979; Hornung 1986). The

*Present address: Thammasat University, Department of Mechanical Engineering, International Institute of Technology, P.O. Box 22, Thammasat Rangit Post Office, Pathumthani 12121, Thailand.

†Present address: 8 Damour Avenue, East Linfield, Sydney NSW 2070, Australia.

object of this paper is to investigate what effect μ and k have on diffraction and (RR $\rightleftharpoons$ IR).

2. Formulation

The Navier-Stokes equations in conservation form may be written

$$\partial_t U + F(U)_x + G(U)_y = R(U)_x + S(U)_y \tag{1}$$

where

$$U^T = (\rho \ \ \rho u \ \ \rho v \ \ \rho E) \tag{2}$$

$$F^T(U) = \left(\rho u \ \ \rho u^2 + p \ \ \rho uv \ \ u(\rho E + p)\right) \tag{3}$$

$$G^T(U) = \left(\rho v \ \ \rho uv \ \ \rho v^2 + p \ \ v(\rho E + p)\right) \tag{4}$$

$$R^T(U) = \left(0 \ \ (\lambda + 2\mu)u_x + \lambda v_y \ \ \mu(u_y + v_x) \ \ uR_2 + vR_3 + \frac{\mu}{P_r}\gamma e_x\right) \tag{5}$$

$$S^T(U) = \left(0 \ \ \mu(u_y + v_x) \ \ (\lambda + 2\mu)v_y + \lambda u_x \ \ uS_2 + vS_3 + \frac{\mu}{P_r}\gamma e_y\right) \tag{6}$$

$$e = E - (u^2 + v^2)/2 \tag{7}$$

where the superscript T indicates the transpose of a row, i.e. U^T for example is a column vector, $(u,\ v)$ are the $(x,\ y)$ components of velocity, ρ is density, p is pressure, E is specific energy, e is the specific internal energy, and subscripting 2, 3 indicates second and third components of a vector. Argon is well described by the perfect gas law

$$p = \rho RT\ . \tag{8}$$

The viscosity was modeled with

$$\mu = \frac{2.01572 \times 10^{-6}\ T^{1.5}}{171.691 + T}\ \mathrm{kg\ m^{-1}\ sec^{-1}}\ . \tag{9}$$

The coefficient of second viscosity was $\lambda = -2/3\mu$ and the thermal conductivity k was obtained from the Prandtl number $P_r = \frac{\mu C_p}{k} = 0.67$, where C_p is the constant-pressure specific heat. The ratio of specific heats γ was taken as $\gamma = 5/3$.

3. Numerics

The algorithm employs a finite-difference, second-order Godunov scheme with Crank-Nicolson correction (Steinthorsson et al. 1994, 1995). It includes an adaptive mesh refinement to five levels where at each level the refinement was by a factor of 4. The cells were square and at the deepest level of refinement 0.4 microns in length, or about 10 mean-free paths.

4. Results

If the ramp angle θ is less than the detachment criterion for RR $\rightleftharpoons$ IR, that is $\theta < \theta_e$, then by experiment some form of Mach reflection (SMR, CMR, DMR) is obtained. There are resolution difficulties as $\theta \to \theta_e$. In our experiments, as described in Virgona (1993), with $\theta = 52°$ which is close to $\theta_e = 53.776°$, the shock system had the appearance of a regular reflection (RR) except that the i and r shocks did not seem to meet on the ramp surface; there was a small gap g between them (Fig. 1a). The numerics have a finer resolution and can resolve it

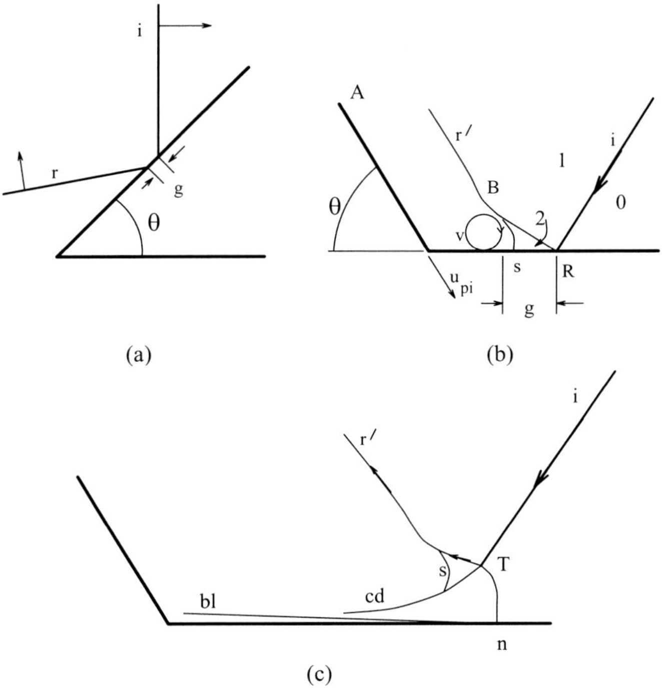

Figure 1. Experimentally observed and numerically resolved diffraction for $\theta = 52$ deg, $\gamma = 5/3$, $M_i = 2.327$: (a) experimental; (b) numerical; (c) after eruption of Mach shock (s is the corner signal and v the vortex).

(Fig. 1b). It comprises a precursor RR with a following and compressive corner signal s. There is a vigorous vortex v downstream of s. The corner signal travels faster than the reflection point and overtakes it, causing the eruption of the Mach shock (Fig. 2). The system then resolves into a self-similar complex Mach reflection (CMR, Fig. 1c).

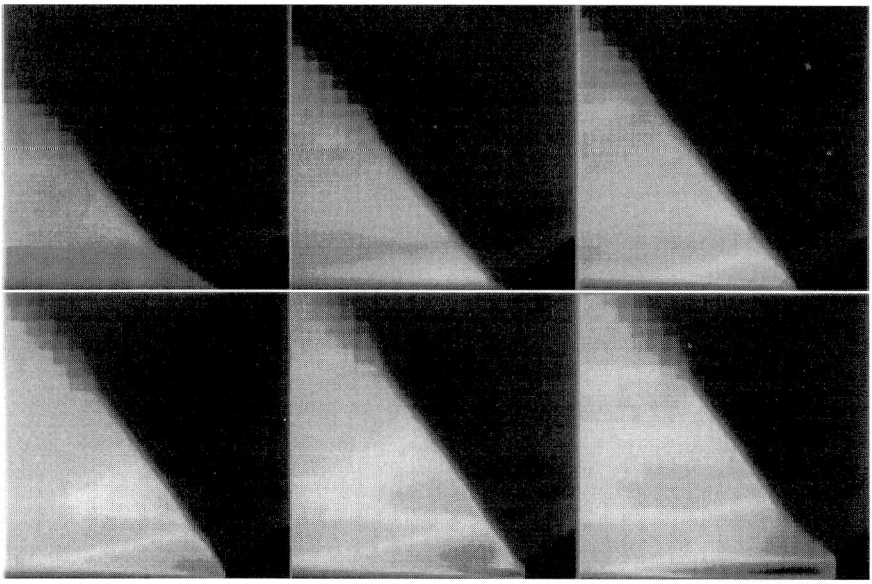

Figure 2. Time sequence of x component of velocity, for $\gamma = 5/3$, $M_i = 2.327$, $\theta = 52°$; ramp surface shown horizontal. Sequence is left to right, top to bottom.

4.1. Shock triple-point trajectory

Some results are presented in Fig. 3a for three values of θ such that $\theta < \theta_e$. Note that the paths are initially curved but become straight with sufficiently large $x(t)$. The straight part corresponds to self-similar CMR, while the curved part corresponds to the non-self-similar evolution. The part which coincides with the x axis ($y = 0$) corresponds to the precursor RR. As shown in Fig. 3b, the initial trajectory leads to a parallax error ϵ in the measurement of the trajectory angle χ, if it is assumed that the trajectory is a straight line passing through the ramp apex, as is sometimes done in experiment. In our case the shocks were photographed after they had traveled about 4 cm along the ramp, so χ depends on a length scale (4 cm). A scale-free trajectory angle ψ is obtained from the slope of the self-similar part of the trajectory. Results for χ are shown in Fig. 4a and for ψ in Fig. 4b. It is remarkable that the χ results support the inviscid θ_e criterion while the ψ results support the von Neumann criterion θ_N. Those results are isothermal with non-slip boundary conditions $T_0 = 273.15$ K and $\mu = 0$ on the ramp surface.

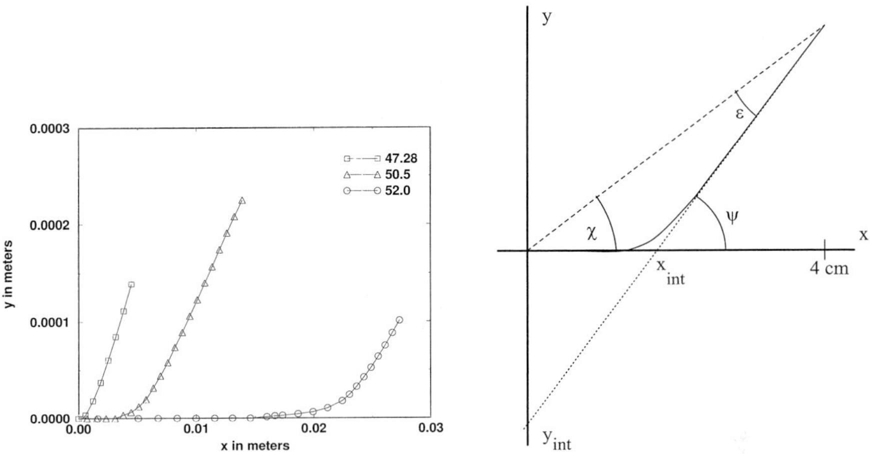

(a) (b)

Figure 3. Shock triple-point trajectories: (a) numerical triple-point trajectories at several wedge angles θ; (b) trajectory angle χ, self-similar trajectory angle ψ, and parallax error angle ϵ.

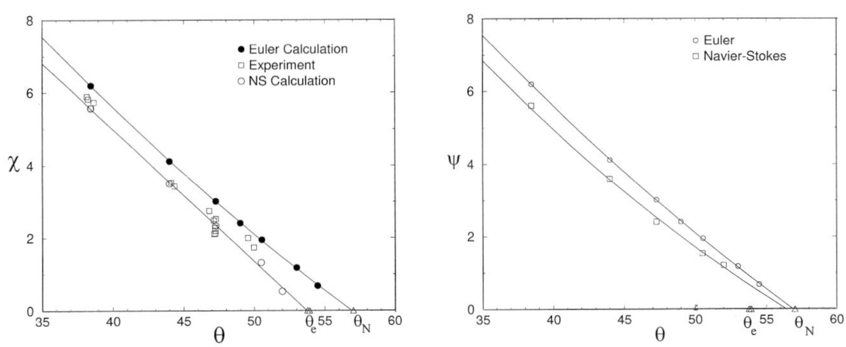

(a) (b)

Figure 4. Trajectory angle versus ramp angle θ for $\gamma = 5/3$, $M_i = 2.327$: (a) χ affected by parallax length scale of 4 cm; (b) scale-free trajectory angle ψ for self-similar flow.

Acknowledgement. This research was supported by the U.S. Department of Energy, Lawrence Livermore National Laboratory, under contract agreement DOE W-7405-ENG-48. We are indebted to Prof. P. Colella for the use of his Euler code data in Fig. 4.

References

Bleakney W, Taub AH (1949) Interaction of shock waves. Rev Mod Phys 21:584–605.

Ben-Dor G (1992) Shock Wave Reflection Phenomena, Springer-Verlag, Berlin.

Henderson LF (1987) Regions and boundaries for reflecting shock wave systems. Z ang Math u Mech (ZAMM) 67:73–86.

Henderson LF, Lozzi A (1975) Experiments on transition of Mach reflexion. J Fluid Mech 68:139–155.

Henderson LF, Lozzi A (1979) Further experiments on transition to Mach reflexion. J Fluid Mech 94:541–560.

Hornung HG (1986) Regular and Mach reflection of shock waves. Ann Rev Fluid Mech 18:33–58.

Kawamura R, Saito H (1956) Reflection of shock waves-1 pseudo-stationary case. J Phys Soc of Japan 11:584–592.

von Neumann J (1943) Oblique reflection of shocks. Explos Res Rep 12, Navy Dept, Bureau of Ordnance, Washington, DC.

Steinthorsson E, Modiano D, Colella P (1994) Computations of Unsteady Viscous Flows Using Solution Adaptive Mesh Refinement in Curvilinear Body-Fitted Grid Systems, AIAA-94-2330; also ICOMP-94-17, NASA TM 106704.

Steinthorsson E, Modiano D, Crutchfield WY, Bell JB, Colella P (1995) An Adaptive Semi-Implicit Scheme for Simulations of Unsteady Viscous Compressible Flows, AIAA-95-1727.

Virgona RJ (1993) A Study of Shock Wave Diffraction from Regular to Irregular Reflection and the Effects of Gas Viscosity, Ph.D. Thesis, The University of Sydney, Sydney, Australia.

Comparison of continuum and molecular approaches for the flow around a perpendicular disk

T. G. Elizarova, I. A. Graur
Institute for Mathematical Modeling, Russian Academy of Sciences
Miusskaya 4a, 125047 Moscow, Russia

A. Chpoun, J. C. Lengrand
Laboratoire d'Aérothermique du CNRS
4 ter, Route des Gardes, 92190 Meudon, France

Abstract: The present work contributes to efforts to define the domain of validity of two continuum approaches, based on Navier-Stokes and quasi-gas-dynamic equations, respectively, by comparing results obtained using each method with those obtained using a direct simulation Monte Carlo method as a reference method. Quasi-gas-dynamic equations are derived from the kinetic equation and include additional diffusion terms. The rarefied flow around a perpendicular disk has been considered for a freestream Mach number varying from 2 to 20.

Key words: Rarefied gas dynamics, Disk, Continuum equations, Monte Carlo simulation

1. Introduction

The interest in defining the domain of validity of continuum approaches for rarefied gas flows is due to the fact that solving continuum equations usually requires less computational resources than methods based on a molecular description. In addition to the usual Navier-Stokes (NS) equations, the quasi-gas-dynamic (QGD) equations were considered. They can be regarded as a new model for the description of viscous gas flows. Results obtained by a direct simulation Monte Carlo (DSMC) method are also presented. The physical modeling is made consistent in all calculations, but DSMC results are free from the near-equilibrium hypothesis that is included in the continuum approaches. Therefore, they can be used as a reference.

The variant of QGD equations used in the present calculations is described in Elizarova et al. (1992, 1994, 1995a, and 1995b). QGD equations may be obtained by averaging the kinetic equation with additional diffusion terms on the left-hand side. They differ from NS equations by the structure of the dissipative terms but they were proved to reduce to NS equations in the limiting case of small Knudsen numbers. In Cartesian coordinates, the QGD system is written as:

$$\begin{aligned}
\rho_t + \nabla_i \rho u^i &= \nabla_i \frac{\mu}{p}(\nabla_j \rho u^i u^j + \nabla^i p), \\
(\rho u^k)_t + \nabla_i \rho u^i u^k + \nabla^k p &= \\
& \nabla_i \frac{\mu}{p} \nabla_j \rho u^i u^j u^k + \nabla_i \frac{\mu}{p} \nabla^i p u^k + \nabla_i \frac{\mu}{p} \nabla^k p u^i + \nabla^k \frac{\mu}{p} \nabla_i p u^i, \\
E_t + \nabla_i (E + p) u^i &= \\
& \nabla_i \frac{\mu}{p} \nabla_j (E + 2p) u^i u^j + \nabla_i \frac{\mu}{p} \nabla^i \frac{p u^k u_k}{2} + \frac{\gamma}{\gamma - 1} \nabla_i \frac{\mu}{\rho} \nabla^i p + \nabla_i \kappa \nabla^i T.
\end{aligned}$$

Considering the perfect-gas equations, one obtains a closed system of equations which describes the space-time evolution of the macroscopic parameters of the gas, where $\mathbf{u}$ = velocity, ρ = density, p = pressure, E = total energy, T = temperature, μ and κ denote viscosity and heat-conductivity coefficients, respectively, and γ denotes the specific heat ratio.

In previous work (Elizarova et al. 1994, 1995a, and 1995b), the flow along a flat plate was considered and QGD equations were found to bring some improvement with respect to NS equations. In contrast with this simple configuration we now consider an infinitely thin circular disk perpendicular to a supersonic freestream. The corresponding axisymmetric flow features shock waves, expansion waves, and possibly recirculation zones. It is expected to be more sensitive to the choice of the equations or methods used for modeling. The problem is also more difficult from a computational point of view due to strong gradients and large differences in the density levels encountered.

2. Computational work

The comparison requires that an identical physical modeling of the gas properties be used in all calculations. This is obtained by considering a monoatomic hard-sphere gas in DSMC, whose transport properties are represented in the continuum approaches by a $\mu \propto T^{1/2}$ viscosity law and a constant Prandtl number ($Pr = 2/3$). Temperature T identifies with the translational temperature and γ is equal to $5/3$. The wall temperature T_w is taken to be equal to the stagnation temperature T_0. Molecules impinging onto the disk surface are assumed to be reflected diffusively with full accommodation at wall temperature T_w. The freestream Mach number Ma_∞ was successively taken equal to 2, 5, 10, and 20. Most calculations were carried out for a freestream Knudsen number $Kn_\infty = \lambda_\infty/D$ equal to 0.1, where λ_∞ is the mean free path and D denotes the diameter of the disk. The corresponding values of the Reynolds numbers $Re = \rho_\infty V_\infty D/\mu_\infty$, $Re_2 = \rho_2 V_2 D/\mu_2 = \rho_\infty V_\infty D/\mu_2$, and $Re_w = \rho_\infty V_\infty D/\mu_w$

Table 1. Flow parameters (Reynolds numbers given for $Kn_\infty = 0.1$).

Ma_∞	Re	Re_2	Re_w	p_{i2}/p_∞	p_{fm}/p_∞	p_2/p_∞	T_2/T_∞	ρ_2/ρ_∞
2	33.0	22.9	21.6	6.34	12.61	4.75	2.078	2.286
5	82.4	28.0	27.0	37.17	67.38	31.0	8.68	3.571
10	164.8	29.1	28.1	147.3	262.5	124.75	32.12	3.883
20	329.6	29.4	28.4	588.0	1043	499.8	125.9	3.970

are given in Table 1; V denotes the velocity,and subscripts $()_\infty$ and $()_2$ refer to conditions in the freestream and behind a Rankine-Hugoniot shock wave, respectively. In addition, the dimensionless pressure on the forward-facing side of the disk is given in Table 1 as estimated by the modified Newtonian approximation (p_{i2}) and by a free-molecule calculation (p_{fm}).

NS calculations were carried out using a commercial code (INCA) based on a finite-volume upwind implicit method with flux vector splitting.

QGD calculations used a code developed at the Institute for Mathematical Modeling (IMM) based on second-order finite-volume space approximation in explicit form and adapted to an axisymmetric geometry.

DSMC calculations were carried out using the code DISIRAF developed at the Laboratoire d'Aérothermique du CNRS based on Bird's ideas (Bird 1976).

For NS, QGD, and DSMC calculations, the grid was refined in the radial ($r-$) direction near the symmetry axis (NS, QGD) and/or near the disk edge (QGD, DSMC). The grid was also refined in the axial ($x-$) direction near the disk plane (NS, QGD, DSMC). A typical mesh consisted of 4000–5000 grid points. The smallest grid intervals were $\delta x \approx \delta r \approx 0.15\,\lambda_\infty$.

Both NS and QGD calculations assumed the gas to be perfect and used the same formulation for velocity slip and temperature jump at the wall.

3. Results and discussion

Due to space limitations, only part of the results will be presented here. Based on grid-sensitivity studies, the results are free from discretization error except for NS results near the edge of the disk. They are given in dimensionless form based on freestream values. Lengths are reduced by the disk radius R. More details can be found in Elizarova et al. (1994) regarding the numerical methods and in Lengrand et al. (1995) for the present calculations.

The distributions of reduced velocity and temperature along the stagnation streamline are given in Fig. 1 for $Kn_\infty = 0.1$ and $Ma_\infty = 2$ and 5. The disk plane was taken as the origin for x. QGD temperature profiles depart

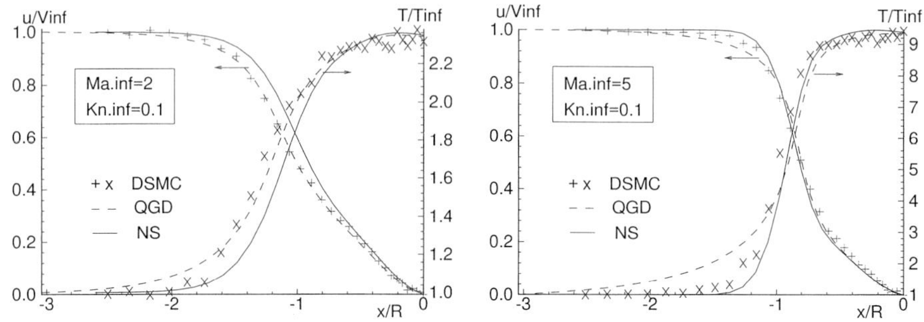

Figure 1. Velocity and temperature profiles ahead of the disk.

from the freestream value earlier than NS and DSMC profiles. Compared with NS results, DSMC and QGD results exhibit differences in the same direction. Thus, additional terms of QGD equations act in the correct direction but they occasionally over-correct the deficiencies of NS equations. The density profiles (not shown) are less sensitive to the model.

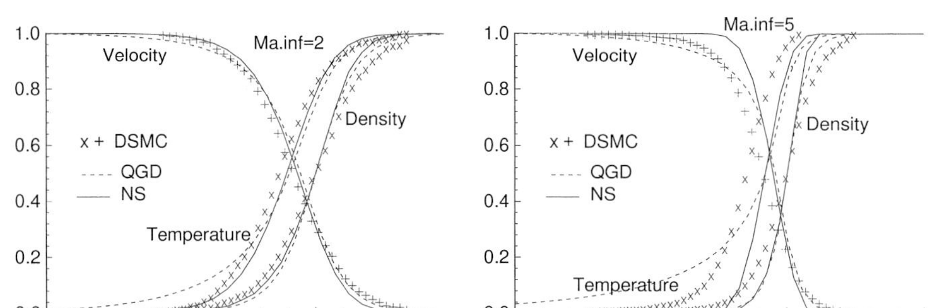

Figure 2. One-dimensional shock-wave structure.

One dimensional shock-wave structures were calculated for the same Mach numbers by the same three methods. However, NS calculations were carried out with a code developed at IMM rather than with the INCA code. Flow parameters are presented (Fig. 2) in the usual dimensionless form based on Rankine-Hugoniot parameter ratios. Abscissas x are divided by λ_∞. The point where $\rho = (\rho_\infty + \rho_2)/2$ is taken as the origin for x. The same observations can be made for the disk stagnation line and for the 1D shock structure.

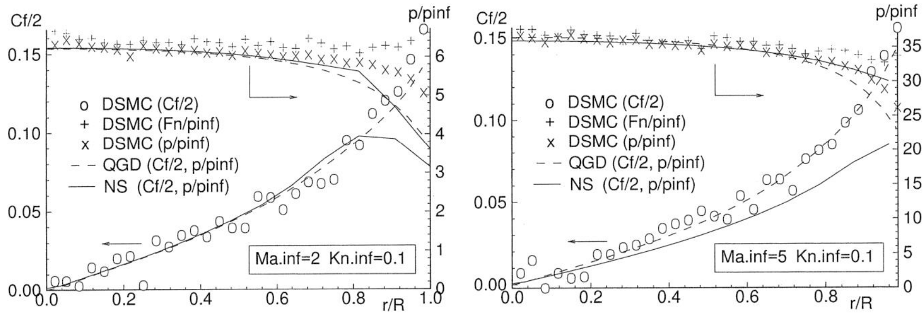

Figure 3. Wall quantities.

Wall quantities (pressure and shear stress F_t) are of interest for practical applications. These quantities are plotted in Fig. 3 for the same flow conditions as in Fig. 1. DSMC makes a difference between normal stress F_n due to molecular normal momentum exchange and pressure p deduced from the equation of state near the wall. Except for near the disk edge, where results are affected by discretization, all pressures are close to one another and close to the Newtonian estimation p_{i2}. However the skin-friction coefficient $C_f/2 = F_t/(\rho_\infty V_\infty^2)$ is much better predicted by QGD than by NS equations.

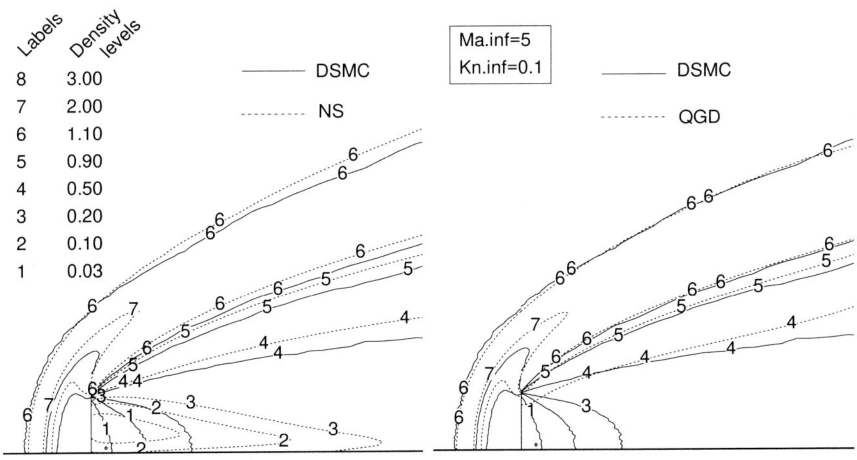

Figure 4. Two-dimensional flow field.

All the above results were obtained from calculations that included also the wake of the disk. Examples of 2D flow fields are given in Fig. 4. The main differences between the results are observed in the near wake, where continuum approaches result in much higher densities and temperatures. This is attributed to a very large local mean free path (10–100 λ_∞) which makes the continuum formulations unrealistic and very sensitive to the model. However the far wake tends to coincide for all three calculations.

Other QGD and NS calculations were carried out for smaller Knudsen numbers. For $Kn_\infty = 0.01$, results coincide except in the near wake for large Mach numbers. For $Kn_\infty = 0.005$, they coincide in the whole flow field.

4. Conclusion

The flow around a disk placed perpendicular to a supersonic rarefied flow was calculated using NS and QGD equations. DSMC calculations were used as a reference. Additional terms in QGD equations act in the right direction to correct (and occasionally over-correct) the differences between NS and DSMC results.

QGD equations are clearly superior to NS ones for calculating the distribution of skin friction along the forward-facing side of the disk under near adiabatic conditions. Subject to later confirmation, one can expect from Reynolds' analogy that the same conclusion will hold for heat transfer in a cold-wall situation. At low Knudsen numbers, QGD equations tend to coincide with NS ones.

More work is needed on other flow configurations (e.g., jet expansion) and other conditions (e.g., cold wall) to generalize the conclusions deduced from the present work and from a previous one where a flat plate was considered.

References

Bird GA (1976) Molecular Gas Dynamics, Clarendon Press, Oxford.

Elizarova TG, Chetverushkin BN, Sheretov YV (1992) Quasi-gas-dynamic equations and computer simulation of viscous gas flows. In: Lecture Notes in Physics 414, pp 421–425.

Elizarova TG, Graur IA, Lengrand JC, Chpoun A (1994) Rarefied Gas Flow Simulation Based on Quasi Gas-Dynamic Equations, Lab. d'Aérothermique du CNRS, RC 94-4.

Elizarova TG, Graur IA, Chpoun A, Lengrand JC (1995a) Comparison of continuum and molecular approaches for rarefied gas flows. In: Proceedings of the 19th Int Symp on Rarefied Gas Dynamics, Oxford (GB), July 25–29, 1994, Oxford Univ Press, Vol 2, p 780.

Elizarova TG, Graur IA, Lengrand JC, Chpoun A (1995b) Rarefied gas flow simulation based on quasi gas-dynamic equations. AIAA J 33(12):2316–2324.

Lengrand JC, Chpoun A, Graur IA, Elizarova TG (1995) Supersonic Rarefied Flow Around a Perpendicular Disk, Lab. d'Aérothermique du CNRS Report R-95-6.

The numerical simulation of unsteady shock-wave/boundary-layer interactions in reacting gas mixtures

Y. S. Weber, Wright Laboratory
Wright-Patterson AFB, OH 45433-7913 USA
E. S. Oran, J. P. Boris, Naval Research Laboratory
Washington DC 20376-5344 USA
J. D. Anderson, Jr., University of Maryland
College Park, MD 20742 USA

Abstract: The reflected-shock/boundary-layer interaction occurring near the end wall in shock tubes has been simulated for both a nonreacting and reacting gas. Specifically, we compare computations in calorically perfect oxygen to dissociating oxygen to assess the effects of endothermic chemical reactions. The results indicate a complex structure consisting of an unstable shear layer in the bifurcation region with embedded vortices. The reduction in specific heat ratio leads to a decrease in the size of the bifurcation angle and reflected-shock velocity. For the conditions examined here, embedded lambda shocks were evident beneath the main bifurcated structure.

Key words: Bifurcation, Reactive, Reflected-shock/boundary-layer interaction

1. Introduction

Numerous analytic and experimental studies on bifurcations of reflected shocks in shock tubes have been performed (Mark 1959, Strehlow et al. 1959, Honda et al. 1975, Brossard et al. 1985). However, due to the high resolution requirements, there have been very few computational simulations. Wilson (1993) simulated reflected-shock interactions for an axisymmetric configuration and assessed the flow quality and impact on test time. Y. Weber (1994a) examined the detailed structure of the separated zone and the impact of the growth of the bifurcation region on the outer inviscid flow for two-dimensional configurations. Details of the separated-flow region show that it is composed of an unstable shear layer with embedded vorticity near the wall. Similar structures were found in simulations of inviscid flows in which reflected shocks interacted with stratified density gradients (Glowacki et al. 1985). In the present work, numerical simulations assess the effect of the chemical dissociation of oxygen on the reflected-shock/boundary-layer interaction. The simulation captures the properties of vortical structures down to the size of a few grid cells.

2. Governing equations and numerical method

Numerical simulations were performed using a reactive Euler solver (Weber J 1994) that was modified to include the viscous transport terms of the Navier-Stokes equations (Weber Y 1994b). For nonreacting O_2, a calorically perfect gas assumption was used to close the equations. For the reacting gas, thermodynamic equilibrium was assumed and the sensible enthalpy was obtained from data in the JANAF tables (Stull 1971). Rate constants for the chemical source terms were obtained from the Dunn and Kang (1973) reaction mechanism. The viscosity and thermal conductivity were computed using Sutherland's law for the nonreactive gas and a mole-weighted form of the Chapman-Enskog equation (Bird et al. 1960) for the reactive gas. Since the objective of these studies is to assess the impact of energy absorption caused by dissociation on the reflected-shock/boundary-layer interaction, diffusion has been neglected.

The convective terms were integrated using the Flux-Corrected Transport (FCT) algorithm (Boris et al. 1993). FCT is a high-order, monotone, positivity preserving algorithm for solving conservation equations. The viscous terms were treated as source terms in the momentum and energy equations. The chemistry integration is time-split from the fluid-dynamic integration. FCT is used to convect the species number densities in the absence of source terms. The resulting species number densities are then used as initial values for the chemical integration which is handled using the Selected Asymptotic Integration Method (SAIM) (Young et al. 1977), which is a robust, efficient second-order ordinary differential-equation solver.

3. Simulation parameters

The effect of chemical dissociation on the reflected-shock/boundary-layer interaction is studied by comparing shock bifurcations in nonreacting O_2 (Case I) and dissociating O_2 (Case II). The initial shock-pressure ratio and temperature are 23.7 and 700 K, respectively. The unit Reynolds number is 2.349×10^3 and is referenced to the conditions behind the incident shock. For Case II, these parameters are based on the initial, frozen state of the gas in Region 2. Based on differences in the specific heat ratio, induced velocity and shock speed are 2% higher in Case II. There were 1024 cells in the x-direction (spaced uniformly) and 256 in the y-direction (clustered near the wall).

4. Results

Comparisons of temperature fields in Fig. 1 show that the reactive-oxygen flow

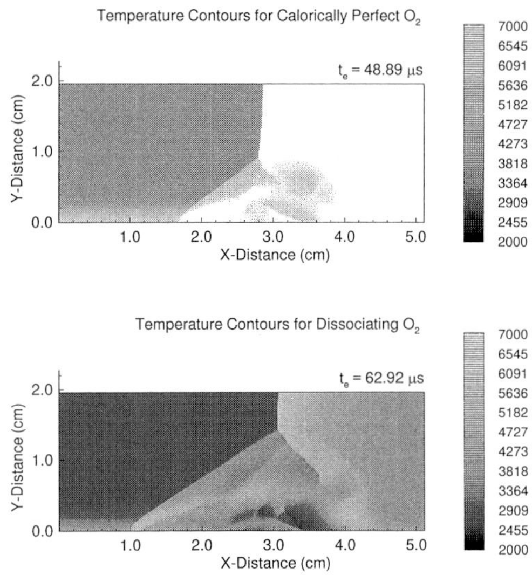

Figure 1. Temperature contours for calorically perfect and reactive O_2.

field is significantly cooler. The maximum temperature is 4800 K compared to the nonreacting value of 8000 K. The trend is opposite for the density field. Fig. 1 also shows the unstable shear layer with embedded vorticity pockets.

The pressure contours of Fig. 2 show qualitatively similar bifurcation structures. The bifurcation angle for Case I is 37° and for Case II is 33°. In both cases, there is a weak shock wave between the bifurcated foot and tail shock, due to the irregular displacement surface over which the inviscid flow travels. As the flow changes direction to remain parallel to the displacement surface, weak shock and expansion waves are generated. The shock located between the bifurcated foot and tail interacts with the tail shock and deforms it. The pressure contours for both cases also show the low-pressure center in the region where the shear layer is reversed. For Case II, the reflected-shock pressure is higher, as shown in the centerline pressure distribution in Fig. 3. This may be attributed to the decrease in specific heat ratio due to the chemical reactions occurring in the flow field.

Small shocklets located between the main bifurcated structure and the lower wall appear only in Case II. They are located near the strong expansion in the bifurcation region. This strong expansion (shown as a white region near the shocklets) expands the flow and causes small lambda shocklets to form. Near the center of the low-pressure field, the specific heat ratio and temperature are low. The lower specific heat leads to a strong expansion which eventually leads

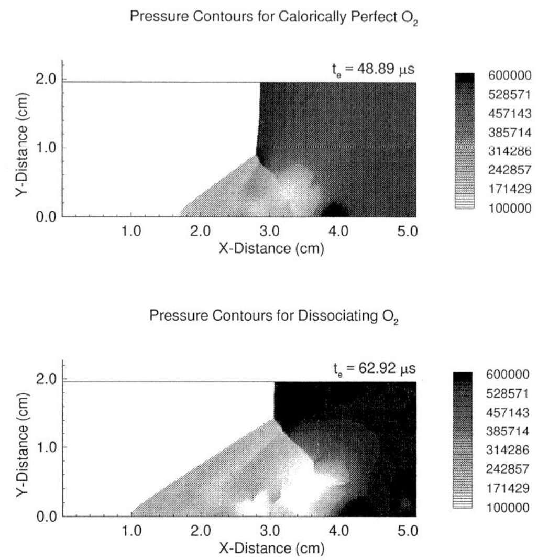

Figure 2. Pressure contours between calorically perfect and reactive O_2.

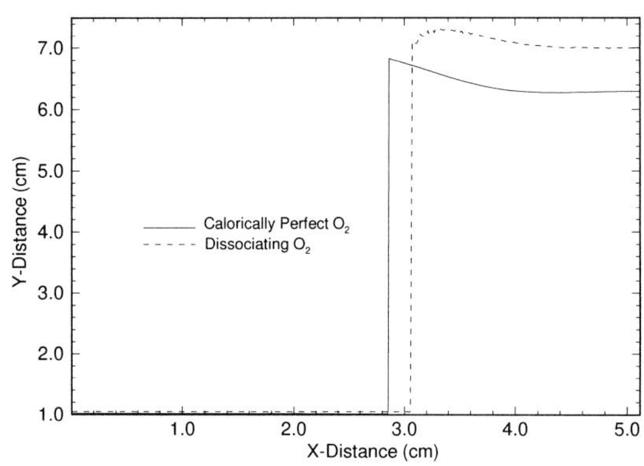

Figure 3. Comparison of centerline pressure profiles for calorically perfect and reactive O_2.

to a shock, or series of shocklets in the case presented here.

Figure 4, the mass-fraction distribution for both O_2 and O in the reactive flow field, shows that there is very little dissociation behind the incident shock, but significant dissociation behind the reflected shock. The reaction zone appears as a gradient in either of the mass fraction contours. The low-energy

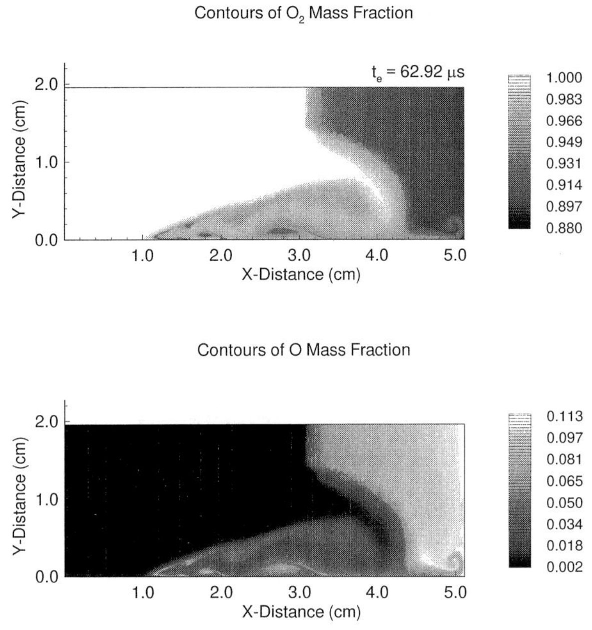

Figure 4. Contours of O and O_2 mass fraction for the reactive flow field.

pockets of vorticity contain monatomic oxygen which has been trapped in there by the fluid mechanics of the interaction.

In the reactive-flow calculation, there is a decrease in shock velocity. This can be attributed to energy absorption by the vibrational energy modes in the gas, as well as an increase in the zero-point energy of the dissociated mixture over the calorically perfect gas.

In conclusion, the chemical reactions lead to a decrease in the specific heat ratio, the reflected-shock velocity, and the bifurcation angle. In addition, the current reactive flow simulation yields embedded shocklets which were not present in previous studies.

Acknowledgement. Funding was provided in part by NRL through ONR and the ARPA Applied and Computational Mathematics Program and the PALACE KNIGHT program of the US Air Force. Computational support was provided by the Army HPC Research Center, the National Center for Supercomputing Applications, the University of Maryland Institute for Advanced Computer Studies, and NRL.

References

Bird RB, Stewart WE, Lightfoot EN (1960) In: Transport Phenomena, John Wiley and Sons, Inc.

Boris JP, Landsberg AM, Oran ES, Gardner JH (1993) LCPFCT — A Flux-Corrected Transport Algorithm for Solving Generalized Continuity Equations, NRL Memorandum Report 6410-93-7192.

Brossard J, Charpentier N (1985) Experimental study of shock wave reflection in a narrow channel. In: 15th Int Symp on Shock Tubes and Shock Waves, D Bershader, R Hanson (eds), Berkeley, CA.

Dunn MG, Kang S (1973) Theoretical and Experimental Studies of Reentry Plasmas, NASA CR-2232.

Glowacki WJ, Kuhl AL, Glaz HM, Ferguson RE (1985) Shock wave interaction with high sound speed layers. In: 15th Int Symp on Shock Tubes and Shock Waves, D Bershader, R Hanson (eds), Berkeley, CA.

Honda M, Takayama K, Onodera O, Kohama Y (1975) Motion of reflected shock waves in shock tube. In: 10th Int Symp on Shock Waves and Shock Tubes, G Kamimoto (ed), Japan.

Mark H (1959) The Interaction of a Reflected Shock Wave with the Boundary Layer in a Shock Tube, NACA TM-1418.

Strehlow RA, Cohen A (1959) Limitations of the reflected shock technique for studying fast chemical reactions and its application to the observation of relaxation in nitrogen and oxygen. J Chem Phys 30:257.

Stull DR (1971) JANAF Thermochemical Tables, National Bureau of Standards NSRDS-NBS 37.

Weber JW (1994) Physical and Numerical Aspects of Two-Dimensional Detonation Simulations Including Detailed Chemical Kinetics on a Massively Parallel Connection Machine, PhD Dissertation, University of Maryland.

Weber YS (1994a) The numerical simulation of shock bifurcation near the end wall in a shock tube. In: 25th AIAA Fluid Dynamics Meeting, AIAA 94-2307.

Weber YS (1994b) The Numerical Simulation of the Reflected-Shock/Boundary-Layer Interaction in Shock Tubes, PhD Dissertation, University of Maryland.

Wilson GJ, Sharma SP, Gillespie WD (1993) Time-dependent simulations of reflected-shock/boundary layer interaction. In: 31st AIAA Aerospace Sciences Meeting, AIAA 93-0480.

Young TR, Boris JP (1977) A numerical technique for solving stiff ordinary differential equations associated with the chemical kinetics of reactive-flow problems. J Appl Phys 81:2424.

Author Index

Keyword Index